Lecture Notes in Computer Science 12742

More information about this subseries at http://www.springer.com/series/7407

Maciej Paszynski · Dieter Kranzlmüller ·
Valeria V. Krzhizhanovskaya ·
Jack J. Dongarra · Peter M. A. Sloot (Eds.)

Computational Science – ICCS 2021

21st International Conference
Krakow, Poland, June 16–18, 2021
Proceedings, Part I

 Springer

Editors
Maciej Paszynski (ORCID)
AGH University of Science and Technology
Krakow, Poland

Valeria V. Krzhizhanovskaya (ORCID)
University of Amsterdam
Amsterdam, The Netherlands

Peter M. A. Sloot (ORCID)
University of Amsterdam
Amsterdam, The Netherlands

ITMO University
St. Petersburg, Russia

Nanyang Technological University
Singapore, Singapore

Dieter Kranzlmüller (ORCID)
Ludwig-Maximilians-Universität München
Munich, Germany

Leibniz Supercomputing Center (LRZ)
Garching bei München, Germany

Jack J. Dongarra (ORCID)
University of Tennessee at Knoxville
Knoxville, TN, USA

ISSN 0302-9743   ISSN 1611-3349   (electronic)
Lecture Notes in Computer Science
ISBN 978-3-030-77960-3   ISBN 978-3-030-77961-0   (eBook)
https://doi.org/10.1007/978-3-030-77961-0

LNCS Sublibrary: SL1 – Theoretical Computer Science and General Issues

This Springer imprint is published by the registered company Springer Nature Switzerland AG
The registered company address is: Gewerbestrasse 11, 6330 Cham, Switzerland

Preface

Welcome to the proceedings of the 21st annual International Conference on Computational Science (ICCS 2021 - https://www.iccs-meeting.org/iccs2021/).

In preparing this edition, we had high hopes that the ongoing COVID-19 pandemic would fade away and allow us to meet this June in the beautiful city of Kraków, Poland. Unfortunately, this is not yet the case, as the world struggles to adapt to the many profound changes brought about by this crisis. ICCS 2021 has had to adapt too and is thus being held entirely online, for the first time in its history.

These challenges notwithstanding, we have tried our best to keep the ICCS community as dynamic and productive as always. We are proud to present the proceedings you are reading as a result of that.

ICCS 2021 was jointly organized by the AGH University of Science and Technology, the University of Amsterdam, NTU Singapore, and the University of Tennessee.

The International Conference on Computational Science is an annual conference that brings together researchers and scientists from mathematics and computer science as basic computing disciplines, as well as researchers from various application areas who are pioneering computational methods in sciences such as physics, chemistry, life sciences, engineering, arts, and humanitarian fields, to discuss problems and solutions in the area, identify new issues, and shape future directions for research.

Since its inception in 2001, ICCS has attracted an increasing number of attendees and higher quality papers, and this year is not an exception, with over 350 registered participants. The proceedings have become a primary intellectual resource for computational science researchers, defining and advancing the state of the art in this field.

The theme for 2021, "**Computational Science for a Better Future**," highlights the role of computational science in tackling the current challenges of our fast-changing world. This conference was a unique event focusing on recent developments in scalable scientific algorithms, advanced software tools, computational grids, advanced numerical methods, and novel application areas. These innovative models, algorithms, and tools drive new science through efficient application in physical systems, computational and systems biology, environmental systems, finance, and other areas.

ICCS is well known for its excellent lineup of keynote speakers. The keynotes for 2021 were given by

- **Maciej Besta**, ETH Zürich, Switzerland
- **Marian Bubak**, AGH University of Science and Technology, Poland | Sano Centre for Computational Medicine, Poland
- **Anne Gelb**, Dartmouth College, USA
- **Georgiy Stenchikov**, King Abdullah University of Science and Technology, Saudi Arabia
- **Marco Viceconti**, University of Bologna, Italy

- **Krzysztof Walczak**, Poznan University of Economics and Business, Poland
- **Jessica Zhang**, Carnegie Mellon University, USA

This year we had 635 submissions (156 submissions to the main track and 479 to the thematic tracks). In the main track, 48 full papers were accepted (31%); in the thematic tracks, 212 full papers were accepted (44%). A high acceptance rate in the thematic tracks is explained by the nature of these tracks, where organisers personally invite many experts in a particular field to participate in their sessions.

ICCS relies strongly on our thematic track organizers' vital contributions to attract high-quality papers in many subject areas. We would like to thank all committee members from the main and thematic tracks for their contribution to ensure a high standard for the accepted papers. We would also like to thank *Springer, Elsevier,* and *Intellegibilis* for their support. Finally, we appreciate all the local organizing committee members for their hard work to prepare for this conference.

We are proud to note that ICCS is an A-rank conference in the CORE classification.

We wish you good health in these troubled times and look forward to meeting you at the conference.

June 2021

Maciej Paszynski
Dieter Kranzlmüller
Valeria V. Krzhizhanovskaya
Jack J. Dongarra
Peter M. A. Sloot

Organization

Local Organizing Committee at AGH University of Science and Technology

Chairs

Maciej Paszynski
Aleksander Byrski

Members

Marcin Łos
Maciej Woźniak
Leszek Siwik
Magdalena Suchoń

Thematic Tracks and Organizers

Advances in High-Performance Computational Earth Sciences: Applications and Frameworks – IHPCES

Takashi Shimokawabe
Kohei Fujita
Dominik Bartuschat

Applications of Computational Methods in Artificial Intelligence and Machine Learning – ACMAIML

Kourosh Modarresi
Paul Hofmann
Raja Velu
Peter Woehrmann

Artificial Intelligence and High-Performance Computing for Advanced Simulations – AIHPC4AS

Maciej Paszynski
Robert Schaefer
David Pardo
Victor Calo

Biomedical and Bioinformatics Challenges for Computer Science – BBC

Mario Cannataro
Giuseppe Agapito

Mauro Castelli
Riccardo Dondi
Italo Zoppis

Classifier Learning from Difficult Data – CLD^2

Michał Woźniak
Bartosz Krawczyk

Computational Analysis of Complex Social Systems – CSOC

Debraj Roy

Computational Collective Intelligence – CCI

Marcin Maleszka
Ngoc Thanh Nguyen
Marcin Hernes
Sinh Van Nguyen

Computational Health – CompHealth

Sergey Kovalchuk
Georgiy Bobashev
Stefan Thurner

Computational Methods for Emerging Problems in (dis-)Information Analysis – DisA

Michal Choras
Robert Burduk
Konstantinos Demestichas

Computational Methods in Smart Agriculture – CMSA

Andrew Lewis

Computational Optimization, Modelling, and Simulation – COMS

Xin-She Yang
Leifur Leifsson
Slawomir Koziel

Computational Science in IoT and Smart Systems – IoTSS

Vaidy Sunderam
Dariusz Mrozek

Computer Graphics, Image Processing and Artificial Intelligence – CGIPAI

Andres Iglesias
Lihua You
Alexander Malyshev
Hassan Ugail

Data-Driven Computational Sciences – DDCS

Craig Douglas

Machine Learning and Data Assimilation for Dynamical Systems – MLDADS

Rossella Arcucci

MeshFree Methods and Radial Basis Functions in Computational Sciences – MESHFREE

Vaclav Skala
Marco-Evangelos Biancolini
Samsul Ariffin Abdul Karim
Rongjiang Pan
Fernando-César Meira-Menandro

Multiscale Modelling and Simulation – MMS

Derek Groen
Diana Suleimenova
Stefano Casarin
Bartosz Bosak
Wouter Edeling

Quantum Computing Workshop – QCW

Katarzyna Rycerz
Marian Bubak

Simulations of Flow and Transport: Modeling, Algorithms and Computation – SOFTMAC

Shuyu Sun
Jingfa Li
James Liu

Smart Systems: Bringing Together Computer Vision, Sensor Networks and Machine Learning – SmartSys

Pedro Cardoso
Roberto Lam

João Rodrigues
Jânio Monteiro

Software Engineering for Computational Science – SE4Science

Jeffrey Carver
Neil Chue Hong
Anna-Lena Lamprecht

Solving Problems with Uncertainty – SPU

Vassil Alexandrov
Aneta Karaivanova

Teaching Computational Science – WTCS

Angela Shiflet
Nia Alexandrov
Alfredo Tirado-Ramos

Uncertainty Quantification for Computational Models – UNEQUIvOCAL

Wouter Edeling
Anna Nikishova

Reviewers

Ahmad Abdelfattah
Samsul Ariffin Abdul
  Karim
Tesfamariam Mulugeta
  Abuhay
Giuseppe Agapito
Elisabete Alberdi
Luis Alexandre
Vassil Alexandrov
Nia Alexandrov
Julen Alvarez-Aramberri
Sergey Alyaev
Tomasz Andrysiak
Samuel Aning
Michael Antolovich
Hideo Aochi
Hamid Arabnejad
Rossella Arcucci
Costin Badica
Marina Balakhontceva

Bartosz Balis
Krzysztof Banas
Dariusz Barbucha
Valeria Bartsch
Dominik Bartuschat
Pouria Behnodfaur
Joern Behrens
Adrian Bekasiewicz
Gebrail Bekdas
Mehmet Belen
Stefano Beretta
Benjamin Berkels
Daniel Berrar
Sanjukta Bhowmick
Georgiy Bobashev
Bartosz Bosak
Isabel Sofia Brito
Marc Brittain
Jérémy Buisson
Robert Burduk

Michael Burkhart
Allah Bux
Krisztian Buza
Aleksander Byrski
Cristiano Cabrita
Xing Cai
Barbara Calabrese
Jose Camata
Almudena Campuzano
Mario Cannataro
Alberto Cano
Pedro Cardoso
Alberto Carrassi
Alfonso Carriazo
Jeffrey Carver
Manuel Castañón-Puga
Mauro Castelli
Eduardo Cesar
Nicholas Chancellor
Patrikakis Charalampos

Bartosz Krawczyk
Dariusz Krol
Valeria Krzhizhanovskaya
Adam Krzyzak
Pawel Ksieniewicz
Marek Kubalcík
Sebastian Kuckuk
Eileen Kuehn
Michael Kuhn
Michal Kulczewski
Julian Martin Kunkel
Krzysztof Kurowski
Marcin Kuta
Bogdan Kwolek
Panagiotis Kyziropoulos
Massimo La Rosa
Roberto Lam
Anna-Lena Lamprecht
Rubin Landau
Johannes Langguth
Shin-Jye Lee
Mike Lees
Leifur Leifsson
Kenneth Leiter
Florin Leon
Vasiliy Leonenko
Roy Lettieri
Jake Lever
Andrew Lewis
Jingfa Li
Hui Liang
James Liu
Yen-Chen Liu
Zhao Liu
Hui Liu
Pengcheng Liu
Hong Liu
Marcelo Lobosco
Robert Lodder
Chu Kiong Loo
Marcin Los
Stephane Louise
Frederic Loulergue
Hatem Ltaief
Paul Lu
Stefan Luding

Laura Lyman
Scott MacLachlan
Lukasz Madej
Lech Madeyski
Luca Magri
Imran Mahmood
Peyman Mahouti
Marcin Maleszka
Alexander Malyshev
Livia Marcellino
Tomas Margalef
Tiziana Margaria
Osni Marques
M. Carmen Márquez
  García
Paula Martins
Jaime Afonso Martins
Pawel Matuszyk
Valerie Maxville
Pedro Medeiros
Fernando-César
  Meira-Menandro
Roderick Melnik
Valentin Melnikov
Ivan Merelli
Marianna Milano
Leandro Minku
Jaroslaw Miszczak
Kourosh Modarresi
Jânio Monteiro
Fernando Monteiro
James Montgomery
Dariusz Mrozek
Peter Mueller
Ignacio Muga
Judit Munoz-Matute
Philip Nadler
Hiromichi Nagao
Jethro Nagawkar
Kengo Nakajima
Grzegorz J. Nalepa
I. Michael Navon
Philipp Neumann
Du Nguyen
Ngoc Thanh Nguyen
Quang-Vu Nguyen

Sinh Van Nguyen
Nancy Nichols
Anna Nikishova
Hitoshi Nishizawa
Algirdas Noreika
Manuel Núñez
Krzysztof Okarma
Pablo Oliveira
Javier Omella
Kenji Ono
Eneko Osaba
Aziz Ouaarab
Raymond Padmos
Marek Palicki
Junjun Pan
Rongjiang Pan
Nikela Papadopoulou
Marcin Paprzycki
David Pardo
Anna Paszynska
Maciej Paszynski
Abani Patra
Dana Petcu
Serge Petiton
Bernhard Pfahringer
Toby Phillips
Frank Phillipson
Juan C. Pichel
Anna
  Pietrenko-Dabrowska
Laércio L. Pilla
Yuri Pirola
Nadia Pisanti
Sabri Pllana
Mihail Popov
Simon Portegies Zwart
Roland Potthast
Malgorzata
  Przybyla-Kasperek
Ela Pustulka-Hunt
Alexander Pyayt
Kun Qian
Yipeng Qin
Rick Quax
Cesar Quilodran Casas
Enrique S. Quintana-Orti

Contents – Part I

ICCS 2021 Main Track

Smoothing Speed Variability in Age-Friendly Urban Traffic Management

José Monreal Bailey[(✉)], Hadi Tabatabaee Malazi, and Siobhán Clarke

University of Dublin Trinity College, Dublin, Ireland
{monrealj,tabatabh}@tcd.ie, Siobhan.Clarke@scss.tcd.ie

Abstract. Traffic congestion has a negative impact on vehicular mobility, especially for senior drivers. Current approaches to urban traffic management focus on adaptive routing for the reduction of fuel consumption and travel time. Most of these approaches do not consider age-friendliness, in particular that speed variability is difficult for senior drivers. Frequent stop and go situations around congested areas are tiresome for senior drivers and make them prone to accidents. Moreover, senior drivers' mobility is affected by factors such as travel time, surrounding vehicles' speed, and hectic traffic. Age-friendly traffic management needs to consider speed variability in addition to drivers' waiting time (which impacts fuel consumption and travel time). This paper introduces a multi-agent pheromone-based vehicle routing algorithm that smooths speed variability while also considering senior drivers during traffic light control. Simulation results demonstrate 17.6% improvement in speed variability as well as reducing travel time and fuel consumption by 11.6% and 19.8% respectively compared to the state of the art.

Keywords: Age-friendly cities · Vehicle routing · Adaptive signal control · Collaborative signal control · Collaborative vehicle routing · Multi-agent systems · Smart city

1 Introduction

The reality of ageing populations requires significant transformations to twenty-first-century living. There are implications for many sectors, including urban services and transportation [1]. Senior drivers feel uncomfortable in hectic traffic, and the risk of being in a car accident has a direct impact on their decision to quit driving [21]. They usually drive slowly and have a high reaction time to traffic events [5,12,16]. An age-friendly smart traffic management system could make senior drivers more confident in traffic, which is important for their well-being.

Smart traffic management systems (STMS) improve traffic flow, minimise traffic congestion and crash rate by applying strategies for vehicle routing and adaptive signal control. While drivers aged over 75 have the highest crash rate of all age groups [6], speed variability is positively correlated with crash frequency

© Springer Nature Switzerland AG 2021
M. Paszynski et al. (Eds.): ICCS 2021, LNCS 12742, pp. 3–16, 2021.
https://doi.org/10.1007/978-3-030-77961-0_1

[19,20]. Therefore, one of the key factors in improving traffic flow for senior drivers is to decrease speed variability. We define the speed variability of a road as the standard deviation of the speed of the vehicles driving on that road at a particular time. A high level of speed variability shows frequent acceleration and deceleration, which leads to an increase in accident risks, especially for senior drivers. Additionally, a vehicle's fuel consumption increases in a *stop and go* mode, with a consequent increase in CO_2 emissions, compared to smooth movement at a constant speed. Moreover, a smooth driving speed is a more pleasant mobility experience even on heavy traffic roads. While much of the related research on STMS focuses on reducing waiting time, the main objective of this work is to reduce speed variability without affecting waiting time.

Researchers have applied a wide range of techniques to mitigate vehicular traffic congestion, such as bee colonies [14], integer linear programming [3], pheromones [4,9,18], collaborative multi-agents [2], and reinforcement learning [22]. Some are centralised and prone to a single point of failure, while others require drivers to share their final destination as well as their planned route. In general, research has concentrated on shortening travel and wait time, reducing fuel consumption and greenhouse emissions [15,18]. There has been limited focus on considering the special needs of senior drivers. In our previous work, we devised a collaborative multi-agent adaptive signal control model to reduce wait time and fuel consumption for senior drivers [2]. However, this was insufficient, as speed variability is also a key factor for such drivers.

In this paper, we propose CoMAAPRAS, which is an age-friendly traffic control system. It adapts to environment changes using a pheromone-based and collaborative multi-agent model. It is based on CoMASig [2], which improves senior drivers' waiting time by extending green light times. CoMAAPRAS extends this model by enabling vehicles to take routes that contribute to reducing speed variability and the accumulative waiting time of senior drivers. Our method uses bio-inspired concepts to reduce speed variability and provide an age-friendly smart traffic light planning system. The model is adaptive, using decentralised, real-time decision-making, without requiring route information from drivers.

2   Related Work

An urban traffic management system can be defined as a large scale, dynamic and decentralised multi-agent system, where vehicles and infrastructure are represented by agents. The interaction among agents can be: (i) homogeneous, where only agents of the same type communicate and make decisions based on the information shared; (ii) heterogeneous, where only agents of different type communicate and take actions; (iii) both, where agents interact with every agent type, share information and make decisions based on that information.

In this section, we review work that addresses different traffic problems, in particular, traffic congestion, and vehicles' fuel consumption and their waiting time. In general, current approaches use vehicle re-routing [3,7,14,18] and urban traffic light control [2,11,22] strategies to address traffic problems. For example,

Cao et al. studied a combination of arrival and total travel times with a semi-decentralised multi-agent based method [3]. Infrastructure agents are responsible for route guidance at each junction by solving a route assignment problem. The method reformulated route assignment as a Mixed-Integer Linear Programming (*MILP*) problem to achieve computation efficiency. A weighted quadratic term is used to minimise the expected travel time according to the potential route assignment. *Arriving on time* is formulated as a probability tail model to maximise the probability of arriving before the deadline. In other words, they incorporate a quadratic term into the objective function of the original arriving on-time problem. *Ng* et al. introduced two Multiple Colonies Artificial Bee Colony (*MC-ABC*) algorithms for online vehicular routing [14]. They provided a flexible re-routing strategy for scheduled logistics to reduce the risk of late delivery. Their methods balance the exploitation and exploration of the original bee colony, and simulation results show that this approach outperforms other artificial bee colony algorithms by avoiding getting trapped in local optima. *Soon* et al. introduced the Eco-friendly Pheromone-based Green Vehicle Routing (*E-PGVR*) strategy [18]. The main focus is on reducing fuel consumption and greenhouse gas emissions by providing green waves to vehicles based on the pheromone intensity of each road. This work assumes fixed traffic light timing and fully connected vehicles. Communication between traffic light agents and intelligent vehicle agents is made through road supervisor agents. The authors argue that similar pheromone-based methods tried to alleviate traffic congestion by rerouting vehicles to routes with lower intensity of pheromone, but the vehicles may still suffer from multiple red signalised junctions. To tackle this problem, the chances for paths with multiple green signalised junctions was increased. To address scalability, the authors also devised a decentralised Hierarchical Multi-Agent Pheromone-based System (*HMAPS*), which uses local dynamic traffic information of m-hops downstream. They also introduced a modified dynamic k-shortest path for path assignment that reduces the computation load. It is worth pointing out that a fixed traffic light program can generate unnecessary waiting times for vehicles at the junctions.

We previously introduced a Collaborative Multi-Agent Signal control algorithm to support senior drivers, called (*CoMASig*) [2]. We focused on senior drivers' special needs and characteristics to improve their driving experiences. Vehicles communicate with traffic signals enabling them to autonomously adjust the phase timing based on the road demand and the proportion of senior drivers. However, our findings were that speed variability was exacerbated in the presence of senior drivers. *Zhou* et al. described an Edge-based Reinforcement Learning algorithm (*ERL*) that optimises traffic lights in real-time by employing edge servers [22]. The low latency edge servers provide a fast computation means for deep neural network training and control feedback. The servers collect local traffic data and aggregate it with the received data from neighbouring edge servers. The approach partitions urban traffic into hierarchical levels of junctions, neighbours, and districts. The modular approach facilitates fine-grained optimisation, with the optimisation processing performed at the three levels in parallel. The extra communication with edge servers and also the cloud may add overhead to the network and affect its performance. Their work limits vehicles' lane-change capability and focuses on the average speeds, which may hide extreme values.

Current research in vehicle traffic re-routing focuses on the reduction of travel time, waiting time, fuel consumption, and greenhouse gas emissions. Different types of drivers (e.g., senior drivers, young drivers) are generally not considered. The current work shows that traffic policies are used to improve traffic congestion, travel times, fuel consumption and greenhouse gas emissions. However, speed variability has not been evaluated as a required feature for an age-friendly urban traffic management system.

3   System Model

We model a traffic network as a grid of 4 by 4 junctions and 16 traffic lights (Fig. 1a). The roads are two way, have 3 lanes each way, and have the same length l. Each *traffic light* ($\text{tl}_{(i,j)}$) controls the incoming traffic flow of the roads to the junction. We refer to these roads as the controlled roads of traffic light $tl_{(i,j)}$. Each traffic light uses the standard NEMA dual-ring, eight-phase structure. There are eight movements at each junction. Figure 1b shows the four phases per ring in a traffic light. The barrier exists to avoid conflicting movements between groups, which means that all phases in one group must end before the next group starts. Each traffic light establishes a neighbourhood composed of all traffic lights at one junction distance. Figure 1a shows an example of neighbourhoods in which a blue dotted circle and a red dotted circle depicts the neighbourhood of TL7 and TL6 traffic lights respectively. Equation 1 shows the set of neighbourhood traffic lights for $tl_{(i,j)}$ which is denoted as $N(tl_{i,j})$. A traffic light is modelled as a collaborative agent, which can send/receive messages

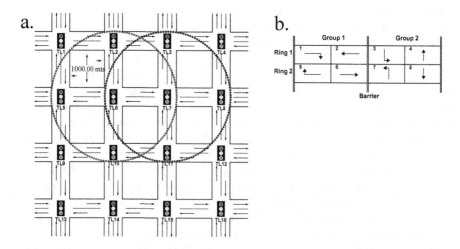

Fig. 1. a. Layout of 4 by 4 grid road network. b. Dual ring, 8 phase controller. (Color figure online)

to/from other traffic lights within its neighbourhood set, and vehicles within the communication range.

$$N(tl_{i,j}) = \{tl_{(k,l)} | (k = i \land (l = j - 1 \lor l = j + 1)) \lor \\ (l = j \land (k = i - 1 \lor k = i + 1))\} \tag{1}$$

A vehicle is denoted by v_x where x is its identifier. A v_x starts its trip at time t_{start} and reaches its destination at t_{end}. The speed of a vehicle is s_x^t where t is the time epoch. Similarly, the location of a vehicle v at time t is denoted by $l(v, t)$, and it is part of a controlled road. A vehicle is also modelled as an agent, and each vehicle agent (v_x) communicates only with the traffic light $(tl_{(i,j)})$ that controls the road on which it is travelling. This process is performed periodically (i.e., every 10 s) to get a significant value while, at the same time, maintaining a good performance. Every v_x communicates with its relevant traffic light through message transfer. Each v_x subscribes to the $tl_{(i,j)}$ that controls the road on which the vehicle is currently driving, and it unsubscribes when it leaves that road.

There are two types of drivers in the system: d_{yd} *young drivers* and d_{sd} *senior drivers*. $d(v_x)$ denotes the driver type of vehicle v_x. Each driver type has a maximum speed. $SMax_{yd}$ and $SMax_{sd}$ are the maximum speed for driver type d_{yd} and d_{sd} respectively. Similarly, each driver type has a reaction time (rt_{yd} and rt_{sd}) defined as the decision-making time of a driver.

The speed variability of a controlled road at time t is defined as the standard deviation of the speed of all vehicles moving on that road at time t. Equation 2 shows the formal definition of speed variability, where N and $\overline{s_{cr}}$ are the number and average speed (Eq. 3) of the vehicles on the controlled road cr at time t.

$$SV_{cr}^t = \sqrt{\frac{1}{N-1} \sum (s_i^t - \overline{s_{cr}})^2} \quad (\forall v_i : l(i, t) \in cr) \tag{2}$$

$$\overline{s_{cr}} = \frac{1}{N} \sum_{\forall i:l(i,t) \in cr} s_i^t \tag{3}$$

The *current waiting time* of vehicle x is defined as the accumulative time that a driver moves below 0.1 m/s from the moment it started its trip to the current time, and voluntary stops (i.e., stops not caused by congestion) are not considered. The current waiting time is denoted by $t_{(tw,x)}$ where x is the vehicle ID (Eq. 4). Similarly, *total waiting time* ($T_{(tw,x)}$) is the accumulative waiting time of vehicle x on the whole trip.

$$t_{(tw,x)} = \sum_{s_x^m < 0.1} |t_m| \quad (\forall t_m : t_{start} \leq t_m \leq now) \tag{4}$$

The *travel time* of a vehicle x is the amount of time the vehicle needed to reach its destination. The *fuel consumption* of a vehicle x corresponds to the total amount of fuel spent by the vehicle on the whole trip.

The objectives of the age-friendly traffic light system are to minimise speed variability (Eq. 5) while reducing total waiting time (Eq. 6) for senior drivers.

$$Min \sum_{\forall cr} SV_{cr}^t \tag{5}$$

$$Min \sum_{\forall x: d(v_x) = d_{sd}} T_{(tw,x)} \qquad (6)$$

4   CoMAAPRAS

The CoMAAPRAS system works with two types of agents: vehicle agents v_x and traffic light agents $tl_{(i,j)}$. This section describes how the system uses pheromones to calculate both the speed variance in a controlled road and waiting time for senior drivers. It also describes the traffic policies that are implemented based on these two values - for example, when to change the lights, or re-route vehicles.

CoMAAPRAS is based on ant-colony optimisation, using the pheromone idea of ants going through different paths to find the shortest one. Vehicle agents communicate their speed and waiting time to traffic light agents, which are in control of the decrease or increase of the pheromones in the road. These pheromones help vehicle agents to find roads with slower speed variance and traffic lights to extend the green light when needed. Traffic lights manage two types of pheromones: a speed variance pheromone ($P_{\hat{v}}$) and a waiting time pheromone (P_{tw}). Each of its controlled roads has a separate value for each pheromone type. For both pheromones, a high value raises a concern about the suitability of a road for senior drivers, while a low value is positive. The values for both pheromones evaporate (i.e., decrease) periodically, to adapt the system to time-related traffic conditions. Pheromone deposits (i.e., increases based on penalties) relating to negative traffic conditions are described in this section.

Speed Variance: Each traffic light updates the speed variance pheromone values for each of its controlled roads from a calculation of each vehicle's speed experience, as communicated when the vehicle is within the communication range. Each v_x records its speed periodically for each control road. When it enters the communication range with the traffic light controlling that road, it sends the full vector containing these measurements and then continues to do so at each recording period. Regardless of any communication from vehicles, every t seconds, each $tl_{(i,j)}$ evaporates (decreases) the values of the speed variance pheromones for all its controlled roads, at a fixed rate (e.g., 10%). This handles the case where there may be no vehicles on the road, which would be positive for senior drivers. The process then assesses whether penalties should be applied to the speed variability pheromone by examining the vehicles' submissions on their periodic speed. Each $tl_{(i,j)}$ increases the $P_{\hat{v}}$ by the speed variance in the controlled road (Eq. 7).

$$P_{\hat{v}} = P_{\hat{v}} + SV_{cr}^t \qquad (7)$$

Waiting Time: When each vehicle agent with a senior driver (v_{sd}) enters communication range with the traffic light controlling the road on which it is travelling, it, like all vehicles, sends the full vector containing the measurements of each speed recording period, and also sends a calculation of the time driving below 0.1 m/s. Each vehicle continues to send these values periodically until it leaves the controlled road. Each traffic light updates its waiting time pheromone values for each of its controlled roads from the reported waiting times. Again, to handle the situation where there are no vehicles on the road, regardless of any communication from senior driver vehicles, every t seconds each $tl_{(i,j)}$ evaporates (decreases) the values of the waiting time pheromones for all its controlled roads, at a fixed rate (e.g., 10%). The process then assesses whether penalties should be applied to the waiting time pheromone, depending on the current waiting time of senior drivers on the road. Each $tl_{(i,j)}$ increases the P_{tw} proportional to $t_{(tw,x)}$ (Eq. 8).

$$P_{tw} = P_{tw} \left(1 + \frac{t_{(tw,x)}}{100} \right) \tag{8}$$

Traffic Policies: Traffic light agents use the pheromones and a set of rules to decide which traffic policy (e.g., light change or rerouting) to apply. Each $tl_{(i,j)}$ communicates with its neighbourhood as shown in Fig. 1, and shares the pheromone information about its controlled roads to its neighbourhood. All neighbourhood traffic lights' information is utilised to decide as to which traffic policy to apply: (i) green light extension, (ii) vehicle re-route, (iii) green light extension and vehicle re-route, or (iv) none, as follows:

i **Green light extension:** Every traffic light compares the pheromone levels in its controlled roads. The waiting time pheromone is increased only by senior drivers. If the controlled roads in green phase have higher waiting time pheromone by a certain percentage (e.g., 10%) than in conflicting phases, then the phase is extended. Otherwise, the traffic light does not extend the green phase and switches to the next phase.

ii **Routing:** Every traffic light compares the speed variance pheromone level on all possible roads that a vehicle may enter after exiting the current road. We refer to these roads as destination roads. If the speed variance pheromone level from any of the possible destination roads differ in more than a certain percentage (e.g., 10%) then the traffic light sends a reroute message to the vehicles in the communication range. The vehicle searches for a new route with the lowest speed variance.

CoMAAPRAS has two steps: *Update Road Pheromone* and *Collaborative Decision Making*, which are explained next.

4.1 Update Road Pheromone

Vehicles send their current speed and waiting time messages periodically to their controlling traffic light when within the communication range. Each traffic light stores all incoming messages and processes them at the end of each time epoch (Algorithm 1). The traffic light processes the information depending on the message type: speed variance or waiting time. Each message contains the vehicle identifier, road identifier, and the set of values stored during its journey until the communication is established. If the message type is speed variance, then the traffic light takes the speed values from the vehicle to calculate its speed variance. Then, the traffic light updates the speed variance pheromone value given the old speed variance and the current one using Eq. 7. The traffic

Algorithm 1. Update Pheromone

```
 1: procedure UPDATEROADPHEROMONE(message)
 2:   if message.type is speed then
 3:     speeds ← message.vehicleSpeeds
 4:     roadId ← message.vehicleRoadId
 5:     newSpeedVar ← speedVariance(speeds)
 6:     pv ← getPvFromRoad(roadId)
 7:     pv ← pv + newSpeedVar
 8:     setPvFromRoad(pv, roadId)
 9:     updateSpeedVar(speeds, roadId)
10:   if message.type is time and message.from is sd then
11:     waitingTime ← message.vehicleWaitingTime
12:     roadId ← message.vehicleRoadId
13:     ptw ← getPtwFromRoad(roadId)
14:     ptw ← ptw * (1 + (waitingTime/100))
15:     setPtwFromRoad(ptw, roadId)
```

light also updates the speed variance for the road with the speed values from the vehicle. If the message type is waiting time, then the traffic light retrieves the accumulative waiting time and road id from the vehicles' message. The traffic light computes the new waiting time pheromone using Eq. 8 and updates its value in the road where the vehicle is driving.

4.2 Collaborative Decision Making

Algorithm 2. Collaborative Decision Making

```
 1: procedure APPLYTRAFFICPOLICY
 2:   greenLightExtension ← TRUE
 3:   routing ← FALSE
 4:   currentPtw ← getPtwFromRoad(currentRoadId)
 5:   for each roadId ∈ tl_{(i,j)} do
 6:     if roadId belongs to a conflicting phase AND greenLightExtension then
 7:       nextRoadPtw ← getPtwFromRoad(roadId)
 8:       if nextRoadPtw > 1.1 * currentPtw then
 9:         greenLightExtension ← FALSE
10:   dRoads ← sort(destinationRoads_{tl_{(i,j)}})
11:   if 1.1 * dRoads[first] <= dRoads[last] then
12:     routing ← TRUE
13:   if greenLightExtension then
14:     tl_{(i,j)} extend green light
15:     for each tl_{(k,l)} ∈ neighbours_{tl_{(i,j)}} do
16:       if can tl_{(k,l)} collaborate? then
17:         tl_{(k,l)} applies green light extension
18:   if routing then
19:     tl_{(i,j)} send re-route message to vehicle
```

Each Traffic light assesses which traffic policy needs to apply from the moment vehicles come into the communication range (Algorithm 2). If a vehicle's current road waiting time pheromone is lower (e.g., by 10%) than in conflicting phases, then the green light extension is disabled. Then, the traffic light evaluates the speed variance pheromones from all possible destination roads for the vehicle. The routing policy is enabled if the speed variances on destination roads differ by a certain percentage (e.g., 10%). If the green light extension traffic policy is enabled, the traffic light will ask the neighbouring traffic lights to collaborate for a green wave, if possible. If the routing policy is enabled the traffic light will send a re-route message to the subscribed vehicle.

5   Experiment Design

In this section, we introduce our evaluation scenario, the baseline methods used for comparison, and the performance metrics.

In our simulation scenario, the signal control plan sequence is fixed and we assume all vehicles can communicate with the infrastructure through a VANET. The evaluation scenario includes three congestion levels for different ratios of senior drivers. Table 1 shows the hourly traffic density in our network for low, medium, and high levels of congestion used in [8] where N, S, W, and E are the cardinal directions of the incoming and outgoing traffic flows. We used two types of driver profiles (younger drivers and senior drivers) and we do not consider any pedestrians in our simulation. The response time (rt_1) for younger drivers is set to the same value as the simulation step (1 s). Senior drivers move at a slower speed ($SMax_2 = 0.6 \times SMax_1$) and their reaction time is 40% slower than young drivers ($rt_2 = 0.6 \times rt_1$) [10]. We consider different ratios of senior drivers ($ratio_{sd} \in [0, 100]$) increasing by 20% each round per execution.

Table 1. Vehicles per hour

		W to E	W to S	W to N	E to W	E to N	E to S	N to S	N to W	N to E	S to N	S to E	S to W	Total
	Low	1668	224	288	1848	144	448	76	560	672	72	912	320	7232
Traffic Levels	Medium	4008	704	512	4440	352	1088	184	1488	1632	176	2208	768	17560
	High	6680	1184	864	7396	592	1824	304	2496	2720	292	3680	1296	29328

The model is developed using Simonstrator, SUMO v1.5, and JDK 8. Simonstrator is a network simulator that facilitates communication between agents [17]. SUMO is an urban traffic simulator that simulates different driver profiles, network layouts and traffic density. Vehicles are distributed using SUMO [13], which generates a normal distribution of the traffic flow. JDK 8 is the Java development kit version 8, which is compatible with both simulators.

The experiments are executed on a Kelvin system, a high performance compute cluster hosted and managed by the Trinity Centre for High-Performance Computing (TCHPC). Each simulation runs on a cluster node that has GNU/Linux, 2.66 GHz Intel processor, and 24 GB of RAM[1]. We execute each of our simulation configurations 11 times which gives a confidence interval of 95%.

We have compared our proposed method against three baselines that aimed at enhancing traffic performance and show improvements in traffic congestion.

Fixed signal control (FSC) is widely used today in urban areas and has a fixed time duration of the lights. The duration has been calculated in advance to improve traffic flow at each junction [8].

---

[1] Kelvin details - https://www.tchpc.tcd.ie/resources/clusters/kelvin.

Eco-friendly Pheromone-based Green Vehicle Routing (E-PGVR) is a pheromone-based routing algorithm that prioritises paths with green waves to reduce fuel consumption, mean travel time, and road congestion [18]. Its performance shows it can improve vehicles journey times.

Collaborative Multi-Agent Signal Control (CoMASig) minimises the waiting time for senior drivers, while it increases the throughput at each junction [2].

The evaluation metrics are speed variability, waiting time, time travel, and fuel consumption defined in Sect. 3.

6   Results

We evaluate the performance of our approach and compare it with the three baselines in 3 traffic level scenarios. We use different percentages of senior drivers (0%, 20%, 40%, 60%, 80%, and 100%) for each of the previous combinations, and present the results from our experiments with different traffic levels and driver types. We present the results for speed variability, waiting time, fuel consumption and travel time in Subsects. 6.1, 6.2, 6.3 and 6.4 respectively.

6.1   Speed Variability

In the low-level traffic scenario, the simulation results show that CoMAAPRAS and CoMASig perform better for senior drivers under low traffic volume (Fig. 2a) since these methods explicitly consider senior drivers in the decision making of their traffic policies. CoMAAPRAS starts to diverge from CoMASig as the percentage of senior drivers in the scenario increases to over 40%. This behaviour is explained as both traffic policies in signal control extension and re-route take place more frequently. The improvements in speed variability are 3.5%, 11.4% and 17.6% for 60%, 80% and 100% of senior drivers respectively.

For the younger drivers in low-level traffic volume, FSC demonstrates the lowest speed variability since the vehicles take advantage of both optimised traffic light timing and the percentage of senior drivers. This approach does not focus on driver types but traffic flow. The results show that the speed variability in FSC starts to improve for every vehicle when the percentage of senior drivers is greater than 40%. In our proposed method, the speed variability of young drivers starts improving when the percentage of senior drivers increases from 40%. Vehicles in CoMAAPRAS are benefited from the number of senior drivers on the road plus the pheromones strategy to re-route and adapt the traffic lights timing. In E-PGVR and CoMASig methods, the speed variability for both young and senior drivers increase as the percentage of seniors drivers rises. In CoMASig, the speed variability decreases by 17.6% between 0% and 20% of senior drivers. This approach considers the overall demand on the road and minimal change in senior drivers' waiting time.

The speed variability for medium and high levels of traffic is completely different from the low-level scenario. Speed variability is not improved in

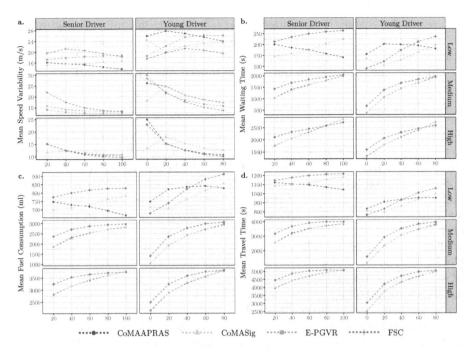

Fig. 2. Plots of performance: a.Speed variability, b.Waiting time, c.Fuel consumption and d.Travel Time, under different traffic levels. For each plot the X-axis corresponds to the percentage of senior drivers, while the Y-axis corresponds to a. the mean speed variability expressed in m/s, b. the mean waiting time in s, the mean fuel consumption expressed in ml and d. the mean travel time in s. Results are separated by driver type (Senior driver, Young driver), and traffic level (Low, Medium and High).

CoMAAPRAS, while E-PGVR shows better performance given the optimised traffic light timing and the pheromones strategy from the ant-colony optimisation algorithm.

6.2 Waiting Time

CoMAAPRAS is the only approach that reduces the waiting time for senior drivers in the low-level traffic scenario, as the percentage of seniors increases (Fig. 2b). The waiting times are improved by 22.6% and 31% when there are 80% and 100% of senior drivers in the scenario. However, CoMAAPRAS does not improve the waiting time in the medium and high-level traffic scenarios, and again, behaves almost the same as CoMASig. Waiting time is directly connected with traffic jams, as it corresponds to the time vehicles driving below 0.1 m/s. The roads controlled by each traffic light have similar speed variability pheromone levels because of the traffic jams. For this reason, CoMAAPRAS only enables the green light extension policy, which is similar to how CoMASig works. E-PGVR and FSC have similar behaviour for the different traffic levels.

The reason for this is the optimised traffic light timing in conjunction with the volume of vehicles per hour entering the roads. The pheromones levels in the roads do not change much to generate a new route, in the case of E-PGVR. The majority of vehicles will drive their predetermined route just like in FSC.

6.3   Fuel Consumption

The simulation results reveal that the fuel consumption in CoMAAPRAS reduces as the number of senior drivers increases for the low-level traffic scenario (Fig. 2c). The reduction in fuel consumption starts from the beginning but improves from approximately 60% of senior drivers. In this case, vehicles drive more smoothly and faster which helps to considerably reduce fuel consumption. The fuel consumption decreases by 1.5%, 9.3% and 19.8% when there are 60%, 80% and 100% of senior drivers. However, CoMAAPRAS does not improve fuel consumption in medium and high-level traffic scenarios. In fact, it behaves almost the same as CoMASig. The reason for this is the traffic congestion generated by the volume of vehicles travelling on the roads. The roads controlled by each traffic light have similar speed variability pheromone levels for which CoMAAPRAS only enables the green light extension policy, which is similar to CoMASig. Likewise, FSC and E-PGVR have similar behaviour, because of the volume of vehicles per hour on the roads and the traffic light optimised timing. In the case of E-PGVR, the pheromone levels do not change vehicles route that often. Most vehicles will drive their predetermined route just like in FSC. Young drivers' fuel consumption with our proposed method does not improve between 20% and 60% of senior drivers, compared to all other approaches.

Despite that, fuel consumption is almost constant between 20% and 80% of senior drivers with CoMAAPRAS. The fuel consumption increases as the number of senior drivers increases in all other approaches. In fact, the fuel consumption improves when there are approximately 80% of senior drivers in the scenario with CoMAAPRAS. In this case, 20% of young drivers benefit from re-routed actions and traffic light time extensions.

6.4   Travel Time

CoMAAPRAS performs promising in reducing the travel time for senior drivers in the low-level traffic scenario, as their number increases (Fig. 2d). This reduction starts from the beginning but it begins to improve when there are approximately 60% of senior drivers in the scenario. The travel time is improved by 2.0%, 7.4%, and 11.6% when there are 60%, 80% and 100% of senior drivers in the scenario. However, CoMAAPRAS does not improve the travel time at all in the medium and high-level traffic scenarios, and again, behaves almost the same as CoMASig. The roads controlled by each traffic light have similar speed variability pheromone levels because of the traffic jams. For this reason, CoMAAPRAS only enables the green light extension policy, which is similar to how CoMASig works. E-PGVR and FSC have similar behaviour for the different traffic levels. The reason for this is the traffic light optimised timing in conjunction with the volume of vehicles per hour entering the roads. In the case of E-PGVR, the pheromones levels in the roads do not change much to generate a new route. The majority of vehicles will drive their predetermined route just like in FSC.

7 Conclusions

The growing number of senior drivers introduces new challenges for an urban traffic management system to meet their needs such as smoothing speed variability. In this paper, we addressed this challenge by devising a collaborative multi-agent age-friendly pheromone-based intelligent traffic signal control that autonomously reroutes vehicles and adapts the signal timings of the traffic controller. We evaluated the performance of our approach in speed variability, waiting time, fuel consumption, and travel time. The results reveal improvements in all of these metrics as the number of senior drivers increases in a low-level traffic scenario, while in the medium and high-level scenarios, similar performance is achieved compared to the state of the art baseline.

We assume connected vehicles in our work, but the mixture of connected and non-connected vehicles can be a topic of future work. Another limitation of our work is the fixed threshold for decision making. An adaptive hyper-parameter setting can be addressed in future work to improve the performance in other traffic levels. Finally, we assume only vehicles in our scenario. A mixture of vehicles and pedestrians can be of interest in future work.

References

1. Peace, dignity and equality on a healthy planet. https://www.un.org/en/sections/issues-depth/ageing/. Accessed 19 Sept 2019
2. Bailey, J.M., Golpayegani, F., Clarke, S.: CoMASig: a collaborative multi-agent signal control to support senior drivers. In: 2019 IEEE Intelligent Transportation Systems Conference (ITSC), pp. 1239–1244 (2019)
3. Cao, Z., Guo, H., Zhang, J.: A multiagent-based approach for vehicle routing by considering both arriving on time and total travel time. ACM Trans. Intell. Syst. Technol. **9**(3), 25:1–25:21 (2017)
4. Cao, Z., Jiang, S., Zhang, J., Guo, H.: A unified framework for vehicle rerouting and traffic light control to reduce traffic congestion. IEEE Trans. Intell. Transp. Syst. **18**(7), 1958–1973 (2017)
5. Doroudgar, S., Chuang, H.M., Perry, P.J., Thomas, K., Bohnert, K., Canedo, J.: Driving performance comparing older versus younger drivers. Traffic Inj. Prev. **18**(1), 41–46 (2017)
6. Ebnali, M., Ahmadnezhad, P., Shateri, A., Mazloumi, A., Heidari, M.E., Nazeri, A.R.: The effects of cognitively demanding dual-task driving condition on elderly people's driving performance; real driving monitoring. Accid. Anal. Prev. **94**, 198–206 (2016)
7. Hamidi, H., Kamankesh, A.: An approach to intelligent traffic management system using a multi-agent system. Int. J. Intell. Transp. Syst. Res. **16**(2), 112–124 (2017). https://doi.org/10.1007/s13177-017-0142-6

8. He, Q., Head, K.L., Ding, J.: Multi-modal traffic signal control with priority, signal actuation and coordination. Transp. Res. Part C Emerg. Technol. **46**, 65–82 (2014)
9. Ho, M.C., Lim, J.M.Y., Soon, K.L., Chong, C.Y.: An improved pheromone-based vehicle rerouting system to reduce traffic congestion. Appl. Soft Comput. **84**, 105702 (2019)
10. Hultsch, D.F., MacDonald, S.W.S., Dixon, R.A.: Variability in reaction time performance of younger and older adults. J. Gerontol. Ser. B **57**(2), 101–115 (2002)
11. Jin, J., Ma, X.: Hierarchical multi-agent control of traffic lights based on collective learning. Eng. Appl. Artif. Intell. **68**, 236–248 (2018). https://www.sciencedirect. com/science/article/pii/S0952197617302658
12. Koppel, S., et al.: The driver behaviour questionnaire for older drivers: do errors, violations and lapses change over time? Accid. Anal. Prev. **113**, 171–178 (2018)
13. Krajzewicz, D., Hertkorn, G., Feld, C., Wagner, P.: Sumo (simulation of urban mobility); an open-source traffic simulation, pp. 183–187 (2002)
14. Ng, K., Lee, C., Zhang, S., Wu, K., Ho, W.: A multiple colonies artificial bee colony algorithm for a capacitated vehicle routing problem and re-routing strategies under time-dependent traffic congestion. Comput. Ind. Eng. **109**, 151–168 (2017)
15. Pan, J., Popa, I.S., Zeitouni, K., Borcea, C.: Proactive vehicular traffic rerouting for lower travel time. IEEE Trans. Veh. Technol. **62**(8), 3551–3568 (2013)
16. Raitanen, T., Törmäkangas, T., Mollenkopf, H., Marcellini, F.: Why do older drivers reduce driving? Findings from three European countries. Transport. Res. F Traffic Psychol. Behav. **6**(2), 81–95 (2003)
17. Richerzhagen, B., Stingl, D., Rückert, J., Steinmetz, R.: Simonstrator: simulation and prototyping platform for distributed mobile applications. In: Proceedings of the 8th International Conference on Simulation Tools and Techniques, SIMUTools 2015, pp. 99–108. ICST (Institute for Computer Sciences, Social-Informatics and Telecommunications Engineering), Brussels, BEL (2015)
18. Soon, K.L., Lim, J.M.Y., Parthiban, R., Ho, M.C.: Proactive eco-friendly pheromone-based green vehicle routing for multi-agent systems. Exp. Syst. Appl. **121**, 324–337 (2019)
19. Stipancic, J., Miranda-Moreno, L., Saunier, N., Labbe, A.: Surrogate safety and network screening: modelling crash frequency using GPS travel data and latent Gaussian spatial models. Accid. Anal. Prev. **120**, 174–187 (2018)
20. Stipancic, J., Miranda-Moreno, L., Saunier, N., Labbe, A.: Network screening for large urban road networks: using GPS data and surrogate measures to model crash frequency and severity. Accid. Anal. Prev. **125**, 290–301 (2019)
21. Sullivan, K.A., Smith, S.S., Horswill, M.S., Lurie-Beck, J.K.: Older adults' safety perceptions of driving situations: towards a new driving self-regulation scale. Accid. Anal. Prev. **43**(3), 1003–1009 (2011)
22. Zhou, P., Braud, T., Alhilal, A., Hui, P., Kangasharju, J.: ERL: edge based reinforcement learning for optimized urban traffic light control. In: 2019 IEEE International Conference on Pervasive Computing and Communications Workshops, pp. 849–854 (2019)

An Innovative Employment of the NetLogo AIDS Model in Developing a New Chain Code for Compression

Khaldoon Dhou[1]([⊠])[iD] and Christopher Cruzen[2][iD]

[1] Texas A&M University Central Texas, Killeen, TX 76549, USA
kdhou@tamuct.edu
[2] University of Missouri St. Louis, St. Louis, MO 63121, USA
christopher.cruzen@mail.umsl.edu

Abstract. In this paper, we utilize the NetLogo HIV model in constructing an environment for bi-level image encoding and employ it in compression. Our model considers converting an image into a virtual environment that comprises female agents testing positive and negative for HIV. Female agents are scattered according to the allocation of the pixels in the original images to be tested. The simulation considers introducing male agents that test positive for HIV, the purpose of which is to track their movements while infecting other HIV- female agents. The progressions of the HIV+ male agents within the simulation take advantage of the relative encoding approach previously used by other image processing and agent-based modeling researchers. That is to say, the simulation allows generating a high proportion of similar movement forms that are similarly encoded regardless of the movements of agents. This is followed up by applying Huffman coding to the obtained chains of movement strings for further reduction. The ultimate results reveal that our product could outperform existing benchmarks using all the images we employed in testing.

Keywords: AIDS · NetLogo · Agent-based model · Chain code · Compression · Huffman coding

1 Introduction

The accelerated expansion of computerized information such as digital documents, and data generated from social media, the Internet of Things (IoT), and Smart Cities (SC), has enlarged the demand for exploring new approaches to encode existing information and decrease its initial size. Among the approaches utilized to encode the data is called 'chain coding' that was first introduced by Freeman in 1961 [18]. His approach considers going along the borders of an image or contour to encode information based on 4 or 8 coding directions. Since then, the Freeman approach was subjective to further developments and enhancements, and researchers are still using it as a foundational method to

© Springer Nature Switzerland AG 2021
M. Paszynski et al. (Eds.): ICCS 2021, LNCS 12742, pp. 17–25, 2021.
https://doi.org/10.1007/978-3-030-77961-0_2

design chain coding applications of unique characteristics. For example, Bons and Kegel [6] introduced a new mechanism called the 'Differential Chain Code' (DCC) that encodes based on the differences between adjacent pixels. Hwang et al. [20] improved the DCC by introducing an operator to eliminate the number of codes. Chain coding is widely used for many purposes such as biometric applications [5], character recognition [28], geometric fusion [1], and compression [13, 37].

The literature on image processing has highlighted several developments of chain coding mechanisms. For example, Bribiesca [7] designed the vertex chain code that allows encoding shapes based on the boundary information and uses one of three characters to represent each code. Similarly, Liu and Srinath [25] explored many chain coding techniques to be used in detecting corner information in images. Siddiqi and Vincent [32] present a group of characteristics that belong to bi-level images that represent textual information and used histograms obtained from chain codes for text recognition. Karczmarek et al. [23] examined a chain coding application that considers organizing particular sets of pixels into chunks that describe their surroundings. Many of the chain coding techniques in the literature are associated with other coding techniques such as Huffman or Arithmetic coding (e.g. [22, 24, 27, 36, 39, 40]). These are used to further compress the chains of codes resulting from the application of a particular chain coding method.

One variation of chain coding is encoding the directions based on how relative they are to their adjacent ones and not the actual directions themselves. This variation started with Liu and Žalik chain code [26] that uses the relative angles between neighboring codes. His method employs eight codes and applies Huffman coding to reduce the number of bits used for character representation. Likewise, Zahir and Dhou [38] created a chain code that considers the relative directions between codes and assigns characters to each code regardless of the paths. Additionally, they had a formula to group, eliminate, and combine characters the purpose of which is to reduce the chain length used for representation and to develop a mechanism that allows lossless and lossy compression. Similarly, Zhao et al. [41] introduced a chain code that consists of six directions and is distortion-free when an image is rotated. The authors concluded that their method offered a theoretical foundation to build a 'bead weaving machine'. Although these previous studies provided efficient chain coding mechanisms, the main issue is that they do not seem to allow simultaneous encoding of multiple parts of an image via the introduction of agents.

A new trend in chain coding is the involvement of agent-based modeling simulations to encode image information. Agent-based modeling is an automated approach that is aimed at simulating the interplay between many entities called agents, over a period of time. It is employed by researchers from various domains including marketing, biology, computing, and social sciences (Examples: [2–4, 30]). A well-known platform to develop agent-based modeling simulations is called 'NetLogo' [34]. It is a programming environment that is embedded with a diverse set of models to be harnessed by numerous research domains. Many

NetLogo models such as the ants, paths, Kermack–McKendrick, and bacteria food hunt models inspired the development of new chain coding applications that include integrating behaviors such as the ones that exist in ants, rabbits, dolphins, beavers, and predator-prey systems [9–11,13,15,17,29] within agent-based modeling simulations. The purpose of integrating these behaviors is to identify agents that can encode image information and use the new strings for image representation and compression.

The advantages of utilizing agent-based modeling simulations to encode image information as opposed to classical approaches are threefold: (a) agent-based modeling approaches allow incorporating biological behaviors to guide the process of information encoding. Examples of existing behaviors that were successfully embedded are the pheromone of ants and the echolocation dolphins [10,29]; (b) employing an agent-based modeling simulation allows assigning multiple agents to encode information instead of being encoded from a single location; (c) agent-based modeling simulations give a researcher the flexibility to explore many parameters and introduce new variations, which can be tested on particular datasets. Utilizing agent-based modeling allows exploring a variety of virtual environments that have been explored by researchers from various domains (e.g. [8,12,14,16]). We believe that these features distinguish agent-based modeling simulations and make them attractive to be employed in various data processing applications.

Although there are few existing agent-based models that were successfully applied in encoding image information, and using the generated representations in compression, further research is needed to investigate new behaviors and biological abstractions. In the present article, we develop a new chain code that is stimulated by the NetLogo HIV model [35] with some modifications and simplifications. The focus of the new model is to build an environment that consists of HIV positively and negatively infected female agents that map to the original distribution of pixels. Additionally, our model introduces HIV-infected male agents that are considered dynamic and move around to infect other HIV-negative female agents in the environment. Furthermore, the model assumes that some female agents can reproduce other HIV-infected male agents that also work on image encoding. The newly generated HIV-positive male agents and the ones that originally exist in the system work on encoding information, while their movements are tracked by the algorithm. The final chains are compressed using Huffman coding, and the results are compared with existing benchmarks used by image processing researchers.

2 Method

The existing method comprises a number of steps that involve the environment creation, agent movements, and other encountered scenarios:

- Step 1: The conversion of an image to a contour representation and the construction of the virtual environment based on the allocation of the pixels in the original image. The virtual environment can be imagined as a two-dimensional grid consisting of cells where each cell corresponds to a pixel. While building the virtual environment, each pixel marked with '1' in the actual image is mapped to a cell with an HIV- female in the grid. On the other hand, each pixel marked with '0' is mapped to a cell with HIV+ female. At the end of this step, the virtual environment will have HIV- and HIV+ female agents (Fig. 1). It is essential to mention that all the female agents within the system are stationary. In other words, they do not change their locations during the life of the simulation.

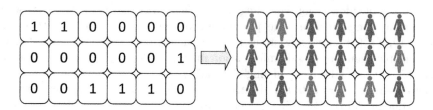

Fig. 1. An example showing the conversion of a sample image to a virtual environment

– Step 2: The inclusion of HIV+ male agents within the environment the purpose of which is to move around and transmit HIV to HIV- females via sexual contacts. These agents highly resemble the one employed in the NetLogo HIV model [35] in terms of their movements and tendency to transmit HIV within a limited human community.

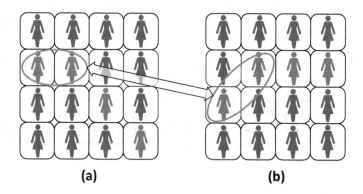

(a)   **(b)**

Fig. 2. Assuming that the first codes (i.e. marked within the blue space) in each of the two sample image segments are similar, the algorithm generates similar codes for each segment regardless of the pixel distribution (Color figure online)

– Step 3: HIV+ male agents begin the search process and their inclination is to identify HIV- female agents. This design serves many purposes: (a) it allows the algorithm to track the needed movements and therefore, generate new chains to be used for representation; (b) it eliminates the movements that do not need to be counted within the string of movements; (c) HIV+ male agents can always consider HIV+ female agents, if necessary (i.e. no HIV- female agents in the neighborhood). Although these movements related to contacts with HIV+ female agents are sometimes necessary to be performed by the algorithm, they are not counted towards the final representation.

– Step 4: The search process involves the employment of a relative coding mechanism that encodes each direction according to the direction that precedes it. The possible encoding directions can be to the right, left, 45° to the right, 45° to the left, and straight movement. In this encoding mechanism, two directions can be different while still encoded similarly (Fig. 2). The advantage of employing a relative coding design lies in the fact that it can generate a large number of similar codes, which can eventually improve the compression ratio as previous studies have shown (e.g. [10, 13, 15]).

(a)   **(b)**

Fig. 3. (a) One connected chain; (b) Two chains and none of them is connected since each contains less than five HIV- female agents that are tied together

– Step 5: The present design assumes that the fifth encountered HIV- female agent within a connected chain has the ability to reproduce an HIV+ male agent to continue working on the encoding process. A connected chain in the present article is defined as five or more HIV- females that are tied together (e.g. Fig. 3).

– Step 6: The Huffman coding algorithm is applied to the final chains that represent the movements of HIV+ male agents for further reduction. The algorithm uses the movements of the agents and the initial coordinates for image reconstruction (i.e. decoding).

3 Results

To assess the effectiveness of the algorithm, we compared our results with standardized benchmarks that are used by image processing researchers: G3, G4, JBIG1, and JBIG2. To this end, we applied our algorithm on eight sample images taken from [42] and compared the size of the chain movements generated by our algorithm with the outcomes of the other algorithms we used for comparison. Table 1 shows the results of a comparison between our algorithm and other existing algorithms. It can be seen from the data in Table 1 that we could outperform all the standardized benchmarks using all the images we utilized in testing. That is to say, for the eight images, our algorithm generated 91052 bits, while G3, G4, JBIG1, and JBIG2 generated 269552, 155584, 112848, and 106904 bits, respectively.

Table 1. The number of bits generated from compressing the chains of movements as opposed to the number of bits generated by other standardized benchmarks: G3, G4, JBIG1, and JBIG2 for a sample of eight images [19,21,31,33,42]

Image	Original size	G3	G4	JBIG1	JBIG2	AIDS
Image 1	35328	10512	4528	4728	4712	**2188**
Image 2	141456	22160	7920	5720	5336	**3430**
Image 3	91008	15888	5792	5024	4888	**2617**
Image 4	414720	102208	81424	62208	58728	**56214**
Image 5	102408	26464	17376	15032	13896	**13718**
Image 6	325120	26704	11936	6312	6104	**3861**
Image 7	544640	45120	16768	7744	7312	**5418**
Image 8	250560	20496	9840	6080	5928	**3606**
Total	**1905240**	**269552**	**155584**	**112848**	**106904**	**91052**

4   Discussion and Conclusion

In this investigation, the aim was to design, and implement an agent-based model of HIV transmission within a social society, to employ it in encoding bi-level image information, and to compress the new chains. The present study extends the previous research studies that explore different behaviors within agent-based modeling environments in encoding image information. The exploration reveals the effectiveness of the current model in encoding image information, and the experimental findings showed that we could outperform well-known standardized benchmarks in bi-level compression.

Our findings suggest that in general, agent-based modeling simulations can be effectively utilized in reducing the size of digital data and change its original representation. The outcomes from the AIDS study are strongly correlated with previous agent-based models that were utilized in image encoding. However, the current study adds a new dimension that shows how to design agent movements that stem from HIV transmission in a society. Furthermore, our investigation offers new insights on agent design and behaviors that can be incorporated for purposes other than building the model itself.

References

1. Al-Asadi, T.A., Witefee, D.M.: Geometrical fusion based on chain code representation. In: IOP Conference Series: Materials Science and Engineering, vol. 928, p. 032033. IOP Publishing (2020)
2. Asgary, A., Cojocaru, M.G., Najafabadi, M.M., Wu, J.: Simulating preventative testing of SARS-CoV-2 in schools: policy implications. BMC Public Health **21**(1), 1–18 (2021)
3. Asgary, A., Najafabadi, M.M., Karsseboom, R., Wu, J.: A drive-through simulation tool for mass vaccination during COVID-19 pandemic. In: Healthcare, vol. 8, p. 469. Multidisciplinary Digital Publishing Institute (2020)
4. Asgary, A., Valtchev, S.Z., Chen, M., Najafabadi, M.M., Wu, J.: Artificial intelligence model of drive-through vaccination simulation. Int. J. Environ. Res. Public Health **18**(1), 268 (2021)
5. Azmi, A.N., Nasien, D., Omar, F.S.: Biometric signature verification system based on freeman chain code and k-nearest neighbor. Multimedia Tools Appl. **76**(14), 15341–15355 (2016). https://doi.org/10.1007/s11042-016-3831-2
6. Bons, J., Kegel, A.: On the digital processing and transmission of handwriting and sketching. Proceedings of EUROCON **77**, 880–890 (1977)
7. Bribiesca, E.: A new chain code. Patt. Recogn. **32**(2), 235–251 (1999)
8. Caci, B., Dhou, K.: The interplay between artificial intelligence and users' personalities: a new scenario for human-computer interaction in gaming. In: Stephanidis, C., et al. (eds.) HCI International 2020 - Late Breaking Papers: Cognition, Learning and Games, pp. 619–630. Springer International Publishing, Cham (2020)
9. Dhou, K., Cruzen, C.: An innovative chain coding technique for compression based on the concept of biological reproduction: An agent-based modeling approach. IEEE Internet Things J. **6**(6), 9308–9315 (2019)
10. Dhou, K., Cruzen, C.: A new chain code for bi-level image compression using an agent-based model of echolocation in dolphins. In: 2020 IEEE 6th International Conference on Dependability in Sensor, Cloud and Big Data Systems and Application (DependSys), pp. 87–91 (2020). https://doi.org/10.1109/DependSys51298.2020.00021
11. Dhou, K.: A novel agent-based modeling approach for image coding and lossless compression based on the wolf-sheep predation model. In: Shi, Y., et al. (eds.) ICCS 2018. LNCS, vol. 10861, pp. 117–128. Springer, Cham (2018). https://doi.org/10.1007/978-3-319-93701-4_9
12. Dhou, K.: Towards a better understanding of chess players' personalities: a study using virtual chess players. In: Kurosu, M. (ed.) HCI 2018. LNCS, vol. 10903, pp. 435–446. Springer, Cham (2018). https://doi.org/10.1007/978-3-319-91250-9_34
13. Dhou, K.: An innovative design of a hybrid chain coding algorithm for Bi-level image compression using an agent-based modeling approach. Appl. Soft Comput. **79**, 94–110 (2019)
14. Dhou, K.: An innovative employment of virtual humans to explore the chess personalities of Garry Kasparov and other class-A players. In: Stephanidis, C. (ed.) HCII 2019. LNCS, vol. 11786, pp. 306–319. Springer, Cham (2019). https://doi.org/10.1007/978-3-030-30033-3_24
15. Dhou, K.: A new chain coding mechanism for compression stimulated by a virtual environment of a predator-prey ecosystem. Future Gener. Comput. Syst. **102**, 650–669 (2020)

16. Dhou, K.: A novel investigation of attack strategies via the involvement of virtual humans: a user study of Josh Waitzkin, a virtual chess grandmaster. In: StephanidisStephanidis, C., et al. (eds.) HCII 2020. LNCS, vol. 12425, pp. 658–668. Springer, Cham (2020). https://doi.org/10.1007/978-3-030-60128-7_48

17. Dhou, K., Cruzen, C.: A highly efficient chain code for compression using an agent-based modeling simulation of territories in biological beavers. Future Gener. Comput. Syst. **118**, 1–13 (2021). https://doi.org/10.1016/j.future.2020.12.016, http://www.sciencedirect.com/science/article/pii/S0167739X20330788

18. Freeman, H.: On the encoding of arbitrary geometric configurations. IRE Trans. Electron. Comput. **2**, 260–268 (1961)

19. Howard, P.G., Kossentini, F., Martins, B., Forchhammer, S., Rucklidge, W.J.: The emerging JBIG2 standard. IEEE Trans. Circ. Syst. Video Technol. **8**(7), 838–848 (1998)

20. Hwang, Y.T., Wang, Y.C., Wang, S.S.: An efficient shape coding scheme and its codec design. In: 2001 IEEE Workshop on Signal Processing Systems, pp. 225–232. IEEE (2001)

21. ISO CCITT Recommend. T.4: Standardization of group 3 facsimile apparatus for document transmission (1980)

22. Jeromel, A., Žalik, B.: An efficient Lossy cartoon image compression method. Multimedia Tools Appl. **79**(1), 433–451 (2020)

23. Karczmarek, P., Kiersztyn, A., Pedrycz, W., Dolecki, M.: An application of chain code-based local descriptor and its extension to face recognition. Patt. Recogn. **65**, 26–34 (2017). https://doi.org/10.1016/j.patcog.2016.12.008, https://www.sciencedirect.com/science/article/pii/S0031320316303971

24. Kim, Y., Kim, K.H., Cho, W.D.: Image compression using chain coding for electronic shelf labels (ESL) systems. IEEE Access **9**, 8497–8511 (2021). https://doi.org/10.1109/ACCESS.2021.3049868

25. Liu, H.C., Srinath, M.: Corner detection from chain-code. Patt. Recogn. **23**(1), 51–68 (1990). https://doi.org/10.1016/0031-3203(90)90048-P, https://www.sciencedirect.com/science/article/pii/003132039090048P

26. Liu, Y.K., Žalik, B.: An efficient chain code with Huffman coding. Patt. Recogn. **38**(4), 553–557 (2005)

27. Lu, C., Dunham, J.G.: Highly efficient coding schemes for contour lines based on chain code representations. IEEE Trans. Commun. **39**(10), 1511–1514 (1991). https://doi.org/10.1109/26.103046

28. Mohamad, M.A., Haron, H., Hasan, H.: Metaheuristic optimization on conventional freeman chain code extraction algorithm for handwritten character recognition. In: Nguyen, N.T., Tojo, S., Nguyen, L.M., Trawiński, B. (eds.) Intell. Inf. Database Syst., pp. 518–527. Springer International Publishing, Cham (2017)

29. Mouring, M., Dhou, K., Hadzikadic, M.: A novel algorithm for bi-level lossless image compression based on ant colonies. In: 3rd International Conference on Complexity, Future Information Systems and Risk, pp. 72–78. Setúbal - Portugal (2018)

30. Najafabadi, M.M.: Modeling an Open Data Ecosystem: The Case of Food Service Establishments Inspection in New York State. State University of New York at Albany (2020)

31. Recommendation T6: Facsimile coding schemes and coding control functions for group 4 facsimile apparatus. International Telecommunication Union, Geneva (1988)

32. Siddiqi, I., Vincent, N.: A set of chain code based features for writer recognition. In: 2009 10th International Conference on Document Analysis and Recognition, pp. 981–985 (2009). https://doi.org/10.1109/ICDAR.2009.136

33. for Standards/International Electrotechnical Commission, I.O., et al.: Progressive bilevel image compression. International Standard 11544 (1993)
34. Wilensky, U.: NetLogo. http://ccl.northwestern.edu/netlogo/, Center for Connected Learning and Computer-Based Modeling, Northwestern University, Evanston, IL (1999). http://ccl.northwestern.edu/netlogo/
35. Wilensky, U.: NetLogo HIV model. Northwestern University, Evanston, IL, Center for Connected Learning and Computer-Based Modeling (1997)
36. Wu, J.T., Ding, J.J.: Improved angle freeman chain code using improved adaptive arithmetic coding. In: 2020 IEEE Asia Pacific Conference on Circuits and Systems (APCCAS), pp. 181–184 (2020). https://doi.org/10.1109/APCCAS50809.2020.9301702
37. Yang, R., Yan, N., Li, L., Liu, D., Wu, F.: Chain code-based occupancy map coding for video-based point cloud compression. In: 2020 IEEE International Conference on Visual Communications and Image Processing (VCIP), pp. 479–482 (2020). https://doi.org/10.1109/VCIP49819.2020.9301867
38. Zahir, S., Dhou, K.: A new chain coding based method for binary image compression and reconstruction. In: Picture Coding Symposium, pp. 1321–1324 (2007)
39. Žalik, B., Mongus, D., Žalik, K.R., Podgorelec, D., Lukač, N.: Lossless chain code compression with an improved binary adaptive sequential coding of zero-runs. J. Vis. Commun. Image Represent. **75**, 103050 (2021). https://doi.org/10.1016/j.jvcir.2021.103050, https://www.sciencedirect.com/science/article/pii/S1047320321000225
40. Žalik, B., Žalik, K.R., Zupančič, E., Lukač, N., Žalik, M., Mongus, D.: Chain code compression with modified interpolative coding. Comput. Electr. Eng. **77**, 27–36 (2019). https://doi.org/10.1016/j.compeleceng.2019.05.001, https://www.sciencedirect.com/science/article/pii/S0045790618327666
41. Zhao, X., Sun, W., Lyu, X., Ma, Z.: On six directions chain code and its application of bead weaving. In: 2018 IEEE 4th International Conference on Computer and Communications (ICCC), pp. 2262–2267 (2018). https://doi.org/10.1109/CompComm.2018.8780823
42. Zhou, L.: A new highly efficient algorithm for lossless binary image compression. ProQuest (2007)

Simulation Modeling of Epidemic Risk in Supermarkets: Investigating the Impact of Social Distancing and Checkout Zone Design

Tomasz Antczak[1], Bartosz Skorupa[1], Mikołaj Szurlej[2],
Rafał Weron[1]([✉]), and Jacek Zabawa[1]

[1] Faculty of Computer Science and Management, Wrocław University of Science
and Technology, 50-370 Wrocław, Poland
{tomasz.antczak,bartosz.skorupa,rafal.weron,jacek.zabawa}@pwr.edu.pl
[2] Faculty of Architecture, Wrocław University of Science and Technology,
50-370 Wrocław, Poland
mikolaj.szurlej@pwr.edu.pl

Abstract. We build an agent-based model for evaluating the spatial and
functional design of supermarket checkout zones and the effectiveness
of safety regulations related to distancing that have been introduced
after the COVID-19 outbreak. The model is implemented in the NetLogo
simulation platform and calibrated to actual point of sale data from
one of major European retail chains. It enables realistic modeling of the
checkout operations as well as of the airborne diffusion of SARS-CoV-2
particles. We find that opening checkouts in a specific order can reduce
epidemic risk, but only under low and moderate traffic conditions. Hence,
redesigning supermarket layouts to increase distances between the queues
can reduce risk only if the number of open checkouts is sufficient to serve
customers during peak hours.

Keywords: Agent-based model · Indoor infection spreading ·
Checkout zone architecture · Decision support · COVID-19 · NetLogo

1 Introduction

The COVID-19 pandemic is unprecedented in contemporary history. In response
to this worldwide threat, governments implemented public health measures
which include isolation/quarantine and social distancing [3]. This has had an
adverse effect on the retail sector. On one hand, it has been badly affected by
the closure of services. On the other, grocery stores have been struggling with
rising demand as consumers tried to stock up for long periods of isolation [8]. To
address concerns, most brick-and-mortar stores have set or have been required
to set restrictions on operating activities (closing of shopping malls, limiting the
number of customers) as well as improved customer safety (installing protective

© Springer Nature Switzerland AG 2021
M. Paszynski et al. (Eds.): ICCS 2021, LNCS 12742, pp. 26–33, 2021.
https://doi.org/10.1007/978-3-030-77961-0_3

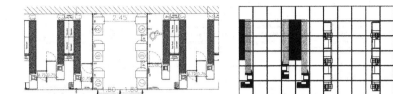

Fig. 1. Blueprint of an actual supermarket checkout zone with six self-service checkouts in the middle and service checkouts to the left and right (*left*), and its representation using NetLogo's square patches of size $1\,m \times 1\,m$ (*right*).

shields and disinfectant dispensers, ensuring distancing). While all these actions lead to incurring costs and negatively impact functionality, their actual value as anti COVID-19 measures is generally unknown.

It is exactly the aim of this study to examine the effectiveness of safety regulations related to distancing that have been introduced in (grocery) supermarkets. These include encouraging customers to keep distance in the queue and closing some checkouts to reduce the congestion in the checkout zone. Although catching the virus from surfaces is plausible, SARS-CoV-2 is predominantly airborne [5]. It is transmitted through the air by people talking, coughing and breathing out small particles called aerosols. To address this problem, we utilize a data-driven agent-based model (ABM) for simulating aligning in queues and movement of customers in the checkout zone introduced in [1] and expand it – by borrowing ideas from [10] – to enable realistic modeling of exhaling, dispersion and absorption of aerosols.

Our model is implemented in the open-source NetLogo simulation platform [11]. Given that ABMs allow designers to evaluate projects to avoid functionally wrong and ineffective solutions [4], our model is not limited to testing the effectiveness of distancing measures for existing checkout zone layouts in supermarkets, but can be also utilized for redesigning them to achieve a desirable customer flow while reducing the risk of infection. This can be achieved by changing the layout, running simulations and measuring if the resulting customer flow has increased and/or the infection risk decreased.

2   Modeling Infection Transmission

The starting point is a recently proposed ABM for simulating customer dynamics in the checkout zone of a (grocery) supermarket [1]. Although customers can be infected at any time, the exposition to potentially hazardous aerosols is the highest at checkout, because of a much higher concentration of people than in other parts of the store. In order to provide a realistic test ground, we calibrate the ABM to actual POS data [2] and consider the checkout zone layout of an existing supermarket. For simplicity and to speed up simulations, the latter is mapped to a square grid with patches of size $1\,m \times 1\,m$, see Fig. 1. Finer spatial granularity can be readily implemented.

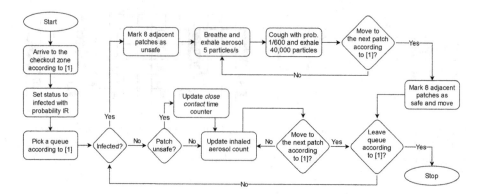

Fig. 2. Flowchart of the simulation steps from the perspective of an agent.

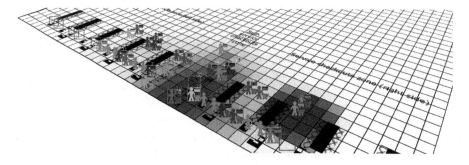

Fig. 3. Visualization of the intensity of aerosol particles at the control height of 1 m. The scale ranges from 0 (white) to 300 particles (dark blue). Two hazardous areas are visible – in the center in the self-service area and in the lower right corner next to one of the service checkouts. Human figures at the checkouts represent cashiers, figures with baskets – customers, and green/red hollow squares – open/closed checkouts. (Color figure online)

Without loss of generality, we assume that agents pick the line with the lowest number of customers in front of them; a scenario labeled #1 in [1]. Again, this can be easily changed to any of the line picking scenarios considered in the cited paper. Moreover, we assume that queues are straight lines and formed along service checkouts or in the axis of the self-service checkout zone.

To evaluate epidemic risk, we expand this ABM to account for the physics of aerosol and droplet dispersion relevant to the hypothesized airborne transmission of SARS-CoV-2. We assume that a hazardous 'close contact' takes place when a susceptible agent is in the immediate proximity of an infected one, i.e., when it occupies one of the 8 squares adjacent to the location of an infected agent, see Fig. 2. Following [10], we let infected individuals release aerosol at a constant rate of 5 particles/s and occasionally cough (on average 6 times/h, each time releasing 40,000 particles). Given that our simulation clock has a granularity of one second, the probability of coughing in each time interval is $6/3600 = 1/600$.

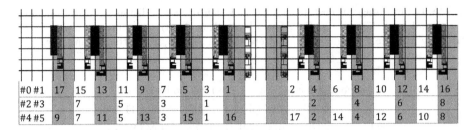

#0 #1	17	15	13	11	9	7	5	3	1		2	4	6	8	10	12	14	16
#2 #3		7		5		3		1				2		4		6		8
#4 #5	9	7	11	5	13	3	15	1	16		17	2	14	4	12	6	10	8

Fig. 4. Opening sequence of the service checkouts in the six considered scenarios. For instance, in scenarios #0 and #1 the checkout to the left of the self-service zone is opened first (1), followed by the checkout to the right (2), then the second checkout to the left (3), etc. Note, that self-service checkouts in the middle are always open.

The aerosol is released at the location of the infected individual into a volume of $1\,\mathrm{m}^3$ at the control height of $1\,\mathrm{m}$. After that, it is immediately diluted via removal and diffusion processes. The removal, e.g., by ventilation or adherence to surfaces, is modeled by a sink term $-c/\tau$ with a timescale of $\tau = 100\,\mathrm{s}$, where c is the amount of aerosol particles. The diffusion is modeled using NetLogo's function 'diffuse' [11], which transfers 5% of aerosol particles within the patch to its eight adjacent patches, see Fig. 3. Finally, susceptible agents accumulate aerosol proportional to their rate of inhalation, on average $0.33\,\mathrm{dm}^3/\mathrm{s}$.

3   Simulation Setup

We consider a supermarket of a large European chain located in southern Poland. The store is equipped with 17 service and 6 self-service checkouts, see Fig. 4, with separate queues to each service and one to the all self-service checkouts. According to standard work rules, all self-service checkouts are constantly open during working hours. The service checkouts are opened dependent on traffic and the availability of cashiers [1]. In the analysed store, as per standard work rules, they were opened in a particular order – starting from the checkout to the left of the self-service zone, followed by the checkout to the right, then the second checkout to the left, etc., see the row labeled '#0 #1' in Fig. 4. Obviously, such rules lead to a congested center of the zone, even under relatively low traffic.

To evaluate the impact of the checkout zone design (here: through the sequence of opening checkouts) and adherence to distancing rules on the risk of infection we consider six scenarios. In the base **scenario #0** standard work rules are in place, all checkouts are available to be opened and safety distances between customers within queues are not respected. **Scenario #1** is a variant of #0 with increased COVID-19 restrictions – markings on the floor make customers keep safety distance within the queues. In **scenarios #2** and **#3**, 50% of service checkouts are permanently closed – every other checkout, including the closest to the self-service checkout zone. In **scenarios #4** and **#5**, the opening sequence first follows the 'every other' rule of #2 and #3, then opens the

Table 1. Percentage of customers with violation of safety distances for all opening hours and peak hours only (i.e., Saturdays 12 pm–4 pm), and three infection rates (IR). By definition there are no violations in scenario #3, hence the missing row.

IR	2.5%	5%	7.5%	2.5%	5%	7.5%
Scenario	All hours			Peak hours		
#0	3.68%	6.76%	9.55%	4.20%	8.18%	11.46%
#1	3.02%	5.67%	7.97%	3.47%	6.45%	9.53%
#2	1.02%	1.92%	2.82%	4.92%	9.40%	13.59%
#4	0.93%	1.75%	2.58%	2.92%	5.57%	8.14%
#5	0.34%	0.72%	1.00%	1.93%	3.78%	5.56%

remaining checkouts starting from the most distant from the center, see Fig. 4. In scenarios #2 and #4 safety distances between customers within queues are not respected, while in #3 and #5 they are.

4   Results

The simulations are conducted for a period of 14 days based on historical POS data regarding customer traffic, basket sizes and cashier availability [2]. Because the number of infected individuals in a population changes dynamically, we consider three levels of the infection rate (IR): 2.5%, 5% and 7.5%. Based on variance stability analysis for the underlying ABM [1, Sect. IV], we perform 10 simulation runs for each scenario and IR level, and average the results.

In Table 1 we report the percentage of customers that violated the safety distance (for any length of time). Note, that by definition, there are no violations in scenario #3, as both horizontal (every second checkout closed) and vertical (markings on the floor) distances between customers exceed one patch or 1.5–2 m measuring from the centers of the occupied patches. This scenario, however, would be difficult to implement in practice due to the significant increase in queue lengths (ca. 1.5 times) and service times (ca. 2 times), and unfeasible during peak hours (i.e., Saturdays 12 pm–4 pm; ca. 7 times longer queues and ca. 15 times longer service times). Out of the remaining ones, scenario #5 yields the lowest values across all infection rates. The percentages are ca. 10 times lower than for the base scenario when all opening hours are considered and over 2 times lower during peak hours. The second best scenario is #4. Scenario #2 trails closely behind, but only for all hours. For peak hours #2 is the worst, even worse than #0. Quite surprising is the advantage of scenario #4 over #1 – both for all and peak hours only – which indicates that avoiding contacts between customers in different queues is more beneficial than contacts within a queue. Of course, only if it is not at the expense of capacity, as in #2.

In the left panels of Fig. 5 we plot the reliability function $R(t)$ of the 'close contact' time t, where $R(t) = 1 - F(t)$ and $F(t)$ is the (empirical) cumulative

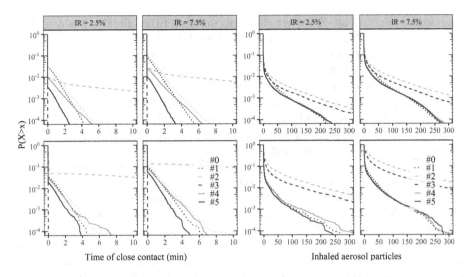

Fig. 5. Reliability function (logarithmic scale) of the 'close contact' time (*left*) and the number of potentially hazardous aerosol particles inhaled by customers (*right*) for all opening hours (*top*) and peak hours only (i.e., Saturdays 12 pm–4 pm; *bottom*) for infection rates IR = 2.5% and 7.5%, and all six scenarios (#0, ..., #5).

distribution function. Clearly, scenario #2 is suboptimal as it yields by far the highest probabilities of very long close contact times, potentially leading to infection. Recall, that by definition, scenario #3 does not allow for 'close contact', hence the vertical reliability function. The drawback is that the much longer queues (see above) lead to larger numbers of inhaled aerosols – scenarios #2 and #3 are much higher than the remaining ones in the right panels of Fig. 5.

Although scenario #5 yields the shortest times of 'close contact', its advantage over the remaining scenarios is not that visible in terms of the number of inhaled particles. For small amounts of absorption (up to ca. 75 particles) and for all opening hours (two upper right panels), strategies that limit close contact between customers in different queues (#4, #5) perform better than the other. However, during peak hours (two lower right panels) the differences nearly disappear as these strategies allow to open adjacent checkouts in case of increased traffic. Interestingly, the numbers of inhaled aerosols are not much higher for peak hours compared to the values obtained for the entire period. This may indicate that – under the considered conditions – the risk of hypothesized aerosol transmission does not increase much with customer traffic.

5 Conclusions

Using an agent-based model able to mimic customer dynamics in the checkout zone [1], expanded to enable realistic modeling of exhaling, dispersion and absorption of potentially hazardous aerosols [10], we evaluate the spatial and functional design of supermarket checkout zones and the effectiveness of safety regulations related to distancing that have been introduced after the COVID-19 outbreak. We find that opening checkouts in a specific order can reduce epidemic risk, but only under low and moderate traffic conditions. During peak hours the differences nearly disappear as strategies #4 and #5 allow to open adjacent checkouts in case of increased traffic. Moreover, scenarios where only every second checkout can be opened turn out to be suboptimal. They increase the social distance and provide a buffer of 1.5–2 m, however, at the cost of significantly increasing the time spent queuing. As a result, the numbers of inhaled aerosol particles are larger than for the other scenarios. Hence, redesigning supermarket layouts to increase distances between the queues can reduce epidemic risk only if the number of open checkouts is sufficient to serve customers during peak hours.

Our model is implemented in the open-source NetLogo simulation platform [11] and can be easily adapted to various situations; the source codes are available from GitHub [1]. In particular, it can be utilized for optimizing supermarket layout, similarly as in [9] for logistic warehouses, or increasing the social-spatial comfort of the customers [7]. We are also aware of the limitations of our study. Our model does not take into account aerodynamic disturbances which may increase the range of airborne droplets [6] nor utilize a realistic collision-free velocity model for pedestrian dynamics [12]. Nevertheless, we believe that it may be used as a decision support tool for architects (re)designing supermarkets and for managers making staffing decisions and planning checkout operations.

References

1. Antczak, T., Weron, R., Zabawa, J.: Data-driven simulation modeling of the checkout process in supermarkets: insights for decision support in retail operations. IEEE Access **8**, 228841–228852 (2020)
2. Antczak, T., Weron, R.: Point of sale (POS) data from a supermarket: transactions and cashier operations. Data **4**(2), 67 (2019)
3. Chakraborty, I., Maity, P.: Covid-19 outbreak: migration, effects on society, global environment and prevention. Sci. Total Environ. **728**, 138882 (2020)
4. Chen, L.: Agent-based modeling in urban and architectural research: a brief literature review. Front. Archit. Res. **1**(2), 166–177 (2012)
5. Lewis, D.: Mounting evidence suggests coronavirus is airborne - but health advice has not caught up. Nature **583**(7817), 510–513 (2020)
6. Li, H., Leong, F., Xu, G., Kang, C., Lim, K., et al.: Airborne dispersion of droplets during coughing: a physical model of viral transmission. Sci. Rep. **11**, 4617 (2021)
7. Nguyen, B., Wang, T.H., Peng, C.: Integration of agent-based modelling of social-spatial processes in architectural parametric design. Archit. Sci. Rev. **63**(2), 119–134 (2020)
8. Pantano, E., Pizzi, G., Scarpi, D., Dennis, C.: Competing during a pandemic? retailers' ups and downs during the COVID-19 outbreak. J. Bus. Res. **116**, 209–213 (2020)
9. Ribino, P., Cossentino, M., Lodato, C., Lopes, S.: Agent-based simulation study for improving logistic warehouse performance. J. Simul. **12**(1), 23–41 (2018)

10. Vuorinen, V., Aarnio, M., Alava, M., Alopaeus, V., Atanasova, N., et al.: Modelling aerosol transport and virus exposure with numerical simulations in relation to SARS-CoV-2 transmission by inhalation indoors. Saf. Sci. **130**, 104886 (2020)
11. Wilensky, U., Rand, W.: An Introduction to Agent-Based Modeling: Modeling Natural, Social, and Engineered Complex Systems with NetLogo. MIT Press, Cambridge (2015)
12. Xu, Q., Chraibi, M.: On the effectiveness of the measures in supermarkets for reducing contact among customers during COVID-19 period. Sustainability **12**, 9385 (2020)

A Multi-cell Cellular Automata Model of Traffic Flow with Emergency Vehicles: Effect of a Corridor of Life

Krzysztof Małecki[1]([✉]) (iD), Marek Kamiński[2], and Jarosław Wąs[3] (iD)

[1] West Pomeranian University of Technology, Żołnierska 52 Street, Szczecin, Poland
kmalecki@wi.zut.edu.pl
[2] Profi-Data Sp. z. o. o, Milczańska 30A Street, Szczecin, Poland
[3] AGH University of Science and Technology, Mickiewicza 30 Street, Krakow, Poland
jarek@agh.edu.pl

Abstract. There are various macroscopic and microscopic road traffic models that allow traffic flow analysis. However, it should be emphasized that standard traffic flow models do not include emergency vehicle traffic. We propose a multi-agent microscopic model for analyzing traffic flow of emergency vehicles with some limitations to the distance between vehicles and their proper distribution (corridor of life) to leave free passage for a privileged vehicle. Real data was used to calibrate and validate the model. Our simulation studies show the importance of certain aspects of road traffic (distance between vehicles, corridor of life, size and type of roadside, friction conflict, etc.) in order to increase/decrease the traffic flow in the aspect of an approaching of emergency vehicle.

Keywords: Agent-Based Modeling (ABM) · Cellular Automata (CA) · Traffic modeling · Emergency vehicles · Friction conflict

1 Introduction

The rapid arrival of emergency services such as the fire brigade, ambulance or police to the places of dangerous incidents is crucial and often seconds decide about the life or health of people. In the urban areas, but also on highways or tunnels, traffic jams and high traffic density often occur. In such situations, leaving the corridor of life is extremely important behavior.

There are many different models of car traffic. Regarding microscopic approach one can point out classical models, including on the one hand, continuous ones like Intelligent Driver Model - a time-continuous car-following model [1]. On the other hand there are discrete models like classical ones: Nagel-Schreckenberg model [2] and Chopard-Luthi-Queloz traffic model [3], where CA paradigm was applied. Although these models are relatively simple they allow for analysis of different relationships [4–8], etc. One can also identify a trend where drivers/cars are represented as agents and their possess different abilities [9].

© Springer Nature Switzerland AG 2021
M. Paszynski et al. (Eds.): ICCS 2021, LNCS 12742, pp. 34–40, 2021.
https://doi.org/10.1007/978-3-030-77961-0_4

The emergency cars simulation was performed in [10]. The author confirmed that the mean speed of the emergency car and its arrival time all depend enormously on the cars density, the route length of the emergency vehicle and the turn capability of the cars. However, following the [11] one can point that the majority of studies based on emergency cars have deals with such problems as: emergency vehicles location problem [12], emergency vehicles dispatching problem [13], and emergency vehicles routing problem which consists of finding a shortest (fastest) path from one location to another [14]. The results of prediction of travel time of the priority vehicles are also available [15, 16].

In this article, the authors proposed a multi-agent multi-cell CA model to study the impact of drivers' behavior on the road on the time of an emergency vehicle moving. The novelty of this work is to develop a CA model that allows vehicles to move sideways. This was achieved by mapping each lane of the road as a grid of small cells that enables this effect. It should be stressed that standard traffic flow models do not take into account the movement of emergency vehicles, however it is highly required to consider such microscopic models in process of stationing deployment of emergency vehicles.

2 Proposed Model

2.1 Road Structure, Types of Vehicles and Real Data

A one-way two-lane road has been mapped. Vehicles move in two directions - forward and sideways. To get high accuracy simulation, the road was divided into many small cells. The grid of a CA consists of two lanes of road divided into cells of 1 m long and 0.25 m wide (Table 1).

Table 1. Road width (according to regulations in Poland [17].

Road width [m]	Road width [cells]	Minimum roadside width [cells]
2.5	10	10
3.0	12	6
3.5	14	2
3.75	15	0

To avoid a situation in which the emergency vehicle traffic would be permanently blocked due to insufficient road width, a roadside was used. The size of the roadside is adapted to the width of the road so that the total width of the road is at least equal to the sum of the width of two trucks and a firefighter car. In the developed system, however, it is possible to widen the roadside in the range of 1 to 5 CA cells on one side of the road.

A few type of vehicles are considered (Table 2). The developed model uses small CA cells to allow the vehicles to move sideways. However, the larger the

Table 2. Types of vehicles (based on real sizes taken from vehicles' catalog).

Type of vehicle	Size (length; width) [m]	Size (length; width) [CA cells]
Ambulance	6; 2	6; 8
Firefighter car	9; 2.5	9; 10
Truck	18; 2.5	18; 10
Small car	3; 1.5	3; 6
Medium car	4; 1.75	4; 7
Big car/Bus	5; 2	5; 8

vehicle is, the more difficult it is to maneuver. Therefore, a number of empirical measurements were carried out, on the basis of which the possible maximum lateral shift after a certain distance forward was determined (Fig. 1).

Fig. 1. Some examples of empirical data taken for different types of vehicles.

2.2 The Corridor of Life

The so-called corridor of life is a free space on the road created for emergency vehicles, by stopping vehicles standing in a traffic jam at the edge of the road. Such solution has been implemented in the developed model (Fig. 2). When there is a situation that the vehicle is forced to reduce its speed, e.g. due to coming to a traffic jam, it will start the descent towards the appropriate edge of the road, thus creating a free passage in the middle of the road. If curb is chosen as the roadside type, then the vehicles will be as close as possible to it first. However, if an emergency vehicle appears and it does not have enough space to continue its travel, then the other vehicles will start to enter the curb at a speed of 1 cell/iteration.

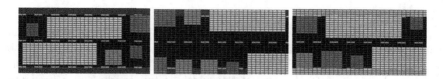

Fig. 2. A traffic jam without a corridor of life (top), with corridor of life (in the middle), and with corridor of life on the road with curbs (bottom)

2.3   Safe Distance Between Cars and the Friction Conflict

It is good practice to maintain such a distance from the vehicle in front of you that, in the event of an unexpected traffic situation, ensure sufficient time and room to react. In the system the safe distance can have values from 0 to 5 CA cells (from 0 to 5 m) and it is one of observed parameter.

In a situation where cars in traffic jams leave a little space for an emergency vehicle, the driver of such vehicle cannot drive at high speed for fear of collision with other vehicles. An emergency vehicle must significantly reduce the speed of the journey. This situation is referred to as a friction conflict, i.e. an emergency vehicle moves between other vehicles, from which it will not be separated even by a single CA cell of free space. Based on observation, we assume that then the maximum speed is automatically limited to 5 cells/iteration (about 18 km/h).

2.4   The Rules of Traffic Movement

The Rules of Emergency Vehicles. In order to properly describe a movement of each type of analysed vehicle, a set of possible actions should be established.

1. Acceleration: $v_i(t) < v_{max} \rightarrow v_i(t+1) = v_i(t) + 1$, where $v_i(t)$ - speed of vehicle in i-cell in time t, v_{max} - the maximum possible speed.
2. The best route - specification of the side shift value (s_{side}), according to the developed algorithm.
3. Breaking - here are two stages:
   (a) based on step2 with formula as follows:

$$s_{side} \neq 0 \rightarrow v_i(t) = \begin{cases} 0 & \text{if } s = 0 \\ max(1, min(s, \sqrt{2sa_b})) & \text{if } s > 0 \end{cases}, \qquad (1)$$

   where $v_i(t)$ - speed of vehicle in i-cell in time t, s - the distance to the nearest obstacle, a_b - a traffic delay factor, s_{side} - the value of side shift.
   (b) The second stage only occurs when the passage is too tight and there is so-called a friction conflict, this is illustrated by the following relationship: $v_i(t) > 5 \rightarrow v_i(t+1) = 5$
4. Identification of the area in which the emergency vehicle's signals are 'heard' or 'visible'.
5. Shift - after completing all previous steps, the vehicle already has a new speed specified and must change its position at $t+1$ according to its value.

The Rules of Classical Vehicles. The subsequent states of classical vehicles are determined as follows:

1. Acceleration: $v_i(t) < v_{max} \rightarrow v_i(t+1) = v_i(t) + 1$, where $v_i(t)$ - speed of vehicle in i-cell in time t, v_{max} - the maximum possible speed.

2. Braking - based on the distance of the car to the nearest vehicle and is expressed by the following formula:

$$v_i(t) = \begin{cases} 0 & \text{if } s = 0 \\ max(1, min(s, \sqrt{2sa_b})) & \text{if } s > 0 \end{cases},$$ (2)

where $v_i(t)$ - speed of vehicle in i-cell in time t, s - the distance to the nearest obstacle, a_b - a traffic delay factor.

3. If the emergency vehicle is approaching, then perform the first and second steps again.
4. Random event - the simulation of unexpected events on the road (pedestrian intrusion, sudden braking in front of an obstacle, etc.). This step is expressed by the following formula: $v_i(t) > 0 \wedge P(t) < p \rightarrow v_i(t+1) = v_i(t) - 1$, where $v_i(t)$ - speed of vehicle in i-cell in time t, $P(t)$ - the value of the random variable at time t, p - the probability of a random event.
5. If it is necessary, try to turn off the road to make place for an emergency vehicle.
6. Consider if being in the signal area of the emergency vehicle
7. Shift - after completing all previous steps, the vehicle already has a new speed specified and must change its position at $t+1$ according to its value.

3   Results

Numerical tests were conducted for two types of emergency vehicles - ambulance and fire brigade. The system was upgraded every 1 s. The results are the arithmetic mean of 100 simulations and show the delay of the emergency vehicle in relation to the time that this vehicle could pass the road while it was traveling with the maximum speed allowed. The road density was assumed in the range from 10% to 20% with a 0.2% step and the distance between vehicles standing in the traffic jam in the range from 0 m to 5 m.

The analysis shows the correlation between the value of delay and the distance between standing vehicles (Fig. 3). The greater the distance left between the vehicles is, the faster the emergency vehicle is able to cross the road with traffic jam. In addition, narrow roads with a single lane width of 2.5 m or 3.0 m generate a much greater delay compared to roads with a single lane width of above 3.5 m. In contrast to the charts for the ambulance, the increase of road density causes a worsening of the travel time, no matter how big the distance between standing cars is. In addition, for lanes with a width of 3 m and 3.5 m, the delay is close to the value of the ambulance delay on the road with a lane width of 2.5 m. This is related to the larger dimensions of the fire brigade vehicle in relation to the ambulance. In the case of a vehicle as large as the fire brigade, the benefits resulting from the increased road width seem to be noticeable only for a road with a width of 3.75 m.

The simulation results show that the lack of free space between vehicles standing in a traffic jam makes it impossible to perform maneuvers to give way to an emergency vehicle. This takes on special importance when the emergency

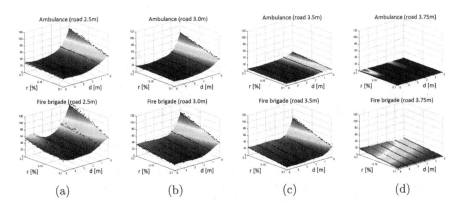

Fig. 3. Ambulance (the first row) and fire brigade (the second row) delays on the road with two lanes (a) 2.5 m wide each (b) 3.0 m wide each, (c) 3.5 m wide each, and (d) 3.75 m wide each, depending on the road density - r and distance between vehicles - d.

vehicle is a fire vehicle, i.e. a vehicle with a much larger size than a police car or ambulance. The biggest differences are noticeable in the vicinity of 20% of the density and distance at the level of 0 m. It can be concluded that in a certain range of values, the negative effect of compaction growth can be offset by the increased distance between vehicles. In this case, it can be assumed that if drivers observe increased traffic flow on the road, they should keep a greater distance between themselves. In the analyzed cases, the fire brigade vehicle again had a greater delay in transit due to the increased its dimensions and the occurring friction conflict.

4   Conclusions

The most important conclusions from the work carried out can be formulated as follows: (1) The proposed microscopic approach in traffic modeling, taking into account the detailed geometry of the roads and the specificity of drivers' behavior in a given area, is useful in the process of stationing deployment of emergency vehicles. (2) The use of small cells of cellular automaton enables efficient modeling of lateral shifts and frictional conflicts. (3) Leaving a greater distance between vehicles immobilized in a traffic jam has a positive effect on better maneuverability and thus less delay in the passage of a priority vehicle. (4) Potential friction conflicts can be predicted on the basis of maps. Geometry of particular roads, as well as behaviors of other road users are crucial factors in planning the arrival time of priority vehicles on a specified route.

Acknowledgement. The authors would like to thank Daniel Budka for helping in the field experiments and MiraTrans Transport i Spedycja Sp. z o.o. for lending a truck with a driver to carry out measurements in order to calibrate the developed model.

References

1. Treiber, M., Hennecke, A., Helbing, D.: Congested traffic states in empirical observations and microscopic simulations. Phys. Rev. E **62**, 1805–1824 (2000)
2. Nagel, K., Schreckenberg, M.: A cellular automaton model for freeway traffic. J. Phys. I France **2** (1992)
3. Chopard, B., Luthi, P.O., Queloz, P.A.: Cellular automata model of car traffic in a two-dimensional street network. J. Phys. A Math. Gen. **29**, 2325–2336 (1996)
4. Schadschneider, A., Schreckenberg, M.: Traffic flow models with 'slow-to-start'rules. Annalen der Physik **509**, 541–551 (1997)
5. Nagel, K., Wolf, D.E., Wagner, P., Simon, P.: Two-lane traffic rules for cellular automata: a systematic approach. Phys. Rev. E **58**, 1425 (1998)
6. Rickert, M., Nagel, K., Schreckenberg, M., Latour, A.: Two lane traffic simulations using cellular automata. Physica A Stat. Mech. Appl. **231**, 534–550 (1996)
7. Liu, M., Shi, J.: A cellular automata traffic flow model combined with a bp neural network based microscopic lane changing decision model. J. Intell. Transp. Syst. **23**, 309–318 (2019)
8. Małecki, K.: The use of heterogeneous cellular automata to study the capacity of the roundabout. In: Rutkowski, L., Korytkowski, M., Scherer, R., Tadeusiewicz, R., Zadeh, L.A., Zurada, J.M. (eds.) ICAISC 2017. LNCS (LNAI), vol. 10246, pp. 308–317. Springer, Cham (2017). https://doi.org/10.1007/978-3-319-59060-8_28
9. Chmielewska, M., Kotlarz, M., Was, J.: Computer simulation of traffic flow based on cellular automata and multi-agent system. In: Wyrzykowski, R., Deelman, E., Dongarra, J., Karczewski, K., Kitowski, J., Wiatr, K. (eds.) PPAM 2015. LNCS, vol. 9574, pp. 517–527. Springer, Cham (2016). https://doi.org/10.1007/978-3-319-32152-3_48
10. Moussa, N.: Modeling evacuation of emergency vehicles by cellular automata models. Int. J. Mod. Phys. C **19**, 947–956 (2008)
11. Fancello, G., Mancini, S., Pani, C., Fadda, P.: An emergency vehicles allocation model for major industrial disasters. Transp. Res. Procedia **25**, 1164–1179 (2017)
12. Bélanger, V., Ruiz, A., Soriano, P.: Recent optimization models and trends in location, relocation, and dispatching of emergency medical vehicles. Eur. J. Oper. Res. **272**, 1–23 (2019)
13. Sudtachat, K., Mayorga, M.E., Chanta, S., Albert, L.A.: Joint relocation and districting using a nested compliance model for ems systems. Comput. Ind. Eng. **142**, 106327 (2020)
14. Zhao, H., Li, J., Nie, C.: Cellular automaton models for traffic flow considering opposite driving of an emergency vehicle. Int. J. Mod. Phys. C **26**, 1550079 (2015)
15. Piórkowski, A.: Construction of a dynamic arrival time coverage map for emergency medical services. Open Geosci. **10**(1), 167–173 (2018)
16. Płokita, I., Piórkowski, A., Lupa, M.: Comparative analysis of algorithms for calculating arrival times of emergency vehicles. Geoinformatica Polonica **2016**, 85–91 (2016)
17. Dz, U.: Announcement of the Minister of Infrastructure and Construction of the Republic of Poland of 23 december 2015 on The Technical Conditions to be met by Public Roads **124**, 9–10 (2015)

HSLF: HTTP Header Sequence Based LSH Fingerprints for Application Traffic Classification

Zixian Tang[1,2], Qiang Wang[1,2], Wenhao Li[1,2], Huaifeng Bao[1,2], Feng Liu[1,2], and Wen Wang[1,2(✉)]

[1] State Key Laboratory of Information Security, Institute of Information Engineering, CAS, Beijing, China
{tangzixian,wangqiang3113,liwenhao,baohuaifeng,liufeng,wangwen}@iie.ac.cn
[2] School of Cyber Security, University of Chinese Academy of Sciences, Beijing, China

Abstract. Distinguishing the prosperous network application is a challenging task in network management that has been extensively studied for many years. Unfortunately, previous work on HTTP traffic classification rely heavily on prior knowledge with coarse grained thus are limited in detecting the evolution of new emerging application and network behaviors. In this paper, we propose HSLF, a hierarchical system that employs application fingerprint to classify HTTP traffic. Specifically, we employ local-sensitive hashing algorithm to obtain the importance of each field in HTTP header, from which a rational weight allocation scheme and fingerprint of each HTTP session are generated. Then, similarities of fingerprints among each application are calculated to classify the unknown HTTP traffic. Performance on a real-world dataset of HSLF achieves an accuracy of 96.6%, which outperforms classic machine learning methods and state-of-the-art models.

Keywords: HTTP header fields · Application traffic classification · Local-sensitive hashing · Traffic fingerprinting

1 Introduction

Network traffic classification is an important task in network management. With the vigorous development of information technology, numerous applications flush into terminal devices, which results in the increasingly complicated network traffic. In order to manage network more effectively, a powerful network traffic classification solution needs to be implemented. Although HTTPS has been popularized in recent years, we found that HTTP is still one of the most mainstream network protocols. Many applications implement client-server communications based on HTTP, so that the importance of the HTTP protocol has not diminished.

HTTP headers play a very important role in application traffic. The headers are flexible to design according to developers' needs. For example, in a POST

© Springer Nature Switzerland AG 2021
M. Paszynski et al. (Eds.): ICCS 2021, LNCS 12742, pp. 41–54, 2021.
https://doi.org/10.1007/978-3-030-77961-0_5

request, Content-Type is often designed to indicate the form of payload submitted by the client. The server will perform different processing strategies for various types of data according to the value of Content-Type. The design of header fields have their specific meanings which directly reflects the developer's thinking in communication module development. Many researchers have conducted in-depth research on HTTP header fields. However, their research on header fields are one-sided that only focuses on limited fields rather than delving into all of them.

In this paper, we propose HSLF, a system that transforms HTTP traffic into fingerprint and classifies application traffic. HSLF generates an HTTP Header Sequence (HHS) based on HTTP headers. The sequence contains all header fields. For each HHS, we perform a Local-sensitive hashing algorithm to convert each sequence into a simple fingerprint. In this way, the sessions sent by various applications are saved as fingerprints with ground-truth labels. Through a large collection of traffic fingerprints, we can gradually describe the features of an application's HTTP traffic, and store fingerprints in a massive database that is continuously updated. We implement application identification by measuring the similarities between labeled fingerprints and fingerprints to be identified. The experimental results indicate that, HSLF outperforms classic machine learning models and two state-of-the-arts in four metrics.

In general, our main contributions can be concluded as follows:

- We comprehensively analyze most of HTTP headers and rationally assessed their importance based on machine learning algorithm.
- We build a dynamically updated fingerprinting database based on LSH algorithm which can be applied for measuring the similarities between HTTP sessions.
- To the best of our knowledge, we are among the first to propose a method that includes all HTTP headers. Evaluated on real-world datasets, HSLF achieves eye-catching performance and is superior to the state-of-the-arts.

2   Background and Related Work

2.1   Traffic Classification Solutions

Many previous traffic classification methods have some limitations that prevent them from being useful in real-time environment. These methods can be divided into rule-based matching and machine learning-based methods. Rule matching methods are mainly based on TCP quintuples (Source IP, Destination IP, Source port, Destination port, protocol) [5] or payload. TCP quintuples based methods may fail for some applications that use ephemeral port allocation or CDN policies. Crotti et al. [3] proposed a flow classification mechanism based on statistical properties of the captured IP packets. However, statistical features based methods are difficult to classify applications accurately. Mainstream payload based methods like Deep Packet Inspection (DPI) [10] and signature seeking are not only infringe on the privacy of end users, but also unsuitable for real-time traffic analysis due to the high resource overhead. Another popular methods are machine learning based models, they train their classifier based on some statistical features in application-generated traffic [12,15,19]. However, due to the limitations of machine learning, it cannot adapt to the growing traffic scale. These methods are far behind network management requirements. In addition to the limitations of the method itself, many traditional classification methods are coarse-grained that include limited categories (e.g. email, news, and games) [8,11]. There are also many methods using fingerprinting algorithm [4,18,22] for traffic classification, and they are proved successful in real environments.

2.2 Research on HTTP Headers

Many researchers have conducted in-depth research on HTTP header fields. Kien Pham et al. [16] studied the $UserAgent$ and identified web crawler traffic based on this field. Fei Xu et al. [20] discovered the $ContentType$ inconsistency phenomenon and used the inconsistency to detect malwares. Arturs et al. [9] and William J. Buchanan1 et al. [1] investigated the use of security headers on the Alexa top 1 million websites to conduct security assessments.

Xu et al. [21] proposed a method that was using the identifier contained in the $UserAgent$ of the HTTP header to distinguish mobile applications. Yao et al. [22] proposed SAMPLES, which automatically collect conjunctive rule corresponds to the lexical context, associated with an application identifier found in a snippet of the HTTP header. However, previous researches pay attention to limited header fields, and few works can adapt to the real environment which contains complex combinations of HTTP header fields.

3 Preprocessing

In this section, we introduce some preprocessing steps. We investigate HTTP headers to determine the weight of each header field. Then we introduce HTTP Header Sequence (HHS), which is used to create a standardized template for an HTTP session.

3.1 Weight Assignment

Different headers contribute differently in application identification. For example, the field $Content - Type$ reflects the MIME type of the subsequent payload, that is, the data processing format, which directly reflects the data type passed in the application; while the field $Date$ represents the time at which the message was originated. It indicates the creation time of the message. Obviously, the contribution of $Content - Type$ to traffic classification is much greater than that of $Date$.

In order to maximize the value of each field, we design the weighting algorithm of the HTTP headers. We perform machine learning methods to train fields to study the influence of headers. Some machine learning models have mechanisms for scoring features which makes them easier to be applied to feature selection tasks. We select 10 typical application including browsing, chat, video etc. By one-hot encoding and standardizing all the header fields, the headers can be learned by the machine learning model.

The machine learning model we choose is Random Forest (RF). RF consists of multiple decision trees. Each node in the tree is a condition about a certain feature, in order to divide the data into two according to different response variables. The node can be determined by the impurity (optimal condition). For classification purposes, Gini Impurity or information divergence is usually used. When training a decision tree, we can calculate how much the tree's impurity is reduced by each feature. For a decision tree forest, it is possible to figure out the average reduction of each feature which can be regarded as the value of feature importance.

The feature score we choose is Gini Impurity (Gini Index). Select a header field F and count the sum of the Gini index (GI) decline degree (or impurity decline degree) of the branch nodes formed by t in each tree of RF. It is defined as:

$$GI(t) = 1 - \sum_{k=1}^{K} p_{k|t}^2 \tag{1}$$

k indicates that there are k categories, and $p_{k|t}$ indicates the proportion of the category k in the node t.

The importance of field F_j at node t ($I_{j|t}$), that is, the amount of Gini index change before and after the node T branch is computed as:

$$I_{j|t} = GI(t) - GI(l) - GI(r) \tag{2}$$

where $GI(l)$ and $GI(r)$ respectively represent the index of F_j on the left and right node. Then, add up the importance of F_i in the whole forest (m trees in) as:

$$I_j = \sum_{i=1}^{m} \sum_{t \in T} I_{j|t} \tag{3}$$

Finally, we normalize all I_j to get each field's final importance (i_j) score. The normalization process is computed as:

$$i_j = \frac{I_j}{\sum_{i=1}^{n} I_i} \tag{4}$$

Importance i indicates the field F's contribution to application classification. We calculate i of all the fields for weight assignment. The results are shown in Table 1. We treat the importance of headers as weights.

Table 1. Importance Ranking list of HTTP headers (Top 20)

Rank	Header	Importance	Rank	Header	Importance
1	User–Agent	0.10407	11	Accept-Language	0.04164
2	Server	0.09036	12	Accept-Encoding	0.0337
3	Path	0.0864	13	Upgrade	0.02585
4	Connection	0.08349	14	Pragma	0.02216
5	Cache-Control	0.06304	15	Date	0.02192
6	Content-Type	0.06136	16	Expires	0.02158
7	Host	0.05998	17	Method	0.01954
8	Accept	0.05816	18	Referer	0.01689
9	Content-Length	0.05671	19	Status-Line	0.01645
10	Cookie	0.04616	20	Last-Modified	0.01463

3.2 HTTP Header Sequence

We define two types of headers: character headers and numeric headers. The character headers are mainly composed of character strings, while a numeric header contains a definite value. We propose different approaches to deal with these two type of headers. For numeric headers, we keep the value directly. For a character header, if the form of characters in this field is relatively simple, such as a Server header "nginx", we retain the original character strings. If the character field is more complicated, such as the *Host* field which will appear with multiple directory separators, we further split the string according to the delimiter. Taking "pan.baidu.com" for example, we split it into "pan", "baidu", "com". Such segmentation is conducive to extracting as many features of long characters as possible.

Meanwhile, there may be multiple request and response exchanges in a session, each header may appear more than once. In most cases, values of a header are constant. However, there are also cases where values are different and cannot be ignored. So we choose to generate a first draft of this session $S = \{F_1 : k_1; F_2 : k_2; ...F_n : k_n\}$ based on the first request and response pair. If the k_i' in the subsequent data is not the same as the k_i in S, then we append k_i' to F_i to update S as $S = \{f_1 : k_1; f_2 : k_2; ...f_i : k_i, k_i'; ...f_n : k_n\}$. Then we allocate the weights achieved in Sect. 3 to each F to form a complete sequence (HHS) for the HTTP session $S = \{F_1(w_1) : k_1; F_2(w_2) : k_2; ...F_i(w_i) : k_i, k_i'; ...F_n(w_n) : k_n\}$.

4   HTTP Header Sequence Based LSH Fingerprints

HSLF is a hierarchical system shown in Fig. 1. In this section, we will present HSLF in detail. We first introduce the proposed fingerprinting algorithm developed from SimHash [2] to generate the fingerprint of HHS. Based on it, we establish a fingerprint database which enable the system to identify applications in HTTP traffic.

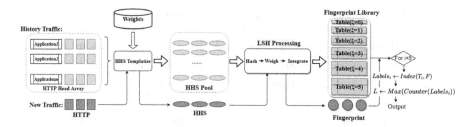

Fig. 1. The overview of HSLF framework.

4.1 LSH Fingerprint

Although we have made predictions through the RF model in Sect. 3 and obtained some successful predictions, general machine learning algorithms cannot adapt to large-scale data classification. In real environments, there are far more than 10 categories. If we use machine learning algorithms for modeling and prediction, we need to learn the characteristics of each application, in which case the consumption of resources will multiply and the increased categories will also decrease the accuracy of model predictions. When we increased the number of applications from 10 to 50, the classification accuracy dropped from 95.8% to 87.2%. Obviously, such accuracy is far from satisfactory for application identification. In addition, it is also difficult to standardize the headers of the string type when modeling machine learning. The efficiency of standardization also directly affects the prediction performance. Therefore, we propose to employ fingerprinting algorithm to describe application traffic.

Local-sensitive hashing (LSH) [6] is an important method with solid theoretical basis for measuring text similarity. It is also widely used in the nearest neighbor search algorithm. The traditional Hash algorithm is responsible for mapping the original content as uniformly and randomly into a signature, which is equivalent to a pseudo-random number generation algorithm in principle. The traditional hash algorithm no longer provides any information except that the original content is not same. The two signatures generated by it may be very different, even if the original content differs by only one byte. Therefore, the traditional hash cannot measure the similarity of the original content in the dimension of the signature. However, the hash signature generated by LSH can represent the similarity of the original content to a certain extent. Its main application is to mine similar data from massive data, which can be specifically applied to text similarity detection, web search and other fields. SimHash is one of LSH algorithms proposed by Charikar [2], and it has been widely used in data retrieval and near-duplicates detection. It maps high-dimensional feature vectors to fixed-length binary bit strings.

The HHS is essentially a high-dimensional vector composed of natural language. Each sequence represents a network communication of an application. In order to identify applications on HTTP traffic, we propose the HHS based

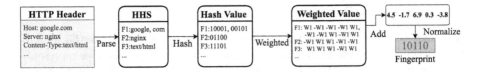

Fig. 2. Example of fingerprint generating.

SimHash algorithm to generate a 64-bit fingerprint for each HTTP session. The details of fingerprinting are as follows.

Templatizing. Templatizing module is to transform an HTTP session into HHS like $\{F_1(w_1) : k_1; F_2(w_2) : k_2; ...; F_i(w_i) : k_i, k_i'; ...; F_n(w_n) : k_n\}$, and saves them in the pool.

Hash. In order to get a fingerprint that can be used to summarize the session and save the characteristics of the session, we utilize the hash algorithm independently on each k_i. Then we add weights and merge these irrelevant hash values to process the dimensionality reduction into the fingerprint format for normal storage. The hash algorithm we choose is MD5 algorithm. The first 64 bits of MD5 are preserved as the result of a hash operation. Since the MD5 algorithm is a classic algorithm, we will not describe it in detail here. k_i's hash result is $h_i = Hash_{MD5}(k_i)$. As we have highly divided the string in the foregoing, the local hash algorithm will not have an avalanche effect on the entire sequence.

Integrate. After calculating the hash for each k_i, we perform a bit-wise accumulation on all h_i: if a bit of $h_{i,j}$ is 0, add the weight value $-w_i$, if it is 1, add the weight value w_i, and finally get the value on each bit weighted value v_i. v_i is computed by:

$$v_i = \sum_{i=1}^{n} (-1)^{1-h_{i,j}} w_i \qquad (5)$$

The s_i of each digit on the final fingerprint is accumulated by all v_i bits and judged according to positive and negative, it is defined as follows:

$$s_i = \begin{cases} 0, & v_i < 0 \\ 1, & v_i \geq 0 \end{cases} \qquad (6)$$

After completing the above steps, we get a fingerprint of 64 bits. The Fig. 2 shows an example flow chart of fingerprinting. For one HTTP session, the first step is to extract headers and generate an HHS. Then, hash each fields' strings and assign weights. Finally, merge all bits with addition to get the fingerprint represents the session (we use a 5-bit value like 10110 for easy explanation).

4.2 Fingerprint Database

We process each HTTP session into a 64-bit fingerprint and measure the similarity between fingerprints by the Hamming distance. Hamming distance represents the number of different bits in two (same length) words. Charikar [2] proposed that when the Hamming distance is less than 3, the text can be considered similar. Our purpose here is not to select all similar texts, but to select the closest one.

First, we need to determine the threshold of the Hamming distance. HSLF and SimHash are substantively different that HHS is not a text composed of natural language. Therefore, we cannot directly use the threshold 3 of natural language text as our threshold for judging similarity. To determine the threshold, we randomly selected 1000 fingerprints and calculated the minimum value of similarities between one of 1000 and all other fingerprints under the same application, which we call internal similarities; and the minimum value of similarities between the one and other fingerprints of different applications, which we call external similarities. The distribution of internal similarities and external similarities is shown in the Fig. 3(a) and Fig. 3(b).

It can be found that most internal similarities are between 0–3, and the internal similarities between 0–5 account for 98.2%; the distribution of external similarities is more scattered that similarities greater than 5 account for 68.7%. So in order to locate a new session to its original application, we consider both the distributions of internal similarities and external similarities and set the threshold to 5. Fingerprints with a similarity of 5 and below are listed as optional.

Gurmeet et al. [14] realized near-duplicate detection at the scale of 8 billion web pages, they found that the problem was how to quickly find fingerprints with a Hamming distance of less than 3, so they proposed an algorithm to speed up the index and successfully used in Google's detection module. Although the amount of fingerprints we achieve is far less large, but if we compare fingerprints one by one to find the most similar fingerprint, the resource overhead is also huge. So we build our method of fingerprint storage and indexing on the basis of Gurmeet's method, which win the time at the cost of space.

In order to facilitate indexing, we divide all fingerprint libraries into multiple levels and store them into 6 tables $\{T_0, T_1, ..., T_5\}$ according to Hamming distance

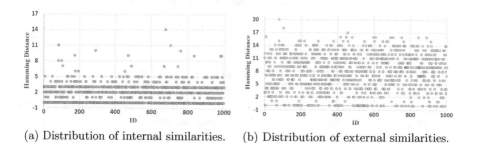

(a) Distribution of internal similarities.   (b) Distribution of external similarities.

Fig. 3. The Similarity distribution of LSH fingerprints

from 0 to 5. The table with Hamming distance 0 is the usual Hash table, a theoretical index structure with a time complexity of $O(1)$. Tables with Hamming distance greater than or equal to 1 are stored as follows:

Copy the entire fingerprint database with the order of 2^d into t sub-tables, each sub-table has an integer p_i and a permutation π_i, the sub-tables $Z_1, Z_2...Z_t$ are constructed by π_i that permutes the p_i bits of all fingerprints in the sub-table to top bits. Then, the fingerprint F to be identified is changed by $\pi_i(F)$. $\pi_i(F)$ will be matched with top p_i bits in each Z_i in parallel. If the fingerprints are all random sequences, there are probably 2^{d-p_i} fingerprints left. Finally, find out fingerprints whose Hamming distance with F are ξ_i. p_i and t are determined according to the actual situation. If p_i is too small, the number of matching fingerprints 2^{d-p_i} is too large, and the search volume is still very high. If p_i is too large, the quantity of sub-tables t will grow a lot.

4.3 Application Identification

The application identification method consists of application fingerprint database pre-work and real-time traffic stream identification. This section describes the algorithm. For a session to be identified, HHS function is used to extract HHS with different fields. $LSHFingerprint$ function calculate session's fingerprint.

The Algorithm 1 describes how an unknown session is identified from a set of requests. The input is fingerprint tables (work as indexable memory data structure) and a new session s. Fingerprint tables $\{T_0, T_1, ..., T_5\}$ are labeled with known application names. Each capture session s will be transformed into an HHS and fingerprinted to get F. Then algorithm queries F from the fingerprint tables T_{ξ_i} in the order of ξ_i from 0 to 5: if a similar fingerprint of F is indexed when $\xi_i = 0$, HSLF will return the application label of the matched fingerprint and end the loop immediately, otherwise continue to find fingerprint with ξ_i self-increasing until ξ_i equal to 5 (the max of pre-work). If multiple fingerprints with different application labels are hit, the return value will be the max number of accumulate labels. If the fingerprint to be predicted does not match any application when $x_i \in [0, 5]$, we will treat such a fingerprint as a new fingerprint and store it in the unknown set for future re-predict. As the size of the fingerprint database continues to increase, fingerprints in the unknown set will be gradually identified.

5 Experiments

In this section, we evaluate our proposed system on real-world dataset to verify the rationality and effectiveness of the system. First, we introduce the public dataset used in the experiments. Then, we detail specific aspects of our approach such as the performance of multi-classification and the superiority compared to machine learning algorithms. Finally, we compare HSLF against some state-of-the-art methods.

Algorithm 1: Application identification

Input: Fingerprint tables $T_0, T_1, ..., T_5$ where $\xi_i \in [0, 5]$,
        A new session s to be identified.
Output: Application identification result L.

1 $S \leftarrow HHS(s)$ $F \leftarrow LSHFingerprint(S)$ $i \leftarrow 0$
2 **while** $i \leq 5$ **do**
3 | $Labels_i \leftarrow Index(T_i, F)$
4 | $num \leftarrow len(Counter(Labels_i))$
5 | **if** $num == 0$ **then**
6 | | $i \leftarrow i + 1$
7 | **else if** $num == 1$ **then**
8 | | return $L \leftarrow Counter(Labels_i)$
9 | **else if** $num > 1$ **then**
10 | | return $L \leftarrow Max(Counter(Labels_i))$
11 |
12 **end**

5.1   Dataset

We collect our original traffic on the lab gateway by using the process log tool PPfilter developed ourselves to mark the application traffic. PPfilter collects network and process log information at the terminal to form a log with a 5-tuple including: IP (source IP and destination IP), port (source port and destination port), port open time, process number (Corresponds to this port), and application-related process name. We use PPfilter to collect their daily traffic for 15 volunteers on campus for one month, and filter out the traffic of each application in the total traffic. During the filtering process, we perform anonymization and privacy protection by replacing or deleting the IP address and some payloads. All collected traffic is marked with application labels. In total, ProcFlow contains 72 GB PCAP files from 35 applications.

5.2   Multi-class Classification Performance

We conduct 10-fold cross validation on ProcFlow and separately verify the performance of HSLF on 10 typical applications. The classification result is shown in Fig. 4. It can be found that each application's three evaluation metrics Precision, Recall, F1-score are satisfactory, and the averages (AVE) are all higher than 0.9. Some applications such as Thunder and Xunfeng can be classified absolutely right.

5.3   Compared with Classic Machine Learning Models

Many previous studies employed machine learning (ML) models for application recognition and proved effective. To prove the superiority of HSLF, we compare it with 4 classic machine learning models: C4.5 (C4.5 decision tree)

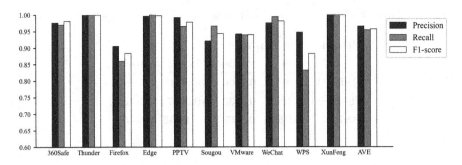

Fig. 4. Precision, recall, F1-score of HSLF on 10 popular Apps.

[7], RF (Random Forests) [17], SVM (Support Vector Machines) [13], kNN (K-NearestNeighbor) [23]. The input of ML models are the one-hot encodings of the headers. To achieve a fair result, all the parameters of machine learning models have been adjusted to the best. We also conduct a 10-fold cross validation on ProcFlow2020 dataset for each model.

The classification results are shown in Table 2. It can be found that the four evaluation metrics of HSLF are: 0.966, 0.926, 0.901, 0.916. Although the accuracy of the four machine learning models are all higher than 85%, the precision, recall, and F1-score are much lower compared to HSLF. It indicates that traditional machine learning classification is unstable, and may have slip in some categories. In contrast, HSLF stay strong performance (higher than 0.9) on precision, recall and F1-score. The comparision result shows that HSLF outperforms machine learning models.

Table 2. Comparison results of HSLF and classic ML methods

Methods	ProcFlow2020			
	ACC	Precision	Recall	F1-score
C4.5	0.8934	0.7075	0.6417	0.6527
Random Forest	0.8946	0.7486	0.6555	0.6806
kNN	0.8777	0.5936	0.5116	0.5310
SVM	0.8699	0.5658	0.4568	0.4848
HSLF	**0.9655**	**0.9257**	**0.9055**	**0.9155**

5.4   Comparison with Other Approaches

In this section, we compare HSLF with two state-of-the-art application traffic classification methods to describe the superiority of HSLF.

TrafficAV [18] is an effective malware identification and classification method. It extracts traffic features, and then uses detection models based on machine

Table 3. Comparison results of Miner-Killer and other methods

Methods	ProcFlow2020			
	ACC	Precision	Recall	F1-score
TrafficAV	0.9020	0.9006	0.9020	0.8995
FlowPrint	0.5007	0.5232	0.5007	0.5063
HSLF	**0.9655**	**0.9257**	**0.9055**	**0.9155**

learning to judge whether the app is malicious or not. TrafficAV contains two detection models namely HTTP detection model and TCP flow detection model. We choose the HTTP detection model for comparison with HSLF. The authors of TrafficAV describe the depiction of the analysis steps and detection solution without open-source implement. Therefore, we faithfully reimplemented the HTTP detection model of TrafficAV feature extraction strategy, and build a classifier based on C4.5 decision tree algorithm the same as TrafficAV.

FlowPrint [4] is a semi-supervised approach for fingerprinting mobile apps from network traffic. It can find temporal correlations among destination-related features of network traffic and use these correlations to generate app fingerprints. FlowPrint calculates the Jaccard distance between two fingerprints and compares with threshold to measure the similarity. The application label of each traffic flow receives the same label as the fingerprint that is most similar to it.

Table 3 shows the identification performance of TrafficAV, FlowPrint and HSLF. It can be found that HSLF outperforms both of them on four metrics. Although the traffic classification purposes of TrafficAV and FlowPrint are different from ours, there are similarities in our solutions. TrafficAV focus on some key information in the HTTP headers, however, TrafficAV only selects 4 headers which ignores the valid information that may be contained in the other headers. In an environment with numerous applications, it is likely to cause collision of feature sequences. Our method HSLF learns from natural language processing to treat the entire headers of the HTTP session as a sequence, and generates a fingerprint of the session to mark the application according to a reasonable weight assignment method. It does not miss any header features and does not rely on specific identifiers. Compared with TrafficAV and FlowPrint, HSLF has better robustness.

6   Conclusion

In this paper, we propose HSLF, a system transforms HTTP sessions into fingerprints for applications traffic classification. HSLF retains almost all the features of HTTP headers and templatizes the headers into HTTP Header Sequence with rational weights. With the help of LSH algorithm, HSLF processes high-dimensional vectors into fingerprints and stores them as tables of different levels. HTTP based applications traffic can be classified by fast index. Experimental results show that HSLF have a satisfactory classification performance and outperforms some classic machine learning methods. Compared with some state-of-the-art approaches, HSLF also has a big improvement. In the future, we will optimize HSLF to fit more network protocols, such as encrypted protocols SSL/TLS.

Acknowledgment. This work was supported by the National Key R&D Program of China with No. 2018YFC0806900 and No. 2018YFB0805004, Beijing Municipal Science & Technology Commission with Project No. Z191100007119009, NSFC No.61902397, NSFC No. U2003111 and NSFC No. 61871378.

References

1. Buchanan, W.J., Helme, S., Woodward, A.: Analysis of the adoption of security headers in http. IET Inf. Secur. **12**(2), 118–126 (2017)
2. Charikar, M.S.: Similarity estimation techniques from rounding algorithms. In: Proceedings of the Thiry-Fourth Annual ACM Symposium on Theory of Computing, pp. 380–388 (2002)
3. Crotti, M., Dusi, M., Gringoli, F., Salgarelli, L.: Traffic classification through simple statistical fingerprinting. ACM SIGCOMM Comput. Commun. Rev. **37**(1), 5–16 (2007)
4. van Ede, T., et al.: Flowprint: semi-supervised mobile-app fingerprinting on encrypted network traffic. In: Network and Distributed System Security Symposium, NDSS 2020. Internet Society (2020)
5. Fraleigh, C., et al.: Packet-level traffic measurements from the sprint IP backbone. IEEE Network **17**(6), 6–16 (2003)
6. Indyk, P., Motwani, R.: Approximate nearest neighbors: towards removing the curse of dimensionality. In: Proceedings of the Thirtieth Annual ACM Symposium on Theory of Computing, pp. 604–613 (1998)
7. Jie, Y., Lun, Y., Yang, H., Chen, L.y.: Timely traffic identification on p2p streaming media. J. China Universities Posts Telecommun. **19**(2), 67–73 (2012)
8. Kaoprakhon, S., Visoottiviseth, V.: Classification of audio and video traffic over http protocol. In: 2009 9th International Symposium on Communications and Information Technology pp. 1534–1539. IEEE (2009)
9. Lavrenovs, A., Melón, F.J.R.: Http security headers analysis of top one million websites. In: 2018 10th International Conference on Cyber Conflict (CyCon), pp. 345–370. IEEE (2018)
10. Li, Y., Li, J.: Multiclassifier: A combination of dpi and ml for application-layer classification in SDN. In: The 2014 2nd International Conference on Systems and Informatics (ICSAI 2014), pp. 682–686. IEEE (2014)
11. Li, Z., Yuan, R., Guan, X.: Accurate classification of the internet traffic based on the SVM method. In: 2007 IEEE International Conference on Communications, pp. 1373–1378. IEEE (2007)
12. Liu, C., He, L., Xiong, G., Cao, Z., Li, Z.: Fs-net: a flow sequence network for encrypted traffic classification. In: IEEE INFOCOM 2019-IEEE Conference on Computer Communications, pp. 1171–1179. IEEE (2019)
13. Liu, C.C., Chang, Y., Tseng, C.W., Yang, Y.T., Lai, M.S., Chou, L.D.: SVM-based classification mechanism and its application in SDN networks. In: 2018 10th International Conference on Communication Software and Networks (ICCSN), pp. 45–49. IEEE (2018)

14. Manku, G.S., Jain, A., Das Sarma, A.: Detecting near-duplicates for web crawling. In: Proceedings of the 16th International Conference on World Wide Web, pp. 141–150 (2007)
15. Moore, A., Zuev, D., Crogan, M.: Discriminators for use in flow-based classification. Technical report (2013)
16. Pham, K., Santos, A., Freire, J.: Understanding website behavior based on user agent. In: Proceedings of the 39th International ACM SIGIR conference on Research and Development in Information Retrieval, pp. 1053–1056 (2016)
17. Raghuramu, A., Pathak, P.H., Zang, H., Han, J., Liu, C., Chuah, C.N.: Uncovering the footprints of malicious traffic in wireless/mobile networks. Comput. Commun. **95**, 95–107 (2016)
18. Wang, S., et al.: Trafficav: an effective and explainable detection of mobile malware behavior using network traffic. In: 2016 IEEE/ACM 24th International Symposium on Quality of Service (IWQoS), pp. 1–6. IEEE (2016)
19. Williams, N., Zander, S.: Evaluating machine learning algorithms for automated network application identification (2006)
20. Xu, F., et al.: Identifying malware with http content type inconsistency via header-payload comparison. In: 2017 IEEE 36th International Performance Computing and Communications Conference (IPCCC), pp. 1–7. IEEE (2017)
21. Xu, Q., Erman, J., Gerber, A., Mao, Z., Pang, J., Venkataraman, S.: Identifying diverse usage behaviors of smartphone apps. In: Proceedings of the 2011 ACM SIGCOMM Conference on Internet Measurement Conference, pp. 329–344 (2011)
22. Yao, H., Ranjan, G., Tongaonkar, A., Liao, Y., Mao, Z.M.: Samples: self adaptive mining of persistent lexical snippets for classifying mobile application traffic. In: Proceedings of the 21st Annual International Conference on Mobile Computing and Networking, pp. 439–451 (2015)
23. Zhang, J., Xiang, Y., Zhou, W., Wang, Y.: Unsupervised traffic classification using flow statistical properties and IP packet payload. J. Comput. Syst. Sci. **79**(5), 573–585 (2013)

Music Genre Classification: Looking for the Perfect Network

Daniel Kostrzewa[✉][ID], Piotr Kaminski, and Robert Brzeski[ID]

Department of Applied Informatics, Silesian University of Technology,
Gliwice, Poland
{daniel.kostrzewa,robert.brzeski}@polsl.pl

Abstract. This paper presents research on music genre recognition. It is
a crucial task because there are millions of songs in the online databases.
Classifying them by a human being is impossible or extremely expensive.
As a result, it is desirable to create methods that can assign a given track
to a music genre. Here, the classification of music tracks is carried out
by deep learning models. The Free Music Archive dataset was used to
perform experiments. The tests were executed with the usage of Convolutional Neural Network, Convolutional Recurrent Neural Networks
with 1D and 2D convolutions, and Recurrent Neural Network with Long
Short-Term Memory cells. In order to combine the advantages of different deep neural network architectures, a few types of ensembles were
proposed with two types of results mixing methods. The best results
obtained in this paper, which are equal to state-of-the-art methods, were
achieved by one of the proposed ensembles. The solution described in the
paper can help to make the auto-tagging of songs much faster and more
accurate in the context of assigning them to particular musical genres.

Keywords: Music information retrieval · Music genre recognition ·
Classification · Deep learning · Convolutional Neural Network ·
Recurrent Neural Networks · Long Short-Term Memory · Ensemble ·
Free music archive dataset

1 Introduction

One of the most vital aspects of music information retrieval is the music genre
recognition. This task aims at classifying pieces of music to their musical genre
by different methods. The issue itself is not trivial because of the similarity of
some genres, e.g., pop and rock or experimental and instrumental genres. Moreover, the given music track can be assigned to many quite different categories. It
turns out that even the distinction of songs made by a human being is not always

This work was supported by Statutory Research funds of Department of Applied
Informatics, Silesian University of Technology, Gliwice, Poland (BKM21 – DK, BK
02/100/BK_21/0008 – RB).

M. Paszynski et al. (Eds.): ICCS 2021, LNCS 12742, pp. 55–67, 2021.
https://doi.org/10.1007/978-3-030-77961-0_6

obvious [7]. However, it can be assumed that music can be divided into individual musical genres. Various machine learning methods can be used for automatic recognition. The classification can be carried out with different classifiers, including those based on deep neural networks. In this research, convolutional and recurrent neural networks, were used for the genre classification task. In the first part of the presented research, different architectures of Convolutional Neural Networks (CNN) [11,13,19], and Convolutional Recurrent Neural Networks with 1D and 2D convolutions (1-D CRNN, 2-D CRNN) [15,27], as well as Recurrent Neural Network with Long Short-Term Memory cells (LSTM) [8,37] were used.

Moreover, to combine the advantages of different deep neural network models, the ensembles were created. Three types of ensembles were proposed, with two methods of results mixing. The first one is merging outcomes by the usage of a single fully connected layer (FCL). The input of meta-classifier is the output of first-level models, transformed by softmax function and combined together into 3-D matrix. The other one is voting (Vote), where the final classification results from majority voting of base models' predictions. All experiments were performed on Free Music Archive Dataset (FMA) [6] – one of the most popular, publicly available databases.

The issue of automatic genre recognition has practical implications in everyday life of many people. It frequently happens that a person's musical taste prefers only specific music genres. Often, one would like to listen to music from a precisely defined genre range. Considering that nowadays, people have access to millions of songs through various music services, it is impossible for a human being to divide them manually. The only solution is the automation of this process. The described practical applications are the reasons why the authors of this paper have researched in this particular direction.

1.1 Related Work

In recent years, the subject of music genre recognition [4,20,29,35] become a vital field of research. The classification of songs can be carried out by various methods, e.g., using classical classifiers [1,14,31]. However, a particular intensity of work in this area can be observed in the deep learning domain [3,10,26]. The most popular solutions employ Convolutional Neural Networks [12,17,18] and Convolutional Recurrent Neural Networks [2,9]. There is also research in which the ensembles of various deep learning models can be found (five CNNs pretrained on ImageNet or Places365 [25], combined outputs from acoustic and visual features [24], and ensemble of one CNN and one RCNN [8]).

Unfortunately, the classification of music genre, performed by other authors, has been made on various databases, with a different number of musical genres, music tracks, and various lengths. The quality of classification [23] also can be measured by many parameters for various criteria and conditions. Therefore, the comparison between the obtained results is difficult, and the conclusions could be ambiguous. Additionally, some of the obtained results, particularly high-performance results, can be deceiving and result from inadequate, flawed experimentation [36,38]. However, the parameters used in this article are one of the most popular and potentially give the best possibilities for comparison with other works.

1.2   Contribution and Paper Structure

The contribution of this article is twofold. Firstly, classification of the musical tracks by the genre using several deep learning models (CNN, CRNN, LSTM) and comparing the obtained results. Secondly, proposing and implementing the classification using several ensembles – various combinations of deep learning models with two types of results' mixing methods. To the best of the knowledge of the authors of this research, there were no created such ensembles yet (with the usage of 2-dimensional convolutions and LSTM cells as well as with different types of results' mixing methods). The comparison of all outcomes, their analysis, and conclusions were made.

The deep neural network architectures created for the purpose of this work are presented in Sect. 2. The used dataset, description of the conducted research, and the obtained results are provided in Sect. 3, while the summary and the final conclusions have been included in Sect. 4.

2   Deep Learning Models

2.1   Convolutional Neural Network

The convolutional neural network (CNN), based on [18], consists of four convolutional layers (Fig. 1). Layer 1 and 2 have both 64 kernels each, whereas layers 3 and 4 have 128 kernels. The kernel size of all layers is equal to 5. After each layer, there is 2-D max pooling applied with kernel size and stride equal 2. In every convolutional layer, $ReLU$ is used as an activation function. Batch normalization is performed afterward. The convolutional layers are followed by one fully connected linear layer with linear activation function and the final output of 8 nodes. The mel-spectrogram is the input of the described architecture. The mel-spectrogram [21] is a type of spectrogram with the Mel scale as its vertical axis. The Mel scale is a result of a non-linear transformation of the frequency scale. The Mel scale is constructed in such a way that sounds at equal distances from each other also for people sound as if they are equidistant from each other. Converting to Mel scale, divides the frequency scale into parts and converts each part into a corresponding Mel scale. To create the mel-spectrogram the song is divided into time windows, then each window is transformed by discrete Fourier transform, from the time domain to frequency domain, then the Mel scale is created using overlapped triangular windows. Finally, the spectrogram can be created by decomposition for each window the magnitude of the signal into its components, corresponding to the frequencies in the Mel scale. In this work, mel-spectrograms with size 128×128 were used.

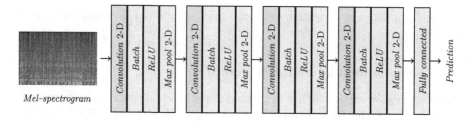

Fig. 1. The architecture of proposed convolutional neural network.

2.2   1-Dimensional Convolutional Recurrent Neural Network

1-Dimensional Convolutional Recurrent Neural Network (1-D CRNN) (Fig. 2) is based on [8], and was created to extract time-dependent features from mel-spectrograms in this model. In this architecture, three 1-D convolutional layers are used. The quantities of kernels are 128, 128, and 64, respectively. In every layer, kernel size equals 5, and stride size equals 1. 1-D batch normalization and *ReLU* are applied after each convolution layer. Following the convolutional layers, there are two stacked Long Short Term Memory (LSTM) cells with a hidden size equal to 128 each. In LSTM cells, dropout is conducted with a probability equal 0.2 to prevent over-fitting. Final prediction is made by one fully connected layer with linear activation.

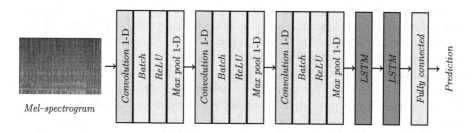

Fig. 2. The architecture of proposed 1-dimensional convolutional recurrent neural network.

2.3   2-Dimensional Convolutional Recurrent Neural Network

2-Dimensional Convolutional Recurrent Neural Network (2-D CRNN) (Fig. 3) in contrary to 1-D CRNN, consists of four 2-D convolutional layers (64, 64, 128, 128 filters respectively) with the kernel size of 3 and stride 1. 2-D batch normalization and *ReLU* are also applied after each convolutional layer. Besides, similar LSTM and fully connected layers are used afterward.

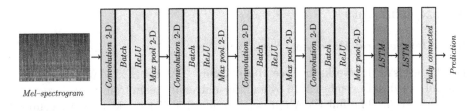

Fig. 3. The architecture of proposed 2-dimensional convolutional recurrent neural network.

2.4 Recurrent Neural Network with Long Short-Term Memory Cells (LSTM)

Recurrent Neural Network with Long Short-Term Memory cells includes three layers with 256 hidden size, and dropout probability equals 0.3. The fully connected layer with linear activation is stacked afterward. The input is the set of MFCC values. The MFCC values are obtained by taking the logarithm of the powers at each of the mel frequencies and do discrete cosine transform [5, 22, 30]. The MFCC values are the amplitudes of the resulting spectrum [40]. In this work, 13 MFCC values were used for each time window.

2.5 Ensembles

In order to combine the advantages of different deep neural network architectures, three types of ensembles were proposed. All types include three networks.

Ensemble 1: Stacked CNNs and 1-D CRNN. Ensemble 1 consists of three different models. Two of them are CNNs (Sect. 2.1), and the third is 1-D CRNN (Sect. 2.2).

Ensemble 2: Stacked CNNs and 2-D CRNN. Similar to ensemble 1, this architecture is built using the two simple CNNs (Sect. 2.1), whereas the third model used is built with 2-D CRNN (Sect. 2.3).

Ensemble 3: Stacked RNN. The third ensemble contains three models with LSTM cells. Each model's cell varies with hyper-parameters such as the cell's hidden size (256, 128, and 256), the number of stacked layers (3, 3, and 2), and dropout probability (0.3, 0.1, and 0.2, respectively).

The output of models in each ensemble is followed by two types of results' mixing methods (to provide the final outcome). The first one is mixing outcomes by usage of the single fully connected layer (FCL). The meta-classifier input is the output results of first-level models transformed by softmax function and combined into a 3-D matrix. Meta-classifier is trained using the validation subset. The other one is voting (Vote), where the final classification results from majority voting of base models' predictions or, if there is no clear winner, meta-classifier's highest score. As a result six different ensembles was proposed (i.e., Ensemble 1 – FCL, Ensemble 1 – Vote, Ensemble 2 – FCL, Ensemble 2 – Vote, Ensemble 3 – FCL, and Ensemble 3 – Vote).

3   Experiments

3.1   Dataset and Hardware Setup

For the needs of this work, the Free Music Archive Dataset (FMA) was chosen. It was published in December 2016 by an inter-university team from Singapore and Switzerland [6]. The full FMA dataset contains 106,574 tracks and 161 genres (including sub-genres) along with its metadata and audio features. Each file has two channels and is sampled with 44.1 kHz. The biggest advantage of the dataset is that it contains a copy-free audio files, is feature-rich, and includes metadata (e.g., genre), and is already split into three subsets (small, medium, and full), which makes it additionally useful for people with lower computational resources.

The small subset of the FMA dataset contains 8000 tracks, each 30 s long, that are categorized into eight top-level genres such as Hip-Hop, Folk, Experimental, International, Instrumental, Electronic, Pop, and Rock. For each genre there, are 1000 tracks assigned, which makes the dataset well-balanced. That is why for training and evaluation purposes in this work, the small FMA dataset has been chosen.

All the experiments were conducted on a notebook with the following hardware configuration: CPU - Intel Core i5-9300HF, memory - 16 GB RAM, GPU - NVIDIA GeForce GTX 1650 4GB GDDR5.

3.2   Additional Settings

Data Normalization
Each of the channel is normalized according to the Eq. 1.

$$I_{normalized}[d] = \frac{I[d] - mean}{std},\qquad(1)$$

where d is an index of dimension in image $d = \{0, 1, 2\}$, I is an input mel-spectrogram, $mean$ and std are arbitrary chosen values equal to 0.5. Normalization can accelerate further computing up to one order of magnitude and decrease error rate [33].

Loss Function
During training as a loss function cross entropy loss is used with the Eq. 2.

$$loss(\hat{y}, class) = -\hat{y}[class] + log\left(\sum_{j=0}^{n-1} exp(\hat{y}[j])\right),\qquad(2)$$

where $\hat{y}$ is a vector consisting classification outputs (in the form of predicted class index of each sample), $class$ is an index of true data class, and n is the size of mini-batch.

Optimizer and Learning Rate
To perform gradient-based optimization in learning phase, Adam (adaptive moment estimation) is used. This algorithm is proven to be memory-efficient. Its another benefit is making rapid progress lowering the cost in the initial stage of training and converging faster than other popular algorithms like AdaGrad or simple stochastic gradient descent (SGD). Learning rate equals 0.001.

Regularization
It is widely believed that nets with a large number of parameters are powerful machine learning systems. Deep networks, however, are very often prone to overfitting. Especially when training set is rather small.

To address this problem dropout regularization method is applied. The term dropout refers to dropping out neurons, both visible and hidden, in a neural network. Dropping a neuron out means temporarily removing it from the network, along with the connections [34]. A single parameter p is passed to a dropout function. It is a fixed probability that a neuron can be abandoned in the training process. For every CNN models p equals 0.2 is used. Values of dropouts for RNN models are presented in Sect. 2.5. Dropouts are not applied while testing.

3.3   Quantitative Results

Several experiments were conducted to compare the performance and efficiency of individual models. The FMA small dataset was divided into three independent subsets in an 80:10:10 ratio. The largest subset was used for training, another two for validation and testing, respectively. After the training phase, results were measured using the following metrics: accuracy, precision, recall, and F1 score.

Table 1 contains quantitative outcomes obtained by first-level models (i.e., CNN, 1-D CRNN, 2-D CRNN, and LSTM). Evaluation is performed on a test subset unseen by the models. The values marked in bold are the highest ones.

Table 1. The results obtained by the first-level models.

Model	Accuracy [%]	Precision [%]	Recall [%]	F1 Score [%]
CNN	**51.63**	**51.81**	**50.47**	48.59
1-D CRNN	49.88	49.63	50.09	**49.43**
2-D CRNN	44.36	45.22	43.84	42.1
LSTM	45.11	45.33	44.59	42.53

The best values were achieved by CNN and 1-D CRNN models, while 2-D CRNN and LSTM were significantly worse. This is surprising because LSTM cells should consider the sequence of the acoustic signal and use it to their advantage.

Table 2 gathers quantitative results for prepared ensembles and their base models. Underlined values are the best outcomes in a particular ensemble, and values marked in bold are the highest values at all.

Table 2. The results obtained by the ensembles.

	Model	Accuracy [%]	Precision [%]	Recall [%]	F1 Score [%]
Ensemble 1	CNN 1	51.63	51.81	50.47	48.59
	CNN 2	50.38	49.61	**55.05**	48.24
	1-D CRNN	49.88	49.63	50.09	49.43
	Ensemble 1 – FCL	55.14	55.22	54.50	53.41
	Ensemble 1 – Vote	**56.39**	**56.25**	54.80	**54.91**
Ensemble 2	CNN 1	50.13	50.62	52.96	48.94
	CNN 2	52.63	52.80	<u>54.09</u>	51.63
	2-D CRNN	44.36	45.22	43.84	42.10
	Ensemble 2 – FCL	53.01	52.92	52.47	50.10
	Ensemble 2 – Vote	<u>53.13</u>	<u>53.28</u>	51.97	<u>51.76</u>
Ensemble 3	LSTM 1	42.98	43.90	43.76	41.20
	LSTM 2	43.86	43.75	43.58	42.39
	LSTM 3	<u>45.11</u>	<u>45.33</u>	<u>44.59</u>	42.53
	Ensemble 3 – FCL	43.36	43.20	42.00	40.84
	Ensemble 3 – Vote	44.74	44.77	43.07	<u>42.86</u>

The analysis of Table 2 leads to some interesting conclusions. As mentioned before, the performance of the 2-D CRNN and LSTM is noticeably lower than the CNNs and 1-D CRNN models. Ensembles 1 and 2 provide significantly better accuracy, precision, and F1 score than the first-level models' values. Moreover, the outcomes show that the voting result's mixing method is slightly better than the fully connected layer. However, in most cases, an ensemble with a fully connected layer can still provide better results than any base model.

3.4   Qualitative Results

Table 3 shows the confusion matrix obtained for Ensemble 1 with the voting model, the best of all developed models. From the results obtained, it can be concluded that not all musical genres are classified with similar effectiveness. For Hip-Hop, almost 80% accuracy was achieved, and for the next four genres (i.e., Rock, Folk, Electronic, and International), the accuracy was between 60% and 70%. This seems to be a satisfactory result. The worst results were for Pop and Experimental, almost 20% and 40% respectively. This is due to the fact that both of these genres combine the features of other genres. Depending on the particular Pop track, it sometimes has Rock, Hip-Hop, Electronic, or other genres' characteristics.

Table 3. Confusion matrix for the best model – Ensemble 1 with Voting [in %].

		Predicted label							
		Hip-Hop	Folk	Experimental	International	Instrumental	Electronic	Pop	Rock
True label	Hip-Hop	79.4	2.1	1.0	2.1	2.1	9.3	3.1	1.0
	Folk	2.0	65.0	5.0	5.0	13.0	2.0	6.0	2.0
	Experimental	5.6	6.5	39.3	4.7	16.8	10.3	5.6	11.2
	International	15.3	4.7	2.4	61.2	1.2	4.7	7.1	3.5
	Instrumental	3.0	10.9	14.9	1.0	56.4	2.0	2.0	9.9
	Electronic	15.6	0.0	3.7	0.9	11.0	62.4	4.6	1.8
	Pop	11.8	15.1	3.2	18.3	7.5	11.8	19.4	12.9
	Rock	0.9	10.4	5.7	4.7	0.0	3.8	7.6	67.0

Table 4. Comparison of different models classifying FMA small with the proposed Ensemble 1 with voting. Recall and F1 score were provided in [39] only. All values are in %.

No.	Model	Accuracy	No.	Model	Accuracy	Recall	F1 score
1	K-Nearest Neighbors [42]	36.4	12	MoER [41]	55.9	–	–
2	Logistic regression [42]	42.3	13	FCN [39]	63.9	43.0	40.3
3	Multilayer perceptron [42]	44.9	14	TimbreCNN [39]	61.7	36.4	35.0
4	Support vector machine [42]	46.4	15	End-to-end [39]	61.4	38.4	34.5
5	Original spectrogram [41]	49.4	16	CRNN [39]	63.4	40.7	40.2
6	Harmonic spectrogram [41]	43.4	17	CRNN-TF [39]	64.7	43.5	42.3
7	Percussive spectrogram [41]	50.9	18	CRNN [32]	53.5	–	–
8	Modulation spectrogram [41]	55.6	19	CNN-RNN [32]	56.4	–	–
9	MFCC [41]	47.1	20	CNN TL [16]	51.5	–	–
10	MoEB [41]	54.1	21	CNN TL [28]	56.8	–	–
11	MoEC [41]	55.6	22	C-RNN [42]	65.2	–	–
23	**Ensemble 1 - vote**	**56.4**	23	**Ensemble 1 - vote**	**56.4**	**54.8**	**54.9**

The assignment of a song to a particular musical genre is often very challenging, even for a skilled listener and music expert. This difficulty increases significantly in vaguely defined genres, such as Pop, Experimental, etc.

3.5   Comparison of the Outcomes

As part of comparing the obtained results with other works' outcomes, a cumulative table was created (Table 4). It contains the results of the works [16, 28, 32, 39, 41, 42], and, at the bottom, in bold, the best results of this research, for Ensemble 1 with voting.

When analyzing Table 4, it can be concluded, that in general, the values obtained by the authors are within the range of those obtained in other similar studies. Lines 1–4 shows (in Table 4) the values achieved by classical classification methods [42], which are lower by 10–20% than those obtained by Ensemble 1. Lines 5–8 presents results of CNNs fed by different spectrograms, line 9 – was treated as traditionally obtained baseline, while lines 10–12 reveals values (all slightly worse than Ensemble 1) achieved by mixtures of CNNs fed by different spectrograms [41]. Lines 13–17 shows the quantitative outcomes of Fully

Convolutional Neural Network, Timbre CNN, End-to-end approach for music auto-tagging, CRNN, and CRNN with Time and Frequency dimensions, respectively [39]. All these methods achieved higher accuracy than Ensemble 1 (up to 8.3%). On the other hand, recall and F1 score are way lower than in presented research (11.3% and 12.6%, respectively). Lines 18–21 presents outcomes for different CRNN models [32] as well as CNN models with the use of transfer learning [16,28]. These results are very close to Ensemble 1 with voting. Moreover, one of them (line 21) is slightly higher (by 0.4%). The last result (line 22) is much higher than obtained by the authors of this work. However, there is a conceptual defect (data leakage from the training of deep learning model to the test data), which leads to overoptimistic outcomes. Summarizing all the results, it can be noticed that the results achieved by the authors of this research are in the range of, or even above the state-of-the-art solutions.

4   Conclusions

We introduced new types of ensembles, consisting of several different deep neural networks as base models, followed by two different results mixing methods, to solve the classification of songs according to musical genres. We have conducted a series of varied experiments for both the developed deep neural network base models (i.e., CNN, 1-D CRNN, 2-D CRNN, and RNN with LSTM cells) and the elaborated ensembles. The quantitative and qualitative studies have shown that the outcomes achieved easily match and in most cases even beat state-of-the-art methods. Moreover, in this research, only mel-spectrogram and MFCCs were used as the input data. Precise enlargement of the number of input signals will likely allow for a further increase in classification quality.

Future work will be centered around conducting experiments on other deep network architectures (possibly with attention mechanism) and other ensemble configurations. A next step, completely different and more demanding research, would be to classify on the assumption that a given work can belong to several musical genres, i.e., a multi-label approach.

The main advantage of using our approach for music genre classification is the novel, automatic application of an ensemble of different deep learning architectures. In case of finding other good base models, they could be easily applied in the ensemble and thus increase the overall quality of the classification.

It has to be remembered that the creating of the ensemble is more labor-intensive, and the classification itself is longer. However, if the main goal is the quality of the assigning song to the proper musical genre process, the possibility of its improving is always valuable.

References

1. Basili, R., Serafini, A., Stellato, A.: Classification of musical genre: a machine learning approach. In: ISMIR (2004)

2. Choi, K., Fazekas, G., Sandler, M., Cho, K.: Convolutional recurrent neural networks for music classification. In: 2017 IEEE International Conference on Acoustics, Speech and Signal Processing (ICASSP), pp. 2392–2396. IEEE (2017)
3. Choi, K., Fazekas, G., Sandler, M., Cho, K.: Transfer learning for music classification and regression tasks. arXiv preprint arXiv:1703.09179 (2017)
4. Costa, Y.M., Oliveira, L.S., Silla, C.N., Jr.: An evaluation of convolutional neural networks for music classification using spectrograms. Appl. Soft Comput. **52**, 28–38 (2017)
5. Davis, S., Mermelstein, P.: Comparison of parametric representations for monosyllabic word recognition in continuously spoken sentences. IEEE Trans. Acoust. Speech Signal Process. **28**(4), 357–366 (1980)
6. Defferrard, M., Benzi, K., Vandergheynst, P., Bresson, X.: FMA: A dataset for music analysis. arXiv preprint arXiv:1612.01840 (2016)
7. Dong, M.: Convolutional neural network achieves human-level accuracy in music genre classification. arXiv preprint arXiv:1802.09697 (2018)
8. Ghosal, D., Kolekar, M.H.: Music genre recognition using deep neural networks and transfer learning. In: Interspeech, pp. 2087–2091 (2018)
9. Gunawan, A.A., Suhartono, D., et al.: Music recommender system based on genre using convolutional recurrent neural networks. Procedia Comput. Sci. **157**, 99–109 (2019)
10. Kereliuk, C., Sturm, B.L., Larsen, J.: Deep learning and music adversaries. IEEE Trans. Multimedia **17**(11), 2059–2071 (2015)
11. Khan, A., Sohail, A., Zahoora, U., Qureshi, A.S.: A survey of the recent architectures of deep convolutional neural networks. Artif. Intell. Rev. **53**(8), 5455–5516 (2020)
12. Kim, T., Lee, J., Nam, J.: Sample-level CNN architectures for music auto-tagging using raw waveforms. In: 2018 IEEE International Conference on Acoustics, Speech and Signal Processing (ICASSP), pp. 366–370. IEEE (2018)
13. Kiranyaz, S., Avci, O., Abdeljaber, O., Ince, T., Gabbouj, M., Inman, D.J.: 1D convolutional neural networks and applications: A survey. arXiv preprint arXiv:1905.03554 (2019)
14. Kostrzewa, D., Brzeski, R., Kubanski, M.: The classification of music by the genre using the KNN classifier. In: Kozielski, S., Mrozek, D., Kasprowski, P., Małysiak-Mrozek, B., Kostrzewa, D. (eds.) BDAS 2018. CCIS, vol. 928, pp. 233–242. Springer, Cham (2018). https://doi.org/10.1007/978-3-319-99987-6_18
15. Labach, A., Salehinejad, H., Valaee, S.: Survey of dropout methods for deep neural networks. arXiv preprint arXiv:1904.13310 (2019)
16. Lee, D., Lee, J., Park, J., Lee, K.: Enhancing music features by knowledge transfer from user-item log data. In: ICASSP 2019-2019 IEEE International Conference on Acoustics, Speech and Signal Processing (ICASSP), pp. 386–390. IEEE (2019)
17. Lee, J., Nam, J.: Multi-level and multi-scale feature aggregation using pretrained convolutional neural networks for music auto-tagging. IEEE Signal Process. Lett. **24**(8), 1208–1212 (2017)
18. Lim, M., et al.: Convolutional neural network based audio event classification. KSII Trans. Internet Inf. Syst. **12**(6), 2748–2760 (2018)
19. Liu, W., Wang, Z., Liu, X., Zeng, N., Liu, Y., Alsaadi, F.E.: A survey of deep neural network architectures and their applications. Neurocomputing **234**, 11–26 (2017)
20. McKay, C., Fujinaga, I.: Musical genre classification: is it worth pursuing and how can it be improved? In: ISMIR, pp. 101–106 (2006)

21. Mermelstein, P.: Distance measures for speech recognition, psychological and instrumental. Pattern Recogn. Artif. Intell. **116**, 374–388 (1976)
22. Mogran, N., Bourlard, H., Hermansky, H.: Automatic speech recognition: an auditory perspective. In: Speech Processing in the Auditory System. Springer Handbook of Auditory Research, vol. 18, pp. 309–338. Springer New York (2004). https://doi.org/10.1007/0-387-21575-1_6
23. Moska, B., Kostrzewa, D., Brzeski, R.: Influence of the applied outlier detection methods on the quality of classification. In: Gruca, A., Czachórski, T., Deorowicz, S., Hareżlak, K., Piotrowska, A. (eds.) ICMMI 2019. AISC, vol. 1061, pp. 77–88. Springer, Cham (2020). https://doi.org/10.1007/978-3-030-31964-9_8
24. Nanni, L., Costa, Y.M., Aguiar, R.L., Silla, C.N., Jr., Brahnam, S.: Ensemble of deep learning, visual and acoustic features for music genre classification. J. New Music Res. **47**(4), 383–397 (2018)
25. Nanni, L., Maguolo, G., Brahnam, S., Paci, M.: An ensemble of convolutional neural networks for audio classification. arXiv preprint arXiv:2007.07966 (2020)
26. Oramas, S., Nieto, O., Barbieri, F., Serra, X.: Multi-label music genre classification from audio, text, and images using deep features. arXiv preprint arXiv:1707.04916 (2017)
27. Pamina, J., Raja, B.: Survey on deep learning algorithms. Int. J. Emerg. Technol. Innov. Eng. **5**(1), 38–43 (2019)
28. Park, J., Lee, J., Park, J., Ha, J.W., Nam, J.: Representation learning of music using artist labels. arXiv preprint arXiv:1710.06648 (2017)
29. Pons, J., Serra, X.: Randomly weighted CNNs for (music) audio classification. In: ICASSP 2019-2019 IEEE International Conference on Acoustics, Speech and Signal Processing (ICASSP), pp. 336–340. IEEE (2019)
30. Sahidullah, M., Saha, G.: Design, analysis and experimental evaluation of block based transformation in MFCC computation for speaker recognition. Speech Commun. **54**(4), 543–565 (2012)
31. Silla, C.N., Koerich, A.L., Kaestner, C.A.: A machine learning approach to automatic music genre classification. J. Braz. Comput. Soc. **14**(3), 7–18 (2008)
32. Snigdha, C., Kavitha, A.S., Shwetha, A.N., Shreya, H., Vidyullatha, K.S.: Music genre classification using machine learning algorithms: a comparison. Int. Res. J. Eng. Technol. **6**(5), 851–858 (2019)
33. Sola, J., Sevilla, J.: Importance of input data normalization for the application of neural networks to complex industrial problems. IEEE Trans. Nucl. Sci. **44**(3), 1464–1468 (1997)
34. Srivastava, N., Hinton, G., Krizhevsky, A., Sutskever, I., Salakhutdinov, R.: Dropout: a simple way to prevent neural networks from overfitting. J. Mach. Learn. Res. **15**(1), 1929–1958 (2014)
35. Sturm, B.L.: A survey of evaluation in music genre recognition. In: Nürnberger, A., Stober, S., Larsen, B., Detyniecki, M. (eds.) AMR 2012. LNCS, vol. 8382, pp. 29–66. Springer, Cham (2014). https://doi.org/10.1007/978-3-319-12093-5_2
36. Sturm, B.L.: The state of the art ten years after a state of the art: future research in music information retrieval. J. New Music Res. **43**(2), 147–172 (2014)
37. Tang, C.P., Chui, K.L., Yu, Y.K., Zeng, Z., Wong, K.H.: Music genre classification using a hierarchical long short term memory (LSTM) model. In: Third International Workshop on Pattern Recognition, vol. 10828, p. 108281B. International Society for Optics and Photonics (2018)
38. Urbano, J., Schedl, M., Serra, X.: Evaluation in music information retrieval. J. Intell. Inf. Syst. **41**(3), 345–369 (2013)

39. Wang, Z., Muknahallipatna, S., Fan, M., Okray, A., Lan, C.: Music classification using an improved CRNN with multi-directional spatial dependencies in both time and frequency dimensions. In: 2019 International Joint Conference on Neural Networks (IJCNN), pp. 1–8. IEEE (2019)
40. Xu, M., Maddage, N.C., Xu, C., Kankanhalli, M., Tian, Q.: Creating audio keywords for event detection in soccer video. In: 2003 International Conference on Multimedia and Expo. ICME2003. Proceedings (Cat. No. 03TH8698), vol. 2, pp. II-281. IEEE (2003)
41. Yi, Y., Chen, K.Y., Gu, H.Y.: Mixture of CNN experts from multiple acoustic feature domain for music genre classification. In: 2019 Asia-Pacific Signal and Information Processing Association Annual Summit and Conference (APSIPA ASC), pp. 1250–1255. IEEE (2019)
42. Zhang, C., Zhang, Y., Chen, C.: SongNet: Real-Time Music Classification. Stanford University Press, Palo Alto (2019)

Big Data for National Security in the Era of COVID-19

Pedro Cárdenas[1]([⊠]), Boguslaw Obara[1], Ioannis Ivrissimtzis[1], Ibad Kureshi[2], and Georgios Theodoropoulos[3]

[1] Department of Computer Science, Durham University, Durham, UK
{pedro.cardenas-canto,boguslaw.obara,ioannis.ivrissimtzis}@durham.ac.uk
[2] Inlecom Systems, Brussels, UK
ibad.kureshi@inlecomsystems.com
[3] Department of Computer Science and Engineering, Southern University of Science and Technology, Shenzhen, China

Abstract. The COVID-19 epidemic has changed the world dramatically as societies adjust their behaviour to meet the challenges and uncertainties of the new normal. These uncertainties have led to instabilities in several facets of society, most notably health, economy and public order. Increasing discontent within societies in response to government mandated measures to contain the pandemic have triggered social unrest, imposing serious threats to national security. Big Data Analytics can provide a powerful force multiplier to support policy and decision makers to contain the virus while at the same time dealing with such threats to national security. This paper presents the utilisation of a big data forecasting and analytics framework to deal with COVID-19 triggered social unrest. The framework is applied and demonstrated in two different disruptive incidents in the United States of America.

Keywords: COVID-19 · Epidemics · Big data · National security · Data analytics · Machine learning

1 Introduction

Global challenges and emergencies such as climate change, epidemics and natural and man-made calamities present unprecedented governance issues. The COVID-19 pandemic has demonstrated how a global challenge can disrupt more than 180 countries. Governments across the globe have taken strict decisions aimed at containing the disease and avoiding massive infections, such as curfews, lockdowns, "stay at home" orders, or compartmentalization of domestic territories according to their infection rates [1].

Such measures represent a meaningful way to control the disease, however, they also have a negative effect on people's lives imposing dramatic changes in the ways of life people had been used to. As a result, containment measures have often be met with varying degrees of social discontent and unrest, from protests

© Springer Nature Switzerland AG 2021
M. Paszynski et al. (Eds.): ICCS 2021, LNCS 12742, pp. 68–82, 2021.
https://doi.org/10.1007/978-3-030-77961-0_7

and non-compliance actions to more violent manifestations such as demonstrations and riots [2]. A state's stability could be seriously undermined by such social instability incidents, which may have a negative effect on national security components such as health, economy, and public order [3].

Policy and decision-makers need to have at their disposal technological tools, acting as force multipliers and enabling insights about disasters and unfolding situations, so that an assessment of the scale of the threat to national and international security can be made [4,5]. Big Data technologies can provide a powerful means in this endeavour [6,7]. As a result the last decade we have witnessed the development of several computational platforms that utilise Big Data analytics to derive insights about disruptive situations that can trigger social unrest [8–13].

Contributing to this effort, in earlier work we have proposed a framework and the associated workflow for the analysis of social media data (Twitter) to derive insights about disruptive events and potential unrest [14–18]. In this paper, this framework is utilised to analyse the COVID-19 pandemic. Our analysis focuses on two geographical areas where acts of social unrest were witnessed as a result of COVID-19 containment measures, namely Michigan and Texas.

The aims and contributions of this paper are twofold. Firstly, to demonstrate the robustness and applicability of our framework for forecasting and analysing important real-world events such as COVID-19 related unrest: would the framework have been able to provide the competent authorities enough notice and insights to deal with the then unfolding crisis? Secondly, to provide interested stakeholders postmortem insights about COVID-19 social crises with the view to contribute to the ongoing global effort to tackle this disruptive situation.

The rest of the paper is organised as follows; Sect. 2 briefly discusses the impact of pandemics such as COVID-19 on National Security; Sect. 3 provides an overview of the framework; Sect. 4 illustrates the operationalisation of the methodology, examining two events regarding COVID-19 related outbreaks; Sect. 5 concludes the paper.

2   COVID-19 and National Security

National security threats refer to those activities that endanger the individuals' physical well-being, or compromise the stability of the state. When we place people at the centre of the analysis of how national security is affected, in which case it is also referred to as human security, we typically distinguish between seven different components: economic security, food security, health security, environmental security, personal security, communal security and political security [3]. Instabilities are generated due to the disruption of one or more these components leading to protests, riots and other forms of violence [17,19]. According to the Global Peace Index [20], civil unrest has doubled over the past decade, and riots, strikes and anti-government protests increased by 244%.

Countries around the world define their domestic security threats based on their internal policies. Pandemics are typically considered national security threats due to their negative social, economic and political impacts [22]. As a global pandemic, COVID-19 represents a serious National Security threat [23]. The negative social and economic effects of lockdowns and curfews have fuelled preexisting social discontent and unrest (e.g. in the Black Lives Matter movement, or the demonstrations in Hong Kong) as well as new anti-lockdown demonstrations. These demonstrations, in turn, act as super-spreader events, further exacerbating the negative impacts of the pandemic [21].

3　An Overview of the Framework

The methodology described in our previous work [14–18], attempts to enrich the security decision-making process scenario. It analyses national security considering its broad spectrum components, including but not limited to health and public order; enabling in such a way to detect timely tipping points and examine a variety of situations as riots, protests or events linked to health issues such as COVID-19.

The framework consists of two main stages (see Fig. 1). An initial phase (Warning Period) continuously analyses data and issues an alert when it identifies that specific societal behavioural characteristics exceed a given threshold (tipping point). The system then gets into its next phase (Crisis Interpretation) by collecting information from numerous sources, such as social networking services or websites, to attempt to zoom in and provide more in-depth insights that unveil data to construe the unfolding crisis and support therefore authorities and other stakeholders into making better decisions.

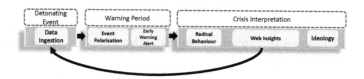

Fig. 1. Conceptual framework, as described in [14–18].

Under the new normal, where COVID-19 tends to modify important behavioural aspects of people's lives, understanding features that directly impact the security of a state becomes crucial. Here, our aim is to use features extracted from our conceptual framework to interpret the health crisis. The characteristics we used here are summarised in Table 1. In the next subsections, we provide a stage by stage overview of the framework.

3.1　Event Polarisation and Early Warning Alert (Q1)

As explained in [17,24,25], a detonating event is an incident that may trigger a disruptive situation that can lead to social unrest and threaten national security

Table 1. Insights derived from the Analytics Framework described in [14–18].

Insights	Stages			
	Early Warning Alert	Radical Behaviour	Ideology	Web Insights
Q1. When do people head towards a situation that evokes that both social stability and national security components can be compromised?	✓	-	-	-
Q2. Which entities are described by people during the crisis?	-	✓	-	-
Q3. What are the radical behavioural traits being conveyed?	-	✓	-	-
Q4. What items are being asked for by individuals in social media?	-	✓	-	-
Q5. Are hostility and authoritarianism traits present during the incident?	-	-	✓	-
Q6. Do embedded web resources in social media texts disclose that the national security components have a horizontal escalation over time?	-	-	-	✓

components. The Warning stage aims to identify which human security components are being affected and, if such impact exceeds a certain threshold, to issue an alert.

More specifically, the alert is issued when the three components of Global Polarisation (GP), Social Media Connectedness (SMC) and Human Security Impact (HSI) have reached a predefined threshold, as described in [14]. First, the GP process performs a sentiment analysis procedure, and when the negative polarisation starts fluctuating above a predefined verge, triggers the next step. The SMC step, based on a Deep Learning model, reveals when individuals are engaged towards the incident. Finally, the HSI step classifies the data corpus into ten human security aspects (health, public order, transport, economy, people, defence, environment, government, information and life), using unsupervised and supervised learning processes. Finally, using a preconfigured scale, it determines if human security components have been compromised.

Once these three steps have been completed, an alert is issued to indicate that the society is heading towards a tipping point, namely a situation where the crisis tends to affect the components that keep a state's stability, which means a point of no return.

3.2 Radical Behaviour (Q2, Q3 and Q4)

Social media is a complex and disarranged milieu in both normal conditions and during crises [26]. Emergencies/incidents are situations where online activity tends to ramp up significantly [27]. Such bursts of activity can provide information linked to various social aspects, but here those aspects that may unbalance the integrity of a state gain special attention.

Based on the analysis of disruptive expressions, the radical behaviour methodology, as detailed in our previous work [16], comprises eight components: (1) Creation of instability scenarios, (2) Identification of affected entities (people, locations, or facilities), (3) Identification of likely affected entities due to their proximity to the incident, (4) Dissection of the intentions expressed towards an entity, (5) Dissemination degree of the crisis (widespread or local incident), (6)

Detection of violent expressions, (7) Classification of violent expressions, and (8) Necessities shared by individuals amid the crisis.

As detailed in [16], the radical behaviour architecture comprises of five stages: Instability Scenarios, Entity Extraction, Wordlists Creation, Content Analytics and Data Interpretation. The sequence of the above mentioned procedures is underpinned by an array of interconnected computational techniques, such as deep learning, natural language processing, supervised and unsupervised learning. It should be noted that the extraction of meaningful insights from the above, depends on the nature of the incident since a violent public disorder or a protest amidst a pandemic have dissimilar roots, and therefore features such as violent expressions cannot be detected with the same frequency.

As part of the communication cycle, individuals use words or specific terms to embody the situation they have to cope with. We note that the pandemic is affecting not only the way people live and work, but also the way they communicate. Indeed, COVID-19 has drawn to the scene numerous new terms that enrich the everyday vocabulary to the level necessary to convey the message. Therefore, the Wordlist Creation step had to be enriched by adding a comprehensive glossary that included terms such as, case fatality rate (CFR), or personal protective equipment (PPE) [28–30].

3.3   Ideology (Q5)

The term ideology usually refers to a set of ideas and beliefs taken from a more complex system of ideas, such as popular sovereignty or nationalism [31]. Another way to approach this concept is by linking it to the processes of giving legitimacy to the power of a dominant group [32]. In both approaches, the term ideology describes fundamental beliefs shared by a social group [33].

During an organised public demonstration or a riot, ideas, beliefs and actions directed against authorities or other groups may emerge [34]. Here, collective emotions denoting appraisals of superiority/inferiority, goal obstruction/injustices, or intolerability, also described in [35] as hostility, play a big role. Protests and other disruptive incidents are two examples of events where hostility may be present and where tints of violence can be a signal of the instability of the state, which in turn may represent the prelude of a crisis [35, 36].

In addition, such disruptive events may evolve due to the fact that people do not empathise with decisions or activities performed by those who hold the "proper authority", which can be perceived as a high level of authoritarianism against them, as measured in an aggression, submission and conventionalism scale [37, 38]. Governments around the globe have introduced various measures to contain the virus. Notwithstanding the differences, the nature of the restrictions may generate traits of hostility and authoritarianism, which is why both ideological characteristics will be used in this work, similarly to the methodology proposed in our work [18].

The ideology traits are spotted following a data analytics procedure that involves the processing of emotions in unstructured data (tweets) to identify the internal components of authoritarianism and hostility, followed by a deep generative model (variational autoencoders) centred on separating the ideological features from the rest of the data. Lastly, such information is compared to a precalibrated model to determine the presence of such ideological characteristics.

3.4 Web Insights (Q6)

While a disruptive event occurs, individuals adopt responses shaped by the nature of the incident, which runs from protests and large-scale mobilisations to violent activity. As a result of such activities, human security components get affected, since a violent incident does not only impact a prima facie component such as "public order", but also "health" as some protesters can be hurt as the situation gets nuanced by more aggressive reactions.

Such an aspect gains importance since national security can be compromised when the number of affected human security components increases over time, which is called horizontal escalation, as described in [39]. Horizontal escalation can be accelerated by the use of web resources (websites, social networking services, independent websites or information outlets), which become a tool to disseminate information and work, for example, as a mouthpiece to organise demonstrations. In the preceding work [15], an architecture aimed at analysing the horizontal escalation of human security components along on-line participative channels was presented. There, a classification model is used to cluster the various human security components from the data corpus, and it yields baseline thresholds to analyse the aforementioned escalation characteristics in future events.

4 Analysing Two COVID-19 Disruptive Events

As stated earlier, the conceptual framework's main objective is to monitor the state of the society at any particular moment and, in case of an alert, to derive deeper insights about the situation and the threat it may constitute to national security. With this goal in mind, two incidents of social unrest, which occurred in April 2020 in Michigan and Texas, are studied, both related to COVID-19 outbreak.

The two events were chosen in consideration of the people's reactions. In both cases, citizens protested after local governments adopted lockdown rules, notwithstanding the strict restrictions imposed to tackle the pandemic.

In the case of Texas, rallies were organised to show disagreement against local restriction measures, and people demanded to reopen the economy [40,41].

By contrast, in Michigan, a convoy of thousands of motorists drove from all over the state to protest the governor's stay-at-home order extension. The protest, known now as Operation Gridlock, involved clogging with their vehicles the streets surrounding the Michigan State Capitol, including the Capitol Loop, and drew national attention [42].

4.1   Data Collection and Cleansing

A data corpus of six million tweets written in English was collected from 10th to 20th April 2020, by considering hashtags such as #covid, #coronavirus, #coronavirusoutbreak and #coronaviruspandemic. Then, two data subsets were extracted from the anterior dataset, each subset containing tweets with a unique combination of specific parameters, such as hashtags that were linked to the studied entities (locations), as depicted in Table 2.

Once these two subsets have been created, tweets appertained to the former clusters were cleansed following the steps described below: (1) URLs were extracted; (2) RT and mention terms were removed; (3) contractions were replaced, for instance, wasn't: was not; (4) punctuation marks were removed; (5) emoticons were replaced by words; (6) Internet slang was replaced by complete expressions using a preconfigured dictionary, for example, AFAIK: "as far as I know", ASAP: "as soon as possible", or BBL: "be back later".

Table 2. Popular hashtags posted on April 2020 linked to two locations, namely, Michigan and Texas. The depicted hashtags in the table involve two tokens, the first one associated with a location and the other with a noun/verb. The two types of tokens are shown in different colours - red and black.

Dataset 1 (Michigan ,USA)		Dataset 2 (Texas, USA)	
#michigan	#michiganprotest	#texas	#reopentexas
#liberatemichigan	#michiganlockdown	#opentexas	#stayhometexas
#freemichigan	#michiganshutdown	#texasstrong	#texans

4.2   Early Warning Alert (Q1)

Once an incident is unfolding, the stability of the state can be compromised due to national security components instability, at which point identifying if an event heads toward a significant disruption scenario becomes a primary task. In light of this, the analysis of three indicators, namely, Global Polarisation, Social Media Connectedness and Human Security Impact, enable the identification of the real nature of the event by triggering an early warning alert, as described in Sect. 3 [14].

Michigan. Figure 2.I shows that the system would generate an alert on 14th April 2020, a day before protests began because the governor's "stay at home" order was declared, and five days before protests escalated (19th April 2020). The triggered alarm suggests that the internal cohesion amongst national security components has been disrupted.

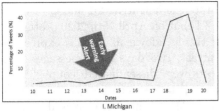
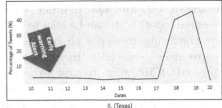

I. Michigan

II. (Texas)

Fig. 2. Early Warning Alert Detection in the events of Michigan and Texas in April 2020.

Texas. As depicted in Fig. 2.II on 11th April 2020, an alert was triggered by the early warning process, eight days before protests against Coronavirus policies intensified (19th April 2020).

4.3   Radical Behaviour (Q2, Q3 and Q4)

The analysis of radical behavioural traits can lead to critical and actionable insights. Table 1 demonstrates how addressing Q2, Q3 and Q4 enables the identification of entities, behavioural traits and required objects; justifying the use of the radical behaviour analysis methodology proposed in our previous work [16] to enrich this part of the analysis.

Table 3. Disruptive Expressions extracted using Word Embeddings and Direct Object (Texas and Michigan).

Location	Dates (2020)							
	April 10	April 11	April 12	April 15	April 16	April 17	April 18	April 19
I. Michigan				Violating -> Lockdown		Stop -> Insanity	Cancel -> Lockdown	Avoid -> Quarantine
				Disagree -> Curfew		Take -> Streets	Protest -> Lockdown	Violating -> Distancing
				Protest -> Rally		Break -> Demand	Take-> Lockdown	Michigan -> Edict
				Protest -> Virus		Protest -> Michigan	Protest -> Distancing	Need -> Lawmaker
						Wear -> Masks	Demand -> Reopening	Break -> Curfew
						Shut -> Now	Liberate -> Lockdown	Want -> Cure
							Block -> Roads	Want -> PPE
							Rally-> Arizona	Take -> Streets
II. Texas		Allow -> Business	Halt -> Covid			Reopen -> Government	Spreading -> Frustation	Hoarding -> PPE
		Avoid -> Corona	Develop -> Diarrhea			Need -> Michigan	See -> Outrage	Open -> Quarantine
		Lifting -> Quarantine	Open -> Employment			Close -> Schools	Wear -> PPE	Observe -> Distancing
		Reopen -> Texas	Help -> Employees				Wear -> Facemask	Protesting -> Coronavirus
		Care -> Lives	Puts -> Halt				Authorizing -> Reopen	Violate -> Lockdown
		Rise -> Deaths						Protest -> Lockdown
		Want -> Nurses						Support -> Boycott
								Make -> Masks
								Rally -> Whattsapp
								Rally -> Austin Texas

Michigan Q2. and Q3. In order to facilitate the narrative, Q2 and Q3 will be presented together. It can be seen in Table 3.I that on April 15th 2020 protesters were conveying messages about violating the lockdown as well as expressing disagreement against the measure. In contrast, two days later messages that expressed an intention to take to the streets were disseminated, coupled with messages that urged people to wear masks while protesting; moreover, messages suggesting the location of the protest, namely Michigan, were conveyed too.

On 18th and 19th April 2020, demands related to the lift of the lockdown and messages urging to continue protesting against the imposed measures were spread. In addition, some other ideas were present, such as demands to reopen, lift the quarantine, and liberate from the lockdown. Incitement to actions affecting various public thoroughfares, such as blocking roads or taking to the streets, were also present.

Q4. Lastly, messages were individuals conveyed their personal needs for PPE (personal protective equipment), or urge for action towards a cure for COVID-19 were shared likewise, see Table 3.I.

Texas Q2. Radical behavioural traits revealed that individuals expressed ideas linked to reopening a specific location, namely, Texas, see Table 3.II. According to the Levin's classification [43], the verb "need" expresses that a person desires something. Following Levin's analysis, on April 17th 2020, messages were posted conveying the desire that a different location, Michigan, would join the incident. As argued in [44], this mention of different locations, suggests that the state is dealing with a widespread event.

Q3. Social media users (Twitter) expressed concepts connected to the demand of allowing business in the city, lifting the quarantine, and contempt towards Coronavirus, as described in Table 3.II. On the other hand, in the following days (17th, 18th and 19th April 2020), messages that instigate violations of the lockdown, urge protest and boycott, close schools, wear PPE, and spread the frustration, were shared.

Q4. Concerns about health were also transmitted, related for example to the need for more nurses, and the rise of deaths.

4.4 Ideology (Q5)

The ideological traits of authoritarianism and hostility reveal important social characteristics. Authoritarianism denotes that individuals do not empathise with decisions or activities performed by those who hold the "proper authority" [37, 38]. Hostility enables the identification of collective emotions which are seen whilst disruptive events take place [35, 36].

In order to begin the dissection of ideology in the COVID-19 datasets, a sentiment analysis process was performed, then tweets with negative polarisation were selected accordingly. In both cases, negative sentiments played the predominant role; Michigan had the highest percentage with 51%, while Texas had 35%. Then authoritarianism and hostility traits were computed using the methodology and thresholds proposed in our previous study [18]. Consistently with our previous approach, when the calculated variables of authoritarianism (aggressiveness, submission and conventionalism) and hostility (anger, contempt and disgust) were above the predefined thresholds, the results were deemed to suggest that the aforementioned ideological traits were present.

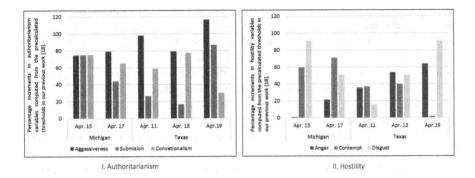

Fig. 3. Ideological traits (Michigan and Texas).

Michigan. A day after the early warning alert was triggered (April 14th 2020), signs of authoritarianism and hostility were detected (April 15th 2020), the same date that the local government imposed the lockdown, see Fig. 3.

Texas. On April 11th 2020, ideological traits were detected, the same date that the early warning alert was triggered, see Fig. 3. This specific point turns into a modular axis, since people were concerned about competing aspects such as the COVID-19 death toll, and lifting the quarantine, see the radical behaviour analysis in Subsect. 4.3 and Table 3.II.

Regarding authoritarianism, it should be noted that in both of the studied cases, aggressiveness was above 60% of the precalculated threshold, while the other two variables showed irregular increments. The consistent increase in aggressiveness suggests that people were conveying messages indicating prejudice/intolerance against a specific topic [38], here a lockdown, a curfew, or a quarantine, see Table 3.

4.5 Web Insights (Q6)

During an incident or a health crisis such as COVID-19, individuals and organisations use digital channels to disseminate information and data such as breaking news, messages or pictures, the analysis of which can help understand whether a crisis is escalating over time. Hence, as described in Section III, the web insights methodology proposed in our previous work [15] enables the study of the horizontal escalation of national security components. Following the methodology there, URLs were classified according to a comprehensive list of entities created over the Wikidata knowledge base. Then, a web scrapping process was conducted to retrieve the content of such web resources.

Michigan. It can be seen in Figs. 4.I and 4.II, that only two media resources were embedded in people's messages while posting a tweet, namely Independent Websites (IW) and Social Networking Services (SNS).

On April 14th 2020, when the early warning alert was triggered, SNS (Instagram and Twitter) were used to convey that one national security component was being affected, in this case, health. One day later, messages posted on those social media sites showed that four national security components were unbalanced (information, defence, business and health). Such increment in the number of affected components (from one to four), demonstrates a horizontal escalation, which according to [39] may represent a disruptive situation, see Fig. 4.I.

It should be noted that both business and government components had the highest intensities, which may complement the behavioural traits previously extracted that referred to violating the lockdown and the disagreement towards that measure (see Table 3.I).

On the other hand, IW showed on April 15th 2020, that three national security components were affected, namely defence, information and government, with government having the highest intensity figure. The result suggests that those web resources were providing a more detailed description of the government's activity (see Fig. 4.II).

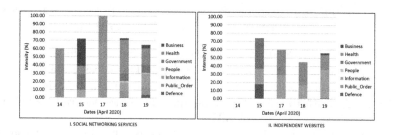

Fig. 4. Horizontal Escalation of the National Security Components during the protests in Michigan (April 2020)

On the following days (17th, 18th and 19th April 2020), both IW and SNS published content affecting the health and information components. The result is relevant since, on April 19th, COVID-19 cases began to spike [42]. By contrast, only SNS revealed information about two other components (people and public order), as displayed in Fig. 4.I.

Texas. Figure 5.I, 5.II and 5.III show that three digital web resources were used by people to disseminate information amidst the protests, namely, IW, SNS and Non-Profit Organisations.

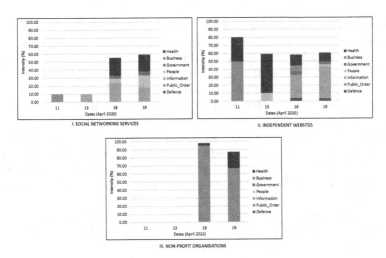

Fig. 5. Horizontal Escalation of the National Security Components during the protests in Texas (April 2020).

As mentioned earlier, the early warning alert and the emergence of ideological traits happened on the same date (11th April 2020). Unlike the previous case, the Independent Websites were used more intensively and they unveil that two national security components were disrupted, business and health; while SNS showed that only the business component was affected, with 80% less intensity than in IW.

Visible changes were displayed between 13th and 18th April 2020, as the IW and SNS showed an increased number of affected components, which exposed a horizontal escalation across the national security factors, which went from two affected factors to five for the IW, and from one to four for SNS.

In addition, Non-profit organisations played a crucial role on the 18th and 19th April 2020, because topics in business and health were affected by them. Moreover, intensity health levels had a considerable increase of 85%, while, by contrast, health levels in SNS and IW showed little change, around 7% on average. Such a difference indicates that Non-profit organisations were stressing issues linked to health.

Finally, it should be noted that on April 19th, when the highest burst of online activity took place (see Fig. 2.II), SNS were used to convey more messages linked to people, as indicated by an increase of 70%; whereas IW were focused on disseminating data related to the information component, which had an increment close to 28%.

5 Conclusions

As COVID-19 has so vividly demonstrated, pandemics constitute a serious National Security threat. Big Data Analytics technologies can provide a powerful force multiplier in the endeavour of competent authorities and stakeholders to manage the pandemic while minimising the security threats it poses.

As a crisis is unfolding, uncertainty is a crucial element and the lack of information is a variable that can obstruct the decision-making process. This paper has discussed the utilisation of a holistic Data Analytics framework for analysing national security aspects in the context of COVID-19. Two real-world cases were considered where authorities' measures to contain the disease via lockdowns resulted in protests and social unrest, namely Michigan and Texas. In both cases the system proved its ability to provide early warning well in advance of the demonstrations (six and eight days respectively). It also demonstrated its capacity to provide insights enabling the better understanding and interpretation of the crisis.

Future work will fully integrate and automate the framework, utilising additional analytics approaches and making the various thresholds involved adaptive to the socio-economic context of deployment.

References

1. Kharroubi, S., Saleh F.: Are Lockdown Measures Effective Against COVID-19? Front Public Health. **8**, 549692 (2020). Accessed 22 Oct 2020, https://doi.org/10.3389/fpubh.2020.549692
2. Peretti-Watel, P., Seror, V., Cortaredona, S., Launay, O., Raude, J., Verger, P., et al.: Attitudes about COVID-19 Lockdown among General Population, France, March 2020. Emerg. Infect. Dis. **27**(1), 301–303 (2021). https://doi.org/10.3201/eid2701.201377
3. Human Development Report: United Nations Development Program, pp. 22–33. Oxford University Press, New York and Oxford (1994)
4. Kureshi, I., Theodoropoulos, G., Magina, E., O'Hare, G., Roche, J.: Towards an info-symbiotic framework for disaster risk management. In: The 19th IEEE/ACM International Symposium on Distributed Simulation and Real Time Applications (DS-RT 2015), Xinhua International Hotel, Chengdu, China, 14–16 October 2015 (2015)
5. Toth, T., Theodoropoulos, G., Boland, S., Kureshi, I., Ghandar, A.: Global challenge governance: time for big modelling? In: 2019 IEEE 18th International Conference on Cognitive Informatics & Cognitive Computing (ICCI*CC), Milan, Italy, pp. 244–253 (2019)
6. Roff, H.: Uncomfortable Ground Truths: Predictive Analytics and National Security. Brookings National Security Report (2020)
7. Akhgar, B., Saathoff, G., Arabnia, H., Hill, R., Staniforth, A., Bayerl, P.: Application of big data for national security: a practitioner's guide to emerging technologies. Butterworth-Heinemann, Paperback ISBN: 9780128019672, eBook ISBN: 9780128019733 (2015)
8. Dolyle, A., et al.: The EMBERS architecture for streaming predictive analytics. In: IEEE International Conference on Big Data, USA (2014)
9. ICEWS Homepage. https://lmt.co/3owTj8L, Accessed 10 Jan 2021
10. Muthiah, S., et al.: EMBERS at 4 years: experiences operating an open source indicators forecasting system. In: Proceedings of the 22nd ACM SIGKDD, New York, NY, USA (2016)

11. Parang, S., Naren, R.: EMBERS AutoGSR: automated coding of civil unrest events. In: Proceedings of the 22nd ACM SIGKDD International Conference on Knowledge Discovery and Data Mining (KDD 2016). Association for Computing Machinery, New York (2016)

12. Zhang, X., Li, J., Zhang, X., Fu, J., Wang, D.: Construction of a geopolitical environment simulation and prediction platform coupling multi-source geopolitical environmental factors. Sci. Technol. Rev. **36**(3) (2018)

13. Van Puyvelde, D., Coulthart, S., Hossain, M.S.: Beyond the buzzword: big data and national security decision-making. Int. Affairs **93**(6) (2017)

14. Cárdenas, P., Theodoropoulos, G., Obara, B., Kureshi, I.: Defining an alert mechanism for detecting likely threats to National Security. In: IEEE International Conference on Big Data, USA (2018)

15. Cárdenas, P., Theodoropoulos, G., Obara, B.: Web insights for national security: analysing participative online activity to interpret crises. In: IEEE International Conference on Cognitive Informatics and Cognitive Computing, Italy (2019)

16. Cárdenas, P., Obara, B., Theodoropoulos, G., Kureshi, I.: Analysing social media as a hybrid tool to detect and interpret likely radical behavioural traits for national security. In: IEEE International Conference on Big Data, USA (2019)

17. Cárdenas, P., Theodoropoulos, G., Obara, B., Kureshi, I.: A conceptual framework for social movements analytics for national security. In: The International Conference on Computational Science (2018)

18. Cárdenas, P., Theodoropoulos, G., Obara, B., Kureshi, I.: Unveiling ideological features through data analytics to construe national security instabilities. In: IEEE International Conference on Big Data, USA (2020)

19. Idris, M.Y.: Social Instability, Policy Uncertainty, and Financial Risk: Evidence from the Egyptian Exchange and Borse de Tunis, Belfer Center for Science and International Affairs Harvard Kennedy School. (2015)

20. Institute for Economics and Peace: Global Peace Index **2020** (2020)

21. Wilson, J.: US lockdown protests may have spread virus widely, cellphone data suggests, The Guardian (2020)

22. Davies, S.E.: National security and pandemics, United Nations, UN Chronicle (2013)

23. Klarevas, L., Clarke, C.: COVID-19 Is a Threat to National Security. Let's Start Treating It as Such, Just Security (2020)

24. Sandoval-Almazan, R., Gil-Garcia, J.R.: Cyberactivism through social media: twitter, youtube, and the Mexican political movement "i'm number 132". In: Hawaii International Conference on System Sciences (2013)

25. Storck, M.: The role of social media in political mobilisation: A case study of the January 2011 Egyptian uprising. University of St Andrews, Scotland (2011)

26. Castillo, C.: Big Crisis Data: Social Media in Disasters and Time-Critical Situations. Cambridge University Press, Cambridge (2016)

27. Bagrow, J.P., Wang, D., Barabási, A.-L.: Collective response of human populations to large-scale emergencies. PLoS ONE **6**(3) (2011)

28. Personal Protective Equipment: Questions and Answers, Centers for Disease Control and Prevention Homepage. https://bit.ly/38wHUQU, Accessed 10 May 2020

29. Coronavirus (COVID-19) Outbreak Glossary, Kaiser Family Foundation. https://bit.ly/3bn4j4R, Accessed 15 Jan 2021

30. Becker, A.: COVID-19 crisis catalog: A glossary of terms, Texas Medical Center Home Page. Accessed 26 May 2020

31. Rudé, G.: Ideology and popular protest. Hist. Reflect./Réflexions Historiques **3**(2), 69–77 (1976)

32. Thompson, J.B.: Studies in the Theory of Ideology. Polity Press, Cambridge (1984)
33. Van Dijk, T.A.: Ideology and discourse analysis. J. Polit. Ideol. **11**(2), 115–140 (2006)
34. Gutting, R.S.: Contentious activities, disrespectful protesters: effect of protest context on protest support and mobilization across ideology and authoritarianism. Polit. Behav. **42**, 865–890 (2019)
35. Matsumoto, D., et al.: The role of intergroup emotions in political violence. Curr. Direct. Psychol. Sci. **24**(5), 369–373 (2015)
36. Troost, D.: Emotions of Protest. Emotions in Politics. Palgrave Studies in Political Psychology series, Palgrave Macmillan, London (2013)
37. Saunders, B.A., Ngo, J.: The right-wing authoritarianism scale. In: Zeigler-Hill, V., Shackelford, T. (eds.) Encyclopedia of Personality and Individual Differences, pp. 71–76. Springer, Cham. (2017). https://doi.org/10.1007/978-3-319-28099-8_1262-1
38. Dunwoody, P., Funke, F.: The aggression-submission-conventionalism scale: testing a new three factor measure of authoritarianism. J. Social Polit. Psychol. **4**(2), 571–600 (2016)
39. Cullen, P.J., Reichborn-Kjennerud, E.: MCDC Countering Hybrid Warfare Project: Understanding Hybrid Warfare. Multinational Capability Development Campaign (2017)
40. Jeffery, A.: Scenes of protests across the country demanding states reopen the economy amid coronavirus pandemic, CNBC. Accessed 18 Apr 2020
41. Hundreds protest COVID-19 orders at Texas Capitol, FOX4News Homepage. https://bit.ly/38IA1rl, Accessed 14 Jan 2021
42. Hutchinson, B.: Operation Gridlock': Convoy in Michigan's capital protests stay-at-home orders, ABCNews. Accessed 16 Apr 2020
43. Levin, B.: English Verb Classes and Alternations: A Preliminary Investigation. University of Chicago Press, Chicago (1993)
44. Reed, T., et al.: Open Source Indicators Program Handbook. The MITRE Corporation, USA (2017)

Efficient Prediction of Spatio-Temporal Events on the Example of the Availability of Vehicles Rented per Minute

Bartlomiej Balcerzak[1]([✉]), Radoslaw Nielek[1,2], and Jerzy Pawel Nowacki[1]

[1] Polish-Japanese Academy of Information Technology, Warsaw, Poland
{b.balcerzak,nielek,nowacki}@pja.edu.pl
[2] Vooom Sp, ul. Jana Czeczota 9, 02-607 Warsaw, Poland

Abstract. This article shows a solution to the problem of predicting the availability of vehicles rented per minute in a city. A grid-based spatial model with use of LSTM network augmented with Time Distribution Layer was developed and tested against actual vehicle availability dataset. The dataset was also made publicly available for researchers as a part of this study. The predictive model developed in the study is used in a multi-modal trip planner.

Keywords: Carsharing · Spatio-temporal events · Machine learning · Prediction · City

1 Introduction

Nowadays, the need for mobility in cities is increasingly met by free-float vehicles sharing systems. Typically, such systems allow vehicles to be rented and returned in any, location within the city area. The specificity of the system is the reason why both from the perspective of service providers and customers, there is a need for efficient prediction of vehicle demand and availability in terms of time and space.

Prediction of spatio-temporal events in urban space is a difficult task due to the number of dimensions that need to be taken into account and high level of non-linearity. The availability of vehicles depends on the day of the week, time of day and location. Next to zones with low and high demand, we also have areas where cars cannot be parked such as closed areas (e.g. military base), parks, ponds, rivers, or pedestrian zones. Moreover, European cities, as opposed to American ones, usually have irregular road networks, what makes the task even more complex. Due to the fact that the prediction results are intended for use in a commercial route planning tool[1], the model, in addition to high precision, must be characterized by high level of adaptability, robustness, and moderate demand for computing power.

---

[1] The prediction module based on results described in this paper is part of the commercial MaaS solution offered by Vooom Inc. available at https://planner.app.vooom.pl/.

© Springer Nature Switzerland AG 2021
M. Paszynski et al. (Eds.): ICCS 2021, LNCS 12742, pp. 83–89, 2021.
https://doi.org/10.1007/978-3-030-77961-0_8

1.1 Contribution

In this article, we present attempts to build an efficient prediction model of the cars availability with the aim to use it in an intelligent multi-modal trip planner. Planning a transfer in intra-city trips using shared free-floating vehicles requires confirmation of the availability of the means of transport in a specific place in the perspective of several dozen minutes (usually in a horizon up to 60 min).

The model we propose is new contribution to the field, since it mostly focuses on the supply side of the car sharing market. Thus, it is based of the trends in availability of cars in a specific city. What is more, by being based on the supply side of things, the model does not require user information for profiling and efficient work. Finally, the model we propose offers means for predicting car availability in long time spans, exceeding the standard time frames for booking cars for short term rent. Creating a city spanning model offers also a challenge for creating efficient architectures for predicting the dynamics of objects sparsely spread within the city grid.

2 Related Work

Prediction of spatio-temporal events in the context of urban transportation has a variety of practical applications, and, therefore, it has been intensively researched in recent years. Li et al. have studied the problem of the availability of free parking lots in NYC [9]. They tested four different spatial units, such as: POIs (Places of Interest), streets, census tracts and 1km grid. It turns out that the best results can be achieved with a combination of 1km grid and random forest algorithm. Robustness of this approach were also confirmed by Goa et al. in a work focused on legality of on-street parking [6], but other methods, such as Graph-Convolutional Neural Networks [15] or Long Short Memory Model (LSTM) [11], were also tested with good results.

Studies done by Schmoeller et al. [10] and Wagner et al. [12] revealed that free-floating carsharing usage concentrates on relatively few hot spots, POIs, such as restaurants and stores, can explain part of spatial variation in car-sharing activities, and the fact that strong imbalance spatial imbalance exists between vehicles demand and supply. As shown by Willing et al. [14], spatial and temporal dimensions are interdependent.

In order to reduce spatio-temporal supply and demand mismatch, various relocation strategies were studied [13]. The majority of studies were focused on solving relocation problem in one-way, station-based systems (see [8] for literature review) or algorithmic aspects of multidimensional optimization problem with constrains (e.g. with use of evolutionary algorithm [7] or operational research [4]). Noteworthy exceptions are works done by Ai et al. [1] (convolutional long short-term memory network applied to prediction of dockless bikes distribution), Bao et al. [2] (hybrid deep learning neural networks used to predict bike availability within 15, 20 and 30 min), Formentin et al. [5] (prediction of the distance to the nearest available vehicle), and Dario et al. [3] (an extensive study on the impact of the type of learning algorithm, time horizon and the specificity of the environment on the accuracy of prediction of vehicle availability in a given area).

3   Model and Problem Definition

Having the presented research in mind, we decided to propose a solution that mainly focuses on the supply side of the car sharing market. Based on the literature review, we decided to use an LSTM model augmented with a Time Distributed Layer with information about car location taken from the last 60 min used as input. This architecture has proven to be the most efficient of those we tested in our work. The location is a 100 m × 100 m square. The total space used in our study (the city of Warsaw and it's direct surroundings) are represented as a matrix of 40 thousands such squares, arranged in 200 by 200 squares shape (thus, covering a shape of 400 squared kilometers). This granularity allows the model to predict car availability at a close proximity of the user. The selection of a 100 m × 100 m cell (henceforth defined as a location) was chosen based on usability of the model. 100 m distance is within a walking distance of a user searching to find available vehicles, and thus our model requires this level of precision.

As a baseline model, we decided to use a simple model which assumes that all cars available at a given time in a given location would remain available at that location. We wanted to test the model's performance at a 15, 30, 45 and 60 min mark.

The model is trained in real time, with each training step occurring on a minute by minute basis. During each training step, the current map of Warsaw (herein designated as t_0 is treated as a target, with 10 past maps serving as inputs. For example, in the case of the 15 min iteration of the model, these 10 inputs are randomly selected from a time span between t_{-15} and t_{-15-60}. This additional randomisation step allows us to minimize the risk of overfitting. The model weights are reset every 24 h at 4 am, since, based on observations of the model behavior, doing so prevents the occurrence of concept drift.

4   Dataset

In our work we used a dataset spanning a period of 30 days between November 1st and November 30th. The data was collected through the use of APIs made available by car rental providers operating in Warsaw, in real time and stored in an SQL database. Due to occasional issues with data availability (instances when API failed to provide data at a given time), the dataset was pruned in order to remove timestamps with a significantly lower number of missing values (these were datastamps that yielded a number of available cars lower than the mean daily value by two standard deviations).

The dataset represents a minute-by-minute availability of cars for rent within the boundaries of the city of Warsaw during daytime hours (between 6 am and 8 pm). Performance comparison within night time (from 1am to 6am) produced a similar pattern to day time hours.Table 1 below represents the basic information about our dataset, while Fig. 1 represents an example of a single observation from the dataset.

Table 1. Basic characteristics of the dataset

Property	Value
Number of observations	25200 (over 50 million records; 4.5 GB raw data)
Size of the observation	200 × 200 binary matrix with each cell representing a 100 × 100 m^2
Coordinates delimiting the analyzed area	20.9°–21.1°, 52.14°–52.35°
Time period represented in the data	11.01.2020–11.30.2020
Hour time span represented in the data	6am–8pm
Average number of locations with available cars	646.58
% of locations where no cars were observed	64.1%

The data collected represents a span of 30 days, thus giving us an opportunity to view how well does the prediction model work during a prolonged period. In addition, the average number of occupied locations at a single moment (around 646), when compared to the size of the grid (40,000 locations) indicates, a very sparse matrix, which provides an additional level of challenge for the classifier. We made the dataset available online[2].

Fig. 1. A heat-map representation of the locations of available cars on December 3rd 2020. The maps were taken for 8 am (left), 11am (right).

---

[2] http://nielek.com/datasets/ICCS2021.html.

4.1   Performance Measures

From a business perspective, there are two instances of incorrect prediction which lead to the most critical outcomes. Firstly, a case when the classifier incorrectly predicts the car to be available at a given location in the future. In such a case the travel planner leaves the user at a place with possibly no available transport, thus increasing user frustration.

Secondly, the classifier may predict a location to have no available vehicles, while such vehicles are present at the location. That may lead to sub optimal routes, thus increasing travel time.

Therefore, our selected measures of classifier performance ought to take these scenarios in to account. With that in mind, we decided to focus on two main measures of classifier performance, i.e. positive class precision and positive class recall.

In order to provide a more general view of the model's performance, we will primarily use F1-score as method for evaluating the feasibility of the proposed model. Since the output of the model is a 200×200, the precision and recall can be calculated on a minute-by-minute basis, thus providing information about the number of correctly predicted locations at a given time.

5   Results

Firstly, we decided to compare results for the entirety of the 30 day period. These aggregated results are presented in Table 2. While the performance measured with the F-1 score shows no improvement over the baseline in the 15 min time frame, the model outperforms the base line in the more distant time frames. One can notice gradual descent of the F-1 score value for the baseline, compared with the more consistent results of the model. What is more, when it comes to the precision score, the model outperforms the baseline in all selected time frames. In the case of the recall, the base line proves superior only in the 15 min mark. These results may be partially explained by the very nature of car sharing. Most providers allow for the booking of a car to stay active for 15 min, thus reducing the mobility of the cars within this time frame. This feature also makes the 15 min time frame less crucial for the performance of the model, while further emphasizing the importance of prediction in the long term, where the user has no means of booking a car for that long.

Table 2. Comparison of the averaged performance measures for the entire month of November. Prediction threshold of 80%

Time frame	Precision/Recall/F1-Score	Baseline precision/Recall/F1-score
15 min	0.9/0.83/0.87	0.86/0.86/0.86
30 min	0.84/0.85/0.84	0.79/0.8/0.79
45 min	0.81/0.87/0.83	0.76/0.76/0.76
60 min	0.77/0.87/0.81	0.72/0.72/0.72

In addition a logistic regression model was also used, the results were below
the baseline level. In order to further review the robustness of our model, we
decided to analyze it's performance during different times of the day, and different days of the week. Results are presented in Fig. 2. The model seems to
be giving the best performance during morning hours and in the early after-
noon, with a significant drop around noon. A similar trend can be observed
when comparing performance by weekdays, with higher scores being observed
on weekends.

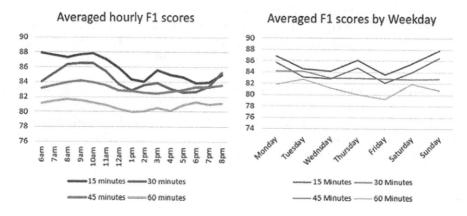

Fig. 2. Averaged F1 scores by hour (left) and by weekday (right)

6 Conclusion

In this article we wanted to propose model allowing for the prediction of the
availability of cars for rent in Warsaw. The model was able to predict the availability of cars within the time spans ranging from 15 to 60 min. The results we
obtained, have shown that our model outperforms the baseline when it comes
to the precision of the prediction. This improvement in performance was mostly
observed when considering time frames such as 45 and 60 min marks. The results
have also shown that the prediction model exceeds the 80% threshold for the
F1 score, regardless of the time of day and time of week; although higher performance was noted in morning hours and during the weekends. This effect,
however, was weaker when predicting longer time spans. The model proposed in
this article proved to be an effective method of predicting car for rent availability
in short time spans. In our future work, we will aim to extend the model to cover
other shared means of transportation, as well as test to the validity of the model
in the case of other cities.

Acknowledgment. Project was partially financed by EU European Regional Developement Fund within the Inteligent Developement Program. Project realised within
the Narodowe Centrum Badań i Rozwoju Szybka Ścieżka program.

References

1. Ai, Y., et al.: A deep learning approach on short-term spatiotemporal distribution forecasting of dockless bike-sharing system. Neural Comput. Appl. **31**(5), 1665–1677 (2019)
2. Bao, J., Yu, H., Wu, J.: Short-term ffbs demand prediction with multi-source data in a hybrid deep learning framework. IET Intell. Transp. Syst. **13**(9), 1340–1347 (2019)
3. Daraio, E., Cagliero, L., Chiusano, S., Garza, P., Giordano, D.: Predicting car availability in free floating car sharing systems: leveraging machine learning in challenging contexts. Electronics **9**(8), 1322 (2020)
4. Folkestad, C.A., Hansen, N., Fagerholt, K., Andersson, H., Pantuso, G.: Optimal charging and repositioning of electric vehicles in a free-floating carsharing system. Comput. Oper. Res. **113**, 104771 (2020)
5. Formentin, S., Bianchessi, A.G., Savaresi, S.M.: On the prediction of future vehicle locations in free-floating car sharing systems. In: 2015 IEEE Intelligent Vehicles Symposium (iv), pp. 1006–1011. IEEE (2015)
6. Gao, S., Li, M., Liang, Y., Marks, J., Kang, Y., Li, M.: Predicting the spatiotemporal legality of on-street parking using open data and machine learning. Ann. GIS **25**(4), 299–312 (2019)
7. Herbawi, W., Knoll, M., Kaiser, M., Gruel, W.: An evolutionary algorithm for the vehicle relocation problem in free floating carsharing. In: 2016 IEEE Congress on Evolutionary Computation (CEC), pp. 2873–2879. IEEE (2016)
8. Illgen, S., Höck, M.: Literature review of the vehicle relocation problem in one-way car sharing networks. Transp. Res. Part B Methodol. **120**, 193–204 (2019)
9. Li, M., Gao, S., Liang, Y., Marks, J., Kang, Y., Li, M.: A data-driven approach to understanding and predicting the spatiotemporal availability of street parking. In: Proceedings of the 27th ACM SIGSPATIAL International Conference on Advances in Geographic Information Systems, pp. 536–539 (2019)
10. Schmöller, S., Weikl, S., Müller, J., Bogenberger, K.: Empirical analysis of free-floating carsharing usage: the Munich and Berlin case. Transp. Res. Part C Emerg. Technol. **56**, 34–51 (2015)
11. Shao, W., Zhang, Yu., Guo, B., Qin, K., Chan, J., Salim, F.D.: Parking availability prediction with long short term memory model. In: Li, S. (ed.) GPC 2018. LNCS, vol. 11204, pp. 124–137. Springer, Cham (2019). https://doi.org/10.1007/978-3-030-15093-8_9
12. Wagner, S., Brandt, T., Neumann, D.: In free float: developing business analytics support for carsharing providers. Omega **59**, 4–14 (2016)
13. Weikl, S., Bogenberger, K.: Relocation strategies and algorithms for free-floating car sharing systems. IEEE Intell. Transp. Syst. Mag. **5**(4), 100–111 (2013)
14. Willing, C., Klemmer, K., Brandt, T., Neumann, D.: Moving in time and space-location intelligence for carsharing decision support. Decis. Supp. Syst. **99**, 75–85 (2017)
15. Yang, S., Ma, W., Pi, X., Qian, S.: A deep learning approach to real-time parking occupancy prediction in transportation networks incorporating multiple spatio-temporal data sources. Transp. Res. Part C Emerg. Technol. **107**, 248–265 (2019)

Grouped Multi-Layer Echo State Networks with Self-Normalizing Activations

Robert Wcisło and Wojciech Czech$^{(\boxtimes)}$ (iD)

Institute of Computer Science, AGH University of Science and Technology,
Kraków, Poland
czech@agh.edu.pl

Abstract. We study prediction performance of Echo State Networks with multiple reservoirs built based on stacking and grouping. Grouping allows for developing independent subreservoir dynamics, which improves linear separability on readout layer. At the same time, stacking enables to capture multiple time-scales of an input signal by the hierarchy of non-linear mappings. Combining those two effects, together with a proper selection of model hyperparameters can boost ESN capabilities for benchmark time-series such as Mackey Glass System. Different strategies for determining subreservoir structure are compared along with the influence of activation function. In particular, we show that recently proposed non-linear self-normalizing activation function together with grouped deep reservoirs provide superior prediction performance on artificial and real-world datasets. Moreover, comparing to standard tangent hyperbolic models, the new models built using self-normalizing activation function are more feasible in terms of hyperparameter selection.

Keywords: Echo state network · Self-normalizing activation ·
Reservoir computing · Deep ESN

1 Introduction

Echo State Networks (ESN), being the leading representative of Reservoir Computing (RC) framework are intensively studied in last years owing to convenient learning paradigm and promising results in modeling dynamic systems. They are classified as a constrained variant of Recurring Neural Networks (RNNs), for which weights of hidden layer are not trained but initialized once, typically randomly [8]. ESNs were successfully applied for univariate/multivariate time series prediction and currently provide proven framework for modeling difficult chaotic systems [11]. Most recent applications of ESNs include medical multivariate time series [3], human activity sensor data [10], electrical load [2] or robot navigation [4].

In recent years, we observe growing number of works addressing the role of reservoir topology. In particular, deep ESNs were introduced as a robust variant of ESN utilizing the effect of layering, which allows to capture multi-scale

M. Paszynski et al. (Eds.): ICCS 2021, LNCS 12742, pp. 90–97, 2021.
https://doi.org/10.1007/978-3-030-77961-0_9

time characteristics of an input signal [7]. Different structural constraints can be applied to the graph of hidden layer links resulting in grouped reservoirs [6], growing subreservoirs organized in blocks [12], small-world reservoirs with segregated input/output nodes [9] or deep reservoirs with simple non-radom structure [5]. The models exhibiting non-trivial, decoupled topology were proven to outperform shallow ESNs in time-series prediction on the most popular benchmark datasets. At the same time, important results regarding non-linear activation function were published in [13]. The authors introduced new, non-linear activation function called Self-Normalizing Activation (SNA). The SNA function projects network pre-activations onto hyperspehere and ensures stable behaviour of ESN, which cannot enter chaotic regime and is less sensitive to perturbations of hyperparameters.

Motivated by recent findings regarding reservoir structure as well as promising properties of SNA, we study multi-layer ESNs with SNA and *tanh* activations. We also propose hybrid ESN architecture with layerying and grouping, then based on benchmark datasets, show that it can achieve superior prediction performance comparing to shallow architectures with the same number of nodes and deep architectures with *tanh* activation function. The key contributions of this work are new multi-layer grouped ESN architecture with SNA activation and their comparison with previously analyzed deep, grouped and shallow ESNs, as well as standard models such as LSTM or moving average. In addition, we present a modular software framework for ESN configuration and testing together with the associated open-source Python library. The library (AutoESN) allows for building different types of ESNs including stacked, grouped, growing and provides implementation of kernel and SNA activation functions.

2   ESN Architectures

Herein, we describe shallow ESNs and provide details on SNA function together with the comment on its parameter *activation radius*. We also present new, grouped multi-layer architecture with SNA activation, which will be tested against standard ESN models with tangent-hyperbolic activation further in experimental section.

2.1   Shallow Echo State Network

Shallow Echo State Network is defined using following state transition equations:

$$\boldsymbol{x}(n) = (1 - \alpha)\boldsymbol{x}(n-1) + \alpha\tilde{\boldsymbol{x}}(n) \tag{1}$$

$$\tilde{\boldsymbol{x}}(n) = f(\boldsymbol{W}_{in}[1; \boldsymbol{u}(n)] + \boldsymbol{W}_x\boldsymbol{x}(n-1)) \tag{2}$$

$$\boldsymbol{y}(n) = \boldsymbol{W}_{out}[1; \boldsymbol{u}(n); \boldsymbol{x}(n)] \tag{3}$$

where $\boldsymbol{u}(n) \in \mathbb{R}^{N_u \times 1}$ is an input signal, $\boldsymbol{y}^t(n) \in \mathbb{R}^{N_y \times 1}$ - an output signal, $n \in \{1, \ldots, T\}$ - discrete time, T - a number of data points in the training dataset, N_x - a number of nodes in the reservoir, N_u - dimensionality of an input, N_y - dimensionality of output, $x(n) \in \mathbb{R}^{N_x \times 1}$ - vector of reservoir node activations, $y(n) \in \mathbb{R}^{N_y \times 1}$ - network output (trained ESN model), $W_{in} \in \mathbb{R}^{N_x \times (N_u+1)}$ - input weight matrix, $W_x \in \mathbb{R}^{N_x \times N_x}$ - recurrent weight matrix, $W_{out} \in \mathbb{R}^{N_y \times (1+N_u+N_x)}$ - readout weight matrix, $f : \mathbb{R}^{N_x} \to \mathbb{R}^{N_x}$ - activation function, $\alpha \in (0, 1]$ - leaking rate and $[\cdot; \cdot]$, $[\cdot; \cdot; \cdot]$ - vertical vector (matrix) concatenation operators. Most typically, W_{in} is fully-connected or dense, while W_x is sparse. Spectral radius ρ is used to scale initial weights of W_x as follows: generating random weights, calculating maximal absolute eigenvalue of W_x, dividing all elements of W_x by this value and scaling obtained matrix with ρ. In case of *tanh* activation function, which currently dominates in practical applications of ESNs, spectral radius and input scaling were identified as the most important hyperparameters affecting non-linearity and memory. Nevertheless, in case of different activation functions such as SNA, the observations are different.

2.2   Self-Normalizing Activation Function on Hyper-Sphere

ESN models were originally built using tangent hyperbolic activation functions, which are applied element-wise on a pre-activation vector. More recently, the self-normalizing activation (SNA) function was proposed as the robust alternative to *tanh*, providing stable predictions and reducing model sensitivity to hyperparameters [13]. It is defined as follows:

$$a(n) = W_{in}[1; u(n)] + W_x x(n-1), \tag{4}$$

$$\tilde{x}(n) = r \frac{a(n)}{\|a(n)\|}, \tag{5}$$

where $a(n)$ is pre-activation vector, $\tilde{x}(n)$ is post-activation vector used in Eq. 1, $\|\cdot\|$ is norm and r is an activation radius - the parameter of SNA function. SNA is the global function (neuron state depends on the states of other neurons), which projects pre-activation vector $a(n)$ into $(N_x - 1)$-dimensional hyper-sphere of radius r. One of interesting SNA properties is that it brings non-linearity and at the same time ensures high memory capacity - on the level of the one exhibited by linear activations. Therefore, it seems to be the best solution for memory non-linearity trade-off. In addition, SNA guarantees that ESN does not display chaotic behaviour as the maximum Lapunov exponent is always zero. Those important characteristics of SNA were proven theoretically and also confirmed by experiments [13]. It can be shown that memory capacity of reservoir with SNA depends on the product $r\rho(W_x)$, therefore in the experimental setup one of those parameters can be fixed. At the same time, scaling factor $\frac{r}{\|a(n)\|}$ (see Eq. 5) significantly reduces the effect of input scaling s. So far, SNA was tested on shallow ESN architectures. Considering its global nature, which can further enrich dynamics reflected by subreservoirs, we decided to study the effects of introducing SNA in more complex ESN architectures.

2.3 Grouped Deep Echo State Network

Combining deep, decoupled ESNs [6] with Self-Normalizing Activation function, we propose new architecture called grouped deep ESN (*gdESN*), which uses two-dimensional organization of subreservoirs. Each subreservoir can be characterized by different hyperparameters: size, leaking rate, activation radius, spectral radius, input scaling and sparsity. The state transitions of *gdESN* are described using equations:

$$\boldsymbol{x}^{(i,j)}(n) = (1 - \alpha^{(i,j)})\boldsymbol{x}^{(i,j)}(n-1) + \alpha^{(i,j)}\tilde{\boldsymbol{x}}^{(i,j)}(n), \qquad (6)$$

$$\tilde{\boldsymbol{x}}^{(i,j)}(n) = f(\boldsymbol{W}_{in}^{(i,j)}[1; \boldsymbol{v}^{(i,j)}(n)] + \boldsymbol{W}_x^{(i,j)}\boldsymbol{x}^{(i,j)}(n-1)), \qquad (7)$$

$$\boldsymbol{v}^{(i,j)}(n) = \begin{cases} \boldsymbol{u}(n) & \text{if } j = 1 \\ \boldsymbol{x}^{(i,j-1)}(n) & \text{if } j > 1, \end{cases} \qquad (8)$$

where N_g is the number of groups ($1 < i \leq N_g$), N_l - the number of layers ($1 < j \leq N_l$) and $\alpha^{(i,j)}$ is leaking rate for i-th group, j-th layer. For *gdESN* with SNA, the function $f \equiv SNA$ but in experimental section we also consider hyperbolic tangent activations.

$$\boldsymbol{y}(n) = \boldsymbol{W}_{out}[1; \boldsymbol{x}^{(1,1)}(n); \ldots; \boldsymbol{x}^{(1,N_l)}(n); \ldots; \boldsymbol{x}^{(N_g,1)}(n); \ldots; \boldsymbol{x}^{(N_g,N_l)}(n)] \quad (9)$$

The node activations $\boldsymbol{x}^{(i,j)}(n)$ from all subreservoirs are concatenated, therefore $\boldsymbol{W}_{out} \in \mathbb{R}^{N_y \times (N_x N_g N_l + 1)}$. The new architecture allows for utilizing decoupled dynamics of grouped ESNs (*gESN*), ordered time-scale organization of deep ESNs (*dESN*) and stability of SNA at the cost of more difficult hyperparameter optimization. Nevertheless, thanks to properties of SNA, the search space can be reduced by fixing input scaling and spectral radius. Moreover, we also gain better computational efficiency. For the fixed total number of neurons N, multiple subreservoirs reduce computational complexity of state updates with $\mathcal{O}(N^2)$ being the cost of matrix-vector multiplication for shallow architecture and $\mathcal{O}(N^2/(N_g N_l))$ - the cost for *gdESN* with N_g groups and N_l layers.

3 AutoESN Library

As a software contribution of this work we created open-source Python library called AutoESN. This library uses PyTorch and provides comprehensive tools for configuring and training decoupled ESN architectures. Comparing to other libraries, like DeepESN [1] our software is based on PyTorch framework, has GPU support and is more flexible in terms of reservoir configuration. Deep-ESN supports deep and shallow ESNs as well as weights initialization and fixed activation functions, which cannot be easily adapted.

In case of AutoESN, we developed the tool for creating different types of decoupled ESNs (including *gdESN*), which enables to mix reservoirs with different kinds of readouts and different custom activation functions. The library consists of two core modules: `Reservoir` and `Readout`. At the core of `Reservoir` module the user can find `GroupedDeepESN` class, which enables creating vanilla ESNs, as well as grouped, deep and grouped deep architectures. Weights initialization and activation functions are separated from reservoir logic. This way, changing regular ESN into subreservoir ESN or into the one having more complex structures of the weight matrices (for example arbitrary graphs) can be achieved with just a few extra lines of code. One can also easily create and exchange activation functions (SNA function is supported natively). The code and documentation of AutoESN library can be accessed on https://github.com/Ro6ertWcislo/AutoESN.

4 Results

Herein, we present experimental comparison of shallow, deep and grouped ESN architectures (including newly proposed *gdESN*) with SNA activation function. The test setup is based on univariate time series prediction for selected benchmark datasets: Mackey Glass (MG) System, Monthly Sunspot Series and Multiple Superimposed Oscillators (MGO). The details regarding each dataset as well as hyperparameter selection can be found under https://github.com/Ro6ertWcislo/AutoESN.

The activation functions considered in this work exhibit different sensitivity to hyperparameters. The behaviour of tangent hyperbolic function highly depends on input scaling, while for SNA its effect is reduced due to normalization factor. Therefore, in experiments we used two different hyperparameter setups as an input to grid search. Each tested model, shallow or decoupled had the total number of 1000 neurons (with the small deviations resulting from subreservoir integer sizes). The best configuration was selected based on minimal NRMSE achieved for validation dataset.

This section describes the results of one step ahead prediction of univariate time series. The experiments on memory capacity and input scaling are documented on the Github page. In addition to ESN models, we also show the best results obtained for simple moving average and LSTM model (with parameters optimized using validation dataset).

Time-Series Prediction. The results of one step ahead prediction obtained for three datasets (MG, Sunspot, MSO) are presented in Table 1. We report average and minimal value of NRMSE achieved for relevant test sets by 8 different models. Each row of the table presents the result obtained by particular model for the best hyperparameter configuration, which was discovered using grid search on validation set. In general, all SNA architectures outperform the ones based on *tanh* activation both in terms of average and minimal prediction error. The only exception is minimal NRMSE for MG achieved by *dESN* with *tanh* activation. Nevertheless, the average NRMSE obtained by the same model is from 2.86 to 4 times worse than the results for SNA models. The primacy of SNA architectures is especially visible for MSO dataset, where the best average

Table 1. Mackey Glass, MSO and Sunspot one step ahead prediction NRMSE results for all architectures on the test set (5 runs). The architectures with smallest NRMSE in terms of both average and minimal value were bolded. ESN denotes regular shallow architecture with 1000 neurons.

Architecture	MG		MSO		Sunspot	
	Min.	Avg.	Min.	Avg.	Min.	Avg.
ESN tanh	0.01785	0.05586	0.01608	0.03113	0.3615	0.3651
gESN tanh	0.02217	0.06678	0.01525	0.02028	0.3450	0.3669
dESN tanh	**0.00579**	0.05240	0.00517	0.01013	0.3902	0.4024
gdESN tanh	0.02035	0.05864	0.00512	0.01128	0.3760	0.4036
ESN SNA	0.00768	0.01749	0.00026	0.00306	0.3318	0.3353
gESN SNA	0.00779	**0.01310**	0.00055	0.00303	0.3430	0.3610
dESN SNA	0.00847	0.01834	0.00023	**0.00163**	0.3321	**0.3332**
gdESN SNA	0.00700	0.01511	**0.00019**	0.00240	**0.3283**	0.3623
LSTM	0.12166	0.10887	0.03816	0.03990	0.34036	0.34428
Moving average	0.21674	–	0.48106	–	0.35459	–

NRMSE *tanh* result is 3.3 times worse than the worst SNA result. The same ratio for MG dataset is 2.86, while for Sunspot it equals 1.01. The selection of hyperparameters for SNA ESNs is cheaper as only the activation radius and regularization factor are adjusted, apart from the layers and groups, which are common for all decoupled models, regardless of activation type. ESNs based on *tanh* are sensitive to hyperparameters, what requires more exhaustive search of parameter space. Comparing the sizes of grid-search spaces we have 960 (*tanh*) vs. 450 (SNA) trials for baseline shallow models, 4800 vs. 2250 for *dESN* and *gESN* models and 9600 vs. 4500 for *gdESN* models. The new *gdESN* model achieved the best minimal NRMSE result for MSO and Sunspot datasets. In case of average NRMSE, it acquired the second best result for MG and MSO. Besides, the Sunspot time-series test was more difficult for *gdESN*, which obtained the worst result from the four decoupled models with SNA. From the perspective of average NRMSE, *dESN* with SNA is the best model, failing only on MG dataset. Overall, the models *gdESN* and *dESN* with SNA are best-suited to predictions on the given benchmark datasets. The hyperparameters selected by grid-search for *gdESN* with SNA are as follows: $N_l = 3$, $N_g = 3$, $r = 50$, $\beta = 0.5$ (MG), $N_l = 3$, $N_g = 2$, $r = 1400$, $\beta = 1.0$ (MSO), $N_l = 3$, $N_g = 2$, $r = 1450$, $\beta = 1.0$ (Sunspot).

5   Conclusion

Combining self normalizing activation function with stacking and grouping of subreservoirs allows for creating robust ESN prediction models, which outperform similar models with *tanh* activation. Grouped Deep SNA architecture proposed in this work can be regarded as a generalization of the previous approaches to reservoir decoupling and forms feasible framework for configuring hyperparameters, which reflect right memory vs. non-linearity balance. The architectures with multiple subreservoirs increase the size of hyperparameter search space per see, but at the same time they enrich reservoir dynamics for the model. Besides, application of SNA function reduces the number of hyperparameters, which have to be adjusted and provides high stability of predictions. In general, for SNA, dividing reservoir into subreservoirs decreases memory capacity (see the results published on Github) but at the same time reduces computational complexity of state activations updates. Our future work will include analysis of interplay between reservoir structure (non-random weighted graphs) and SNA activation function. We also plan to extend our analysis of *gdESN* SNA architecture to different benchmark time-series.

Acknowledgement. The research presented in this paper was financed from the funds assigned by Polish Ministry of Science and Higher Education to AGH University of Science and Technology.

References

1. Deepesnpy. https://github.com/lucapedrelli/DeepESN
2. Bianchi, F.M., De Santis, E., Rizzi, A., Sadeghian, A.: Short-term electric load forecasting using echo state networks and PCA decomposition. IEEE Access **3**, 1931–1943 (2015)
3. Bianchi, F.M., Scardapane, S., Løkse, S., Jenssen, R.: Reservoir computing approaches for representation and classification of multivariate time series. IEEE Trans. Neural Netw. Learn. Syst. **32**, 2169–2179 (2020)
4. Chessa, S., Gallicchio, C., Guzman, R., Micheli, A.: Robot localization by echo state networks using RSS. In: Bassis, S., Esposito, A., Morabito, F.C. (eds.) Recent Advances of Neural Network Models and Applications. SIST, vol. 26, pp. 147–154. Springer, Cham (2014). https://doi.org/10.1007/978-3-319-04129-2_15
5. Gallicchio, C., Micheli, A.: Reservoir topology in deep echo state networks. In: Tetko, I.V., Kůrková, V., Karpov, P., Theis, F. (eds.) ICANN 2019. LNCS, vol. 11731, pp. 62–75. Springer, Cham (2019). https://doi.org/10.1007/978-3-030-30493-5_6
6. Gallicchio, C., Micheli, A., Pedrelli, L.: Deep reservoir computing: a critical experimental analysis. Neurocomputing **268**, 87–99 (2017)
7. Gallicchio, C., Micheli, A., Pedrelli, L.: Design of deep echo state networks. Neural Netw. **108**, 33–47 (2018)
8. Jaeger, H., Haas, H.: Harnessing nonlinearity: predicting chaotic systems and saving energy in wireless communication. Science **304**(5667), 78–80 (2004)
9. Kawai, Y., Park, J., Asada, M.: A small-world topology enhances the echo state property and signal propagation in reservoir computing. Neural Netw. **112**, 15–23 (2019)
10. Palumbo, F., Gallicchio, C., Pucci, R., Micheli, A.: Human activity recognition using multisensor data fusion based on reservoir computing. J. Amb. Intell. Smart Environ. **8**(2), 87–107 (2016)
11. Pathak, J., et al.: Hybrid forecasting of chaotic processes: using machine learning in conjunction with a knowledge-based model. Chaos Interdisc. J. Nonlinear Sci. **28**(4), 041101 (2018)

12. Qiao, J., Li, F., Han, H., Li, W.: Growing echo-state network with multiple sub-reservoirs. IEEE Trans. Neural Netwo. Learn. Syst. **28**(2), 391–404 (2016)
13. Verzelli, P., Alippi, C., Livi, L.: Echo state networks with self-normalizing activations on the hyper-sphere. Sci. Rep. **9**(1), 1–14 (2019)

SGAIN, WSGAIN-CP and WSGAIN-GP: Novel GAN Methods for Missing Data Imputation

Diogo Telmo Neves[1,2,3(✉)], Marcel Ganesh Naik[2], and Alberto Proença[3]

[1] Intelligent Analytics for Mass Data (IAM), German Research Center for Artificial Intelligence (DFKI), Berlin, Germany
[2] Charité – Universitätsmedizin, and Berlin Institute of Health, Berlin, Germany
{diogo-telmo.neves,marcel.naik}@charite.de
[3] Centro ALGORITMI, Universidade do Minho, Braga, Portugal
{dneves,aproenca}@di.uminho.pt

Abstract. Real-world datasets often have missing values, which hinders the use of a large number of machine learning (ML) estimators. To overcome this limitation in a data analysis pipeline, data points may be deleted in a data preprocessing stage. However, an alternative better solution is data imputation.

Several methods based on Artificial Neural Networks (ANN) have been recently proposed as successful alternatives to classical discriminative imputation methods. Amongst those ANN imputation methods are the ones that rely on Generative Adversarial Networks (GAN).

This paper presents three data imputation methods based on GAN: SGAIN, WSGAIN-CP and WSGAIN-GP. These methods were tested on datasets with different settings of missing values probabilities, where the values are missing completely at random (MCAR). The evaluation of the newly developed methods shows that they are equivalent or outperform competitive state-of-the-art imputation methods in different ways, either in terms of response time, the data imputation quality, or the accuracy of post-imputation tasks (e.g., prediction or classification).

Keywords: Missing data · Data imputation · Generative Adversarial Network

1 Introduction

Real-world datasets often have missing values, a large number of those can be found at the website of OpenML [23] or at the Machine Learning (ML) repository maintained by the University of California at Irvine [6]. TBase [14], one of the largest transplant databases in Europe, is another example of a large collection

M. G. Naik—The second author is a participant in the BIH Charité Digital Clinician Scientist Program funded by the Charité – Universitätsmedizin Berlin, the Berlin Institute of Health and the German Research Foundation (DFG).

© Springer Nature Switzerland AG 2021
M. Paszynski et al. (Eds.): ICCS 2021, LNCS 12742, pp. 98–113, 2021.
https://doi.org/10.1007/978-3-030-77961-0_10

of data with missing values. A few reasons for the missing values in TBase are: a sensor malfunction, a nurse that forgot to register the weight of a patient, or an attribute (of the database) that accepts NULL entries. Yet, domain experts often find incorrect values for a given attribute (aka variable or feature), due to quantitative or qualitative data errors [1,2]. Incorrect values may also be marked as missing values, turning a complete dataset into an incomplete one or an already incomplete dataset may end up with more missing values.

Datasets with missing values will hinder the use of a large number of machine learning (ML) estimators and/or will impair the ML model quality. As a consequence, the conclusions and insights that could be extracted from the data may be of no use [9,12,15,17,22]. Handling missing values is a common task during the data preprocessing stage of a data analysis pipeline [1,2], which has challenged many researchers. One possibility is to apply *listwise deletion* (i.e., remove the data points that have at least one missing value) as a way to get a complete dataset from an incomplete one. However, listwise deletion is not always an adequate solution, since it may lead to a high decrease in the amount of data points and, ultimately, to an empty dataset, which would turn impracticable or meaningless the planned data analysis [9,12,15,17,22].

A substantial effort has been dedicated to devise new algorithms, methods, libraries and frameworks for robust data imputation, from univariate to multivariate techniques, from basic imputation (e.g., mean, median and mode) to regression-based algorithms (e.g., linear, logistic, or stochastic regression), from discriminative to generative imputation methods. Among the latter, the Generative Adversarial Networks (GAN) became popular for its extraordinary capability of capturing the data distribution. GAN is the base for our novel data imputation methods, and also used in *purify* for synthetic data generation[1].

The mechanisms of missingness are typically classified as missing at random (MAR), missing completely at random (MCAR) and missing not at random (MNAR). A precise definition of these terms can be found in [18]. This work only addresses the MCAR mechanism, mainly due to: (i) modelling MAR or MNAR data requires domain knowledge as well as a deep insight on every detail of the data, hardly feasible with third-party datasets, while in MCAR the missingness does not depend on the observed data, nor on the unobserved data; (ii) modelling MCAR data requires less complex ML models, since no feature (attribute or variable) dependency needs to be modeled; (iii) literature describing the main imputation methods (discriminative or generative) usually assume that missing data is MCAR; (iv) by assuming the same missingness setting, fair comparisons of our work against competitive methods can be established.

---

[1] The first author used synthetic data to develop and test ML models for the kidney disease pilot (see https://www.bigmedilytics.eu/pilot/kidney-disease/) of BigMedilytics (an EU-funded project supported by the European Union's Horizon 2020 research and innovation programme under grant agreement No. 780495).

Current references in generative imputation methods are GAIN [24], and a Wasserstein GAIN (WGAIN) [7]. GAIN achieved excellent results due to the exceptional ability of its GAN-based architecture to generate a model data distribution close to the real data distribution [8,24]; however, it has some caveats, namely the optimization process is delicate and unstable, as theoretically shown in [3], and the training phase is a computationally expensive task [11,20].

The key contributions of this work are the three novel generative imputation methods, all improved versions of the GAIN implementation: a *Slim* GAIN (SGAIN), a *Wasserstein Slim* GAIN *with Clipping Penalty* (WSGAIN-CP) and a *Wasserstein Slim* GAIN *with Gradient Penalty* (WSGAIN-GP).

Other contributions include:

- concrete and scientific evidence (in Sect. 2) that the algorithm described in Sect. 3.3, is a trustworthy implementation of a Wasserstein GAN and has a better overall quality when compared to the one described in [7];
- an empirical experimental demonstration (in Sect. 4) that the novel methods outperform state-of-the-art imputation methods in different aspects, namely in response time, on data imputation quality, or in the accuracy of post-imputation tasks (e.g., prediction or classification).

2   Imputation Methods Based on GAN

Imputation methods can be classified as discriminative or generative methods. Classical imputation methods typically fall in the former class, while advanced methods based on ANN fall in the latter. Classical methods, such as KNNImpute [21], MICE [5] and MissForest [19] are mature and have a wide acceptance, being used in several domains or applications. Typically, the use of these methods is grounded on assumptions about the data itself, namely on data distributions, correlations, skewness, and dependencies. On the other hand, generative methods rely on a different approach, no matter if for data imputation or not: to learn from data samples.

Recently, Goodfellow *et al.* devised a novel deep learning architecture, GAN, based on two ANN that contest each other, a *generator* and a *discriminator* [8], which became known as *a minimax two-player game*. Despite their impressive capability to learn from data samples and to capture the data distribution, they have undesired features, namely mode collapse and vanishing gradients.

To mitigate these GAN caveats, a Wasserstein GAN was proposed [4], which impose bounds on the weights of the discriminator ANN (renamed as *critic*). Few months later an improvement to the training of a Wasserstein GAN was also proposed [10]. A little after, Soon Yoon *et al.* proposed GAIN [24], a simple but yet very effective and accurate data imputation method based on a GAN. However, since GAIN is based on a vanilla GAN, it still has the aforementioned caveats (besides the ones mentioned in Sect. 1).

To reduce the impact of those limitations, we decided to develop a *slim* Wasserstein GAIN and explore variants with two types of penalties: a *weight clipping penalty* (WSGAIN-CP) and a *gradient penalty* (WSGAIN-GP). These data imputation methods exhibit better execution times than GAIN.

Very recently, Friedjungová *et al.* published a work [7] that claims to have implemented a data imputation method using a Wasserstein GAN (WGAN) [7]. Unfortunately, there is no public repository with that implementation, which hinders a thoroughly comparison against such work. Furthermore, in our opinion, the algorithm described in [7] cannot be considered a WGAN [4] since, for instance, the critic ANN is not trained more times than the generator ANN. Additionally, the results presented in [7] do not promote a fair comparison with those published in [24] and several inconsistencies and discrepancies are noticed. Therefore, we consider that our WSGAIN-CP data imputation method is a trustworthy implementation of a Wasserstein GAN and has a better overall quality when compared to the one described in [7].

3   Novel Generative Imputation Methods

This section introduces our novel generative imputation methods: SGAIN, derived from the implementation of GAIN [24] and grounded on the seminal work of Goodfellow *et al.* [8]; WSGAIN-CP and WSGAIN-GP, Wasserstein variations of the baseline data imputation method, SGAIN. Before introducing these three generative imputation methods, we start paving the way with some notation and giving a brief problem formulation to ease the understanding of each algorithm.

3.1   Notation and Problem Formulation

In terms of linear algebra operations, let us consider $\odot$, $\oplus$, and $\ominus$ as the element-wise multiplication, addition, and subtraction operations, respectively.

Let us consider:

- $\mathcal{X}$ an $\mathbb{R}^d$ data space: $\mathcal{X} = \mathcal{X}_1 \times \cdots \times \mathcal{X}_d$;
- d independent continuous and/or discrete random variables $\mathbf{X} = (X_1, \ldots, X_d)$ that take values in $\mathcal{X}$ and whose distribution is denoted by $P(\mathbf{X})$;
- $\mathbf{M} = (M_1, \ldots, M_d)$ as being a mask of the values in $\mathbf{X}$, where *a missing value* in $\{X_i\}_{i=1}^d$ is represented by a zero and any non-missing value is represented by a one; thus, $\{M_i\}_{i=1}^d$ assume values in $\{0, 1\}$.

The goal of the imputation process is to estimate values for all missing values in each $\{X_i\}_{i=1}^d$. However, it could happen that the estimation is done for every element, no matter if it corresponds to a missing value or not. Generative imputation methods (e.g., [24]) do exactly this, since they attempt to model the distribution of the (existing) data (i.e., $P(\mathbf{X})$).

Let us denote $\bar{\mathbf{X}}$ as the output vector of an abstract imputation function that produces estimations for all elements in each $\{X_i\}_{i=1}^d$. Then, the estimation final vector, which we denote as $\hat{\mathbf{X}}$, can be assembled as follows:

$$\hat{\mathbf{X}} = \mathbf{M} \odot \mathbf{X} \oplus (1 \ominus \mathbf{M}) \odot \bar{\mathbf{X}} \tag{1}$$

3.2 Slim GAIN

This section introduces the *Slim GAIN* imputation method (SGAIN). The architecture of SGAIN (in Fig. 1) is derived from the counterpart of GAIN [24].

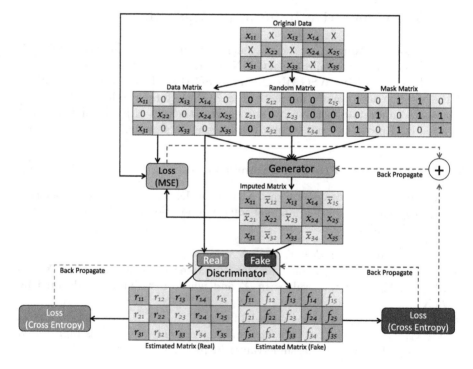

Fig. 1. The architecture of SGAIN.

In contrast to the architecture of GAIN, in SGAIN there is no Hint Generator and, consequently, no Hint Matrix is generated. The architecture of SGAIN is even slimmer, since both the generator and the discriminator neural networks have only two layers, whereas in GAIN each of them has three layers.

SGAIN uses the hyperbolic tangent activation function (aka `tanh`) in the output layers of the generator and the discriminator. The rationale to use the `tanh` and not the `sigmoid` activation function is twofold: (i) the convergence of the optimizer used in a neural network is usually faster if its inputs are linearly transformed to have zero means and unit variances, and decorrelated, as discussed in [13]; and (ii) the derivatives of the `tanh` activation function are larger than the derivatives of the `sigmoid`, which means that the optimizer can converge faster when `tanh` is used.

The architecture of SGAIN also invokes twice the discriminator, one for the *real* data and the other for the *fake* data. This brings the architecture of SGAIN closer to the one used by Goodfellow *et al.* in [8].

Generator. The generator (G) inputs are $\mathbf{Z}$ and $\mathbf{M}$, and the output is $\bar{\mathbf{X}}$. $\mathbf{Z}$ is a d-dimensional variable, $\mathbf{Z} = (Z_1, \ldots, Z_d)$, in which each $\{Z_i\}_{i=1}^d$ has the non-missing values of $\{X_i\}_{i=1}^d$ and the missing values in $\{X_i\}_{i=1}^d$ are replaced by random values (aka noise). We denote $\mathbf{N} = (N_1, \ldots, N_d)$ as the output of a function that draws out random values from a continuous uniform distribution (a common configuration is to use the interval $[-0.01, +0.01]$).

Formally, the $\mathbf{Z}$ vector is assembled as follows:

$$\mathbf{Z} = \mathbf{M} \odot \mathbf{X} \oplus (1 \ominus \mathbf{M}) \odot \mathbf{N} \tag{2}$$

$\bar{\mathbf{X}}$ is formally defined as:

$$\bar{\mathbf{X}} = G(\mathbf{Z}, \mathbf{M}) \tag{3}$$

Discriminator. The discriminator (D) is the other player of the *minimax* game described in [8]. In SGAIN, the discriminator input is either the *real* data (i.e., $\mathbf{X}$) or the *fake* data (i.e., $\bar{\mathbf{X}}$); the fake data is produced by G (Eq. 3).

Using back-propagation, the different outputs of D are used to compute the losses of the generator and of the discriminator.

Algorithm 1: Pseudo-code of SGAIN

    Input: X   `// dataset w/ missing values`

1   **Parameter:** mb   `// mini-batch size`

2   **Parameter:** α `// hyper-parameter of Generator loss`

3   **Parameter:** n_iter   `// number of iterations`

4   **Result:** $\hat{\mathbf{X}}$ `// imputed dataset`

5   $\mathbf{M} \leftarrow mask(\mathbf{X})$ `// each miss. value is 0, otherwise 1`

6   **for** $iter \leftarrow 1$ to n_iter **do**

7   Draw mb samples from $\mathbf{X}$: $\{\tilde{\mathbf{x}}(j)\}_{j=1}^{mb}$

8   Draw mb samples from $\mathbf{M}$: $\{\tilde{\mathbf{m}}(j)\}_{j=1}^{mb}$

9   Draw mb i.i.d. samples of $\mathbf{N}$: $\{\tilde{\mathbf{n}}(j)\}_{j=1}^{mb}$ `// random noise`

10   **for** $j \leftarrow 1$ to mb **do**

11          $\tilde{\mathbf{z}}(j) \leftarrow \tilde{\mathbf{m}}(j) \odot \tilde{\mathbf{x}}(j) \oplus \left(1 \ominus \tilde{\mathbf{m}}(j)\right) \odot \mathbf{n}(j)$ `// Equation 2`

12          $\bar{\mathbf{x}}(j) \leftarrow G\left(\tilde{\mathbf{z}}(j), \tilde{\mathbf{m}}(j)\right)$ `// Equation 3`

13   **end**

        `// (1) Discriminator Optimization`

14   Update D using Adam or RMSprop or SGD

15      $\nabla_{\theta_D} - \frac{1}{mb}\sum_{j=1}^{mb} \mathcal{L}_D\left(D\left(\tilde{\mathbf{x}}(j)\right), D\left(\bar{\mathbf{x}}(j)\right), \tilde{\mathbf{m}}(j)\right)$

        `// (2) Generator Optimization`

16   Update G using Adam or RMSprop or SGD

17      $\nabla_{\theta_G} - \frac{1}{mb}\sum_{j=1}^{mb} \mathcal{L}_G\left(D\left(\bar{\mathbf{x}}(j), \tilde{\mathbf{m}}(j)\right)\right) + \frac{\alpha}{mb}\sum_{j=1}^{mb} \mathcal{L}_{MSE}\left(\tilde{\mathbf{x}}(j), \bar{\mathbf{x}}(j), \tilde{\mathbf{m}}(j)\right)$

18 **end**

19 $\mathbf{Z} \leftarrow \mathbf{M} \odot \mathbf{X} \oplus (1 \ominus \mathbf{M}) \odot \mathbf{N}$ `// Equation 2`

20 $\bar{\mathbf{X}} \leftarrow G(\mathbf{Z}, \mathbf{M})$ `// Equation 3`

21 $\hat{\mathbf{X}} \leftarrow \mathbf{M} \odot \mathbf{X} \oplus (1 \ominus \mathbf{M}) \odot \bar{\mathbf{X}}$ `// Equation 1`

SGAIN Algorithm. A common step during the training of a GAN is to draw samples from the (training) data, which form a mini-batch of data used in an iteration. The SGAIN algorithm, as shown in Algorithm 1, also has this step. To fully understand the algorithm let us denote $\tilde{x}$, $\tilde{m}$, and $\tilde{n}$ as samples draw from $\mathbf{X}$, $\mathbf{M}$, and $\mathbf{N}$, respectively, and for the same data points.

In a nutshell, the SGAIN algorithm follow these steps:

- draws the $\tilde{x}$, $\tilde{m}$, and $\tilde{n}$ samples, assembles $\tilde{z}$ with them (Eq. 2) and computes $\bar{x}$, which is the output of the generator G (Eq. 3);
- optimizes the discriminator (in line 14) with the loss function (line 15) given by Expression 4;
- optimizes the generator (in line 16) with the loss function (line 17) given by Expression 5;
- repeats these steps for the given number of iterations (n_iter);
- after training SGAIN, estimates the whole set of missing values (in lines 19, 20 and 21).

$$\nabla_{\theta_D} \frac{1}{mb}\sum_{j=1}^{mb}\left[\tilde{\mathbf{m}}(j) \odot D\left(\tilde{\mathbf{x}}(j)\right)\right] - \frac{1}{mb}\sum_{j=1}^{mb}\left[\left(1 \ominus \tilde{\mathbf{m}}(j)\right) \odot D\left(\bar{\mathbf{x}}(j)\right)\right] \quad (4)$$

$$\nabla_{\theta_G} - \frac{1}{mb} \sum_{j=1}^{mb} \left[\left(1 \ominus \tilde{\mathbf{m}}(j) \right) \odot D\left(\bar{\mathbf{x}}(j) \right) \right] + \frac{\alpha}{mb} \sum_{j=1}^{mb} \left[\tilde{\mathbf{m}}(j) \odot \left(\tilde{\mathbf{x}}(j) \ominus \bar{\mathbf{x}}(j) \right)^2 \right] \quad (5)$$

For the sake of brevity, the parameter that allows to select a specific optimizer (the options are: Adam, RMSProp, and SGD) as well as the hyper-parameters of that optimizer, are elided in Algorithm 1.

3.3   Wasserstein Slim GAIN with Clipping Penalty

This Section introduces the *Wasserstein Slim GAIN with Clipping Penalty* imputation method (WSGAIN-CP). This method was inspired in the work described in [4,8,24] and it aims to reduce the main caveats that affect a vanilla GAN [8,24], such as mode-collapse and vanishing gradients [4]. The architecture of WSGAIN-CP remains almost identical to that of SGAIN (Fig. 1).

Generator. The generator of WSGAIN-CP is identical to the counterpart of SGAIN (Sect. 3.2); they even share the same loss function (see Expression 6 and Expression 5).

$$\nabla_{\theta_G} - \frac{1}{mb} \sum_{j=1}^{mb} \left[\left(1 \ominus \tilde{\mathbf{m}}(j) \right) \odot C\left(\bar{\mathbf{x}}(j) \right) \right] + \frac{\alpha}{mb} \sum_{j=1}^{mb} \left[\tilde{\mathbf{m}}(j) \odot \left(\tilde{\mathbf{x}}(j) \ominus \bar{\mathbf{x}}(j) \right)^2 \right] \quad (6)$$

Critic. In a Wasserstein GAN, as it is WSGAIN-CP, the discriminator is named *critic* (C) and is trained more times than the generator. A common configuration is to train the critic 5 times more per each train of the generator. Moreover, for each train of the critic its weights are kept within a predefined interval (usually, that interval is $[-0.01, +0.01]$), a technique known as *(weight) clipping penalty*. It deserves to be noticed that, this penalty is not a component of the loss function of the critic, as shown in Expression 7 (which is identical to Expression 4) and in Lines 18 and 19 of Algorithm 2.

$$\nabla_{\theta_C} \frac{1}{mb} \sum_{j=1}^{mb} \left[\tilde{\mathbf{m}}(j) \odot C\left(\tilde{\mathbf{x}}(j) \right) \right] - \frac{1}{mb} \sum_{j=1}^{mb} \left[\left(1 \ominus \tilde{\mathbf{m}}(j) \right) \odot C\left(\bar{\mathbf{x}}(j) \right) \right] \quad (7)$$

WSGAIN-CP Algorithm. For the sake of brevity, we elide to fully describe the WSGAIN-CP algorithm (Algorithm 2). However, one should notice that the loss function of the critic (in line 18) is given by Expression 7, whereas the loss function of the generator (in line 22) is given by Expression 6. In line 19, the weights of the critic network are kept within an interval whose bounds are determined by the *clip* parameter. Again, the parameter that allows to select a specific optimizer (the options are: Adam, RMSProp, and SGD), as well as the hyper-parameters of that optimizer, are elided in Algorithm 2.

Algorithm 2: Pseudo-code of WSGAIN-CP

	Input: X	// dataset w/ missing values
1	**Parameter**: mb	// mini-batch size
2	**Parameter**: α	// hyper-parameter of Generator loss
3	**Parameter**: clip	// clip value of Critic weights
4	**Parameter**: n_iter	// number of iterations
5	**Parameter**: n_critic	// additional times to train the Critic
6	**Result:** $\hat{\mathbf{X}}$	// imputed dataset
7	$\mathbf{M} \leftarrow mask(\mathbf{X})$	// each miss. value is 0, otherwise 1

8 **for** $iter \leftarrow 1$ to n_iter **do**

9   **for** $extra \leftarrow 1$ to n_critic **do**

10   Draw mb samples from $\mathbf{X}$: $\{\tilde{\mathbf{x}}(j)\}_{j=1}^{mb}$

11   Draw mb samples from $\mathbf{M}$: $\{\tilde{\mathbf{m}}(j)\}_{j=1}^{mb}$

12   Draw mb i.i.d. samples of $\mathbf{N}$: $\{\tilde{\mathbf{n}}(j)\}_{j=1}^{mb}$ // random noise

13   **for** $j \leftarrow 1$ to mb **do**

14              $\tilde{\mathbf{z}}(j) \leftarrow \tilde{\mathbf{m}}(j) \odot \tilde{\mathbf{x}}(j) \oplus \left(1 \ominus \tilde{\mathbf{m}}(j)\right) \odot \mathbf{n}(j)$ // Equation 2

15              $\bar{\mathbf{x}}(j) \leftarrow G\left(\tilde{\mathbf{z}}(j), \tilde{\mathbf{m}}(j)\right)$ // Equation 3

16   **end**

         // (1) Critic Optimization

17   Update C using Adam or RMSprop or SGD

18          $\nabla_{\theta_C} \frac{1}{mb} \sum_{j=1}^{mb} \mathcal{L}_C \left(C\left(\tilde{\mathbf{x}}(j)\right), C\left(\bar{\mathbf{x}}(j)\right), \tilde{\mathbf{m}}(j)\right)$

19          $w_c \leftarrow clip\_critic\_weights(w_c, -clip, +clip)$

20   **end**

     // (2) Generator Optimization

21   Update G using Adam or RMSprop or SGD

22      $\nabla_{\theta_G} - \frac{1}{mb} \sum_{j=1}^{mb} \mathcal{L}_G \left(C\left(\bar{\mathbf{x}}(j), \tilde{\mathbf{m}}(j)\right)\right) + \frac{\alpha}{mb} \sum_{j=1}^{mb} \mathcal{L}_{MSE}\left(\tilde{\mathbf{x}}(j), \bar{\mathbf{x}}(j), \tilde{\mathbf{m}}(j)\right)$

23 **end**

24	$\mathbf{Z} \leftarrow \mathbf{M} \odot \mathbf{X} \oplus (1 \ominus \mathbf{M}) \odot \mathbf{N}$	// Equation 2
25	$\bar{\mathbf{X}} \leftarrow G(\mathbf{Z}, \mathbf{M})$	// Equation 3
26	$\hat{\mathbf{X}} \leftarrow \mathbf{M} \odot \mathbf{X} \oplus (1 \ominus \mathbf{M}) \odot \bar{\mathbf{X}}$	// Equation 1

3.4   Wasserstein Slim GAIN with Gradient Penalty

This section introduces the *Wasserstein Slim GAIN with Gradient Penalty* imputation method (WSGAIN-GP). This method was inspired in the work described in [4,8,10,24] and the motivation behind it is to reduce the main caveats that affect a vanilla GAN [8,24] as well as the ones that affect a WGAN [4], namely the undesired behaviour that can arise due to weight clipping [10]. The architecture of WSGAIN-GP remains almost identical to that of WSGAIN-CP.

Generator. The generator of WSGAIN-GP is identical to the counterpart of WSGAIN-CP (Sect. 3.3).

Critic. The critic (C) of WSGAIN-GP is almost identical to the counterpart of WSGAIN-CP (Sect. 3.3). However, since WSGAIN-GP aims to get rid of the undesired behaviour that can arise due to weight clipping [10], there is no weight clipping. Instead, to improve its training the critic uses a technique known as *gradient penalty.* The gradient penalty is a component of the loss function, which is the only difference between Expression 7 and Expression 9.

Algorithm 3: Pseudo-code of WSGAIN-GP

    Input: X   // dataset w/ missing values

1   **Parameter:** mb   // mini-batch size

2   **Parameter:** α // hyper-parameter of Generator loss

3   **Parameter:** *lambda*   // hyper-parameter of Critic loss

4   **Parameter:** n_iter   // number of iterations

5   **Parameter:** n_critic   // additional times to train the Critic

6   **Result:** $\hat{\mathbf{X}}$ // imputed dataset

7    $\mathbf{M} \leftarrow mask(\mathbf{X})$ // each miss. value is 0, otherwise 1

8   **for** $iter \leftarrow 1$ *to* n_iter **do**

9   **for** $extra \leftarrow 1$ *to* n_critic **do**

10   Draw mb samples from $\mathbf{X}$: $\{\tilde{\mathbf{x}}(j)\}_{j=1}^{mb}$

11   Draw mb samples from $\mathbf{M}$: $\{\tilde{\mathbf{m}}(j)\}_{j=1}^{mb}$

12   Draw mb i.i.d. samples of $\mathbf{N}$: $\{\tilde{\mathbf{n}}(j)\}_{j=1}^{mb}$ // random noise

13   Draw mb i.i.d. samples of $\mathbf{N}$: $\{\tilde{\epsilon}(j)\}_{j=1}^{mb}$ // random noise

14   **for** $j \leftarrow 1$ *to* mb **do**

15              $\tilde{\mathbf{z}}(j) \leftarrow \tilde{\mathbf{m}}(j) \odot \tilde{\mathbf{x}}(j) \oplus \left(1 \ominus \tilde{\mathbf{m}}(j)\right) \odot \mathbf{n}(j)$ // Equation 2

16              $\bar{\mathbf{x}}(j) \leftarrow G\left(\tilde{\mathbf{z}}(j), \tilde{\mathbf{m}}(j)\right)$ // Equation 3

17              $\dot{\mathbf{x}}(j) \leftarrow \tilde{\mathbf{m}}(j) \odot \left(\tilde{\epsilon}(j) \odot \tilde{\mathbf{x}}(j)\right) \oplus \left(\left(1 \ominus \tilde{\mathbf{m}}(j)\right) \odot \left(1 \ominus \tilde{\epsilon}(j)\right) \odot \bar{\mathbf{x}}(j)\right)$

18   **end**

          // (1) Critic Optimization

19   Update C using Adam or RMSprop or SGD

20          $\nabla_{\theta_C} \frac{1}{mb} \sum_{j=1}^{mb} \mathcal{L}_C\left(C\left(\tilde{\mathbf{x}}(j)\right), C\left(\bar{\mathbf{x}}(j)\right), \tilde{\mathbf{m}}(j)\right) + \frac{\lambda}{mb} \sum_{j=1}^{mb} \mathcal{L}_{GradPen}\left(C\left(\dot{\mathbf{x}}(j)\right)\right)$

21   **end**

    // (2) Generator Optimization

22   Update G using Adam or RMSprop or SGD

23    $\nabla_{\theta_G} - \frac{1}{mb} \sum_{j=1}^{mb} \mathcal{L}_G\left(C\left(\bar{\mathbf{x}}(j), \tilde{\mathbf{m}}(j)\right)\right) + \frac{\alpha}{mb} \sum_{j=1}^{mb} \mathcal{L}_{MSE}\left(\tilde{\mathbf{x}}(j), \bar{\mathbf{x}}(j), \tilde{\mathbf{m}}(j)\right)$

24 **end**

25 $\mathbf{Z} \leftarrow \mathbf{M} \odot \mathbf{X} \oplus (1 \ominus \mathbf{M}) \odot \mathbf{N}$ // Equation 2

26 $\bar{\mathbf{X}} \leftarrow G(\mathbf{Z}, \mathbf{M})$ // Equation 3

27 $\hat{\mathbf{X}} \leftarrow \mathbf{M} \odot \mathbf{X} \oplus (1 \ominus \mathbf{M}) \odot \bar{\mathbf{X}}$ // Equation 1

$$\dot{\mathbf{x}}(j) = \tilde{\mathbf{m}}(j) \odot \left(\tilde{\epsilon}(j) \odot \tilde{\mathbf{x}}(j)\right) \oplus \left(\left(1 \ominus \tilde{\mathbf{m}}(j)\right) \odot \left(1 \ominus \tilde{\epsilon}(j)\right) \odot \bar{\mathbf{x}}(j)\right) \qquad (8)$$

$$\nabla_{\theta_C} \frac{1}{mb} \sum_{j=1}^{mb} \left[\tilde{\mathbf{m}}(j) \odot C\left(\tilde{\mathbf{x}}(j)\right)\right] - \frac{1}{mb} \sum_{j=1}^{mb} \left[\left(1 \ominus \tilde{\mathbf{m}}(j)\right) \odot C\left(\bar{\mathbf{x}}(j)\right)\right]$$
$$+ \frac{\lambda}{mb} \sum_{j=1}^{mb} \left(\left\|\nabla_{\dot{\mathbf{x}}(j)} C\left(\dot{\mathbf{x}}(j)\right)\right\|_2 \ominus 1\right)^2 \qquad (9)$$

WSGAIN-GP Algorithm. For the sake of brevity, we elide to fully describe the WSGAIN-GP algorithm (Algorithm 3). However, four details deserve to be noticed: (i) the new random noise (line 13) is generated as described in Sect. 3.2; (ii) the statement in line 17 corresponds to Eq. 8 and is crucial to compute the norm of the gradients (line 20 and Expression 9); (iii) the loss function of the critic (line 20) is given by Expression 9; as aforementioned, the gradient penalty component is the only difference between Expression 7 and Expression 9; and (iv) the loss function of the generator (line 23) is exactly the same of that in WSGAIN-CP Algorithm 2 (line 22) and is given by Expression 6. Again, the parameter that allows to select a specific optimizer (the options are: Adam, RMSProp, and SGD) as well as the hyper-parameters of that optimizer, are elided in Algorithm 3.

4   Experimental Results

This section evaluates the novel generative imputation methods on 10 real-world datasets from the ML repository maintained by the University of California at Irvine [6], as presented in Table 1.

Table 1. Short description of datasets.

Name	Area	Instances	Attributes	Continuous	Discrete	Target	Model
Breast Cancer	Life	569	31 (32)	30	0	Diagnosis	LR
Credit Card	Business	30000	24 (25)	14	9	Def. Pay. Next Month	LR
EEG Eye State	Life	14980	15 (15)	14	0	Eye Detect.	KNN
Iris	Life	150	5 (5)	4	0	Class	KNN
Letter Recognition	Computer	20000	17 (17)	16	0	Letter	KNN
Online News Popularity	Business	39644	60 (61)	56	3	Shares	LR
Spambase	Computer	4601	58 (58)	57	0	Spam	LR
(Red) Wine Quality	Business	1599	12 (12)	11	0	Quality	KNN
(White) Wine Quality	Business	4898	12 (12)	11	0	Quality	KNN
Yeast	Life	1484	9 (10)	6 (7)	2 (1)	Local. Site	KNN

During the data preprocessing several data transformations were applied: drop one (non-relevant) attribute (aka variable or feature) in Breast Cancer, Credit Card, Online News Popularity, and Yeast datasets; perform one-hot encoding of discrete (aka categorical) attributes; and scale the continuous (aka numerical) attributes to fit inside the interval $[-1.00, +1.00]$, using the scikit-learn [16] `MinMaxScaler`[2]. In the Yeast dataset we considered one continuous

---

[2] https://scikit-learn.org/stable/modules/generated/sklearn.preprocessing. MinMaxScaler.html.

attribute as being discrete, thus, it has two discrete attributes instead of just one. The missing values are introduced under an MCAR setting and, unless otherwise stated, the amount of missing values is 20% of all data points. Moreover, the amputation is evenly distributed by all attributes. However, it is highly likely for an amputated dataset to have rows with just missing values, we did nothing to prevent such cases.

We obtained GAIN from its GitHub repository[3] and for every run we kept unchanged the batch size (128), the hint rate (0.9), the hyper-parameter alpha (100), and the number of iterations (9000). To promote a fair comparison, we also used these values to run SGAIN, WSGAIN-CP, and WSGAIN-GP (we recall that our methods do not have the hint generator). However, since the critic of WSGAIN-CP and WSGAIN-GP was trained five times more than the generator, we decided to divide the number of iterations by three, in these cases. For WSGAIN-CP and WSGAIN-GP the generator was trained 3000 times, whereas the critic was 15000 times, which sums up to as many times the generator and the discriminator of GAIN are trained (each 9000 times).

Three types of experimental results are discussed in the next sections: (i) response times, which measure how fast the methods can present results; (ii) the quality of the results, measured by the root mean square error (RMSE) between the imputed values and the original deleted values; and (iii) the area under the receiver operating characteristics (AUROC), which in this study is used to measure a model accuracy of post-imputation prediction.

4.1 Response Times

Figure 2 shows the mean execution time, taken from ten executions, of GAIN, SGAIN, WSGAIN-CP and WSGAIN-GP on each dataset. The exact same amputated datasets were used by each algorithm to perform the imputation of missing values. In all cases, SGAIN outperformed GAIN, with improvements ranging from (roughly) 20 to 30% less than the execution time of GAIN. WSGAIN-CP and WSGAIN-GP also outperformed GAIN, the only exception is that GAIN is marginally faster than WSGAIN-CP on the Letter Recognition dataset. In general, SGAIN, WSGAIN-CP and WSGAIN-GP are considerable faster than GAIN, in particularly SGAIN. This is an extremely significant result since each one of our novel generative imputation methods are trained faster than GAIN.

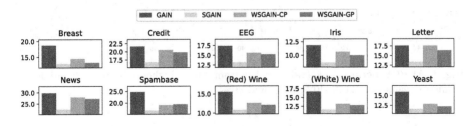

Fig. 2. Response times in seconds (lower is better).

---

[3] https://github.com/jsyoon0823/GAIN.

4.2   RMSE Performance

A common way to compare the quality of the imputed data is to measure how close are the imputed data points to the counterpart data points in the original (i.e., complete) dataset. Usually, this is achieved by computing the *root mean square error* (RMSE). Figure 3 reports on the RMSE achieved by GAIN, SGAIN, WSGAIN-CP, and WSGAIN-GP on all datasets and from ten executions. The exact same amputated datasets were used by each algorithm to perform the imputation of missing values.

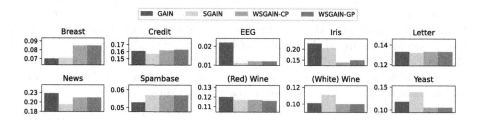

Fig. 3. RMSE performance (values range from 0 to 1, lower is better).

The main observation that can be derived from the RMSE results is that our novel generative imputation methods have competitive performance to that of GAIN. This is again a very significant result, since it shows, besides the observed consistency, that the changes and optimizations of the architecture of SGAIN, from which WSGAIN-CP and WSGAIN-GP were derived, do not impair the quality of the missing data imputation.

4.3   AUROC Performance

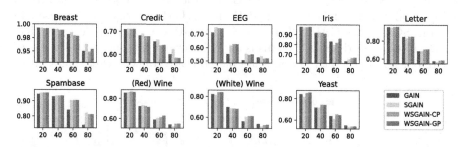

Fig. 4. AUROC performance (values range from 0 to 1, higher is better); the amounts of missing values are 20%, 40%, 60%, and 80%.

In this study, the areas under the receiver operating characteristics (AUROC) are used to measure the models accuracies of post-imputation predictions. The models are the *Logistic Regression* (LR)[4] and the *K Neighbours Classifier* (KNN)[5] (see rightmost column of Table 1). We do not present results for the (Online) News (Popularity) dataset since the post-imputation predictions took too longer to complete due to the size of the dataset.

The main observation that can be derived from the AUROC results shown in Fig. 4 is that both new WSGAIN imputation methods present the higher values in almost all datasets and for all settings of missing values. This is again a very significant result, since it shows, besides the observed consistency, that the changes and optimizations of the architecture of SGAIN, from which WSGAIN-CP and WSGAIN-GP were derived, do not impair the models accuracies of post-imputation predictions.

5 Conclusions

This paper presented and discussed three novel generative imputation methods: SGAIN, WSGAIN-CP, and WSGAIN-GP. SGAIN is a slimmer and optimized GAN derived from GAIN, whereas, whereas WSGAIN-CP and WSGAIN-GP are variations of a Wasserstein GAN. These methods are available online in a GitHub repository[6].

The performed experimental work comparatively evaluated our novel methods with the competition, using real-world datasets from different domains, with distinct characteristics and under various settings. Results explicitly showed that our methods outperformed the reference GAN-based imputation method (GAIN) and implicitly showed that they outperformed other imputation methods (e.g. MICE, MissForest, Matrix Completion, Auto-Encoder, and EM) [7,24]. This latter conclusion is derived from the fact that our methods outperform GAIN and GAIN outperforms those data imputation methods, as shown in [24]. The measured response times also showed that the newly developed methods are considerably faster than GAIN and require significant less time to be trained.

Acknowledgements. The first author thanks the ALGORITMI research centre, Universidade do Minho, where he conducts part of his research as an external collaborator. The third author developed his work at ALGORITMI research centre, Universidade do Minho, supported by FCT – Fundação para a Ciência e Tecnologia within the R&D Units Project Scope: UIDB/00319/2020.

References

1. Abedjan, Z., et al.: Detecting data errors: where are we and what needs to be done? Proc. VLDB Endowment **9**(12), 993–1004 (2016)

---

[4] https://scikit-learn.org/stable/modules/generated/sklearn.linear_model. LogisticRegression.html.

[5] https://scikit-learn.org/stable/modules/generated/sklearn.neighbors. KNeighborsClassifier.html.

[6] https://github.com/dtneves/ICCS_2021.

2. Abedjan, Z., Golab, L., Naumann, F.: Profiling relational data: a survey. VLDB J. **24**(4), 557–581 (2015)
3. Arjovsky, M., Bottou, L.: Towards principled methods for training generative adversarial networks. arXiv preprint arXiv:1701.04862 (2017)
4. Arjovsky, M., Chintala, S., Bottou, L.: Wasserstein gan. arXiv preprint arXiv:1701.07875 (2017)
5. Buuren, S.v., Groothuis-Oudshoorn, K.: mice: Multivariate imputation by chained equations in r. J. Stat. Softw., 1–68 (2010)
6. Dua, D., Graff, C.: UCI machine learning repository (2017). http://archive.ics.uci.edu/ml
7. Friedjungová, M., Vašata, D., Balatsko, M., Jiřina, M.: Missing features reconstruction using a wasserstein generative adversarial imputation network. In: Krzhizhanovskaya, V.V., et al. (eds.) ICCS 2020. LNCS, vol. 12140, pp. 225–239. Springer, Cham (2020)
8. Goodfellow, I., et al.: Generative adversarial nets. Adv. Neural Inf. Process. Syst. **27**, 2672–2680 (2014)
9. Graham, J.W.: Missing data analysis: making it work in the real world. Ann. Rev. Psychol. **60**, 549–576 (2009)
10. Gulrajani, I., Ahmed, F., Arjovsky, M., Dumoulin, V., Courville, A.C.: Improved training of wasserstein gans. In: Advances in Neural Information Processing Systems, pp. 5767–5777 (2017)
11. Huqqani, A.A., Schikuta, E., Ye, S., Chen, P.: Multicore and GPU parallelization of neural networks for face recognition, pp. 349–358 (2013)
12. Lall, R.: How multiple imputation makes a difference. Polit. Anal. **24**(4), 414–433 (2016)
13. LeCun, Y.A., Bottou, L., Orr, G.B., Müller, K.-R.: Efficient BackProp. In: Montavon, G., Orr, G.B., Müller, K.-R. (eds.) Neural Networks: Tricks of the Trade. LNCS, vol. 7700, pp. 9–48. Springer, Heidelberg (2012)
14. Liefeldt, L., et al.: Donor-specific hla antibodies in a cohort comparing everolimus with cyclosporine after kidney transplantation. Am. J. Transplant. **12**(5), 1192–1198 (2012)
15. Little, R.J., Rubin, D.B.: Statistical Analysis with Missing Data, vol. 793. John Wiley & Sons, Hoboken (2019)
16. Pedregosa, F., et al.: Scikit-learn: machine learning in python. J. Mach. Learn. Res. **12**, 2825–2830 (2011)
17. Rubin, D.B.: Multiple Imputation for Nonresponse in Surveys, vol. 81. John Wiley & Sons, Hoboken (2004)
18. Schafer, J.L., Graham, J.W.: Missing data: our view of the state of the art. Psychol. Methods **7**(2), 147 (2002)
19. Stekhoven, D.J., Bühlmann, P.: Missforest–non-parametric missing value imputation for mixed-type data. Bioinformatics **28**(1), 112–118 (2012)
20. Strigl, D., Kofler, K., Podlipnig, S.: Performance and scalability of GPU-based convolutional neural networks. In: 2010 18th Euromicro Conference on Parallel, Distributed and Network-based Processing, pp. 317–324. IEEE (2010)
21. Troyanskaya, O., et al.: Missing value estimation methods for DNA microarrays. Bioinformatics **17**(6), 520–525 (2001)
22. Van Buuren, S.: Flexible Imputation of Missing Data. CRC Press, Boca Raton (2018)

23. Vanschoren, J., van Rijn, J.N., Bischl, B., Torgo, L.: Openml: networked science in machine learning. SIGKDD Explor. **15**(2), 49–60 (2013)
24. Yoon, J., Jordon, J., Van Der Schaar, M.: Gain: Missing data imputation using generative adversarial nets. arXiv preprint arXiv:1806.02920 (2018)

Deep Learning Driven Self-adaptive Hp Finite Element Method

Maciej Paszyński[1(✉)], Rafał Grzeszczuk[1], David Pardo[2,3,4],
and Leszek Demkowicz[5]

[1] AGH University of Science and Technology, Kraków, Poland
{paszynsk,grzeszcz}@agh.edu.pl
[2] The University of the Basque Country, Bilbao, Spain
[3] Basque Center for Applied Mathematics, Bilbao, Spain
[4] IKERBASQUE, Bilbao, Spain
[5] Oden Institute, The University of Texas at Austin, Austin, USA
leszek@oden.utexas.edu

Abstract. The finite element method (FEM) is a popular tool for solving engineering problems governed by Partial Differential Equations (PDEs). The accuracy of the numerical solution depends on the quality of the computational mesh. We consider the self-adaptive hp-FEM, which generates optimal mesh refinements and delivers exponential convergence of the numerical error with respect to the mesh size. Thus, it enables solving difficult engineering problems with the highest possible numerical accuracy. We replace the computationally expensive kernel of the refinement algorithm with a deep neural network in this work. The network learns how to optimally refine the elements and modify the orders of the polynomials. In this way, the deterministic algorithm is replaced by a neural network that selects similar quality refinements in a fraction of the time needed by the original algorithm.

Keywords: Partial differential equations · Finite element method · Adaptive algorithms · Neural networks

1 Introduction

The self-adaptive hp-Finite Element Method (FEM) has been developed for many years by the community of applied mathematicians working in the field of numerical analysis [3–6,9]. They require extremely high numerical accuracy, which is difficult to obtain by other numerical methods. In this paper, we refer to the iterative Algorithm 1 proposed by [3], and we introduce the simplified, one-step, Algorithm 2 as a kernel for the selection of the optimal refinements for the interiors of elements. The edge refinements are adjusted by taking the minimum of the corresponding orders of interiors. We further propose how to replace Algorithm 2 with a Deep Neural Network (DNN).

The DNN can make similar quality decisions about mesh refinements as Algorithm 2, while the online computational time is reduced. The main motivation

© The Author(s) 2021
M. Paszynski et al. (Eds.): ICCS 2021, LNCS 12742, pp. 114–121, 2021.
https://doi.org/10.1007/978-3-030-77961-0_11

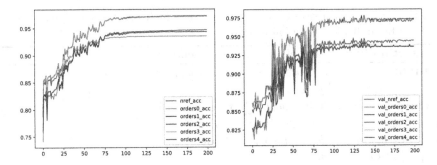

Fig. 1. The convergence of accuracy of training (left) and validation (right) datasets.

for this work is the following observation. We have noticed that making random of 10% of the decision about element refinements made by the self-adaptive hp-FEM algorithm does not disturb the algorithm's exponential convergence. Thus, the possibility of teaching the deep neural network making decisions optimal up to 90% is enough to keep the exponential convergence.

Algorithm 1: Self-adaptive hp-FEM algorithm

    Input: Initial mesh, PDE, boundary conditions, error
    Output: Optimal mesh
1   coarse mesh = initial mesh
2   Solve the coarse mesh problem
3   Generate fine mesh
4   Solve the fine mesh problem
5   **if** *maximum relative error > accuracy* **then**
6 | **return** *fine mesh solution*
7   **end**
8   Select optimal refinements for every hp finite element from the coarse mesh (**Call algorithm 2**)
9   Perform all required h refinements
10   Perform all required p refinements
11   coarse mesh = actual mesh
12   **goto 2**

---

2 Self-adaptive hp-FEM with Neural Network

We focus on the L-shape domain model problem [5,6] to illustrate the self-adaptive applicability hp-FEM algorithm for the solution of a model problem with a singular point. The gradient of the solution tends to infinity, and intensive mesh refinements are needed to approximate this behavior properly.

We describe in Algorithm 1 the self-adaptive hp-algorithm, initially introduced by [3]. It utilizes Algorithm 2 for the selection of the optimal refinements over element K. This algorithm delivers exponential convergence of the numerical error with respect to the mesh size, which has been verified experimentally by multiple numerical examples [3, 4].

Algorithm 2: Selection of optimal refinements over K

Input: Element K, coarse mesh solution $u_{hp} \in V_{hp}$, fine mesh solution $u_{\frac{h}{2},p+1} \in V_{\frac{h}{2},p+1}$

Output: Optimal refinement V_{opt}^K for element K

1   **for** *coarse mesh elements* $K \in T_{hp}$ **do**
2   **for** *approximation space* $V_{opt} \in K$ **do**
3       $rate_{min} = \infty$
4   Compute the projection based interpolant $w|_K$ of $u_{\frac{h}{2},p+1}|_K$
5   Compute the error decrease rate

$$rate(w) = \frac{\left| u_{\frac{h}{2},p+1} - u_{hp} \right|_{H^1(K)} - \left| u_{\frac{h}{2},p+1} - w \right|_{H^1(K)}}{\Delta\text{nrdof}(V_{hp}, V_{opt}^K, K)}$$

6   **if** $rate(w) < rate_{min}$ **then**
7         $rate_{min} = rate(w)$
8   Select V_{opt}^K corresponding to $rate_{min}$ as the optimal refinement for element K
9   **end**
10   **end**
11 **end**
12 Select orders of approximation on edges as minimum of corresponding orders from neighboring interiors

---

Our goal is to replace Algorithm 2 with a deep neural network. The left column in Fig. 2 presents the optimal distribution of refinements, as provided by the deterministic algorithm. We can see that all the h refinements (breaking of elements) are performed towards the point singularity. We also see that the p refinements are surrounding the singularity as layers with a different color. They change from red, light and dark pink ($p = 6, 7, 8$), through brown ($p = 5$), yellow ($p = 4$), green ($p = 3$), blue ($p = 2$) and dark blue ($p = 1$) close to the singularity.

The refinements performed by the iterative Algorithm 1 are executed first closer to the singularity. With the iterations, the differences between the coarse and fine mesh solution tend to zero [3].

Dataset. We propose the following samples to train the DNN:

Input variables: coarse mesh solution $u_{hp} \in V_{hp}$ for element K, the element sizes and coordinates, the norm of the fine mesh solution over element K, the maximum norm of the fine mesh solution over elements
Output variables: Optimal refinement V_{opt}^K for element K.

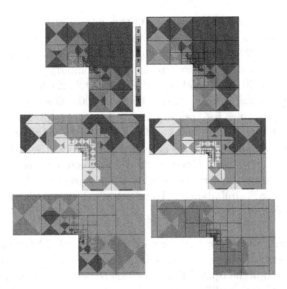

Fig. 2. The mesh provided by the deterministic hp-FEM algorithm and by the deep learning-driven hp-FEM algorithm. Different colors denote different polynomial orders of approximation on element edges and interiors. The original L-shape domain. Zoom $1\times$, $1000\times$, $100000\times$ towards the center. The sequence of hp refined meshes generated by deterministic algorithm (left panel) and DNN driven algorithm (right panel).

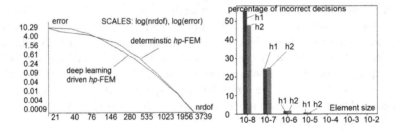

Fig. 3. Let panel: The comparison of deterministic and DNN hp-FEM on original L-shape domain. **Right panel:** The sizes (horizontal h1/vertical h2 directions) from 10^{-2} (right) down to 10^{-8} (left) of the elements where MPL network made incorrect decisions during verification.

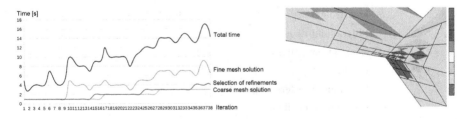

Fig. 4. Left panel: The execution times of the parts of the self-adaptive hp-FEM algorithm. **Right panel:** The refinements generated by DNN for a distorted mesh.

We construct the dataset by executing the deterministic Algorithm 1 for the model L-shape domain problem. We perform 50 iterations of the hp-adaptivity, generating over 10,000 deterministic element refinements, resulting in 10,000 samples. We repeat this operation for rotated boundary conditions (4) by the following angles: 10, 20, 30, 40, 50, 60, 70, 80, and 90°. Each rotation changes the solution and the samples. We obtain a total of 100,000 samples. We randomly select 90% of the samples for training and use the remaining 10% as a test set. We further sample the training data and use 10% of training data as a validation set. After one-hot encoding the categorical variables, each sample is represented by a 136-dimensional vector. Since it is much more common for the deterministic algorithm to make specific h refinement decisions ($nref$) for the L-shape domain problem, the dataset is imbalanced. To mitigate this, we apply supersampling of underrepresented $nref$ classes.

DNN Architecture. We use a feed-forward DNN[1] with 12 fully-connected layers. After 8 layers, the network splits into 6 branches, 4 layers each: the first branch decides about the optimal $nref$ parameter - h refinement, the remaining branches decide about modifying the polynomial orders - p refinement. Experiments have shown that further expanding of the network makes it prone to overfitting [8]. Splitting the network into branches assures sufficient parameter freedom for each variable. This approach also simplifies the model: there is no need to train a DNN for each variable. Since all possible decisions are encoded as categorical variables, we use cross-entropy as the loss function. We encoded the input data as a 136-dimensional normalized vector, as detailed in Table 1. We assume that the polynomial degree will not exceed $n = 11$. We train the network for up to 200 epochs with validation loss-based early stopping on an Nvidia Tesla v100 GPGPU with 650 tensor cores available in ACK Cyfronet PROMETHEUS cluster [2]. To minimize the loss function, we use the Adam optimizer [7], with the learning rate set to $10e^{-3}$. We apply kernel L2-penalty throughout the training as a means of regularization and dropout [10] with probability 0.5. The network converges after approximately 110 epochs.

Table 1. Dimensionality of specific input features to the DNN, encoded in a single 136-dimensional vector. Polynomial coefficients that do not exist in a given polynomial order are always 0.

Feature name	Data dimensionality
Polynomial degree	1
Element coordinates	2
H1 norms	2
Polynomial orders (one-hot)	10
Polynomial coefficients	121

DNN Performance. The network achieved over 92% accuracy on the test set. We run three tests to assess whether such a network can be used in the *hp*-FEM.

First numerical experiment is to reproduce the deterministic Algorithm 2 for the original L-shape domain problem, presented in Figs. 1, 2 and left panel in Fig. 3. Both deterministic and DNN-driven algorithms provide exponential convergence. The verification phase shows that the DNN makes up to 50% of incorrect decisions when the element sizes go down to 10^{-7} and less, see the right panel in Fig. 3. Thus, at the zoom of 100,000 times, we see some differences in Fig. 2. Despite that, the algorithm still converges exponentially.

Second Numerical Experiment. We run the self-adaptive *hp*-FEM algorithm, and we provide zeros as the coarse mesh solution degrees of freedom. We get the same convergence. This second test shows that the DNN is not sensitive with respect to the coarse mesh solution and that the norm of the fine mesh solution, the maximum norm, and the coordinates and dimensions of the elements are enough to make proper decisions. The DNN looks at the fine mesh solution's norms at the given and neighboring elements and, based on these data in decides whether the element is to be broken and how it should be broken. Thus, we can replace Algorithm 2 and the coarse mesh solution phase with the DNN. Left panel in Fig. 4 presents the execution times of particular parts of the *hp*-FEM algorithm. The removal of the coarse mesh solution phase and replacing Algorithm 2 by the DNN saves up to 50% of the execution times.

Third Numerical Experiment. The third test, illustrated in Fig. 5, concerns the L-shape domain algorithm with boundary conditions rotated 45° (no samples for this case were provided in the training set). The DNN *hp*-FEM also provides exponential convergence in this case.

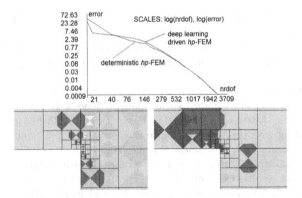

Fig. 5. The convergence for deterministic and DNN hp-FEM algorithms for the L-shape with b.c. rotated by 45°. The meshes of the deterministic *hp*-FEM algorithm and by DNN driven hp-FEM algorithm for the L-shape domain with b.c. rotated by 45° Zoom $10^5 \times$ times.

Fourth Numerical Experiment. The last test, illustrated in the right panel in Fig. 4 concerns randomly disturbed mesh, different from the training set. The DNN captures both top and bottom singularities. It produces hp refinements towards the bottom singularity and p refinements towards the top singularity. The resulting accuracy was 1% of the relative error after ten iterations.

3   Conclusions

We replaced the algorithm selecting optimal refinements in the self-adaptive hp-FEM by a deep neural network. We obtained over 92% of correct answers, the same accuracy of the final mesh, and exponential convergence of the mesh refinement algorithm. A very interesting observation is that DNN requires coordinates of elements (to recognize the adjacency between elements), the dimensions of elements (to recognize the refinement level), the H^1 norm of the solution over the element, and the maximum norm of the solutions over elements. The DNN by "looking" at the norms over adjacent elements, recognizes with 92% accuracy the proper p-refinement of the element. The replacement of the coarse mesh solver (line 2 in Algorithm 1) and the optimal refinements selection (Algorithm 2) by the DNN allows for a 50% reduction of the computational time.

The DNN used is available at

http://home.agh.edu.pl/paszynsk/dnn_hp2d/dnn_hp2d.tar.gz

Acknowledgement. This work was partially supported by National Science Centre grant no. 2016/21/B/ST6/01539. The visit of Maciej Paszyński at the Odens Institute has been supported by J. T. Oden Research Faculty Fellowship. Additionally, this work was supported by National Science Centre, Poland grant no. 2017/26/M/ST1/ 00281.

References

1. Bebis, G., Georgiopoulos, M.: Feed-forward neural networks. IEEE Potent. **13**(4), 27–31 (1994)
2. Bubak, M., Kitowski, J., Wiatr, K. (eds.): eScience on Distributed Computing Infrastructure. LNCS, vol. 8500. Springer, Cham (2014). https://doi.org/10.1007/978-3-319-10894-0
3. Demkowicz, L.: Computing with hp-Adaptive Finite Elements, vol. 1. Chapman & Hall/CRC Applied Mathematics & Non-linear Science (2006)
4. Demkowicz, L., Kurtz, J., Pardo, D., Paszyński, M., Rachowicz, W., Zdunek, A.: Computing with hp-Adaptive Finite Elements, vol. 2. Chapman & Hall/CRC Applied Mathematics & Non-linear Science (2007)
5. Guo, B., Babuška, I.: The hp version of the finite element method, part i: the basic approximation results. Comput. Mech. **1**(1), 21–41 (1986)
6. Guo, B., Babuška, I.: The hp version of the finite element method, part ii: general results and applications. Computat. Mech. **1**(1), 203–220 (1986)
7. Kingma, D.P., Ba, J.: Adam: a method for stochastic optimization. arXiv preprint arXiv:1412.6980 (2014)

8. Salman, S., Liu, X.: Overfitting mechanism and avoidance in deep neural networks. arXiv preprint arXiv:1901.06566 (2019)

9. Schwab, C.: p-and hp-Finite Element Methods. The Clarendon Press, Oxford University Press, New York (1998)

10. Srivastava, N.: Improving neural networks with dropout. Univ. Toronto **182**(566), 7 (2013)

Machine-Learning Based Prediction of Multiple Types of Network Traffic

Aleksandra Knapińska[(✉)] [iD], Piotr Lechowicz[iD], and Krzysztof Walkowiak[iD]

Department of Systems and Computer Networks,
Wrocław University of Science and Technology, Wrocław, Poland
`aleksandra.knapinska@pwr.edu.pl`

Abstract. Prior knowledge regarding approximated future traffic requirements allows adjusting suitable network parameters to improve the network's performance. To this end, various analyses and traffic prediction methods assisted with machine learning techniques are developed. In this paper, we study on-line multiple time series prediction for traffic of various frame sizes. Firstly, we describe the gathered real network traffic data and study their seasonality and correlations between traffic types. Secondly, we propose three machine learning algorithms, namely, linear regression, k nearest neighbours, and random forest, to predict the network data which are compared under various models and input features. To evaluate the prediction quality, we use the root mean squared percentage error (RMSPE). We define three machine learning models, where traffic related to particular frame sizes is predicted based on the historical data of corresponding frame sizes solely, several frame sizes, and all frame sizes. According to the performed numerical experiments on four different datasets, linear regression yields the highest accuracy when compared to the other two algorithms. As the results indicate, the inclusion of historical data regarding all frame sizes to predict summary traffic of a certain frame size increases the algorithm's accuracy at the cost of longer execution times. However, by appropriate input features selection based on seasonality, it is possible to decrease this time overhead at the almost unnoticeable accuracy decrease.

Keywords: Traffic prediction · Machine learning · Application-aware network

1 Introduction

With a constant growth of internet traffic, its analysis and prediction can be beneficial for the network operators in various scenarios. Intuitively, they can be applied for resources planning during network migration or redimensioning. In the case of any budget limitations, the most congested links can be properly

This work was supported by National Science Centre, Poland under Grant 2019/35/B/ST7/04272.

M. Paszynski et al. (Eds.): ICCS 2021, LNCS 12742, pp. 122–136, 2021.
https://doi.org/10.1007/978-3-030-77961-0_12

maintained in the first place. Another application for traffic analysis and prediction could be proactive traffic routing and virtual topology adaptation [14]. The biggest role is played here by the real-time or on-line models, which adjust their predictions by analysing live network traffic. In the case of congestion or an unexpected traffic spike, the network can adjust and reconfigure quickly. Another noteworthy application of traffic prediction could be energy efficiency. Unused network links or transponders can be forced into a low power consumption state for varying intervals to save energy [4].

The traffic in the network's optical layer is aggregated from the traffic in the packet layer fulfilled by various services and applications that can have different requirements, e.g., regarding resilience, latency, or security. Multilayer application-aware networks can identify these various traffic types in the optical layer and apply suitable optimization methods to mitigate their requirements [11,12]. The knowledge about the amount and general patterns of different network traffic types would be a substantial help to the resources planning and allocation processes. However, the information about the exact distribution of the network traffic generated by different applications and services may not be available for the network operator or its amount may not be effectively processed in real-time. In such a case, the frame size distribution can be studied as a good representation of the network traffic diversity. Different frame sizes can be an indication of different types of traffic. For example, frames bigger in size are used when a higher amount of data needs to be sent [7], including content traffic, like video [2]. Some in-depth studies indicate also more detailed examples: the maximal DNS packet size when UDP is used is 580 bytes [7] and signaling traffic in P2P IPTV uses packets of size up to 127 bytes [2]. More sporadically, other smaller frame sizes are used, usually representing residual parts of traffic otherwise using larger frames. However, less utilized frame sizes are not necessarily less important: they might be also used by some crucial protocols. Studying the number of frames in different sizes and the amount of traffic using them, and then the most common traffic patterns can help better utilize the network resources. For that reason, the analysis and prediction of the network traffic in different frame sizes can be seen as a step into multilayer application-aware network planning and optimisation.

The main contribution of this work is the analysis and prediction of the network traffic distinguishing frame sizes. In more detail, we present the data preparation process along with a seasonality and correlation analysis. Following that, we investigate different models and input features sets in three machine learning algorithms for the best compromise between prediction quality and the time of execution for an on-line traffic prediction. We repeat the experiments on three additional datasets from different time periods and places, with varying data granularity to confirm the generalization of the presented methodology and findings.

The rest of the paper is organized as follows. Literature review can be found in Sect. 2. The data analysis is presented in Sect. 3, followed by the description of the proposed models and algorithms in Sect. 4. Numerical experiments can be found in Sect. 5. Finally, Sect. 6 concludes the paper.

2 Related Work

The topic of network traffic analysis and prediction has been comprehensively analysed in several papers in the last five years, both stand-alone [9] and as a chapter of more general surveys on machine learning in optical networks [3,8,14,17]. These machine learning algorithms can be divided into two categories, namely, supervised and unsupervised learning, and applied to achieve various goals [14]. On the one hand, in supervised learning, algorithms during the training phase are aware of expected results and can be applied, e.g., for traffic forecast based on historical data, quality of transmission estimation [13] or routing [19]. On the other hand, in unsupervised learning, there is no prior knowledge of the expected results and it can be used to find patterns (similarities) and structures in the traffic or to extract features, e.g., traffic anomaly detection [5] or attack detection [6]. There are several traffic prediction methods present in the literature, the majority being based either on autoregressive moving average (ARIMA) or long short-term memory (LSTM) recurrent neural networks [9]. However, pure time-series forecasting approaches have recently been indicated as limiting, and in terms of non-time-series forecasting methods, the results highly depend on the datasets used [3]. Therefore, there is a need for more creative approaches, with respect to computational overhead and accuracy.

An interesting way to improve the prediction accuracy is using data analysis methods to create additional input features to the algorithms, on top of the amount of network traffic at consecutive points in time. In [15], the authors add an autocorrelation coefficient to an LSTM-based traffic prediction model to improve its accuracy. That way, the information about seasonality in the time series can be captured. Further, in [10], the use of daily and weekly seasonal patterns is also explored and three data-driven LSTM methods are proposed.

The problem addressed in this work, namely, prediction network traffic for different frame size ranges separately is a simultaneous multiple time series prediction. Such a problem, for example was addressed in [16] for forecasting the demand for thousands of products across multiple warehouses. The authors use a model based on linear regression with additional spike analyser and safety stock rule engine. Seasonality analysis and information about annual events are used for designing a self-updating model successfully forecasting multiple time series in a short time. Encoding events in a multiple time series model was also used for example in [1] for forecasting UK electricity demand. Categorising the day being forecasted as a working day or a non-working day and using it as an additional feature in a kNN model showed advantages over conventional kNN. In [18], it is shown that not only the seasonality, but also correlations between predicted time series can be a valuable piece of information added to the model predicting methane outbreaks in coal mines. In this case, a model based on random forest is successfully used, taking into account additional parameters derived from cross correlations (including autocorrelations) between selected pairs of time series.

To the best of our knowledge, the prediction of network traffic for different frame sizes separately has not been studied in the literature. To fill this research gap, in this work we analyse the gathered real network data for various frame sizes and study their seasonality and correlations. Next, we propose appropriate machine learning models and algorithms to efficiently forecast traffic based on the historical data.

3 Data Analysis

Data analysis and further experiments were conducted on multiple datasets containing real data. The first dataset is composed of the Seattle Internet Exchange Point (SIX) data from a four-week period, between the 1st and the 28th of November 2020. The second dataset consists of SIX data from the succeeding four-week period – from 28th of November to 26th of December 2020. Both datasets have 5-minute sampling. To create the third dataset, the SIX dataset from November was resampled using a maximum of 1 h aggregation. The obtained dataset represented the same period as the first one, being significantly smaller at the same time. The data were collected weekly from the SIX website[1], so that raw 5 min sampling values are available for the whole investigated period. Two of the databases published in RRD format are used, namely, aggregated traffic in bits and frame size distribution. The numerical values were extracted from the databases using RRDtool[2]. There are 13 frame size ranges in the original data. For simplicity, in this paper, they are represented by letters, as shown in table 1. In Fig. 1, the frame size distribution of the gathered input data is plotted for a sample week.

Table 1. Frame sizes - letter representation

Frame size in bytes	64	65–128	129–256	257–384	385–512	513–640	641–768	769–896	897–1024	1025–1152	1153–1280	1281–1408	1409–1536
Letter represen-tation	a	b	c	d	e	f	g	h	i	j	k	l	m

The bit value of traffic in different frame sizes was calculated from the collected databases as follows. Let n be the total number of frames in a given time point, x_i the size of a frame of i-th size in bits, y_i the percentage of frames of i-th size divided by 100, s the aggregate traffic in given time point in bits per second. From the data, we know the values of x_i, y_i and s in a given time point. The aggregated traffic s can be expressed as $s = \sum_{i=1}^{13} x_i \cdot y_i \cdot n$. Thus, the total number of frames n and the traffic in frames of i-th size can be easily calculated.

---

[1] https://www.seattleix.net/statistics/
[2] https://oss.oetiker.ch/rrdtool/

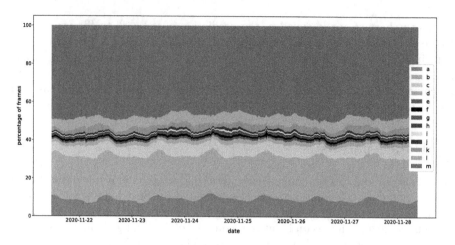

Fig. 1. Input data - frame size distribution

In Fig. 2, we present the calculated traffic for different frame sizes in the investigated 4 week period. As can be observed, the vast majority of traffic is transmitted in the biggest frames, denoted as size m. Interestingly, the second most used frame size b, does not carry much traffic, because of its small size.

To test if the conducted methodology can be generalized to use on data from other sources, we obtained our fourth dataset from a European provider, containing 1 month worth of data with different frame size ranges. Instead of SIX's 13 even ranges, the European dataset uses 7 uneven ones: 64–127 bytes (a), 128–255 bytes (b), 256–511 bytes (c), 512–1023 bytes (d), 1024–1513 bytes (e), 1514 bytes (f) and <1515 bytes (g). The received dataset covers a period from the 1st of November to the 1st of December 2020 and is aggregated to a 3-hour average. In Fig. 3, we present the calculated traffic in different frame sizes in European dataset.

Further data analysis and modelling were performed in Python, using standard machine learning, statistical, and plotting packages: Scikit-learn, statsmodels, pandas, SciPy, NumPy, seaborn, and matplotlib.

When briefly observing the traffic plots, it can be suspected that the data has strong seasonality. It can be further explored by checking the autocorrelation for all the frame sizes. In Fig. 4, we present the autocorrelation function values for all the frame sizes for 5 different lag values in the SIX November dataset with 5-minute sampling. As can be observed, there is a very strong autocorrelation for the 5 min lag. It means, that with 5-minute sampling, the amount of traffic in every timestamp is highly correlated with the previous one. The autocorrelation after 1h lag is also very high. Examining the autocorrelation values in the succeeding columns, it can be stated that there is daily, two-day and weekly seasonality in the data. The same dependence was observed for all the other datasets. The exact values of the autocorrelation function vary for specific frame sizes but these fluctuations are not radical. That information can be later used

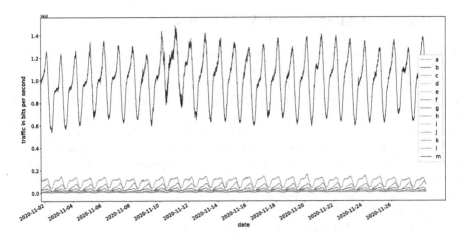

Fig. 2. Internet traffic in different frame sizes in SIX in November 2020

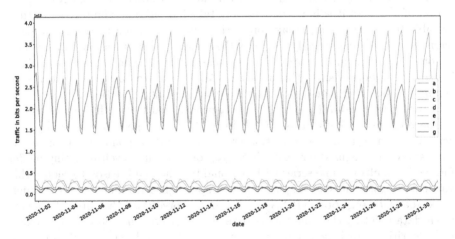

Fig. 3. Internet traffic in different frame sizes in the European dataset

for creating additional features for the machine learning algorithms for better traffic prediction.

Despite large differences in traffic values for different frame sizes, it can be suspected, that there are correlations between them, especially since the traffic for all the frame sizes has similar seasonality. Figure 5 presents a correlation plot between the traffic in all the frame sizes and the aggregate traffic in SIX November dataset. Several correlations with a value higher than 0.9 can be found between the traffic for some frame sizes. Similar relationships can be observed in all considered datasets. These correlations are an important piece of information and will be used as a help for the models and algorithms to better predict the traffic in specific frame sizes.

4   Proposed Models and Algorithms

Taking into account information obtained from the conducted data analysis, we propose three models for network traffic prediction in specific frame sizes. That is, predicting the traffic in frames of size x based on historical traffic in all the frame sizes (model 1), predicting the traffic in frames of size x based on historical traffic in three less correlated frame sizes (model 2) and predicting the traffic in frames of size x based on historical traffic only in frames of size x (model 3).

To help the models make better predictions, we use additional input features, which we chose based on the seasonality in the data proved by calculating auto-correlations. That means, that on top of the amount of traffic in a considered point in time, we add extra features indicating the amount of traffic in important points in the past, e.g., 5 min before, 24 h before, 1 week before. We further discuss the choice of specific additional features in Sect. 5.

Having the models and additional features prepared we need to choose the machine learning (ML) algorithms. In this work, we forecast a relatively short period of 5 min, so it is important to only take into account the regressors that are able to be trained and make predictions fast. Although methods like deep learning tend to be extremely accurate, they require a lot of training time, which is not suitable for our scenario. For that reason, after trying several algorithms including Support Vector Machines and AdaBoost, we chose three relatively simple and fast regressors described in the subsequent subsections.

Linear regression (LR) is a simple approach that tries to fit a linear model to the relationship between observed linear data. The goal is to find the best generalization so that the prediction error for new data points is the smallest. This approach was used for example in [16] for designing a self-updating model forecasting multiple time series with seasonality. The main reason to choose this particular algorithm is its simplicity, which implies fast training and forecasting.

k nearest neighbours (kNN) is a method of predicting the output for a new input data point by checking the outputs of its k nearest neighbours. Therefore, this algorithm can handle non-linearity in the data well, since it predicts the values of new data points only by checking the most similar (nearest) ones. This approach was used for example in [1] for creating a multiple time series model with additional features identifying weekdays and weekends, which have different seasonality patterns. Again, the main reason to use this algorithm is its simplicity, which implies speed.

Random forest (RF) is an example of an ensemble method based on decision trees. Each decision tree uses a random subset of features and their decisions are averaged to improve the overall accuracy and prevent over-fitting. This is an example of combining a set of weak models to create one strong model. This approach was used successfully, for example, in [18] for creating a prediction model for multiple time series with additional features, including correlations between the considered time series'. RF proved to make the most accurate predictions among the algorithms considered in [18]. This method can be time and resource-consuming in large datasets, however, in our case, the number of features is low, which means that a small number of trees is sufficient for the prediction.

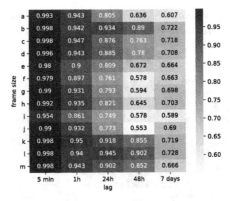

Fig. 4. Autocorrelation for traffic in all the frame sizes for different lag values

5   Numerical Experiments

In this project, we use Scikit-learn implementation of the ML algorithms. Their chosen parameters were tuned by grid search and their tested and selected values are presented in Table 2. In all the algorithms, individual data points represent the network traffic in specific points in time.

Table 2. Tuning chosen parameters for the algorithms

Algorithm	Parameter	Tested values	Chosen value
kNN	weights	'uniform', 'distnace'	'uniform'
	n_neighbours	1, 3, 5, 8, 10	8
RF	n_estimators	3, 5, 10, 15, 20, 50	10

To evaluate and compare the predictions made by different algorithms, a suitable error metric is needed. Because the amount of traffic in different frame sizes varies significantly, so do the absolute error metrics. In order to directly compare the performance of chosen regressors for all considered frame sizes, a percentage error metric is the most reasonable choice. For that reason, we decided to use the root mean squared percentage error (RMSPE) for the evaluation.

Multiple experiments were run to find the best model. As the main goal is to create an on-line traffic prediction model that is able to quickly respond to changing network conditions, there are two important factors to be considered: prediction quality and time. For that reason, we use additional input features which we briefly described in Sect. 4. We initially add four extra features indicating the amount of traffic 5 min before (*previous_timestamp*), 1 h before (*an_hour_ago*), 24 h before (*yesterday*) and 1 week before (*last_week*). The

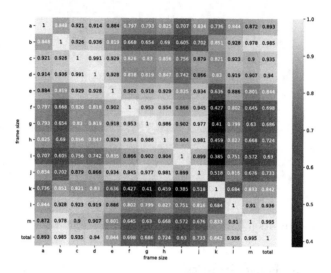

Fig. 5. Correlations between traffic in different frame sizes and aggregate traffic

use of additional features enables us to directly teach the models important similar datapoints, which makes it possible not to consider them in order. For that reason, in all the models and algorithms we use 10-fold cross validation.

Table 3. RMSPE comparison in different models and algorithms, SIX November dataset, 5-minute sampling; best model for each algorithm highlighted

Alg.	Frame size												
	a	b	c	d	e	f	g	h	i	j	k	l	m
Model 1													
LR	0.0126	0.0094	0.0115	0.0188	0.0466	0.0444	0.0274	0.0261	0.0781	0.0383	0.0165	0.0163	0.0107
kNN	0.0305	0.0179	0.0364	0.0409	0.0627	0.0566	0.0493	0.0500	0.0952	0.0685	0.0479	0.0267	0.0138
RF	0.0160	0.0128	0.0149	0.0209	0.0580	0.0519	0.0295	0.0284	0.0872	0.0405	0.0194	0.0198	0.0141
Model 2													
LR	0.0148	0.0123	0.0161	0.0231	0.0517	0.0479	0.0325	0.0319	0.0866	0.0443	0.0199	0.0171	0.0136
kNN	0.0302	0.0141	0.0185	0.0287	0.0706	0.0692	0.0619	0.0619	0.1071	0.0794	0.0196	0.0182	0.0143
RF	0.0166	0.0132	0.0163	0.0233	0.0648	0.0494	0.0342	0.0338	0.0867	0.0478	0.0207	0.0213	0.0148
Model 3													
LR	0.0164	0.0131	0.0184	0.0253	0.0555	0.0491	0.0354	0.0351	0.0898	0.0478	0.0225	0.0177	0.0142
kNN	0.0164	0.0131	0.0167	0.0245	0.0579	0.0489	0.0333	0.0337	0.0860	0.0434	0.0212	0.0185	0.0143
RF	0.0175	0.0137	0.0187	0.0261	0.0630	0.0499	0.0353	0.0357	0.0937	0.0489	0.0231	0.0205	0.0152

In Table 3, we present the RMSPE values for three models described in Sect. 4, for the SIX November dataset with 5 min sampling. As can be concluded, the choice of the model depends on the choice of the regressor - LR and RF get their lowest RMSPE values in model 1, and kNN - in model 3. However, the overall lowest RMSPE values are obtained by LR. As an illustration,

Table 4. RMSPE comparison in different models and algorithms, SIX December dataset, 5-minute sampling; best model for each algorithm highlighted

Alg.	Frame size												
	a	b	c	d	e	f	g	h	i	j	k	l	m
Model 1													
LR	0.0151	0.0110	0.0109	0.0177	0.0336	0.0304	0.0247	0.0258	0.0609	0.0329	0.0155	0.0182	0.0111
kNN	0.0284	0.0181	0.0368	0.0353	0.0501	0.0532	0.0483	0.0485	0.0841	0.0629	0.0448	0.0287	0.0135
RF	0.0180	0.0146	0.0152	0.0206	0.0379	0.0347	0.0268	0.0276	0.0718	0.0356	0.0189	0.0213	0.0144
Model 2													
LR	0.0153	0.0140	0.0129	0.0205	0.0374	0.0343	0.0312	0.0321	0.0713	0.0407	0.0174	0.0191	0.0137
kNN	0.0317	0.0150	0.0206	0.0238	0.0606	0.0669	0.0649	0.0663	0.0999	0.0774	0.0186	0.0191	0.0142
RF	0.0185	0.0155	0.0158	0.0226	0.0414	0.0390	0.0323	0.0332	0.0741	0.0419	0.0192	0.0210	0.0148
Model 3													
LR	0.0183	0.0152	0.0180	0.0237	0.0411	0.0372	0.0341	0.0352	0.0731	0.0440	0.0215	0.0198	0.0145
kNN	0.0184	0.0148	0.0166	0.0229	0.0414	0.0366	0.0318	0.0328	0.0717	0.0407	0.0208	0.0211	0.0144
RF	0.0196	0.0158	0.0179	0.0247	0.0433	0.0394	0.0345	0.0348	0.0750	0.0439	0.0217	0.0206	0.0152

Table 5. RMSPE comparison in different models and algorithms, SIX November dataset, 1-hour maximum aggregation; best model for each algorithm highlighted

Alg.	Frame size												
	a	b	c	d	e	f	g	h	i	j	k	l	m
Model 1													
LR	0.0236	0.0205	0.0254	0.0315	0.1219	0.0918	0.0424	0.0384	0.1411	0.0706	0.0315	0.0427	0.0225
kNN	0.0598	0.0379	0.1064	0.0996	0.1491	0.1543	0.1437	0.1338	0.2084	0.1986	0.1364	0.0768	0.0406
RF	0.0299	0.0286	0.0344	0.0381	0.1345	0.1347	0.0491	0.0429	0.1450	0.0788	0.0475	0.0556	0.0334
Model 2													
LR	0.0249	0.0350	0.0579	0.0635	0.1349	0.1394	0.0643	0.0652	0.1492	0.0909	0.0746	0.0516	0.0394
kNN	0.0477	0.0315	0.0523	0.0647	0.1598	0.1716	0.1580	0.1551	0.2247	0.2325	0.0562	0.0375	0.0343
RF	0.0290	0.0337	0.0490	0.0592	0.1119	0.1289	0.0711	0.0679	0.1714	0.0966	0.0666	0.0355	0.0363
Model 3													
LR	0.0411	0.0380	0.0679	0.0688	0.1539	0.1189	0.0816	0.0797	0.1754	0.1084	0.0675	0.0536	0.0517
kNN	0.0419	0.0342	0.0515	0.0599	0.1521	0.1270	0.0781	0.0720	0.1629	0.0966	0.0596	0.0449	0.0452
RF	0.0457	0.0356	0.0452	0.0585	0.1500	0.1477	0.0804	0.0709	0.1793	0.0926	0.0603	0.0541	0.0462

in Fig. 6, we present the comparison of the prediction results between model 1 and 3 obtained by LR for a sample frame size. Some differences between the models can be spotted in the presented zoomed-in fragment, showing that the prediction based on the historical traffic from all the frame sizes rather than a single one is generally closer to the real values.

The same trends were observed in all the remaining datasets. Table 4 presents the results for the SIX December dataset, while Table 5 presents for the SIX November dataset with 1h aggregation, with additional features *an_hour_ago*, *yesterday* and *last_week*. In Table 6, we present the RMSPE values obtained for the European dataset. Because of the 3-hour average aggregation, after calculating the autocorrelation function values we decided to use the following additional input features: the amount of traffic 24 h before (*yesterday*), 48 h

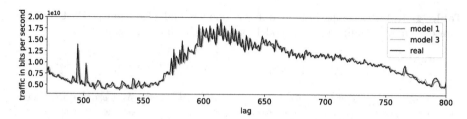

Fig. 6. Prediction results for traffic in frames of size i, LR regressor, SIX November dataset, 5 min sampling - zoomed-in fragment

before (*two_days_ago*) and 1 week before (*last_week*). As can be seen, similarly to the other datasets, the prediction quality is higher taking into account the historical traffic in all the frame sizes simultaneously, with LR being the most accurate among considered regressors.

Table 6. RMSPE comparison in different models, European dataset, 3-hour average aggregation; best model for each algorithm highlighted

Algorithm	Frame size						
	a	b	c	d	e	f	g
Model 1							
LR	0.0156	0.0225	0.0193	0.0247	0.0139	0.0165	0.0306
kNN	0.0262	0.1084	0.0895	0.1220	0.0351	0.0281	0.0332
RF	0.0202	0.0333	0.0299	0.0314	0.0250	0.0257	0.0345
Model 2							
LR	0.0236	0.0369	0.0331	0.0386	0.0276	0.0287	0.0350
kNN	0.0323	0.1275	0.1075	0.1283	0.0357	0.0324	0.0487
RF	0.0296	0.0421	0.0411	0.0378	0.0331	0.0340	0.0384
Model 3							
LR	0.0300	0.0441	0.0422	0.0370	0.0343	0.0345	0.0422
kNN	0.0307	0.0760	0.0591	0.0618	0.0359	0.0350	0.0398
RF	0.0318	0.0460	0.0440	0.0372	0.0365	0.0382	0.0410

In Table 7, we present the mean percentage advantage of the RMSPE values obtained by the best regressor, LR, in model 1 over model 3 for all considered datasets. As can be concluded, the prediction of the amount of traffic in a specific frame size is better considering the historical data of the traffic for all the frame sizes when compared to the prediction based only on one frame size.

Table 8 presents average time of execution for considered regressors for tested models (note, model 1a and 1b are described further). The measurements were performed on a machine with an Intel Core i5-1038NG7 processor with 16 GB RAM. As can be observed, all the algorithms are the fastest in model 3 because of the smallest dataset size. Nevertheless, at this stage, the prediction quality is more important than the time of execution, especially considering the very short

Table 7. Mean percentage advantage of RMSPE of model 1 over model 3 for the LR regressor in considered datasets

Dataset	Frame size												
	a	b	c	d	e	f	g	h	i	j	k	l	m
SIX November, 5 min sampling	22%	28%	36%	25%	16%	10%	22%	25%	12%	18%	25%	9%	24%
SIX November, 1 h aggregation	43%	46%	63%	54%	21%	23%	48%	52%	20%	35%	53%	20%	56%
SIX December, 5 min sampling	18%	26%	38%	24%	18%	18%	26%	25%	17%	24%	27%	9%	22%
European, 3 h aggregation	48%	49%	54%	33%	60%	52%	27%	–	–	–	–	–	–

Table 8. Time of execution in seconds, SIX November dataset, 5-min sampling

Algorithm	Model 1	Model 2	Model 3	Model 1a	Model 1b
LR	0.0084	0.0043	0.0033	0.0095	0.0062
kNN	0.0379	0.0198	0.0155	0.0485	0.0277
RF	0.3931	0.1370	0.0588	0.4790	0.2880

time of execution for the best regressor - LR. For that reason, we chose model 1 for further analysis.

In Table 9 we present different choices of input features for model 1 in two datasets with 5-minute sampling for the best algorithm - LR. In model 1a, compared to model 1, we change the last input feature: from *last_week* to *two_days_ago*. As can be seen, for some frame sizes the RMSPE values are marginally lower in model 1a, however, the times of execution are higher, as can be seen in Table 8. It can be explained by the size of the dataset - the amount of network traffic "a week before" cannot be obtained for the first week worth of data while the amount of traffic "two days before" cannot be obtained for

Table 9. RMSPE comparison in different versions of model 1, SIX November and December datasets with 5-minute sampling

Dataset	Frame size												
	a	b	c	d	e	f	g	h	i	j	k	l	m
Model 1													
SIX November	0.0126	0.0094	0.0115	0.0188	0.0466	0.0444	0.0274	0.0261	0.0781	0.0383	0.0165	0.0163	0.0107
SIX December	0.0151	0.0110	0.0109	0.0177	0.0336	0.0304	0.0247	0.0258	0.0609	0.0329	0.0155	0.0182	0.0111
Model 1a													
SIX November	0.0142	0.0091	0.0115	0.0190	0.0455	0.0431	0.0272	0.0261	0.0827	0.0383	0.0165	0.0156	0.0106
SIX December	0.0149	0.0113	0.0115	0.0177	0.0338	0.0300	0.0244	0.0258	0.0580	0.0338	0.0166	0.0192	0.0111
Model 1b													
SIX November	0.0128	0.0094	0.0117	0.0189	0.0466	0.0442	0.0275	0.0265	0.0789	0.0394	0.0168	0.0161	0.0108
SIX December	0.0150	0.0113	0.0111	0.0180	0.0336	0.0305	0.0252	0.0263	0.0609	0.0332	0.0157	0.0180	0.0112

only the first two days worth of data. For that reason, the size of the dataset is smaller in model 1.

Model 1b has the smallest both dataset and number of features. Comparing to model 1, we delete the feature *an_hour_ago*, because the autocorrelation values for the lag of 5 min and 1 h are both extremely high, so the information provided by features obtained from both of them are similar. Indeed, the differences in RMSPE values between model 1 and model 1b are marginal and the time gains from using a smaller set of features in model 1b are significant. For that reason, we propose model 1b as the best trade-off between prediction quality and time of execution. We conducted the analysis described above on the remaining algorithms and the same trends were observed.

Analysing the predictions made by the models, it can be seen that traffic in some of the frame sizes is easier to predict than in the other ones. The lowest RMSPE values are achieved by the models predicting the traffic in frames of size a, b, c and m, while the highest RMSPE values are obtained by the models predicting the traffic in frames of size e, f and i - k. That might have been caused by the traffic in the middle frame sizes being less regular. In Fig. 7 and 8, we present zoomed-in fragments of the actual and predicted amounts of the network traffic – an easier to predict traffic in frames of size m and more difficult to predict traffic in frames of size e accordingly.

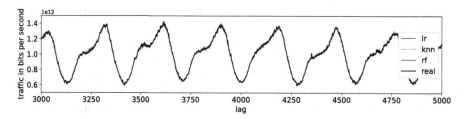

Fig. 7. Prediction results for traffic in frames of size m, model 1b, SIX November dataset, 5 min sampling - zoomed-in fragment

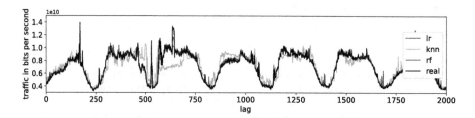

Fig. 8. Prediction results for traffic in frames of size e, model 1b, SIX November dataset, 5 min sampling - zoomed-in fragment

6   Conclusions and Future Work

In this paper, we focus on the prediction and analysis of network traffic composed of various frame sizes. In more detail, the developed model is able to forecast traffic of a certain type based on the historical data. Firstly, we described gathered real network traffic data and its preparation process required for feature extraction and further analysis. After that, we detected seasonality patterns by calculating autocorrelations for different lag values showing similar patterns for different frame sizes. Moreover, the correlations between different traffic types were investigated, indicating similarities between traffic patterns in certain frame sizes. Next, we proposed three machine learning algorithms and ran extensive numerical experiments on four datasets to evaluate their efficiency. According to the results, linear regression yields the highest accuracy having its RMSPE values on average 50% lower than kNN and 15% lower than random forest.

Additionally, we investigated the impact of different models and input features choices, finding the best compromise between prediction quality and time of execution.

In future work, we plan to focus on the prediction of traffic for various applications and services to improve the performance of multilayer application-aware networks.

References

1. Al-Qahtani, F.H., Crone, S.F.: Multivariate k-nearest neighbour regression for time series data-a novel algorithm for forecasting UK electricity demand. In: The 2013 International Joint Conference on Neural Networks (IJCNN), pp. 1–8. IEEE (2013)
2. Biernacki, A., Krieger, U.R.: Session level analysis of P2P television traces. In: Zeadally, S., Cerqueira, E., Curado, M., Leszczuk, M. (eds.) FMN 2010. LNCS, vol. 6157, pp. 157–166. Springer, Heidelberg (2010). https://doi.org/10.1007/978-3-642-13789-1_15
3. Boutaba, R., Salahuddin, M.A., Limam, N., Ayoubi, S., Shahriar, N., Estrada-Solano, F., Caicedo, O.M.: A comprehensive survey on machine learning for networking: evolution, applications and research opportunities. J. Internet Serv. Appl. 9(1), 1–99 (2018). https://doi.org/10.1186/s13174-018-0087-2
4. Cenedese, A., Tramarin, F., Vitturi, S.: An energy efficient ethernet strategy based on traffic prediction and shaping. IEEE Trans. Commun. 65(1), 270–282 (2016)
5. Chen, X., Li, B., Shamsabardeh, M., Proietti, R., Zhu, Z., Yoo, S.J.B.: On real-time and self-taught anomaly detection in optical networks using hybrid unsupervised/supervised learning. In: 2018 European Conference on Optical Communication (ECOC), pp. 1–3 (2018)
6. Furdek, M., Natalino, C., Lipp, F., Hock, D., Giglio, A.D., Schiano, M.: Machine learning for optical network security monitoring: a practical perspective. J. Lightwave Technol. 38(11), 2860–2871 (2020)
7. Garsva, E., Paulauskas, N., Grazulevicius, G.: Packet size distribution tendencies in computer network flows. In: 2015 Open Conference of Electrical, Electronic and Information Sciences (eStream), pp. 1–6. IEEE (2015)
8. Gu, R., Yang, Z., Ji, Y.: Machine learning for intelligent optical networks: a comprehensive survey. J. Netw. Comput. Appl. 157 (2020)

9. Joshi, M., Hadi, T.H.: A review of network traffic analysis and prediction techniques. arXiv preprint arXiv:1507.05722 (2015)
10. Krishnaswamy, N., Kiran, M., Singh, K., Mohammed, B.: Data-driven learning to predict wan network traffic. In: Proceedings of the 3rd International Workshop on Systems and Network Telemetry and Analytics, pp. 11–18 (2020)
11. Lehman, T., Yang, X., Ghani, N., Gu, F., Guok, C., Monga, I., Tierney, B.: Multilayer networks: an architecture framework. IEEE Commun. Mag. **49**(5), 122–130 (2011)
12. Lopez, V., Konidis, D., Siracusa, D., Rozic, C., Tomkos, I., Fernandez-Palacios, J.P.: On the benefits of multilayer optimization and application awareness. J. Lightwave Technol. **35**(6), 1274–1279 (2017)
13. Mata, J., et al.: A SVM approach for lightpath QoT estimation in optical transport networks. In: 2017 IEEE International Conference on Big Data (Big Data), pp. 4795–4797 (2017)
14. Musumeci, F., et al.: An overview on application of machine learning techniques in optical networks. IEEE Commun. Surv. Tutor. **21**(2), 1383–1408 (2019)
15. Shihao, W., Qinzheng, Z., Han, Y., Qianmu, L., Yong, Q.: A network traffic prediction method based on LSTM. ZTE Commun. **17**(2), 19–25 (2019)
16. Wagner, N., Michalewicz, Z., Schellenberg, S., Chiriac, C., Mohais, A.: Intelligent techniques for forecasting multiple time series in real-world systems. Int. J. Intell. Comput. Cybern. (2011)
17. Xie, J., Yu, F.R., Huang, T., Xie, R., Liu, J., Wang, C., Liu, Y.: A survey of machine learning techniques applied to software defined networking (sdn): Research issues and challenges. IEEE Commun. Surv. Tutor. **21**(1), 393–430 (2018)
18. Zagorecki, A.: Prediction of methane outbreaks in coal mines from multivariate time series using random forest. In: Yao, Y., Hu, Q., Yu, H., Grzymala-Busse, J.W. (eds.) RSFDGrC 2015. LNCS (LNAI), vol. 9437, pp. 494–500. Springer, Cham (2015). https://doi.org/10.1007/978-3-319-25783-9_44
19. Zhong, Z., Hua, N., Yuan, Z., Li, Y., Zheng, X.: Routing without routing algorithms: an AI-based routing paradigm for multi-domain optical networks. In: 2019 Optical Fiber Communications Conference and Exhibition (OFC), pp. 1–3 (2019)

Scalable Handwritten Text Recognition System for Lexicographic Sources of Under-Resourced Languages and Alphabets

Jan Idziak[1], Artjoms Šeļa[2], Michał Woźniak[2], Albert Leśniak[2], Joanna Byszuk[2], and Maciej Eder[2(✉)]

[1] Singapore, Singapore
[2] Institute of Polish Language, Polish Academy of Sciences, al. Mickiewicza 31, 31-120 Krakow, Poland
{artjoms.sela,michal.wozniak,albert.lesniak,joanna.byszuk,
maciej.eder}@ijp.pan.pl

Abstract. The paper discusses an approach to decipher large collections of handwritten index cards of historical dictionaries. Our study provides a working solution that reads the cards, and links their lemmas to a searchable list of dictionary entries, for a large historical dictionary entitled the *Dictionary of the 17th- and 18th-century Polish*, which comprises 2.8 million index cards. We apply a tailored handwritten text recognition (HTR) solution that involves (1) an optimized detection model; (2) a recognition model to decipher the handwritten content, designed as a spatial transformer network (STN) followed by convolutional neural network (RCNN) with a connectionist temporal classification layer (CTC), trained using a synthetic set of 500,000 generated Polish words of different length; (3) a post-processing step using constrained Word Beam Search (WBC): the predictions were matched against a list of dictionary entries known in advance. Our model achieved the accuracy of 0.881 on the word level, which outperforms the base RCNN model. Within this study we produced a set of 20,000 manually annotated index cards that can be used for future benchmarks and transfer learning HTR applications.

Keywords: Handwritten text recognition · Index cards archives · Lexicography · Neural network · Convolutional neural network · Recurrent neural network · Connectionist temporal classification · Keras OCR · ResNet · Spatial transformer networks · Synthetic dataset

1 Introduction

Decades of lexicographic work that was done before the popularization of machine-readable texts provide rich lexicographic and/or linguistic data that

---

J. Idziak—Independent Scholar.

M. Paszynski et al. (Eds.): ICCS 2021, LNCS 12742, pp. 137–150, 2021.
https://doi.org/10.1007/978-3-030-77961-0_13

is extremely hard to reuse today or to be integrated into modern databases, corpora and collections. Not only are these original resources handwritten, but they are also unstructured, or at best their structure is limited to an alphabetical order of the respective items. Card files served as tools of lexicographic description and, when collected into catalogues, allowed random access to vast bodies of lexical information. These cards were building blocks of lexicons and dictionaries, long before corpus linguistics that relied on digitized texts appeared [18]. The lexicographic resources in question involve millions of handwritten cards for various historical dictionaries, ranging from Latin (with the archetypical *Thesaurus Linguae Latinae*, one of the first initiatives of this kind), to medieval and modern language varieties (Middle Dutch, Old Czech, Old Norse Prose, or Middle High German to name but a few). In most cases, the index cards are acquired throughout several decades – sometimes dating back to the 19th century – and they contain comprehensive documentation for all known words of respective language varieties.

The lexicographic collections held at the Institute of Polish Language of the Polish Academy of Sciences are no exception, with its extensive card catalogues of *The Old Polish Dictionary, The Dictionary of the 17th- and 18th-century Polish, The Dictionary of Polish Dialects, The Great Dictionary of Polish*, as well as a few onomastic dictionaries of proper nouns. A single index card contains a lemma (a base word in a header) followed by its context excerpted from actual historical documents. Stored in dedicated boxes and alphabetized, the index cards are used by lexicographers to compose subsequent dictionary entries. Since most of the aforementioned dictionaries are not completed yet, the handwritten index cards serve as a work-in-progress source of information, and, even in the case of the dictionaries that are already published, the index cards are still valid as their supplementary materials. The biggest challenge, however, is that they are not machine-readable, and not linked to the searchable lexicographic databases.

While recent years saw a development of numerous approaches to optical and handwritten text recognition (HTR) also in the relation to humanistic data, e.g. Transkribus [16,20] which provides excellent performance, such solutions are better suited to longer texts, fewer scribal hands and require significant amounts of training data [8], also posing limitations as to the number of pages that can be annotated. Meanwhile, the index cards contain short excerpts, typically no more than a sentence of context next to the lemma, followed by source description, and are produced by numerous lexicographers, often showing inconsistent handwriting style that can be understood only by themselves or other team members. In fact, while the second poses a significant challenge, also in the case of the need to prepare manually annotated training set, computer vision methods which rely on less easy to observe patterns hold great promise of perhaps outperforming human reading of more illegible scribblings.

This project sets up an operational workflow for retrieving lexical data from handwritten card catalogues, followed by matching their lemmas to the list of dictionary entries. To build and test a prototype, we have chosen *The Dictionary of the 17th- and 18th-century Polish*, which is a good example of a lexicographic source that combines traditional materials of 2.8 million handwritten index cards with modern technologies – the dictionary itself is a fully digital database, and it is linked to an annotated corpus of the 17th-century Polish [4,21]. Moreover, in a pilot study a small selection of the index cards was manually mapped onto a list of dictionary entries [3]. This preliminary work, however, clearly shows that manual mapping is hardly feasible in real-scale setups and would involve an immense effort expressed in thousands of working hours. Our project aims at overcoming this limitation by using an automated approach that would simplify the work of lexicographers in preparing the digital entries. The main goal of the project, however, goes beyond the prototype applied to the 17th-century lexicographic sources. The diverse range of the obtained outcomes makes this study potentially interesting both from a computer sciences point of view, as well as from a digital humanities perspective.

The research presented in this paper provides the following contributions:

1. We propose a unified modular workflow that is adjustable to any language, since the model relies solely on a synthetic dataset; our workflow can be easily extended to other under-resourced languages, including languages with extended Latin alphabets or non-Latin scripts.
2. We provide a working HTR detection and recognition prototype (also as a deployed demo web application) that outperforms baseline models significantly.
3. We provide a synthetic dataset of artificially generated 500,000 words in Polish, supplemented by another set of 30,000 random strings with uniform distribution of Polish diacritics.
4. We offer a manually labelled set of 20,000 words in Polish to be used as ground truth in future applications and model evaluation settings.

2   Related Work

Modern HTR heavily leans towards solutions based on Artificial Neural Networks and it was shown that various architectures of deep learning improve performance in the handwriting recognition task [6,9,10,12,15,22,27–29] compared to Hidden Markov models [25,26]. Handwritten text imposes severe challenges for a machine vision technologies that we counter by combining several known methods: (1) the problem of significant variation in writing style, shapes, sizes and possible deformations of characters is solved through spatial deformation rectification achieved with the Spatial Transformation Network that learns translation-invariant features [15]. (2) Since source material in HTR tasks is often based on a large and diverse lexicon which adds to the difficulty of decision making of the model, Word Beam Search is often used to decode CTC layer and constrain prediction errors [24]. Many best-performing post-OCR error correction systems depend on contextual awareness provided by a language model [23], which cannot be adapted to our case of isolated index words. Instead we rely on lexicon-aided decoding of the model's predictions that broadly follows the character-level error correction framework [7]. (3) It is often very difficult to achieve a system generalizability in HTR in a specific domain because of the lack of a large amount of ground truth handwriting samples. Recent studies propose to compensate this by generating vast amounts of synthetic images from character strings [14]. It was even shown that CNN algorithm trained only on the synthetic data outperformed other methods on the text detection task [12]. (4) Finally, the level at which text segmentation for HTR is deployed was subjected to discussion. Early approaches [19] employed line segmentation in the HTR context, while word segmentation or character region awareness methods [12,13] are gaining more popularity recently.

3   Data

3.1   Original Data

Our main focus was the *The Dictionary of the 17th- and 18th-century Polish* (https://sxvii.pl/), a partially-completed lexicographic database that is based on roughly 2.8 million index cards manually filled by different hands (in rare cases typewritten) in the years 1954–1995 [3,4]. The cards are stored in 836 alphabetized boxes, and, after a digitization project conducted in the years 2010 –2015, they are now freely accessible in bitmap image format, saved under names linking to their respective boxes (e.g. *Egzekucja –Ekspediowanie*). For the sake of this study, we drew a sample of 100,000 cards taken uniformly across all boxes, which made up our primary dataset (denoted as `PL-100k-main` hereafter).

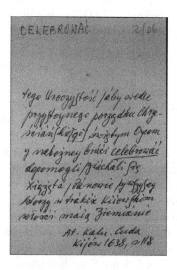

Fig. 1. A handwritten index card from *The Dictionary of the 17th- and 18th-century Polish*.

Almost all of the cards – rare exceptions being phonetic variant pointers – followed a conventional scheme for encoding a given word's attestation (see Fig. 1): (1) a lemma served as an index which ensured navigation and accessibility, while (2) the body of the card registered immediate context of the word occurrence, followed by (3) a bibliographical reference to sources and sometimes including other information (grammatical form, alternate spelling, etc.). Our lemma detection procedure at times encountered difficulties when the additional information about grammatical form was written next to the header lemma, rather than within the card's body. A significant part of the cards from the *The Dictionary of the 17th- and 18th-century Polish* had their index word written in "handwritten capitals" (majuscule alphabet) to keep the words recognizable, which limits the variability of shapes and maintains somewhat clear boundaries between letters. A mix of minuscule and all-majuscule handwriting meant that the HTR model should be able to perform simultaneously on both of these modes of writing, which added an additional step to the transfer learning workflow (as discussed below).

The list of all the 86,000 dictionary entries of the *The Dictionary of the 17th- and 18th-century Polish* is publicly accessible (https://sxvii.pl/). We used the list to serve as a constrained set of possible words to match them against the resulting predictions from our HTR system.

3.2 Synthetic Data for Training

The majority of already existing training datasets for HTR are limited to English. Since Polish language uses an extended Latin script with additional 16 diacritics (8 lowercase and 8 capitalized), we had to make sure that Polish is at least partially represented [11]. Our transfer-learning approach involved the already existing large dataset of English words CVIT [17] that we further enhanced with an artificially generated set of Polish words. To this end, we randomly excerpted 500,000 actual Polish words from the corpus of 17th-century Polish texts *Korba* [21], and generated their bitmap representations using a variety of fonts that mimicked the handwriting both in lowercase and uppercase (this is our PL-500k-synthetic dataset). Not only were the words and the font shapes picked at random, but also the final bitmap images were slightly distorted using different augmentation steps. The augmentation distortion included (1) posterization – which maximizes the image contrast, (2) equalization of the image histogram, (3) solarization – which inverts all pixel values above a threshold, (4) affine geometric transformation. Consequently, the set we obtained consisted of artificial bitmap representations of actual Polish words. This allowed for training the representation of the Polish diacritics, while the proportion of diacritics to standard characters followed their natural distribution.

Apart from the above PL-500k-synthetic dataset, we also prepared an additional set of 30,000 randomly generated strings containing solely Polish diacritical marks – with uniform distribution – in order to improve performance for rare polish diacritics. We denote this set as PL-30k-diacritics.

3.3   Manually Labelled Subsets

To obtain a carefully curated set for evaluation purposes, we drew a small sample from the original dataset, namely 20,000 bitmap images, that were then manually corrected for their bounding boxes, and manually labeled by the project members. This set (`PL-20k-hand-labelled`) was our primary evaluation set, since it provided ground truth for both neatly cropped images and corrected labels. The set has been made publicly available for future benchmarking and applications. In order to facilitate the tedious annotation work, we applied the annotation tool Prodigy (https://prodi.gy/).

Yet another manually annotated subset was prepared (`PL-3k-boundaries`) to optimize the performance of word detection. We drew at random 3,000 images from the original dataset, applied the word detection module as discussed above, and manually-corrected the resulting bounding boxes around the detected lemmas (Fig. 2).

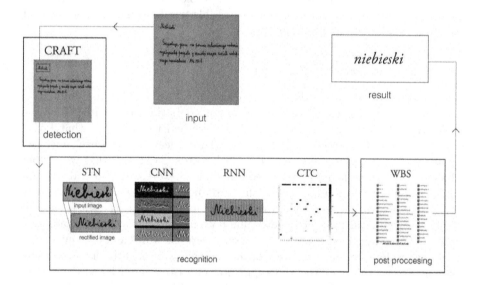

Fig. 2. The HTR workflow, including the detection, recognition, and post-processing steps.

4   Methodological Workflow

Our general workflow involved three main parts: (1) detection, (2) recognition, and (3) postprocessing. The first two steps are based on deep learning and convolutions, while the third step is an optimization technique aimed at improving the final predictions of the neural network.

4.1 Detection of the Index Word

Since all cards had a header lemma, usually placed distinctly from the body text, it is relatively easy to access it: we cut a card to top 300 pixels in order to optimize for computation time, and used Keras OCR Craft model for text detection and bounding box assignment [2]. Then for each card, we identified the first bounding box located in the top-left area, which in the vast majority of cases contains the word in question. We selected 3,000 hand-labeled images to optimize the position of the bounding boxes (the `PL-3k-boundaries` set).

4.2 Recognition

The recognition stage can be broken down to four consecutive components based on TPS-ResNet-BiLSTM-CTC architecture proposed by Baek et al. [1]. We used already existing pre-trained model that we further improved:

1. Input text image was first rectified with the help of the Spatial Transformer Network, or STN [15] with Thin Plate Spline (TPS) transformation [29]. The aim of this step is to ensure that the images are consistent in terms of contrast, saturation and so forth, and thus easier to process at the feature extraction step.
2. Feature extraction using a Convolutional Neural Network (CNN) setup. It extracted relevant features from the image and focused on attributes that are characteristic to particular characters. After a few preliminary tests, we chose a ResNet backbone, because it provided a clear improvement in accuracy.
3. The features extracted in step 2 are fed sequentially into the Bidirectional LSTM layer (BiLSTM).
4. Finally, the Connectionist Temporal Classification (CTC) layer was involved. The benefit of using it is at least two-fold. Firstly, the predictions show the ability to overcome a variable size of the input sequence, even if the number of features is fixed. Secondly, the CTC layer accounts for words with repeated letters, thus helping to differentiate between, say, the words "to" and "too". Also, because most of the publicly available datasets used English, a standard Attention layer would not guarantee sufficiently good performance for other languages. The CTC layer, on the contrary, provided a matrix of all possible predictions which could be further generalized beyond English (Fig. 3).

4.3 Transfer Learning

In our approach we first took an existing model pre-trained on two datasets: MJSynth (MJ) [14] and SynthText (ST) [15]. These datasets are not designed for HTR problems and do not provide Polish characters, therefore we involved a few additional datasets to enhance the model. Firstly, we added the CVIT database with 9 million images of handwritten words based on English corpus. Subsequently, further domain improvement was done using Polish words from a

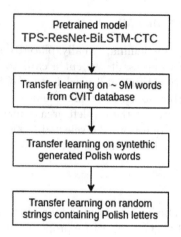

Fig. 3. Transfer learning workflow.

synthetic dataset `PL-500k-synthetic` as discussed above. The last step involved enhancing the performance specifically for the recognition of Polish diacritics. To achieve this, we used the randomly generated strings containing all characters with uniform distribution (`PL-30k-diacritics`).

Initially, we also used the IAM handwritten database containing carefully-labelled handwritten words [19]. After several rounds of transfer learning tests, however, it became clear that the dataset representation and the sample images significantly differ from texts produced in natural conditions. This is caused by the fact that the words in the IAM handwritten database are cut closely around the word outlines, which confuses the recognition system. Our observations suggest that using this dataset leads to overfitting of neural network models with a large number of parameters. Consequently, any applications based on word representation of the IAM handwritten dataset do not seem suitable in real life situations. We would even argue that this dataset should not be used as the main dataset for performance evaluation as well.

4.4   Postprocessing

The predictions as produced by our workflow will always contain some wrongly reconstructed words. However, some of these mistakes are relatively easy to correct, due to the non-random nature of the language. For instance, a human would instinctively guess that the string "cvolution" resembles the word "evolution". The string "ancl" would require an additional split second to decipher, because it could be reconstructed as "and", as "ant" and perhaps even as "uncle" or "anele". Since the list of possible words is limited, an optimization algorithm can be used that matches the input sequence with the nearest element from the closed set of words known in advance. Real-life situations might be more challenging, given the fact that new words emerge at times, e.g. the algorithm would disregard the string "covfefe" as a valid word, and would replace it either with "coffee" or with "coverage".

In the case of our project, however, a vast majority of words written down on index cards match the close set of words stored as a list of 86,000 dictionary entries. Additionally, we took advantage of the fact that the index cards are alphabetized (as is the list of dictionary entries), and stored in 836 boxes with known ranges of the alphabet ascribed to each box. Consequently, in our constrained Word Beam Search (WBC) approach not only did we take into consideration the CTC match of the predictions and the expected dictionary entries, but we also assumed that it is very unlikely to match an index card from a given box to a word belonging to a distant part of the alphabetized list.

In the first round of our procedure, we additionally involved a quality check step. To this end, we prepared the aforementioned `PL-20k-hand-labelled` dataset: we randomly selected 20,000 index cards, manually checked the predictions, and corrected all the misrecognized words. The aim of this step was two-fold: firstly, to allow for re-training the model with the clean ground-truth dataset (while keeping in mind that for the sake of this study we didn't apply it, for the sake of a clear-cut separation of the training set and the evaluation set) and secondly, to prepare 20,000 manually corrected cards for public access for future benchmarking and improving HTR models (Fig. 4).

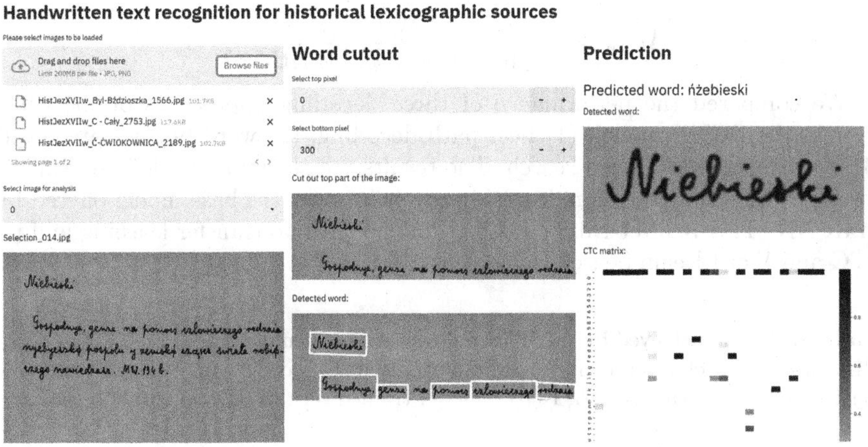

Fig. 4. Online demo application of the HTR system.

5   Application

Alongside our full-size system that is capable of consecutively processing millions of cards, we also designed an online application for testing our workflow and visually presenting how a given index card is being processed. After the file is selected, the app automatically performs cutting, detection and recognition of the word. In post-processing, the app optimizes for the best path CTC encoding. Both vanilla prediction and the CTC matrix are available for inspection.

6   Results

The results of the system are based on accuracy scores achieved on a subset of the original data (the `PL-100k-main` datset). The detection component of the workflow achieved 0.93 of intersection over union on 3,000 hand labeled images (the `PL-3k-boundries` dataset). The results for the recognition model are best presented in comparison to a baseline model. We have excluded 20,000 index cards (the `PL-20k-manual`) to serve as a test set, the average word length was 5.99 characters. Accuracy is calculated as the number of words that were classified correctly divided by the number of all words in the dataset:

$$Acc = \frac{1}{n} \sum_{i}^{n} \mathbb{1}_{y_i = \hat{y}_i}$$

We use Levenshtein distance as the edit distance. Average edit distance is calculated as the sum of the edit distance divided by the number of words. Average normalized edit distance is calculated as averaged edit distance per number of characters in the word:

$$\Delta_{Lnorm}(i) = \frac{\Delta_L(y_i, \hat{y}_i)}{\max(\text{length}(y_i), \text{length}(\hat{y}_i))}$$

We compared the performance of three decoding approaches of the CTC layer in the RCNN model: (1) best path decoding, (2) word beam search, and (3) constrained word beam search. The results are shown in Table 1. As can be observed, we achieved a significant improvement over the base model offered by an RCNN (TPS-ResNet-BiLSTM-CTC) model with no transfer learning and no CTC and Word Beam Search refinement.

Table 1. Results achieved by the HTR models. All of them are based on TPS-ResNet-BiLSTM-CTC architecture. BP – best path, WBS – word beam search, WBS-C – constrained word beam search, POL – model trained on the Polish synthetic set.

Model	Word accuracy	Normalised edit distance	Edit distance	Average edit on misclassified
BP	0.3755	0.1898	1.0871	**1.7484**
WBS	0.0995	0.6194	6.1763	6.8711
BP-POL	0.4332	0.2246	1.2120	2.0945
WBS-POL	0.6655	0.2033	1.0993	2.7063
WBS-C-POL	**0.8810**	**0.0479**	**0.3165**	3.2125

Since our pipeline is, among other things, aimed to aid lexicographers in linking index cards to existing databases, the model that minimizes edit distance on *wrong* predictions sometimes could be more useful than the WBS model that is designed for word-level accuracy and makes use of external information (alphabet range of a box). Average edit distance on wrongly recognized words (Table 1) shows this effect: both best path decoding models have less edits than WBS decoding (1.75 and 2.09), suggesting that in a realistic setting of manual work with the full-scale collection, less greedy models could provide more "useful" predictions despite the drop in word-level accuracy. In addition, the best-performing constrained WBS model has a higher rate of edits in misclassified words than unconstrained WBS which implies that sometimes words fall out of the box's alphabetical range but then are still forcefully fitted to that range by WBS-C. Most probably, these mistakes happen because index word position could be recognized incorrectly in a situation when a card has multiple words on the top.

7   Discussion

Our results show a trade-off between a straightforward word-level accuracy and amount of noise captured by a model. High values of average edit distance on the misclassified words for the Word Beam Search models show that the algorithm is likely to pick up noise in the activation map and predict long words even though the best path encoding would ignore such activation. If the label is "a" and the predicted word is "aproksymacja", the edit distance would be eleven. This is also visible for the vanilla base model with the Word Beam Search encoding: it achieves the accuracy of less than 0.1, while the base model is able to get to 0.37. Our best-performing model achieves the word-level accuracy of 0.88 due to highly constrained output, limited by the alphabetical range of a box for a source word, but at the same time its tendency to aggressively fit predictions to longer or unrelated words remains an issue.

The vanilla RCNN model had lower edit distances than the vanilla model with the knowledge of the full Polish alphabet. This happened because Polish diacritics extend possibilities of decoding and are simultaneously not frequently encountered. Thus, a model that is aware of Polish diacritics, could predict a as $ą$ or e as $ę$, while in real life $ę$ and $ą$ are infrequent. At the same time, the vanilla base model would not make these errors as it does not have $ą$ or $ę$ available for a prediction at all.

Our work highlights the importance and possibilities of automated information extraction from a historical archive, on the example of index card catalogues of dictionaries of historical variants of Polish developed in the Institute of Polish Language of the Polish Academy of Sciences. The way in which index cards are organized, facilitates the recovery of some parts of the source structure (e.g. index lemma, body, references) and invites further processing, such as knowledge linking, that goes beyond the undiscriminating plain text recognition from a given image. That recognition of already structured information could be further extended with additional layers of layout analysis of a card image.

8   Conclusion

In this paper, we presented a HTR solution tailored for processing large collections of handwritten index card catalogues. Although primarily designed to deal with the Polish language and lexicographic sources, our solution also expands HTR applicability to under-resourced languages and alphabets, providing domain-specific dataset that could be further reused for a wide array of tasks. The linguistic archives around the world present a wide variety of historical, dialectal, onomastic and other lexicographic data. Such linguistic projects were often undertaken decades before the creation of corpus linguistic tools and digital databases. HTR pipelines could contribute greatly to quickening the pace of work on turning index cards into dictionaries. In the future, a similar approach could be used to globally solve the problem of linking data of the past to the existing resources and linguistic platforms which today massively inhabit the digital space and utilize its affordances. The recognition of different handwriting and definition building patterns can also serve as an invaluable resource for examining the conventions of work in dictionary project as well as the study of individual contributions.

Software and Data

Code, models and data: https://github.com/perechen/htr_lexicography Web demo application: http://149.156.30.114:8503/

The elements of the system implemented in (or inspired by) other solutions: synthetic data generation [5], detection model [2], recognition model [1], CTC and Word Beam Search [24].

The 17th-century index cards (the original dataset): https://rcin.org.pl/dlibra/publication/20029.

Acknowledgements. This research was partly conducted as a result of a project supported by Poland's National Science Centre (project number UMO-2013/11/B/HS2/02795). The authors are grateful to Bartłomiej Borek for his IT support, which included setting up the server and the environment for conducting our experiments.

References

1. Baek, J., et al.: What is wrong with scene text recognition model comparisons? dataset and model analysis. In: International Conference on Computer Vision (ICCV) (2019)
2. Baek, Y., Lee, B., Han, D., Yun, S., Lee, H.: Character region awareness for text detection. In: Proceedings of the IEEE Conference on Computer Vision and Pattern Recognition, pp. 9365–9374 (2019)
3. Bilińska-Brynk, J., Rodek, E.: Paper quotation slips to the Electronic Dictionary of the 17th-and 18th-Century Polish - digital index and its integration with the Dictionary. In: EURALEX XIX Proceedings, pp. 465–470 (2020)

4. Bronikowska, R., Majdak, M., Wieczorek, A., Żółtak, M.: The Electronic Dictionary of the 17th-and 18th-century Polish - towards the open formula asset of the historical vocabulary. In: EURALEX XIX Proceedings pp. 471–475 (2020)
5. Chu, W.: Text renderer (2021). https://github.com/Sanster/text_renderer
6. Doetsch, P., Kozielski, M., Ney, H.: Fast and robust training of Recurrent Neural Networks for offline handwriting recognition. In: 2014 14th International Conference on Frontiers in Handwriting Recognition, pp. 279–284 (2014). https://doi.org/10.1109/ICFHR.2014.54
7. Farra, N., Tomeh, N., Rozovskaya, A., Habash, N.: Generalized character-level spelling error correction. In: Proceedings of the 52nd Annual Meeting of the Association for Computational Linguistics. vol. 2, pp. 161–167. Association for Computational Linguistics, Baltimore, Maryland (2014). https://doi.org/10.3115/v1/P14-2027. http://aclweb.org/anthology/P14-2027
8. Franzini, G., et al.: Attributing authorship in the noisy digitized correspondence of Jacob and Wilhelm Grimm. Frontiers Digital Humanities 5 (2018). https://doi.org/10.3389/fdigh.2018.00004
9. Graves, A., Fernández, S., Schmidhuber, J.: Multi-dimensional recurrent neural networks. In: de Sá, J.M., Alexandre, L.A., Duch, W., Mandic, D. (eds.) ICANN 2007. LNCS, vol. 4668, pp. 549–558. Springer, Heidelberg (2007). https://doi.org/10.1007/978-3-540-74690-4_56
10. Graves, A., Schmidhuber, J.: Offline handwriting recognition with multidimensional Recurrent Neural Networks. Adv. Neural Inf. Process. Syst. 21, 545–552 (2008). https://proceedings.neurips.cc/paper/2008/hash/66368270ffd51418ec58bd793f2d9b1b-Abstract.html
11. Grzelak, D., Podlaski, K., Wiatrowski, G.: Analyze the effectiveness of an algorithm for identifying Polish characters in handwriting based on neural machine learning technologies. J. King Saud University - Comput. Inf. Sci. (2019). https://doi.org/10.1016/j.jksuci.2019.08.001
12. Gupta, A., Vedaldi, A., Zisserman, A.: Synthetic data for text localisation in natural images. In: Proceedings of the IEEE Conference on Computer Vision and Pattern Recognition, pp. 2315–2324 (2016) https://openaccess.thecvf.com/content_cvpr_2016/html/Gupta_Synthetic_Data_for_CVPR_2016_paper.html
13. He, K., Zhang, X., Ren, S., Sun, J.: Deep residual learning for image recognition (2015) http://arxiv.org/abs/1512.03385
14. Jaderberg, M., Simonyan, K., Vedaldi, A., Zisserman, A.: Synthetic data and artificial neural networks for natural scene text recognition (2014). http://arxiv.org/abs/1406.2227
15. Jaderberg, M., Simonyan, K., Zisserman, A., Kavukcuoglu, K.: Spatial Transformer Networks (2016), http://arxiv.org/abs/1506.02025
16. Kahle, P., Colutto, S., Hackl, G., Mühlberger, G.: Transkribus: A service platform for transcription, recognition and retrieval of historical documents. In: 2017 14th IAPR International Conference on Document Analysis and Recognition (ICDAR), vol. 04, pp. 19–24 (2007). https://doi.org/10.1109/ICDAR.2017.307
17. Krishnan, P., Jawahar, C.V.: Matching handwritten document images. In: Leibe, B., Matas, J., Sebe, N., Welling, M. (eds.) ECCV 2016. LNCS, vol. 9905, pp. 766–782. Springer, Cham (2016). https://doi.org/10.1007/978-3-319-46448-0_46
18. Landau, S.I.: Dictionaries: The art and craft of lexicography. Cambridge University Press, 2 edn. (2001)
19. Marti, U.V., Bunke, H.: The IAM-database: an English sentence database for offline handwriting recognition. Int. J. Doc. Anal. Recogn. 5(1), 39–46 (2002). https://doi.org/10.1007/s100320200071

20. Muehlberger, G., et al.: Transforming scholarship in the archives through handwritten text recognition: Transkribus as a case study. J. Documentation **75**(5), 954–976 (2019). https://doi.org/10.1108/JD-07-2018-0114

21. Ogrodniczuk, M., Gruszczyński, W.: Connecting Data for Digital Libraries: The Library, the Dictionary and the Corpus. In: Jatowt, A., Maeda, A., Syn, S.Y. (eds.) ICADL 2019. LNCS, vol. 11853, pp. 125–138. Springer, Cham (2019). https://doi.org/10.1007/978-3-030-34058-2_13

22. Pal, A., Singh, D.: Handwritten English character recognition using neural network. Int. J. Comput. Sci. Commun. **1**(2), 141–144 (2010)

23. Rigaud, C., Doucet, A., Coustaty, M., Moreux, J.P.: ICDAR 2019 competition on Post-OCR text correction. 15th International Conference on Document Analysis and Recognition, pp. 1588–1593 (2019). https://hal.archives-ouvertes.fr/hal-02304334/document

24. Scheidl, H., Fiel, S., Sablatnig, R.: Word beam search: a connectionist temporal classification decoding algorithm. In: 2018 16th International Conference on Frontiers in Handwriting Recognition (ICFHR), pp. 253–258 (2018). https://doi.org/10.1109/ICFHR-2018.2018.00052

25. Sánchez, J.A., Romero, V., Toselli, A.H., Vidal, E.: Icfhr 2014 competition on handwritten text recognition on Transcriptorium datasets (HTRtS). In: 2014 14th International Conference on Frontiers in Handwriting Recognition, pp. 785–790 (2014). https://doi.org/10.1109/ICFHR.2014.137

26. Sánchez, J.A., Romero, V., Toselli, A.H., Vidal, E.: Icfhr 2016 competition on handwritten text recognition on the READ dataset. In: 2016 15th International Conference on Frontiers in Handwriting Recognition (ICFHR) pp. 630–635 (2016). https://doi.org/10.1109/ICFHR.2016.0120

27. Voigtlaender, P., Doetsch, P., Ney, H.: Handwriting recognition with large multidimensional long short-term memory recurrent neural networks. In: 2016 15th International Conference on Frontiers in Handwriting Recognition (ICFHR), pp. 228–233 (2016). https://doi.org/10.1109/ICFHR.2016.0052

28. Xiao, S., Peng, L., Yan, R., Wang, S.: Deep network with pixel-level rectification and robust training for handwriting recognition. In: 2019 International Conference on Document Analysis and Recognition (ICDAR), pp. 9–16 (2019). https://doi.org/10.1109/ICDAR.2019.00012

29. Yin, X., Yin, X., Huang, K., Hao, H.: Robust text detection in natural scene images. IEEE Trans. Pattern Anal. Mach. Intell. **36**(5), 970–983 (2014). https://doi.org/10.1109/TPAMI.2013.182

Out-Plant Milk-Run-Driven Mission Planning Subject to Dynamic Changes of Date and Place Delivery

Grzegorz Bocewicz[1]([⊠]) [iD], Izabela Nielsen[3] [iD], Czeslaw Smutnicki[2] [iD], and Zbigniew Banaszak[1] [iD]

[1] Faculty of Electronics and Computer Science,
Koszalin University of Technology, Koszalin, Poland
{grzegorz.bocewicz,zbigniew.banaszak}@tu.koszalin.pl
[2] Department of Materials and Production, Aalborg University, Aalborg, Denmark
czeslaw.smutnicki@pwr.edu.pl
[3] Faculty of Electronics, Wroclaw University of Science and Technology, Wroclaw, Poland
izabela@mp.aau.dk

Abstract. We consider a dynamic vehicle routing problem in which a fleet of vehicles delivers ordered services or goods to spatially distributed customers while moving along separate milk-run routes over a given periodically repeating time horizon. Customer orders and the feasible time windows for the execution of those orders can be dynamically revealed over time. The problem essentially entails the rerouting of routes determined in the course of their proactive planning. Rerouting takes into account current order changes, while proactive route planning takes into account anticipated (previously assumed) changes in customer orders. Changes to planned orders may apply to both changes in the date of services provided and emerging notifications of additional customers. The considered problem is formulated as a constraint satisfaction problem using the ordered fuzzy number (OFN) formalism, which allows us to handle the fuzzy nature of the variables involved, e.g., the timeliness of the deliveries performed, through an algebraic approach. The computational results show that the proposed solution outperforms the commonly used computer simulation methods.

Keywords: Out-plan milk-run system · Dynamic planning · Delivery uncertainty

1 Introduction

In the paper an out-plant Dynamic Milk-run Routing Problem (DMRP), which consists of designing vehicle routes in an online fashion as orders executed in supply networks are revealed incrementally over time, is considered. In real-life settings, the Out-plant Operating Supply Networks (O^2SNs) [12], apart from randomly occurring disturbances (changes in the execution of already planned requests/orders and the arrival of new ones, traffic jams, accidents, etc.), an important role is played by the imprecise nature of the parameters which determine the timeliness of the services/deliveries performed [18].

© Springer Nature Switzerland AG 2021
M. Paszynski et al. (Eds.): ICCS 2021, LNCS 12742, pp. 151–167, 2021.
https://doi.org/10.1007/978-3-030-77961-0_14

This is because the time of carrying out the operations of transport and service delivery depends on both the transport infrastructure and the prevailing weather conditions as well as on human factors. The imprecise nature of these parameters is implied due to the operator's psychophysical disposition (e.g., manifested in various levels of stress, distraction, fatigue), disturbances in the flow of traffic, etc. resulting in delays in deliveries and unloading/loading operations. Therefore, the time values of the operations performed vary and are uncertain.

The non-stationary nature of the uncertainty of the parameters mentioned, and the usually small set of available historical samples in practice, limits the choice of a formal data model to a fuzzy-numbers-driven one. It means the uncertainty of O^2SN data connected with traffic disturbances as well as the changes in service delivery dates require the use of a model based on the formalism of fuzzy sets. However, it is worth noting that the specificity of the process involved in the course of the services' delivery schedule planning results in the need to determine the sequentially cumulative uncertainty in the performance of the operations involved in it. The question that arises concerns the method for avoiding additional uncertainty introduced in the combinations of summing up uncertainties of cyclically executed operations, e.g., in cyclic production [3] or distribution [2]. In this context, in contrast to standard fuzzy numbers, the support of a fuzzy number obtained by algebraic operations performed on the Ordered Fuzzy Numbers (OFNs) domain does not expand. In turn, however, the possibility of carrying out algebraic operations is limited to selected domains of the computability of these supports. This is a reason why this contribution focuses on the development of sufficient conditions implying the calculability of arithmetic operations that guarantee the interpretability of the results obtained. Consequently, the objective is to develop an algebraic model aimed at fast calculation of fuzzy schedules for vehicles as well as for planning of time buffers enabling the adjustment of currently fuzzy schedules to baseline schedules assuming the deterministic nature of operation times (i.e., their crisp values).

The present study is a continuation of our previous work [1–3, 20] on methods for fast online prototyping of supply schedules and transport routes of a tugger train fleet making adjustments for the tradeoff points between fleet size and storage capacity [21] as well as problems regarding the planning and control of production flow in departments of automotive companies [3]. Its main contribution is threefold:

- An OFN algebra allows the possibility to plan vehicle fleet scenarios aimed at requests while taking into account the uncertainty of the deliveries' operation times.
- In contrast to standard fuzzy numbers, the support of a fuzzy number does not expand as a result of algebraic operations performed on the OFN domain.
- The objective is to maximize the number of new transportation requests, which are inserted dynamically throughout the assumed time horizon.

The remainder of this paper is organized as follows. Section 2 reviews the literature. Section 3 provides the OFNs' framework. Section 4 contains the problem formulation supported by illustrative examples. The model, the methodology used in solving the problem and the conclusions are provided in Sects. 5, 6, and 7, respectively.

2 Literature Review

Most of the problems appearing in O^2SNs are aimed at searching for an optimal distribution policy, i.e., a plan of whom to serve, how much to deliver, and what routes to travel by what fleet of vehicles. Examples of such problems [4, 21] include both simple ones, such as the Mix Fleet VRP, Multi-depot VRP, Split-up Delivery VRP, Pick-up and Delivery VRP, VRP with Time Windows, and more complex ones. Many works are devoted to the Periodic Vehicle Routing Problems (PVRP) aimed at searching for an optimal periodic distribution policy providing a set of routes assigning customers to vehicles that minimize the total travel cost while satisfying vehicle capacity and the time periods when customers should be visited [9]. In turn, the Multi-Depot PVRP with Due Dates and Time Windows [6] being its extension while determining regularly repeated routes to travel by each vehicle in order to satisfy the customer demands can be seen as a kind of the Milk-run Vehicle Routing problem with Time Windows [11]. Since milk-run routing and scheduling problems are NP-hard VRPs, they are solved by using heuristic methods [10, 16]. Regardless of problems typical for in-plant or out-plant milk-run systems [12] or problems that accentuate the dynamic or static character of vehicle routing [10, 15], the goal is to search for optimal solutions.

The Dynamic VRPs that arise when new customers appear in the tours after the starting visit are among the more important and more challenging extensions of VRP [8, 19]. Solving such problems comes down to setting proactive and/or reactive routing strategies [14]. Proactive routing strategies are based on a certain knowledge about demand and are used to anticipate a possible order from a new customer. In turn, reactive routing strategies try to reschedule used fleets due to the occurrence of a new order when a new one arrives. In focusing on the search for such solutions, it is usually assumed that planned routings and schedules are robust to assumed disturbances [15, 16].

It is worth noting that relatively few studies are devoted to the problems of out-plant milk-run dynamic routing and systems in which services are provided by appointment. In systems of this type, the dynamic multi-period VRP is solved, which involves scheduling services in a rolling horizon fashion, in which new requests received, but unfulfilled, during the first period together with the set of customer requests preplanned for the next period constitute a new portfolio of orders [7]. An exhaustive review of VRP taxonomy for milk-run systems can be found in [4].

In many real-life situations, due to the uncertainty of DMRP data caused by traffic disturbances (uncertain travel times, daily changes in traffic intensity, etc.) the service provided cannot be estimated in a precise way. However, the majority of models of the so-called Fuzzy VRP only assume vagueness for fuzzy customer demands to be collected and fuzzy travel times. The literature on these issues is very scarce [6], despite the rapidly growing demand for predictive oriented service providers. The quickly developing enterprise servitization indicates a growing need for this type of service [13].

To summarize, the review shows a gap in the literature in terms of analytical approaches for the assessment of possible rerouting and rescheduling scenarios. The current paper aims to fill this gap by providing the method combining the declarative modeling paradigm with OFN algebra.

3 An Ordered Fuzzy Numbers Framework

The milk-run routing and scheduling problems developed so far have limited use due to the data uncertainty observed in practice. The values describing parameters such as transport time or loading/unloading times depend on the human factor, which means they cannot be determined precisely. It is difficult to account for data uncertainty by using fuzzy variables due to the imperfections of the classic fuzzy numbers algebra [2]. Equations which describe the relationships between fuzzy variables using algebraic operations do not meet the conditions of the Ring. This means that no matter what algebraic operations are used, the support of the fuzzy number, which is the result of these operations, expands. An example of imperfections of the classic fuzzy numbers algebra is shown in Fig. 1, where the uncertainty increases with successive cycles.

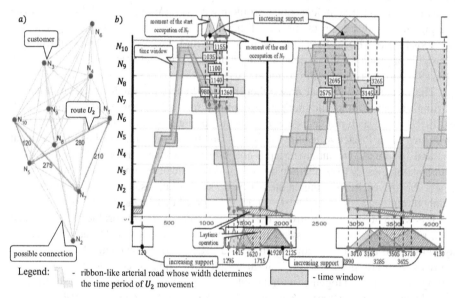

Fig. 1. Route connecting customers N_1, N_5, N_{10}, N$_7$ served by deliveries a), fuzzy schedule b) (Color figure online)

Figure 1 distinguishes the fuzzy values of the start/end moments of service operations carried out on nodes N_7 and N_1 located along the route selected for U_2 (orange line). The attainable values of these moments are characterized by an increasing support in subsequent cycles (the level of uncertainty increases).

In the case of classic fuzzy numbers $\hat{a}$, $\hat{b}$, $\hat{c}$ (marked with the symbol $\wedge$), it is impossible to solve a simple equation $\hat{A} + \hat{X} = \hat{C}$. This fact significantly hinders the application of approaches based on declarative models. Therefore, we propose a model based on OFN algebra in which it is possible to solve algebraic equations. OFNs can be defined [17] as a pair of continuous real functions (f_A – "up"; g_A – "down") i.e.:

$$\hat{A} = (f_A, g_A), \text{ where:} f_A, g_A : [0, 1] \to \mathbb{R} \tag{1}$$

Assuming that f_A is increasing and g_A is decreasing as well as that $f_A \leq g_A$, the membership function μ_A of the OFN $\hat{A}$ is as follows (see OFN rows in Table 1):

$$\mu_A(x) = \begin{cases} f_A^{-1}(x) & when \ x \in UP_A \\ g_A^{-1}(x) & when \ x \in DOWN_A \\ 1 & when \ x \in CONST_A \\ 0 & in \ the \ remaining \ cases \end{cases} \tag{2}$$

where, $UP_A = (l_{A0}, l_{A1})$, $CONST_A = (l_{A1}, p_{A1})$ and $DOWN_A = (p_{A1}, p_{A0})$. OFNs are two types of orientation [17]: **positive**, when $\hat{A} = (f_A, g_A)$; **negative**, when $\hat{A} = (g_A, f_A)$. The algebraic operations used in the proposed model are as follows:

Definition 1. Let $\hat{A} = (f_A, g_A)$ and $\hat{B} = (f_B, g_B)$ be OFNs. $\hat{A}$ is a number equal to $\hat{B}$ ($\hat{A} = \hat{B}$), $\hat{A}$ is a number greater than $\hat{B}$ or equal to or greater than $\hat{B}$ ($\hat{A} > \hat{B}$; $\hat{A} \geq \hat{B}$), $\hat{A}$ is less than $\hat{B}$ or equal to or less than $\hat{B}$ ($\hat{A} < \hat{B}$, $\hat{A} \leq \hat{B}$) if: $\forall_{x \in [0,1]} f_A(x) * f_B(x) \wedge g_A(x) * g_B(x)$, where: the symbol $*$ stands for: $=, >, \geq, <,$ or $\leq$. ∎

Definition 2. Let $\hat{A} = (f_A, g_A)$, $\hat{B} = (f_B, g_B)$, and $\hat{C} = (f_C, g_C)$ be OFNs. The operations of addition $\hat{C} = \hat{A} + \hat{B}$, subtraction $\hat{C} = \hat{A} - \hat{B}$, multiplication $\hat{C} = \hat{A} \times \hat{B}$ and division $\hat{C} = \hat{A}/\hat{B}$ are defined as follows: $\forall_{x \in [0,1]} f_C(x) = f_A(x) * f_B(x) \wedge g_C(x) = g_A(x) * g_B(x)$, where: the symbol $*$ stands for $+, -, \times,$ or $\div$; The operation of division is defined for $\hat{B}$ such that $|f_B| > 0$ and $|g_B| > 0$ for $x \in [0, 1]$. ∎

The ordered fuzzy number $\hat{A}$ is a proper OFN [QW] when one of the following conditions is met: $f_A(0) \leq f_A(1) \leq g_A(1) \leq g_A(0)$ (for positive orientation) or $g_A(0) \leq g_A(1) \leq f_A(1) \leq f_A(0)$ (for negative orientation). They allow us to specify the conditions which guarantee that the result of algebraic operations is a proper OFN [2]:

Theorem 1. Let $\hat{A}$ and $\hat{B}$ be proper OFNs with different orientations: $\hat{A}$ (positive orientation), $\hat{B}$ (negative orientation). If one of the following conditions holds:

- $(|UP_A| - |UP_B| \geq 0)$ ∧ $(|CONST_A| - |CONST_B| \geq 0)$ ∧ $(|DOWN_A| - |DOWN_B| \geq 0)$,
- $(|UP_B| - |UP_A| \geq 0)$ ∧ $(|CONST_B| - |CONST_A| \geq 0)$ ∧ $(|DOWN_B| - |DOWN_A| \geq 0)$,

then the result of the operation $\hat{A} + \hat{B}$ is a proper OFN $\hat{C}$.

where: UP_X – an image (codomain) of function f_X, $CONST_X = \{x \in X : \mu_X = 1\}$, $DOWN_X$ – an image of function g_X, $|a|$ – length of the interval a. ∎

The fulfillment of the conditions underlying the above theorem may lead to a reduction in the fuzziness of the sum of OFNs with different orientations. This is because algebraic operations (in particular sums) take values which are proper OFNs, i.e., are fuzzy numbers which are easy to interpret.

4 Problem Formulation

Let us consider the graph $G = (N, E)$ modeling an O^2SN. The set of nodes $N = \{N_1, \ldots, N_\lambda, \ldots, N_n\}$, (where $n = |N|$) includes one node representing distribution center N_1 and $\{N_2, \ldots, N_n\}$ nodes representing customers. The set of edges $E = \{(N_i, N_j) | i, j \in \{1, \ldots, n\}, i \neq j\}$ determines the possible connections between nodes. Given is a fleet of vehicles $\mathcal{U} = \{U_1, \ldots, U_k, \ldots, U_K\}$. The customers $\{N_2, \ldots, N_n\}$ are cyclically visited (with period T) by vehicles U_k traveling from node N_1. Variable Q_k denotes the payload capacity of vehicles U_k. Execution of the ordered delivery z_λ by the customer N_λ takes place in the fuzzy period $\widehat{t_\lambda}$ (represented by an OFN). The moment when the vehicle U_k starts delivery to the customer N_λ is indicated by fuzzy variable $\widehat{y_\lambda^k}$ (represented by an OFN). The deliveries ordered by the customer N_λ are carried out in the delivery time interval (the time window for short) $TW_\lambda = [ld_\lambda; ud_\lambda]$, i.e., $\widehat{y_\lambda^k} \geq ld_\lambda$ and $\widehat{y_\lambda^k} + \widehat{t_\lambda} \leq ud_\lambda$. It is assumed that the fuzzy variable $\widehat{d_{\beta, \lambda}}$ (taking the form of an OFN) determines traveling time between nodes N_β, N_λ, where: $(N_\beta, N_\lambda) \in E$. The routes of U_k are represented by sequences: $\pi_k = (N_{k_1}, \ldots, N_{k_i}, N_{k_{i+1}}, \ldots, N_{k_\mu})$, where: $k_i \in \{1, .., n\}$, $(N_{k_i}, N_{k_{i+1}}) \in E$. Moreover, the following assumptions are met: Z denotes a sequence of required amounts of goods z_λ ($\lambda = 1, \ldots, n$); Π denotes a set of routes π_k, $k = 1, \ldots, K$; node N_1 representing the distribution center occurs only once in each route of the set Π; node representing the customer N_λ ($\lambda > 1$) occurs only once in the route belonging to the set Π; the amount of goods transported by U_k cannot exceed payload capacity Q_k, deliveries are being made over a given periodically repeating time horizon with period T.

In that context, typical proactive planning of goods distribution fundamentally involves the question: *Given a fleet $\mathcal{U}$ providing deliveries to the customers allocated in a network G (ordering assumed amounts of goods z_λ). Does there exist the set of routes Π guaranteeing timely execution of the ordered services following time windows TW_λ?*

For the purpose of illustration, let us consider network G shown in Fig. 2a), where 10 nodes (1 distribution center and 9 customers) are serviced by fleet $\mathcal{U} = \{U_1, U_2, U_3\}$.

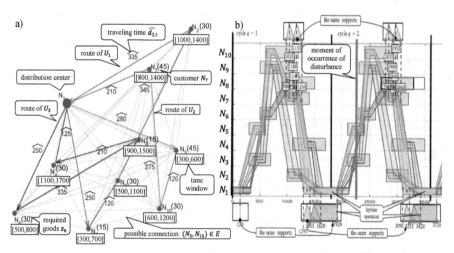

Fig. 2. Graph G modeling the considered O^2SN a) and corresponding fuzzy cyclic schedule b)

Table 1. Assumed traveling time values

$\widehat{d_{1,3}}$	$\widehat{d_{3,9}}$	$\widehat{d_{9,2}}$	$\widehat{d_{2,1}}$	$\widehat{w_1^1}$	$\widehat{d_{1,5}}$	$\widehat{d_{5,10}}$	$\widehat{d_{10,7}}$	$\widehat{w_1^2}$	$\widehat{d_{1,6}}$	$\widehat{d_{6,8}}$	$\widehat{d_{8,4}}$	$\widehat{d_{4,1}}$	$\widehat{w_1^3}$
				115 270 435				300 435 505					330 400 465

The following routes: $\pi_1 = (N_1, N_3, N_9, N_2)$ (green line), $\pi_2 = (N_1, N_5, N_{10}, N_7)$ (orange line), $\pi_3 = (N_1, N_6, N_8, N_4)$ (blue line) guarantee the delivery of the required services related to the fulfillment of the ordered amount of goods to all customers cyclically (within the period $T = 1800$). The solution was determined assuming that the vehicle payload capacity Q_k is equal to 120 and required amounts of goods are equal $Z = (0, 30, 15, 30, 45, 30, 45, 15, 30, 30)$. The corresponding fuzzy cyclic schedules are shown in Fig. 2b). This solution assumes that travelling times $\widehat{d_{\beta,\lambda}}$ (in Table 1) are represented by an OFNs and times of node occupation $\widehat{t_\lambda}$ are singletons ($\widehat{t_\lambda} = 120$, $\lambda = 1, \ldots, 10$). The implemented routes are determined in the process of proactive planning. However, other disruptions may occur in the process of implementing such planned routes. An example of such a disturbance IS concerns changes of delivery time windows or new order notification from a customer located outside of a given route.

Such a disturbance is presented in Fig. 2b) where the dispatcher receives information on $TW_8^* = [2900; 3500]$ changing the delivery time window being located at the node N_8 (from $TW_8 = [2700; 3300]$ to) – see the second ($q = 2$) cycle of schedule (moment $t^* = 2300$ when U_1 occupies N_3, U_2 occupies N_5 and U_3 is moving to N_7). Due to this change, the adopted routes do not guarantee the implementation of maintenance services on the set dates – the handling of N_8 according to the new $TW_8^* = [2900; 3500]$ prevents the timely handling of the customer N_8 and vice versa. In such a situation, it becomes necessary to answer to the following question:

Given a vehicle fleet U providing deliveries to the customers allocated in a network G. Vehicles move along a given set of routes Π according to a cyclic fuzzy schedule $\widehat{\mathbb{Y}}$. Given is a disturbance $IS(t^*)$ related either to changing from TW_λ to TW_λ^* or occurrence of a new order from the customer located in the place N_λ outside of a given route. Does there exist a rerouting $^*\Pi$ and rescheduling $^*\widehat{\mathbb{Y}}$ of vehicles, which guarantee timely execution of the ordered amounts of goods but not at the expense of the already accepted orders?

The possibility of reactive (dynamic) planning of vehicle missions in the event of a disturbance occurrence is the subject of the following chapters.

5 OFN-Based Constraint Satisfaction Problem

In general the problem under consideration can be formulated in the following way:

Given a fleet U providing deliveries to the customers allocated in a network G (customers are serviced by prescheduled time windows TW). Vehicles move along a given set of routes Π according to the cyclic fuzzy schedule $\widehat{\mathbb{Y}}$. Assuming the appearance of the disturbance $IS(t^*)$ (which changes TW to TW^* and/or location of customer N_λ to

N_λ^* at the moment t^*), a feasible way of rerouting (*Π) and rescheduling ($\widehat{^*\mathbb{Y}}$) of MSTs, guaranteeing timely execution of the ordered services, is sought.

Parameters

G :   graph of a transportation network $G = (N, E)$,

$\mathcal{U}$:   set of vehicles: $\mathcal{U} = \{U_1, \ldots, U_k, \ldots, U_K\}$, U_k is the k-th vehicle,

K: size of vehicle fleet,

TW: set of delivery time windows: $TW = \{TW_1, \ldots, TW_\lambda, \ldots, TW_n\}$, where $TW_\lambda = [ld_\lambda; ud_\lambda]$ is a deadline for service at the customer N_λ (see example in Fig. 2),

$IS(t)$:   state of vehicle fleet mission at the moment t: $IS(t) = \left(M(t), {}^*TW(t), {}^*E(t) \right)$ where:
   $M(t)$ is an allocation of vehicles at the moment t: $M(t) = \left(N_{a_1}, \ldots, N_{a_k}, \ldots, N_{a_K} \right)$, where $a_k \in \{1, .., n\}$ determines the node N_{a_k} occupied by U_k (or the node the U_k is headed to).
   ${}^*TW(t)$ is the set of time windows TW_λ^* at the moment t : ${}^*TW(t) = \{TW_1^*, \ldots, TW_\lambda^*, \ldots, TW_n^*\}$, where $TW_\lambda^* = [ld_\lambda^*; ud_\lambda^*]$.
   ${}^*E(t)$ is a set of edges (with traveling times $\widehat{d_{\beta,\lambda}}$) of network G (customer location) at the moment t.
   The state $IS(t^*)$ such that the following condition $\left[{}^*TW(t^*) \neq TW \right] \vee \left[{}^*E(t^*) \neq E \right]$ holds is called the **disturbance** occurring at the moment t^*.

T: period in which all customers should be serviced (see Fig. 2b) – $T = 1800$),

Π: set of routes π_k for the network G, when there is no disturbance, where π_k is a route of U_k:
   $\pi_k = \left(N_{k_1}, \ldots, N_{k_i}, N_{k_{i+1}}, \ldots, N_{k_\mu} \right)$, where: $x_{k_i, k_{i+1}}^k = 1$ for $i = 1, \ldots, \mu - 1$

$$x_{\beta,\lambda}^k = \begin{cases} 1 & \text{if } U_k \text{ travels from node } N_\beta \text{ to node } N_\lambda \\ 0 & \text{otherwise} \end{cases},$$

z_λ :   customer's N_λ demand,

U_k: maximum loading capacity of U_k,

c_λ^k: weight of goods delivered to N_λ by U_k, when there is no disturbance,

$\widehat{d_{\beta,\lambda}}$:   fuzzy traveling time along the edge $\left(N_\beta, N_\lambda \right)$ – defined as positive OFNs,

$\widehat{t_\lambda}$:   fuzzy time of node N_λ occupation (represented by an OFN),

$\widehat{\mathbb{Y}}$: fuzzy schedule of fleet $\mathcal{U}$, $\widehat{\mathbb{Y}} = \left(\hat{Y}, \hat{W}, C \right)$ when there is no disturbance:

$\hat{Y}$: set of $\widehat{Y^k}$, where $\widehat{Y^k}$ is a sequence of moments $\widehat{y_\lambda^k}$: $\widehat{Y^k} = \left(\widehat{y_1^k}, \ldots, \widehat{y_\lambda^k}, \ldots, \widehat{y_n^k} \right)$, $\widehat{y_\lambda^k}$ is fuzzy time at which U_k arrives at node N_λ,

$\hat{W}$: set of $\widehat{W^k}$, where $\widehat{W^k}$ is a sequence of laytimes $\widehat{w_\lambda^k}$: $\widehat{W^k} = \left(\widehat{w_1^k}, \ldots, \widehat{w_\lambda^k}, \ldots, \widehat{w_n^k} \right)$, $\widehat{w_\lambda^k}$ is laytime at node N_λ for U_k.

C: set of C^k, where C^k is a sequence of delivered goods c_λ^k: $C^k = \left(c_1^k, \ldots, c_\lambda^k, \ldots, c_n^k \right)$, c_λ^k is weight of goods delivered to node N_λ by U_k.

Variables

$*x_{\beta,\lambda}^k$: binary variable indicating the travel of U_k between nodes N_β, N_λ after occurrence of the disturbance $IS(t^*)$:

$$*x_{\beta,\lambda}^k = \begin{cases} 1 \text{ if } U_k \text{ travels from node } N_\beta \text{ to node } N_\lambda \\ 0 \qquad\qquad\qquad \text{otherwise} \end{cases},$$

$*c_\lambda^k$: weight of goods delivered to node N_λ by U_k, after occurrence of $IS(t^*)$,

$\widehat{*y_\lambda^k}$: fuzzy time at which U_k arrives at node N_λ, after occurrence of $IS(t^*)$,

$\widehat{*w_\lambda^k}$: laytime at node N_λ for U_k, after occurrence of the disturbance $IS(t^*)$,

$\widehat{*s^k}$: take-off time of U_k.

$*\pi_k$: route of U_k, after occurrence of the disturbance $IS(t^*)$,: $*\pi_k = \left(N_{k_1}, \ldots, N_{k_i}, N_{k_{i+1}}, \ldots, N_{k_\mu} \right)$, where: $*x_{k_i,k_{i+1}}^k = 1$ for $i = 1, \ldots, \mu - 1$ and $*x_{k_\mu,k_1}^k = 1$; $*\Pi$ is a set of routes $*\pi_k$.

$*C^k$: set of $*c_\lambda^k$, payload weight delivered by U_k; $*C$ is family of $*C^k$.

$\widehat{*W^k}$: sequence of laytimes $\widehat{*w_\lambda^k}$: $\widehat{W^k} = \left(\widehat{*w_1^k}, \ldots, \widehat{*w_n^k} \right)$; $\widehat{*W}$ is a set of $\widehat{*W^k}$,

$\widehat{*Y^k}$: sequence of moments $\widehat{*y_\lambda^k}$: $\widehat{*Y^k} = \left(\widehat{*y_1^k}, \ldots, \widehat{*y_n^k} \right)$; $\widehat{*Y}$ is a set of $\widehat{*Y^k}$,

$\widehat{*\mathbb{Y}}$: fuzzy schedule of fleet $\mathcal{U}$, after occurrence of $IS(t^*)$: $\widehat{*\mathbb{Y}} = \left(\widehat{*Y}, \widehat{*W}, *C \right)$.

Constraints

$$\widehat{*s^k} \geq 0; k = 1 \ldots K \tag{3}$$

$$\left(\widehat{s^k} \leq t^*\right) \Rightarrow \left(\widehat{*s^k} = \widehat{s^k}\right); k = 1 \ldots K \tag{4}$$

$$\left(\widehat{y_j^k} \leq t^*\right) \Rightarrow \left(*x_{i,j}^k = x_{i,j}^k\right); j = 1 \ldots n; i = 2 \ldots n; k = 1 \ldots K \tag{5}$$

$$\left(\widehat{y_j^k} \leq t^*\right) \Rightarrow \left(\widehat{*y_j^k} = \widehat{y_j^k}\right); j = 2 \ldots n; k = 1 \ldots K \tag{6}$$

$$\left(\widehat{y_j^k} \leq t^*\right) \Rightarrow \left(\widehat{*w_j^k} = \widehat{w_j^k}\right); j = 2 \ldots n; k = 1 \ldots K \tag{7}$$

$$\sum\nolimits_{j=1}^{n} *x_{1,j}^k = 1; k = 1 \ldots K \tag{8}$$

$$\left(*x_{1,j}^k = 1\right) \Rightarrow \left(\widehat{*y_j^k} = \widehat{*s^k} + \widehat{d_{1,j}}\right); j = 1 \ldots n; k = 1 \ldots K \tag{9}$$

$$\left(*x_{i,j}^k = 1\right) \Rightarrow \left(\widehat{*y_j^k} = \widehat{*y_i^k} + \widehat{d_{i,j}} + \widehat{t_i} + \widehat{*w_i^k}\right); j = 1 \ldots n; i = 2 \ldots n; k = 1 \ldots K \tag{10}$$

$$\left(\widehat{y_i^k} \leq t^*\right) \Rightarrow \left(*c_i^k = c_i^k\right); i = 1 \ldots n; k = 1 \ldots K \tag{11}$$

$$*c_i^k \leq Q \times \sum\nolimits_{j=1}^{n} *x_{i,j}^k; i = 1 \ldots n; k = 1 \ldots K \tag{12}$$

$$\left(*x_{i,j}^k = 1\right) \Rightarrow *c_j^k \geq 1; k = 1 \ldots K; i = 1 \ldots n; j = 2 \ldots n \tag{13}$$

$$\sum\nolimits_{k=1}^{K} *c_i^k = z_i; i = 1 \ldots n \tag{14}$$

$$\widehat{*s^k} + T = \widehat{*y_1^k} + \widehat{t_1} + \widehat{*w_1^k}; k = 1 \ldots K \tag{15}$$

$$\widehat{*y_j^k} \geq 0; i = 1 \ldots n; k = 1 \ldots K \tag{16}$$

$$\sum\nolimits_{j=1}^{n} *x_{i,j}^k = \sum\nolimits_{j=1}^{n} *x_{j,i}^k; i = 1 \ldots n; k = 1 \ldots K \tag{17}$$

$$\widehat{*y_i^k} \leq T, i = 1 \ldots n; k = 1 \ldots K \tag{18}$$

$$*x_{i,i}^k = 0; i = 1 \ldots n; k = 1 \ldots K \tag{19}$$

$$\widehat{*y_i^k} + \widehat{t_i} + c \times T \leq ud_\lambda^*, i = 1 \ldots n; k = 1 \ldots K \tag{20}$$

$$\widehat{{}^*y_i^k} + c \times T \geq ld_\lambda^*, \ i = 1 \ldots n; k = 1 \ldots K \tag{21}$$

It is assumed that the arithmetic operations contained in the above constraints meet the conditions of Definition 1 and Theorem 1. The rescheduling and rerouting of the vehicles then resulting in a new plan of delivery are the result of the disturbance $IS(t^*)$. In that context, when disturbance $IS(t^*)$ occurs, the new set of routes $^*\Pi$ and a new schedule $\widehat{{}^*\mathbb{Y}}$, which guarantee timely servicing of customers, are determined by solving the following Ordered Fuzzy Constraint Satisfaction (OFCS) Problem (22):

$$\widehat{FCS}\left(\widehat{\mathbb{Y}}, \Pi, IS(t^*)\right) = \left(\left(\hat{\mathcal{V}}, \widehat{\mathcal{D}}\right), \hat{\mathcal{C}}\left(\widehat{\mathbb{Y}}, \Pi, IS(t^*)\right)\right), \tag{22}$$

where:

$\hat{\mathcal{V}} = \left\{\widehat{{}^*\mathbb{Y}}, {}^*\Pi\right\}$ – a set of decision variables: $\widehat{{}^*\mathbb{Y}}$ – a fuzzy cyclic schedule guaranteeing timely provision of service to customers in the case of disturbance IS, and $^*\Pi$ – a set of routes determining the fuzzy schedule $\widehat{{}^*\mathbb{Y}}$;

$\widehat{\mathcal{D}}$ – a finite set of decision variable domains: $\widehat{{}^*y_\lambda^k}, \widehat{{}^*w_\lambda^k} \in \mathcal{F}$ ($\mathcal{F}$ is a set of OFNs (1)), $^*x_{\beta,\lambda}^k \in \{0, 1\}, {}^*c_\lambda^k \in \mathbb{N}$;

$\hat{\mathcal{C}}$ – a set of constraints which take into account the set of routes Π, fuzzy schedule $\widehat{\mathbb{Y}}$ and disturbance $IS(t^*)$, while determining the relationships that link the operations occurring in vehicles fleet cycles (5)–(21).

To solve $\widehat{FCS}$ (22), the values of the decision variables from the adopted set of domains for which the given constraints are satisfied must be determined.

6 Dynamic Mission Planning

The idea standing behind the proposed reaction to occurring disruption $IS(t^*)$ can be reduced to dynamic adaptation (i.e., rerouting and rescheduling) of previously adopted routes Π, and schedules $\widehat{\mathbb{Y}}$, i.e., adjusting them (if possible) to the changes in time windows TW^* or occurrence of a new order from the customer located outside of a given route E^*. Let $\widehat{\mathbb{Y}}(q)$ denote the output fuzzy schedule of the q-th cycle defined as:

$$\widehat{\mathbb{Y}}(q) = \left(\widehat{Y}(q), \widehat{W}(q), C(q)\right) \tag{23}$$

where $\widehat{Y}(q), \widehat{W}(q), C(q)$ are families of following sets ($q = 1, 2 \ldots, Q$):

$$\widehat{Y^k}(q) = \left(\widehat{y_1^k}(q), \ldots, \widehat{y_\lambda^k}(q), \ldots, \widehat{y_n^k}(q)\right) \text{ and } \widehat{y_\lambda^k}(q) = \widehat{y_\lambda^k} + (q-1) \times T,$$

$$\widehat{W^k}(q) = \left(\widehat{w_1^k}(q), \ldots, \widehat{w_\lambda^k}(q), \ldots, \widehat{w_n^k}(q)\right) \text{ and } \widehat{w_\lambda^k}(q) = \widehat{w_\lambda^k} + (q-1) \times T,$$

$$C^k(q) = \left(c_1^k(q), \ldots, c_\lambda^k(q), \ldots, c_n^k(q)\right) \text{ and } c_\lambda^k(q) = c_\lambda^k + (q-1) \times T.$$

In that context the schedule's sequence: $\widehat{\mathbb{Y}}(1), \widehat{\mathbb{Y}}(2), \ldots, \widehat{\mathbb{Y}}(Q)$ determined by routes Π, presents the execution of adopted delivery plans in subsequent $q = 1, 2 \ldots, Q$. It is assumed that disturbance $IS(t^*)$ can occur in any time $t^* \in \{T \times (q-1), \ldots, T \times q\}$.

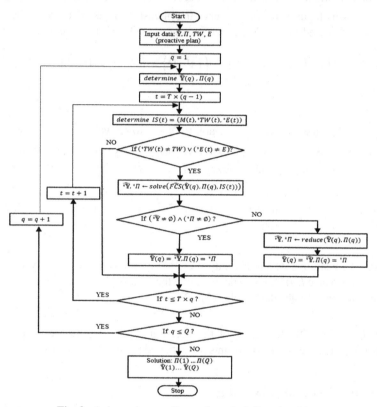

Fig. 3. A dynamic rerouting and rescheduling algorithm

An algorithm that supports dynamic planning, based on the proposed concept of $\widehat{FCS}$ (23), is shown in Fig. 3. The algorithm processes the successive customer service cycles $q = 1, 2, \ldots, Q$. If there is a disturbance $IS(t^*)$ (i.e., the condition $\left(^*TW(t) \neq TW\right) \vee \left(^*E(t) \neq E\right)$ holds) in a given cycle q (at moment t^*), then problem $\widehat{FCS}$ is solved (*solve* function). The function *solve* represents algorithms implemented in declarative programming environments (responsible for the search for admissible solutions to the problems considered). The existence of an admissible solution (i.e., $\left(^*\widehat{\mathbb{Y}} \neq \emptyset\right) \wedge \left(^*\Pi \neq \emptyset\right)$) means that there are routes which ensure that customers are serviced on time when disturbance $IS(t^*)$ occurs in the cycle q. If an admissible solution does not exist, then the currently used routes and the associated vehicle schedule should be modified (*reduce* function) in such a way that it removes the delivery operation at node N_λ at which disturbance $IS(t^*)$ occurs.

The presented algorithm generates in reactive mode (in situations of occurrence $IS(t^*)$) alternative corrected versions of the assumed customer delivery plan. The computational complexity of the algorithm depends on methods used to solve the problem $\widehat{FCS}$ (function *solve*). This problem was implemented in the IBM ILOG environment.

7 Computational Experiments

Consider the network from Fig. 2a), in which the three vehicles $\mathcal{U} = \{U_1, U_2, U_3\}$ cyclically service customers N_2-N_{10} with period $T = 1800$ [u.t]. The fuzzy traveling times between nodes $\widehat{d_{\lambda,\beta}}$ and the time windows TW are as shown in Table 1 and Fig. 2a), respectively. Routes $\pi_1 = (N_1, N_3, N_9, N_2)$, $\pi_2 = (N_1, N_5, N_{10}, N_7)$, $\pi_3 = (N_1, N_6, N_8, N_4)$ determine the fuzzy schedule $\widehat{\mathbb{Y}}$, as shown in Fig. 4a). It is easy to see (Fig. 4a) that in the second cycle ($q = 2$) of the fuzzy schedule (at the moment $t^* = 2300$ for the location $_1M(t^*) = (N_3, N_5, N_6)$), the disturbance $_1IS(t^*)$ concerning unplanned changes in time window $TW_8^* = [2900; 3500]$ (instead $TW_8 = [2700; 3300]$) on node N_8 is announced (customer locations remain unchanged, i.e., $_1^*E(t^*) = E$).

Given this, an answer to the following question is sought: Does there exist a set of routes $_1^*\Pi$ operated by vehicles U_1, U_2 and U_3 for which the fuzzy cyclic schedule $_1^*\widehat{\mathbb{Y}}$ will guarantee that all customers are serviced on time when disturbance $_1IS(t^*) = \left(_1M(t^*)\right), _1^*TW(t^*), _1^*E(t^*)\right)$ occurs?

In order to find the answer to this question, we used the algorithm shown in Fig. 3. The problem (22) $\widehat{FCS}$ was then implemented in the constraint programming environment IBM ILOG (Windows 10, Intel Core Duo2 3.00 GHz, 4 GB RAM). The solution time for problems of this size does not exceed 30 s. The following routes were obtained: $_1^*\pi_1 = (N_1, N_3, N_7, N_4)$, $_1^*\pi_2 = (N_1, N_5, N_9, N_2)$, $_1^*\pi_3 = (N_1, N_6, N_{10}, N_8)$ (see Fig. 4c). It should be noted that the route change occurs when the vehicles U_1, U_2 and U_3 are serving customers N_3, N_5, N_7, respectively. New routes allow for timely delivery to the rest of customers despite changing the time window at the customer N_8.

An example of another type of disturbance $_2IS(t^*)$ is presented in the third cycle ($q = 3$). At the moment $t^* = 4200$ (location $_2M(t^*) = (N_7, N_9, N_6)$) customer's N_{10} location change is signaled – see Fig. 3d). In this case the adopted plan does not guarantee timely deliveries. Similar to the previous case, an attempt was made to designate the routes $_2^*\Pi$ guaranteeing the timely servicing of all customer caused by $_2IS(t^*) = \left(_2M(t^*), _2^*TW(t^*), _2^*E(t^*)\right)$ occurrence.

The following routes $_2^*\pi_1 = (N_1, N_3, N_7, N_8)$, $_2^*\pi_2 = (N_1, N_5, N_9, N_2)$, $_2^*\pi_3 = (N_1, N_6, N_{10}, N_4)$ (see Fig. 4d) were obtained as a solution to the problem (22) $\widehat{FCS}$. In the case under consideration, the change of routes only applies to U_1 and U_3 vehicles. Taking over the customer N_8 by U_1 (see Fig. 4d) allows for timely handling of N_{10} despite changing its location.

In fuzzy schedule $^*\widehat{\mathbb{Y}}$ (Fig. 4a), the operations are represented as ribbon-like "arterial roads", whose increasing width shows the time of vehicle movement resulting from the growing uncertainty. It is worth noting that the uncertainty is reduced at the end of each time window as a result of the operation of vehicles waiting at node N_1. The increasing uncertainty is not transferred to the subsequent cycles of the system. Uncertainty is reduced as a result of the implementation of the OFN formalism. The vehicles' waiting time at node N_1 has a negative orientation (laytimes $*\widehat{w_1^1}$, $*\widehat{w_1^2}$ and $*\widehat{w_1^3}$- Table 1). Taking the above factors into account, the proposed method of dynamic planning vehicle missions in cyclic delivery systems is unique, due to the possibility of taking into account the reduction of uncertainty in subsequent work cycles of the considered system.

Moreover, the routes $*\pi_1$, $*\pi_2$, $*\pi_3$ remain unchanged (see routes $*\pi_1$, $*\pi_2$, $*\pi_3$ in Fig. 4a) until a disturbance occurs, and then they are rerouted, rescheduled and finally synchronized again so that all customers are serviced on time.

In addition to the above experiments, the effectiveness of the proposed approach was evaluated for distribution networks of different sizes (different numbers of customers and vehicles). The results are presented in Table 2. To summarize, the experiments were carried out for networks containing 6–18 nodes in which services were made by sets consisting of 1–4 vehicles. This means that the problems considered can be solved in online mode (<900 s) when the size of service distribution network does not exceed 16

Table 2. Results of computational experiments carried out for selected instances of O^2SN.

Number of nodes n	Number of vehicles K	Disturbance IS	
		Time window change for one customer	Relocation of one customer
		Calculation time [s]	Calculation time [s]
6	1	<1	<1
6	2	<1	<1
6	3	1	<1
6	4	5	3
10	1	10	9
10	2	25	20
10*	3	30	27
10	4	67	60
14	1	150	132
14	2	321	294
14	3	554	410
14	4	>900	>900
18	1	>900	>900
18	2	>900	>900
18	3	>900	>900
18	4	>900	>900

* - solution from Fig. 4

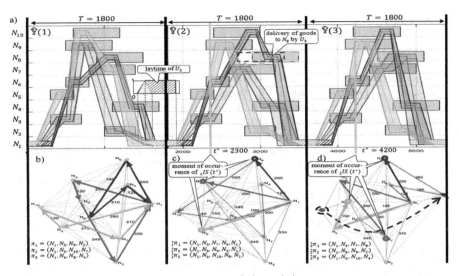

Fig. 4. Response schedules to disturbances $_1IS(t^*)$, $_2IS(t^*)$ in the network G from Fig. 2a)

nodes. In the case of larger networks, the effect of combinatorial explosion limits the practical use of this method.

8   Conclusions

The main contribution of this work in the field of dynamic vehicle routing and scheduling involves demonstrating the possibility of using of OFNs' algebra framework in the course of decision making following dynamic changes of date and place delivery. Besides of the capability to handle fuzzy nature of variables, the proposed approach can be used for rapid prototyping of time buffers' sizing and allocation underlying proactive and/or reactive routing strategies.

The results of experiments confirm the competitiveness of the analytical approach in relation to the computer-simulation-based solutions aimed at rerouting and rescheduling of a vehicle fleet following the milk-run manner. In this context a compromise between the sizes of delivery cycles and the size of time buffers, taking into account the uncertainty of the data characterizing the O^2SN will be recorded and streamlined into the proposed approach. Moreover, the related issues, concerning vehicle mission planning aimed at dynamic planning of multi-period outbound delivery-driven missions being implemented in a rolling horizon approach, will be the subject of future work.

Acknowledgments. This research was carried out under the internship: Declarative models of the vehicles fleet mission planning (Aalborg University).

References

1. Bocewicz, G., Banaszak, Z., Nielsen, I.: Multimodal processes prototyping subject to grid-like network and fuzzy operation time constraints. Ann. Oper. Res. **273**(1–2), 561–585 (2017). https://doi.org/10.1007/s10479-017-2468-5
2. Bocewicz, G., Banaszak, Z., Rudnik, K., Smutnicki, C., Witczak, M., Wójcik, R.: An ordered-fuzzy-numbers-driven approach to the milk-run routing and scheduling problem. J. Comput. Sci. **49**, 101288 (2021). https://doi.org/10.1016/j.jocs.2020.101288
3. Bocewicz, G., Nielsen, I., Banaszak, Z.: Reference model of a milk-run delivery problem. In: Hamrol, A., Kujawińska, A., Barraza, M.F.S. (eds.) MANUFACTURING 2019. LNME, pp. 150–160. Springer, Cham (2019). https://doi.org/10.1007/978-3-030-18789-7_14
4. Braekers, K., Ramaekers, K., Nieuwenhuyse, I.V.: The vehicle routing problem: state of the art classification and review. Comput. Ind. Eng. **99**, 300–313 (2016)
5. Brito, J., Moreno-Pérez, J.A., Verdegay, J.L.: Fuzzy optimization in vehicle routing problems. In: Proceedings of the Joint 2009 International Fuzzy Systems Association, World Congress and European Society of Fuzzy Logic and Technology, pp. 1547–1552 (2009)
6. Cantu-Funes, R., Salazar-Aguilar, M.A., Boyer, V.: Multi-depot periodic vehicle routing problem with due dates and time windows. J. Oper. Res. Soc. (2017). https://doi.org/10.1057/s41274-017-0206-7
7. Xi, C.: Multi-period dynamic technician routing and scheduling problems with experience-based service times and stochastic customers. Ph.D. (Doctor of Philosophy) thesis, University of Iowa (2016). https://doi.org/10.17077/etd.g8ozrc3x
8. Euchi, J., Yassine, A., Chabchoub, H.: The dynamic vehicle routing problem: solution with hybrid metaheuristic approach. Swarm Evol. Comput. **21**, 41–53 (2015). https://doi.org/10.1016/j.swevo.2014.12.003
9. Francis, P.M., Smilowitz, K.R., Tzur, M.: The period vehicle routing problem and its extensions. In: Golden, B., Raghavan, S., Wasil, E. (eds.) The Vehicle Routing Problem: Latest Advances and New Challenges. Operations Research/Computer Science Interfaces, vol. 43. Springer, Boston (2008). https://doi.org/10.1007/978-0-387-77778-8_4
10. Hanshar, F.T., Ombuki-Berman, B.M.: Dynamic vehicle routing using genetic algorithms. Appl. Intell. **27**(1), 89–99 (2007). https://doi.org/10.1007/s10489-006-0033-z
11. Huang, M., Yang, J., Teng, M.A., Li, X., Wang, T.: The modeling of milk-run vehicle routing problem based on improved C-W algorithm that joined time window. Transp. Res. Procedia **25**, 716–728 (2017)
12. Kilic, H.S., Durmusoglu, M.B., Baskak, M.: Classification and modeling for in-plant milk-run distribution systems. Int. J. of Adv. Manuf. Technol. **62**, 1135–1146 (2012)
13. Kryvinska, N., Kaczor, S., Strauss, C.: Enterprises' servitization in the first decade—retrospective analysis of back-end and front-end challenges. Appl. Sci. **10**(8), 2957 (2020). https://doi.org/10.3390/app10082957
14. Orozco, J.-A., Barceló, J.: Reactive and proactive routing strategies with real-time traffic information. Procedia. Soc. Behav. Sci. **39**, 633–648 (2012)
15. Pavone, M., Bisnik, N., Frazzoli, E., Isler, V.: A stochastic and dynamic vehicle routing problem with time windows and customer impatience. Comput. Sci. Eng. **14**, 350–364 (2009). https://doi.org/10.1007/s11036-008-0101-1
16. Sáeza, D., Cortésb, C.E., Núñez, A.: Hybrid adaptive predictive control for the multi-vehicle dynamic pick-up and delivery problem based on genetic algorithms and fuzzy clustering. Comput. Oper. Res. **35**, 3412–3438 (2008)
17. Prokopowicz, P., Ślęzak, D.: Ordered fuzzy numbers: definitions and operations. In: Prokopowicz, P., Czerniak, J., Mikołajewski, D., Apiecionek, Ł., Ślęzak, D. (eds.) Theory and Applications of Ordered Fuzzy Numbers, pp. 57–79. Springer, Cham (2017).https://doi.org/10.1007/978-3-319-59614-3_4

18. Polak-Sopinska, A.: Incorporating human factors in in-plant milk run system planning models. In: Ahram, T., Karwowski, W., Taiar, R. (eds.) IHSED 2018. AISC, vol. 876, pp. 160–166. Springer, Cham (2018). https://doi.org/10.1007/978-3-030-02053-8_26

19. Wangham, M., Adriano, D.D., Montez, C., Novaes, A.G.N.: DMRVR: dynamic milk-run vehicle routing solution using fog-based vehicular ad hoc networks. Electronics **2020**, 9 (2010). https://doi.org/10.3390/electronics9122010

20. Wójcik, R., Bocewicz, G., Bożejko, W., Banaszak, Z.: Milk-run routing and scheduling subject to a trade-off between vehicle fleet size and storage capacity. Manage. Prod. Eng. Rev. **10**(3), 41–53 (2019)

21. Yang, J., Huang, M., Ma, T., Li, X., Wang, T.: The modeling of milk-run vehicle routing problem based on improved C-W algorithm that joined time window. Transp. Res. Procedia **25**, 716–728 (2017)

An Efficient Hybrid Planning Framework for In-Station Train Dispatching

Matteo Cardellini[1]🆔, Marco Maratea[1]🆔, Mauro Vallati[2](✉)🆔, Gianluca Boleto[1], and Luca Oneto[1]🆔

[1] University of Genoa, Genova, Italy
{matteo.cardellini,marco.maratea,gianluca.boleto,luca.oneto}@unige.it
[2] University of Huddersfield, Huddersfield, UK
m.vallati@hud.ac.uk

Abstract. In-station train dispatching is the problem of optimising the effective utilisation of available railway infrastructures for mitigating incidents and delays. This is a fundamental problem for the whole railway network efficiency, and in turn for the transportation of goods and passengers, given that stations are among the most critical points in networks since a high number of interconnections of trains' routes holds therein. Despite such importance, nowadays in-station train dispatching is mainly managed manually by human operators.

In this paper we present a framework for solving in-station train dispatching problems, to support human operators in dealing with such task. We employ automated planning languages and tools for solving the task: PDDL+ for the specification of the problem, and the ENHSP planning engine, enhanced by domain-specific techniques, for solving the problem. We carry out a in-depth analysis using real data of a station of the North West of Italy, that shows the effectiveness of our approach and the contribution that domain-specific techniques may have in efficiently solving the various instances of the problem. Finally, we also present a visualisation tool for graphically inspecting the generated plans.

Keywords: Railway station · In-station train dispatching · Automated planning · PDDL+

1 Introduction

Railways play a significant economical role in our society for transporting either goods or passengers, but the increasing volume of people and freight transported on railways is congesting the networks [2]. In-station train dispatching is the problem of optimising the effective utilisation of available railway infrastructures for mitigating incidents and delays, and their impact. This is a fundamental problem for the whole railway network efficiency, and in turn for the transportation of goods and passengers, given that stations are among the most critical points in networks since a high number of interconnection of trains' routes

© Springer Nature Switzerland AG 2021
M. Paszynski et al. (Eds.): ICCS 2021, LNCS 12742, pp. 168–182, 2021.
https://doi.org/10.1007/978-3-030-77961-0_15

holds therein. Despite such importance, nowadays in-station train dispatching is mainly managed manually by human operators, while automatic support would be of great help for the wide railway network efficiency. The nature of real-world applications often necessitates the representation of the dynamics of the application in terms of mixed discrete/continuous behaviour, including effects, processes, exogenous events, and continuous activities.

In this paper we present a framework (shown in Fig. 2 of Sect. 3) for solving in-station train dispatching problems, so to support human operators in dealing with such task. We employ automated planning languages and tools: in automated planning, knowledge is made explicit in a symbolic problem instance models. When using the dominant family of planning knowledge representation languages – PDDL [13], a problem is specified via an initial state and goal state, that need to be reached. States, represented via fluents, can be changed by applying actions, that are specified via preconditions and effects, represented as a Boolean combination of predicates. A *solution (plan)* is a sequence of actions such that a consecutive application of the actions in the plan (starting in the initial state) results in a state that satisfies the goal [7]. Such classical planning model has been recently extended, in order to handle mixed discrete/continuous behaviours, resulting in PDDL+ [5], which thus allows to model hybrid domains. In addition to actions, PDDL+ introduces *continuous processes* and *exogenous events*, that are triggered by changes in the environment. Processes are used to model continuous changes. PDDL+ applications in real-world domains include real-time UAV manoeuvring [14], efficient battery load management [6], and urban traffic control [12].

Our framework, that extends a research prototype [3], is organised in three main elements: a preprocessor, that takes as input a description of the in-station train dispatching problem, and outputs a PDDL+ instance model; a planning engine, able to solve the PDDL+ problem and computing a plan, and a visualisation tool. We present all such elements and focus on the planning engine, its input/output, and its performance when employing the ENHSP planning engine [15,16], enhanced by domain-specific techniques. We carry out a detailed analysis using real data of a station of the North West of Italy, provided by Rete Ferroviaria Italiana (RFI), that shows the effectiveness of our approach and the contribution that the implemented domain-specific techniques may have in efficiently solving the various instances of the problem.

The paper is structured as follows. Section 2 introduces our target problem. Then, Sect. 3 presents our framework and its elements. The experimental analysis of our modified planning engine on the real data is shown in Sect. 4. The paper ends in Sect. 5 by discussing related work and drawing conclusions.

2   The In-Station Train Dispatching Problem

This section describes the in-station train dispatching problem, following the formalisation introduced in [9].

A railway *station* can be represented as a tuple $\mathscr{S} = \langle G, I, P, E^+, E^- \rangle$. G is an undirected graph, $G = \langle S, C \rangle$, where S is the set of nodes representing

Fig. 1. A subsection of the real-world railway station considered in the experimental analysis. Flags are used to delimit itineraries; shorter segments represent track segments; bold indicates platforms.

the track segments, i.e., the minimal controllable rail units, and C is a set of edges that defines the connections between them. Their status can be checked via track circuits, that provide information about occupation of the segment and about corresponding timings. Track segments are grouped in itineraries I. Each itinerary $i = \langle s_{m1}, s_{m2}, ..., s_{mn} \rangle \in I$ is a path graph, subset of the graph G, representing a sequence of connected track segments. Track segments are grouped in itineraries manually by experts of the specific railway station, and a track segment can be included in more than one itinerary. While track segments are the minimal controllable units of a station, itineraries describe paths that the trains will follow in order to move inside the station.

P is a set of platforms that can be used to embark/disembark the train. Each platform $p \in P$ has an associated set of track segments $\{s_i, \ldots, s_j\}$ indicating that trains stopping at those track segments will have access to the platform.

E^+ is a set of entry points to the station from outside railway network; similarly, E^- is a set of exit points of the station that allow the train to leave the station and enter the outside railway network. For the sake of this formalisation, entry points and exit points behave like buffers of infinite size which are connected to a single track segment that is the first (last) track segment the train will occupy after (before) entering (exiting) the station.

Figure 1 provides a schematic representation of part of the Italian railway station we use in our analysis. In the figure, track segments and platforms are easily recognisable, and flags are used to indicate initial and end points of itineraries. A track segment can be occupied by a single train at the time. For safety reasons, a train is required to *reserve* an itinerary, and this can be done only if the itinerary is currently not being used by another train. While a train is navigating the itinerary, the track segments left by the train are released. This is done to allow trains to early reserve itineraries even if they share a subset of the track segments. A train t going through the controlled railway station is running a route R_t in the station graph G, by reserving an itinerary and moving through the corresponding track segments. A route $R_t \subseteq I$ is a sequence of *connected disjointed itineraries* which the train t will travel in the station. To simplify the notation we define with $h(R_t) \in S$ the first track segment of the route and with $l(R_t) \in S$ the last.

Considering a single railway station, there are four possible types of train:

- *Origin*: the train originates at the controlled station. It is initially at a platform p, and it is required to leave the controlled station via a specified exit point.
- *Destination*: the controlled station is the final destination of the train. The train is expected to reach the destination at time t_a via a given entry point, and is required to reach a platform to end its trip.
- *Transit*: the train is moving through the controlled station without making a stop. The train is expected to reach the destination at time t_a via a given entry point, and is required to reach a specified exit point without stopping at a platform.
- *Stop*: the train is making an intermediate stop at the controlled station. The train is expected to reach the destination at time t_a via a given entry point, and is required to reach a specified exit point after a stop at a platform for embarking/disembarking passengers.

The sequence of itineraries $R_t = (i_1, i_2, ..., i_n)$, with $i_k \in I$, which the train t will run across have to be connected in the graph G.

A *timetable* is the schedule that includes information about when trains should arrive at the controlled station, when they arrive at a platform, and the time when they leave a platform.

We are now in the position to define the in-station train dispatching problem as follows: Given a railway station $\mathscr{S}$, a set of trains T and their current position within the station or their time of arrival at the controlled station, find a route for every train that allows to respect the provided timetable, as much as possible.

3 Description of the Framework

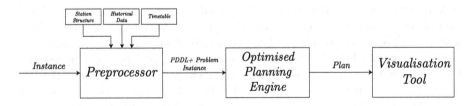

Fig. 2. Architecture of the proposed framework.

The architecture of the proposed framework is shown in Fig. 2. An instance can be defined using a provided interface, allowing a human operator to specify the trains that need to be controlled, their type (i.e., *Origin, Destination, Transit* or *Stop*) and the expected arrival/departure times, together with all the features needed to cluster the historical tracks' travel times (as explained in Sect. 3.1).

The prepocessor generates a PDDL+ problem instance, that can then be parsed and reasoned upon by a dedicated optimised planning engine, by relying also on information about the structure of the controlled station, timetables, and historical data. The preprocessor uses the morphology of the station to compute all the possible routes that a train can move through in order to reduce the combinations that a plan may follow.

The optimised planning engine is based on the well-known ENHSP [15,16] system, enhanced with domain-specific optimisations to improve performance on in-station train dispatching instances. Finally, the generated plans can be graphically shown by a visualisation tool, that aims at supporting a human operator in checking and simulating the planned movements of trains inside the station.

3.1   Preprocessor

The preprocessor is in charge of translating a given problem instance into a PDDL+ problem instance, so that it can be solved by a PDDL+ planning engine. We can define a PDDL+ problem model as a t-uple $\langle I, G, A, O \rangle$. I is a description of the initial state, that characterises the initial state of the world. G is the list of goals that must be achieved. A is the set of actions, events, and processes that are used to model the evolution of the world. Actions are under the control of the planning engine, while events and processes are exogenous and are triggered automatically when the corresponding preconditions are met. Processes are used to model continuous changes, while events are instantaneous changes of the state of the world. Finally, O is the list of objects involved in the problem.

To perform its task, the preprocessor relies on additional data provided by Rete Ferroviaria Italiana (RFI). Here we introduce the provided data, and describe the structure of the PDDL+ model that is generated.

Data Provided. RFI provided the access to the data of 5 months (January to May 2020) of train movements of one medium-sized railway station of the Liguria region, situated in the North West of Italy. A subsection of the station is shown in Fig. 1: the station has been anonymised, and the complete structure can not be provided due to confidentiality issues. The modelled station includes 130 track segments (out of which 34 are track switches), 107 itineraries, 10 platforms, 3 entry points, and 3 exit points. The average number of trains per day was 130 before COVID-19-related lockdown and 50 after movement restrictions were enforced in Italy. The travel times information needed by the PDDL+ model, e.g., the time needed by each train to complete a track segment, to leave a platform, maximum times, etc., were calculated by leveraging on historical data provided by RFI. The whole dataset of historical tracks' travel times were clustered by a variety of features, such as train characteristics (e.g., passengers, freight, high speed, intercity, etc.), station transit characteristics (i.e., overall train trip inside the rail-network, and entry and exit points), weekdays, and weather conditions. According to the characteristics of the problem at hand, it is then possible to estimate the travel times of every train by assigning the average value of the times inside the corresponding cluster.

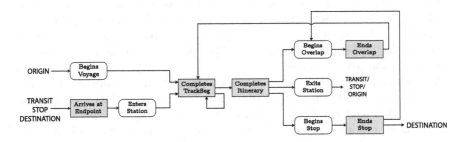

Fig. 3. A flow chart showing the movements of the train inside the station based on its type. Rectangles represent events and squared rectangles represent actions. Processes are omitted for clarity.

Times. The travel times information are encoded in the PDDL+ model using the following fluents:

- `arrivalTime(t)`: the time a_t at which the train t arrives at the controlled station.
- `segmentLiberationTime(t,s,i)`: the amount of time it takes for train t to free segment $s \in i_n$ since it starts moving on itinerary i.
- `timeToRunItinerary(t,i)`: the amount of time it takes for train t to move through itinerary i.
- `stopTime(t,p)`: the expected amount of time it takes for train t to embark/disembark passengers or goods at platform p.
- `timeToOverlap(t,i_n,i_m)`: the time it takes for train t to completely leave itinerary i_n and move to itinerary i_m.
- `timetableArrivalTime(t)` (`timetableDepartureTime(t)`): the time in which the train t should arrive at (departure from) a platform, according to the official timetable.

Movement of Trains. In this section we focus on the PDDL+ structures used to model the movement of trains per type. An overview of the events and actions used to model trains' movement is provided in Fig. 3. For the sake of readability, we say that a Boolean predicate is activated (de-activated) when its value is made True (False).

Transit Train. If a train is of type Transit it will traverse the station without stopping to embark/disembark passengers or goods at any of the platforms. The movement of the train through the station is regulated using the following PDDL+ constructs (listed in application's order):

- A process `incrementTime` encodes an explicit notion of passing time by increasing the fluent `time`.
- An event `arrivesAtEntryPoint(t,e⁺)` is triggered when the fluent `time` reaches the value of the predicate `arrivalTime(t)`, and it requires that the predicate `trainEntersFromEntryPoint(t,e⁺)`, indicating the entry point for the train, is active. The event makes true the predicate `trainHasArrivedAtStation(t)` signalling that the train is at the gateway of the station ready to enter.
- An action `entersStation(t,e⁺,i)` can be used by the planning engine to let the train t enter the station from the entry-point e^+, using the itinerary i. The precondition that must hold in order for this action to be taken are: (i) the entry point is connected to the itinerary i, and (ii) the itinerary i is free, i.e., none of its track segments are occupied by a train. As a result of this action i is reserved by t by blocking all the track segments in it via dedicated predicates `trackSegBlocked(s)`, and by the predicate `trainInItinerary(t,i)`.
- A process `incrementTimeReservedItinerary(i)` keeps track of how many seconds have passed since the reservation of an itinerary i by any train, updating the fluent `tReservedItinerary(i)` accordingly.
- The event `completesTrackSeg(t,s,i)` is triggered when a train t has left the track segment s. This is triggered when `tReservedItinerary(i)` has reached the value specified in the fluent `segmentLiberationTime(t,s,i)`, and as a result the segment s is freed by de-activating the predicate `trackSegBlocked(s)`.
- As soon as the train has reached, with its head, the end of the itinerary (so when `tReservedItinerary(i)` is greater than `timeToRunItinerary(t,i)`), the event `completesItinerary(t,i)` is triggered. It activates a predicate `trainHasCompletedItinerary(t,i)`.
- An action `beginsOverlap(t,iₙ,iₘ)` can be used by the planning engine to encode the movement of train t from itinerary i_n to itinerary i_m. This action can be used only if `trainHasCompletedItinerary(t,iₙ)` is active and all the track segments $s \in i_n$ are not occupied by another train.
- A process `incrementOverlapTime(t,iₙ,iₘ)` keeps track of the time a train t is taking to move all its carriages through the joint that connects i_n to i_m by increasing the value of the fluent `timeElapsedOverlapping(t,iₙ,iₘ)`.
- An event `endsOverlap(t,iₙ,iₘ)` is triggered when the fluent that counts the time of overlap `timeElapsedOverlapping(t,iₙ,iₘ)` is greater than the fluent `timeToOverlap(t,iₙ,iₘ)`. The event signals that the train is no longer on itinerary i_n deactivating the predicates `trainInItinerary(t,iₙ)` and resetting the value of function `tReservedItinerary(iₙ)` to 0.
- Finally, when a train has completed its routes of itineraries, and with the last itinerary i has reached the segment leading to an exit point of the station e^-, the action `exitsStation(t, i, e⁻)` allows the train to exit the station activating the predicate `trainHasExitedStation(t)`.

Part of the described constructs are shown in Fig. 4. The goal for a train t of type *Transit* is to reach the state in which the predicate `trainHasExitedStation(t)` is active.

```
(:action entersStation
:parameters
 (T1 - train EP1 - entryP
 I1-2 itinerary)
:precondition (and
 (trainIsAtEndpoint T1 EP1)
 ...
 (not (trainHasExitedStation T1))
 (not (trainHasEnteredStation T1))
 (not (trackSegBlocked cdb1))
 (not (trackSegBlocked cdb2))
 ...
 (not (trackSegBlocked cdb6)))
:effect (and
 (not (trainIsAtEndpoint T1 EP1))
 (itineraryIsReserved I1-2)
 (trainInItinerary T1 I1-2)
 (trainHasEnteredStation T1)
 (trackSegBlocked cdb1)
 (trackSegBlocked cdb2)
 ...
 (trackSegBlocked cdb6)))
```

```
(:event arrivesAtEntryPoint
:parameters (T1 - train EP1 - entryP)
:precondition (and
 (>= time (ArrivalTime T1))
 (not (trainHasEnteredStation T1))
 (trainEntersFromEntryPoint T1 EP1))
:effect (and (trainIsAtEndpoint T1 EP1)
 (trainHasArrivedAtStation T1)
 (assign (trainEntryIndex T1)
 (trainsArrivedAtEndpoint EP1))
 (increase (trainsArrivedAtEndpoint EP1) 1 )))

(:event completeTrackSeg
:parameters
 (T1 - train cb1 - trackSeg I1-2 itinerary)
:precondition (and
 (>= (timeReservedIt I1-2) 29)
 (trainInItinerary T1 I1-2)
 (trackSegBlocked cdb1))
:effect (and
 (not (trackSegBlocked cdb1))))

(:process incrementTimeReservedItinerary
:parameters(I1-2 - itinerary)
:precondition (itineraryIsReserved I1-2)
:effect (increase
 (timeElapsedReservedItinerary I1-2) #t ))
```

Fig. 4. The partially grounded action entersStation(t,e$^+$,i$_n$) (left), events arrivesAtEntryPoint(t,e$^+$) and completesTrackSeg(t,s,i$_n$), and process incrementTimeReservedItinerary(i$_n$) (right).

Stop Train. A train of this type needs to stop at a platform before exiting the station, as it is making an intermediate stop at the controlled station. Three other PDDL+ constructs are added with respect to a train of type Transit in order to model this behaviour:

- The action beginStop(t,i,p) is used to allow the planning engine to stop the train t at a platform p after having completed itinerary i, and having the platform at the end of the itinerary. The effect of this action is to signal that the train is at the platform by activating the predicate trainIsStoppingAtStop(t,p) and all the track segments $s \in p$ are blocked in order to achieve mutual exclusion on the platform.
- A process increaseTrainStopTime(t) keeps track of the time spent by the train at a platform. This is done by increasing the value of the function trainStopTime(t) over time.
- The train t can begin its route towards the exit point only after a time stopTime(t) has passed – this it to allow passengers or goods to embark / disembark. A train cannot leave the platform before the timetabled departure, set via the function timetableDepartureTime(t). For this reason the event endStop(t,i,p) that signals that the train is ready to leave the platform can be triggered only if the fluent trainStopTime(t) has reached the value set in the predicate stopTime(t) and the fluent time is greater than timetableDepartureTime(t). This event activates the predicate trainHasStopped(t).

```
(:action beginStop                          (:process increaseTrainStopTime
:parameters                                 :parameters(T1 - train)
(T1 - train I3 - itinerary                  :precondition (trainIsStopping T1)
 S1 - platform)                             :effect
:precondition (and                           (increase (trainStopTime T1) #t ))
 (trainHasCompletedItinerary T1 I3)
 (trainInItinerary T1 I3)                   (:event endStop
 (not (trainIsOverlapping T1))              :parameters
 (not (stopIsOccupied S1))                   (T1 - train I3 - itinerary S1 - platform)
 (not (trainIsStopping T1))                 :precondition (and
 (not (trainHasStoppedAtStop T1 S1))         (>= (trainStopTime T1) 5 )
 ...)                                         (>= time (DepartTime T1))
:effect (and                                 (trainInItinerary T1 I3)
 (trainIsStoppingAtStop T1 S1)               (trainIsStoppingAtStop T1 S1)
 (trainIsStopping T1)                        (stopIsOccupied S1))
 (assign (trainStopTime T1) 0 )            :effect (and
 (stopIsOccupied S1)                         (not (trainIsStoppingAtStop T1 S1))
 (not (itineraryIsReserved I3))              (not (trainIsStopping T1))
 (not (trackSegBlockedtrackSegC))            (trainHasStoppedAtStop T1 S1)
 ...                                         (trainHasStopped T1)
 (not (trackSegBlockedtrackSegL))))          (not (stopIsOccupied S1))))
```

Fig. 5. The partially grounded action beginStop(t,i_n,p_k) (left), process increaseTrainStopTime(t) and event endStop(t,i_n,p_k) (right).

The train t will then begin its trip towards the exit point with the action beginsOverlap(t,i_n,i_m) where i_n is the itinerary where the platform is located, and i_m is a subsequent connected itinerary. Part of the described constructs are shown in Fig. 5. For a train of type *Stop* the goal is to reach a state in which the predicates trainHasExitedStation(t) and trainHasStopped(t) are both active.

Origin Train. A train of type *Origin* needs to specify the platform the train is parked at. For this reason the predicates trainIsStoppingAtStop(t,p) and trainHasStopped(t) are activated at the initial state of the problem, indicating the train t is departing from platform p. This type of train resemble the train of type *Stop* but without the possibility to enter the station, since it is already inside it at the beginning of the plan. Only one action is introduced in order to model the behaviour of an *Origin* train:

– An action beginVoyage(t,p,i) can be used by the planning engine to allow the departure of train t from platform p via itinerary i. This action can not be executed before the timetabled departure time timetableDepartureTime(t).

The goal of a train of type *Origin* are the same of a train of type *Transit*: it must have exited the station.

Destination Train. A train of type *Destination* behaves like an *Stop* train but, after having reached the platform, it will remain parked there. For this reason no additional PDDL+ constructs are needed. A train of type *Destination* will simply have in its goal the need to reach a state in which the predicate trainIsStopping(t) is active, meaning so that the train t has reached its stop.

3.2 Optimised Planning Engine

The considered domain-independent PDDL+ planning engine ENHSP is based on a forward search approach on the discretised PDDL+ planning instance. In other words, the planning engine deals with time-related aspects by discretising it using a given delta value. We enhanced ENHSP in three main ways: adaptive delta, domain-specific heuristic, and constraints.

Adaptive Delta. Traditionally, PDDL+ planning engines exploit a fixed time discretisation step for solving a given instance. In the proposed model, it is possible to know a-priori when events will be triggered, so it is possible to exactly predict when the world will change. It is therefore possible to exploit a dynamic time discretisation. When nothing happens, there is no need to discretise time. This is similar in principle to the approach exploited by decision-epoch planning engines for dealing with temporal planning problems [4].

Specialised Heuristic. Following the traditional A* search settings, the cost of a search state z is calculated as $f(z) = g(z) + h(z)$, where $g(z)$ represents the cost to reach z, while $h(z)$ provides an heuristic estimation of the cost needed to reach a goal state from z. In our specialisation, $g(z)$ is calculated as the elapsed modelled time from the initial state to z. $h(z)$ is a domain-specific heuristic calculated according to the following equation:

$$h(z) = \sum_{t \in T(z)} \rho_t(z) + \pi_t(z) \tag{1}$$

where $T(z)$ is the set of trains of the given problem that did not yet achieve their goals at z. $\rho_t(z)$ is a quantity that measures the time that, starting from the current position, the considered train needs to reach its final destination and is computed as follows:

$$\rho_t(z) = \max_{R \in \mathcal{R}_t(z)} \sum_{i_n \in R} \texttt{timeToRunItinerary(t, } i_n\texttt{)} \tag{2}$$

where $\mathcal{R}_t(z)$ is the set containing all the possible routes R, as sequences of itineraries, that a train t can run across in order to reach its final destination (i.e., an exit point or a platform) from the state z. Since the initial and final position of every train is known a-priori and based upon its type (*Origin, Destination, Stop, Transit*) the set can be computed beforehand and used in the search phase.

The penalisation method $\pi_t(z)$ gives a very high penalisation value P for each goal specified for the considered train t that has not yet been satisfied at state z. For instance, if a train of type *Stop* has not yet entered the station, a penalisation of $2 \times P$ is given to the heuristic since there are two goals related to the train that has not been satisfied in the initial state, i.e., stopping at a platform and leaving the station from an exit-point.

Constraints. The planning engine has been extended by explicitly considering 3 set of constraints. The first set limits the time a train is allowed to stay in the controlled station. The other sets limit, respectively, the time passed from the arrival of the train in the station, and the time spent stopping at a platform. The idea behind such constraints is to avoiding situations where trains are left waiting for long periods of time, occupying valuable resources. The maximum times are calculated a-priori, according to historical data, and depends on the structure of the railway station.

3.3   Visualisation Tool

The last element of the framework shown in Fig. 2 is the visualisation tool. This is a pivotal tool, that allows to visually inspect the generated plans and can be used by domain experts to compare alternative solutions, and by AI experts to validate the plans.

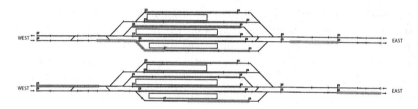

Fig. 6. Two screenshots from the visualisation tool capturing the itineraries occupied by three trains at different time steps. Top: The red train is moving from the West entry point to platform 1 (counting from the bottom), the blue train is stopped at platform 4 and the green train is approaching the East exit point. Bottom: The red train has stopped at the platform and is moving towards the East exit point, the blue and the green trains are leaving the station from the West and East exit points, respectively. (Color Figure online)

Figure 6 shows two screenshots from the visualisation tool capturing the itineraries occupied by three trains (red, green, blue) in two subsequent time instants. The tool can parse a solution plan, represented using the PDDL+ standard, and extracts the movements of trains and their position at each time step. The itineraries are then coloured based on the train that is currently moving on them. The tool provides an interface that allows the user to move forward or backward in time, to see how the network conditions evolve. In the example shown in Fig. 6, the red train is moving from the West entry point to platform 1 (counting from the bottom), the blue train is stopped at platform 4 and will soon move to the West exit point and the green train is approaching the East exit point.

4   Experimental Analysis

The aim of this experimental analysis is to assess the importance of the domain-specific techniques implemented in the planning engine. As a first step, to understand that the proposed framework can accurately model the dynamics of the in-station train dispatching problem, we validated on historical data.

To perform this analysis, we considered the 5 months of historical data, and encoded all the registered movements of trains under the form of PDDL+ plans – as if the movements were generated using our approach. We then validated the plans against our PDDL+ model with a PDDL+ validator. This step checks if the PDDL+ model presented in the paper is capable of representing the observed movements. Remarkably, the validation process showed that all the movements observed in the 5 months of historical data were correctly validated. This result also indicates that the proposed modelisation, and the overall framework, can provide an encompassing framework for comparing different strategies to deal with recurrent issues, and for testing new train dispatching solutions.

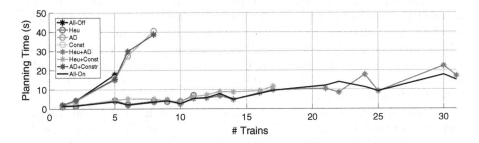

Fig. 7. The CPU-time needed for planning instances with an increasing number of trains using different combinations of domain-specific optimisation techniques.

We can now turn our attention to assessing the usefulness of the improvements proposed in Sect. 3.2. To perform this analysis, we selected the day – in February 2020, before the start of the COVID-19 lockdown in Italy – with the minimum mean squared deviation of recorded train timings from the official timetable. This was done to guarantee that no emergency operations were executed by the human operators. We considered the evening peak hour of the day, 17:00–20:00, when commuters return at home. During the selected time-slot 31 trains move through the station. We considered different time windows of the considered 3-h period to generate instances with a increasing number of trains to be controlled. These instances are then tested using the planning engine with different combinations of the domain-specific optimisations techniques. Figure 7 shows the CPU planning time required by the 8 combinations to deal with increasingly large instances. The combination labelled *All-Off* considers ENHSP run using the default settings, and none of the optimisation techniques introduced in this paper. The *All-On* label indicates instead the planning engine run using all the optimisations. Finally, labels *AD, Heu,* and *Const* are used to indicate the use of, respectively, adaptive delta, domain-specific heuristic, and constraints. The analysis was performed with a cut-off time of 60 CPU-time seconds, on a 2.5 GHz Intel Core i7 Quad-processors with 16 GB of memory made available and a Mac OS operating system.

According to the results in Fig. 7, the *All-Off* combination can plan no more than 5 trains. Inspecting the instances with more that 5 trains it can be seen that, while the first 5 trains are close to each other, the 6*th* train arrives at the station a couple of minutes after the others. This gap of time, in which nothing happens, causes the planning engine to wast a significant amount of time in trying to identify actions to apply to reach the goal, at every discretisation delta. The use of the adaptive delta allows the planning engine to skip between times where nothing happens, therefore it significantly reduces the amount of computations made by the planning engine, solving up to 8 trains. The use of the domain-specific heuristic alone can solve instances up to 11 trains included. Notably, there is a significant synergy between the heuristic and the adaptive delta. When these two optimisations are activated, the planning engine performance are significantly boosted, given that with these two techniques enabled we are able to solve all instances, as for the *All-On* label. On the other hand, the use of the constraints can provide some performance improvement, but limited. With regards to the shape of the generated solutions, we observed that all the approaches lead to similar solutions: there are no major differences from this perspective. We further experimented with other peak hours (07:30–10:00 and 12:00–14:30), and results are similar to the ones shown.

Summarising, the performed experiments indicate that the use of the specialised heuristic is the single most important component of the domain-specific planning engine. The adaptive delta plays an important role as well, but their synergic combination allows to solve all evaluated instances. The use of constraints provides some improvement, but not as significant as the other elements.

5   Related Work and Conclusions

Automated approaches have been employed for solving variants of the in-station train dispatching problem. Mixed-integer linear programming models have been introduced in [11] for controlling a metro station. The experimental analysis demonstrated the ability of the proposed technique to effectively control a metro station but also highlighted scalability issues when it comes to control larger and more complex railway stations. More recently, constraint programming models have been employed in [8] for performing in-station train dispatching in a large Indian terminal: this approach is able to deal with a large railway station, but only for very short time horizons, i.e., less than 10 min. In [1] a hybrid approach that extends Answer Set Programming (ASP) [10] is used to tackle real-world train scheduling problems, involving routing, scheduling, and optimisation.

In this paper, we presented a framework for solving in-station train dispatching, focused on modelling the problem with the PDDL+ language, able to specify mixed discrete/continuous behaviour of railway systems, and solving with the ENHSP planning engine enhanced with domain-specific techniques. Results on real data of a station of the North West of Italy, provided by Rete Ferroviaria Italia, show that our solution is efficient, scalable with regards to the number of trains and can handle large time horizons, for the station at hand. Future work will focus on modelling additional stations of the North of Italy, and to run field tests of the proposed framework.

Acknowledgements. This work has been partially funded by Hitachi Rail STS through the RAIDLab (Railway Artificial Intelligence and Data Analysis Laboratory), a joint laboratory between Hitachi Rail STS and University of Genoa. This work has been supported by RFI (Rete Ferroviaria Italiana) who provided the data for the analysis (we sincerely thank Renzo Canepa for his support). Mauro Vallati was supported by a UKRI Future Leaders Fellowship [grant number MR/T041196/1].

References

1. Abels, D., Jordi, J., Ostrowski, M., Schaub, T., Toletti, A., Wanko, P.: Train scheduling with hybrid answer set programming. In: Theory and Practice of Logic Programming, pp. 1–31 (2020)
2. Bryan, J., Weisbrod, G.E., Martland, C.D.: Rail freight solutions to roadway congestion: final report and guidebook. Transp. Res. Board (2007)
3. Cardellini, M., Maratea, M., Vallati, M., Boleto, G., Oneto, L.: In-station train dispatching: a PDDL+ planning approach. In: Proceedings of ICAPS (2021)
4. Cushing, W., Kambhampati, S., Weld, D.S.: When is temporal planning really temporal? In: Proceedings of IJCAI, pp. 1852–1859 (2007)
5. Fox, M., Long, D.: Modelling mixed discrete-continuous domains for planning. J. Artif. Intell. Res. **27**, 235–297 (2006)
6. Fox, M., Long, D., Magazzeni, D.: Plan-based policies for efficient multiple battery load management. J. Artif. Intell. Res. **44**, 335–382 (2012)
7. Ghallab, M., Nau, D., Traverso, P.: Automated Planning: Theory & Practice. Morgan Kaufmann Publishers, Burlington (2004)
8. Kumar, R., Sen, G., Kar, S., Tiwari, M.K.: Station dispatching problem for a large terminal: a constraint programming approach. Interfaces **48**(6), 510–528 (2018)
9. Lamorgese, L., Mannino, C.: An exact decomposition approach for the real-time train dispatching problem. Oper. Res. **63**(1), 48–64 (2015)
10. Lifschitz, V.: Answer set planning. In: Gelfond, M., Leone, N., Pfeifer, G. (eds.) LPNMR 1999. LNCS (LNAI), vol. 1730, pp. 373–374. Springer, Heidelberg (1999). https://doi.org/10.1007/3-540-46767-X_28
11. Mannino, C., Mascis, A.: Optimal real-time traffic control in metro stations. Oper. Res. **57**(4), 1026–1039 (2009)
12. McCluskey, T.L., Vallati, M.: Embedding automated planning within urban traffic management operations. In: Proceedings of ICAPS, pp. 391–399 (2017)
13. Mcdermott, D., et al.: PDDL - The Planning Domain Definition Language. Technical report, Yale Center for Computational Vision and Control (1998)

14. Ramírez, M., et al.: Integrated hybrid planning and programmed control for real time UAV maneuvering. In: Proceedings of AAMAS, pp. 1318–1326 (2018)
15. Scala, E., Haslum, P., Thiébaux, S., Ramírez, M.: Interval-based relaxation for general numeric planning. In: Proceedings of ECAI, pp. 655–663 (2016)
16. Scala, E., Haslum, P., Thiébaux, S., Ramírez, M.: Subgoaling techniques for satisficing and optimal numeric planning. J. Artif. Intell. Res. **68**, 691–752 (2020)

Evaluating Energy-Aware Scheduling Algorithms for I/O-Intensive Scientific Workflows

Tainã Coleman[1](✉), Henri Casanova[2], Ty Gwartney[2],
and Rafael Ferreira da Silva[1](✉)

[1] USC Information Sciences Institute, Marina del Rey, CA, USA
{tcoleman,rafsilva}@isi.edu
[2] Information and Computer Sciences, University of Hawaii, Honolulu, HI, USA
{henric,tyg}@hawaii.edu

Abstract. Improving energy efficiency has become necessary to enable sustainable computational science. At the same time, scientific workflows are key in facilitating distributed computing in virtually all domain sciences. As data and computational requirements increase, I/O-intensive workflows have become prevalent. In this work, we evaluate the ability of twopopular energy-aware workflow scheduling algorithms to provide effective schedules for this class of workflow applications, that is, schedules that strike a good compromise between workflow execution time and energy consumption. These two algorithms make decisions based on a widely used power consumption model that simply assumes linear correlation to CPU usage. Previous work has shown this model to be inaccurate, in particular for modeling power consumption of I/O-intensive workflow executions, and has proposed an accurate model. We evaluate the effectiveness of the two aforementioned algorithms based on this accurate model. We find that, when making their decisions, these algorithms can underestimate power consumption by up to 360%, which makes it unclear how well these algorithm would fare in practice. To evaluate the benefit of using the more accurate power consumption model in practice, we propose a simple scheduling algorithm that relies on this model to balance the I/O load across the available compute resources. Experimental results show that this algorithm achieves more desirable compromises between energy consumption and workflow execution time than the two popular algorithms.

Keywords: Scientific workflows · Energy-aware computing · Workflow scheduling · Workflow simulation

1 Introduction

Scientific workflows have become mainstream for the automated execution of computational workloads on parallel and distributed computing platforms. Over the past two decades, workflow applications have become more complex as the

© Springer Nature Switzerland AG 2021
M. Paszynski et al. (Eds.): ICCS 2021, LNCS 12742, pp. 183–197, 2021.
https://doi.org/10.1007/978-3-030-77961-0_16

volume of data and processing needs have substantially increased. At the same time, computational platforms have increased their processing capacity and developed mechanisms for efficient workload management. An integral aspect of the workload processing lifecycle is energy management by which the system makes scheduling and resource provisioning decisions so as to maximize workload throughput while reducing or bounding energy consumption [9]. In the past few years, energy management for scientific workflows has gained traction, especially due to their singular contributions to major scientific discoveries. As a result, several works have proposed energy-aware workflow scheduling algorithms for a range of computing environments, and in particular cloud computing platforms [8,11,12,22].

These scheduling algorithms have all been designed and evaluated using the traditional model for application power consumption [5,7,10,12,13,21]: power consumption is linear in CPU utilization and does not account for I/O operations. In previous work [15,17], we have experimentally quantified the accuracy of this model and have identified sources of inaccuracy. A key finding was that power consumption on multi-socket, multi-core compute nodes is not linearly related to CPU utilization. Another finding was that, unsurprisingly, I/O operations (i.e., disk reads and writes) significantly impact power consumption. Since many workflows are I/O-intensive, accounting for the power consumption of I/O is crucial for power management of their executions. In that same work, we then proposed a power consumption model that accounts for (i) computations that execute on multi-socket, multi-core compute nodes; and (ii) I/O operations and the idle power consumption caused by waiting for these operations to complete. Experimental results show that this model, unlike the traditional simpler model, has high accuracy with respect to real-world workflow executions on production platforms.

Given the inaccuracies of the traditional power consumption model, it is unclear whether previously proposed energy-aware workflow scheduling algorithms that use this model can be effective in practice. Published evaluations show good results, but these results report on power consumption computed based on that very same, inaccurate, model. One of the goals of this work is to assess the effectiveness on these algorithms in practice. To this end, we evaluate two popular (i.e., most cited according to Google Scholar) energy-aware workflow scheduling algorithms for cloud platforms [11,22]. We perform this evaluation both when assuming the traditional power consumption model and when assuming the accurate model in [15,17]. We then propose a simple I/O-aware workflow scheduling algorithm that uses the accurate model, and compare this algorithm to the algorithms in [11,22]. Specifically, this work makes the following contributions:

1. An evaluation of two popular energy-aware workflow scheduling algorithms [11,22] when used for executing I/O-intensive workflow applications, using the traditional power model in the literature;

2. An analysis of how the results from the above evaluation differ when the energy consumption of the schedules produced by the two algorithms is estimated using the accurate power model proposed in [15,17].
3. A simple I/O- and energy-aware workflow scheduling algorithm that leverages the accurate power consumption model proposed in [15,17] to judiciously allocate threads to cores in multi-socket, multi-core platforms and to reduce the power consumption due to I/O operations.
4. An evaluation of this algorithm, which shows that it achieves a preferable compromise between energy consumption and workflow execution time when compared to the two algorithms in [11,22].

2   Background

2.1   Scientific Workflows

Scientific workflows are a cornerstone of modern scientific computing, and they have underpinned some of the most significant discoveries of the last decade [4,6]. They are used to describe complex computational applications that require efficient and robust management of large volumes of data, which are typically stored/processed on heterogeneous, distributed resources. In many cases, a workflow can be described as a directed acyclic graph, where the vertices represent tasks and the edges represent data or control dependencies. As workflows continue to be adopted by scientific projects and user communities, they are becoming more complex. Today's production workflows can be composed of millions of individual tasks that execute for milliseconds to hours, and that can be single-threaded programs, multi-threaded programs, tightly coupled parallel programs (e.g., MPI programs), or loosely coupled parallel programs (e.g., MapReduce jobs), all within a single workflow [14]. Many of these workflows are used to analyze terabyte-scale datasets obtained from streams, from files, or object stores. As a result, most workflows comprise I/O-intensive tasks, and many workflows are mostly composed of such tasks [3]. These are the workflows we specifically target in this work.

2.2   Power and Energy Consumption

In [15,17], we have investigated the impact of CPU utilization and disk I/O operations on the energy usage of I/O-intensive workflow executions on platforms that comprises multi-socket, multi-core compute nodes. In contrast to the traditional power consumption model used in the energy-aware workflow scheduling literature, we find that power consumption is impacted non-linearly by CPU utilization and depends on the way in which workflow tasks are allocated to cores and sockets. Our experimental results also show that I/O operations, as well as the idling due to waiting for these operations to complete, have significant impact on overall power consumption. Based on these results, we proposed a power consumption model for I/O-intensive workflows that accounts for the above phenomena. Experimental evaluation of this model showed that it accurately captures real-world behavior, whereas the traditional model used in the literature can be inaccurate by up to two orders of magnitudes. Below, we briefly describe both models, which are used for the experiments conducted in this work. (Note that neither model accounts for energy consumption due to RAM usage.)

Traditional Power and Energy Model. Energy-aware workflow scheduling studies [5,7,10,12,13,21,22] typically assume that the dynamic power (i.e., not accounting for the host's idle power consumption) consumed by the execution of a task at time t, $P(t)$, is linearly related to the task's CPU utilization, $u(t)$, as:

$$P(t) = (P_{\max} - P_{\min}) \cdot u(t) \cdot \tfrac{1}{n}, \tag{1}$$

where $P_{\max}$ is the power consumption when the compute node is at its maximum utilization, $P_{\min}$ is the idle power consumption (i.e., when there is no or only background activity), and n is the number of cores. Note that $u(t)$ can be computed by benchmarking the task on a dedicated compute node.

Given this power consumption model, the energy consumption of a task, E, is modeled as follows:

$$E = r \cdot P_{\min} + \int_0^r P(t)dt, \tag{2}$$

where r denotes the task's execution time.

I/O-Aware Power Consumption Model. The model proposed in [15,17] models $P(t)$, the power consumption of a compute node at time t, as:

$$P(t) = P_{\text{CPU}}(t) + P_{\text{I/O}}(t), \tag{3}$$

where $P_{\text{CPU}}(t)$, resp. $P_{\text{I/O}}(t)$, is the power consumption due to CPU utilization, resp. I/O operations.

Let s denote the number of sockets on the compute node, and n the number of cores per socket, so that the total number of cores on the compute node is $s \cdot n$. Let K denote the set of tasks that use at least one core on the compute node. $P_{\text{CPU}}(t)$ is then defined as follows:

$$P_{\text{CPU}}(t) = \sum_{k,i,j} P_{\text{CPU}}(k,i,j,t), \tag{4}$$

where $P_{\text{CPU}}(k,i,j,t)$ is the power consumption of CPU utilization at time t due to the execution of task k ($k \in K$) on socket i ($0 \le i < s$) at core j ($0 \le j < n$) on the compute node. Experiments on real-world systems show that power consumption does not linearly increase as cores on sockets are allocated to workflow tasks, and that the behavior depends on the scheme used to allocate each additional core on a socket. We consider two such schemes: (i) *unpaired* – cores are allocated to tasks in sequence on a single socket until all cores on that socket are allocated, and then cores on the next socket are allocated in sequence; and (ii) *pairwise* – cores are allocated to tasks in round-robin fashion across sockets (i.e., each core is allocated on a different socket than the previously allocated core). Both core allocation schemes can be supported by configuring the hardware/software infrastructure accordingly. Based on experimental results, the following model for $P_{\text{CPU}}(k, i, j, t)$ is derived:

$$
P_{\text{CPU}}(k,i,j,t) = \begin{cases} (P_{\max} - P_{\min}) \cdot \frac{u(t)}{s \cdot n} & \text{if } j = 0 \text{ (first core on a socket)} \\ 0.881 \cdot P_{\text{CPU}}(k,i,j-1,t) & \text{if } j > 0 \text{ and } pairwise \\ 0.900 \cdot P_{\text{CPU}}(k,i,j-1,t) & \text{if } j > 0 \text{ and } unpaired \end{cases} \tag{5}
$$

where $u(t)$ is the task's CPU utilization at time t. The model is written recursively as the power consumption due to allocating a task to a core on a socket depends on the power consumption due to previously allocated cores on that socket. The 0.881 and 0.900 coefficients above are obtained from linear regressions based on measurements obtained on real-world platforms [15, 17].

Similarly to the definition of P_{CPU}, we have:

$$
P_{\text{I/O}}(t) = \sum_{k,i,j} P_{\text{I/O}}(k, i, j, t), \tag{6}
$$

where $P_{\text{I/O}}(k, i, j, t)$ is the power consumption of I/O operations at time t due to the execution of task k $(k \in K)$ on socket i $(0 \leq i < s)$ at core j $(0 \leq j < n)$ on the compute node. $P_{\text{I/O}}(k, i, j, t)$ is modeled as follows:

$$
P_{\text{I/O}}(k,i,j,t) = \begin{cases} 0.486 \cdot (1 + 0.317 \cdot \omega(t)) \cdot P_{\text{CPU}}(k,i,j,t) & \text{if } pairwise \\ 0.213 \cdot (1 + 0.317 \cdot \omega(t)) \cdot P_{\text{CPU}}(k,i,j,t) & \text{otherwise} \end{cases} \tag{7}
$$

where the 0.486 and 0.213 values above come from linear regressions [15, 17], and $\omega(t)$ is 0 if I/O resources are not saturated at time t, or 1 if they are (i.e., idle time due to IOWait). More precisely, $\omega(t)$ is equal to 1 whenever the volume of I/O requests placed by concurrently running tasks exceeds some platform-dependent maximum I/O throughput. In Eq. 7, $\omega(t)$ is weighted by an application-independent single factor (0.317).

A detailed description and evaluation of the above model is available in [15, 17]. In this work, we limit our analysis to the *unpaired* scheme, as it yields the lowest energy consumption of the two schemes. For simplicity, in the rest of this paper we denote the above model for the unpaired scheme as the **realistic** model (in contrast to the *traditional* model described earlier).

3   Analysis of Energy-Aware Workflow Scheduling Algorithms

In this section, we describe two widely-used energy-aware workflow scheduling algorithms that leverage the traditional power consumption model for making scheduling decisions described in the previous section. We then evaluate the energy consumption for schedules computed by these two algorithms using both the traditional and the realistic models. We do so by using a simulator that can simulate the power consumption of a workflow execution on a compute platform for either model. We perform these simulations based on real-world execution traces of three I/O-intensive workflow applications. The specific scheduling problem that these algorithms aim to solve is as follows.

Scheduling Problem Statement. Consider a workflow that consist of single-threaded tasks. This workflow must be executed on a cloud platform that comprises homogeneous, multi-core compute nodes. Initially, all compute nodes are powered off. A compute node can be powered on at any time. Virtual machine (VM) instances can be created at any time on a node that is powered on. Each VM instance is started for an integral number of hours. After this time expires, the VM is shutdown. A node is automatically powered off if it is not running any VM instance. The cores on a node are never oversubscribed (i.e., a node runs at most as many VM instances as it has cores). A VM runs a single workflow task at a time, which runs uninterrupted from its start until its completion. The metrics to minimize are the workflow execution time, or makespan, and the total energy consumption of the workflow execution.

3.1   SPSS-EB

The Static Provisioning-Static Scheduling under Energy and Budget Constraints (SPSS-EB) algorithm in [11] computes a static resource provisioning and task schedule at the onset of application execution. It considers tasks in topological order (i.e., respecting task dependencies). For each task to be scheduled, with earliest start time t (computed based on the completion times of its already scheduled parent tasks), the algorithm considers the three options below in sequence:

1. If possible, schedule the task to run at the earliest time $t' \geq t$ on a VM instance that is already scheduled to be running at time t' and that will be able to complete the task before this VM instance expires.
2. Otherwise, if possible, schedule a new VM instance to start at time t on a node that has already been scheduled to be powered on and will have at least one idle core at that time, and schedule the task on that instance at time t.
3. Otherwise schedule a new node to be powered on at time t, schedule a new VM instance to be started on that node at time t, and schedule the task to execute on that VM instance at time t.

For each option above, if multiple VM instances or nodes are possible, pick the one that will complete the task the earliest, breaking ties by picking the VM instance or node that will lead to the highest energy saving.

We refer the reader to [11] for a more detailed description of and pseudo-code for the algorithm. Note that the algorithm therein also considers the monetary cost of running VM instances (as charged by the cloud provider), which is used to break ties and also used to evaluate the goodness of the schedule. In this work, we ignore monetary cost and only consider energy consumption and execution time.

3.2 EnReal

Like SPSS-EB, the Energy-aware Resource Allocation (EnReal) algorithm [22] computes a static schedule by considering tasks in topological order. For each task to be scheduled, with earliest start time t (computed based on the completion times of its already scheduled parent tasks), the algorithm follows the following steps:

1. If possible, schedule the task to run at the earliest time $t' \geq t$ on a VM instance that is already scheduled to be running at time t' or shortly after time t' (defined by "time partitions" – computed based on the number of overlapping tasks) and that will be able to complete the task before this VM instance expires. Ties are broken by picking the VM instance that would consume the least amount of energy, and then by picking the VM instance that would lead to a more balanced allocation of tasks on compute nodes.
2. Otherwise, if possible, start a new VM at time $t' \geq t$ on a node that is already scheduled to be on at time t'. Ties are broken by picking the node with the highest number of already scheduled VM instances.
3. Otherwise, schedule a new node to be powered on at time t, schedule a new VM instance to be started on that node at time t, and schedule the task to execute on that VM instance at time t.

We refer the reader to [22] for a more detailed description of and pseudocode for the algorithm. Note that the algorithm therein also considers migration, which relocates a VM instance from a compute node to another. The objective is to save energy by co-locating computations so as to reduce the number of nodes that are powered-on. In this work, we ignore migration as the energy savings it provides for relatively short-running tasks is marginal [20].

3.3 Workflow Energy Consumption Analysis

The analysis presented in this work is based on the simulated execution of real-world execution traces of scientific workflow applications executed on the Chameleon Cloud [2] platform, an academic cloud testbed. These traces are distributed as part of the WfCommons project [18] and represent a number of different configurations, in which the number of tasks and their characteristics (e.g., input data size, number of I/O operations, flops) vary. Therefore, we argue that these traces form a representative set of small- and large-scale workflow configurations. Specifically, we consider three I/O-intensive workflows:

1. *1000Genome* – A bioinformatics workflow that identifies mutational overlaps using data from the 1000 genomes project in order to provide a null distribution for rigorous statistical evaluation of potential disease-related mutations. We consider 15 instances of 1000Genome, with between 260 and 902 tasks.
2. *Montage* – An astronomy workflow for generating custom mosaics of the sky. The workflow re-projects images to correct orientation, and rectifying a common flux scale and background level. We consider 9 instances of Montage, with between 59 and 2122 tasks.

3. *SoyKB* – A bioinformatics workflow that resequences soybean germplasm lines selected for desirable traits such as oil, protein, soybean cyst nematode resistance, stress resistance, and root system architecture. We consider 9 instances of SoyKB, with between 96 and 676 tasks.

Simulator. We have developed a simulator for our experimental evaluation and validation purposes. The simulator is based on WRENCH [1], a framework for implementing simulators of workflow management systems, with the goal of producing simulators that are accurate and can run scalably on a single computer while requiring minimal software development effort. In [1], we have demonstrated that WRENCH achieves these objectives, and provides high simulation accuracy for workflow executions using state-of-the-art workflow systems. To ensure accurate and coherent results, the simulations conducted here use the same platform description as for the evaluation of the power model developed in our previous work [15,17]: 4 compute nodes each with 2 hexacore processors. The simulator code and experimental scenarios used in the rest of this paper are all publicly available online [19].

Evaluation Results. Figure 1-*top* shows the simulated energy consumption of the schedules computed by the SPSS-EB and EnReal algorithms, as computed with both the *traditional* and the *realistic* models, for all Montage, SoyKB, and 1000Genome workflow application instances. Recall that both algorithms make scheduling decisions assuming that the traditional model holds in practice. Energy consumption does not necessarily increase monotonically with the number of workflow tasks due to irregular workflows structures. Comparing the "SPSS-EB/traditional" to the "EnReal/traditional" results would thus correspond to comparisons traditionally done in the literature. Instead, comparing the "SPSS-EB/realistic" to the "EnReal/realistic" results corresponds to a realistic comparison. We can see that, in some cases, results vary significantly. For instance, for the 364-task 1000Genome workflow, the traditional comparison gives a clear advantage to SPSS-EB, while the realistic comparison gives a larger advantage to EnReal. In total, such "reversals" are observed for 6 of the 9 Montage executions, none of the SoyKB executions, and 6 of the 15 1000Genome executions. When no reversals occur, the magnitude of the advantage of one algorithm over the other can be largely overestimated when assuming that the traditional model holds. For instance, consider the 1000Genome execution with 820 and 920 tasks. A traditional comparison would indicate that EnReal consumes marginally less energy than SPSS-EB, while a realistic comparison shows that, in fact, SPSS-EB consumes about twice as much energy as EnReal. Overall, the traditional model can lead to misleading results. We conclude that published results evaluating these and other energy-aware workflow scheduling algorithms do not allow for an accurate quantitative comparison of how algorithms would perform in practice, and in particular for I/O-intensive workflows.

The results in Fig. 1-*top* show smaller discrepancies between the traditional and the realistic models for EnReal than for SPSS-EB. We term the absolute difference between the energy consumption computed based on the two models

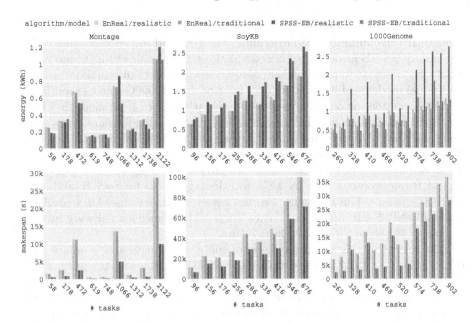

Fig. 1. Simulated workflow energy consumption (top) and makespan (bottom) for each workflow application instance for the SPSS-EB and EnReal algorithms. Results are shown when simulating energy consumption with the traditional power model and the realistic power model.

Table 1. Energy consumption error (maximum, mean, and standard deviation) for both algorithms and for each set of workflow instances.

Workflow	Algorithm	Energy error (KWh)					
		Max		Mean		Stand. deviation	
Montage	SPSS-EB	0.45	(160.71%)	0.09	(39.31%)	0.17	(52.07%)
	EnReal	0.22	(17.01%)	0.06	(7.81%)	0.01	(1.89%)
SoyKB	SPSS-EB	0.66	(33.76%)	0.13	(5.37%)	0.37	(15.72%)
	EnReal	0.12	(10.46%)	0.11	(9.17%)	0.02	(1.65%)
1000Genome	SPSS-EB	5.88	(360.57%)	1.95	(187.46%)	1.80	(120.50%)
	EnReal	0.35	(26.29%)	0.11	(14.65%)	0.10	(7.08%)

the "error". Table 1 summarizes the results in Fig. 1 and shows the maximum error, the mean error, and the standard deviation of the error for both algorithms computed for all workflow instances for each application. The maximum error is up to three orders of magnitude for SPSS-EB, but only up to half an order of magnitude for EnReal. The mean error for SPSS-EB can also be much larger than that of EnReal, especially for the 1000Genome workflows. One may wonder how come, for some workflows, EnReal results are much less sensitive to the choice of the power consumption model. We analyzed the schedules computed by EnReal. When multiple options are possible for scheduling a task,

EnReal balances the distribution of tasks among cores and computing nodes. As a result, it "involuntarily" also distributes the I/O load, which saves energy. By contrast, SPSS-EB favors early task completions, thus leading to I/O contention, which can translate to much higher energy consumption than would occur if the traditional model held in practice (e.g., for 8 of the 15 1000Genome executions).

When considering only the results obtained with the realistic model in Fig. 1-*top*, there are significant reductions in energy consumption for most cases when using the EnReal algorithm relative to using the SPSS-EB algorithm. However, these energy savings come at the cost of higher makespans. Makespan results are shown in Fig. 1-*bottom*. EnReal consistently leads to higher makespans than SPSS-EB (on average higher by 4.35x for Montage, 1.71x for SoyKB, and 3.81x for 1000Genome). This is because EnReal fosters resource re-use. More precisely, and unlike SPSS-EB, it does not create a new VM instance if an already running instance will become idle in the near future.

The results in this section make it possible to evaluate and compare algorithms using our realistic power model, but these algorithms make their scheduling decisions based on the traditional model. In practice, they can make very suboptimal decisions, such as execute I/O-intensive tasks concurrently on the same compute node in an attempt to save energy. Such decisions can be particularly harmful for the overall energy consumption since the time waiting for I/O operations to complete, as seen in [15,17], can significantly increase idle power consumption. In the next section, we investigate whether is possible to design a simple algorithm that makes good decisions based on the realistic model.

4   Energy-Aware Scheduling of I/O-Intensive Workflows

In this section, we present an energy-aware workflow scheduling algorithm that accounts for the energy cost of I/O by using the power consumption model described in Sect. 2.2 as a basis for making scheduling decisions. Our goal is to show that it is possible, and in fact straightforward, to design an algorithm that compares favorably to previously proposed algorithms that rely on the traditional power model.

4.1   I/O- and Energy-Aware Scheduling

We propose IOBalance, an I/O- and energy-aware workflow scheduling algorithm that aims at minimizing energy consumption of I/O-intensive workflows by reducing I/O contention and data movement operations. Contention is lessened by distributing tasks that perform high number of I/O operations among available (running) nodes; data movement reduction is achieved by assigning tasks to nodes in which most of the tasks' input data are available (i.e., has been produced by a previous task).

Like SPSS-EB and EnReal, IOBalance computes a static schedule, deciding when to power hosts on and when to start VM instances. Tasks are scheduled in topological order, marking a task ready whenever all its parent tasks have been scheduled (initially all entry tasks are marked ready). Among all ready tasks to be scheduled, the algorithm first schedules the task with the highest volume of I/O operations, breaking ties based on the task's earliest start time (i.e., picking the task that can be started the earliest). Given a task to be scheduled with earliest start time t, IOBalance considers three options below in sequence:

1. If possible, schedule the task on a VM instance that is already scheduled to be running at time t or later, and that can complete the task before expiration. For each such VM instance the algorithm determines: (i) the energy cost of the task's execution (computed using the power consumption model in Eq. 3); and (ii) the earliest time at which the task could start on this VM instance. The algorithm picks the VM instance with the lowest energy cost. If multiple instances have the same energy cost, then it picks the one that can start the task the earliest.
2. Otherwise, if possible, schedule a new VM instance to start at time t on a host that has already been scheduled to be powered on and will have at least one idle core at that time, and schedule the task on that instance at time t. If multiple such hosts exist, pick the host that already stores the largest amount of data needed by the task (so as to reduce data movements). VM instances on the same host are allocated to cores in round-robin fashion.
3. Otherwise, schedule a new host to be powered on at time t, schedule a new VM instance to be started on that host at time t, and schedule the task to execute on that VM instance at time t. If multiple such hosts exist, pick the host that already stores the largest amount of data needed by the task.

4.2   Experimental Evaluation

To evaluate the effectiveness of our algorithm and compare it to SPSS-EB and EnReal, we implemented it in the simulator used for the experiments in Sect. 3.3. Hereafter, we only show results for the realistic power consumption model. That is, the power consumed by the execution of the workflow on the compute platform is simulated based on the realistic model. However, recall that SPSS-EB and EnReal make scheduling decisions assuming the traditional model.

Figure 2-*top* shows the simulated energy consumption of the schedules computed by the SPSS-EB, EnReal, and IOBalance algorithms, for all Montage, SoyKB, and 1000Genome workflow application instances. Overall, IOBalance saves significant energy when compared to SPSS-EB for all Montage and SoyKB workflow instances. Energy savings are up to 53% and on average 32% for Montage, up to 44% and on average 18% for SoyKB, and up to 64% and on average 36% for 1000Genome. When compared to EnReal, our proposed algorithm leads to schedules that consume more energy for most workflow instances. Specifically, consumed energy is higher than that of EnReal by up to 52% and on average 30% for SoyKB workflows, and up to 48% and on average 18% for 1000Genome workflows. For Montage workflows, however, IOBalance leads to lower energy consumption than EnReal for 6 of the 9 workflow instances, by up to 47%.

These energy results must be put in perspective with the makespan results shown in Fig. 2-*bottom*. For 30 of the 33 workflow instances IOBalance leads to

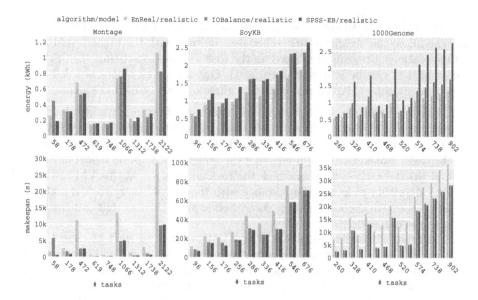

Fig. 2. Simulated workflow energy consumption (top) and makespan (bottom) for each workflow application instance for the SPSS-EB, EnReal, and IOBalance algorithms. Results are shown when simulating energy consumption with the realistic power model.

makespan that is within 5% of the makespan achieved by SPSS-EB (the exceptions are the 58- and 178-task Montage workflows, and the 176-task SoyKB workflow). IOBalance schedules tasks in a way that reduces potential I/O contention, thus saving on energy, while rarely increasing the makespan when compared to SPSS-EB. For instance, the 1000Genome workflow is composed of a large number of I/O-intensive tasks that run for a couple of minutes and process high volumes of data (O(GB)) [16]. These tasks, if all scheduled on the same node (or a small number of nodes), suffer from local and remote I/O contention, which increases power consumption as defined by $\omega(t)$ in $P_{\mathrm{I/O}}$ (Eq. 7). The distribution of I/O-intensive tasks among cores/nodes counterbalanced by CPU-bound tasks prevents waiting for I/O operations to complete, which saves on energy by avoiding idle power consumption, but allows CPU-intensive tasks to benefit from the idle CPU cycles, which reduces makespan. IOBalance leads to makespan shorter, by up to 333% and on average by 94%, than EnReal for all workflow instances. So although EnReal schedules consume less energy, as noted when comparing SPSS-EB and EnReal in Sect. 3.3, this energy saving comes at the cost of 2x longer makespan on average.

We conclude that even a simple algorithm like IOBalance can improve on the state of the art because it makes scheduling decisions based on the realistic power consumption model. SPSS-EB and EnReal achieve different compromises between energy and makespan, with SPSS-EB consuming more energy but leading to shorter makespans and EnReal consuming less energy but leading to much longer makespans. IOBalance achieves a compromise that is strictly preferable to that achieved by SPSS-EB, saving significant energy while achieving similar makespans. For most workflow instances it leads to higher energy consumption than EnReal but achieves much shorter makespans.

5   Related Work

The need to manage energy consumption in large cloud data-centers has received significant attention in this past decade [9]. At the application and system levels, researchers have investigated techniques and algorithms to enable energy-efficient executions. In the scientific workflows literature, a range of energy-aware workflow task scheduling or resource provisioning algorithms [5,7,10,12,13,21] have been proposed. In [7], only dynamic energy consumption is considered and an algorithm composed of five sub-algorithms (VM selection, sequence tasks merging, parallel tasks merging, VM reuse, and task slacking algorithm) is proposed. Similarly, [12] also only considers dynamic energy consumption, but uses reinforcement learning for a more budget oriented analysis. Other works consider both static and dynamic energy consumption [5,10,13,21]. All these works make the strong assumption that power consumption is linearly correlated with CPU utilization and equally divided among virtual machines (or CPU cores within a compute node), and ignore power consumption due to I/O. The work in [15,17] shows that, at least for I/O-intensive workflow executions, these assumptions do not hold in practice, and proposes an accurate, if more complex, power consumption model. We use this model in this work.

Dynamic Voltage and Frequency Scaling (DVFS) is a well-known power management technique [7,10,12,21]. It is used to decrease processor frequency to save energy and is often paired with slack time optimization, a method that takes advantage of idle slots resulting from early-completing tasks. Algorithms that implement this combined approach generally succeed in reducing power consumption (see the comparison to EnReal in [5]). This reduction comes at a sharp increase in workflow makespan. The algorithms considered in this work (the two algorithms in [11,22] and the algorithm proposed in Sect. 4.1) do not use DVFS, but could conceivably be augmented to do so.

6   Conclusion and Future Work

In this work, we evaluate two popular energy-aware workflow scheduling algorithms when used for executing I/O-intensive workflow applications. We quantify the energy consumption of the schedules they compute using both the traditional, but inaccurate, power consumption model used in the literature and the accurate power model developed in [15,17]. We show that comparing these algorithms under the traditional model leads to misleading results. Furthermore, as both algorithms make scheduling decisions based on the inaccurate traditional model, it is unclear how effective they can be in practice. For this reason, we propose a simple I/O-aware workflow scheduling algorithm that uses the accurate power consumption model, and compare this algorithm to the above two algorithms. Experimental results show that this algorithm achieves a strictly preferable tradeoff between makespan and energy than one of its two competitors. Although it often leads to higher energy consumption than the other competitor, it also achieves a 2x shorter makespan on average. In future work, we plan to broaden our analysis to consider the use of DVFS.

Acknowledgments. This work is funded by NSF contracts #1923539, #1923621, #2016610, and #2016619. Results presented in this paper were obtained using the Chameleon testbed supported by the National Science Foundation.

References

1. Casanova, H., et al.: Developing accurate and scalable simulators of production workflow management systems with wrench. Future Gener. Comp. Sy. **112**, 162–175 (2020)
2. Chameleon cloud. https://chameleoncloud.org (2021)
3. De Oliveira, D.C., et al.: Data-intensive workflow management: for clouds and data-intensive and scalable computing environments. Synth. Lect. Data Manage. **14**(4), 1–179 (2019)
4. Deelman, E., et al.: Pegasus, a workflow management system for science automation. Futur. Gener. Comput. Syst. **46**, 17–35 (2015)
5. Ghose, M., et al.: Energy efficient scheduling of scientific workflows in cloud environment. In: IEEE HPCC (2017)
6. Klimentov, A., et al.: Next generation workload management system for big data on heterogeneous distributed computing. J. Phys: Conf. Ser. **608**, 012040 (2015)
7. Li, Z., et al.: Cost and energy aware scheduling algorithm for scientific workflows with deadline constraint in clouds. IEEE Trans. Serv. Comput. **11**(4), 713–726 (2018)
8. Ma, X., et al.: An IoT-based task scheduling optimization scheme considering the deadline and cost-aware scientific workflow for cloud computing. EURASIP J. Wirel. Commun. Netw. **2019**(1), 1–19 (2019)
9. Orgerie, A.C., et al.: A survey on techniques for improving the energy efficiency of large-scale distributed systems. ACM Comput. Surv. (CSUR) **46**(4), 1–31 (2014)
10. Pietri, I., Sakellariou, R.: Energy-aware workflow scheduling using frequency scaling. In: International Conference on Parallel Processing Workshops (2014)
11. Pietri, I., et al.: Energy-constrained provisioning for scientific workflow ensembles. In: International Conference on Cloud and Green Computing (CGC) (2013)
12. Qin, Y., Wang, H., Yi, S., Li, X., Zhai, L.: An energy-aware scheduling algorithm for budget-constrained scientific workflows based on multi-objective reinforcement learning. J. Supercomput. **76**(1), 455–480 (2019). https://doi.org/10.1007/s11227-019-03033-y
13. Shepherd, D., et al.: Workflow scheduling on power constrained VMs. In: IEEE/ACM 8th International Conference on Utility and Cloud Computing (2015)
14. Ferreira da Silva, R., et al.: A characterization of workflow management systems for extreme-scale applications. Future Gener. Comput. Syst. **75**, 228–238 (2017)
15. Ferreira da Silva, R., Orgerie, A.-C., Casanova, H., Tanaka, R., Deelman, E., Suter, F.: Accurately simulating energy consumption of I/O-intensive scientific workflows. In: Rodrigues, J., et al. (eds.) ICCS 2019. LNCS, vol. 11536, pp. 138–152. Springer, Cham (2019). https://doi.org/10.1007/978-3-030-22734-0_11

16. Ferreira da Silva, R., et al.: Using simple pid-inspired controllers for online resilient resource management of distributed scientific workflows. Future Gener. Comp. Sy. 95 (2019)
17. Ferreira da Silva, R., et al.: Characterizing, modeling, and accurately simulating power and energy consumption of i/o-intensive scientific workflows. Journal of Computational Science 44, 101157 (2020)
18. Ferreira da Silva, R., et al.: Workflowhub: Community framework for enabling scientific workflow research and development. In: IEEE WORKS Workshop (2020)
19. Energy-aware simulator. https://github.com/wrench-project/energy-aware-simulator (2021)
20. Wang, X., et al.: Delay-cost tradeoff for virtual machine migration in cloud data centers. J. Netw. Comput. Appl. **78**, 62–72 (2017)
21. Wu, T., et al.: Soft error-aware energy-efficient task scheduling for workflow applications in DVFS-enabled cloud. J. Syst. Architect. **84**, 12–27 (2018)
22. Xu, X., et al.: EnReal: an energy-aware resource allocation method for scientific workflow executions in cloud environment. IEEE Trans. Cloud Comput. **4**(2), 166–179 (2015)

A Job Shop Scheduling Problem with Due Dates Under Conditions of Uncertainty

Wojciech Bożejko$^{1(\boxtimes)}$ ⓘ, Paweł Rajba^2 ⓘ, Mariusz Uchroński^1 ⓘ, and Mieczysław Wodecki^3 ⓘ

^{1} Department of Automatics, Mechatronics and Control Systems, Wrocław University of Technology, Janiszewskiego 11-17, 50-372 Wrocław, Poland
{wojciech.bozejko,mariusz.uchronski}@pwr.edu.pl
^{2} Institute of Computer Science, University of Wrocław, Joliot-Curie 15, 50-383 Wrocław, Poland
pawel@cs.uni.wroc.pl
^{3} Department of Telecommunications and Teleinformatics, Wrocław University of Science and Technology, Wybrzeże Wyspiańskiego 27, 50-370 Wrocław, Poland
mieczyslaw.wodecki@pwr.edu.pl

Abstract. In the work we consider a job shop problem with due dates under conditions of uncertainty. Uncertainty is considered for operation execution times and job completion dates. It is modeled by normal and Erlang random variables. We present algorithms whose constructions are based on the tabu search method. Due to the application of the probabilistic model, it was possible to obtain solutions more resistant to data disturbances than in the classical approach.

Keywords: Scheduling · Job shop · Weighted tardiness · Uncertainty

1 Introduction

The paper deals with the job-shop problem, widely regarded as one of the most difficult combinatorial problems. Due to many of its practical applications, it has been intensively researched for many years. There are few works that dealt with the deterministic job shop problem with additional parameters and constraints. One of the few is the work of Balas et al. [1], which deals with the problem of the job shop with time windows, penalties for untimely completion of jobs and machine setups. A review of different variants of multi-machine problems with machine setups is contained in Sharma and Janin [8]. There are definitely fewer works devoted to the job shop problem with uncertain parameters. Kim et al. [5] presented a modification of classical construction algorithms solving deterministic scheduling problems. Shoval and Efatmaneshnik [9] consider the problem in which jobs execution times are random variables with a normal distribution. The method of determining the set of feasible schedules is presented, as well as the criteria for selecting an appropriate sub-optimal schedule from this set.

© Springer Nature Switzerland AG 2021
M. Paszynski et al. (Eds.): ICCS 2021, LNCS 12742, pp. 198–205, 2021.
https://doi.org/10.1007/978-3-030-77961-0_17

There are several practical reasons for introducing due dates. First, there might be dependencies to other departments or contracted delivery dates as a part of the over supply chain processes. The other reason is that there might be time window when machines are not available e.g., because of some maintenance activities. In this paper we consider a job shop problem with due dates for jobs completion and penalties for tardy jobs, applying total weighted tardiness criterion. Uncertain parameters, i.e., processing times and due dates are random variables with the normal or Erlang distributions. Normal distribution is the natural choice when we expect a standard behavior of some phenomenon, obviously we can also leverage very useful mathematical properties in the modeling transformations. Erlang distribution was investigated as, for set of parameters, it introduces a possibility to model events which diverge from the standard behavior. This distribution also offers some useful mathematical properties which have been applied in the modeling activities.

2 Problem Formulation

The literature presents various ways of defining the job shop problem, nevertheless in this paper we will use the definitions and symbols from [4]. Let $\mathcal{J} = \{1, 2, \ldots, n\}$ be a set of jobs, $\mathcal{M} = \{1, 2, \ldots, m\}$ - a set of machines. By $\mathcal{O} = \{1, 2, \ldots, o\}$ we denote a set of operations, where an operation is equal to an action of job execution on a machine in a *machine order*. Therefore, we can partition the set of $\mathcal{O}$ operations into sequences corresponding to individual jobs, i.e., define the j job as sequence of o_j operation, which will be indexed sequentially with the numbers $(l_{j-1} + 1, \ldots, l_j)$. The operations should be performed in a given order (*technological order*), where $l_j = \sum_{i=0}^{j} o_i$ ($j \in \mathcal{J}$, $l_0 = 0$, $\sum_{i=1}^{n} o_i = o$). The operation $k \in \mathcal{O}$ should be performed on the $\mu_k \in \mathcal{M}$ machine in time (duration) $p_k > 0$. For each job j there is a desired *completion date* d_j and designated a *penalty factor* w_j for exceeding it. The considered variant of the problem consists in determining the moments of starting the execution of jobs on machines so that the technological order is maintained, and minimize the sum of penalties for untimely execution of jobs (*tardiness*).

Over the years of research on the job shop problem, there have emerged many ways of modeling it. In the following part, we will present one of them, the so-called disjunctive graph. A disjunctive graph with weights in vertices can be defined as a pair of sets $G = (V, K \cup D \cup R)$ [4], defined as follows:

1. The set of the vertices of the graph $V = \{1, 2, \ldots, o\} \cup \{s, t\}$, where the vertex i ($i = 1, 2, \ldots, n$) corresponds to the i-th operation. The weight of a vertex is equal to the execution time of the operation it represents. The remaining (extra) vertices s and t have a weight zero.
2. The set of arcs is the union of the sets:
   (i) *conjunctive arcs* K between successive operations within the job (representing the technological sequence): $K = \bigcup_{j=1}^{n} \bigcup_{i=l_{j-1}+1}^{l_j - 1} \{(i, i + 1)\}$,
   (ii) *disjunctive arcs* D between operations performed on the same machine: $D = \bigcup_{i,j \in \mathcal{O}, i \neq j, \nu_i = \nu_j} \{(i, j), (j, i)\}$,

(iii) arcs R, in the form of $(s, l_{j-1}+1)$ and (l_j, t), where $l_{j-1}+1$ is the first and l_j the last operation of the job $j \in \mathcal{J}$.

The subset $D' \subset D$ containing exactly one arc from each pair of arcs is called a *representation* of disjunctive arcs. It can be easily shown that the subgraph $G' = (V, K \cup D' \cup R)$ of the disjunctive graph G represents an feasible solution if and only if G' does not contain cycles.

For a determined feasible representation of D', the time of completion of i-th job C_i should be calculated as the length of the longest path in the graph $G' = (V, K \cup D' \cup R)$ from the vertex s to l_i (l_i – the last operation of job i). Then $T_i = \max\{0, C_i - d_i\}$ is the job's delay (*tardiness*), and $\mathcal{F} = \sum_{i=1}^{n} w_i T_i$ the sum of the costs of tardiness (*penalties*).

3   Uncertainty

We will consider a job shop problem with random parameters, where the job completion times C_i and the due dates d_i are expressed as random variables $\tilde{C}_i$ and $\tilde{d}_i$ with normal or Erlang distributions. More specifically, we will consider the following four cases: (**A**) $\tilde{C}_i \sim N(C_i, \alpha \cdot C_i)$, (**B**) $\tilde{d}_i \sim N(d_i, \beta \cdot d_i)$, (**C**) $\tilde{C}_i \sim \mathcal{E}(C_i, 1)$, (**D**) $\tilde{d}_i \sim \mathcal{E}(d_i, 1)$, where the α, β parameters will be specified at the stage of computational experiments. Additionally, we define a random variable $\tilde{T}_i$ representing the size of delay (tardiness) of i-th job. After randomization, to compare solutions (just like at paper [2]), we use the following comparative criteria: $\mathcal{W}_E = \sum_{i=1}^{n} w_i E(\tilde{T}_i)$, $\mathcal{W}_{ED} = \sum_{i=1}^{n} w_i (E(\tilde{T}_i) + \theta \cdot D^2(\tilde{T}_i))$, where $E(\tilde{T}_i)$ is an expected value of $\tilde{T}_i$, $D^2(\tilde{T}_i)$ is an variation and a parameter $0 < \theta < 1$. Criteria similar to $\mathcal{W}_E$ and $\mathcal{W}_{ED}$ are considered in papers [2,3,7].

Normal Distribution. Here we will consider the following two cases: (**A**) $\tilde{C}_i \sim N(C_i, \alpha \cdot C_i)$ and (**B**) $\tilde{d}_i \sim N(d_i, \beta \cdot d_i)$. If X is a random variable, then by F_X and f_X we will denote its cumulative distribution function and density.

   Case A ($\tilde{C}_i \sim N(C_i, \alpha \cdot C_i)$). Then, tardiness is a random variable defined as $\tilde{T}_i = \max\{0, \tilde{C}_i - d_i\}$. For shortening of the notation, let $\mu_i = (d_i - C_i)/\sigma_i$.

Theorem 1. *If the job completion times are independent random variables with a normal distribution* $\tilde{C}_i \sim N(C_i, \sigma_i)$ *($\sigma_i = \alpha \cdot C_i$, then the expected value*

$$E(\tilde{T}_i) = (1 - F_{\tilde{C}_i}(d_i)) \left(\frac{\sigma_i}{\sqrt{2\pi}} e^{\frac{-(\mu_i)^2}{2}} + (C_i - d_i) \left(1 - F_{N(0,1)}(\mu_i)\right) \right).$$

For calculating the value of the criterion function $\mathcal{W}_{ED}$ it is necessary to know the standard deviation of the random variable $\tilde{T}_i$. Because $D^2(\tilde{T}_i) = E(\tilde{T}_i^2) = (E(\tilde{T}_i))^2$, we will present a method for determining the random variable $\tilde{T}_i^2$.

Theorem 2. *If the times of execution of jobs are independent random variables with a* $\tilde{C}_i \sim N(C_i, \sigma_i)$ *($\sigma_i = \alpha \cdot C_i$, $i \in \mathcal{J}$), then the expected value of the random* *variable $\tilde{T}_i^{\,2}$ constituting the square of delay:*

$$E(\tilde{T}_i^{\,2}) = \left(1 - F_{N(0,1)}\left(\nu_i\right)\right)\left(\frac{(C_i - d_i)\sigma_i}{\sqrt{2\pi}}e^{\frac{-(\mu_i)^2}{2}}\right.$$
$$\left. + (C_i^2 + \sigma_i^2 + d_i^2 - 2d_i\sigma_i)\left(1 - F_{N(0,1)}\left(\mu_i\right)\right)\right).$$

Both proven theorems allow quick calculation of the first two central moments for random dates of jobs' completion.

Case B ($\tilde{d}_i \sim N(d_i, \beta \cdot d_i)$). Tardiness of job $i \in \mathcal{J}$ is $\tilde{T}_i = \max\{0, C_i - \tilde{d}_i\}$. We will now present two theorems that allow us to calculate the value of $\mathcal{W}_E$ and $\mathcal{W}_{ED}$. We omit the proofs of theorems because they are similar to the proofs of Theorems 1 and 2.

Theorem 3. *If the requested due dates for completing jobs are independent random variables with a normal distribution $\tilde{d}_i \sim N(d_i, \sigma_i)$ ($\sigma_i = \beta \cdot d_i$, $i \in \mathcal{J}$), then the expected tardiness value*

$$E\left(\tilde{T}_i\right) = F_{N(0,1)}\left(-\mu_i\right)\left(C_i F_{N(0,1)}\left(-\mu_i\right) + \frac{\sigma_i}{\sqrt{2\pi}}e^{-(-\mu_i)^2}2 - d_i F_{N(0,1)}\left(-\mu_i\right)\right).$$

Theorem 4. *If the latest due dates for completing jobs are normally distributed independent random variables $\tilde{d}_i \sim N(d_i, \sigma_i)$ ($\sigma_i = \beta \cdot d_i$, $i \in \mathcal{J}$), then the expected value of the random variable $\tilde{T}_i^{\,2}$*

$$E(\tilde{T}_i^{\,2}) = F_{N(0,1)}\left(-\mu_i\right)\left((d_i^2 + \sigma_i^2 + C_i^2 - 2C_i d_i)F_{N(0,1)}\left(-\mu_i\right)\right.$$
$$\left. + \frac{\sigma_i(C_i - d_i)}{\sqrt{2\pi}}e^{\frac{-(\mu)^2}{2}}\right).$$

Erlang Distribution. It results directly from the definition of the Erlang distribution that the density function of the random variable $\tilde{C}_i \sim \mathcal{E}(\nu_i, \lambda_i)$

$$f_{\tilde{C}_i}(x) = \begin{cases} \frac{1}{(\nu_i - 1)!}\lambda_i^{\nu_i}x^{\nu_i - 1}e^{-\lambda_i x}, & \text{if } x > 0, \\ 0, & \text{if } x \le 0. \end{cases} \tag{1}$$

We will consider two cases: **(C)** $\tilde{C}_i \sim \mathcal{E}(C_i, 1)$ and **(D)** $\tilde{d}_i \sim \mathcal{E}(d_i, 1)$.

Case C ($\tilde{C}_i \sim \mathcal{E}(C_i, 1)$). In this case, the tardiness in the execution times $\tilde{T}_i = \max\{0, \tilde{C}_i - d_i\}$. Similarly, as in the case of normal distribution, we present two theorems claim that is used when calculating the value of the criterion comparing the solution.

Theorem 5. *If the job completion times $\tilde{C}_i$ are independent random variables with the Erlang distribution $\mathcal{E}(C_i, 1)$, then the expected value of the job tardiness*

$$E(\tilde{T}_i) = (1 - F_{\mathcal{E}(C_i,1)}(d_i))\left(C_i\left(1 - F_{\mathcal{E}(C_i+1,1)}(d_i)\right) - d_i\left(1 - F_{\mathcal{E}(C_i,1)}(d_i)\right)\right).$$

Theorem 6. *If the job completion times $\tilde{C}_i$ are independent random variables with the Erlang distribution $\mathcal{E}(C_i, 1)$, $i \in \mathcal{J}$, then the expected value*

$$E(\tilde{T}_i^{\,2}) = \left(1 - F_{\mathcal{E}(C_i,1)}(d_i)\right)\left(C_i(C_i+1)\left(1 - F_{\mathcal{E}(C_i+2,1)}(d_i)\right)\right.$$
$$\left. - 2d_iC_i\left(1 - F_{\mathcal{E}(C_i+1,1)}(d_i)\right) + d_i^2\left(1 - F_{\mathcal{E}(C_i,1)}(d_i)\right)\right).$$

In summary, when the job completion times arerandom variables with the Erlang distribution, then the values of the function $\mathcal{W}_E$ and $\mathcal{W}_{ED}$ are calculated using the Theorem 5 and 6.

Case D ($\tilde{d}_i \sim \mathcal{E}(d_i, 1)$). Similarly, as for case **C**, we determine the cumulative distribution function of the density of job tardiness $i \in \mathcal{J}$ given as $\tilde{T}_i = \max\{0, C_i - \tilde{d}_i\}$. Then, we can proceed to calculate the expected values of random variables $\tilde{T}_i$ and $\tilde{T}_i^{\,2}$.

Theorem 7. *If the deadlines are independent random variables with the Erlang distribution $\tilde{d}_i \sim \mathcal{E}(d_i, 1)$, $i = 1, 2, \ldots n$, then the expected value of tardiness $E(\tilde{T}_i) = F_{\mathcal{E}(\delta,1)}(C_i)(C_iF_{\mathcal{E}(\delta,d_i)}(C_i) - \frac{\delta}{d_i}F_{\mathcal{E}(\delta+1,1)}(C_i))$, where $\delta = d_1 + \ldots + d_i$.*

Theorem 8. *If the due dates for completing jobs are independent random variables with the Erlang distribution $\tilde{d}_i \sim \mathcal{E}(d_i, 1)$, then the expected value of the square of tardiness for $\delta = d_1 + \ldots + d_i$ is*

$$E(\tilde{T}_i^{\,2}) = F_{\mathcal{E}(\delta,1)}(C_i)\left((\delta+1)F_{\mathcal{E}(\delta+2,1)}(C_i) - 2C_i\delta F_{\mathcal{E}(\delta+1,1)}(C_i) + C_i^2F_{\mathcal{E}(\delta,1)}(C_i)\right).$$

Equalities (Theorem 7 and 8) will be used in $\mathcal{W}_E$ and $\mathcal{W}_{ED}$ calculation.

4   Computational Experiments

In order to carry out computational experiments, a simplified version of the tabu search algorithm for solving the job shop problem presented in the work [4] was adopted. Later in this section the deterministic algorithm will be denoted by $\mathcal{AD}$, whereas probabilistic one by $\mathcal{AP}$, where $\mathcal{AP}_E$ denoted probabilistic algorithm with criterion $\mathcal{W}_E$ and $\mathcal{AP}_{ED}$ – with $\mathcal{W}_{ED}$ criterion.

Test instances of 82 examples of deterministic data were taken from OR-Library [6]. Additionally, in accordance with the uniform distribution, the coefficients of the penalty function for the delay of the jobs w_i were drawn from the set $\{1, 2, \ldots, 10\}$. Similarly, the requested completion dates for the d_i jobs were drawn from the set $[P_i, (3/2)P_i]$, where $P_i = \sum_{j=1}^m p_{i,j}$, $i = 1, 2, \ldots, n$. The set of these (deterministic) data was marked with Ω. Let F be the solution value determined by the tested algorithm, and F^* the value of the reference solution. Relative error of the solution F given by $\delta = \frac{F - F^*}{F^*} \cdot 100\%$ indicates by how many percent the solution of the algorithm is worse/better than reference one.

Let $\tilde{\eta}$ be an instance of probabilistic data, $\mathcal{Z}(\tilde{\eta})$ data set generated from $\tilde{\eta}$ by disturbance of task execution times according to the assumed schedule. We have further: A_{ref} – algorithm designating reference solutions, A – algorithm whose resistance are tested (in our case $A_{\mathcal{P}}$ or $A_{\mathcal{P}+\mathcal{B}}$), $\pi_{(A,x)}$ – solution designated by the algorithm A for data x, $F(\pi_{(A,x)}, y)$ – value of criterion function of solution $\pi_{(A,x)}$ for the instance y. Then $\Delta(A, \tilde{\eta}, \mathcal{Z}(\tilde{\eta})) = \dfrac{\sum_{\varphi \in \mathcal{Z}(\tilde{\eta})} F(\pi_{A,\tilde{\eta}}, \varphi) - \sum_{\varphi \in \mathcal{Z}(\tilde{\eta})} F(\pi_{(A_{ref}, \varphi)}, \varphi)}{\sum_{\varphi \in \mathcal{Z}(\tilde{\eta})} F(\pi_{(A_{ref}, \varphi)}, \varphi)}$,

we call *resistance of solution* $\pi_{(A,\tilde{\eta})}$ (designated by the algorithm A for the instance $\tilde{\eta}$) on the set of distributed data $\mathcal{Z}(\tilde{\eta})$.

If $\tilde{\mathcal{P}}$ is a set of instances of probabilistic data of the examined problem, then the expression $\mathcal{S}(A, \tilde{\mathcal{P}}) = \frac{1}{|\tilde{\mathcal{P}}|} \sum_{\tilde{\eta} \in \tilde{\mathcal{P}}} \Delta(A, \tilde{\eta}, \mathcal{Z}(\tilde{\eta}))$ we call a *resistance coefficient* of the algorithm A on the set $\tilde{\mathcal{P}}$. For one deterministic instance, we generate three instances of probabilistic data from the set $\mathcal{D}$. In total, *probabilistic data set* $\tilde{\mathcal{P}}$ has 1125 instances.

The most important results of the computational experiments are presented in Tables 1 and 2. The $\mathcal{AD}$ column contains the results of the deterministic algorithm. Columns labeled $\mathcal{AP}_E$ and $\mathcal{AP}_{ED}$ contain the results of the probabilistic algorithm based on the expected value or the expected value and variance, respectively. In addition to the mean errors, additional columns were added with the percentage of cases for which the probabilistic algorithm gave results not worse than the deterministic algorithm (column %NG) and the corresponding percentage for which the algorithm gave better results (in tables this column is marked as %L).

Table 1. The robustness of the $\mathcal{AD}$ and $\mathcal{AP}$ algorithms expressed as a percentage for random times with **normal distribution.**

Random	α	$\mathcal{AD}$	$\mathcal{AP}_E$	%NG	%L	$\mathcal{AP}_{ED}$	%NG	%L
Durations	0.1	60.93	68.34	70	51	73.17	56	49
	0.2	69.91	78.58	66	62	88.81	49	44
	0.3	71.13	81.12	62	60	88.65	46	45
	0.4	80.41	85.68	63	62	92.16	52	52
	Mean	**70.60**	**78.43**	**65**	**59**	**85.70**	**51**	**48**
Due dates	0.1	24.49	23.82	72	65	24.40	65	61
	0.2	7.28	5.80	66	63	5.61	67	66
	0.3	3.06	2.16	77	76	2.17	72	71
	0.4	1.74	1.14	80	80	1.14	78	78
	Mean	**9.15**	**8.23**	**74**	**71**	**8.33**	**70**	**69**

Part 'due dates' of Table 1 presents the results of computational experiments for the case in which the requested due dates for completing jobs are random variables with a normal distribution. This time we note that the resistance coefficients are better for the probabilistic algorithm and this is the case for all β parameter values. Moreover, also the percentage of cases for which the probabilistic algorithm was not worse or better is clearly in favor of the probabilistic algorithm.

Table 2. The robustness of the $\mathcal{AD}$ and $\mathcal{AP}$ algorithms expressed as a percentage for random times with **Erlang distribution.**

Random	$\mathcal{AD}$	$\mathcal{AP}_E$	%NG	%L	$\mathcal{AP}_{ED}$	%NG	%L
Durations	57.20	57.14	90	54	268.32	45	38
Due dates	51.86	52.20	90	56	53.50	78	55

The jobs execution times in 'durations' part of Table 2 are random. We note that the resistance coefficients are slightly better for the probabilistic algorithm. Also, the percentage of cases for which the probabilistic algorithm was not worse or better than the deterministic algorithm is significantly in favor of the probabilistic algorithm. In turn, 'due dates' part of Table 2 presents test results for the case in which the required due dates for completing jobs have the Erlang distribution. We note that the resistance coefficients are slightly better for the deterministic algorithm. On the other hand, however, if we look at the percentage of cases for which the probabilistic algorithm was not worse or better than the deterministic algorithm, there is a noticeable advantage of the probabilistic algorithm.

5 Summary

The performed computational experiments have shown that the solutions determined by the probabilistic version of algorithm are more robust to disturbances (that is: stable) than the ones determined by the deterministic algorithm, for majority of considered variants. Thus, the concept of uncertainty modeling and comparing the solutions (being random variables) for solving the job shop problem with uncertain parameters were confirmed.

Acknowledgments. The paper was partially supported by the National Science Centre of Poland, grant OPUS no. 2017/25/B/ST7/02181.

References

1. Balas, E., Simonetti, N., Vazacopoulos, A.: Job shop scheduling with setup times, deadlines and precedence constraints. J. Sched. **11**(4), 253–262 (2008). https://doi.org/10.1007/s10951-008-0067-7
2. Bożejko, W., Rajba, P., Wodecki, M.: Stable scheduling of single machine with probabilistic parameters. Bull. Pol. Acad. Sci. Tech. Sci. **65**(2), 219–231 (2017)

3. Bożejko, W., Rajba, P., Wodecki, M.: Robustness of the uncertain single machine total weighted tardiness problem with elimination criteria applied. In: Zamojski, W., Mazurkiewicz, J., Sugier, J., Walkowiak, T., Kacprzyk, J. (eds.) DepCoS-RELCOMEX 2018. AISC, vol. 761, pp. 94–103. Springer, Cham (2019). https://doi.org/10.1007/978-3-319-91446-6_10
4. Grabowski J., Wodecki M.: A very fast tabu search algorithm for job shop problem. In: Rego, C., Alidaee, B. (eds.) Metaheuristic Optimization Via Memory and Evolution, pp. 117–144. Kluwer Academic Publishers (2005)
5. Kim, J.G., Jun, H.B., Bang, J.Y., Shin, J.H., Choi, S.H.: Minimizing tardiness penalty costs in job shop scheduling under maximum allowable tardiness. Processes 8(11), 1398 (2020)
6. OR-Library. http://www.brunel.ac.uk/~mastjjb/jeb/info.html
7. Rajba, P., Wodecki, M.: Sampling method for the flow shop with uncertain parameters. In: Saeed, K., Homenda, W., Chaki, R. (eds.) CISIM 2017. LNCS, vol. 10244, pp. 580–591. Springer, Cham (2017). https://doi.org/10.1007/978-3-319-59105-6_50
8. Sharma, P., Jain, A.: A review on job shop scheduling with setup times. J. Eng. Manuf. 230(3), 517–533 (2016)
9. Shoval, S., Efatmaneshnik, M.: A probabilistic approach to the Stochastic Job-Shop Scheduling problem. Procedia Manuf. 21, 533–540 (2018)

New Variants of SDLS Algorithm for LABS Problem Dedicated to GPGPU Architectures

Dominik Żurek, Kamil Piętak[✉], Marcin Pietroń,
and Marek Kisiel-Dorohinicki

AGH University of Science and Technology, al. Adama Mickiewicza 30,
30-059 Krakow, Poland
{dzurek,kpietak,pietron,doroh}@agh.edu.pl

Abstract. Low autocorrelation binary sequence (LABS) remains an open hard optimisation problem that has many applications. One of the promising directions for solving the problem is designing advanced solvers based on local search heuristics. The paper proposes two new heuristics developed from the steepest-descent local search algorithm (SDLS), implemented on the GPGPU architectures. The introduced algorithms utilise the parallel nature of the GPU and provide an effective method of solving the LABS problem. As a means for comparison, the efficiency between SDSL and the new algorithms is presented, showing that exploring the wider neighbourhood improves the results.

Keywords: LABS · GPGPU · Steepest-descent local search

1 Introduction

This paper concentrates on solving the low autocorrelation binary sequence problem using efficient parallel computations on the GPGPU. It introduces a new variant of local search heuristics for LABS together with very efficient realisations designed for the GPGPU architectures. LABS, one of the hard discrete problems despite wide research, remains an open optimisation problem for long sequences. It has wide range of applications including communication engineering [13,14], statistical mechanics [2,10] and mathematics [7,8].

LABS is an NP-hard combinatorial problem with simple a formulation. It consists of finding a binary sequence $S = \{s_0, s_1, \ldots, s_{L-1}\}$ with length L, where $s_i \in \{-1, 1\}$ which minimises the energy function $E(S)$:

$$C_k(S) = \sum_{i=0}^{L-k-1} s_i s_{i+k} \quad \text{and} \quad E(S) = \sum_{k=1}^{L-1} C_k^2(S) \tag{1}$$

The simplest method of solving LABS is exhaustive enumeration that provides the best results, but can be applied only to small values of L.

© Springer Nature Switzerland AG 2021
M. Paszynski et al. (Eds.): ICCS 2021, LNCS 12742, pp. 206–212, 2021.
https://doi.org/10.1007/978-3-030-77961-0_18

Some researchers use partial enumeration, choosing so-called skew symmetric sequences [11] that are the most likely solutions for many lengths (e.g. for $L \in [31, 65]$ 21 best sequences are skew symmetric). Enumerative algorithms are obviously limited to small values of L by the exponential size of the search space. Therefore, a lot of various heuristic algorithms have been developed. They use some plausible rules to locate good sequences more quickly. A well-known method for such techniques is *steepest descend local search* (SDLS) [1] or tabu search [6]. In recent years, a few modern solvers based on the self-avoiding walk concept have been proposed. The most promising solvers are *lssOrel* [3] and *xLostavka* [4], which are successfully used for finding skew-symmetric sequences of lengths between 301 and 401 [5]. Another direction of research is using evolutionary multi-agents systems with local optimisation algorithms [9].

In this paper, we propose two new algorithms that are derived from basic SDLS and are implemented on the GPGPU. The first algorithm, SDLS-2, extends the notion of the sequence neighbourhood to a 2-bit distance. The second, SDLS deep through (SDLS-DT), introduces the recurrent exploration of sequences in both the 1-bit and 2-bit neighbourhood. In this paper, we compare them to the SDLS algorithm implemented on the GPGPU described in [15].

2   New Variants of SDLS Search Algorithms for LABS

The new approach to resolve the LABS problem based on SDLS algorithm relies upon increasing the search area of searching during each single iteration.

The implementation of the proposed algorithms utilizes the notion of the neighbourhood of a sequence S with length L obtained by flipping one symbol in the sequence: $N(S) = \{flip(S, i), i \in \{1, .., L\}\}$, where $flip(s_1 \ldots s_i \ldots s_L, i) = s_1 \ldots s_i \ldots s_L$ [6].

All computed products can then be stored in a $(L-1) \times (L-1)$ table $T(S)$, such that $T(S)_{ij} = s_j s_{i+j}$ for $j \leq L - i$, and saving the values of the different correlations in a $L-1$ dimensional vector $C(S)$, defined as $C(S)_k = C_k(S)$ for $1 \leq k \leq L - 1$. Cotta observed that flipping a single symbol s_i multiples by -1 the value of cells in $T(S)$ where s_i is involved, the fitness of sequence $flip(S, i)$ can be efficiently recomputed in time $O(L)$.

2.1   The SDLS-2 Algorithm with Extended Neighbourhood

The SDLS-2 algorithm extends the notion of neighbourhood to sequences that differ by up to 2 bits. In this case, besides searching for the solution in the neighbourhood with a distance equal to 1, the best results in a single iteration are explored among sequences that differ on two bits with regard to the input sequence. If the best sequence has grater energy than the best sequence from the current iteration, the last one becomes the reference sequence for the next iterations. If the best sequence founded in the current iteration is worse than the input sequence, the algorithm is stopped and the actual reference sequence is returned as a result. In the case of the sequence with the length L, in the single step, this algorithm implies the necessity of looking through $\frac{L(L-1)}{2}$ more solutions than the traditional SDLS algorithm [15].[1]

2.2 Sequential Version of the SDLS-DT Algorithm

In this approach, it is not possible to estimate the number of generated solutions in a single step of the algorithm. The searching of solutions in locality one and two is done in the single step in this case. One step of described algorithm should be defined as a single run of the external loop (Algorithm 1, line 3). Inside the body of this loop, the energy after changing the value on the position with the index with the same value as the current loop's counter is calculated first of all (Algorithm 1, line 4). L energies, based on the sequences which are different on two bits comparison to the original sequence are then calculated (Algorithm 1, line 7). The lowest of the obtained energies is chosen (Algorithm 1, line 8) and if its value is lower then the current reference energy (Algorithm 1, line 9), update of the reference energy E_r and the input sequence corresponding to it occurs (Algorithm 1, line 10). The new run of the external loop then begins. In the following loop runs, there is no possible way to estimate on how many bits the searching space differs comparison to the input sequence because of its dependence upon the winning sequence, which in this case can have a different value on one or two positions. The each single step of the algorithm is done in case of when improvement is not observed (Algorithm 1, line 9) (which is equivalent with breaking the *while* loop). The number of that steps is equal to the length of the input sequence. Figure 1 illustrates two first steps of the algorithm on a sample sequence.

Algorithm 1. Sequential version of SDLS-DT algorithm

1: **function** SEQUNTIALSDLS-DT(S)
2:     $E_r = compute\_reference\_energy(S)$
3:     **for** $i := 0$ **to** $len(L)$ **do**
4:         $E[i] = compute\_single\_energy\_by\_mutation\_i_{th}\_bit(S)$
5:         $improvement := true$
6:         **while improvement do**
7:             $E_{local\_II} = compute\_energies\_by\_mutation\_of\_two\_bits(S)$
8:             $E_{best} = compute\_lowest\_energy(E[i], E_{local\_II})$
9:             **if** $E_r < E_{best}$ **then** $improvement := false$
10:                 $update\_reference\_sequence\_and\_energy()$
        end while
    end for
11:     **return** E_{best}
    end function

---

[1] The GPGPU implementation of discussed approach is not able to check a large space so as a result it provides the worst results. For that reason the detailed description and analysis of this algorithm will be conducted as a future work.

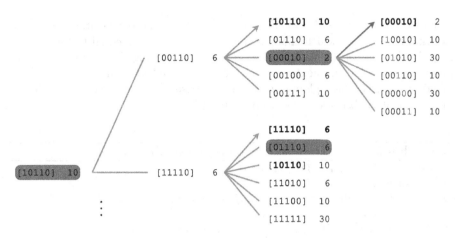

Fig. 1. The example of iteration in SDLS-DT algorithm

2.3   Parallel Version of SDLS-DT for GPGPU

In the GPGPU implementation, the first step is the creation of the external loop with the length L (Algorithm 2, line 2). The first operation in this loop changes the value to the opposite value on the position with the index equal to the loop's counter, which is marked as *bit*. The change is realised by thread with that index (Algorithm 2, line 3). For that sequence, the structures $C(S)$ and $T(S)$ are created (Algorithm 2, line 4). Based on the actual sequence, the first energy which provides the first solution in the neighbourhood with distance one is then calculated. This energy becomes the current reference energy (Algorithm 2, line 7). The sequence which is different on one position comparison with the original sequence, becomes the input sequence for the second step of the algorithm. The second step of the proposed method is looking for a solution until the next attempts are unable to find a better solution than the current solution which is realised by the internal *while* loop (Algorithm 2, line 6). In each iteration of this loop L energies are calculated according to the SDLS algorithm (Algorithm 2, lines 7–8). In this case, the thread with ID equal to the counter loop is not blocked because according to this method this thread calculates the energy for the original sequence. From the energies calculated in that way the energy with the lowest value E_{best} is chosen. If this value is lower than the value of the current reference energy E_r (Algorithm 2, line 10) it then becomes the reference energy and its sequence becomes the input sequence to the next iteration of the *while* loop (Algorithm 2, line 11). In this case, the winning thread actualises the $C(S)$ and $T(S)$ (Algorithm 2, line 12). It should be noted that the search could be in a space with a distance higher than two bits. In the second iteration, the input to the internal loop could be the sequence with two different bits than the original sequence. In each internal iteration, the distance between its input sequence and the original sequence could increase. This means that it is not possible to verify which sequence was checked, so there is no sense in blocking any thread. In the case of the occurrence of a *break* in the *while* loop, the second iteration of the external loop begins which changes the bit on the position equal to the counter loop. In the case of the *ith* iteration, the *ith* bit is changed and it becomes the input sequence for the next iteration of the internal loop. In each run of the external loop, the best global energy is actualised in cases in which an in the improvement is observed (Algorithm 2, line 13).

Algorithm 2. Parallel version of SDLS-DT on GPGPU

1: **function** PARALLELSDLS-DT(S)
2:     **for** $bit := 0$ **to** $len(L)$ **do**
3:         **if** $threadId == bit$ **then** $S_{block}[threadId]* = -1$ **end if**
4:         $create\_T(S)\_and\_C(S)()$
5:         $E_r := compute\_energy\_with\_local\_one(S_{block})$
6:         **while** $improvement$ **do**
7:             $mutation\_of\_threadId\_bit(S_{block})$
8:             $ParallelValueFlip(S_{block}, T', C')$
9:             $E_{best} := compute\_lowest\_energy()$
10:             **if** $E_r < E_{best}$ **then** $improvement := false$ **end if**
11:             $update\_reference\_sequence\_and\_energy()$
12:             $update\_T(S)\_and\_C(S)()$
        end while
13:         **if** $E_{best_{global}} < E_{best}$ **then** $E_{best_{global}} := E_{best}$ **end if**
    end for
14:     **return** $E_{best_{global}}$
    end function

---

3   Effectiveness of the Proposed Algorithms

In order to measure the effectiveness of the proposed algorithms, each parallel version of them was performed on the Nvidia Tesla V100-SXM2-32GB[2]. Each algorithm was seeking the optimal solution for three different input lengths $(128, 256)$. In the first iteration, the processor randomises 128 different sequences, one for each GPU block. With data generated in such a way, each kernel starts searching the minimum value of energy. The moment that the best energy from each block is found, it is stored as the current best energy $E_{global\_optimum}$. The processor generates a new set of 128 sequences for which the algorithm repeats the search process and if applicable, the global energy is updated. The entire process was run 10 times for one hour each.

Table 1 contains the number of searched solutions along with the average of searched solutions for the single thread block and the number of individual kernels that are run. Figure 2 presents the efficiency of the algorithms. As could be observed for each size of the problem, the best solution was obtained by

---

[2] https://www.nvidia.com/en-us/data-center/v100/.

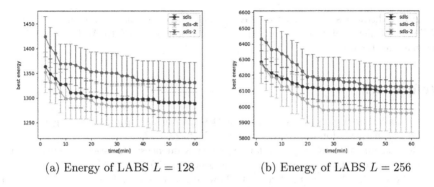

(a) Energy of LABS $L = 128$ (b) Energy of LABS $L = 256$

Fig. 2. Energies achieved by basic SDLS, SDLS-2 and SDLS-DT algorithms

Table 1. The number of explored solutions during one minute computations

Sequence length	Method	Number of searched solutions	Average	Standard deviation	Number of running kernels
128	**SDLS**	20 740 434 073	2807	531.8	57773
128	**SDLS-2**	16 865 830 144	224 089	41837.6	593
128	**SDLS-DT**	31 176 794 592	356 752	46302.4	703
256	**SDLS**	22 380 865 408	11159	2 836	15 661
256	**SDLS-2**	15 975 437 994	1 657 780	257 819	74
256	**SDLS-DT**	24 978 089 130	3 019 130	292 121	68

the *SDLS-DT* algorithm. This fact is due to the parallel implementation of this algorithm was able to successively explore significantly larger space than SDLS and SDLS-2 algorithms. Despite the results being similar to those achieved with SDLS, the efficiency of the SDLS-2 algorithm is weaker. This fact proves that the efficiency of the algorithm is high but the proposed implementation is not effective. The improvement of this implementation is considered as future work.

4   Conclusions and Further Work

This paper is the next step of our research related to efficient algorithms for LABS realised using GPGPU architectures. The presented new SDLS algorithms show a significant improvement in effectiveness compared to the traditional SDLS approach. They can be further combined with meta-heuristics such as evolutionary algorithms, which constitute a basis for the concept of a hybrid environment in the master-slave model that was proposed in [12].

In the near future the authors plan to implement parallel variants of the self-avoiding walk, *lssOrel* or *xLostavka* solvers. We also consider extending the search neighbourhood in tabu search heuristics and propose new variants of the algorithm dedicated to the GPGPU. The next interesting and promising direction of research is to design and implement an evolutionary multi-agent system with new local optimisation on the GPGPU units.

212 D. Żurek et al.

Acknowledgments. The research presented in this paper was realized thanks to funds of Polish Ministry of Science and Higher Education assigned to AGH University of Science and Technology.

References

1. Bartholomew-Biggs, M.: The Steepest Descent Method, pp. 1–8. Springer, Boston (2008). https://doi.org/10.1007/978-0-387-78723-7_7
2. Bernasconi, J.: Low autocorrelation binary sequences: statistical mechanics and configuration space analysis. J. De Physique **48**, 559–567 (1987)
3. Bošković, B., Brglez, F., Brest, J.: Low-Autocorrelation Binary Sequences: On Improved Merit Factors and Runtime Predictions to Achieve Them. arXiv e-prints arXiv:1406.5301 (Jun 2014)
4. Brest, J., Bošković, B.: A heuristic algorithm for a low autocorrelation binary sequence problem with odd length and high merit factor. IEEE Access **6**, 4127–4134 (2018). https://doi.org/10.1109/ACCESS.2018.2789916
5. Brest, J., Bošković, B.: In searching of long skew-symmetric binary sequences with high merit factors (2020)
6. Gallardo, J.E., Cotta, C., Fernández, A.J.: Finding low autocorrelation binary sequences with memetic algorithms. Appl. Soft Comput. **9**(4), 1252–1262 (2009)
7. Günther, C., Schmidt, K.U.: Merit factors of polynomials derived from difference sets. J. Comb. Theor. Ser. A **145**, 340–363 (2016)
8. Jedwab, J., Katz, D.J., Schmidt, K.U.: Advances in the merit factor problem for binary sequences. J. Comb. Theor. Ser. A **120**(4), 882–906 (2013)
9. Kowol, M., Byrski, A., Kisiel-Dorohinicki, M.: Agent-based evolutionary computing for difficult discrete problems. Procedia Comput. Sci. **29**, 1039–1047 (2014)
10. Leukhin, A.N., Potekhin, E.N.: A Bernasconi model for constructing ground-state spin systems and optimal binary sequences. J. Phys. Conf. Ser. **613**, 012006 (2015)
11. Packebusch, T., Mertens, S.: Low autocorrelation binary sequences. J. Phys. A Math. Theor. **49**(16), 165001 (2016)
12. Piętak, K., Żurek, D., Pietroń, M., Dymara, A., Kisiel-Dorohinicki, M.: Striving for performance of discrete optimisation via memetic agent-based systems in a hybrid CPU/GPU environment. J. Comput. Sci. **31**, 151–162 (2019)
13. Zeng, F., He, X., Zhang, Z., Xuan, G., Peng, Y., Yan, L.: Optimal and z-optimal type-ii odd-length binary z-complementary pairs. IEEE Commun. Lett. **24**(6), 1163–1167 (2020)
14. Zhao, L., Song, J., Babu, P., Palomar, D.P.: A unified framework for low autocorrelation sequence design via majorization-minimization. IEEE Trans. Signal Process. **65**(2), 438–453 (2017)
15. Żurek, D., Piętak, K., Pietroń, M., Kisiel-Dorohinicki, M.: Toward hybrid platform for evolutionary computations of hard discrete problems. Procedia Comput. Sci. **108**, 877–886 (2017). International Conference on Computational Science, ICCS 2017, 12–14 June 2017, Zurich, Switzerland

Highly Effective GPU Realization of Discrete Wavelet Transform for Big-Data Problems

Dariusz Puchala$^{(\boxtimes)}$ and Kamil Stokfiszewski

Institute of Information Technology, Lodz University of Technology,
Ul. Wolczanska 215, 90-924 Lodz, Poland
{dariusz.puchala,kamil.stokfiszewski}@p.lodz.pl

Abstract. Discrete wavelet transform (DWT) is widely used in the tasks of signal processing, analysis and recognition. Moreover it's practical applications are not limited to the case of one-dimensional signals but also apply to images and multidimensional data. From the moment of introduction of the dedicated libraries that enable to use graphics processing units (GPUs) for mass-parallel general purpose calculations the development of effective GPU based implementations of one-dimensional DWT is an important field of scientific research. It is also important because with use of one-dimensional procedure we can calculate DWT in multidimensional case if only the transform's separability is assumed. In this paper the authors propose a novel approach to calculation of one-dimensional DWT based on lattice structure which takes advantage of shared memory and registers in order to implement necessary inter-thread communication. The experimental analysis reveals high time-effectiveness of the proposed approach which can be even 5 times higher for Maxwell architecture, and up to 2 times for Turing family GPU cards, than the one characteristic for the convolution based approach in computational tasks that can be classified as big-data problems.

Keywords: Discrete wavelet transform · GPU computations · Mass-parallel computations · Lattice structure · Time effectiveness

1 Introduction

Discrete wavelet transform (DWT) is one of the basic tools of digital signal processing. It finds wide applications in such areas as processing of multidimensional data [1,2], image compression [3,4], image watermarking [5,6], analysis and clustering of high dimensional data [7,8], data mining [9] and many more (see [10,11]). It should be noted that even in case of images (or multidimensional data) one-dimensional wavelet transform procedure is a basic tool used in calculations due to the fact that two-dimensional (or respectively many dimensional) DWT can be calculated in a row-column scheme with use of one-dimensional

© Springer Nature Switzerland AG 2021
M. Paszynski et al. (Eds.): ICCS 2021, LNCS 12742, pp. 213–227, 2021.
https://doi.org/10.1007/978-3-030-77961-0_19

transform. For this reason it is very important to develop time-effective real-izations of one-dimensional DWT. In the recent years we could observe an intense research on improvement of such algorithms (e.g. see [12]). This trend is especially visible thanks to the growing popularity of graphics processing units (GPUs) for which massively parallel algorithms for calculation of DWT were also constructed (c.f. [13–21]).

The solutions found in the literature are focused mainly on the case of two-dimensional data. For example in paper [18] the authors proposed the approach to calculation of 2D DWT based on row-column approach and the lifting scheme (see [10]) which takes advantage of shared memory and registers. In paper [16] the problem of calculation of 2D and 3D DWT based on row-column approach (and its extension to three dimensions) using shared memory and registers is addressed wherein the authors put a great effort on the issues of effective memory access. Finally in paper [17] we can find an approach based on lifting scheme where row and column passes are merged into inseparable transform. In selected papers we can find simple solutions to calculation of 2D DWT based on Haar wavelets and row-column scheme (e.g. see papers [14] and [15]). The problem of calculation of one-dimensional DWT was widely addressed in papers [19–21]. The authors of these papers analyzed convolution based approach and the approach taking advantage of lattice structure with multiple calls of kernel functions (we refer to this approach as a naive lattice structure in the remaining part of the paper). To the best knowledge of the authors of this paper the problem of effective calculation of one-dimensional DWT (i.e. using shared memory and registers) is not solved.

In this paper we propose a novel approach to calculation of one-dimensional wavelet transform. The proposed approach is based on the lattice structure and requires only one call of the kernel function. It also takes advantage of the shared memory and registers for communication between threads. Thanks to the mentioned features it can be characterized by high time-effectiveness. The experimental analysis shows that the proposed approach can be even 5 times faster for Maxwell architecture GPUs (NVIDIA GTX960), and even 2 times for Turing architecture (NVIDIA RTX2060), than the approach based on the convolution. Such acceleration is achieved mainly in case of big-data problems which makes the proposed approach particularly useful in modern computational tasks.

2   Calculation of One-Dimensional DWT

In practical applications discrete one-dimensional wavelet transform is implemented as a two-channel filter bank with the structure shown in Fig. 2 (see e.g. [22]). Here an input signal $X(z) = X_0(z^2) + z^{-1} X_1(z^2)$ (with $X_0(z)$ and $X_1(z)$ describing even/odd numbered samples of input data) is in the first place decimated at blocks ($\downarrow 2$), which gives $[X_1(z), X_0(z)]^T$ vector, and then the same vector enters the block of filters defined by polyphase matrix $E(z)$ of the form:

$$E(z) = \begin{bmatrix} H_0(z) \ H_1(z) \\ G_0(z) \ G_1(z) \end{bmatrix},$$

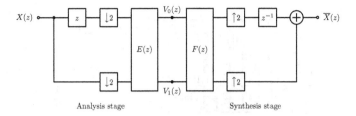

Fig. 1. The structure of two-stage filter bank in polyphase representation

where $H_0(z)$, $G_0(z)$ and $H_1(z)$, $G_1(z)$ represent, respectively, even/odd numbered coefficients of impulse responses of the filters. Both filters $H(z)$ and $G(z)$, together with decimators ($\downarrow 2$), form the analysis stage. At the output of this stage we get the signal in the form of $[V_0(z), V_1(z)]^T$ components, where $V_0(z)$ and $V_1(z)$ are obtained in the result of low-pass and high-pass filtering respectively (we refer here to the typical configuration of filters). The synthesis stage consists of filters described by the polyphase matrix $F(z)$. The synthesis and analysis filters are selected in a way satisfying the perfect reconstruction (PR) condition $F(z)E(z) = I$, with I being an identity matrix. With that assumption, and with use of upsampling blocks ($\uparrow 2$), it is possible to restore the input signal at the output of the bank, i.e. $\overline{X}(z) = X(z)$, if only there was no interference with the components obtained at the output of the analysis stage.

The practical implementations of the block of filters from Fig. 1 can be based on: (i) convolution approach, (ii) lattice structure, (iii) lifting scheme (see [10]). Lattice structure and lifting scheme approaches allow for almost two-fold reduction in the number of additions and multiplications and can be characterized by the smallest number of parameters required to describe both orthogonal and biorthogonal wavelet transforms, see e.g. [10]. This feature can be crucial from the point of view of adaptive parametric structures. In this paper we consider convolution approach, which is widely used in the literature (see [10]), and lattice structure approach, which was chosen as a basis for the proposed effective and mass-parallel solution. Moreover, in the rest of the paper we assume N and M parameters to describe the size of input data and the length of the impulse responses of the filters respectively.

2.1   Calculations Based on the Convolution

The basic approach to calculation of DWT is based directly on the formula $[V_0(z), V_1(z)]^T = E(z)[X_1(z), X_0(z)]^T$. In practice it can be realized as the convolution of input signal with the impulse responses of both filters $H(z)$ and $G(z)$. Taking into account the polyphase representation of the bank and the decimation operations ($\downarrow 2$) it can be easily verified that the convolution should be calculated only at the position of even numbered samples of input signal. Such computational scheme offers good possibilities for mass parallel calculation of DWT. The approach adopted in this paper assumes the calculation of

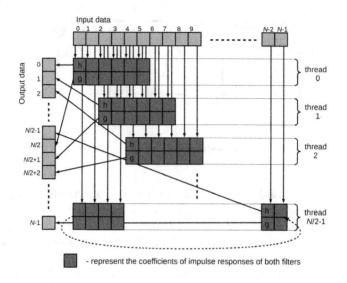

- represent the coefficients of impulse responses of both filters

Fig. 2. Mass parallel computation of DWT based on one-dimensional convolution in case of periodic signals ($M = 6$).

convolution in separate threads starting from each even numbered sample of input data. Moreover in order to reduce the number of memory transactions each thread realizes calculations for both filters $H(z)$ and $G(z)$ operating on the same data samples (see Fig. 2). This gives the number of $N/2$ threads required to process the data. The results of computations must be written to the additional output table. Moreover all the calculations can be done fully independently, which means that no additional synchronization mechanism is required. As a result the implementation of mass-parallel convolution based calculations of DWT requires one execution of the kernel function. It should be noted that Fig. 2 shows the case where the periodic repetition of input data is assumed. This is the boundary condition considered in this paper. The additional analysis of the computational structure reveals that it can be characterized by a number of approximately 4 arithmetical operations performed for a single data transaction. It is enough to consider operations within one thread, where we have: $4M - 2$ arithmetical operations (additions and multiplications) and $M + 2$ data transactions (data read and write operations). The total numbers of additions ($\mathcal{L}_{ADD}$) and multiplications ($\mathcal{L}_{MUL}$) required by the convolution approach can be described by the formulas:

$$\mathcal{L}_{ADD} = N(M - 1), \quad \mathcal{L}_{MUL} = NM. \tag{1}$$

In turn the total number of memory transactions ($\mathcal{L}_{MEM}$), a number of sequential steps ($\mathcal{L}_{SEQ}$) (considers arithmetical operations when calculations within all threads can be realized at the same time) and a number of kernel function calls equal respectively:

$$\mathcal{L}_{MEM} = \frac{1}{2}N(M + 2), \quad \mathcal{L}_{SEQ} = 4M - 2, \quad \mathcal{L}_{KER} = 1. \qquad (2)$$

It should be noted that the number of sequential steps that must be performed is a measure of complexity of parallel algorithm.

2.2 Calculations Based on the Lattice Structure

Although the computations within a filter bank are effective and can be described by linear complexity $\mathcal{O}(MN)$, where N is the size of input signal, and M describes the size of filters, its efficiency can be increased even more with use of lattice structures. Lattice structures allow for almost twofold reduction in the number of multiplications and additions and allow for accurate parametrization in the sense of a number of free parameters (see [10]). The construction of a lattice structure requires factorization of the polyphase matrix $E(z)$ into a product of simple matrices. Following [10] such factorization can be described as:

$$E(z) = BD\Lambda(z^{-1}) \left(\prod_{i=2}^{\frac{M}{2}-1} A_{\frac{M}{2}-i} D\Lambda(z^{-1}) \right) A_0, \qquad (3)$$

where the matrices used in formula (2) can be defined as:

$$B = \begin{bmatrix} a & b \\ c & d \end{bmatrix}, \quad A_i = \begin{bmatrix} 1 & t_i \\ s_i & 1 \end{bmatrix}, \quad \Lambda(z) = \begin{bmatrix} 1 & 0 \\ 0 & z \end{bmatrix}, \quad D = \begin{bmatrix} 0 & 1 \\ 1 & 0 \end{bmatrix}.$$

Matrices B and A_i for $i = 0, 1, \ldots, \frac{M}{2} - 2$ represent basic operations within the lattice structure which are parametrized with the values of a, b, c, d and t_i, s_i for $i = 0, 1, \ldots, \frac{M}{2} - 2$ parameters. It should be noted that in case of the orthogonal bank of filters the following relations take place: $s_i = -t_i$ for $i = 0, 1, \ldots, \frac{M}{2} - 2$ and $d = \pm a$, $c = \mp b$. For example orthogonal Daubechies 4 (db4) wavelet can be decomposed into the set of parameters: $t_0 = -0.322276$, $t_1 = -1.23315$, $t_2 = -3.856628$ and $a = 0.15031$, $b = 0.006914$ and $d = a$, $c = -b$. In case of biorthogonal bank of filters all of the parameters are required in general, though the well-known postulate of multi-resolution analysis, i.e. $H_0(-1) = 0$ and $H_1(1) = 0$ (c.f. [10]), which must be met by any wavelet transform, makes the values of b and c dependent on the values of the remaining parameters. Then the formula (3) describes the accurate factorization of matrix $E(z)$ where the number of free parameters is the smallest possible depending on the considered family of wavelet transforms. In Fig. 3 we can see the data flow diagram of the lattice structure based on the decomposition formula (3).

   In case of the lattice structure parallelization of calculations is possible due to the independent character of computations realized within A_i matrices for $i = 0, 1, \ldots, M/2 - 2$ and B matrix at each stage of calculations (see Fig. 3). The operations described by A_i and B matrices are represented graphically as

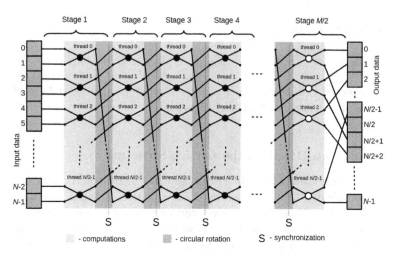

Fig. 3. Mass-parallel computation of DWT based on one-dimensional lattice structure in case of periodic signals ('•' - operations described by matrices A_i for $i = 0, 1, \ldots, M/2 - 2$; 'o' - operations described by matrix B).

butterfly operators '•' and 'o' respectively. It should be noted that at each stage the operations within butterfly operations are independent, since each butterfly operator operates on an individual pair of input data. If we assume that each butterfly operator is implemented by one thread then the total required number of threads will be $N/2$. However, it is necessary to synchronize the calculations between consecutive stages which is possible with use of global synchronization mechanisms. Hence, the mass-parallel realization of the lattice structure based directly on the data flow diagram from Fig. 3 requires $M/2$ calls of the kernel function. Since the time required to call a kernel function is a thousand times longer than a time of a single mathematical operation we will call this approach naive. Further analysis allows to say that the lattice structure described by the factorization formula (3) can be characterized by a number of 1 (for A_i operators) and 3/2 (for B operators) arithmetical operations for one data transaction. The total numbers of additions and multiplications required by the lattice structure can be expressed respectively as:

$$\mathcal{L}_{ADD} = \frac{1}{2}NM, \quad \mathcal{L}_{MUL} = \frac{1}{2}N(M + 2). \tag{4}$$

Moreover the total number of memory transactions, sequential steps and the calls of kernel function can be described by:

$$\mathcal{L}_{MEM} = NM, \quad \mathcal{L}_{SEQ} = 2(M + 1), \quad \mathcal{L}_{KER} = \frac{1}{2}M. \tag{5}$$

Summarizing, the direct comparison of both described approaches indicates approximately twofold reduction of the number of arithmetic operations in case of lattice structure which translates into twofold reduction of the sequential steps required by the mass parallel realizations of both approaches. However, the convolution based approach can be characterized by almost twice smaller number of memory transaction than the lattice structure. Moreover the naive lattice structure requires much higher number of kernel function calls. It should be noted that a single kernel function call may take even several thousands of clock cycles while a single mathematical operation requires around 20 cycles. Hence such an approach may be characterized by poor time-effectiveness.

3   The Proposed Lattice Structure

In this paper we propose a novel approach to calculation of DWT based on the lattice structure. We choose the lattice structure as the starting point since it can be characterized by almost two times smaller number of arithmetic operations than the convolution based approach. This feature can be crucial in practical scenarios when operating on big data sets because GPU computations have both sequential and parallel character. The main goal of introducing a new approach is to increase the efficiency of DWT calculations. The increase in efficiency is possible if we reduce a number of kernel function calls and take advantage of shared memory and effective local synchronization mechanism. The mentioned objectives can be achieved only if the computations within the entire lattice structure can be partitioned into separate and independent sets of operations which can be executed within separate blocks of threads. However, if we look at the structure from Fig. 3, it is difficult to distinguish separate groups of base operations due to the specific permutation (cyclic rotation) of data between suc-

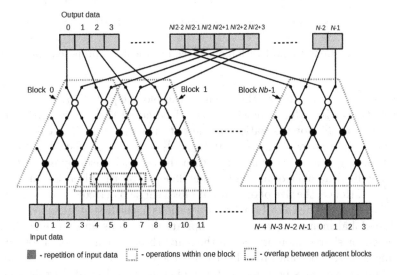

Fig. 4. The organization of calculations in case of the proposed approach to mass-parallel realization of DWT ($M = 6$).

cessive stages. Such partitioning is possible, however, if we accept the possibility of repeating some basic operations. The proposed approach is shown in Fig. 4.

In the proposed approach, the base operations are grouped into independent blocks according to the presented partition scheme (see Fig. 4). Each block contains the set of base operations from successive stages that create trapezoidal structures (indicated with orange dotted lines). Although adjacent blocks overlap by $K = M - 2$ elements of the input table, computations across all blocks are independent and may be performed at the same time. In addition, operations within a single block can take advantage of shared memory as well as efficient synchronization mechanism in the form of the __syncthreads() function. In Fig. 5 we show the organization of calculations within a single block.

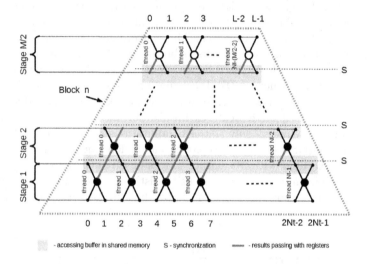

Fig. 5. Operations realized within one block in case of the proposed approach.

The size of input data at the input of the block equals $2N_t$, where N_t is a number of threads within a block. It should be noted that every thread in the block is responsible for realization of one butterfly operator, i.e. realizations of operations required by A_i for $i = 0, 1, \ldots, \frac{M}{2} - 2$ and B matrices. However, in the following stages, the number of threads used decreases by one with each stage. The final number of active threads depends on the filter size and equals $N_t - \frac{M}{2} - 1$. Hence, the size of output data for a single block equals $L = 2N_t - K$ with $K = M - 2$. The role of the base operations in the first stage is to read data from the global memory, perform appropriate operations and write the results to registers or shared memory. Since the same threads realize operations in the following stages, it becomes possible to pass the results directly through the registers (blue connections between butterfly operations in Fig. 5). In this way, the base operations within internal stages read data from registers or shared memory and write the results of calculations at the same localizations. The operations at the last stage write the results to the global memory. Synchronization between stages is realized with use of the fast intra-block synchronization.

It should be noted that with such organization of calculations the realization of the assumed boundary condition requires to extend the input table by $M - 2$ elements and fill them with the first $M - 2$ elements of the input data (see the green elements of input table in Fig. 4). In this way the size of input table for the whole structure from Fig. 4 equals $N + (M - 2)$, where N is the size of input data. The size of input data, which is at the same time the size of the output of the whole structure, can be calculated as $N = N_b L$, where N_b is the number of blocks of threads. The further analysis of the proposed approach allows to derive the formulas for the number of additions $(\mathcal{L}_{ADD})$ and multiplications $(\mathcal{L}_{MUL})$:

$$\mathcal{L}_{ADD} = \frac{1}{2}NM + \left(\frac{1}{4}N_b MK\right), \mathcal{L}_{MUL} = \frac{1}{2}N(M+2) + \left(\frac{1}{2}N_b K\left(\frac{1}{2}K+1\right)\right),$$
(6)

and also the numbers of kernel calls $(\mathcal{L}_{KER})$ and the sequential steps $(\mathcal{L}_{SEQ})$ that must be realized within the algorithm:

$$\mathcal{L}_{KER} = 1, \quad \mathcal{L}_{SEQ} = 2(M + 1) .$$
(7)

The comparison of formulas (4) and (6) allows to state that the proposed approach can be characterized by the higher number of arithmetical operations resulting from the overlaps between the blocks of threads. However, the comparison of formulas (5) and (7) allows to see that the number of sequential steps stays the same but the number of kernel function calls in the proposed approach equals 1. Also the analysis of the number of global memory transactions, where in case of the proposed approach we have:

$$\mathcal{L}_{GMEM} = 2N + N_b K, \quad \mathcal{L}_{SMEM} = \frac{1}{2}N_b K\left(2N_t - \frac{1}{2}K\right)$$

with $\mathcal{L}_{GMEM}$ and $\mathcal{L}_{SMEM}$ describing the numbers of global and shared memory transactions respectively, allows to state that the proposed approach reduces significantly the number of global memory transactions for shared memory transactions and register based data passing.

In the proposed implementation (see Listing 2) the kernel function `proposed()` is called only once. The arguments of the kernel function include `input[]` and `output[]` buffers intended for input data and the results of computations respectively, as well as the `coeff[]` table holding the parameters required by the operators at the following stages of the lattice structure. The remaining three parameters are: `N2` $= \frac{N}{2}$, `L` – the filter length and `m2` $= \frac{M}{2}$. It should be noted that buffer `sbuf[]` of size N_t elements is located in the shared memory and also the passing of the intermediate results of computations between stages through registers is realized with use of `fVal[]` variable.

In order to avoid a situation in which the elements of input data are first read by the threads operating on even indexes and next operating on odd indexes, thus making the memory references not optimally organized, we improve the efficiency of memory transfers by introducing `ll_to_ff()` function (see Listing 1), which reads two `float` elements as one 64-bit reference to `long long` type. Then using dedicated CUDA API function and bit-shift operation it is possible to extract two 32-bit `float` variables from the value of one 64-bit variable.

```
1   __device__ inline void ll_to_ff(long long* d,float& a, float& b) {
2       long long bb;
3       bb=*((long long*)d);
4       a=__uint_as_float((unsigned int)bb);
5       b=__uint_as_float((unsigned int)(bb>>32));
6   }
```

Listing 1. Function reading two `float` variables as one 64-bit reference.

Using the function `ll_to_ff()` form Listing 1 we come up with the following kernel code implementing the proposed DWT computation method.

```
1   __global__ void proposed(float* input,float* output,
2                            float* coeff,int N2,int L,int m2) {
3       __shared__ float sbuf[Nt];   // Shared memory buffer
4       int iInd0,iInd1,iInd2,iInd3; // Additional variables in registers
5       float a,b,c,d,fX,fY,fVal;
6       // Initialization and mapping of thread coordinates to data index
7       iInd0=1;
8       iInd1=threadIdx.x;
9       iInd2=L2*blockIdx.x+threadIdx.x;
10      iInd3=blockDim.x-1;
11      // First stage: reading input data and values of operation parameters
12      ll_to_ff((long long*)coeff,a,b);
13      ll_to_ff((long long*)input+iInd2,fX,fY);
14      // Computations within the first stage
15      sbuf[iInd1]=fX+a*fY;
16      fVal=b*fX+fY;
17      __syncthreads();   // Synchronization
18      while(iInd0<m2) { // Internal stages
19          iInd1+=1;
20          if (threadIdx.x<iInd3) { // Filtering of unneeded threads
21              // Reading operation parameters
22              ll_to_ff((long long*)coeff+iInd0,a,b);
23              // Computations within stage
24              fY=sbuf[iInd1];
25              sbuf[iInd1]=fVal+a*fY;
26              fVal=b*fVal+fY;
27          }
28          iInd0+=1;
29          iInd3-=1;
30          __syncthreads();      // Synchronization
31      }
32      if (threadIdx.x<iInd3) { // Last stage
33          // Reading operator parameters
34          iInd1+=1;
35          ll_to_ff((long long*)coeff+iInd0,a,b);
36          iInd0+=1;
37          ll_to_ff((long long*)coeff+iInd0,c,d);
38          // Computations within stage and storing results to global memory
39          fY=sbuf[iInd1];
40          output[iInd2]=a*fVal+b*fY;
41          output[iInd2+N2]=c*fVal+d*fY;
42      }
43  }
```

Listing 2. Kernel function for the proposed approach.

4   Experimental Analysis

In order to verify the effectiveness of the proposed approach a series of experiments was performed including various sizes of filters and input data. During the experiments we considered filter sizes M changing in a range from 4 to 32 coefficients and the experimental data of sizes between 1024 and $1024 \cdot 10^5$ elements. In Fig. 6 we present selected experimental results obtained with NVIDIA GTX960 GPU card (Maxwell family) which is characterized by the total number of 1024 computing cores and 4 GB of global memory (128-bit memory bus and 112.2 GB/s of memory bandwidth). The second GPU card used was NVIDIA RTX2060. It is a representative of the Turing architecture of NVIDIA GPU cards with a number of 1920 computing cores and 6 GB of global memory (192-bit memory bus and 336 GB/s of memory bandwidth). The selected results obtained for the second card are presented in Fig. 6.

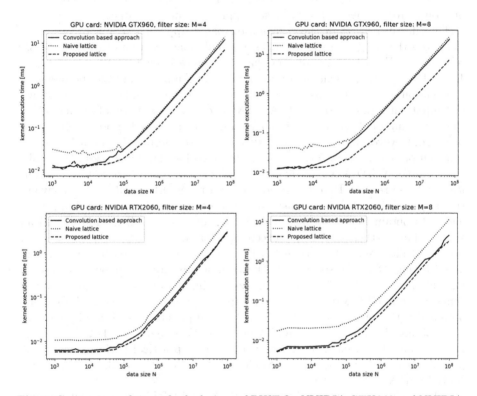

Fig. 6. Comparison of time of calculations of DWT for NVIDIA GTX960 and NVIDIA RTX2060 cards and filter sizes $M = 4$ and $M = 8$.

In Fig. 6 we present the results obtained with NVIDIA GTX960 card for the popular filter sizes, i.e. $M \in \{4, 8\}$. The analysis of results indicates a huge advantage of the convolution approach over the approach using the naive lattice for data sizes up to around 10^5 elements. The ratio of execution times (convolution time/lattice time)[1] lies between 0.15 and 1.0 depending on the filter and input data sizes. For example for $M = 4$ the ratio is below 0.50 for smaller data sizes ($N \leq 0.5 * 10^5$), and grows up to 1 at $N \approx 10^5$ to stay close to 0.9 for the remaining data sizes. It shows that the naive lattice can be more than 2 times slower than the convolution based approach. For $M = 8$ the ratio changes between 0.3 and 0.93 depending on the size of input data. The most favorable results for the naive lattice are obtained with filters of size $M = 32$ and indicate only 15% advantage of the convolution approach for data sizes $N \geq 0.25 * 10^5$. On this basis we state that naive lattice allows to obtain much worse results than the results possible to obtain with use of convolution. Hence, the naive lattice is characterized by significantly lower time-efficiency than the convolution approach.

In case of the proposed approach the situation looks differently. For filter sizes $M \in \{4, 8\}$ (highly practical sizes of filters), and starting from data size of around 10^4 elements, the proposed approach can be characterized by much higher effectiveness than the convolution based approach. For the considered filter sizes the obtained ratios of times of calculations are around 1.9, 2.7 and 3.4 respectively. Even for sizes smaller than 10^4 the ratio changes between 0.9 and 1.1. In Fig. 7(a) we present the values of ratios of computation times between the convolution and the proposed approach for all of the considered sizes of filters and input data. In case of NVIDIA GTX960 card the data size of around 10^4 is the lower limit of effectiveness of application of the proposed approach. It should be noted, however, that the highest advantage of the proposed approach over the convolution one is above 5 times and is obtained for high filter sizes ($M \geq 24$).

In case of NVIDIA RTX2060 card, see Fig. 6, the naive lattice obtains significantly worse results. Here the ratios are in a range 0.1 to 0.3. So the convolution approach is in the worst case even 10 times faster. The proposed approach allows to obtain better results than the convolution for all data sizes and all sizes of filters (there are a few exceptions for $M \in \{4, 8\}$ and $N \approx 10^8$). For small filter sizes, like $M \in \{4, 8\}$, the proposed approach is up to 1.25 and 1.5 times faster respectively. The highest advantage can be observed for long filters $M \in \{30, 32\}$ and high sizes of input data and it equals even 2 times, see Fig. 7(b). Moreover, if we compare results of the proposed approach obtained with both considered GPUs, then the possible speed-up of computations obtained with NVIDIA RTX2060 is in the range [1.75, 4.12], while the average value is around 2.68 times. The averaged practical performance of the proposed method is 83 GFlops/s for NVIDIA GTX960, and 233 GFlops/s for NVIDIA RTX2060.

The results of the proposed approach for NVIDIA RTX2060 card where compared to sequential convolution and lattice CPU implementations. The CPU variants allowed to obtain better results for small amount of processed data,

---

[1] We measured times of kernel launch and calculations with NVIDA's nvprof profiler. During experiments we used Intel i7-9700, 12 MB cache, 32 GB RAM, Windows 10 platform, and CUDA 10, C++ implementations.

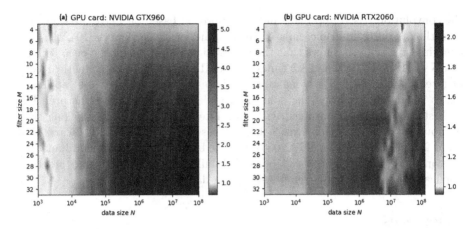

Fig. 7. Ratios of times of calculations (convolution time/proposed lattice time) of DWT for NVIDIA's (a) GTX960 and (b) RTX2060 cards.

i.e. $MN \leq 16384$ for convolution, or $MN \leq 32768$ for lattice approach. Here, the computation time ratios between CPU and GPU were within the range $[0.20, 0.99]$, e.g. for lattice approach with $M = 4$, and $N = 1024$, the CPU implementation was 5 times faster than the proposed approach for GPU. However, for longer filters and input data sizes the maximum speed-up of computations obtained with the proposed approach was more than 200 or 120 times, respectively, when compared to convolution and lattice based CPU implementations.

On the basis of the conducted experimental analysis we can state that the proposed approach to calculation of one-dimensional wavelet transform allows to obtain better results than CPU lattice approach, and the popular and widely used convolution approach for GPU. The possible advantage is several hundred times over CPU, and several times in case of GPU, and it can be observed for large data sets. Thus, we conclude that the proposed approach is highly effective for big-data problems.

5   Conclusions

In this paper the authors propose a novel approach to calculation of one-dimensional discrete wavelet transform. The presented solution takes advantage of shared memory and registers to implement communication between threads. Moreover only one call of a kernel function is required. The computations within the proposed approach are based on the lattice structure which is an effective tool for calculation of one-dimensional wavelet transform. The experimental analysis involving two models of GPU cards (NVIDIA GTX960 and NVIDIA RTX2060) reveals high time-effectiveness of the proposed solution. It is significantly faster than popular solutions dedicated for GPUs and known from the literature, i.e.

the convolution based approach and the approach based on naive lattice structure that requires multiple calls of kernel functions (see papers [19–21]). The experimental analysis shows that the obtained acceleration can be even 3 times (for NVIDIA GTX960) and 1.5 times (for NVIDIA RTX2060) for practical sized of filters ($M \in \{4, 6, 8\}$) when compared to the convolution based approach which highly outperforms the solution based on naive lattice structure. The highest acceleration is obtained with long filters (i.e. $M = 32$) and long sizes of input data and it equals even 5 times in case of NVIDIA GTX960 card (and up to 2 times for NVIDIA RTX2060). On the basis of the experimental analysis we conclude that the proposed approach fulfills its role and can find practical applications with particular emphasis on big-data problems.

References

1. Nakonechny, A., Veres, Z.: The wavelet based trained filter for image interpolation. In: Proceedings of IEEE First International Conference on Data Stream Mining and Processing, pp. 218–221 (2016)
2. Galletti, A., Marcellino, L., Russo, L.: A GPU parallel algorithm for image denoising based on wavelet transform coefficients thresholding. In: 14th International Conference on Signal-Image Technology and Internet-Based Systems (SITIS), pp. 485–492 (2018)
3. Nashat, A.A., Hussain Hassan, N.M.: Image compression based upon wavelet transform and a statistical threshold. In: International Conference on Optoelectronics and Image Processing (ICOIP), pp. 20–24 (2016)
4. Paul, A., Khan, T.Z., Podder, P., Ahmed, R., Rahman, M.M., Khan, M.H.: Iris image compression using wavelets transform coding. In: 2nd International Conference on Signal Processing and Integrated Networks (SPIN), pp. 544–548 (2015)
5. Stolarek, J., Lipiński, P.: Improving watermark resistance against removal attacks using orthogonal wavelet adaptation. In: Bieliková, M., Friedrich, G., Gottlob, G., Katzenbeisser, S., Turán, G. (eds.) SOFSEM 2012. LNCS, vol. 7147, pp. 588–599. Springer, Heidelberg (2012). https://doi.org/10.1007/978-3-642-27660-6_48
6. Liu, J.: An image watermarking algorithm based on energy scheme in the wavelet transform domain. In: IEEE 3rd International Conference on Image, Vision and Computing (ICIVC), pp. 668–672 (2018)
7. Babichev, S.: Optimization of Information Preprocessing in Clustering Systems of High Dimension Data. Radio Electr. Comp. Sci. Control. (2), 135–142 (2014)
8. Kim, B., McMurry, T., Zhao, W., Wu, R., Berg, A.: Wavelet-based functional clustering for patterns of high-dimensional dynamic gene expression. J. Comput. Biol. **17**(8), 1067–1080 (2010)
9. Dong, J.: Data mining of time series based on wave cluster. In: International Forum on Information Technology and Applications, IFITA 2009 (2009)
10. Nguyen, T., Strang, G.: Wavelets and Filter Banks. Cambr. Press, Welleslay (1996)
11. Mallat, S.G.: A theory for multiresolution signal decomposition: the wavelet representation. IEEE Trans. Pattern Anal. Mach. Intell. **11**(7), 674–693 (1989)
12. Yatsymirskyy, M.: Lattice structures for synthesis and implementation of wavelet transforms. J. Appl. Comput. Sci. **17**(1), 133–141 (2009)
13. Jeannot, E., Namyst, R., Roman, J. (eds.): Euro-Par 2011. LNCS, vol. 6853. Springer, Heidelberg (2011). https://doi.org/10.1007/978-3-642-23397-5

14. Angaji, E.T., Ebrahimi, S.A.R.: Accelerating Haar wavelet transform with CUDA-GPU. In: 13th International Conference on Natural Computation, Fuzzy Systems and Knowledge Discovery (ICNC-FSKD), pp. 791–796 (2017)

15. Aslan, S., Badem, H., Karaboga, D., Basturk, A., Ozcan, T.: Accelerating discrete haar wavelet transform on GPU cluster. In: 9th International Conference on Electrical and Electronics Engineering (ELECO), pp. 1237–1240 (2015)

16. Quan, T.M., Jeong, W.: A fast discrete wavelet transform using hybrid parallelism on GPUs. IEEE Trans. Parallel Distrib. Syst. **27**(11), 3088–3100 (2016)

17. Barina, D., Kula, M., Matysek, M., Zemcik, P.: Accelerating discrete wavelet transforms on GPUs. In: IEEE International Conference Image Processing (ICIP), pp. 2707–2710 (2017)

18. Enfedaque, P., Auli-Llinas, F., Moure, J.C.: Implementation of the DWT in a GPU through a register-based strategy. IEEE Trans. Parallel Distrib. Syst. **26**(12), 3394–3406 (2015)

19. Puchala, D., Szczepaniak, B., Yatsymirskyy, M.: Lattice structure for parallel calculation of orthogonal wavelet transform on GPUs with CUDA architecture. Przeglad Elektrotechniczny **91**(7), 52–54 (2015)

20. Puchala, D., Stokfiszewski, K., Szczepaniak, B., Yatsymirskyy, M.: Effectiveness of fast Fourier transform implementations on GPU and CPU. Przeglad Elektrotechniczny **92**(7), 69–71 (2016)

21. Puchala, D., Stokfiszewski, K., Wieloch, K., Yatsymirskyy, M.: Comparative study of massively parallel GPU realizations of wavelet transform computation with lattice structure and matrix-based approach. In: IEEE Second International Conference on Data Stream Mining and Processing (DSMP), pp. 88–93 (2018)

22. Cooklev, T.: An efficient architecture for orthogonal wavelet transforms. IEEE Signal Process. Lett. **13**(2), 77–79 (2006)

A Dynamic Replication Approach for Monte Carlo Photon Transport on Heterogeneous Architectures

Ryan Bleile[1,2](✉), Patrick Brantley[1], Matthew O'Brien[1], and Hank Childs[2]

[1] Lawrence Livermore National Laboratory, Livermore, CA 94550, USA
[2] University of Oregon, Eugene, OR 97403, USA
bleile1@llnl.gov

Abstract. This paper considers Monte Carlo photon transport applications on heterogenous compute architectures with both CPUs and GPUs. Previous work on this problem has considered only meshes that can fully fit within the memory of a GPU, which is a significant limitation: many important problems require meshes that exceed memory size. We address this gap by introducing a new dynamic replication algorithm that adapts assignments based on the computational ability of a resource. We then demonstrate our algorithm's efficacy on a variety of workloads, and find that incorporating the CPUs provides speedups of up to 20% over the GPUs alone. Further, these speedups are well beyond the FLOPS contribution from the CPUs, which provide further justification for continuing to include CPUs even when powerful GPUs are available. In all, the contribution of this work is an algorithm that can be applied in real-world settings to make more efficient use of heterogeneous architectures.

Keywords: Monte Carlo · Photon transport · Load balance · GPU

1 Introduction

Monte Carlo transport is an important computational technique for nuclear science applications, including applications in physics, nuclear reactors, medical diagnostics, and more. The technique involves simulating phenomena by calculating how a sample of particles moves through and interacts with a background medium. A key consideration for the approach is how many particles to employ, as adding more particles provides more accuracy but comes at the expense of

NOTICE: This manuscript has been authored by Lawrence Livermore National Security, LLC under Contract No. DE-AC52-07NA2 734-I with the US. Department of Energy. The United States Government retains, and the publisher, by accepting the article for publication, acknowledges that the United States Government retains a non-exclusive, paid-up, irrevocable, world-wide license to publish or reproduce the published form of this manuscript, or allow others to do so, for United States Government purposes. LLNL-CONF-817536

M. Paszynski et al. (Eds.): ICCS 2021, LNCS 12742, pp. 228–242, 2021.
https://doi.org/10.1007/978-3-030-77961-0_20

increased computational requirements. In many cases, achieving sufficient accuracy requires a large number of particles, which in turn creates significant computational requirements.

Supercomputers are often used to simulate computationally-expensive Monte Carlo photon transport problems. These machines can calculate many more operations per second than a normal computer, which in turn enables many workloads to be simulated on feasible time scales. That said, these machines create significant challenges, as they require both parallel coordination within a compute node and across compute nodes. Further, the architectures of their compute nodes are increasingly heterogeneous, often containing both multi-core CPUs and one or more GPUs.

A typical strategy for a heterogenous supercomputer is to use the CPUs only for management and communication with other compute nodes and to use the GPUs to transport particles. This approach usually pairs each GPU with one CPU core to drive the application, and leaves the rest of the CPU cores idle. Based on the relative FLOPS, utilizing the CPU only for management tasks would appear to be an acceptable strategy. Using Lawrence Livermore's RZAnsel [1] supercomputer as an example, the GPUs make up 1,512 TFLOPS, while the CPUs make up 58 TFLOPS, for a total system GPU+CPU count of 1,570 TFLOPS. This means that GPUs and CPUs make up 96.3% and 3.7% of the total FLOPS, respectively—the programmer effort to engage the CPU may not be viewed as worthwhile. However, CPUs have other benefits, including increased memory size and reduced latency to access memory. Further, many operations for Monte Carlo photon transport are not FLOP-bound. In all, engaging CPUs to carry out computation has the potential to add benefits beyond their FLOPS contributions (e.g., beyond 3.7%).

O'Brien et al. [8] were the first to demonstrate benefits from incorporating CPUs alongside GPUs to carry out Monte Carlo photon transport. That said, their algorithm was limited in utility, because it could only be applied to meshes that could fit entirely within GPU memory. This limitation is crucial in the context of supercomputers, since typical simulations at large scale use computational meshes that exceed GPU memory. Such meshes are decomposed into domains (or blocks), with each block small enough to fit within memory and each compute node working on one (or more) blocks. This domain decomposition complicates execution, as each compute node can only transport particles where it has valid data. In this paper, we expand upon the work by O'Brien et al. to deal with domain-decomposed meshes. We accomplish this by introducing two new algorithms: one for load balancing and one for building communication graphs. We also analyze the effects of domain decomposition on the performance of hybrid heterogeneous approaches. In all, the contribution of this work is a practical algorithm that translates the potential demonstrated by O'Brien et al. into a real world setting.

2 Background

Monte Carlo photon transport problems divide their spatial domains amongst its compute resources (i.e., MPI Ranks) in a non-traditional manner. In many physics simulations, there is a one-to-one mapping between compute resources and spatial domains—a physics simulation with N compute resources has N spatial domains, and each compute resource has its own unique spatial domain. With Monte Carlo photon transport problems, the full mesh is often too large to fit into one compute resource's memory, but not so large that it must be fully partitioned across the total memory of all the compute nodes. Saying it another way, there are often fewer spatial domains than compute resources, and so multiple computational resources can operate on the same domain at the same time. Consider a simple example with two spatial domains (D_0 and D_1) and four compute resources (P_0, P_1, P_2, and P_3). One possible assignment is for D_0 to be on P_0, P_1, and P_2 and D_1 to be on P_3, another possible assignment is for D_0 to be on P_0 and P_1 and D_1 to be on P_2 and P_3, and so on. Overall, domain assignment is an additional component for optimizing performance.

In the Monte Carlo community, the mapping of spatial domains to compute resources is referred to as "replication," as the mapping will replicate some domains across the resources. There are two main strategies for replication: static and dynamic. Static replication makes assignments when the program first begins and uses those assignments throughout execution. Dynamic replication changes assignments as the algorithm executes, in order to maintain load balancing. Both replication strategies aim to improve efficiency—they operate by replicating the spatial domains that have more particles, in order to distribute the workload more evenly across compute resources.

Dynamic replication is part of an overall approach for Monte Carlo transport. Each cycle of a Monte Carlo approach consists of three phases: initialization, tracking, and finalization. When incorporating dynamic replication, the initialization phase executes the dynamic replication algorithm. The tracking phase does a combination of particle transport and communicating particles. Particle transport operates mostly in an embarrassingly parallel fashion, up until particles move from one spatial domain to another (and thus need to be re-assigned to a compute resource that has that spatial domain) and thus MPI communication is required. The finalization phase processes the distributed results of the tracking phase. Importantly, the initialization phase determines the performance of the tracking phase—if the domain assignments from the dynamic replication algorithm create balanced work for each compute resource, then all compute resources should complete the tracking phase at the same time, ensuring parallel efficiency.

Tracking is the computationally dominant portion of the algorithm. During tracking, each particle makes small advancements for short periods of time, and each advancement is referred to as a "segment." The type of activity within a segment can vary, which affects the computational cost and duration of the advancement for a segment. In this paper the three relevant activities are: (1) collisions with the background material, (2) moving between mesh elements, and (3) moving to the end of the time step. Tracking concludes when each particle has advanced for a period equal to the overall cycle duration—if the overall cycle takes ΔT seconds, if a given particle advances via N segments for that cycle, and if each segment i advances for some time t_i seconds, then $\sum_{i=0}^{N-1} t_i = \Delta T$.

3   Related Works

Many works have studied spatial domain decomposition methods for Monte Carlo particle transport. The method was introduced by Alme et al. [2], as they split a problem into a few parts, allowing for replications of spatial domains in order to parallelize the workloads while maintaining processor independence. Their proposed method was adopted by the Mercury simulation code and implemented in a production environment; this team then provided empirical evidence for its efficacy [11]. Spatial domain decomposition methods were further analyzed by Brunner et al. [4,5], who also contributed improvements for increasing scalability and improving performance overall. One of their important improvements for scalability was to add point-to-point communication, allowing processors in different spatial domains to communicate directly with one another. This was a change from a model where each spatial domain had a single processor which was in charge of all communication for that group of processors.

Work by O'Brien et al. [9] introduced dynamic replication. Their scheme performed regular evaluation of parallel efficiency and then performed load balancing when efficiency dropped below a specified threshold. O'Brien et al. [10] extended this work by adding a communication graph, which defined which processors can perform point-to-point communication during a cycle. Using these new communication graph algorithms, O'Brien was able to successfully scale Mercury on LLNL's Sequoia supercomputer to over one million processors while maintaining good parallel performance. This work showed that keeping the load balance during particle communication within a cycle is important for scaling parallel performance. When particles were communicated to neighbors without considering load balance, a single processor could become bogged down with significantly more work—work which potentially could have been shared. Additional extensions to this work can be seen in other groups as well, such as with Ellis et al. who looked into additional mapping algorithms under specific conditions in the Monte Carlo transport code, Shift [6]. Their work extends the communication graph concept by combining it with Monte Carlo variance reduction techniques to improve the overall efficiency for their use-cases.

While many works have focused on algorithmic improvements, many others have focused on evaluating load imbalance effects. In his PhD thesis, Paul Romano expanded upon the concept of domain decomposition algorithms by providing new analytical understanding [12]. In particular, Romano provided a basis for understanding the importance of load imbalance and being able to determine analytically the benefit of this method. Wagner et al. [13] took a more empirical approach when studying load imbalance of reactor physics problems. They considered the problem of load imbalance stemming from spatial decomposition, and proposed new decomposition methods for handling this issue. Similarly, Horlik et al. [7] explored several spatial domain decomposition methods and analyzed their effect on load imbalance. In summary, each of these groups identified load imbalance as a problem and proposed analysis and solutions that fit their specific needs.

As noted in the introduction, our closest comparator is a separate work from O'Brien et al. [8]. This work considered the problem of balancing particles in a given spatial domain among processors of varying speeds, but it did not consider domain decomposed meshes. As domain replication strategy is an important aspect to achieve performant algorithms, developing an algorithm that supports both heterogenous computing and domain decomposition is non-trivial and requires fresh investigation. This gap is the focus of our work.

4   Our Method

This section describes our novel dynamic replication algorithm for Monte Carlo transport. Our algorithm is optimized for heterogenous architectures—it assumes that individual computational resources will have different levels of compute power, and makes assignments based on that knowledge. Our algorithm consists of three steps:

1. **Assignment** (Sect. 4.1): identify how many times to replicate each spatial domain, and then assign those domains to compute resources.
2. **Distribution** (Sect. 4.2): partition the particles across compute resources.
3. **Mapping** (Sect. 4.3): build a communication graph between compute resources in order to communicate particles that have exited their current spatial domain during tracking.

4.1   Step 1: Assignment

This step produces an assignment of compute resources to spatial domains, with the goal of making an assignment that minimizes execution time. In particular, the number of particles per spatial domain varies, and so the goal is to replicate the domains with the most particles in order to assign a commensurate level of compute to each domain. The algorithm works by considering work and compute as proportions—if a domain has 10% of the particles, then that domain should be replicated so that it gets 10% of the compute resources. Further, if the assignments are effective, then all compute resources should complete at the same time during the tracking phase.

To make assignments, our algorithm needs to understand (1) how much work needs to be performed and (2) how capable the compute resources are. In both cases, we use results from the previous cycle, which we find to be a good representation for what work to expect in the next cycle. Explicitly, the total work for each domain is the number of segments to execute. We consider the per-domain work from the previous cycle as our estimate for the upcoming cycle. For compute rate, we consider how many segments per second each type of resource achieved. That is, we measure the average number of segments per second over all of the CPUs and the same for GPUs. Using past performance automatically accounts for variation in translating FLOPS to segments across hardware; where the FLOP ratio between a GPU and CPU may be 100:1, the ratio in average number of segments per second may be much lower, like 20:1.

Our algorithm depends on considering both work and compute in proportion to the whole, and we define three terms for ease of reference. Let PW_i be the proportion of work within spatial domain i. For example, if domain i has 10% of the total estimated work, then $PW_i = 0.1$. Further, let $PC\text{-}GPU$ and $PC\text{-}CPU$ be the proportion of total compute for a GPU and a CPU, respectively. For example, if a GPU can do 100 million segments per second, if a CPU can do 5 million segments per second, and if there are 4 GPUs and 20 CPUs, then the total capability is 500 million segments per second, and $PC\text{-}GPU = 0.2$ and $PC\text{-}CPU = 0.01$.

At the beginning of program execution, we assign each domain one GPU and one CPU. This ensures that every domain has "surge" capability in case the work assignment estimates are incorrect (which can happen when particles migrate from one domain to another at a high rate). Such surge capability prevents the worst case scenario—one compute resource takes a long time to complete its work, and the others sit idle. Further, one of these compute resources (either the CPU or GPU) can act as a "foreman" for its spatial domain. These foremen are bound to a spatial domain throughout program execution. When a compute resource is assigned a new spatial domain, it can get that domain from the appropriate foreman. The remaining compute resources can then be assigned to work on spatial domains dynamically.

Our assignment algorithm works in two phases. The first phase decides how many compute resources should be assigned to each spatial domain, and what type they should be. The second phase uses this information to make actual assignments to specific compute resources, being careful to minimize communication by keeping the same spatial domains on the same compute resources when possible.

The first phase employs a greedy algorithm, and is described in pseudocode below labeled "MakeGreedyAssignments." It begins by setting up an array variable that tracks how much work is remaining for each spatial domain ("RemainingWork") using the predicted work (PW_i) and taking into account the pre-allocated resources (one CPU and one GPU for each of the M spatial domains). The final step is to assign the remaining compute resources to spatial domains. $NGPU$ is the number of GPUs, it begins by assigning the $NGPU - M$ available GPUs to spatial domains, one at a time. Each time, the algorithm first finds the spatial domain d with most remaining work, i.e., its evaluations takes into account that resources have been assigned previously. After the GPUs, it then makes assignments for each of the $NCPU - M$ available CPUs in a similar manner.

function MAKEGREEDYASSIGNMENTS(M, NGPU, NCPU, PW, PCGPU, PCCPU)
    for i in range(M) **do**
        WorkRemaining[i] = PW[i]
    end for
    for i in range(M) **do**   ▷ Account for preallocated resources
        WorkRemaining[i] -= (PCGPU+PCCPU)
    end for
    NGPU -= M

NCPU -= M
for i in range(NGPU) **do**   ▷ Replicate remaining resources greedily
    d = FindDomainWithMostWork(WorkRemaining)
    WorkRemaining[d] -= PCGPU
    AssignGPUToSpatialDomain(d)
end for
for i in range(NCPU) **do**
    d = FindDomainWithMostWork(WorkRemaining)
    WorkRemaining[d] -= PCCPU
    AssignCPUToSpatialDomain(d)
end for
end function

All replication schemes nearly always have some load imbalance. Consider a problem with two spatial domains with equal amounts of particles ($PW_0 = PW_1 = 0.5$) and three GPU compute resources, C_0, C_1, C_2 where $PC\text{-}GPU = 0.333$. Then C_0 and C_1 will be foremen, and the only question is whether to replicate domain 0 or 1 on C_2. Whatever the outcome, one domain will have a $WorkRemaining$ value of 0.167. In this example, it would be up to the foreman to carry out this extra work and it would be likely that the extra compute resources would be idle as it does so. Fortunately, these effects get smaller as concurrencies get larger. Also, the heterogeneous nature of compute helps on this front, as there are more resources (the CPUs) that are smaller (i.e., smaller values of $PC\text{-}CPU$) leading $WorkRemaining$ values being closer to 0 on the whole.

The second phase assigns specific compute resources. Every time a compute resource is assigned a new domain, it must retrieve this domain from its corresponding foreman, incurring a communication cost. So the goal of this phase is to repeat assignments between compute resources and domains. For example, if the output of the first phase indicates that domain d should have 3 GPUs, then the second phase checks to see if there are 3 GPUs that had d in the previous cycle. If so, then those GPUs should be assigned to d again for the current cycle, as this prevents unnecessary communication. Of course, as the number of compute resources applied to a domain increases, new compute resources must be located and communication costs are inevitable.

4.2 Step 2: Distribution

This step partitions the particles across compute resources. This partitioning must honor the spatial domain assignments, i.e., if particle P lies within spatial domain D, then the particle can only be to assigned to compute resources that were assigned D. In our approach, we perform this partitioning relative to performance—GPU compute resources get more particles and CPU compute resources get less, and the proportion between them corresponds to $\frac{PC\text{-}GPU}{PC\text{-}CPU}$. The remainder of implementation details follow trivially from previous work [8].

4.3   Step 3: Mapping

Mapping refers to establishing a communication graph between compute resources. This mapping is needed when particles exit their spatial domain. When this happens, they need to be sent from their current compute resource to another compute resource that is operating on their new spatial domain.

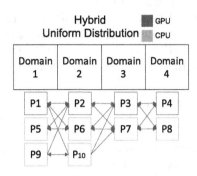

Fig. 1. Result of our Map step with 4 spatial domains, 4 GPU compute resources, and 6 CPU compute resources. The square boxes show which domains neighbor (1–2, 2–3, 3–4).

In a domain replication environment, a poor communication graph can affect overall performance. For example, assume that domain d is replicated by K compute resources— C_0, C_1, ... $C_{(K-1)}$. One possible communication graph could instruct all other compute resources to send their particle entering d to C_0. This is bad: C_0 would spend more time doing communication than the other C_i resources and it also will end up with more particles to transport. Instead, a better mapping would lead to an even spread of particles between the C_i's.

For a given compute resource, our Map algorithm makes connections to all neighboring domains. It uses a round robin algorithm to prevent load imbalance, specifically:

$$index_A \; mod \; size_B = index_B \; mod \; size_A$$

where A is a list of resources from one domain and B is a list of resources from a second domain (see Fig. 1). Our Map algorithm makes two connections for each neighboring domain d—one to a CPU compute resource that contains d and one to a GPU compute resource that contains d. Each connection also has a weighting which dictates the proportion of particles communicated. For our experiments, we set the weights to be proportional to their compute abilities (*PC-GPU* and *PC-CPU*), i.e., a GPU resource would be sent many more particles than a CPU resource. That said, exploring different weights would be interesting future work, in particular weights where CPUs get more particles.

5   Experiment Overview

This section provides an overview of our experiments, and is organized into three subsections. Subsection 5.1 describes the hardware and software used for our experiments. Subsection 5.2 describes the factors we vary to form our set of experiments. Finally, Subsect. 5.3 describes the measurements we use to evaluate our results.

5.1 Hardware and Software

Our experiments were run on LLNL's RZAnsel supercomputer. This platform has two Power 9 CPUs (22 cores per CPU, of which 20 are usable), 4 Nvidia Volta GPUs (84 SMs per GPU), and NVLink-2 Connections between the sockets on each node. In addition, there are a total of 256 GB of CPU memory and 64 GB of GPU memory per node [1]. For software, we used Imp [3], a Monte Carlo code that solves time-dependent thermal x-ray photon transport problems.

5.2 Experimental Factors

Our experiments vary two factors: workload (11 options) and Hardware configuration (3 options). We ran the cross product of experiments, meaning 33 experiments overall.

Workloads: our 11 unique workloads consisted of three distinct problems ("Crooked Pipe," "Holhraum," and "Gold Block"), with one of those problems ("Gold Block") having nine different variations. Details for each of the three distinct problems are as follows:

- **Crooked Pipe:** a problem that simulates transport through an optically thin pipe with a U-shaped kink surrounded by an optically thick material. The Crooked Pipe problem is load imbalanced since particles are sourced into the leading edge of the pipe, causing spatial domains that contain this region to have a much higher amount of work per cycle than the others. This is a common test problem in the Monte Carlo photon transport community as well as an excellent driver for testing load balancing methods.
- **Hohlraum:** a problem that simulates the effects of Lawrence Livermore's NIF laser on a gold hohlraum. Particles in this problem start in an incredibly hot gold wall and then propagate throughout the mostly hollow interior, colliding with a central obstruction as well as the surrounding gasses. This problem starts out very load imbalanced with most work in the hot region.
- **Gold Block:** a homogenous test problem that simulates a heated chunk of gold. This problem is a solid cylinder of gold with reflecting boundary conditions. Since this problem is a homogeneous material with reflecting boundary conditions, we can modify the length scale of the problem in order to change the ratio of the number of collision segments with the number of total segments by changing the number of mesh element crossing segments and leaving all else fixed. We use this length scaling to create a total of 9 configurations, with an unscaled version at the center we refer to as the Base Gold Block. Specifically, we took our Base Gold Block problem and halved the length scale 4 times consecutively, and similarly doubled the length scale 4 times consecutively to create these configurations. The goal with this scaling is to understand performance with respect to the percent of time performing collisions segments versus other segment types.

Hardware configurations: we ran each of the workloads with:

- **Hybrid:** our algorithm, scheduled with both GPUs and CPUs.
- **CPU-Only:** scheduled using only CPU resources
- **GPU-Only:** scheduled using only GPU resources

For the **CPU-Only** and **GPU-Only** tests, we were able to perform experiments using our algorithm, since our algorithm simplifies to be the same as predecessor work when the resources are homogeneous. Further, all experiments were run on 4 nodes, meaning we used: for CPU-Only 160 CPU resources, for GPU-Only 16 GPU resources, and for Hybrid 144 CPU resources + 16 GPU resources. Each of these experiments were run with a total of 80 million particles.

5.3   Measurements

To analyze our results, we considered three types of measurements:

- **Throughput** defines the number of segments, on average, that a processor will be able to process in one second. This metric is used to compare application performance in a consistent manner, regardless of hardware or software configuration.
- **Segment Counters** divide the segments into the three different activity types considered in this paper (see Sect. 2). Specifically, these counters count the total number of times each type of segment has occurred across all segments in the simulation. Segment counters are useful for understanding how performance varies with respect to different segment types.
- **Efficiency** determines the success of a load balance algorithm. For each domain i, we calculate the ratio of the compute resource applied to that domain (sums of $PC\text{-}GPU$ and $PC\text{-}CPU$ for the assigned C_i's) and work for that domain (PW_i). For example, a given domain may have 8% of the compute resources and 10% of the total work, for a ratio of 0.8 or an efficiency of 80%. Our efficiency metric is the minimum of these ratios over all domains, meaning 1.0, 100%, is a perfect score (compute resources applied perfectly in proportion to work for all domains) and less than 1.0 indicates the inefficiency—a score of 0.5 indicates that one domain has been given half the resources it needs, i.e. it has an efficiency of 50%.

6   Results

Our results are organized into three parts:

- Section 6.1 evaluates the performance of our overall heterogenous algorithm.
- Section 6.2 evaluates the efficacy of our load balancing algorithm.
- Section 6.3 evaluates the importance of our algorithm's surge capability.

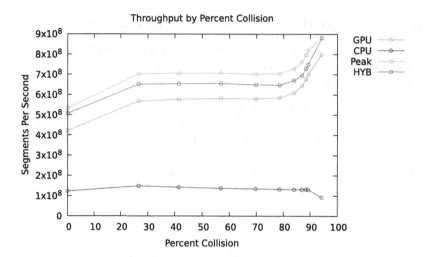

Fig. 2. Plot showing throughput (segments per second) as a function of what proportion of the segments were of the type "collision," as given by the segment counters. It plots four lines, one for each of our three hardware configurations, and one for a theoretical "peak" configuration (described in Sect. 6.1). Each of the dots come from our workloads—the left-most (∼0% collisions) come from the Crooked Pipe problem, the right-most (∼100% collisions) come from the Hohlraum problem, and the remainder come from variations of the Gold Block problem.

6.1   Algorithm Performance

Figure 2 shows the performance results for our 33 experiments. This figure contains a line for "peak" performance. This does not represent actual experiments, but rather a theoretical analysis of the potential peak speedup from using both CPUs and GPUs. This line was calculated by taking the GPU performance and adding 90% of the CPU performance (since some CPUs are needed to manage GPUs in a heterogeneous environment, specifically 36 of the 40 CPU cores were used for computation, while the remaining 4 managed GPUs). This peak line should be viewed as a "guaranteed-not-to-exceed" comparator.

One important finding is on the potential of heterogenous computing for this problem. While the CPUs have only 3% of the FLOPs of the GPUs, their performance (i.e., throughput) is much better than 3%. CPUs have 26.2% of the throughput for the Crooked Pipe problem, 20.4% for the Base Gold Block problem, and 10.4% for the Hohlraum problem. In all, this provides important evidence that including CPUs can be much more beneficial than a basic FLOP analysis. Of course, this potential can only be leveraged with an effective algorithm.

With respect to actual achieved performance, our heterogenous algorithm ("Hybrid") performed quite well. It was 20.4% faster than GPU-only for the Crooked Pipe problem, and approximately 10% faster for the other problems. Relative to the peak line, our algorithm achieved 95.1% for Crooked Pipe, 91.8%

for Base Gold Block, and 99.6% for Hohlraum. The performance is greatest where the amount of collision segments is dominant, which is also where there is a larger amount of compute used by the resources. In this region, the CPUs are less valuable, but still more valuable than the hardware specification predicts. On the other end of the spectrum, where there are less collisions and more mesh element crossing segments, the compute is lower, and the GPUs are less performant. This enables the CPUs to provide an even greater benefit overall.

Table 1. This table shows the efficiency for our three workloads over a full program execution. Minimum efficiency represents the worst assignment over all compute resources and cycles: for one cycle of the Crooked Pipe problem, there was a compute resources which had about 20% too much work to finish on time with the other compute resources. Maximum efficiency speaks to the best cycle: for one cycle of the Crooked Pipe problem, the most underpowered compute resource had only 0.1% too much work. Average efficiency speaks to the behavior across cycles: (1) for each cycle, identify the most underpowered compute resource and calculate how much extra work it has, and (2) take the average over all cycles of the extra work amounts. For the Crooked Pipe problem, the average efficiency is 99.55%, meaning that the average slowdown for completing a cycle due to load balancing was <0.5%.

Problem	Minimum efficiency	Maximum efficiency	Average efficiency
Crooked pipe	81.76%	99.92%	99.55%
Base gold block	81.76%	99.9%	99.90%
Hohlraum	81.76%	86.16%	85.80%

6.2   Load Balance Efficiency

Table 1 plots efficiency results for our three workloads. On the whole, the minimum efficiency values for these workloads are low. That said, these conditions occur in the first few cycles, as these cycles do not have a history of performance to base their load balancing decisions on. For Crooked Pipe and Base Gold Block, the average efficiencies indicate that good load balance is achieved quickly and maintained throughout the run. The Hohlraum problem had worse efficiency. This was because one domain was a "hot spot"—it had much more work than the other domains. This topic is explored further in the following subsection.

6.3   Surge Capability

The Hohlraum workload demonstrates the value in our "surge capability" (ensuring that one GPU and CPU are assigned to each domain). This workload had 4 domains, and domain 1 had the majority of particles, to the point meriting assignment of every non-foreman compute resource. That said, during a compute cycle, domain 1's particles stream out rapidly into neighboring domains. Our surge capability ensured extra compute resources were allocated, and this made a 3× performance improvement for this case. Figure 3 has more details on this comparison, with Gantt charts that show behavior within a cycle. Finally, the "surge" allocation had no impact on the other two problems since their work was more balanced, and they would have received those resources anyway.

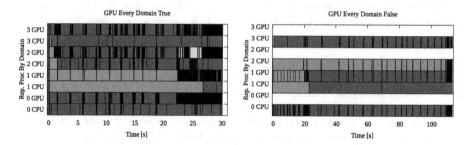

Fig. 3. Gantt charts for a single cycle of the Hohlraum workload. The left Gantt chart corresponds to our algorithm, and completes in 30 s. The right Gantt chart is a variant of our algorithm where there is no minimum compute allocation (i.e., the "surge capability" is disabled). This variant took 113 s, 3.8× slower. The Gantt plots show an evolution over time per compute resource, with red representing "idle" time, yellow representing communication time ("MPI Send/Recv"), and green representing time spent tracking particles. The blank spaces in the right chart occur because there is no compute resource assigned to that domain, for example no GPU resource for domain 3. Finally, these Gantt charts show only the first CPU and first GPU for a domain, and the other compute resources are not plotted. In particular, the remaining compute resources in the left Gantt chart (our algorithm) are doing tracking (green) at a high rate, consistent with the overall efficiency of 85%—some of the worst performers for this workload (domains 0, 2, and 3) are being plotted.

7   Conclusion

In this paper, we introduce a novel load balancing algorithm which can efficiently partition heterogeneous compute resources across domains. We demonstrate results using this algorithm, in a production Monte Carlo photon transport code, running a variety of workloads. This work was motivated by the performance difference seen in practice between Monte Carlo transport codes running on CPUs and GPUs when compared with the ratio of the available FLOPs. Our algorithm demonstrated up to a 20% performance benefit, which is much greater than the 3.7% predicted by solely looking at the ratio of FLOPs. Additionally, our algorithm achieves 85% to 99% load balancing efficiency on the problems demonstrated.

In terms of future work, we wish to study more workloads, such as neutron transport, and to run larger problems using more compute resources. We also plan to look more at the surge capability, and whether better predictions can be made about when it is needed. In particular, the assignment of at least one GPU to every domain can be wasteful in some configurations, possibly limiting performance by as much as 15% in extreme cases. Further, while domain replication algorithms assume that each MPI rank has a single domain for a given cycle, more adaptive approaches, including switching a domain mid-cycle and splitting a domain into pieces could have performance benefits. Another improvement would be adapting assignments based on current loads, in case the target resource already is overloaded. Finally, we plan to extend our algorithm to work with even more heterogeneous architectures in the future, including multiple types of accelerators on a node and even nodes with different types of compute power. This work will happen as such machines come online, as most supercomputers are using strictly CPUs and GPUs at this time.

References

1. Livermore computing center high performance computing: Rzansel. https://hpc.llnl.gov/hardware/platforms/rzansel. Accessed 9 Dec 2020
2. Alme, H.J., Rodrigue, G.H., Zimmerman, G.B.: Domain decomposition models for parallel Monte Carlo transport. J. Supercomputing **18**(1), 5–23 (2001)
3. Brantley, P., et al.: A new implicit Monte Carlo thermal photon transport capability developed using shared Monte Carlo infrastructure. In: The International Conference on Mathematics and Computational Methods Applied to Nuclear Science and Engineering (M and C 2019), Portland, Oregon, pp. 25–29. (2019)
4. Brunner, T.A., Brantley, P.S.: An efficient, robust, domain-decomposition algorithm for particle Monte Carlo. J. Comput. Phys. **228**(10), 3882–3890 (2009)
5. Brunner, T.A., et al.: Comparison of four parallel algorithms for domain decomposed implicit Monte Carlo. J. Comput. Phys. **212**(2), 527–539 (2006)
6. Ellis, J.A., et al.: Optimization of processor allocation for domain decomposed Monte Carlo calculations. Parallel Comput. **87**, 77–86 (2019)
7. Horelik, N., Siegel, A., Forget, B., Smith, K.: Monte Carlo domain decomposition for robust nuclear reactor analysis. Parallel Comput. **40**(10), 646–660 (2014)
8. O'Brien, M., et al.: Hybrid CPU-GPU load balancing for Monte Carlo particle transport. In: Proceedings of the 26th International Conference on Transport Theory (ICTT-26), Sorbonne Univeristy, Paris, France (2019)
9. O'Brien, M.J., Brantley, P.S., Joy, K.I.: Scalable load balancing for massively parallel distributed Monte Carlo particle transport. In: Proceedings of International Conference on Mathematics and Computational Methods Applied to Nuclear Science and Engineering (M and C 2013), Sun Valley, Idaho, vol. 45, pp. 647–658 (2013)
10. O'Brien, M.: Dynamic load balancing of parallel Monte Carlo transport calculations via spatial redecomposition. In: Proceedings of the Joint International Topical Meeting on Mathematics and Computation and Supercomputing in Nuclear Applications, pp. 16–19 (2007)
11. Procassini, R., O'Brien, M., Taylor, J.: Load balancing of parallel Monte Carlo transport calculations. In: Proceedings of the 2005 ANS Topical Meeting in Mathematics and Computation, Avignon, France, 12–15 September 2005 (2005)

12. Romano, P.K.: Parallel algorithms for Monte Carlo particle transport simulation on exascale computing architectures. Ph.D. thesis, Massachusetts Institute of Technology (2013)
13. Wagner, J.C., et al.: Hybrid and parallel domain-decomposition methods development to enable Monte Carlo for reactor analyses. Prog. Nucl. Sci. Technol. **2**(1), 815–820 (2011)

Scientific Workflow Management on Hybrid Clouds with Cloud Bursting and Transparent Data Access

Bartosz Baliś[1]([📧]) [iD], Michał Orzechowski[1], Łukasz Dutka[2][iD],
Renata G. Słota[1][iD], and Jacek Kitowski[1,2][iD]

[1] Department of Computer Science, AGH University of Science and Technology,
Krakow, Poland
{balis,morzech,rena,kito}@agh.edu.pl
[2] AGH University of Science and Technology, ACK Cyfronet AGH, Krakow, Poland

Abstract. Cloud bursting is an application deployment model wherein additional computing resources are provisioned from public clouds in cases where local resources are not sufficient, e.g. during peak demand periods. We propose and experimentally evaluate a cloud-bursting solution for scientific workflows. Our solution is *portable* thanks to using Kubernetes for deployment of the workflow management system and computing clusters in multiple clouds. We also introduce *transparent data access* by employing a virtual distributed file system across the clouds, allowing jobs to use a POSIX file system interface, while hiding data transfer between clouds. To *balance load distribution* and *minimize the communication volume* between clouds, we leverage graph partitioning, while ensuring that the algorithm distributes the load equally at each parallel execution stage of a workflow. The solution is experimentally evaluated using the HyperFlow workflow management system integrated with the Onedata data management platform, deployed in our on-premise cloud in Cyfronet AGH and in the Google Cloud.

Keywords: Cloud bursting · Scientific workflow management · Scientific data management · Hybrid cloud · Kubernetes

1 Introduction

Cloud bursting is an application deployment model leveraging *hybrid cloud* [17], in which the application normally runs in a private cloud, while during peak demand periods it is partially deployed on additional resources allocated in a public cloud [10]. Hybrid computing approach based on on-premise computing infrastructure and resources allocated on-demand from a public cloud is increasingly adopted in scientific computing [19]. The motivations to move scientific computations to public clouds include lack of appropriate local resources [4], insufficient computing infrastructure due to growing user base [1], and better turnaround times achieved in the cloud due to long queue wait time in a local

© Springer Nature Switzerland AG 2021
M. Paszynski et al. (Eds.): ICCS 2021, LNCS 12742, pp. 243–255, 2021.
https://doi.org/10.1007/978-3-030-77961-0_21

cluster, even if it is better in terms of raw performance [3,16]. Consequently, cloud bursting is definitely an attractive model for scientific applications.

In this paper, we propose a cloud bursting approach for scientific workflows leveraging graph partitioning. In our solution we address the following problems: (1) *Portability*: we deploy the Workflow Management System and computing clusters on the private and public clouds using Kubernetes, taking benefit from its cloud-agnostic application deployment model [20]. Consequently, we support any cloud in which a Kubernetes cluster can be deployed. (2) *Transparent data access*: we employ a virtual distributed file system across two clouds, so that workflow jobs use a POSIX file system interface, while data is transparently transferred regardless of its location (local or remote), and underlying storage technology. (3) *Execution optimizations*: we introduce an algorithm for partitioning of the workflow in order to balance task distribution between partitions (load balancing in space) and in individual parallel execution phases of the workflow (load balancing in time); the algorithm allows uneven partitions (cluster sizes), and optimizes the volume of communication between clouds. We estimate the required size of the cluster rented in the public cloud by taking into account the level of parallelism of the workflow and resource demands of workflow jobs. The solution is experimentally evaluated with the HyperFlow workflow management system [2] integrated with the Onedata data management platform [8].

The paper is organized as follows. Section 2 presents related work. Section 3 describes the proposed solution. Section 4 contains experimental evaluation of the solution. Section 5 concludes the paper.

2   Related Work

The usefulness of hybrid infrastructures in scientific computing have been recognized since the early days of clouds [5,21,24]. Hybrid clouds have also been the subject of research specifically in the context of scientific workflows. Liu and others [15] introduce an algorithm for dynamic placement of large scientific workflow data sets in hybrid clouds. In [13] and [6], the authors propose algorithms for scheduling of workflows on hybrid clouds, with deadline and cost constraints. These algorithms heavily rely on accurate predictions of task execution times on specific computing resources which can be difficult [23].

In [22] graph partitioning is used to guide scheduling of scientific workflows in a cluster of nodes, but not in the context of a hybrid cloud. Cloud bursting is supported in the Galaxy system dedicated for life science workflows [9]. The key component in this solution is a job mapper which decides to which cloud a given job should be submitted. This decision-making is based on simple criteria, e.g. the resource availability, location of specific tools, or location of data [1].

Overall, no approaches investigate graph partitioning for hybrid execution of scientific workflows with cloud bursting, combined with architectural aspects enabling portability and transparent data access. All these issues are addressed by the research presented in this paper.

3 Cloud Bursting Solution for Scientific Workflows

3.1 Architecture

Figure 1 shows a conceptual architecture of the proposed solution. The private cloud is running Kubernetes cluster 1 – the *home cluster*. We assume that the user is assigned a certain quota of N virtual CPUs (vCPU) in the home cluster, distributed across a certain number of nodes. The user decides to allocate another Kubernetes cluster in a public cloud in order to speed up the computations. The second cluster has size of M vCPUs.

The workflow graph is then divided into two partitions, one for each of the clusters. The relative sizes of the partitions are, respectively, $\frac{N}{N+M}$ and $\frac{M}{N+M}$. The partitioning should minimize the volume of the inter-cluster communication and maximize task parallelism, see Sect. 3.2 for more details.

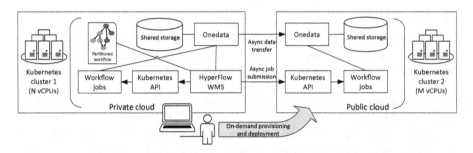

Fig. 1. High-level concept of the proposed cloud bursting architecture for scientific workflows. The user has a cluster with N virtual CPUs in the private cloud and allocates another cluster with M vCPUs in the public cloud. The workflow graph is divided into two partitions proportional to cluster sizes. The partitioning should minimize inter-cloud data transfer without reducing task parallelism.

How to estimate the desired size (in virtual CPUs) of the second cluster? The upper limit on M can be estimated as follows:

$$M = \sum_{p_i \in P_{LoP}(w)} cpuReq(p_i) - N_{vcpu}(cluster_1) + 0.5 * N_{nodes} \qquad (1)$$

where $P_{LoP}(w)$ is the subset of workflow tasks that belong to the largest graph level (execution phase) of the workflow graph. In other words, a cluster of its size would accommodate for the largest level of parallelism (LoP) in the workflow and avoid task waits. The factor $0.5 * N_{nodes}$ represents the fact that about 0.5 of vCPU is reserved on each node for the Kubernetes middleware. Let us note that the Level of Parallelism in a Directed Acyclic Graph (DAG) of tasks is not necessarily equal to the size of the largest level graph. A better estimation can be done by simulating the workflow 'execution wave' [12]. $cpuReq(p_i)$ denotes the CPU request (in vCPUs) of task p_i, while $N_{vcpu}(cluster_1)$ is the number of vCPUs in Cluster 1.

Figure 1 also shows a simplified deployment of components on both clusters. The HyperFlow engine [2] runs in the home cluster and executes the workflow. Based on the information on graph partitioning, HyperFlow creates workflow jobs either in the home or the remote Kubernetes cluster. The Onedata data management platform [8] is responsible for synchronizing data between the clusters. The input data of the workflow is initially located on the storage node in the home cluster. On the remote cluster, Onedata creates a virtual POSIX file system and synchronizes metadata about the input files, so that the files are visible locally, even though they physically exist in a remote file system. Onedata transfers the files only when they are accessed, and only these blocks that were actually read which is efficient for very large files whose only small parts are read by remote processes.

3.2   Workflow Partitioning

To map workflow tasks to different clusters (clouds), we use graph partitioning [18]. Graph vertices represent jobs while edges denote communications between them. Vertex *weight* denotes the *computational cost* of a job, in our case the requested amount of *vCPU*, e.g. 0.5. Vertex *size*, on the other hand, represents the *communication cost* due to the execution of the job, i.e., transfer of input files from a storage node and transfer of output files to the storage node. A typical partitioning algorithm will try to minimize the *edge cut* metric, i.e. the number of edges that connect different partitions, while balancing the load (vertex weights) between the partitions. Figure 2a shows an example workflow partitioned in this way which results in a suboptimal mapping of workflow tasks onto clusters. First, because the volume of communication between clusters is not minimized. Second, because job parallelism is reduced – jobs run only in the first cluster, then only in the second one. In the case of scientific workflows, it is important to divide the computational tasks equally between the partitions at each individual *execution phase* of the workflow [22]. The execution phase of a task $Ep(p_i)$ corresponds to its level in the graph, i.e. it is an integer $1..L$ defined as follows:

– If *task* p_i has no *predecessors*, $Ep(p_i) = 1$.
– Otherwise $Ep(p_i) = max([Ep(predecesors(p_i)]) + 1$.

While the edge cut metric often minimizes the communication volume, for certain types of graphs this is not the case. For example, if a vertex has multiple edges that connect to vertices from another partition, the inter-partition communication would occur once, but the edge cut metric would count each edge separately. For such graphs, a metric optimizing the number of boundary vertices works better, because it actually optimizes the communication volume [18]. Figure 2b shows a partitioning of the same workflow according to this metric. This partitioning not only reduces the inter-partition communication, but also results in a better task parallelism. As shown in the figure, the two parallel phases of the workflow are distributed across two partitions.

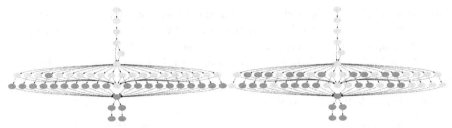

(a) Minimizing the edgecut.   (b) Minimizing the communication volume.

Fig. 2. Workflow partitioning between clusters. Colors denote partitions to which tasks are assigned. (Color figure online)

In our case, we assume that the input data of the workflow is located on a storage node in the home cluster. Therefore, we add an additional, special *storage node* to the graph that has weight of 0 (no computational cost) and represents the transfer of workflow input data to workflow jobs. The storage node is a special *fixed node* which is always pre-assigned to the the partition that represents the cluster where the data is located.

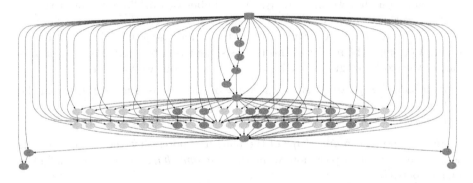

Fig. 3. Workflow graph with the special storage node and unequal partitions (cluster sizes): cluster $1 = 0.4$, cluster $2 = 0.6 *$ total $vCPU$. Partitioning optimized for total communication volume. Colors denote partitions to which tasks are assigned. (Color figure online)

A partitioning must also address the problem of different cluster sizes and allow unequal partitions while minimizing communication volume and maximizing task parallelism. Figure 3 shows a partitioning where the storage node is depicted and the sizes of the partitions vary, such that partition 1 has weight of 0.4, while partition 2 of 0.6.

While we found that the communication volume metric often results in desirable task parallelism, the addition of the storage node may easily disrupt it. Therefore we have decided to use multiple weights for vertices to ensure that the load is equally distributed at each parallel execution phase of the workflow [22]. A vertex has now n weights, where n is equal to the number of levels in the workflow graph. For the sake of example, let us assume that a workflow graph has 7 levels. Job p, which belongs to graph level 4, and whose computational cost $C_p = 0.5 \, vCPU$ will have the following weights vector:

$$w_p = [\, 0 \; 0 \; 0 \; 0.5 \; 0 \; 0 \; 0 \,]$$

In other words, $w_i = C_p$ for $i = level(p)$, 0 otherwise. As a result, the partitioning algorithm will perform multi-constraint optimization, trying to balance each of the weights separately, leading to balanced task parallelism.

4   Evaluation

This section presents evaluation of the proposed solution. Section 4.1 presents analysis of the workflow partitioning algorithm for different synthetic workflow graphs, while Sect. 4.2 describes the setup of the experimental evaluation and shows its results.

4.1   Workflow Partitioning Analysis

In order to study the impact of different parameters of the algorithm on the partitioning quality, we have analyzed a number of workflow graphs from the Pegasus workflow gallery [7]: Cybershake 1000 vertices, Epigenomics 997 vertices, LIGO 1000 vertices, Sipht 968 vertices, and Montage 2.0 6448 vertices. Figure 4 presents the results of this analysis, depicting total communication volume of the workflow graph partitioning depending on two variables:

- *distribution of partition sizes*: the value on the x axis denotes the size of the first partition in relation to the entire graph.
- *allowed load imbalance between partitions* (ufactor): 1%, 5%, and 10%.

In addition, the *round shape* of the marker denotes that the storage node is assigned to the bigger partition, while the *diamond shape* means that it is in the smaller partition.

Several observations can be made from these results. In general, not surprisingly, allowing load imbalance among the partitions results in a better communication volume. However, the gain heavily depends on the workflow structure. For Cybershake, for example, it is not significant in any of the partitioning variants, while in the case of Montage 2.0, the difference can be substantial.

Table 1 shows the distribution of workflow tasks between partitions individually for workflow execution phases and for various workflow examples. In all cases partitions sizes were equal and the partitioning was optimized for the communication volume, except for Sipht where the result for the edge cut optimization is also shown for comparison. As one can see, the optimization for the communication volume results in approximately equal distribution of tasks among partitions, but some divergences are inevitable due to the structure of the workflows, as is in the case of the Montage 2.0 workflow which is perfectly divided when using 3 partitions.

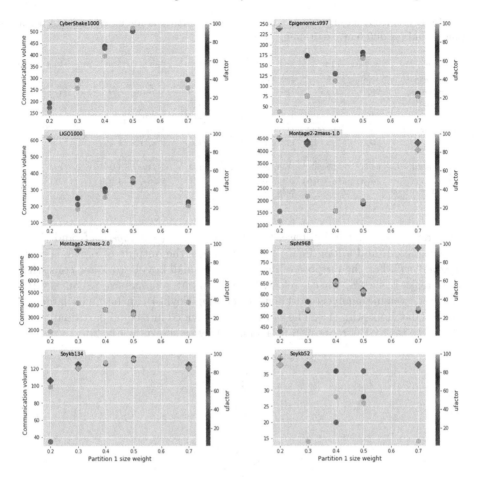

Fig. 4. Workflow partitioning results for eight different workflow graphs showing communication volume for different partition size distribution (0.2/0.8, 0.3/0.7, 0.4/0.6, 0.5/0.5 and 0.7/0.3 and allowed load imbalance between partitions (1%, 5%, 10%, denoted by parameter *ufactor* of value 10, 50 and 100, respectively).

4.2 Experimental Runs

For the purpose of experimental runs, we have set up two small Kubernetes clusters: one in the home cloud in Cyfronet AGH consisting of 4 nodes with 6 vCPU and 21 GB of RAM each, and a remote one in Google Cloud comprising 6 nodes, each with 4 vCPU and 16 GB of RAM. Consequently, both clusters were approximately equal in terms of the number of virtual CPUs and the amount of RAM per vCPU. For shared storage, we used NFS in the home cluster and Ceph in the remote one. In both cases, these served as a backed storage for Onedata that exposed a POSIX virtual file system for the workflow jobs.

In the experimental runs we used the genomic workflow Soykb [14], whose structure is shown in Fig. 5. The workflow is characterized by a large input file

Table 1. Distribution of workflow tasks between partitions for individual execution phases and for different partitioning strategies.

Workflow	# Tasks	Ph1	Ph2	Ph3	Ph4	Ph5	Ph6	Ph7	Ph8	Ph9
Cybershake	1000	5/5	240/256	240/254						
Epigenomics	997	3/4	119/126	119/126	119/126	119/126	3/4	1/0	1/0	1/0
Sipht	968	353/359	44/20	68/60	21/11	12/20				
Sipht (edge cut)	968	265/447	41/23	128/0	32/0	32/0				
Ligo	1000	123/ 106	123/106	6/14	123/128	123/ 128	6/14			
Montage 2.0 (dss)	6448	102/90	3148/2900	2/1	1/2	64/128	1/2	1/2	1/3	
Montage 2.0 (dss), 3 partitions	6448	64/64/64	2016/2016/2016	1/1/1	1/1/1	64/64/64	1/1/1	1/1/1	2/1/1	

– the reference genome database, two parallel stages (*genotype_gvcfs* and *filtering_snp* tasks), and a long final task that merges the outputs of previous tasks (*merge_gcvf*). For the purpose of illustration, a small Soykb workflow consisting of 52 tasks is shown, with a relatively small input genome size (about 1 GB). However, genome analysis workflows (including Soykb) can be much larger in size and data footprint [11]. The workflow was divided into equal partitions (with respect to task CPU requests), in a similar way as shown in Fig. 2b.

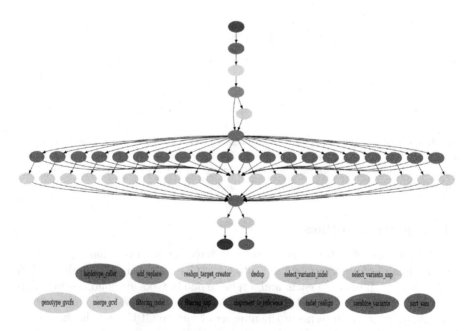

Fig. 5. Structure of the Soykb workflow.

The visualization of the execution of the workflow in the hybrid cloud is depicted in Fig. 6. Labels on the Y axis denote nodes on which tasks were executed, with k8s*/gke* denoting nodes in Cyfronet/Google Cloud, respectively. A given node may occur multiple times (e.g. k8s3-0 and k8s3-1) if tasks were running in parallel on this node. Note that in general more tasks were running in parallel on nodes in Cyfronet since they contained more virtual CPUs per node.

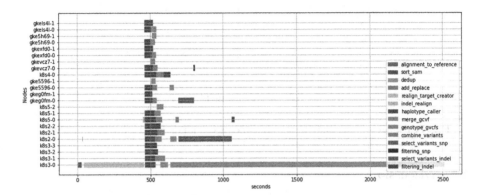

Fig. 6. Example visualization of the Soykb workflow execution in two clouds.

Figure 7 shows the data distribution among the two experimental clouds after the execution of the workflow. As one can see, only about 19% of all input and output files of the workflow were either transferred to or created in the remote (Google) cloud. On the other hand, almost all files were needed in the home cloud, where the final merge tasks were executed. Figure 7b shows that only about 50% of the largest input file was transferred from the home cloud to the remote cloud during the execution. This can be explained by looking at the data access pattern of this file, shown in Fig. 8. The chart shows which blocks of the file are accessed by individual jobs of the workflow. The type of jobs is denoted by different colors. The graph reveals that the tasks from the two parallel stages of the workflow (*genotype_gvcfs* and *filtering_snp*) read only small portions of the genome database, clearly showing a data-parallelism pattern. This is an opportunity for optimization since it is not necessary to transfer the entire file to the remote cloud. On the other hand, the two tasks with the 'gather' pattern (merge_gcvf and combine_variants) read the entire file which hints that these tasks should run in the home cluster in order to avoid the transfer of the entire file. To guide the partitioning algorithm into this optimization, the weights of the graph edges – denoting the cost of communication – should be based on the **number of file blocks read, not the entire input file size**. However, collecting such data on low-level data access patterns is not always straightforward, requiring instrumentation of low-level file IO subroutines. Because the Onedata system employs block-level data transfer optimization, and the load is divided equally into the two clouds, it is expected that about half of the input file will be transferred to the remote cloud.

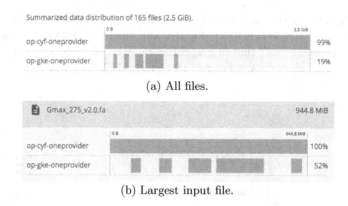

(a) All files.

(b) Largest input file.

Fig. 7. Runtime data distribution between two clouds for the Soykb workflow.

4.3 Discussion

The presented method has several limitations. The graph partitioning package that we have used (Metis) does not allow for defining fixed nodes. As a result, the obtained communication volume is only reliable when we assume that the partition to which the storage node was assigned represents the home cluster. Consequently, it could be difficult to optimize partitioning where the home cluster is significantly smaller (e.g. 0.2/0.8) because the storage node may tend to be assigned to the larger partition, depending on the workflow structure.

The method could be improved by introducing information about scheduling of jobs in which their allocation to time slots is taken into account. Such a scheduling algorithm could, rather than mapping jobs to a fixed set of nodes, predict the requirements for resources (vCPUs) in a given point of workflow execution. Consequently, the required size of the public cluster could be estimated better. Moreover, combined with auto-scaling capabilities, the remote cluster could be scaled up and down as needed.

We have focused on balancing the load due to requested vCPUs. The method could be extended to take into account also other resources, e.g. memory and storage. Supporting balancing of memory consumption would be relatively easy to implement – it would require adding memory-request weights to the partitioning algorithm. The Kubernetes scheduler already supports memory requests and assigns jobs to nodes based on them.

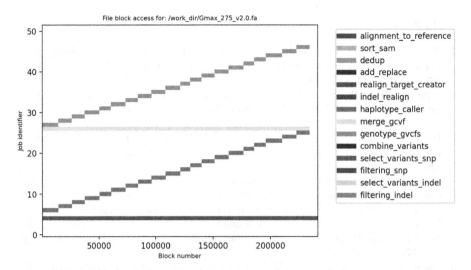

Fig. 8. Data access pattern of the largest input file of the Soykb workflow. The graph shows which blocks of the file were read by individual jobs with colors denoting different job types. The pattern reveals that the tasks from the two parallel stages of the workflow (*genotype_gvcfs* and *filtering_snp*) read only small portions of the file. (Color figure online)

5   Conclusions and Future Work

We have shown a solution for execution of scientific workflows in hybrid clouds with cloud bursting and transparent data access. Our solution enables portable deployment of scientific workflows into two or more clouds as long as they support Kubernetes clusters, as most large public cloud providers currently do. Transparent data access hides data transfer between clouds and allows for block-level optimization wherein only the required file blocks are transferred. We have used graph partitioning to ensure load balancing of the workload across two clouds while minimizing the communication volume. These capabilities were evaluated experimentally using the HyperFlow workflow management system integrated with the Onedata data management platform and deployed in a hybrid cloud comprising two Kubernetes clusters located in the Cyfronet AGH's private cloud and the Google Cloud. Future work involves more detailed analysis and experiments to investigate the impact of several factors on the execution performance and quality of workflow partitioning, including the accuracy of job resource requests, and accurate graph vertex and edge weights, reflecting data access patterns in scientific workflows.

Acknowledgements. The research presented in this paper has been partially supported by the funds of Polish Ministry of Science and Higher Education assigned to the AGH University of Science and Technology.

References

1. Afgan, E., Coraor, N., Chilton, J., Baker, D., Taylor, J., Team, G.: Enabling cloud bursting for life sciences within galaxy. Concurrency Comput. Pract. Experience **27**(16), 4330–4343 (2015)
2. Balis, B.: Hyperflow: a model of computation, programming approach and enactment engine for complex distributed workflows. Future Gener. Comput. Syst. **55**, 147–162 (2016)
3. Balis, B., Figiela, K., Jopek, K., Malawski, M., Pawlik, M.: Porting HPC applications to the cloud: a multi-frontal solver case study. J. Comput. Sci. **18**, 106–116 (2017)
4. Belgacem, M.B., Chopard, B.: A hybrid HPC/cloud distributed infrastructure: coupling EC2 cloud resources with HPC clusters to run large tightly coupled multiscale applications. Future Gener. Comput. Syst. **42**, 11–21 (2015)
5. Bicer, T., Chiu, D., Agrawal, G.: A framework for data-intensive computing with cloud bursting. In: 2011 IEEE International Conference on Cluster Computing, pp. 169–177. IEEE (2011)
6. Chang, Y.S., Fan, C.T., Sheu, R.K., Jhu, S.R., Yuan, S.M.: An agent-based workflow scheduling mechanism with deadline constraint on hybrid cloud environment. Int. J. Commun. Syst. **31**(1), e3401 (2018)
7. Da Silva, R.F., Chen, W., Juve, G., Vahi, K., Deelman, E.: Community resources for enabling research in distributed scientific workflows. In: 2014 IEEE 10th International Conference on e-Science, vol. 1, pp. 177–184. IEEE (2014)
8. Dutka, Ł., et al.: Onedata - a step forward towards globalization of data access for computing infrastructures. Procedia Comput. Sci. **51**, 2843–2847 (2015). International Conference On Computational Science, ICCS 2015
9. Goonasekera, N., Mahmoud, A., Chilton, J., Afgan, E.: Galaxycloudrunner: enhancing scalable computing for galaxy. BioRxiv (2020)
10. Guo, T., Sharma, U., Shenoy, P., Wood, T., Sahu, S.: Cost-aware cloud bursting for enterprise applications. ACM Trans. Internet Technol. (TOIT) **13**(3), 1–24 (2014)
11. Hazekamp, N., et al.: Combining static and dynamic storage management for data intensive scientific workflows. IEEE Trans. Parallel Distrib. Syst. **29**(2), 338–350 (2017)
12. Ilyushkin, A., Ghit, B., Epema, D.: Scheduling workloads of workflows with unknown task runtimes. In: 2015 15th IEEE/ACM International Symposium on Cluster, Cloud and Grid Computing, pp. 606–616. IEEE (2015)
13. Lin, B., Guo, W., Lin, X.: Online optimization scheduling for scientific workflows with deadline constraint on hybrid clouds. Concurrency Comput. Pract. Experience **28**(11), 3079–3095 (2016)
14. Liu, Y., et al.: PGen: large-scale genomic variations analysis workflow and browser in SoyKB. In: BMC Bioinformatics, BioMed Central, vol. 17, p. 337 (2016)
15. Liu, Z., et al.: A data placement strategy for scientific workflow in hybrid cloud. In: 2018 IEEE 11th International Conference on Cloud Computing (CLOUD), pp. 556–563. IEEE (2018)
16. Marathe, A., et al.: A comparative study of high-performance computing on the cloud. In: Proceedings of the 22nd International Symposium on High-Performance Parallel and Distributed Computing, pp. 239–250 (2013)
17. Mell, P., Grance, T.: The NIST definition of cloud computing (2011)
18. Moulitsas, I., Karypis, G.: Architecture aware partitioning algorithms. In: Bourgeois, A.G., Zheng, S.Q. (eds.) ICA3PP 2008. LNCS, vol. 5022, pp. 42–53. Springer, Heidelberg (2008). https://doi.org/10.1007/978-3-540-69501-1_6

19. Netto, M.A., Calheiros, R.N., Rodrigues, E.R., Cunha, R.L., Buyya, R.: HPC cloud for scientific and business applications: taxonomy, vision, and research challenges. ACM Comput. Surv. (CSUR) **51**(1), 1–29 (2018)
20. Orzechowski, M., Balis, B., Pawlik, K., Pawlik, M., Malawski, M.: Transparent deployment of scientific workflows across clouds-kubernetes approach. In: 2018 IEEE/ACM International Conference on Utility and Cloud Computing Companion (UCC Companion), pp. 9–10. IEEE (2018)
21. Parashar, M., AbdelBaky, M., Rodero, I., Devarakonda, A.: Cloud paradigms and practices for computational and data-enabled science and engineering. Comput. Sci. Eng. **15**(4), 10–18 (2013)
22. Tanaka, M., Tatebe, O.: Workflow scheduling to minimize data movement using multi-constraint graph partitioning. In: 2012 12th IEEE/ACM International Symposium on Cluster, Cloud and Grid Computing (CCGRID 2012), pp. 65–72. IEEE (2012)
23. Tchernykh, A., Schwiegelsohn, U., Alexandrov, V., Talbi, E.: Towards understanding uncertainty in cloud computing resource provisioning. Procedia Comput. Sci. **51**, 1772–1781 (2015)
24. Wu, H., et al.: Automatic cloud bursting under fermicloud. In: 2013 International Conference on Parallel and Distributed Systems, pp. 681–686. IEEE (2013)

Scaling Simulation of Continuous Urban Traffic Model for High Performance Computing System

Mateusz Najdek[1(✉)], Hairuo Xie[2], and Wojciech Turek[1(✉)]

[1] AGH University of Science and Technology, Krakow, Poland
{najdek,wojciech.turek}@agh.edu.pl
[2] University of Melbourne, Melbourne, Australia

Abstract. Urban traffic simulation of extensive areas with complex driver models poses a significant computational challenge. Developing highly scalable parallel simulation algorithms is the only feasible way to provide useful results in this case. In this paper, we present extensions of the SMARTS system, a traffic simulation tool, which provides efficient scalability with a large number of parallel processes. The presented extensions enabled its scalability for HPC-grade systems. The extended version has been thoroughly tested in strong and weak scalability scenarios for up to 2400 computing cores of a supercomputer. The satisfactory scalability has been achieved by introducing several significant improvements, which have been discussed in details.

Keywords: Urban traffic simulation · Simulation scalability · Simulation metrics · HPC

1   Introduction

The domain of Agent-based Modeling has been attracting more and more attention in recent years, providing means for explaining phenomena observed in complex social situations [10]. Microscopic urban traffic is a very interesting real-life case within the domain, where a large number of autonomous agents (drivers) coexist and interact in a common system (road network). The simulation of this phenomenon is a significant computational challenge. Simulation algorithms repetitively modify one large data structure, which, when performed in parallel, must be properly synchronized. Providing parallelization methods for such a task, which could efficiently utilize High-Performance Computing (HPC) hardware, remains an open problem.

The work presented in this paper is the result of cooperation between the creators of the SMARTS traffic simulator [11] from University of Melbourne and the HPC team from AGH University of Science and Technology. The main contribution is the extended version of the SMARTS system, which allows efficient simulation of the IDM [4] model on 2400 cores of a supercomputer. We also describe introduced improvements and provide a discussion of typical issues with creating or migrating spatial simulations for HPC-grade systems.

© Springer Nature Switzerland AG 2021
M. Paszynski et al. (Eds.): ICCS 2021, LNCS 12742, pp. 256–263, 2021.
https://doi.org/10.1007/978-3-030-77961-0_22

2   Scalable Traffic Simulations

The need for parallel execution of traffic simulation has been identified at the end of XX century [7]. At least a few attempts to creating a centrally synchronized parallel traffic simulation were tried later on. Work presented in [6] or [9] report scalability at a few nodes. The concept of increasing the time between global synchronization, presented in [2], led to positive impact on scalability and significant errors in the simulation results. Another approach, presented in [13], introduced a complex protocol for correcting the errors resulting from rare synchronizations.

In [12] authors suggested that the global synchronization in parallel traffic simulation algorithm might not be necessary. Removing centralized element from the computation is crucial in terms of HPC-grade scalability, however, achieving this in traffic simulations was not straightforward. Probably the first reports on a method which overcame this problem were presented in [18]. The authors presented an Asynchronous Synchronization Strategy, which assumed that each computing process communicates with limited number of processes – only those responsible for adjacent fragments of the modeled environment. The authors achieved linear scalability with 32 parallel workers. The concept has been extended in [14,15], where a simulation of discrete traffic model scaled linearly to 19200 parallel processes running on 800 nodes of a supercomputer, which is probably the largest setup tested so far in this type of simulations.

The work presented in this paper is the result of merging the experiences from building HPC-scalable traffic simulations with the SMARTS tool, which supports continuous traffic models and manages realistic maps. *Scalable Microscopic Adaptive Road Traffic Simulator (SMARTS)* [11] is a free, light-weight and versatile simulation tool developed at the School of Computing and Information Systems, University of Melbourne, Australia. The simulator has been used in many research projects [8,16,17]. SMARTS builds road networks based on freely-available OpenStreetMap data, which allows researchers to simulate traffic in real road networks around the world. During a simulation, the position of vehicles on roads is updated based on a continuous model (a car-following model), *Intelligent Driver Model (IDM)* [4], which computes the acceleration of the vehicle based on several factors such as the distance to an impeding object. A simulated vehicle is also controlled by a lane-changing model, *Minimizing Overall Braking Induced by Lane Changes (MOBIL)* [3], which is used to determine whether it is safe and beneficial for the vehicle to change to specific traffic lanes. SMARTS also models traffic lights and a number of traffic rules.

SMARTS is a distributed system that has one *server* and one or more *workers*. It divides the simulation area into multiple sub-areas, assigned to different workers that run in parallel and exchange information only with workers responsible for adjacent sub-areas. The workers do not need to communicate with the server during simulation, which helps prevent a communication bottleneck at the server. SMARTS also adopts a Priority Synchronous Parallel model, which uses a two-phase simulation approach to reduce the impact of the slowest worker [5].

3   Towards HPC-Grade Traffic Simulation

The concept of the simulation method presented in SMARTS is suitable for distributed computing, but for a high level of parallelization at High-Performance Computing (HPC) systems several modifications were required. Considering massive HPC parallelism, where the number of workers can be orders of magnitude larger than in the case of small clusters, one should always focus on minimizing the non-parallel part of process or possibly getting rid of it entirely. Original elements of SMARTS, like workers initialization, task division, communication, and results collecting could significantly reduce overall efficiency and even crash the entire system. To overcome these problems the following improvements were required.

3.1   Initialization of the Simulation

SMARTS imposes on each worker to know metadata of its neighboring workers with which it must communicate. To obtain this information, a connection must be established with a centralized server, shared among all running workers. The server builds a dedicated connection with each worker synchronously. This sequentiality overhead cannot be easily avoided, but the effect is relatively reduced in long simulations. It also does not influence the proper simulation stage.

Another much greater sequential part of initialization, referred to as *setup* in Sect. 4, requires all workers to obtain simulation configuration from the server by message pushing model. Conceptually, it is better to aggregate data dedicated to all workers within a single node and distribute them locally, involving their resources for parallelizing the process. In HPC practice, a far more efficient solution is to use the common filesystem for shared data distribution. Memory extensive part of shared message content is then loaded locally in parallel.

3.2   Data Scalability

To be able to simulate large scenarios within the available memory, one has to analyze the volume of data and divide them properly between available computing resources to ensure data-scalability. Generally, two types of data can be distinguished: static and dynamic. The static, such as the model of the environment, should be divided among all processing workers. Currently, SMARTS system balances the workload by generating vehicles with random routes on-the-fly, using the whole map at every worker. The optimal solution is to use predefined routes and remove the need for global planning from workers. This will reduce the setup time and also memory usage within a single node. The second data type (dynamic) can increases at demand and correspond mainly to the mobile entities. To prevent complete memory consumption, a safe estimation of maximal volume for each worker should be performed. Another method of prevention is online monitoring of transmitted data and dynamic load balancing – which is a challenge for further research.

3.3   Massive Parallelism Issues

All operations in a distributed environment which use randomness require special attention (or should be avoided), as they can cause complications visible only in large scale. Defining internal identifiers of individual processes based on a randomly generated string may seem a sufficient approach. However, being a variant of *the Birthday problem*, it can, with an increasing number of processes cause problems, when generated the same identifier. A better approach is to associate an internal identifier with a uniquely defined entity in the cluster topology as it guarantees unambiguity. Another thing to be taken into consideration is access to shared system resources, such as random socket number selection by different processes. It can manifest itself when a larger number of workers is involved, causing not obvious race conditions between separate processes within a computing node. Incoming messages were processed so far by dedicated, emerging threads in a thread-per-message approach. This solution is quickly worn-out at a larger scale and causes a massive decrease in efficiency with more incoming messages. To overcome this problem Thread Pooling based solution was introduced where the creation time of thread is avoided and its total amount is restricted. The messages are stored in a queue and handled by reusable processing threads.

3.4   Simulation Results Processing

The required details of the collected metrics should be carefully considered since fined-grained metrics cause larger sizes of exchanged messages and also more frequent communication. Where every timestamped vehicle positions are not required or are in final post-processing aggregated anyway, then it is better to perform the aggregation ahead at each worker. In such a case instead of gathering trajectories directly, it is better to transform them into more coarse-grained metrics such as density and flow per road. The original architecture, where trajectories were collected by the server, is not feasible with more than 240 workers for the reason of the limited amount of memory. In order to support this feature in the HPC-grade version, we introduced a solution to bind the trajectory metrics with the corresponding vehicle state. It is gradually growing, burdening workers instead of the server. The proposed solution allows for independent serialization of such results at the end of simulation, fully in parallel, instead of massively time-consuming sequential serialization. It totally eliminates the sequential part of this process and greatly speeds it up by a factor of the number of workers used.

4   Evaluation of Traffic Simulation Scalability

Different test scenarios were prepared with an increasing number of processing cores. The first scenario of strong scalability included a fixed minimum number of 2.4 million vehicles in total. The second one (weak scalability) increased from 24 thousand up to 2.4 million vehicles in proportion to the number of workers.

Routes were randomly generated using A-star algorithm. Both scenarios used 59 km × 55 km area of Beijing. The road graph contained approximately 400 thousand directed edges, which correspond to streets and 220 thousand nodes, that represent intersections and points of road division into smaller sections. The average length of a road is about 85 m and the total length is 58.000 km.

All tests were performed using the Prometheus supercomputer, a part of the PL-Grid infrastructure[1]. Currently, Prometheus is ranked as the 324th computer in the worldwide top 500 list[2]. It contains 2232 computing nodes, HP Apollo 8000 with Xeon E5-2680v3 CPUs working at 2.5 GHz. Each node has 24 physical cores – the total number of computing cores is 53,568. Nodes are connected by the Infiniband FDR 56 Gb/s network.

For evaluation purposes, two main categories of performance metrics were collected depending on the source. The first describes the server perspective. Figures 1 and 3 shows execution time of 4 components: global time of initialization, model setup and loading, simulation, serialization of obtained results. The second type describes simulation from the worker's side. Figures 2 and 4 present the average execution step time from 50 steps for data collected over the last 1000 out of 1500 steps. The length of a single step was 0.2 s, which is equivalent to 6 min of real-time traffic. Synchronization is based on decentralized communication and result are collected in a decentralized manner. All tests were repeated 3 times to confirm the stability of the obtained results. Each node has been configured to run 24 workers and the server was run on a separate node.

4.1 Strong Scalability

As Amdahl [1] pointed out in 1967, for fixed-sized problem overall speedup is limited by the factor of sequential part that cannot be subject to parallelization. Based on that, strong scalability is defined as the variance of the solution's execution time to the utilized number of computing processors. In this experiment, each worker minimally simulated the number of vehicles equivalent to roughly a fraction of the fixed number of 2.4 million vehicles by the number of workers. Since the size of the problem is large, it was not feasible to run scenario using less than 4 nodes. Figure 1 shows results of global execution times for each component of the process.

The whole process for the same task size benefits greatly from adding up to 1440 workers and further increase causes it to suffer due to the initialization part. In Fig. 2 the median of simulation step time is reduced and distribution becomes more convergent with an increasing number of workers.

---

[1] http://www.plgrid.pl/.

[2] https://www.top500.org/lists/top500/list/2020/11?page=4.

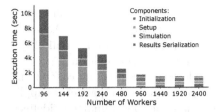

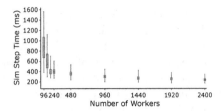

Fig. 1. Global times of process components with increasing number of workers for growing problem size.

Fig. 2. Simulation step times for increasing problem size in proportion to the number of workers.

4.2   Weak Scalability

As Gustafson [1] noticed, in practice predominantly size of the problem scales with the number of available resources. The observation is that if the problem is small, there is no benefit in using large amounts of resources, and a more sensible approach is to use resources according to the problem size. Weak scaling is concerned with how the execution time for solving a scaled size problem varies with the number of CPUs used. The computational costs depend on the number of vehicles and the results are presented in figures below (Figs. 3 and 4), where each worker initially simulated a minimal number of 1000 vehicles.

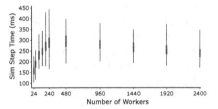

Fig. 3. Global times of process components with increasing number of workers for growing problem size.

Fig. 4. Simulation step times for increasing problem size in proportion to the number of workers.

The main factor extending the duration of the entire process is the initialization part. As shown in Fig. 4 results confirm algorithm scalability as the simulation time of each task remains within a constant range. The conclusion is by increasing the number of workers communication overhead is not strongly affecting this method.

5   Conclusions and Further Work

The presented solution to the problem of efficient urban traffic simulation is most likely the first to efficiently utilize HPC-grade hardware for simulation of continuous traffic model with real-life scenarios. The simulation algorithm itself exposes proper features both in strong and weak scalability tests. Achieved overall scalability of the simulation system is not perfect, however, in all of the analyzed cases addition of hardware improved the simulation efficiency.

Extensions introduced to the SMARTS platform resulted in conclusions regarding HPC-grade software design, which goes beyond the particular application. Identified issues and proposed solutions may become valuable guidelines for researchers willing to scale spatial simulations to HPC-grade. The extended versions of SMARTS system, used for conducting the presented experiments, is available at: https://github.com/mateusznajdek/SMARTS-extension.

A few aspects still can and should be improved. More efficient strategies for initialization are definitely needed. Also, scalable mechanisms for model division should be introduced. These will be the subject of further research.

Acknowledgments. The research presented in this paper was funded by the National Science Centre, Poland, under the grant no. 2019/35/O/ST6/01806. The research was supported in part by PL-Grid Infrastructure.

References

1. Gustafson, J.: Reevaluating amdahl's law. Commun. ACM **31**(5), 532–533 (1988)
2. Kanezashi, H., Suzumura, T.: Performance optimization for agent-based traffic simulation by dynamic agent assignment. In: Proceedings of 2015 Winter Simulation Conference, WSC 2015, pp. 757–766. IEEE Press, Piscataway (2015)
3. Kesting, A., Treiber, M., Helbing, D.: General lane-changing model MOBIL for car-following models. J. Transp. Res. Board **1999**(1), 86–94 (2007)
4. Kesting, A., Treiber, M., Helbing, D.: Enhanced intelligent driver model to access the impact of driving strategies on traffic capacity. Trans. Royal Soc. London A **368**(1928), 4585–4605 (2010)
5. Khunayn, E.B., Karunasekera, S., Xie, H., Ramamohanarao, K.: Straggler mitigation for distributed behavioral simulation. In: 2017 IEEE 37th International Conference on Distributed Computing Systems (ICDCS), pp. 2638–2641. IEEE (2017)
6. Klefstad, R., Zhang, Y., Lai, M., Jayakrishnan, R., Lavanya, R.: A distributed, scalable, and synchronized framework for large-scale microscopic traffic simulation. In: Proceedings of 2005 IEEE Intelligent Transportation Systems, 2005, pp. 813–818 (2005)
7. Nagel, K., Schleicher, A.: Microscopic traffic modeling on parallel high performance computers. Parallel Comput. **20**(1), 125–146 (1994)
8. Nguyen, U.T., et al.: A randomized path routing algorithm for decentralized route allocation in transportation networks. In: ACM SIGSPATIAL, pp. 15–20 (2015)
9. O'Cearbhaill, E.A., O'Mahony, M.: Parallel implementation of a transportation network model. J. Parallel Distrib. Comput. **65**(1), 1–14 (2005)
10. Railsback, S.F., Grimm, V.: Agent-Based and Individual-Based Modeling: A Practical Introduction. Princeton University Press, Princeton (2019)
11. Ramamohanarao, K., et al.: SMARTS: scalable microscopic adaptive road traffic simulator. ACM Trans. Intell. Syst. Technol. (TIST) **8**(2), 1–22 (2016)

12. Rickert, M., Nagel, K.: Dynamic traffic assignment on parallel computers in transims. Futur. Gener. Comput. Syst. **17**(5), 637–648 (2001)
13. Toscano, L., D'Angelo, G., Marzolla, M.: Parallel discrete event simulation with erlang. In: Proceedings of the 1st ACM SIGPLAN Workshop on Functional High-performance Computing, FHPC 2012, pp. 83–92. ACM, New York (2012)
14. Turek, W.: Erlang-based desynchronized urban traffic simulation for high-performance computing systems. Futur. Gener. Comput. Syst. **79**, 645–652 (2018)
15. Turek, W., Siwik, L., Byrski, A.: Leveraging rapid simulation and analysis of large urban road systems on HPC. Transp. Res. Part C: Emerging Technol. **87**, 46–57 (2018)
16. Xie, H., Karunasekera, S., Kulik, L., Tanin, E., Zhang, R., Ramamohanarao, K.: A simulation study of emergency vehicle prioritization in intelligent transportation systems. In: IEEE VTC Spring, pp. 1–5 (2017)
17. Xie, H., Tanin, E., Karunasekera, S., Qi, J., Zhang, R., Kulik, L., Ramamohanarao, K.: Quantifying the impact of autonomous vehicles using microscopic simulations. In: ACM SIGSPATIAL, pp. 1–10 (2019)
18. Xu, Y., Cai, W., Aydt, H., Lees, M., Zehe, D.: An asynchronous synchronization strategy for parallel large-scale agent-based traffic simulations. In: Proceedings of the 3rd ACM SIGSIM Conference on Principles of Advanced Discrete Simulation, SIGSIM PADS 2015, pp. 259–269. ACM, New York (2015)

A Semi-supervised Approach for Trajectory Segmentation to Identify Different Moisture Processes in the Atmosphere

Benjamin Ertl[1,2]([📧]) [ID], Matthias Schneider[2] [ID], Christopher Diekmann[2] [ID],
Jörg Meyer[1] [ID], and Achim Streit[1] [ID]

[1] Steinbuch Centre for Computing (SCC), Karlsruhe Institute of Technology (KIT),
Karlsruhe, Germany
`{benjamin.ertl,joerg.meyer2,achim.streit}@kit.edu`
[2] Institute for Meteorology and Climate Research (IMK-ASF),
Karlsruhe Institute of Technology (KIT), Karlsruhe, Germany
`{matthias.schneider,christopher.diekmann}@kit.edu`

Abstract. Different moisture processes in the atmosphere leave distinctive isotopologue fingerprints. Therefore, the paired analysis of water vapour and the ratio between different isotopologues, for example $\{H_2O, \delta D\}$ with δD as the standardized HDO/H_2O isotopologue ratio, can be used to investigate these processes. In this paper, we propose a novel semi-supervised approach for trajectory segmentation to extract information that enables us to identify atmospheric moisture processes. While our approach can be transferred to a variety of domains as well, we focus our evaluation on Lagrangian air parcel trajectories and modelled $\{H_2O, \delta D\}$ fields. Our final aim is to understand the free tropospheric $\{H_2O, \delta D\}$ pair distribution that is observable by satellite sensors of the latest generation. Our method adopts a recently developed density-based clustering algorithm with constrained expansion, *CoExDBSCAN*, which identifies clusters of temporal neighbourhoods that are only expanded with regards to a priori constraints in defined subspaces. By formulating a constraint for the correlation of $\{H_2O, \delta D\}$, we can segment trajectories into multiple phases and extract the regression coefficients for each phase. Grouping segments with similar coefficients and comparing them to theoretical values allows us to find interpretable structures that correspond to atmospheric moisture processes. The experimental evaluation demonstrates that our method facilitates an efficient, data-driven analysis of large-scale climate data and multivariate time series in general.

Keywords: Semi-supervised clustering · Multivariate time-series · Time-series segmentation · Climate research.

1 Introduction

With advances in technology that translates into an increasing amount of data, researchers across many disciplines face new challenges analysing and gaining

© Springer Nature Switzerland AG 2021
M. Paszynski et al. (Eds.): ICCS 2021, LNCS 12742, pp. 264–277, 2021.
https://doi.org/10.1007/978-3-030-77961-0_23

knowledge from massive volumes of data. For example, the U.S. National Oceanic and Atmospheric Administration (NOAA) cites for their Big Data Program that tens of terabytes of data are generated from satellites, radars, ships, weather models, and other sources a day [13]. Unsupervised learning methods such as cluster analysis are particularly useful in analyzing large amounts of data since it allows domain experts to consider groups of objects rather than individual objects and to focus on a higher level representation of the data [19]. While many advances in cluster algorithms have been made for spatiotemporal data [11,19], such as atmospheric model data, many proposed methods lack the exploitation of available a priori knowledge that might improve the output quality [5]. Especially semi-supervised learning clustering algorithms, which incorporate a prior knowledge into the clustering process, can improve the quality of the results [3]. In this paper, we propose a novel semi-supervised approach for subsequence time series clustering based on our recently developed density-based clustering algorithm with constrained expansion called *CoExDBSCAN* [4]. By applying *CoExDBSCAN* we can segment trajectories of $\{H_2O, \delta D\}$ pair distributions into multiple phases which can be associated with atmospheric moisture processes. Identifying such processes is an important scientific task to infer the dynamics of cloud-circulation systems. Investigating the atmosphere from a cloud-circulation system perspective is essential to address the significant uncertainty of climate predictions [1].

The two main contributions of our work presented here are:

- Adaptation of our *CoExDBSCAN* algorithm for trajectory segmentation by formulating a constraint on the $\{H_2O, \delta D\}$ pair distribution to differentiate multiple phases of a trajectory.
- Extracting information about the regression coefficients for each phase and comparing the distribution of coefficients to theoretical values to identify corresponding atmospheric moisture processes.

The remainder of this paper is organized as follows. Section 2 introduces related work and background knowledge about our data, the theory behind the relation of $\{H_2O, \delta D\}$ and atmospheric moisture processes. In Sect. 3 we present the experimental evaluation of our proposed approach and provide a discussion of the results in Sect. 4. We conclude the paper in Sect. 5 and provide an outlook on future research.

All dataset together with the code for this paper are publicly available in the supplementary GitHub repository[1].

2   Background

2.1   Trajectory Clustering

Trajectory clustering and subsequence time series clustering are well established research fields. Trajectories can be described as sets of measurements which

---

[1] https://github.com/bertl4398/iccs2021.

are measured as a function of an independent variable, typically time, where each individual trajectory measures a possible multidimensional response variable [7]. One of the first comprehensive methods for this type of data has been introduced by Gaffney et al. [7], who proposed a probabilistic mixture regression model applying the Expectation Maximization (EM) algorithm to cluster trajectories and demonstrated their approach analysing extratropical cyclones [8]. Since clustering whole trajectories can overlook common behaviour in partial segments of the trajectories, Lee et al. [10] proposed a partition-and-group framework and a trajectory clustering algorithm called *TRACLUS*, which they demonstrated among others in the field of climate research for hurricanes landfall forecasts. Following the given notion of trajectory data, there is no distinction to time series data in general, however the data records per individual trajectory can frequently be too short to be amenable to conventional time series modelling techniques, which requires specialized approaches [7]. Zolhavarieh et al. [21] compiled a well received survey about subsequence time series clustering algorithms and applications. More recently there have also been subsequence time series cluster algorithms proposed that are model-based [9], completely unsupervised [20] or semi-supervised [4]. We have recently developed the semi-supervised algorithm *CoExDBSCAN*, that utilizes the original *DBSCAN* algorithm proposed by Ester et al. [6], to find density-connected clusters in a defined subspace of features and restricts the expansion of clusters to a priori constraints. Because we can formulate an a priori constraint on the $\{H_2O, \delta D\}$ pair distribution of our data based on expert knowledge, *CoExDBSCAN* is a suitable choice to our problem. *CoExDBSCAN* has been demonstrated to be especially suited for spatiotemporal data, where one subspace of features defines the spatial extent of the data and another subspace the correlations between features. We can apply the algorithm to differentiate multiple phases in our trajectory data. However, by focusing on the temporal aspect of the trajectories, i.e. considering our data as multivariate time series, in distinction to the original algorithm we define the time space of the data as subspace for the distance based density computations. In this way we are able to find subsequences that follow our formulated constraint on the $\{H_2O, \delta D\}$ pair distribution. For a detailed explanation of the original *DBSCAN* and *CoExDBSCAN* algorithms as well as the pseudo code of the algorithms, we refer to the original papers by Ester et al. [6] and Ertl et al. [4] respectively.

2.2   Research Data

Our final research aim is to link the global $\{H_2O, \delta D\}$ pair distribution observed in the MUSICA IASI satellite-based remote sensing data set [2,17] to different moisture processes that occurred prior to the observation. This dataset offers well-documented $\{H_2O, \delta D\}$ pair data from the year 2014 to 2020 with high quality and global coverage. The generation of this unique dataset has become only recently possible through advances in satellite sensor technology and retrieval theory. In this dataset and throughout this paper, H_2O indicates the water

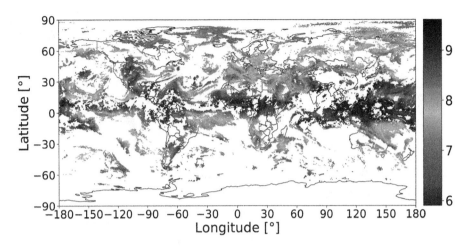

Fig. 1. Example global MUSICA IASI H_2O data for morning satellite overpasses at 2016-06-08. H_2O values are in parts per million by volume (ppmv) in logarithmic scale. The depicted data are limited to cloud free observations and have been filtered for best quality (retrievals with good sensitivity and low errors). The red square indicates our area of interest for the trajectory clustering.

vapour concentration measured in parts per million by volume (ppmv); δD corresponds to the standardised ratio value between light and heavy water i.e. H_2O and HDO [17]. Figure 1 and Fig. 2 illustrate the characteristics of the MUSICA IASI dataset.

Figure 1 shows the global H_2O observations retrieved from the infrared atmospheric sounding interferometer (IASI) onboard the EUMETSAT (European Organisation for the Exploitation of Meteorological Satellites) Metop-A and Metop-B (Meteorological Operational) platforms for morning overpasses at the 8th June 2016 for about five kilometers altitude. For this single day 183,036 individual observations are available after filtering out cloudy and partly cloudy observations as well as observations with bad quality.

Figure 2 depicts the $\{H_2O, \delta D\}$ pair distribution starting from the same date at the 8th June 2016 until the 30th June 2016 for the area of interest (red rectangle in Fig. 1). All MUSICA IASI $\{H_2O, \delta D\}$ data are shown as gray dots and the contours are at 2.5%, 10% and 50% levels, meaning the percentage of data lying outside the indicated area.

Different water cycle processes affect the isotopic composition of atmospheric water differently, for example lighter isotopes evaporate preferentially while heavier isotopes condense preferentially. The red lines in Fig. 2 illustrate the theoretical dependencies of δD as a function of H_2O. Noone et al. differentiate between five processes that leave a distinct trace in the $\{H_2O, \delta D\}$ value space [14]:

1. **Rayleigh pseudoadiabatic** process in which the liquid water that condenses is assumed to be removed as soon as it is formed, by idealized instantaneous precipitation (red dotted line in Fig. 2)

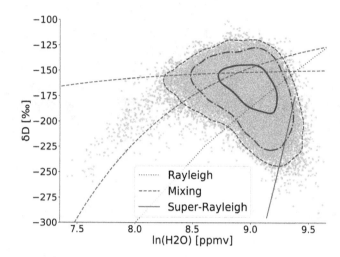

Fig. 2. Example MUSICA IASI $\{H_2O, \delta D\}$ pair distribution. The 61,283 observations are for morning and evening satellite overpasses from 2016-06-08 to 2016-07-30 with H_2O in logarithmic scale in the area of interest (see red rectangle in Fig. 1); red lines indicate theoretical lines. (Color figure online)

2. **Super-Rayleigh** remoistening associated with isotopic exchange as rain-drops evaporate into a subsaturated layer (red solid line in Fig. 2)
3. **Reversible moist adiabatic** process with a transition to a Rayleigh process when condensation is to ice and irreversible (not shown, would be a line with a weaker slope as the dotted line in Fig. 2)
4. **Mixing** of two different mixing members having a specific $\{H_2O, \delta D\}$ char-acteristic (red dashed lines in Fig. 2 show two examples)
5. **Terrestrial transpiration** mixing with land source (not shown)

Noone's work establishes a theoretical basis for using isotope ratio observa-tions paired with the water vapor mixing ratio to identify different water sources, condensation processes, and transport pathways in the troposphere. Moreover, Noone et al. were able to derive slope and intercept of the linear relationship between H_2O and δD from measurements of the isotope ratio of water vapor at the Mauna Loa Observatory [15].

In this paper, we apply our segmentation algorithm to the $\{H_2O, \delta D\}$ pairs modelled along Lagrangian air parcel trajectories. As model data we use the high-resolution data from the regional isotope-enabled atmospheric model COSMO-iso [12] and the trajectories are determined with the tool LAGRANTO [18]. The trajectories' calculation setup is oriented towards the overpass times and altitudes representative for the MUSICA IASI data. Analysing the model data enables us to reveal the kind of moisture processes that can be observed in the MUSICA IASI $\{H_2O, \delta D\}$ pair data. We utilize the theoretical and observa-tional findings by Noone in our experimental evaluation to identify atmospheric moisture processes that correspond to different $\{H_2O, \delta D\}$ pair distribution for different segments of our trajectory data. Our focus will be on rain events, where Rayleigh pseudoadiabatic, Super-Rayleigh, and reversible moist adiabatic processes affect the $\{H_2O, \delta D\}$ pair distribution, in contrast to non-rain events, where air mass mixing processes are dominating the $\{H_2O, \delta D\}$ pair distribution. Our semi-supervised approach with the adaptation of the *CoExDBSCAN* algorithm with the appropriate constraint formulation for our research objective is detailed in the next section.

3 Experimental Evaluation

For our experimental evaluation, we organize our Lagrangian air parcel trajectory dataset as multivariate time series, e.g. backward trajectories for individual air parcels, and focus on an area of interest with the arrival of all trajectories above West Africa at pressure levels 575 and 625 hectopascal (hPa). The trajectories are calculated daily for local morning (9 am) and evening (9 pm) times during the period from June 8, 2016 to July 30, 2016, resulting in 12,720 individual trajectories (11,853 after filtering) with 169 time steps each with a time delta of one hour; each trajectory comprises a time frame of 7 d. Figure 3a illustrates our trajectory dataset, depicting 3,194 individual trajectories out of 11,853 that have been colored according to their similarity, using *DBSCAN* on a precomputed distance matrix. For the precomputed distance matrix, each trajectory has been converted to a $4 \cdot 169 = 676$ dimensional vector; the latitudinal (1) and longitudinal (2) difference for each time point to the arrival coordinates, the bearing (3) for each consecutive point and the scaled height difference (4) for each consecutive point, 169 points each. This initial clustering is only done to associate and compare individual segments as a result of our *CoExDBSCAN* segmentation, see Fig. 3b, and is completely independent from our trajectory clustering.

For the trajectory segmentation we adopt the *CoExDBSCAN* algorithm by defining the temporal order of the data points, the time dimension, as the spatial subspace. Following the definition of the *CoExDBSCAN* ϵ-neighbourhood, see [4] and Definition 1 for reference, with the time dimension as the spatial subspace the ϵ-neighbourhood describes a neighbourhood of lagged points, similar to a time window, where the maximum lag in time for the initial data points is defined by the ϵ parameter and the minimal amount of data points that are required to form a cluster is defined by the *minPts* parameter.

Definition 1. *Let DB be a database of points. The CoExDBSCAN ϵ-neighbourhood of a point p, denoted by $N_\epsilon(p)$, is defined by*

$$N_\epsilon(p) = \{q \in DB | dist(p_S, q_S) \leq \epsilon \wedge constraints(p_R, q_R)\} \quad (1)$$

where p_S, q_S are the subspace representations of point p and q of the user-defined spatial subspace S, p_R, q_R are the subspace representations of point p and q of the user-defined constraint subspace R and the **constraints** *function evaluates* **true** *for each constraint C_i in a user-defined set of constraints $C = \{C_1, C_2, ..., C_n\}$.*

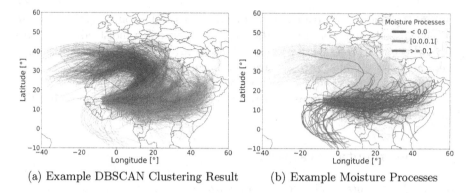

(a) Example DBSCAN Clustering Result   (b) Example Moisture Processes

Fig. 3. Illustration of 3,194 trajectories out of 11,853 that have been colored according to their geographical closeness, bearing similarity and height difference along each individual trajectory (a). (b) illustrates the association of individual trajectories to different moisture processes as a result of the *CoExDBSCAN* segmentation.

In general, we consider a trajectory as a time series T of size m as an ordered sequence of real-value data, $T = (t_1, t_2, ..., t_m)$, and a subsequence of length n as $T_{i,n} = (t_i, t_{i+1}, ..., t_{i+n-1})$, where $1 \leq i \leq m - n + 1$; a subsequence is considered an arranged sequence of data that omits some elements without changing the order of the remaining element [21]. The algorithm starts with an initial time point and considers all temporal neighbours to form a cluster following Definition 1. Our constraint on the $\{H_2O, \delta D\}$ pair distribution to differentiate multiple phases of individual trajectories has been conceptualized together with domain experts and the constraint parameter δ empirically determined, see Definition 2. We constrain the expansion of clusters e.g. subsequences by including a neighbouring point in the ϵ-neighbourhood only if the residuals of an ordinary least squares linear regression for the current cluster points without the neighbouring point deviates only by a certain factor δ from the ordinary least squares linear regression including the neighbouring point, formulated in Definition 2.

Definition 2. *A point t_j belongs to a subsequence of points $T_{i,n} = (t_i, t_{i+1}, ..., t_{i+n-1})$, where $1 \leq i \leq m - n + 1$, iff*

$$（Y_{t_j} - \hat{Y}_{t_j}）^2 < \delta \cdot \frac{1}{n} \sum_{k=i}^{(i+n-1)} （Y_{t_k} - \hat{Y}_{t_k}）^2 \tag{2}$$

where Y and $\hat{Y}$ are the dependent variable and fitted value of the linear regression respectively.

Algorithm 1 gives a pseudo code representation of our semi-supervised trajectory segmentation approach. The algorithm takes a set of trajectories as input, as well as the parameters and subspace selections for *CoExDBSCAN*. To exemplify our approach and demonstrate the flexibility to segment likewise only a subset of individual trajectories, we focus on rain events within trajectories without any loss of generality. These events have been identified by retrieving the time points where the moving average over three consecutive time points for the relative humidity are above a certain threshold and extending the event to at least three time points if necessary.

Algorithm 1: Semi-Supervised Trajectory Segmentation

---

    input : trajectories T
    input : time radius ϵ
    input : density threshold $minPts$
    input : residual threshold δ
    output: point labels per trajectory $label_t$ initially $undefined$

1 **foreach** *trajectory t in trajectories T* **do**
2   timePoints = sortByTime(t, ascending);
3   phasesAscending = CoExDBSCAN(timePoints.time, timePoints.{ln(H2O),ln(deltaD/1000 + 1)}, ϵ, $minPts$, δ);
4   timePoints = sortByTime(t, descending);
5   phasesDescending = CoExDBSCAN(timePoints.time, timePoints.{ln(H2O),ln(deltaD/1000 + 1)}, ϵ, $minPts$, δ)
6   **foreach** *phaseAscending in phasesAscending* **do**
7   **foreach** *phaseDescending in phasesDescending* **do**
8   **if** $sum(OrdinaryLeastSquares(phaseAscending).residuals^2) < sum(OrdinaryLeastSquares(phaseAscending).residuals^2)$ **then**
9               $label_t \leftarrow phaseAscending$;
10   **else**
11               $label_t \leftarrow phaseAscending$;

The identified trajectories with a subset of rain events are sorted by time in ascending and descending order, see Line 2 and 4 in Algorithm 1. For each time ordering we compute the labels using *CoExDBSCAN* with the time dimension, timePoints.time, as the spatial subspace and the natural logarithm of the water vapour values together with the natural logarithm of the isotopologue ratio value divided by 1,000 plus one, timePoints.{ln(H2O),ln(deltaD/1000 + 1)}, as the constraint subspace, see Line 3 and 5. The remaining parameters have been empirically determined and proposed by domain experts and set to $\epsilon = 2$, $minPts = 3$ and $\delta = 2$ in this example. Computing the segmentation in both temporal orders is necessary, because the outcome of the linear regression in our constraint depends on the deviation of the residuals from the current cluster points, which can be different following the trajectory points in ascending or descending temporal order. The final segmentation of the trajectory, or the subset of time points within a trajectory, is the selection of phases from the ascending and descending *CoExDBSCAN* run where the outcome with the squared residual sum is lowest, Line 6 to 11 in the algorithm.

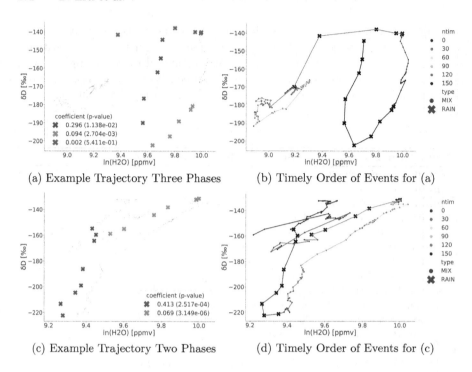

(a) Example Trajectory Three Phases   (b) Timely Order of Events for (a)

(c) Example Trajectory Two Phases   (d) Timely Order of Events for (c)

Fig. 4. Example trajectories with two rain sequences, which can be identified by the crosses. (a) with additional segmentation of a continuous sequences based on the defined constraint for the regression coefficients and (c) without additional segmentation. (b) and (d) illustrate the timely order of events according to the number of hours before arrival from 168 to 0 (dark blue to dark red). (Color figure online)

Figure 4 shows the segmentation result of two example trajectories. Figure 4a and Fig. 4c depict the results of our semi-supervised segmentation approach with dots indicating non-rain events (air mass mixing) and crosses indicating rain events, which have been segmented into different coloured phases with different regression coefficients. Figure 4b and Fig. 4d visualize the temporal order of the corresponding trajectories' events coloured according to the number of hours before arrival from 168 to 0 (dark blue to dark red). The first example trajectory in Fig. 4a has two rain events with three distinct phases, where the first event from hour 168 to hour 159 before arrival (see Fig. 4b) has been segmented into two different phases with different regression coefficients with statistical significance (p-value < 0.05, blue crosses and orange crosses); the second event starts at hour 124 and ends at hour 119 before arrival with a steady regression coefficient, however not statistically significant (green crosses). The second example trajectory in Fig. 4c shows again two rain events, the first from hour 140 to hour 133 before arrival and the second from hour 38 to hour 32 before arrival. The regression coefficients are unvarying with statistical significance (blue crosses and orange crosses). We clearly observe different slopes of the linear regression lines fitted to the two different rain events.

Comparing the rain segments from both example trajectories with the theoretical evolution of δD as a function of H_2O, see Sect. 2.2, we can interpret the segmented event in Fig. 4a (blue and orange crosses) and the two timely separated phases in Fig. 4c (blue and orange crosses) as different kind of rain processes. For example, the first phase in Fig. 4c (blue crosses) is in line with super-Rayleigh processes, i.e. there is some interchange between condensed moisture and vapour. The second phase (orange crosses) is close to a Rayleigh process, i.e. can be explained by condensation and direct rainout, without significant interchange between condensed moisture and vapour.

These two example trajectories demonstrate that our approach can **(1) identify timely separated events** as well as **(2) split timely connected events** with varying regression coefficients. The latter gives novel opportunities for identifying details of the related rain processes. Being able to distinguish these fine-grained structures in large volumes of data emphasizes the benefit of applying our method, which we discuss in the following section.

The histogram in Fig. 5a outlines the distribution of regression coefficients with statistical significance (p-value < 0.05) for all rain segments as a result of our semi-supervised trajectory segmentation, analysing all 11,853 trajectories, clipped to the interval $[-1.0, 1.0]$ (omitting four segments < -1.0 and three segments > 1.0). Of all rain events 7.6% have a negative slope, 58.6% have a slope between 0.0 and 0.1, and 33.8% have a slope greater than 0.1.

Figure 5b depicts the distribution of the $\{H_2O, \delta D\}$ pairs modelled in the area of interest at the altitudes that are representative for the MUSICA IASI data. The grey dots show all model data, i.e. represent the same as the grey dots in Fig. 2, but modelled instead of measured data. The lines are 50% contour lines. The grey contour line is for the whole data set, see Fig. 2. The blue, orange, and green contour lines are for data points representing air masses that experienced rain events during the last five days prior to arrival. The colours are according to the coefficients that characterise the rain events (i.e. they are in line with Fig. 5a). If an air mass experienced rain events with two different characteristics, the respective data point belongs to both groups. We can clearly identify air masses that experienced a Super-Rayleigh process (green contour line is almost completely below the theoretical Rayleigh line).

4   Discussion

As our experimental evaluation demonstrates, our approach can successfully segment time series, e.g. trajectories or subsequences of trajectories, into sequences that contain temporal close points which follow a priori constraints. With our constraint formulated in Definition 2 each segment is differentiated by the deviation from the ordinary least squares regression residuals, which in effect splits sequences if their individual linear regression is a better fit than their combined linear regression. This constraint is particularly useful for our research objective

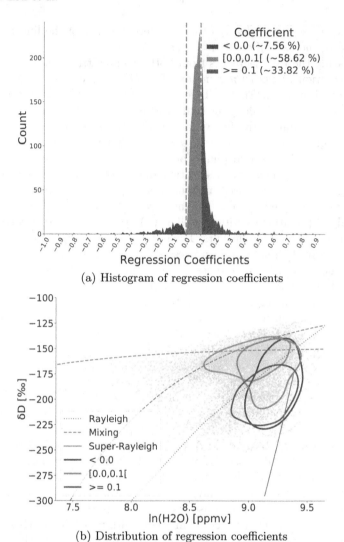

(a) Histogram of regression coefficients

(b) Distribution of regression coefficients

Fig. 5. Histogram (a) of regression coefficients (slope of linear regression line) for all rain segments as a result of our semi-supervised trajectory segmentation with statistical significance (p-value < 0.05); and $\{H_2O, \delta D\}$ distributions (b) of the model data in the area of interest for all data points (grey dots and contour line) and for data points representing air masses that experienced rain events (blue, orange and green colours are as in (a) and represent rain events having different regression coefficients); contour levels are at 50%. (Color figure online)

to analyse $\{H_2O, \delta D\}$ pair distributions that follow theoretical linear relations. However, in general this constraint restricts the cluster expansion to the correlation of time point values and can be transferred to different domains and datasets. For example motion data in 2D or 3D space can be segmented with our approach and the same constraint formulation as well, identifying recurring, similar motions that follow similar linear correlations.

Segmenting our Lagrangian air parcel trajectories enables us to better understand the complex dynamics that cause the $\{H_2O, \delta D\}$ pair distribution observable in the MUSICA IASI dataset. Since our approach is able to distinguish fine-grained structures in large volumes of data, it is an effective data-driven analysis method for this purpose. For instance, with the grouping of trajectories we can draw conclusions on atmospheric moisture transport patterns: Fig. 3b suggests that if we observe a $\{H_2O, \delta D\}$ data pair below the Rayleigh line, the air mass has very likely been transported from East to West Africa.

Ongoing research in the area of observational atmospheric data orientated towards the combination of different sensors might soon allow the global detection of the vertical distribution of $\{H_2O, \delta D\}$ pairs (a method for such synergetic combination has recently been demonstrated using CH_4 as an example, Schneider et al. [16]). By using state-of-the-art reanalysis datasets, for calculating the trajectories our approach could then be used to directly identify different moisture processes in the measured $\{H_2O, \delta D\}$ fields.

5   Conclusion

In this paper we propose a novel semi-supervised approach for trajectory segmentation and demonstrate our approach to identify different processes in the atmosphere. We adopt our recently developed algorithm *CoExDBSCAN*, density-based clustering with constrained expansion, for trajectory segmentation by formulating a constraint on the deviation from the ordinary least squares regression residuals of combined or split segments. By further extracting information about the regression coefficients for each segment and comparing the distribution of coefficients to theoretical values we are able to identify corresponding atmospheric processes. This approach is demonstrated in our experimental evaluation for Lagrangian air parcel trajectories together with data from the regional isotope-enabled atmospheric model COSMO-iso [12]. We are able to successfully extract segments from subsequences of temporal continuous events and compare these segments with the theoretical evolution of our dependent variable δD as a function of H_2O. By orienting the trajectory calculations towards the overpass times and altitudes represented by the satellite data, we can use this model evaluation to better understand the satellite data.

For future work, we plan to further improve the extraction of specific $\{H_2O, \delta D\}$ sequences in the model domain and use the model data of rain, cloud etc. to clearly relate specific $\{H_2O, \delta D\}$ distributions to distinct moist processes. For instance we want to demonstrate that our method can be used to distinguish between air masses that experienced shallow convection or deep convection. The long-term perspective is to generate multisensor $\{H_2O, \delta D\}$ observation data, i.e. to create observation data that offer $\{H_2O, \delta D\}$ pair information at different altitudes. Then use trajectories calculated from state-of-the-art reanalysis fields and apply the method directly to the observational data. Data-driven analysis of observation data could significantly improve the insight into the dynamics of such cloud-circulation systems and would help to improve the significant uncertainty of climate predictions.

References

1. Bony, S., et al.: Clouds, circulation and climate sensitivity. Nature Geosci. **8**(4), 261–268 (2015). https://doi.org/10.1038/ngeo2398
2. Borger, C., et al.: Evaluation of MUSICA IASI tropospheric water vapour profiles using theoretical error assessments and comparisons to GRUAN vaisala rs92 measurements. Atmos. Measur. Techn. **11**(9), 4981–5006 (2018). https://doi.org/10.5194/amt-11-4981-2018
3. Dinler, D., Tural, M.K.: A survey of constrained clustering. In: Celebi, M.E., Aydin, K. (eds.) Unsupervised Learning Algorithms, pp. 207–235. Springer, Cham (2016). https://doi.org/10.1007/978-3-319-24211-8_9
4. Ertl., B., Meyer., J., Schneider., M., Streit., A.: CoExDBSCAN: density-based clustering with constrained expansion. In: Proceedings of the 12th International Joint Conference on Knowledge Discovery, Knowledge Engineering and Knowledge Management - Volume 1: KDIR, pp. 104–115. INSTICC, SciTePress (2020). https://doi.org/10.5220/0010131201040115
5. Ertl., B., Meyer., J., Streit., A., Schneider., M.: Application of mixtures of gaussians for tracking clusters in spatio-temporal data. In: Proceedings of the 11th International Joint Conference on Knowledge Discovery, Knowledge Engineering and Knowledge Management - Volume 1: KDIR, pp. 45–54. INSTICC, SciTePress (2019). https://doi.org/10.5220/0007949700450054
6. Ester, M., Kriegel, H.P., Sander, J., Xu, X.: A density-based algorithm for discovering clusters in large spatial databases with noise. In: Proceedings of the Second International Conference on Knowledge Discovery and Data Mining, KDD 1996, pp. 226–231. AAAI Press (1996)
7. Gaffney, S., Smyth, P.: Trajectory clustering with mixtures of regression models. In: Proceedings of the Fifth ACM SIGKDD International Conference on Knowledge Discovery and Data Mining, KDD 1999, pp. 63–72. Association for Computing Machinery, New York (1999). https://doi.org/10.1145/312129.312198
8. Gaffney, S.J., Robertson, A.W., Smyth, P., Camargo, S.J., Ghil, M.: Probabilistic clustering of extratropical cyclones using regression mixture models. Climate Dyn. **29**(4), 423–440 (2007). https://doi.org/10.1007/s00382-007-0235-z
9. Hallac, D., Vare, S., Boyd, S., Leskovec, J.: Toeplitz inverse covariance-based clustering of multivariate time series data. In: Proceedings of the 23rd ACM SIGKDD International Conference on Knowledge Discovery and Data Mining, KDD 2017, pp. 215–223. Association for Computing Machinery, New York (2017)
10. Lee, J.G., Han, J., Whang, K.Y.: Trajectory clustering: a partition-and-group framework. In: Proceedings of the 2007 ACM SIGMOD International Conference on Management of Data, pp. 593–604. Association for Computing Machinery, New York (2007). https://doi.org/10.1145/1247480.1247546
11. Maciąg, P.S.: A survey on data mining methods for clustering complex spatiotemporal data. In: Kozielski, S., Mrozek, D., Kasprowski, P., Małysiak-Mrozek, B., Kostrzewa, D. (eds.) BDAS 2017. CCIS, vol. 716, pp. 115–126. Springer, Cham (2017). https://doi.org/10.1007/978-3-319-58274-0_10

12. Miltenberger, A.K., Pfahl, S., Wernli, H.: An online trajectory module (version 1.0) for the nonhydrostatic numerical weather prediction model COSMO. Geosci. Model Dev. **6**(6), 1989–2004 (2013)
13. NOAA: National Oceanic and Atmospheric Administration Big Data Program. https://www.noaa.gov/organization/information-technology/big-data-program. Accessed 30 Nov 2020
14. Noone, D.: Pairing Measurements of the Water Vapor Isotope Ratio with Humidity to Deduce Atmospheric Moistening and Dehydration in the Tropical Midtroposphere. J. Climate **25**(13), 4476–4494 (012). https://doi.org/10.1175/JCLI-D-11-00582.1
15. Noone, D., et al.: Properties of air mass mixing and humidity in the subtropics from measurements of the D/H isotope ratio of water vapor at the Mauna Loa Observatory. J. Geophys. Res. Atmo. **116**(D22) (2011). https://doi.org/10.1029/2011JD015773
16. Schneider, M., et al.: Synergetic use of IASI and TROPOMI space borne sensors for generating a tropospheric methane profile product. submitted to Atmospheric Measurement Techniques (2021)
17. Schneider, M., et al.: Accomplishments of the MUSICA project to provide accurate, long-term, global and high-resolution observations of tropospheric $H_2O,\delta D$ pairs - a review. Atmos. Measure. Tech. **9**(7), 2845–2875 (2016). https://doi.org/10.5194/amt-9-2845-2016
18. Sprenger, M., Wernli, H.: The LAGRANTO Lagrangian analysis tool - version 2.0. Geosci. Model Dev. **8**(8), 2569–2586 (2015). https://doi.org/10.5194/gmd-8-2569-2015
19. Wang, S., Cai, T., Eick, C.F.: New spatiotemporal clustering algorithms and their applications to ozone pollution. In: 2013 IEEE 13th International Conference on Data Mining Workshops, pp. 1061–1068 (2013). https://doi.org/10.1109/ICDMW.2013.14
20. Zhang, Q., Wu, J., Zhang, P., Long, G., Zhang, C.: Salient subsequence learning for time series clustering. IEEE Trans. Pattern Anal. Mach. Intell. **41**(9), 2193–2207 (2019). https://doi.org/10.1109/TPAMI.2018.2847699
21. Zolhavarieh, S., Aghabozorgi, S., Teh, Y.W.: A review of subsequence time series clustering. Sci. World J. **2014** (2014). https://doi.org/10.1155/2014/312521

mRelief: A Reward Penalty Based Feature Subset Selection Considering Data Overlapping Problem

Suravi Akhter[1(✉)], Sadia Sharmin[3], Sumon Ahmed[1], Abu Ashfaqur Sajib[2(✉)],
and Mohammad Shoyaib[1]

[1] Institute of Information Technology, University of Dhaka, Dhaka, Bangladesh
bsse0827@iit.du.ac.bd, {sumon,shoyaib}@du.ac.bd
[2] Department of Genetic Engineering and Biotechnology, University of Dhaka,
Dhaka, Bangladesh
abu.sajib@du.ac.bd
[3] Department of Computer Science and Engineering, Islamic University of Technology,
Gazipur, Bangladesh
sharmin@iut-dhaka.edu

Abstract. Feature selection plays a vital role in machine learning and data mining by eliminating noisy and irrelevant attributes without compromising the classification performance. To select the best subset of features, we need to consider several issues such as the relationship among the features (interaction) and their relationship with the classes. Even though the state-of-the-art, Relief based feature selection methods can handle feature interactions, they often fail to capture the relationship of features with different classes. That is, a feature that can provide a clear boundary between two classes with a small average distance may be mistakenly ranked low compared to a feature that has a higher average distance with no clear boundary (data overlapping). Moreover, most of the existing methods provide a ranking of the given features rather than selecting a proper subset of the features. To address these issues, we propose a feature subset selection method namely modified Relief (mRelief) that can handle both feature interactions and data overlapping problems. Experimental results over twenty-seven benchmark datasets taken from different application areas demonstrate the superiority of mRelief over the state-of-the-art methods in terms of accuracies, number of the selected features, and the ability to identify the features (gene) to characterize a class (disease).

Keywords: Feature selection · mRelief · Data overlapping

1 Introduction

Feature selection is the process of selecting a feature subset S from the original feature set F such that S includes the most informative and relevant features for classification. Through this reduction process, the noisy, redundant and irrelevant features are eliminated which in turn improves the classification accuracy, reduces over-fitting as well as

© Springer Nature Switzerland AG 2021
M. Paszynski et al. (Eds.): ICCS 2021, LNCS 12742, pp. 278–292, 2021.
https://doi.org/10.1007/978-3-030-77961-0_24

the complexity of the model. In general, the existing feature selection methods can be divided into three main categories: wrapper, embedded, and filter methods [28]. Among them, filter based methods are most popular as they are computationally less expensive and not biased to any classification algorithm. Over the years different filter criteria have been introduced such as Correlation [8], Mutual Information (MI) [23,27,28], and Distance [10,12,17].

Earlier methods of feature selection use correlation metric to assess the quality of features. Pearson's correlation coefficient (PCC) is a well-known method in this regard [8]. However, these methods cannot capture the non-linear relationship between features and class variable. MI based selection methods overcome these problems and can detect the nonlinear relationship both for categorical and numerical data containing multiple classes [21]. One of the state-of-the-art methods in this regard, Maximum Relevance Minimum Redundancy (mRMR) [23] selects feature incrementally by maximizing the relevancy between a feature and class variable as well as minimizing the redundancy among the features. However, mRMR discards some features which provides additional information about class despite its redundancy. This problem is solved by Joint Mutual Information (JMI) [20] to some extent. Recently, Gao et al. [7] introduces a new feature selection method called Min-Redundancy and Max-Dependency (MRMD) where a new feature redundancy term is proposed for better approximation of the dependency between the features and class variable. Another MI based hybrid method namely Information-Guided Incremental Selection (IGIS+) is proposed in [22] for gene selection where they employ interaction information for ranking the features and then select the final subset utilizing classifier performance. However, high classification accuracy does not always ensure that the selected genes are relevant to a particular disease identification. Again, it is also well-known that a single feature alone can not predict the class properly without considering its interaction with the other features. Thus, we need to identify the inter-relationship among the features properly to select a better subset of features. However, MI based methods often fail to capture the higher order interaction among the features.

Relief based methods (RBM) such as Relief [17] can capture feature interactions better than the MI based method. Instead of searching through feature combinations, it uses the concept of k nearest neighbors (NN) to derive feature statistics that indirectly account for interactions. RBMs are particularly interesting because they can perform better even if the feature dimension increases. Relief was originally designed for binary classification and extended to ReliefF [14] to handle multiple classes. Both in Relief and ReliefF, we need to choose the value of k appropriately otherwise, it might become difficult to capture the informative information. Instead of fixing the value of k, it will be more advantageous if one can determine a volume from where reliable information can be extracted. Inspired by this idea, SURF [12] define a volume with a radius considering the average distance of all the instances and use the *hit* and *miss* within that radius (near *hit/miss*). It helps to capture the informative information even if the interaction is small. RelielfF shows low success rate in this case. Again, along with the *near* instance, other instances might contain important information and should be used for selecting the features. SURF* [11] was designed to capture two way interaction in feature weighting using near and far instance weighting. Even though it improves the performance of SURF* but requires more computation. To reduce the computational complexity retaining the performance, MultiSURF* [10] is proposed. However, SURF, SURF* [11], and MultiSURF* are mainly designed to handle genomics data (usually features contain few discrete value such as 0, 1 or 2) and might fail to achieve better performance for other problems having different type of data. MultiSURF [29], one of the recent methods in RBM group, solve this problem with less computational time compared to MultiSURF*.

Despite the importance of RBMs, most of them only rank the features and does not consider redundancy. Moreover, they may suffer if data overlapping exists and fail to capture the relationship between a feature and class. To understand this issue, let us consider an example as given in Fig. 1 where F_1 and F_2 are two features having instances C_+^1 to C_+^3 for one class and C_-^1 to C_-^4 other class. C_+^T is the target instance. Relief gives more priority to feature F_1 than F_2 though F_2 have better separability. (for details, see illustrative example in Sect. 3.3).

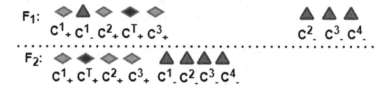

Fig. 1. Data overlapping among different classes

To solve the aforementioned problems, we propose a new feature selection method namely modified Relief (mRelief). The main contributions of this paper are as follows: First, it can capture feature interaction with the help of kNN considering all the features. Second, a $reward - penalty$ scheme is introduced here to detect the data overlapping among different classes and thus, establish a relationship between feature and class properly. Third, a redundancy criteria is proposed to remove the redundant features. In addition to superior performance in different datasets, mRelief can identify the important genes for disease identification.

The rest of the paper is organized in the following four sections. Section 2 describes the preliminaries required to understand the paper, our proposed mRelief is presented in Sect. 3, dataset description, implementation details and experimental results are described in Sect. 4. Finally, Sect. 5 concludes the paper.

2   Preliminaries

Several relief based methods namely Relief, ReliefF, SURF, SURF*, MultiSURF, and MultiSURF* are described in this section.

2.1   Relief

Relief [17] is a distance based feature selection method that uses 1NN to find the nearest two instances of two different classes from a target instance. The distance is called *hit* $(\delta_h(x_n^{f_i}))$ when the target and NN belong to the same class and *miss* $(\delta_m(x_n^{f_i}))$ otherwise. The score (J_{relief}) of Relief for a feature f_i is updated using Eq. 1.

$$J_{relief}(f_i) = J_{relief}(f_i) - \frac{\delta_h(x_n^{f_i})}{N} + \frac{\delta_m(x_n^{f_i})}{N} \tag{1}$$

here, N is the total number of samples. For nominal feature values, both δ_h and δ_m are '0' if the values are same and '1' otherwise. For numerical values, normalized feature values are used to calculate distances.

2.2 ReliefF

ReliefF [14] is an extension of Relief to multi-class problem where instead of 1NN, kNN is used to calculate *hit* and *miss* distances. Like Relief, it also calculates kNN for the same class (hits) and for each of other classes (misses). Finally, the score $(J_{ReliefF})$ is updated using Eq. 2 where $p(y_n)$ is the miss and $p(c_n)$ is the hit class probability.

$$J_{ReliefF}(f_i) = J_{ReliefF}(f_i) - \sum_{i=1}^{K} \frac{\delta_{h_i}(x_n^{f_i})}{m*k} + \sum_{c_n \neq y_n} \frac{p(y_n)}{1 - p(c_n)} \sum_{i=1}^{K} \frac{\delta_{m_i}(x_n^{f_i})}{m*k} \tag{2}$$

2.3 SURF and SURF*

SURF [12] considers the nearest instances that have small distances than a threshold (T) using Eq. 3, where, 'T' is defined by taking the average distance of all instance pairs.

$$\delta(x_i, x_j) < T \tag{3}$$

here, x_i is the target instance and x_j is the *hit/miss* instance. Considering these *hit* and *miss*, SURF score, J_{surf} is calculated as using Eq. 2.

SURF* [11] utilizes both the near and far instances (outside T) and its score, (J_{surf*}) is calculated by combining near and far instances. Near score is calculated using the traditional ReliefF scoring method (Eq. 2). Far scoring is the opposite of near scoring that is this sign of the second (*hit*) and third term (*miss*) of the right hand side of Eq. 2 become opposite.

2.4 MultiSURF* and MultiSURF

In MultiSURF* [10], threshold (T) for a target instance is identified by taking average distance of all instances from the target instance defined in Eq. 4. Note that, T remains same in SURF*, but it varies in MultiSURF* for the target instance.

$$T_i = \frac{\sum_{j=1}^{N-1} \delta(x_j, x_i)}{N - 1} \tag{4}$$

In case of MultiSURF*, a dead-band zone (D_i) is defined using the standard deviation (σ_i) of the distances during the calculation of T_is. A x_n is near when $\delta(x_i, x_j) < (T_i - D_i)$ and far when $\delta(x_i, x_j) > (T_i + D_i)$. MultiSURF considers only the *near* score instead of both *near* and *far* as in MultiSURF*.

3   Modified Relief (mRelief)

We propose modified Relief (mRelief) that includes a *reward* − *penalty* scheme to select a subset of features. The process of selecting features using mRelief can be divided into two parts: Candidate Feature Selection and Final Feature Subset Selection, which are described as follows:

3.1   Candidate Feature Selection

In this step, we filter the feature set F by removing the irrelevant and noisy features that do not contribute to the classification. A feature is irrelevant if it does not possess the ability to discriminate among the different classes properly. To measure this ability, we use paired t-test considering two sets of distances: one for *hit* and another for *miss* for a particular feature f_i. This test helps us to determine if the means of *hit* (μ_h) and *miss* (μ_m) distances are significantly different from each other under the null hypothesis, $H_0 : \mu_h = \mu_m$. When f_i accept (H_0) on α confidence interval, it is considered as irrelevant feature and removed from the feature set. Based on this hypothesis testing, we define feature irrelevance as given in Definition 1. Following this process, we remove the irrelevant and noisy features and obtain a candidate feature set S^c.

Definition 1. *Feature Irrelevance: A feature f_i is called irrelevant if there is no significant difference between the distances of hit and miss.*

3.2   Final Feature Subset Selection

Here, we first rank the candidate features S^c based on their individual relevance and then, select a subset of features S that jointly maximizes the relevancy as well as minimizes redundancy among them which is described as follows:

Ranking the Candidate Features: To rank the features in S^c, we assign a value for each feature f_i based on its score J_1 as defined in Eq. 5.

$$J_1(f_i) = \frac{1}{N} \sum_{n=1}^{N} \exp\left(-\frac{\delta_m(x_n^{f_i})}{\delta_h(x_n^{f_i})}\right) \tag{5}$$

here, $x_n^{f_i}$ is the nth instance of f_i, N is the total number of instances, $\delta_h(x_n^{f_i})$ and $\delta_m(x_n^{f_i})$ are the average distances of k nearest hit and miss respectively from a particular target instance (x_n). J_1 represents the class discrimination ability of f_i and it decreases if the target instance differs less from the nearby instances of the same class than the nearby instances of the other classes, and increases otherwise. It is well-known that the most discriminatory features have the least uncertainties. By minimizing the score J_1, we expect to identify the features that have better discrimination capability among the classes which in turn reduces the uncertainty. With the help of derivation given in [31], the relation between *score* and uncertainty is specified in Theorem 1.

Theorem 1. *Score minimization is equivalent to entropy minimization.*

Proof. It is straightforward to show that the distance (δ_h) from x_n to one of its NN ($h(x_n^{f_i})$) within the same class (*hit*) of a feature f_i is proportional to the volume (V) of the hyperspheres having the radius ($x_n^{f_i}$-$h(x_n^{f_i})$). At the same time, the posterior probability, $p(c_n|x_n^{f_i})$ of x_n to be class c_n is equivalent to $\frac{1}{NV}$. Therefore, it is evident that $\delta_h \propto p(c_n|x_n^{f_i})$. From the set of probabilities calculated for all the instances of f_i, one can calculate the uncertainties of the instances to be the member of that class by using entropy $H(x^{f_i}) = \sum_{n=1}^{N} p(x_n^{f_i}) \log p(x_n^{f_i})$. Similarly, for the other classes (*miss*), it can be shown that $\delta_m \propto p(y_n|x_n^{f_i})$ where $c_n \neq y_n$. Combining both *hit* and *miss* distances as given in Eq. 5 and their associated probabilities, it can easily be shown that minimizing J_1 also minimizes the entropy. This proves the theorem.

Note that, entropy minimization does not always ensure *score* minimization, but *score* minimization always ensure entropy minimization as shown in Theorem 1. However, only minimization of J_1 in Eq. 5 is not reliable enough due to data overlapping among different classes as shown in Fig. 1. To solve this problem, we introduce *reward-penalty* scheme. The definition of *reward* is given in Definition 2.

Definition 2. *Reward: Reward is a distance based measure that indicate the clear class separability of hit and miss instances and can be calculated using Eq. 6.*

$$dR = \frac{max(\delta_{h_k})}{min(\delta_{MoH})} * \frac{k - n_{MoH}}{k} \tag{6}$$

here, δ_{h_k} are the distances of the k nearest hit, δ_{MoH} is the distances of *miss* outside *hit*, and n_{MoH} is the number of *miss* outside *hit*. On the other hand, *penalty* term is defined in Definition 3.

Definition 3. *Penalty: Penalty is a distance based measure that indicates the amount of mixing of hit and miss instances and can be calculated using Eq. 7*

$$dP = \frac{min(\delta_{MiH})}{max(\delta_{h_k})} * \frac{k - n_{MiH}}{k} \tag{7}$$

here, n_{MiH} is the number of *miss* inside of *hit*. The values of dR and dP ranges from 0 to 1. The lower the value of dR and dP, the more the reward and penalty. Incorporating dR and dP to Eq. 5, we propose a new score J_2 defined in Eq. 8 to calculate the relevance of each feature $f_i(\in S^c)$ and sort them in ascending order to get the ranking of the features.

$$J_2(f_i) = \frac{1}{N} \sum_{n=1}^{N} \exp\left(-\frac{\delta_m(x_n^{f_i})}{\delta_h(x_n^{f_i})} - \frac{dP(x_n^{f_i})}{dR(x_n^{f_i})} \right) \tag{8}$$

Generating a Criteria for Subset Selection. The ranking based on J_2 does not confirm the best combination of features set for classification. Moreover, redundant features may exist in the ranking that need to be eliminated despite its relevancy. For this, let us define redundancy (δ_{red}) in term of distance which is given in Definition 4.

Definition 4. *Redundancy:* *A feature f_i is fully redundant in term of distance with the selected feature subset S when it has same discrimination ability as f_j ($f_j \in S$) and can be calculated using Eq. 9.*

$$\delta_{red}(f_i, S) = \sum_{f_j \in S} \frac{1}{N} \sum_{n=1}^{N} \exp\left(-\frac{\delta_m(x_n^{\{f_j, f_i\}})}{\delta_h(x_n^{\{f_j, f_i\}})}\right) - J_1(f_i) \qquad (9)$$

here, $\delta_h(x_n^{\{f_j, f_i\}})$ represents distance considering the target instance from f_i and calculating $J_1(f_i)$ for *hit* and *miss* instance for f_j. Lower value of δ_{red} indicates high redundancy with the selected feature.

Equation 8 calculates J_2 for a single feature f_i. This equation can also be used for a set of features S using kNN for that S. By incorporating this redundancy term with Eq. 8 that calculates distances considering all the features in S and f_i, we propose our final criteria (J_3) for feature subset selection as given in Eq. 10.

$$J_3(f_i, S) = \frac{1}{N} \sum_{n=1}^{N} \exp\left(-\frac{\delta_m(x_n^{S, f_i})}{\delta_h(x_n^{S, f_i})} - \frac{dP(x_n^{S, f_i})}{dR(x_n^{S, f_i})} - \delta_{red}\right) \qquad (10)$$

Search Strategy: We adopt a greedy forward search strategy to select the best subset of features without having low redundancy among them. At first, we include the top ranked (obtained using J_2) feature to our final subset S as the first selected feature. After that, we consider the subsequent features ($S^c \setminus f_1$) in the ranked list one by one and evaluate their goodness in combination with the selected subset S using the score J_3. The overall process of mRelief is shown in Algorithm 1. If the feature f_i minimizes the J_3, it shows that f_i gives additional information with the selected feature and is added f_i to S otherwise discarded. Following this process, we select the final subset of feature.

Algorithm 1. mRelief

Input: Dataset (D): all instances, $X =\{x_1, x_2, x_3, ...x_n\}$ and features, $F=\{f_1, f_2, f_3, ...f_p\}$
Parameter: Number of neighbour (k)
Output: Subset of features, S $\subseteq$ F

1: Select the candidate feature set S^c performing t-test
2: Calculate relevance score J_2 of each feature $f_i \in S^c$ using Eq-(8)
3: Sort S^c based on their score J_2 in ascending order
4: select f_1 with minimum score J_2 value
5: $S \leftarrow f_1; S^c \leftarrow S^c \setminus f_1; T \leftarrow J_2(f_1)$
6: **for all** i = 2: $|S^c|$ **do**
7:    Calculate score $J_3(f_i, S)$ using Eq-(10)
8:    **if** $J_3(f_i, S) < T$ **then**
9:        $S \leftarrow S \cup f_i; T \leftarrow J_3(f_i, S)$
10:   **end if**
11:   $S^c \leftarrow S^c \setminus f_i$
12: **end for**
13: **return** S

3.3 An Illustrative Example

To understand the impact of $reward\text{-}penalty$ scheme, let us consider an example where F_1 and F_2 are two features having the instance values shown in Table 1. Here C_x^y represents instance y (1 to m) belong to class x (+ and $-$) and T represents the target instance. Considering Eq. 2, the score of ReliefF for F_1 and F_2 are 0.0342 and 0.0225 respectively. According to ReliefF, F_1 gets higher value compared to F_2. However, it is evident that the instances of F_2 are more clearly separable than F_1 and thus, F_2 should get higher priority. This is because, for F_1, an instance of '-ve' class resides within the '+ve' class (also shown in Fig. 1. In case of mRelief (Eq. 8), the score of F_1 and F_2 are $0.1368e^{-38}$ and $\simeq 0$ respectively. As mRelief gives priority to the minimized score, it ranks F_2 better than F_1 which is desired.

Table 1. Sample dataset for the illustrative example

Instance	C_+^1	C_-^1	C_+^2	C_+^T	C_+^3	C_-^2	C_-^3	C_-^4
F_1	0.13	0.15	0.17	0.18	0.19	0.6	0.62	0.65
F_2	0.11	0.3	0.14	0.12	0.15	0.31	0.35	0.40
class	+	$-$	+	+	+	$-$	$-$	$-$

4 Result Analysis and Discussions

To demonstrate the experimental result, we first describe the datasets and then, present the experimental setup of different methods along with the proposed one and their evaluation process. Finally, mRelief is compared with other state-of-the-art methods from different aspects.

4.1 Dataset Description

To compare mRelief with other state-of-the-art methods, we use twenty benchmark datasets collected from UCI machine learning repository[1]. We also use seven cancer related datasets namely (Lung [2], CNS [25], SRBCT [16], Colon [1], Leukemia [9], MLL [24]) and GDS3341 [5] collected from different sources. The characteristics of the datasets are presented in column 2 to 4 of Table 2.

4.2 Implementation Detail

We conduct 10 fold cross-validation (10-CV) for the datasets with a large number of samples (>250), ten runs of 5-CV for the datasets having their samples between 100 and 250; and Leave-One-Out (LOO) otherwise. For fair comparison, the same strategy

---

[1] https://archive.ics.uci.edu/ml.

Table 2. Dataset overview and performance comparison among different algorithms in terms of Accuracy. Significant win is marked as * and loss as ° means mRelief significantly perform better/worse in the comparison of existing methods.

Dataset	Features	Class	Instance	MRMD	MultiSURF	MultiSURF*	SURF*	SURF	ReliefF	mRelief
yeast	8	10	1484	0.517*	0.532	0.522*	0.562	0.497*	0.560	**0.563(7)**
wine	13	3	178	0.963*	0.982	0.983	0.984	0.987	0.984	**0.989(12)**
heart	13	2	270	0.819	0.833	0.804	0.800*	0.833	0.833	**0.837(11)**
segment	19	7	2310	0.894*	0.919	0.800*	0.809*	0.914	0.906	**0.926(10)**
steel	27	7	1941	**0.684**	0.649	0.676	0.677	0.637	0.662	0.668(18)
ionosphere	33	2	351	0.833	0.822	0.831	0.819	0.836	0.831	**0.856(17)**
dermatology	34	6	366	0.948	0.815*	0.838*	0.868*	0.800*	0.840*	**0.950(13)**
appendicitis	7	2	106	0.850	0.858	0.833	0.842	0.875	0.842	**0.875(6)**
german	20	2	1000	**0.763**	0.747	0.742	0.740	0.749	0.753	0.752(15)
sonar	60	2	208	0.755	0.764	0.790	0.750	0.764	0.759	**0.791(17)**
libras	91	15	360	**0.744°**	0.571*	0.469*	0.527*	0.567*	0.584*	0.702(29)
page-blocks	10	5	5472	0.912*	0.923	0.901*	0.901*	0.921	0.921	**0.921(5)**
saheart	9	2	462	0.710	0.681	0.668*	0.685	0.672*	0.685	**0.717(6)**
southgerman	21	2	1000	**0.771**	0.753	0.752	0.753	0.750	0.754	0.767(15)
page-blocks0	10	2	5472	0.926*	0.918*	0.901*	0.901*	0.928	0.906*	**0.933(7)**
vehicle0	18	2	846	0.819*	0.822*	0.921	0.921	0.809*	0.833*	**0.926(6)**
ecoli3	7	2	336	**0.926°**	0.886	0.886	0.886	0.886	0.886	0.886(6)
new-thyroid1	5	2	215	0.941	0.945	0.818*	0.818*	0.945	0.945	**0.950(2)**
musk	166	2	476	0.783	0.790	0.775	0.775	**0.810**	0.763*	0.796(45)
semeion	256	10	1593	0.874*	0.820*	0.750*	0.759*	0.867*	0.855*	**0.895(99)**
Win/ Tie/ Loss	-	-	-	15/0/5	19/1/0	19/1/0	18/2/0	17/2/1	18/1/1	-
Sig. Win/ Loss	-	-	-	7/2	5/0	9/0	8/0	6/0	6/0	-

is followed for all other methods used in this paper. The results of mRelief are compared with other RBMs such as ReliefF, SURF, SURF*, MultiSURF*, and MultiSURF. We also compare mRelief with two MI based methods namely MRMD and IGIS+. Features are normalized to the interval [0,1] for all RBMs along with mRelief as suggested in [14]. For MRMD and IGIS+, we follow the suggested discretization as given in [7] and [22]. There are various type of classifier such as SVM, KNN, XGBoost [4] and for imbalance data PEkNN [15], kENN [18] can be used to measure the accuracy. The average accuracy of n-fold cross-validation is calculated using SVM (linear kernel) to measure the performance of the compared methods. We use the same number of features that mRelief selects for all other compared methods (except IGIS+) as they are all ranking methods.

4.3 Comparison of mRelief with the State-of-the-Art Methods

Tables 2 and 3 present the average accuracies of mRelief in comparison with other state-of-the-art methods along with the number of selected features given in the parenthesis. For these tables, win/tie/loss is calculated using the accuracies and shown in the second last row of the tables. To evaluate the significance of the improvements in accuracies among different methods, paired t-tests (at 95% significance level) are performed and shown in the last row of the tables. Here, Win/Tie/Loss indicates the number of datasets for which mRelief performs better/equally-well/worse than other aforementioned methods. The best performing method is presented in boldfaced. Analyzing these tables, we find that for most of the datasets, mRelief outperforms the other methods with less number of features. The reason behind such performance is that mReleif is able to detect feature interaction, data overlapping and redundant features. The detailed discussion of experimental results is presented as follows.

Table 3. Performance comparison among different algorithms on high Dimensional Datasets in terms of Accuracy. Significant win is marked as * and loss as ° means mRelief significantly perform better/worse in the comparison of existing methods.

Dataset	Features	Class	Instance	IGIS+	MultiSURF	MultiSURF*	SURF*	SURF	ReliefF	mRelief
Lung	12600	5	203	0.902(9)	0.852*	0.892*	0.899*	0.871*	0.918*	**0.931(77)**
CNS	7129	2	60	0.618(3)	0.582*	0.603*	0.629	0.605	0.628	**0.631(7)**
SRBCT	2308	4	83	0.923(9)*	0.987*	0.942*	0.928*	0.982*	0.994*	**0.997(51)**
Colon	2000	2	62	0.727(3)*	0.829	0.695*	0.649*	**0.845**	0.835	0.843(21)
Leukemia	7129	2	72	0.903(3)	0.932	0.865*	0.861*	0.937	0.937	**0.939(37)**
MLL	12582	3	72	0.843(5)*	0.944*	0.873*	0.844*	0.953	0.951*	**0.965(64)**
GDS-3341	30802	2	41	0.862(4)	**1.000**	0.780	0.927	0.976	**1.000**	**1.000(26)**
Win/ Tie/ Loss	-	-	-	7/0/0	6/1/0	7/0/0	7/0/0	6/0/1	6/1/0	-
Sig. Win/ Loss	-	-	-	3/0	4/0	5/0	5/0	2/0	3/0	-

Impact of Feature Interaction and Data Overlapping. To understand the impact of feature interaction and data overlapping, let us consider the data visualization for *Semeion* dataset shown in Fig. 2. We observe a major data overlapping between class 3 and 10 in this figure. As an MI based method, MRMD can handle data overlapping, its overall accuracy for this dataset is 87.4% which is close to mRelief (89.5%) (mRelief significantly wins in this case). Note that, MRMD can classify class 3 and 10 more accurately than RBMs (with 80% and 87% accuracy respectively), mRelief achieve 88.7% and 89.07% accuracy in this case. It shows although MRMD performs better than other RBMs, it can not exceed the performance of mRelief due to its lower capability of approximating the higher-order feature interaction. Instead of MRMD, we use IGIS+ (an MI based method) for the genomics datasets which is mainly designed to detect genes (features) for such datasets. This reason helps mRelief to win for all genomics dataset compared to IGIS+ shown in Table 3. On the other hand, RBMs (Table 3) can capture feature interaction but they confuse among the classes if data overlapping exists. Therefore, even the best performing RBM (SURF) does not perform well and the accuracies of class 3 and 10 are 85.2%, and 82.9% respectively. These results demonstrate the superiority of mRelief over the other methods in terms of capturing interactions and the ability to handle data overlapping.

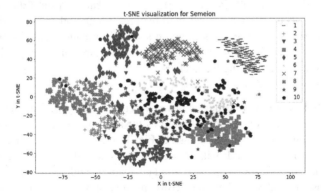

Fig. 2. Data Visualization using tSNE[19]. Here, (1–10) represents different classes.

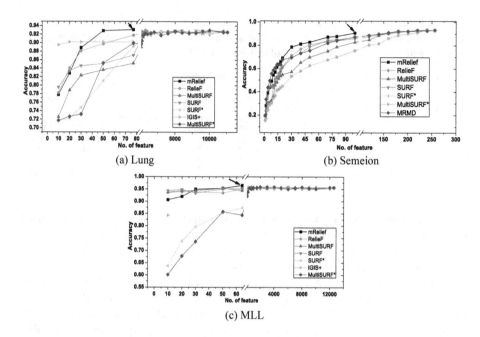

(a) Lung   (b) Semeion

(c) MLL

Fig. 3. Comparison of accuracies for different number of selected features

Impact of Feature Selection. mRelief uses Eq. 10 to select the best set of features. Note that, among all the compared methods, mRelief is the only feature selection method. To demonstrate the capability of mRelief in this regard, we plot the accuracies of different methods for different number of selected features using $Lung$, $Semeion$ and MLL dataset as shown in Fig. 3 (arrow mark indicates the performance of total selected features using mRelief). From these figures, we observe that mRelief can identify the relevant features properly and reach the highest point of accuracy for $Lung$, and MLL dataset and very close to the highest accuracies for $Semeion$ dataset. Note that, for $Lung$ dataset, mRelief selects 7089 candidate features from 12600 features and achieves 93.2% accuracy using 77 selected features. On the other hand, other methods require considerably more features to attain such performance. This justifies the purpose of feature selection methods that the selected features are highly relevant to the class and less redundant among themselves.

Impact of Redundancy. mRelief achieves a better ranking with a small set of features as it can identify non-redundant features for most of the datasets. For this, let us consider $Lung$ and MLL datasets shown in Fig. 3. In MLL dataset, mRelief performs better than the existing methods with less number of selected features, because mRelief has the mechanism of removing the redundant features and selects the features having better class separability. However, RBMs achieve similar accuracies with a higher number of features as they can not eliminate the redundant ones. In MLL dataset, IGIS+ obtains satisfactory performance (though these accuracies are lower compared

Table 4. Ranking of the selected gene

Method	Pathway	Rank	FDR
Using both score and frequency			
mRelief	Proteasome	1	4.41E−42
	Epstein-Barr virus infection	2	1.62E−32
	Cell cycle	3	1.40E−23
	Viral carcinogenesis	4	7.13E−18
	Hepatitis B	5	1.09E−15
	Kaposi's sarcoma-associated herpesvirus infection	6	3.19E−14
	Pathways in cancer	8	4.95E−13
Using the respective method's score			
mRelief	Viral carcinogenesis	2	6.94E−10
	Hepatitis B	3	8.11E−10
	Kaposi's sarcoma-associated herpesvirus infection	4	1.17E−08
	Epstein-Barr virus infection	5	2.58E−07
ReliefF	Pathways in cancer	4	3.06E−11
	Kaposi's sarcoma-associated herpesvirus infection	5	3.29E−11
SURF	N/A		
MultiSURF	Pathways in cancer	1	0.000716
SURF*	Hepatitis B	5	6.07E−12
	Viral carcinogenesis	7	7.88E−12
MultiSURF*	Viral carcinogenesis	2	1.31E−10
	Cell cycle	6	2.06E−09
IGIS+	Pathways in cancer	4	0.387

to mRelief) with a fewer number of selected features in comparison with mRelief. We use IGIS+ only for gene datasets as the proposal is only for these types of datasets. However, the selected features often fail to identify the relevant genes for a disease whereas the selected features of mRelief are truly relevant for classification as well as disease identification which is described in the following subsections.

Impact of Important Gene Identification. To demonstrate that mRelief can select relatively a small subset of best performing features (gene) to identify a particular class (disease related gene), we use $GDS3341$ [5], gene expression dataset. Both qualitative and quantitative changes in gene expression contribute to the development of different diseases. Therefore, we investigate the biological significance of the top fifty genes selected by different methods, except IGIS+. IGIS+ can select only ten different genes, which are used as input in further analysis. To explore the biological importance of the top selected genes, we use NetworkAnalyst 3.0 [33] for identifying the cellular pathways.

The $GDS3341$ dataset includes nasopharyngeal carcinoma tissue samples. Epstein-Barr virus (EBV) is well known to cause nasopharyngeal carcinoma (NPC), which is an epithelial cancer prevalent in Southeast Asia [3, 32]. Hepatitis B virus (HBV) infection plays a role in the development of NPC [30]. Epstein-Barr virus (EBV) and human herpesvirus, which is also known as Kaposi sarcoma-associated herpesvirus (KSHV), belong to the human gammaherpes virus family [6]. EBV manipulates the ubiquitin-proteasome system in EBV-associated malignancies [13]. Table 4 shows the pathways that have been identified based on protein-protein interaction networks. The list of pathways are ranked based on false discovery rates (FDR), which are adjusted p-values used in the analysis of large datasets generated in high-throughput experiments in order to correct for random events that falsely appear significant [26]. Only the top ten most significant pathways are considered. Table 4 shows only those pathways relevant to NPC. Based on the ranks of the NPC associated pathways, mRelief unequivocally perform better than the other methods. Although the selected genes of IGIS+ can identify "Pathways in cancer" with only ten selected genes, it is not significant statistically $(FDR > 0.05)$.

5   Conclusion

In this paper, we propose mRelief that selects a better feature subset with higher accuracies compared to the state-of-the-art methods over a large set of benchmark datasets. Moreover, it identifies a set of features that is highly representative of a particular class and thus can be used in different applications including gene selection for disease identification. However, mRelief can be adapted using SURF or MultiSURF instead of fixed k, which we will address in the future.

Acknowledgement. This research is supported by a grant (19IF12116, 2019–2020) from the Innovation fund of the ICT Division, Ministry of Posts, Telecommunications and Information Technology, Bangladesh. The authors are thankful for the support.

References

1. Alon, U., et al.: Broad patterns of gene expression revealed by clustering analysis of tumor and normal colon tissues probed by oligonucleotide arrays. Proc. Natl. Acad. Sci. **96**(12), 6745–6750 (1999)
2. Bhattacharjee, A., et al.: Classification of human lung carcinomas by mrna expression profiling reveals distinct adenocarcinoma subclasses. Proc. Natl. Acad. Sci. **98**(24), 13790–13795 (2001)
3. Cao, Y.: Ebv based cancer prevention and therapy in nasopharyngeal carcinoma. NPJ Precision Oncol. **1**(1), 1–5 (2017)
4. Chen, T., Guestrin, C.: Xgboost: a scalable tree boosting system. In: Proceedings of the 22nd ACM SIGKDD, pp. 785–794 (2016)
5. Dodd, L.E., et al.: Genes involved in dna repair and nitrosamine metabolism and those located on chromosome 14q32 are dysregulated in nasopharyngeal carcinoma. Cancer Epidemiol. Prev. Biomarkers **15**(11), 2216–2225 (2006)

6. Frere, C., et al.: Therapy for cancer-related thromboembolism. Semin. oncol. **41**(3), 319–338 (2014)
7. Gao, W., Hu, L., Zhang, P.: Feature redundancy term variation for mutual information-based feature selection. Appl. Intell. **50**(4), 1272–1288 (2020)
8. Goh, L., Song, Q., Kasabov, N.: A novel feature selection method to improve classification of gene expression data. In: Proceedings of the Second Conference on Asia-Pacific Bioinformatics-Volume 29, pp. 161–166. Australian Computer Society, Inc. (2004)
9. Golub, T.R., et al.: Molecular classification of cancer: class discovery and class prediction by gene expression monitoring. Science **286**(5439), 531–537 (1999)
10. Granizo-Mackenzie, D., Moore, J.H.: Multiple threshold spatially uniform ReliefF for the genetic analysis of complex human diseases. In: Vanneschi, L., Bush, W.S., Giacobini, M. (eds.) EvoBIO 2013. LNCS, vol. 7833, pp. 1–10. Springer, Heidelberg (2013). https://doi.org/10.1007/978-3-642-37189-9_1
11. Greene, C.S., Himmelstein, D.S., Kiralis, J., Moore, J.H.: The informative extremes: using both nearest and farthest individuals can improve relief algorithms in the domain of human genetics. In: Pizzuti, C., Ritchie, M.D., Giacobini, M. (eds.) EvoBIO 2010. LNCS, vol. 6023, pp. 182–193. Springer, Heidelberg (2010). https://doi.org/10.1007/978-3-642-12211-8_16
12. Greene, C.S., Penrod, N.M., Kiralis, J., Moore, J.H.: Spatially uniform relieff (surf) for computationally-efficient filtering of gene-gene interactions. BioData Mining **2**(1), 1–9 (2009)
13. Hui, K.F., Tam, K.P., Chiang, A.K.S.: Therapeutic strategies against epstein-barr virus-associated cancers using proteasome inhibitors. Viruses **9**(11), 352 (2017)
14. Kononenko, I.: Estimating attributes: analysis and extensions of RELIEF. In: Bergadano, F., De Raedt, L. (eds.) ECML 1994. LNCS, vol. 784, pp. 171–182. Springer, Heidelberg (1994). https://doi.org/10.1007/3-540-57868-4_57
15. Kadir, M.E., Akash, P.S., Sharmin, S., Ali, A.A., Shoyaib, M.: A proximity weighted evidential k nearest neighbor classifier for imbalanced data. In: Lauw, H.W., Wong, R.C.-W., Ntoulas, A., Lim, E.-P., Ng, S.-K., Pan, S.J. (eds.) PAKDD 2020. LNCS (LNAI), vol. 12085, pp. 71–83. Springer, Cham (2020). https://doi.org/10.1007/978-3-030-47436-2_6
16. Khan, J., Wei, J., Ringner, M., Saal, L., Ladanyi, M., Westermann, F.: Classification and diagnostic prediction of cancers using gene expression profiling 300 and artificial neural networks. Nature Med. **7**, 673–679 (2001)
17. Kira, K., Rendell, L.A.: The feature selection problem: traditional method and a new algorithm. AAAI **2**, 129–134 (1992)
18. Li, Y., Zhang, X.: Improving k nearest neighbor with exemplar generalization for imbalanced classification. In: Huang, J.Z., Cao, L., Srivastava, J. (eds.) PAKDD 2011. LNCS (LNAI), vol. 6635, pp. 321–332. Springer, Heidelberg (2011). https://doi.org/10.1007/978-3-642-20847-8_27
19. Van der Maaten, L., Hinton, G.: Visualizing data using t-sne. Journal of machine learning research **9**(11), (2008)
20. Moody, J., Yang, H.: Data visualization and feature selection: New algorithms for nongaussian data. Adv. Neural Inf. Process. Syst. **12**, 687–693 (1999)
21. Naghibi, T., Hoffmann, S., Pfister, B.: A semidefinite programming based search strategy for feature selection with mutual information measure. IEEE Trans. Pattern Anal. Mach. Intell. **37**(8), 1529–1541 (2014)
22. Nakariyakul, S.: A hybrid gene selection algorithm based on interaction information for microarray-based cancer classification. PloS One **14**(2), e0212333 (2019)
23. Peng, H., Long, F., Ding, C.: Feature selection based on mutual information criteria of max-dependency, max-relevance, and min-redundancy. IEEE Trans. Pattern Anal. Mach. Intell. **27**(8), 1226–1238 (2005)

24. Pollack, J.R., et al.: Genome-wide analysis of dna copy-number changes using cdna microarrays. Nature Genet. **23**(1), 41–46 (1999)
25. Pomeroy, S., et al.: Gene expression-based classification and outcome prediction of central nervous system embryonal tumors. Nature **415**(24), 436–442 (2002)
26. Rouam, S.: False discovery rate (fdr). encyclopedia of systems biology. Cancer Epidemiol. Prevention Biomarkers **36**, 731–732 (2013)
27. Roy, P., Sharmin, S., Ali, A.A., Shoyaib, M.: Discretization and feature selection based on bias corrected mutual information considering high-order dependencies. In: Lauw, H.W., Wong, R.C.-W., Ntoulas, A., Lim, E.-P., Ng, S.-K., Pan, S.J. (eds.) PAKDD 2020. LNCS (LNAI), vol. 12084, pp. 830–842. Springer, Cham (2020). https://doi.org/10.1007/978-3-030-47426-3_64
28. Sharmin, S., Shoyaib, M., Ali, A.A., Khan, M.A.H., Chae, O.: Simultaneous feature selection and discretization based on mutual information. Pattern Recogn. **91**, 162–174 (2019)
29. Urbanowicz, R.J., Olson, R.S., Schmitt, P., Meeker, M., Moore, J.H.: Benchmarking relief based feature selection methods for bioinformatics data mining. J. Biomed. Inform. **85**, 168–188 (2018)
30. Weng, J.-J., et al.: Effects of hepatitis b virus infection and antiviral therapy on the clinical prognosis of nasopharyngeal carcinoma. Cancer Med. **9**(2), 541–551 (2020)
31. Yang, S.H., Hu, B.G.: Discriminative feature selection by nonparametric bayes error minimization. IEEE Transa. Knowl. Data Eng. **24**(8), 1422–1434 (2012)
32. Young, L.S., Dawson, C.W.: Epstein-barr virus and nasopharyngeal carcinoma. Chinese J. Cancer **33**(12), 581 (2014)
33. Zhou, G., et al.: Networkanalyst 3.0: a visual analytics platform for comprehensive gene expression profiling and meta-analysis. Nucleic Acids Res. **47**(W1), W234–W241 (2019)

Reconstruction of Long-Lived Particles in LHCb CERN Project by Data Analysis and Computational Intelligence Methods

Grzegorz Gołaszewski[1], Piotr Kulczycki[1,2(✉)], Tomasz Szumlak[1], and Szymon Łukasik[1,2]

[1] Faculty of Physics and Applied Computer Science,
AGH University of Science and Technology, Kraków, Poland
{ggolaszv,kulpi,szumlak.slukasik}@agh.edu.pl
[2] Systems Research Institute, Polish Academy of Sciences, Warsaw, Poland

Abstract. LHCb at CERN, Geneva is a world-leading high energy physics experiment dedicated to searching for New Physics phenomena. The experiment is undergoing a major upgrade and will rely entirely on a flexible software trigger to process the data in real-time. In this paper a novel approach to reconstructing (detecting) long-lived particles using a new pattern matching procedure is presented. A large simulated data sample is applied to build an initial track pattern by an unsupervised approach. The pattern is then updated and verified by real collision data. As a performance index, the difference between density estimated by nonparametric methods using experimental streaming data and the one based on theoretical premises is used. Fuzzy clustering methods are applied for a pattern size reduction. A final decision is made in a real-time regime with rigorous time boundaries.

1 Introduction

Particle physics represents one of the hottest areas for the applications of Machine Learning research. Computational intelligence, understood as a group of methods reacting to the environment in new ways, making useful decisions by imitating intelligent organisms or social mechanisms is being used to tackle tasks difficult to deal with, using standard statistical apparatus [3]; see also [6,7].

The LHCb (Large Hadron Collider beauty) [8] experiment is one of eight particle physics detector experiments collecting data at the Large Hadron Collider (LHC), CERN, Geneva. It has been collecting proton-proton collision data during Run 1 and Run 2 data taking periods from 2010 to 2012 and from 2015 to 2018 respectively. Its primary physics mission is to search for New Physics and provide precise measurement of the charge-parity violation in the heavy quark sector [12]. Thanks to its extraordinary tracking system, in time, the physics programme has been extended and the detector became a general purpose in forward direction with a high performance high level trigger system [1]. As the experiment is currently undergoing a major modernisation increasing its instantaneous luminosity (and for the preparation for Run 3 that will start in 2021)

© Springer Nature Switzerland AG 2021
M. Paszynski et al. (Eds.): ICCS 2021, LNCS 12742, pp. 293–300, 2021.
https://doi.org/10.1007/978-3-030-77961-0_25

the entire tracking system must be exchanged [11] and new readout electronics capable of on-detector raw data processing must be designed [10]. Most of the physics analyses in LHCb use, so called long tracks, i.e. trajectories of particles traversing the whole active volume of the LHCb tracking system.

The goal of this contribution is to propose a new algorithm, based on principles of data mining, for detecting and reconstructing so called long-lived particles in the revised detector scheme under new, more strict time and accuracy constraints. The problem of detecting such particles will be treated here as the instance of the classification task, that is assigning a tested element $\tilde{x}$ to one of the designated classes with known sets of representative elements. The reconstruction of long-lived particles for each analyzed track classifier should be assigned one of the label: 1 if the track is associated with the long-lived particle, and 0, if otherwise. The proposed algorithm relies on non-parametric construction of training patterns representing tracks of long-lived particles, with online adaptation of this training set. In addition to that, the algorithm uses fuzzy clustering for obtaining repeatedly a reduced representation of the training set.

2   Long-Lived Particles in LHCb

The most important, from the point of view of this paper, sub-systems of the upgraded spectrometer are those comprising the tracking system: vertex detector (VELO), upstream tracker (UT) and scintillating fibre tracker (FT); see Fig. 1. The VELO is a pixel detector located around the proton-proton interaction point and dedicated to precise measurement of charged particle trajectories. The UT detector provides additional measurements upstream of the magnetic field and is capable of reconstructing up to two space points. Finally, the FT is designed to sample trajectories after the magnet and provides up to three space points. The overall design goals of this system are to provide track reconstruction with efficiency greater than 95% and average momentum resolution of 0.5% for charged particles with momenta $p < 20$ GeV. Most of the physics analyses in LHCb use long tracks, which are trajectories of particles traversing the whole active volume of the LHCb tracking system.

The tracks reconstructed in the LHCb detector are divided into types depending on the sub-detectors in which they are observed. VELO tracks are defined as those which have hits only in the VELO detector. Upstream tracks are defined as those which have measurements only in the VELO and UT detectors. Upstream tracks are also referred to as Velo-UT tracks. FT tracks are defined as those which have hits only in the FT stations. Downstream tracks, which are the subject of this paper, must have hits both in the UT and FT trackers. Particles that are reconstructed as long tracks must traverse all of the tracking detectors. These tracks provide the best momentum resolution.

Downstream tracks, that are the signature of long-lived particles, are currently being reconstructed at LHCb using a state-of-art *PatLongLivedTracking* algorithms. The input of the algorithm constitutes a track segment container and hits reconstructed in the UT detector. The FT tracks are prepared by a stand-alone code that is not a part of the long-lived tracks reconstruction. In order to

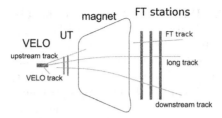

Fig. 1. A simplified diagram of the LHCb tracker system

increase purity of the input FT tracks they are processed by a multivariate classifier (binned Boosted Decision Tree) to discard bad candidates, e.g. segments that do not represent a real particle. Selected candidates are then propagated through the magnetic field to the UT tracker, assuming that they originate at the proton-proton interaction point. The UT detector comprises four layers of silicon micro-strip detectors divided into two stations UTa and UTb. Each station is capable of providing one space point (for each reconstructed trajectory) by probing x and y coordinates. The latter is measured by tilted by 5° and –5° layers UTau and UTbv also known as stereo layers, probing u and v coordinates respectively. The algorithm is searching for one hit in either of the x-probing layers which gives significant constraints on the flight trajectory of a particle. Next, an attempt is made to match a hit in the remaining x layer. Finally, hits are being searched for in the stereo layers. The hits are subsequently fitted with a parabola and χ^2 metric is calculated.

3   Proposed Approach

The algorithm being proposed here is divided into three separate phases. The initial stage includes data preparation and parameter adjustment. It is carried out before the experiment starts. This is followed by the on-line stage – the main phase from the point of view of the entire LHCb experiment. It is implemented in real time in the trigger circuit of the detector. Finally the calibration stage involves the modification of parameters and updating the on-line algorithm.

3.1   Initial Stage

Generation of Mass Density Training Patterns. The algorithm starts with the creation of the f mass distribution density of detected long-lived particles for each of the two kinds of particles which are K_S^0 i Λ. Due to the large number of factors affecting the measurements, the normal distribution is assumed in the form $f(m) = \frac{1}{\hat{\sigma}\sqrt{2\pi}} e^{-\frac{(m-\hat{m})^2}{2\hat{\sigma}^2}}$, where the value of the parameter $\hat{m}$ is established using physical premises. To determine the value of $\hat{\sigma}$, the Monte Carlo simulated events, prepared by the LHCb collaboration, is used (with the sample large enough for the estimator $\hat{\sigma}$ to be stable with assumed accuracy).

Construction of the Pattern of Pairs of Long-Lived Particle Tracks.
The next step will be to create an n-element set W of unique reference pairs using
the Monte Carlo detector model. Such a set is created for each kind of particle
and all described procedures is applied separately in regard of particle kind. Each
pair $(w^i \in W, i = 1, 2, ..., n)$ is described by the following parameters: particle
mass m_i; coordinates (V_x^i, V_y^i, V_z^i) of the vertex of primary particle decay; the set
of points $(x_{a,1}^i, v_1^i, u_1^i, x_{b,1}^i), (x_{a,2}^i, v_2^i, u_2^i, x_{b,2}^i)$ belonging to the subsequent hits on
UT measurement planes, for the 1st and 2nd track from the pair, respectively;
FT-seed state-vectors $s_1^i = (x_1^i, y_1^i, t_{x,1}^i, t_{y,1}^i, \frac{q}{p}_1^i)$ and $s_2^i = (x_2^i, y_2^i, t_{x,2}^i, t_{y,2}^i, \frac{q}{p}_2^i)$
associated with the appropriate tracks. The pattern elements generated in this
way are then sorted relative to the coordinate $x_{1,1}$ (hit from the first UT measurement plane, and the first track of each pair) and divided into sectors of
length resulting from measuring accuracy. This allows indexing of the pattern
elements, which will significantly speed up the search for appropriate pairs at a
later stage. Each pair is also initially assigned the fuzzy membership $\mu_i = 1$ in
the pattern set [4]. The pattern prepared in this way will then be transferred to
the LHCb trigger environment.

3.2   Online Stage

The following section describes the procedure for finding pairs of long-lived particle traces from the data set. Such a set consists of hits from all UT detector
measurement planes and FT-seed state vectors from T1, T2, T3 detectors created in earlier procedures. Due to the high frequency of intersection of proton
beams and because this procedure is only one of many carried out during the
experiment, it is necessary that the algorithm uses the least amount of both time
and memory resources. It must also be constructed with a perspective of parallel
implementation (in case of the execution comparing created tracks at this stage
of the algorithm is lost in practice). It consists of the following steps.

1. Hits from the UTa_x plane are assigned to the respective sectors created in
   the initial stage, which allows them to be associated with pairs of tracks from
   the pattern, assigned to a given sector or directly adjacent sectors.
2. For each pair of reference traces (w_i) to which at least one hit has been
   assigned, all UT measurement planes are searched in order to find the corresponding hits from these planes and the first trace from the pair. The similarity criteria used result from technical measurement accuracy.
3. If at least 7 out of 8 possible hits on the UT detector were found, the best-
   matched elements are found in the set of FT-seeds according to the corresponding similarity criterion.
4. When the FT-seeds found in this way are sufficiently similar to the reference
   ones for both pair of tracks, hits from the UT detector with matched FT-
   seeds are marked as true tracks and forwarded for further analysis, which is
   beyond the scope of the algorithm presented.

3.3   Calibration Stage

As part of the experiment, calibration procedures are carried out, improving the effectiveness of the procedure and adapting it to changing conditions. What's more, in the period between experiment breaks, in parallel to the on-line algorithm running, an update of the track pair patterns is launched and during the next break, an improved pattern is loaded into the system. Data for modifying the pattern are read from the buffer. Because the collected particles come from the very end of the reconstruction process, they are assigned a mass of m_i, along with all parameters that describe pattern pairs.

Adding New Elements to the Pattern. For all new w'_j elements, their similarity d to the most similar element from the W pattern is determined: $\Delta_j = \min\limits_{w_i \in W} \{d(w_i, w'_j)\}$, on the basis of which the decreasing function of membership of an element to a pattern is determined. If the element most similar w_i comes from the experiment, then their degrees of belonging become equal to $\mu_j = \mu_i$, while if w_i comes from the Monte Carlo model, the membership is

$$\mu_j = \begin{cases} 1 - 2\frac{\Delta_j^2}{\epsilon_\mu^2} & \text{, for } 0 \le \Delta_j \le \frac{\epsilon_\mu}{2} \\ 2\frac{(\epsilon_\mu - \Delta_j)^2}{\epsilon_\mu^2} & \text{, for } \frac{\epsilon_\mu}{2} < \Delta_j \le \epsilon_\mu \\ 0 & \text{, for } \Delta_j > \epsilon_\mu \end{cases} \tag{1}$$

where $\epsilon_\mu > 0$ describes the maximum similarity deciding element's acceptance.

Removing Redundant Elements. As the pattern increases in size, it is necessary to remove excess elements. For this purpose, the elements of the pattern are sorted according to the degree of the membership μ_i, and then $k < n$ elements with the lowest membership μ_i such that $\hat{f}(m_i) - f(m_i) > 0$ are removed from the pattern. In the above formula, f denotes the reference mass distribution created at the beginning of the procedure, while $\hat{f}$ describes nuclear mass distribution given in the form of the kernel estimator $\hat{f}(m) = \frac{1}{nh}\sum_{n=1}^{n} K(\frac{m-m_i}{h})$ where $K : \mathbb{R} \to [0, \infty)$ denotes the assumed kernel function and $h > 0$ means the smoothing parameter calculated by optimizing criteria [5].

Fuzzy Pattern Clustering. If the maximum acceptable size of the pattern is exceeded, while maintaining high values of μ, it is grouped by the fuzzy c-means clustering procedure [9]. Members of each of the K clusters are replaced by a representative (in the form of a centroid of a given cluster) with the weight $\psi_i = \sum_{j=1}^{n} c_{i,j} s_j$, where $c_{i,j}$ is a membership of the element j to cluster i. In addition to reducing the size of the pattern, this procedure also allows the gradual removal of Monte Carlo elements from the pattern, by replacing them with elements designated by means of experiments. The set of patterns for particular particle kinds is clustered separately.

4   Evaluation Criteria and Preliminary Results

The most important figures used to assess the performance of the downstream tracking algorithm are reconstruction efficiency and ghost fractions. Both are measured using simulated data samples by counting the number of correctly reconstructed tracks and compared to the number of tracks that would be reconstructed by a perfect algorithm. The latter tracks are called reconstructible. This corresponds to the concept of classification accuracy [2].

The following definitions are used within the LHCb experiment. A particle is reconstructible as a downstream track if it is reconstructible as a FT track and has at least one hit in both UTa and UTb stations. A particle is reconstructible as a FT track if it has at least one x and one stereo hit associated to it in each of the three FT stations. A particle is considered reconstructed as a downstream track if at least 70% of the FT-station hits on a track are associated to it and the track has no more than one wrongly associated UT hit. Based on these definitions one can construct two types of efficiencies: the overall tracking efficiency $\epsilon_{rec,tot}$ and the efficiency related to the algorithm itself $\epsilon_{rec,alg}$. The former is useful for physics studies and represents the total reconstruction efficiency, i.e. corresponds to the number of downstream reconstructed tracks divided by the number of downstream reconstructible tracks. The latter also depends on the efficiency of the FT tracking and is calculated by dividing the number of downstream reconstructed and FT reconstructed tracks by the number of downstream reconstructible and FT reconstructed tracks. At the same time, it is worth mentioning that in our analyses electrons are discarded, since they undergo multiple scattering and more energy loss than other heavier particles. In addition to the efficiency functions ghost tracks must be also considered. Ghosts are a consequence of large particle occupancies and finite detector granularity. We define the fraction of the ghost tracks as R_{ghost} which corresponds to the number of ghost tracks related to the total number of downstream reconstructed tracks.

The presented algorithm has been preimplemented and integrated with the LHCb test environment. Tests were carried out on data generated in a Monte Carlo simulation and from real data from previous experiments. Because the actual data consists solely of measurements, it is not known which of them represents true particle tracks and there is no real possibility to fully reliably determine the algorithm efficiency. A set of training patterns was generated first. It consisted of about 250000 elements. A testing dataset, which was separately generated, consisted of the following particles: UT+T – 3815 particles, UT+T>GeV – 2327, UT+T_strange – 273, UT+T_strange>5GeV – 122, noVelo+UT+T_strange – 186, noVelo+UT+T_ strange>5GeV – 73 and 65621 particles of other types excluding noise. The tracking efficiency ($\epsilon_{rec,tot}$) was calculated with regards to the different particle types. The performance of the algorithm was demonstrated in Table 1. Taking into account the preliminary stage of the research it was found to be very promising. In all cases the majority of particles belonging to each of the groups were correctly identified.

Table 1. Tracking efficiency $\epsilon_{rec,tot}$ for tested particle types.

Particle type	Tracking efficiency
UT+T	62.6
UT+T>GeV	75.6
UT+T_strange	58.8
UT+T_strange>5GeV	77.8
noVelo+UT+T_strange	54.5
noVelo+UT+T_strange>5GeV	66.7

5 Final Comments

This paper presents an algorithm for detecting traces of long-lived particles using elements of data analysis and computational intelligence. The aforementioned procedure is subject to numerous quality requirements as well as restrictions, especially the time needed for its implementation. This implies great possibilities to modify the concept during its implementation, depending on the partial results obtained. One of the planned research directions is the addition of a data validity mechanism known from data stream mining to the calibration stage, which would allow, in the long term, to remove pattern elements for which similar tracks are not found. In addition, this could improve the ability of the algorithm to adapt to variable physical parameters of the detector.

Acknowledgments. We acknowledge support from CERN and LHCb and from the national agency: MEiN and National Science Centre (Poland) UMO-2018/31/B /ST2/03998. The work was also supported by the Faculty of Physics and Applied Computer Science AGH UST statutory tasks within subsidy of MEiN.

References

1. Aaij, R., et al.: Tesla: an application for real-time data analysis in high energy physics. Comput. Phys. Commun. **208**, 35–42 (2016)
2. Cady, F.: The Data Science Handbook. Wiley (2017)
3. Da, R.: Computational Intelligence in Complex Decision Making Systems. Atlantis Computational Intelligence Systems. Atlantis Press, Paris (2010)
4. Valdez, F.: Bio-inspired optimization methods. In: Kacprzyk, J., Pedrycz, W. (eds.) Springer Handbook of Computational Intelligence, pp. 1533–1538. Springer, Heidelberg (2015). https://doi.org/10.1007/978-3-662-43505-2_81
5. Kulczycki, P.: Kernel estimators for data analysis. In: Ram, M., Davim, J. (eds.) Advanced Mathematical Techniques in Engineering Sciences, pp. 177–202. CRC/Taylor and Francis (2018)
6. Ballová, D.: Trend analysis and detection of change-points of selected financial and market indices. In: Kulczycki, P., Kacprzyk, J., Kóczy, L.T., Mesiar, R., Wisniewski, R. (eds.) ITSRCP 2018. AISC, vol. 945, pp. 372–381. Springer, Cham (2020). https://doi.org/10.1007/978-3-030-18058-4_30

7. Mesiar, R., Kolesárová, A.: On some recent construction methods for bivariate copulas. In: Kulczycki, P., Kóczy, L.T., Mesiar, R., Kacprzyk, J. (eds.) CITCEP 2016. AISC, vol. 462, pp. 243–253. Springer, Cham (2017). https://doi.org/10.1007/978-3-319-44260-0_15

8. LHCb collaboration: LHCb detector performance. Int. J. Mod. Phys. A **30**(7), 1530022 (2015)

9. Miyamoto, S., Ichihashi, H., Honda, K.: Algorithms for Fuzzy Clustering: Methods in c-Means Clustering with Applications. Studies in Fuzziness and Soft Computing. Springer, Heidelberg (2008). https://doi.org/10.1007/978-3-540-78737-2

10. Steinkamp, O.: LHCb upgrades. J. Phys. Conf. Ser. **1271**(1), 012010 (2018)

11. Szumlak, T.: Events reconstruction at 30 MHz for the LHCb upgrade. Nucl. Instrum. Methods Phys. Res. Sect. A **936**(1), 356–357 (2019)

12. Vecchi, S.: Overview of recent LHCb results. EPJ Web Conf. **192**(24), 8 (2018)

Motion Trajectory Grouping for Human Head Gestures Related to Facial Expressions

Maja Kocoń[(⊠)] [ID]

West Pomeranian University of Technology Szczecin, Sikorskiego 37, Szczecin, Poland
maja.kocon@zut.edu.pl

Abstract. The paper focuses on human head motion in connection with facial expressions for virtual-based interaction systems. Nowadays, the virtual representation of a human, with human-like social behaviour and mechanism of movements, can realize the user-machine interaction. The presented method includes the head motion because head gestures transmit additional information about the interaction's situational context. This paper presents head motion analysis based on the rotation of rigid objects technique for virtual-based interaction systems. First, we captured the head gestures of a human subject, expressing three basic facial expressions. The proposed motion model was described using three non-deformable objects, which reflect the neck and head skeleton movement's character. Based on the captured actions, the motion trajectories were analyzed, and their characteristic features were distinguished. The obtained dependencies were used to created new trajectories using piecewise cubic Hermite interpolating polynomial (PCHIP). Furthermore, the trajectories assigned to the rigid model have been grouped according to their similarities for a given emotional state. This way, using a single master trajectory and a set of coefficients, we were able to generate the whole set of trajectories for joint rotations of the head for the target emotional state. The resulting rotation trajectories were used to create movements on the three-dimensional human head.

Keywords: Trajectory grouping · Emotion intensity · Rigid object · Motion similarity

1 Introduction

Nowadays, intuitive and robust interaction between human and machine is still a challenging problem. One of the methods to improve traditional communication and to humanize machine is the use of some virtual reality elements. First of all, we need to focus on the face and head motion aspects. Facial expression and head gestures provide information about emotions, intentions and mood, therefore it is the main communication channel in interaction [1]. Although the face mainly transmits the person's emotional state, it does not adequately describe the situational context. Therefore, in the analysis and synthesis of human motion, it

© Springer Nature Switzerland AG 2021
M. Paszynski et al. (Eds.): ICCS 2021, LNCS 12742, pp. 301–315, 2021.
https://doi.org/10.1007/978-3-030-77961-0_26

is essential to consider head movement aspects. The most popular perception studies about the meaning of the head motion in interpersonal communication have been described in [8,11], where the significant role of the movement in nonverbal communication was emphasized. For example, the essential head gesture such as tilting, nodding or shaking is important in active listening, where the head nodding can substitute verbal information like "yes" or "no" [2,13,23] or can be used instead "this one"/"that one" when you point at something [28]. Additionally, the head movements determine the meaning of the words [9,27] simplify the verbal expressions and specify the intensity of the emotions [6].

The importance of the head in emotion perception is described in [12], where Hess et al. proved that head position strongly influences reactions to negative emotions like fear and anger. Head movement with anger and fear expression strengthens the recognition of these emotions. In [21] Mignault et al. demonstrated that gesture called extension is correlated with joy expression, and motion called flexion is associated with "inferiority emotions" like guilt. Additionally, researchers have observed that during the conversation, the head movements are not random; these movements are used to influence interaction [10]. Head pose affects communication, complete verbal message and can improve the virtual person's realism, which is essential for many applications. The 3D model of the human head is increasingly used in many applications related to HMI [30], such as serious games [4], user-friendly interfaces [5], personalized agents in telepresence systems [29], driver assistance systems [26], or social robotics [3,18].

2   Related Works

There are many different methods for generating head movements. Motion can be generated based on captured characteristic points from video sequences, based on spoken words, or generated randomly. Most human head studies can be categorized as rule-based or data-driven frameworks. The first methods define rules for head gestures that build semantic labels, such as shaking, nodding, and tilting. In this case, the set of head movements are limited, which results in repeatability sequences. In comparison, data-driven methods use motion capture sequences and based on given head motion trajectories, new realizations of head movements are generate [24].

One popular idea used for motion synthesis is the head motion prediction based on the audio signals. Marsella et al. [20] proposed an approach for synthesizing a three-dimensional character movement based on the acoustic signal's prosodic analysis. They used the acoustic analysis to select the essential category of behaviour correlated with emotional state and words. Their proposed system can synthesize the head's different facial expressions and movements by transitioning from one gesture to another using co-articulation with other animations. Their method consists of semantic and prosody, and they can generate more appropriate virtual human motion than only the prosody method.

Based on the analysis of the relation between head gestures and conversation actions, Liu et al. [19] proposed an approach to generate head nodding and tilting where motion rules were obtained from human interaction features. In the first step, they labelled the sentence in the conversation database with selected head gestures, and then they created a correlation between sentence and proper head movement. They found an association between head nodding and the last syllable of phrase limits and head tilting when the subject was thinking or embarrassed. They used fixed shapes trajectories for head nods and head tilts, where for example, the for nodding and tilting intensity and the duration was kept the same; therefore, the effects of timing can be estimated. To assess the naturalness of the head motion on human-like robots, they used perceptual evaluation. They find that the naturalness is improved when head nodding and tilting are incorporated into the structure.

Lee et al. [17] describes a framework to generate head motion and other motion like eyelid and eye gaze motion. Proposed approach based on Gaussian Mixture Models (GMM) and gradient descent optimization algorithm to create head motion from speech attributes. Nonlinear Dynamic Canonical Correlation Analysis model determines the eye gaze from head motion synthesis. To obtain the current head gesture, they need two previous frames and prosodic features for the actual frame. Then head postures are calculated by maximization the last joint distribution exploiting gradient descent.

In [25] for head poses parallel generation frame with emotional, synthetic speech aligned with the real speech from motion capture sequences is creating. This frame is used to train the initial models or adapt the previous models obtained with natural speech. The main advantage of this solution is the reduction of mismatch. Besides, head movement with speech-driven methods can ignore the message context, even when speech is synchronized with head movements.

Another idea used for head motion synthesis based on the audio signals is described in [9]. Greenwood et al. have explained the concept of generating head movements from voice. For several conversational scenarios, six hours of natural and expressive speech were captured. Then, head motion and speech have been connected using deep two-way Long Short Term Memory networks, which allows analyzing the language's long structure. Finally, they have obtained and extended the model by conditioning with previous movements using a Conditional Variational Autoencoder.

3   Proposed Approach

Human motion synthesis is an essential topic in man-machine interaction systems. In this case, a machine definition can refer to a human-like robot or avatar – a virtual representation of the human displayed on the screen. HMI researchers' main goal is to design the interaction to be more like interpersonal communication, closer to human-like perception and understanding. For this purpose, non-verbal signals such as facial expressions and head gestures are widely used.

The method presented in this paper is preliminary and can be extended with additional rigid elements for more head gestures correlated with different emotional intensity.

3.1 Head Gestures

Head gestures were analyzed for three universal emotional states [7]: joy, sadness and fear. The selected emotions were characterized by the most significant facial expression muscles' activity and the most active head movements. Emotions can also be expressed with different intensity. It depends on the situational context and the person's character or mood. Therefore, we took the intensity of selected emotions into account, and we have used three intensities of emotions: weak, medium, strong. Head movement is based on rotation, and possible rotations for the human head are shown in Fig. 1. They are based on our previous work [15], where for each subtype of expression, we have estimated ranges of movements.

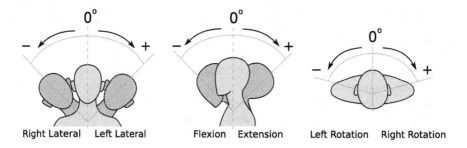

Fig. 1. Selected gestures of the human head.

3.2 Rigid Head Model

In contrast to the face, where we have elastic movements of mimic muscles, the human head can be described as a solid body. Therefore, we have used the set of non-deformable, rigid elements connected by joints for the head's action. Based on the anatomical structure of the head, we have selected three rigid elements that correspond to the head and neck. In this way proposed head consists of three segments that correspond to the neck and skull, which indicates the natural movements of the head [16]: C_1, C_2, C_3, where element C_1 indicate pitch motion, element C_2 roll rotation and element C_3 yaw motion (as depicted in Fig. 2). We used a kinematic chain with three degrees of freedom because head gestures generation with three degrees of freedom is more useful than only one or two angles [22]. The full description of the motion for three rigid elements for one of the emotion called joy with high intensity is shown in the Fig. 3, where C_1, C_2, C_3 refers to the rigid elements, and x, y, z rotations around the axis, as shown in Fig. 2. Analyzing the individual rotational motion of components, we decided to examine the relations between rotations of specific joints.

To determine the properties of human head movements, we have captured trajectories of joints for every element of the rigid model. The data describes three emotional states: joy, sadness and fear with various intensities grades: four for joy, seven for sadness and six for fear. Since three coordinates represent every joint position, the resulting number of trajectories is equal to $3 \cdot 3 \cdot 17 = 153$.

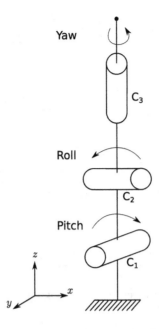

Fig. 2. The chain of rigid elements defined for the head.

4   Rotation Trajectories Analysis

Analysis of the movements of rigid elements was made for three primary emotions with different intensities. In this case, we have used three rigid items in three-dimensional space for head movements modelling, as shown in Fig. 2. Motion trajectories were obtained for all rigid elements in all axes. Motion data were obtained from captured characteristic points of the head, and from the collected motion data, a trajectory was obtained using piecewise cubic Hermite interpolating polynomial [14].

4.1   Ranges of Rotation Angles

In the first stage, an analysis of rotation ranges was performed, and we were able to determine the maximum deflection of a rigid element that can be achieved during rotation. The range of rotation angles in trajectory was calculated using the following formula:

$$R = |max[p(n)] - min[p(n)]|, \tag{1}$$

where $p(n)$ - trajectory of angle values $n = 0, \ldots, N - 1$, and N is the number of points in the trajectory.

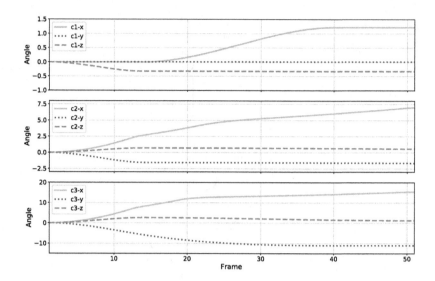

Fig. 3. Motion trajectories for strong intensity joy emotion.

Based on the analysis, we observed that no rotation actions occur in 35% cases with different emotions, rigid elements and axes. On the other hand, the participation of individual rigid elements in the rotation is shown in Table 1.

Table 1. Cases with lack of rotation actions for each rigid element.

Element	X axis	Y axis	Z axis
C_1	10%	95%	95%
C_2	0%	29%	41%
C_3	23%	41%	23%

According to the measurements, the most extensive range of motion was happen for C_2 and C_3 elements for pitch and roll head movements, while the smallest occurs for the yaw head movement. Element C_1 has the lowest range of rotations for all actions. In the next step, we have computed ranges for all trajectories in all emotional states. Table 2 presents a small subset of trajectories with the highest and the lowest values of R (constant trajectories with $R = 0$ were ignored). For all rigid elements and coordinates, the most extensive range was observed for strong joy and strong sadness in pitch movement. The C_2 and C_3 elements are responsible for the upper part of the head rotation and are very active for sadness state. However, the component C_1 is responsible for the lower part of the neck shows the smallest range for the fear state. The roll movements are the most active part when sadness and joy expressions occur. The C_2 and C_3 elements are responsible for the movement of the upper part of the head and demonstrate the significant activity. In the case of yaw movement, the highest activity was observed for C_2 and C_3 elements.

Table 2. Example cases with the highest and lowest R values.

R	Emotion	Element	Rotation
28.6059	Strong sadness	C_2	Pitch
28.6059	Strong sadness	C_2	Pitch
18.6531	Medium joy	C_2	Pitch
16.7949	Strong sadness	C_2	Roll
15.4474	Strong joy	C_3	Pitch
15.0925	Strong sadness	C_2	Pitch
15.0925	Medium sadness	C_2	Pitch
...	...	...	...
0.2767	Medium sadness	C_2	Roll
0.2401	Weak sadness	C_3	Yaw
0.2207	Weak fear	C_1	Pitch
0.0275	Strong sadness	C_3	Roll
0.0112	Strong joy	C_1	Roll
0.0078	Weak joy	C_3	Roll

4.2 Trajectories Similarity

A preliminary analysis of obtained trajectories for considered emotional states indicates a noticeable similarity between them. Therefore, we have decided to compare them to create groups where all trajectories but one can be obtained by scaling and translating the single selected trajectory. For this purpose, we have compared all combinations of trajectories in pairs and compute the mean squared error (MSE) since the length of every trajectory is the same:

$$M = \frac{1}{N} \sum_{n=0}^{N-1} [p_1(n) - p_2(n)]^2, \qquad (2)$$

where: $p_1(n)$ represents the first trajectory, $p_2(n)$ denotes the second trajectory and N is the number of points in single trajectory. The case when $M = 0$ shows that $p_1(n) = p_2(n)$ for every n. Thus, the value M can be treated in this study as a value of similarity.

To calculate the similarity between trajectories, we create a set of pairs for all cases in our dataset. The number of combinations without repetitions was equal to $\binom{2}{17.9}$ resulting in $\binom{2}{153} = 11628$ pairs to compare. An interesting situation occurs when $M = 0$ and $R > 0$, which means that both trajectories are identical and represents the change over time (for $R = 0$ both trajectories are constant). We have calculated all combinations between trajectories and measured their ranges for the whole set. In the result, we found seven pairs satisfying such condition and Table 3 contains the found pairs.

Table 3. Trajectories that satisfy the similarity criterion.

Set	Emotion	Element	Rotation
1	Medium sadness	C_1	Pitch
	Strong sadness	C_1	Pitch
2	Medium sadness	C_2	Pitch
	Strong sadness	C_2	Pitch
3	Medium sadness	C_2	Roll
	Strong sadness	C_2	Roll
4	Medium sadness	C_2	Yaw
	Strong sadness	C_2	Yaw
5	Medium sadness	C_3	Pitch
	Strong sadness	C_3	Pitch
6	Medium sadness	C_3	Roll
	Strong sadness	C_3	Roll
7	Medium sadness	C_3	Yaw
	Strong sadness	C_3	Yaw

The connections between trajectories concern the sadness emotion with various intensities. In the Fig. 4, the first trajectory is connected with the C_3 element in yaw motion for the medium intensity, which is similar to the high intensity of sadness (for the same element). The same situation can be found for C_1 element in sadness state with medium and strong intensities for pitch motion. In the evaluated comparison with the MSE criterion, we have concluded that compared trajectories have to the same phase when it reaches its target angle of rotation which is marked in that figure with the dashed line.

The duration of a single-phase animation for all emotional states was determined based on the most extended emotional reaction for an event. In our case, the longest period of movement phase is equal to 2.125 s (assuming the animation frame rate equal to 24 frames per second). For all trajectories in our dataset, we have calculated the time spans for rotation phases (where the angle is changing in time), and the results are presented in Fig. 5. It is visible that the occurrence of short periods of rotations is frequent in the considered dataset.

4.3 Trajectories Grouping

An essential part of the rotation trajectory is the phase when the angle remains constant up to the end of the animation. Having this phase in mind and considering

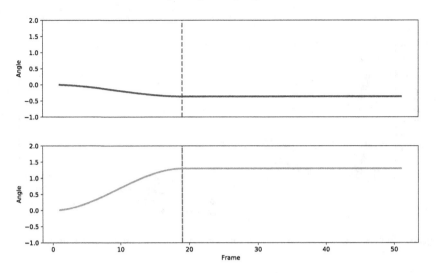

Fig. 4. Comparison of trajectories showing the similarity between elements C_3 in yaw and C_1 in pitch rotations.

all trajectories, a set of the group may be created with the same phases. In any of such groups, the trajectories are interrelated, which means that selecting one parent trajectory are interrelated from the group, the remaining trajectories can be computed by scaling and translating the parent trajectory. In the result of grouping for our dataset, twelve groups were obtained. Having selected parent trajectory $p_1(n)$, the rest of the trajectories can be described by the following formula:

$$p_j(n) = c \cdot [p_1(n) - d_1] \cdot \frac{r_2}{r_1} + d_2, \tag{3}$$

where $j = 2, \ldots, J - 1$ and J denotes the number of trajectories and $p_j(n)$ is the j-th child trajectory. The remaining parameters determine the transformation properties. The selection of $p_1(n)$ from the group is arbitrary.

The first parameter $c \in \{-1, 1\}$ defines if the child trajectory has to be mirrored vertically and can be determined as follows:

$$c = \begin{cases} 1 \text{ if } s_x = s_y \\ -1 \text{ if } s_x \neq s_y \end{cases}, \tag{4}$$

where: $s_x = sgn(p_1(0) - p_1(N - 1))$, $s_y = sgn(p_j(0) - p_j(N - 1))$ and $sgn(x)$ is signum function. The d_1, d_2 parameters are used to align both trajectories:

$$d_1 = max[p_1(n)],$$

$$d_2 = \begin{cases} min[p_j(n)] \text{ if } c = -1 \\ max[p_j(n)] \text{ if } c = 1 \end{cases}. \tag{5}$$

Finally, the adaptation of ranges for both trajectories is realized by scaling the $p_1(n)$ trajectory by the ratio of ranges r_2/r_1, where r_1 is the range of parent and r_2 child trajectory:

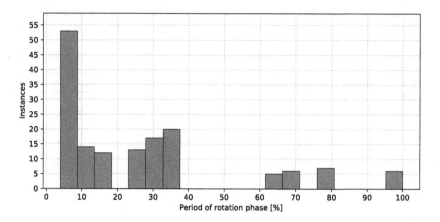

Fig. 5. The length of rotation actions for the length of the single phase of animation.

$$r_1 = |max[p_1(n)] - min[p_1(n)]|,$$
$$r_2 = |max[p_j(n)] - min[p_j(n)]|. \qquad (6)$$

Since the ratio is always positive, its value determine if the dynamic range of parent trajectory is extended ($r_2/r_1 > 1$) or compressed ($r_2/r_1 < 1$).

4.4   Final Rotation of the Head

As an example of using the proposed mechanism, we have selected a group of trajectories with rotation phase in 19 frames. We have chosen a first trajectory (intense sadness emotional state) in the group depicted in Fig. 6, where $d_1 = 0$ and $r_1 = 1.3564$. Obtained rotations based on the trajectory are presented in Fig. 7.

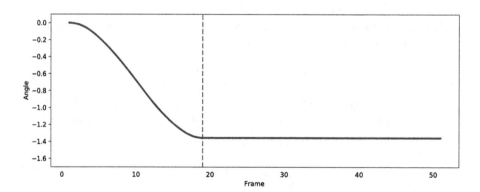

Fig. 6. The parent trajectory of child trajectories in the group for strong sadness state rotations.

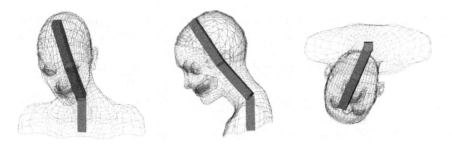

Fig. 7. Example configuration of rigid elements in strong sadness state for the front, side and top view.

Next, using it as the parent trajectory, we have calculated the parameters for the rest trajectories in the group. Table 4 summarizes the coefficients obtained for one group where $p_1(n)$ represents weak joy state, element C_1 with pitch movement. Selecting all trajectories for intense sadness emotional state, we have rendered a few frames of animation for this case presented in Fig. 8.

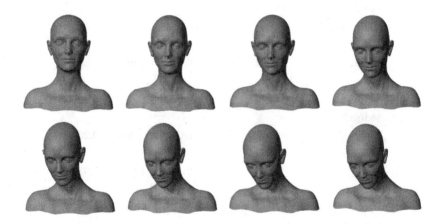

Fig. 8. Example phases of head animation for strong sadness emotional state.

The final data for animation can be expressed as a single reference trajectory with set parameters to generate the rest of the trajectories for a specific emotional state.

Table 4. The coefficients of child trajectories obtained for one group with rotation phase performed in 19 frames.

Emotion	Element	Rotation	c	d_2	r_2	r_2/r_1
Weak Joy	C_2	Pitch	−1	0	4.8811	3.5986
Weak Joy	C_3	Pitch	−1	0	5.1591	3.8035
Weak Joy	C_3	Roll	1	0	0.0078	0.0058
Weak Joy	C_3	Yaw	−1	0	4.5821	3.3781
Medium Sadness	C_1	Pitch	−1	0.0104	1.2895	0.9507
Medium Sadness	C_2	Pitch	1	−0.1220	15.0925	11.1269
Medium Sadness	C_2	Roll	−1	0.0777	9.6130	7.0871
Medium Sadness	C_2	Yaw	1	−0.0144	1.7840	1.3152
Medium Sadness	C_3	Pitch	1	−0.0168	2.0816	1.5347
Medium Sadness	C_3	Roll	1	−0.0002	0.0275	0.0203
Medium Sadness	C_3	Yaw	1	−0.0029	0.3605	0.2658
Strong Sadness	C_1	Pitch	−1	0.0104	1.2895	0.9507
Strong Sadness	C_2	Pitch	1	−0.1220	15.0925	11.1269
Strong Sadness	C_2	Roll	−1	0.0777	9.6130	7.0871
Strong Sadness	C_2	Yaw	1	−0.0144	1.7840	1.3152
Strong Sadness	C_3	Pitch	1	−0.0168	2.0816	1.5347
Strong Sadness	C_3	Roll	1	−0.0002	0.0275	0.0203
Strong Sadness	C_3	Yaw	1	−0.0029	0.3605	0.2658

5   Conclusions

The principal aim of our work was to create relations between the facial expressions and head movements for the better virtual human-machine interaction. For this purpose, we analyzed motion trajectories for three basic emotions for three different intensities of these emotions. The obtained results were used to create movement on the three-dimensional head of a human. The results indicate that based on the similarity of the trajectories, the rotations can be performed using a limited set of the original trajectories in the database. We have presented a technique on how to group a set of trajectories using similarity measure between them and how to restore them inside the group using a single trajectory and additional coefficients. We have used the mean square error measure to determine the similarity level between pairs of trajectories. The mean absolute error, root mean squared error or other measurements can also be used for this task. However, selecting a proper similarity measure taking the problem's specificity into mind is not a trivial task and needs further investigation. The presented work can be extended by adding new rigid elements and increasing the number of head gestures correlated with different emotional intensities. In future work, a scheme for selecting a base trajectory in groups using optimization techniques is planned.

References

1. Boker, S.M., Cohn, J.F., Theobald, B.J., Matthews, I., Brick, T.R., Spies, J.R.: Effects of damping head movement and facial expression in dyadic conversation using real-time facial expression tracking and synthesized avatars. Philos. Trans. R. Soc. Lond. B Biol. Sci. **364**(1535), 3485–3495 (2009)
2. Boker, S.M., et al.: Something in the way we move: motion dynamics, not perceived sex, influence head movements in conversation. J. Exp. Psychol. Hum. Percept. Perform. **3**(37), 874–891 (2011)
3. Breazeal, C., Thomaz, A.L.: Learning from human teachers with socially guided exploration. In: ICRA, pp. 3539–3544. IEEE (2008)
4. Cai, Y., van Joolingen, W., Walker, Z. (eds.): VR, Simulations and Serious Games for Education. GMSE. Springer, Singapore (2019). https://doi.org/10.1007/978-981-13-2844-2
5. Clavel, C., Plessier, J., Martin, J.C., Ach, L., Morel, B.: Combining facial and postural expressions of emotions in a virtual character. In: Ruttkay, Z., Kipp, M., Nijholt, A., Vilhjalmsson, H.H. (eds.) Intelligent Virtual Agents. Lecture Notes in Computer Science, vol. 5773, pp. 287–300. Springer, Heidelberg (2009). https://doi.org/10.1007/978-3-642-04380-2_31
6. Cohn, J.F., Reed, L.I., Moriyama, T., Xiao, J., Schmidt, K.L., Ambadar, Z.: Multimodal coordination of facial action, head rotation, and eye motion during spontaneous smiles. In: Proceedings of the 6th IEEE International Conference on Automatic Face and Gesture Recognition, pp. 129–135 (2004)
7. Ekman, P.: Emotions Revealed: Recognizing Faces and Feelings to Improve Communication and Emotional Life. Owl Books (2007)

8. Graf, H.P., Cosatto, E., Strom, V., Huang, F.J.: Visual prosody: facial movements accompanying speech. In: 5th IEEE International Conference on Automatic Face and Gesture Recognition, pp. 396–401 (2002)

9. Greenwood, D., Laycock, S., Matthews, I.: Predicting head pose from speech with a conditional variational autoencoder. In: Proceedings of the Interspeech, pp. 3991–3995 (2017)

10. Gunes, H., Pantic, M.: Dimensional emotion prediction from spontaneous head gestures for interaction with sensitive artificial listeners. In: Allbeck, J., Badler, N., Bickmore, T., Pelachaud, C., Safonova, A. (eds.) IVA 2010. LNCS (LNAI), vol. 6356, pp. 371–377. Springer, Heidelberg (2010). https://doi.org/10.1007/978-3-642-15892-6_39

11. Harrigan, J., Rosenthal, R., Scherer, K.R.: The New Handbook of Methods in Nonverbal Behavior Research. Series in Affective Science. Oxford University Press (2005)

12. Hess, U., Adams, R.B., Kleck, R.E.: Looking at you or looking elsewhere: the influence of head orientation on the signal value of emotional facial expressions, motivation and emotion. Motiv. Emot. **31**(2), 137–144 (2007). https://doi.org/10.1007/s11031-007-9057-x

13. Heylen, D.: Challenges ahead: head movements and other social acts during conversations. In: Halle, L., Wallis, P., Woods, S., Marsella, S., Pelachaud, C., Heylen, D.K. (eds.) Proceedings of the Joint Symposium on Virtual Social Agents, pp. 45–52. The Society for the Study of AI and the Simulation of Behaviour (2005)

14. Kahaner, D., Moler, C., Nash, S.: Numerical Methods and Software. Prentice-Hall Inc., USA (1989)

15. Kocoń, M.: Influence of facial expressions on the human head movements. In: 41st International Conference on Telecommunications and Signal Processing, TSP 2018, Athens, Greece, 4–6 July 2018, pp. 1–5. IEEE (2018)

16. Kocoń, M.: Head movements in the idle loop animation. IADIS Int. J. Comput. Sci. Inf. Syst. **15**(2), 137–147 (2020)

17. Le, B.H., Ma, X., Deng, Z.: Live speech driven head-and-eye motion generators. IEEE Trans. Vis. Comput. Graph. **18**, 1902–1914 (2012)

18. Ligthart, M., Hindriks, K., Neerincx, M.A.: Reducing stress by bonding with a social robot: towards autonomous long-term child-robot interaction. In: Companion of the 2018 ACM/IEEE International Conference on Human-Robot Interaction, HRI 2018, pp. 305–306. ACM (2018)

19. Liu, C., Ishi, C.T., Ishiguro, H., Hagita, N.: Generation of nodding, head tilting and eye gazing for human-robot dialogue interaction. In: 2012 7th ACM/IEEE International Conference on Human-Robot Interaction (HRI), pp. 285–292 (2012)

20. Marsella, S., Xu, Y., Lhommet, M., Feng, A., Scherer, S., Shapiro, A.: Virtual character performance from speech. In: Proceedings of the 12th ACM SIGGRAPH/Eurographics Symposium on Computer Animation, SCA 2013, pp. 25–35. Association for Computing Machinery (2013)

21. Mignault, A., Chaudhuri, A.: The many faces of a neutral face: head tilt and perception of dominance and emotion. J. Nonverbal Behav. **27**, 111–132 (2003). https://doi.org/10.1023/A:1023914509763

22. Mukherjee, S., Robertson, N.: Deep head pose: gaze-direction estimation in multimodal video. IEEE Trans. Multimedia **17**, 1 (2015)

23. Munhall, K.G., Jones, J.A., Callan, D.E., Kuratate, T., Vatikiotis-Bateson, E.: Visual prosody and speech intelligibility: head movement improves auditory speech perception. Psychol. Sci. **15**(2), 133–137 (2004)

24. Sadoughi, N., Busso, C.: Head motion generation. Handbook of Human Motion, pp. 2177–2200. Springer, Cham (2018). https://doi.org/10.1007/978-3-319-14418-4_4

25. Sadoughi, N., Liu, Y.P., Busso, C.: Meaningful head movements driven by emotional synthetic speech. Speech Commun. **95**, 87–99 (2017)

26. Schwarz, A., Haurilet, M., Martinez, M., Stiefelhagen, R.: Driveahead - a large-scale driver head pose dataset. In: 2017 IEEE Conference on Computer Vision and Pattern Recognition Workshops (CVPRW), pp. 1165–1174 (2017)

27. Sun, X., Truong, K.P., Pantic, M., Nijholt, A.: Towards visual and vocal mimicry recognition in human-human interactions. In: IEEE International Conference on Systems, Man, and Cybernetics, pp. 367–373 (2011)

28. Tojo, T., Matsusaka, Y., Ishii, T., Kobayashi, T.: A conversational robot utilizing facial and body expressions. In: 2000 IEEE International Conference on Systems, Man, and Cybernetics, vol. 2, pp. 858–863 (2000). https://doi.org/10.1109/ICSMC.2000.885957

29. Vidrascu, L., Devillers, L.: Real-life emotion representation and detection in call centers data. In: Tao, J., Tan, T., Picard, R.W. (eds.) Affective Computing and Intelligent Interaction. Lecture Notes in Computer Science, vol. 3784, pp. 739–746. Springer, Heidelberg (2005). https://doi.org/10.1007/11573548_95

30. Wang, K., Zhao, R., Ji, Q.: Human computer interaction with head pose, eye gaze and body gestures. In: 2018 13th IEEE International Conference on Automatic Face Gesture Recognition, p. 789 (May 2018). https://doi.org/10.1109/FG.2018.00126

DenLAC: Density Levels Aggregation Clustering – A Flexible Clustering Method –

Iulia-Maria Rădulescu$^{(\boxtimes)}$ ⓘ, Alexandru Boicea ⓘ, Ciprian-Octavian Truică ⓘ, Elena-Simona Apostol ⓘ, Mariana Mocanu ⓘ, and Florin Rădulescu ⓘ

University Politehnica of Bucharest, Bucharest, Romania
{iulia.m.radulescu,alexandru.boicea,ciprian.truica,elena.apostol,
mariana.mocanu,florin.radulescu}@upb.ro

Abstract. This paper introduces DenLAC (Density Levels Aggregation Clustering), an adaptable clustering algorithm which achieves high accuracy independent of the input's shape and distribution. While most clustering algorithms are specialized on particular input types, DenLAC obtains correct results for spherical, elongated and different density clusters. We also incorporate a simple procedure for outlier identification and displacement. Our method relies on defining clusters as density intervals comprised of connected components which we call density bins, through assembling several popular notions in data mining and statistics such as Kernel Density Estimation, the density attraction and density levels theoretical concepts. To build the final clusters, we extract the connected components from each density bin and we merge adjacent connected components using a slightly modified agglomerative clustering algorithm.

Keywords: Clustering · Kernel density estimation · Probability density function · Hierarchical clustering

1 Introduction

Clustering is the process of grouping a set of points into clusters such that the points in the same cluster have high similarity but are significantly dissimilar from the points in other clusters. Even though a large number of clustering algorithms have been developed over time, most of them perform well only on certain datasets, as we show in Sect. 2 and Sect. 6. Therefore, we propose Den-LAC, which achieves high accuracy for various types of datasets regardless of the input's shape or distribution.

Relying on the cluster definitions proposed in [14] and [4,13], we demonstrate that clusters consist of density intervals containing connected components, which we call density bins, the key concept of our method. We then extract the connected components employing a nearest neighbour approach: we iterate through each density bin's objects, trying to expand clusters by recurrently merging the nearest neighbors for each object. The DenLAC algorithm sequentially applies

© Springer Nature Switzerland AG 2021
M. Paszynski et al. (Eds.): ICCS 2021, LNCS 12742, pp. 316–329, 2021.
https://doi.org/10.1007/978-3-030-77961-0_27

the following steps: i) evaluate the input's probability density function; ii) remove the outliers with the help of the probability density function; iii) assign the input objects to the corresponding density bins; iv) split the density bins into adjacent connected components; v) aggregate the connected components from the previous step in order to build the final clusters.

We evaluate DenLAC alongside several well-known clustering algorithms on eight synthetic bi-dimensional datasets and on the Iris and Tetragonula real datasets in order to demonstrate its flexibility. The employed quality functions reveal that DenLAC obtains higher accuracy as compared to the other considered algorithms.

2   Related Work

Clustering algorithms fall into five broad categories: partitional (representative based), hierarchical, density-based, probabilistic and spectral.

Representative based algorithms rely on the intuitive notion of distance to cluster the data points. The objects within the dataset are partitioned around a subset of chosen representatives. K-means [18] and CLARANS (Clustering Large Applications based on Randomized Search) [22] are two examples of these clustering methods. K-means is sensitive to outliers because it relies on the mean computation. It is also a randomized algorithm, therefore it considerably depends on the parameters initialization. Also, K-means uses the square-error to measure the quality of the clustering results and therefore works well when clusters are compact clouds that are rather well separated from one another [12]. CLARANS is also randomized, thus, it is sensitive to the parameters initialization; moreover it only relies on the distances between the objects to be clustered.

Hierarchical algorithms rely on computing the pairwise distances between the input dataset's instances. There are two main types of hierarchical clustering methods: agglomerative (bottom-up) clustering and divisive (top-down) clustering. Hierarchical algorithms do not scale well with large data sets due to the quadratic computational complexity of calculating all the pairwise distances and suffer from problems such as chaining. BIRCH (Balanced Iterative Reducing and Clustering using Hierarchies) [31] is a more efficient hierarchical clustering algorithm used for large datasets. Because BIRCH structures the data as a tree, the clustering result depends on the input objects order. In addition to this, BIRCH does not obtain accurate results for arbitrarily shaped clusters as it uses the diameter of the cluster to determine its boundaries. CURE (Clustering Using Representatives) [12] is an agglomerative hierarchical algorithm which can identify non-spherical clusters. CURE fails to take into account special characteristics of individual clusters, thus making incorrect merging decisions when the underlying data does not follow the assumed model [16].

Density based clustering methods rely on the consideration that the typical density of points within a cluster is considerably higher than outside of the cluster [7]. DBSCAN (Density Based Spatial Clustering of Applications with Noise) [7] measures density as the number of points within the radius ε of a point within the analyzed dataset and is accurate for arbitrarily shaped groups of objects. However, it fails to cluster datasets with varying density accurately.

The Gaussian Mixture [2] is a probabilistic clustering algorithm. It assumes that the input objects were drawn from k multivariate normal distributions, where k is the number of clusters and relies on Expectation Maximization to determine each distribution's specific parameters. The algorithm works well only on datasets which follow the normal distribution (elliptical clusters) and since it uses Expectation Maximization for the mixture model's parameters computation, it is sensitive to the parameters initialization.

Spectral Clustering [21] relies on computing the eigenvectors associated with the Laplacian matrix derived from the dataset's similarity graph and then clusters them using any algorithm, for example K-means. Spectral clustering cannot successfully cluster datasets that contain structures at different scales of size and density [20].

3   Methodology

In this section we present the theoretical concepts on which DenLAC is based on. We first introduce the notion of density attraction, defined in [14]. Then, we reproduce Chaudhuri et al.'s density levels based cluster definition [4]. Finally we explain how our method utilizes the aforementioned concepts.

3.1   DENCLUE (DENsity-Based CLUstEring) Algorithm

The DENCLUE (DENsity-based CLUstEring) algorithm [14] identifies clusters using a hill climbing procedure which is guided by the probability density's function gradient, assigning the objects approaching the local maximas of the probability density function to the same cluster. [14] characterize center-defined clusters and arbitrary-shaped clusters in terms of the local maximas of the probability density function, called "density attractors": i) the center-defined clusters consist of the objects which are "density attracted" by a density attractor whose density is greater than a given threshold ξ and ii) the arbitrary shaped clusters consist of objects which are "density attracted" by a set of density attractors, provided that there exists a path between the density attractors with density above a given threshold ξ.

We display the formal density attractor-based cluster definitions proposed in [14] below, using the following notations: i) D represents the d-dimensional input dataset, ii) $x \in D$ denotes an object in the input dataset iii) we note the probability density function with f iv) $x*$ represents a density attractor, more specifically a local maxima of the probability density function f

Definition 1 (Center-Defined Clusters).   *A center-defined cluster given a density attractor $x*$ is a subset C of the input dataset D with $x \in C$ being attracted by $x*$ and $f(x*) \geq \xi$. Points $x \in D$ are called outliers if they are density attracted by a local maximum $x*_0$ with $f(x_0*) < \xi$.*

Definition 2 (Arbitrary-Shape Clusters). *An arbitrary-shape cluster given the set X of density attractors is a subset C of the input dataset D where: i) $\forall x \in C \exists x* \in X : f(x*) \geq \xi$, x is density attracted to $x*$ and ii) $\forall x*_1, x*_2 \in X : \exists$ a path $P \subset F^d$ from $x*_1$ to $x*_2$ with $\forall p \in P : f(p) \geq \xi$, where F^d is a d-dimensional feature space.*

3.2 Clusters as Connected Components at Different Density Levels

[4,13] define clusters as connected components at each density level λ of the dataset D. The density levels are regions for which the density value is equal to or above a particular level and can be estimated using Kernel Density Estimation [5]. The formal definition is:

Definition 3 (Density Levels). *Given a level λ, the λ-density level is: $L_\lambda = \{x : P(x) \geq \lambda\}$.*

The connected components at a lower density level incorporate the connected components at higher density levels. Thus, a cluster of density f is any connected component of $x : f(x) \geq \lambda$, for any $\lambda > 0$. The collection of all such clusters forms an (infinite) hierarchy called the cluster tree [4]. Finding consistent and reliable approximations of the cluster tree has been studied extensively in [4,13].

3.3 Clusters as Adjacent Density Bins

From Definition 1 we infer that the members of a center-defined cluster tend to the local maxima of the probability density function within the cluster: the estimated probability density values of the cluster members decrease linearly as they move away from the density attractor. From Definition 2 we infer that the probability density values for the members of an arbitrary shaped cluster are greater than a given threshold ξ. Assuming that the probability density function is estimated using KDE employing a smooth kernel function, such as the Gaussian kernel, we infer that the estimate's values do not rise or drop suddenly.

Considering the deductions above, we view clusters as sets of adjacent density intervals consisting of connected components, which we call density bins. We formally define density bins with the help of Definition 3 as the set differences between adjacent density levels described in Definition 4.

Definition 4 (Density Bins). *For a given dataset D with the probability density function $f(\overline{D})$ and n density levels L_{λ_i}, where $i \in (0, n)$, we define density bins as: $B_i = L_{\lambda_i} \setminus L_{\lambda_{i-1}}$.*

For example, bn bins, each consisting of cm_n connected components, are structured as follows:

B_1: $L_{\lambda_1} \approx C_{1,1} \cup C_{1,2} \cup ... \cup C_{1,m_1}$
B_2: $L_{\lambda_2} \setminus L_{\lambda_1} \approx C_{2,1} \cup C_{2,2} \cup ... \cup C_{2,m_2}$
...

B_{bn}: $L_{\lambda_{bn}} \setminus L_{\lambda_{bn-1}} \approx C_{bn,1} \cup C_{bn,2} \cup ... \cup C_{bn,m_{bn}}$

Where C_{i,m_i} is a connected component for the i-th bin, with $i = \overline{1, bn}$.

4   Algorithm Pipeline

The proposed algorithm pipeline, which is graphically displayed in Fig. 1, consists of the following modules: i)*Density estimation*: we estimate the density of the input dataset using Kernel Density Estimation ii)*Outlier identification and displacement*: we eliminate the outliers using the probability density function; iii)*Density based partitioning*: we identify the density levels in the input dataset and extract compact sets of objects that are positioned at large distances one to another which represent the differences between adjacent density levels; iv)*Distance based splitting*: we split the compact sets of object using the distance between them; v)*Merging the final partitions*: we merge the closest partitions and return the final clusters in order to minimize the distance function [31].

Fig. 1. DenLAC algorithm pipeline

In the following sections we describe each step in detail, using the following notations: i) $D = d_1, ..., d_n$ represents the d-dimensional, input dataset of size n ii) $B = \bigcup_{i=1}^{bn} B_i$ represents the density bins set of size bn, defined in the previous section.

4.1   Density Estimation. Outlier Identification and Displacement

We identify the dense regions of the input set D by estimating the probability density function using Kernel Density Estimation (also known as KDE). We choose KDE due to its independence of the input dataset distribution's underlying form. Out of the variety of available kernel functions (cosine, exponential, linear, epanechnikov, gaussian, etc.) we prefer the gaussian kernel function since it produces the smoothest estimate, thus being the most suitable for the density levels partitioning. We employ Scott's rule of thumb [25] for determining the optimal bandwidth, as it is a classic, well-established method.

DenLAC utilizes the probability density function to partition the input dataset according to its density. However, the probability density function is also helpful for removing potential outliers by detecting probability density function's unusually low values.

To determine these values we used the interquartile range, noted IQR [26], defined as the difference between the 25th and the 75th percentiles. This method is more robust than Z-score for the reason that Z-score relies on the mean and the standard deviation, which are affected by outliers. The kth percentile can be defined as a value x_j from a given data set $X = x_1, x_2, ..., x_n$ having the property that k percent of the values are less or equal than x_j. The probability density function values of the outliers, associated with regions with extremely low density, are more than $1.5 \times IQR$ below the 25th percentile. Therefore, we remove the objects with probability density function values in this interval.

4.2 Density Based Partitioning

We partition the input dataset D into bn density bins by allocating each object d_i to a density bin B_i according to its estimated probability density function, as follows:

Given a dataset instance $d_i \in D$ and a density bin B_i delimited by the boundaries B_i^{min} and B_i^{max}, we assign d_i to B_i if $f(d_i) \geq B_i^{min}$ and $f(d_i) < B_i^{max}$, where $f : D \to \mathbb{Z}$ is the estimated probability density function. The number of density intervals, bn, must be supplied at runtime as parameter. Section 5, Subsect. 5.1 offers several guidelines on choosing the algorithm's parameters. The density bins set and the density based split result for the Aggregation [11] synthetic dataset can be visualized in Fig. 2. The density bins before and after allocating the input dataset's objects are graphically represented in Fig. 2b (the regions colored with the same shade blue) and Fig. 2c (the groups of points with the same colour) respectively.

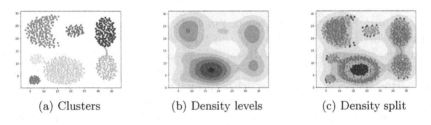

(a) Clusters   (b) Density levels   (c) Density split

Fig. 2. Density based splitting

4.3 Splitting the Density Bins into Distinct Connected Components

The resulting density bins consist of distinct connected components (Definition 4) which must be separated. For example, all the points marked with red in Fig. 2c are assigned to the same density interval according to their estimated probability density. Therefore, the set of points must be divided into distinct connected components. We split the connected components using a nearest-neighbour approach. We examine each object within a density bin and start extending a cluster from it. We extend a cluster by adding the object's nearest neighbours whose distances to the selected object are smaller than a specific threshold.

We determine the correct threshold for each bin using the same approach as [17] and [10]. The above-mentioned papers address the problem of DBSCAN's radius threshold ϵ automatic computation by laying out a k-distance plot representing the density levels as interconnected smooth curves. Similar to [17], we compute the distances between each point and its k-th neighbour, where k depends on the input dataset's D dimension d and is automatically determined as follows: $k = 2 \cdot d - 1$. In order to deduce k we used the same rule of thumb for deducing the nearest neighbour as the one mentioned in [24]. Then, we use these values to draw the k-distance plot; the threshold value, $threshold_{kDistance}$, corresponds to the sharp change point of the outlined curve. We formally define our k-distance plot as $kDistance\colon B_{i_{sortedK}} \rightarrow D_k$, where $B_{i_{sortedK}}$ represents the set of points belonging to bin B_i sorted by their distance to their k-th neighbour and D_k represents the set of the actual distances. We note that because a bin consists of points belonging to the same density interval, the k-distance plot will usually incorporate a single sharp change point.

If the clusters are not well separated either because of noise or chains, we need to over partition the bins in order to better separate and emphasise the high density regions. This is why we introduce the expansion factor ef, a custom parameter utilized to weight the threshold. We cover ef's value selection in detail in Sect. 5, Subsect. 5.1. Therefore, we formally define the threshold using the inflection point of the k-distance plot, $threshold_{kDistance}$ and the expansion factor ef: $threshold = ef \cdot threshold_{kDistance}$.

4.4 Merging the Adjacent Connected Components

The resulting adjacent connected components must ultimately be merged in order to build the final clusters. For this purpose we use an agglomerative approach: we compute the pairwise distances between all partitions, we merge the closest partitions into a cluster, we recompute the distances that have changed due to the merge and continue with the following closest partitions. In the majority of cases, where no over-partitioning is needed, the process is considerably faster than plain hierarchical agglomerative clustering, because it starts with already clustered large groups of points instead of just points. The linkage method is configurable trough the input parameters. We offer two options: smallest pairwise distance (the default option) and centroid linkage. For elongated clusters or well separated gaussian clusters, we use the smallest pairwise distance linkage. This is the most suitable linkage criterion because regardless the input dataset's shape, the continuous regions previously obtained, which we want to join, are elongated. For noisy datasets or gaussian clusters connected by chains, we use centroid linkage - so the noise or the chains do not affect the quality of the result.

5   Algorithm

5.1   Choosing the Hyperparameters

The number of clusters, the bn number of bins, the expansion factor ef and the method to compute the inter cluster linkage must be provided as parameters. Determining the optimal number of clusters is a comprehensively documented problem; the elbow and silhouette methods represent popular solutions. The number of bins bn is generally equal to the number of density levels of the optimal cluster tree (as defined in [4]). However, if the input dataset contains chains (see Aggregation [11]), we must choose a greater value for bn in order to force the separation of the chained bins. Therefore, $bn \in [3, 9]$. The expansion factor ef is generally equal to one. For elongated and well separated spherical clusters there is no need to apply any weight on the closest mean (defined in Sect. 4, Subsect. 4.3). However, if the input dataset is noisy, ef should be lower than 1 in order to "prune" the noisy points. Therefore, $ef \in [0.1, 2]$. We support two linkage methods: single linkage (smallest pairwise distances), which is the default option, and centroid linkage. For elongated and well separated spherical clusters single linkage performs best. This is because the intermediary connected components which we join in the aggregation step are usually elongated, regardless the input dataset's structure. For noisy or chained datasets, centroid linkage obtains best results.

5.2   Implementation

We start by estimating the probability density function. Then, we group the input objects into partitions representing bn density bins. Afterwards, we iterate trough each partition and split its connected components. Lastly, we merge the adjacent connect components in order to build the final clusters. The density bins splitting procedure consists of the following steps: i) we start with an arbitrary object and retrieve its $k = d + 1$ nearest neighbours computed as all the objects whose distance to the specified object is smaller than the threshold defined in Subsect. 4.3; ii) we start a new cluster from the current object, add the object's neighbors to the cluster and try to expand it; iii) when we cannot further expand the current cluster we return to Step 1 and repeat the process.

The full algorithm is presented in pseudocode 1, whereas the distance based split is displayed in pseudocode 2. The implementation is freely available online[1], alongside run instructions and graphical results for all synthetic bidimensional datasets.

6   Experimental Results

6.1   Clustering Evaluation Methods

In order to evaluate DenLAC's accuracy we use two different classes of evaluation methods: i) evaluation methods based on the mutual information (specifically the Entropy [29] and ii) evaluation methods based on counting pairs of objects distributed either to the same cluster or to different clusters by a given clustering algorithm versus the ground truth [29] (specifically the Rand Index [23] and the Adjusted Rand Index [27]).

---

[1] https://github.com/IuliaRadulescu/DENLAC.

Algorithm 1. DenLAC algorithm

1: **procedure** DENLAC(*inputDataset, numberOfClusters, bn, ef, linkageMethod*)
2: *pdf* ← the probability density estimate of *inputDataset*;
3: *outliers* ← the outliers, identified employing the IQR method on the *pdf*;
4: *inputDataset* = *inputDataset* \ *outliers*;
5: Recompute *pdf* for the updated *inputDataset*;
6: Split *pdf* into *bn* intervals: $B_1, ..., B_{bn}$;
7: **for** B_k in $B_1, ..., B_{bn}$ **do**
8:   **for** d_i in *inputDataset* **do**
9:    **if** $d_i >= B_k^{min}$ and $d_i < B_k^{max}$ **then**
10:     $B_k = B_k \cup d_i$;
11:    **end if**
12:   **end for**
13: **end for**
14: *connectedComponents* ← ∅;
15: **for** B_k in $B_1, ..., B_{bn}$ **do**
16:   *connectedComponentsBk* = separateConnectedComponents(B_k, ef);
17:   *connectedComponents*   =   *connectedComponents*   ∪
   connectedComponentsBk;
18: **end for**
19: *clusters* ← the connectedComponents aggregation result using *linkageMethod*
   and *numberOfClusters*;
20:   **return** *clusters*;
21: **end procedure**

The *Entropy* [29] measures the uncertainty regarding the cluster membership of a randomly chosen object, assuming that all elements of the input dataset have the same probability of being picked. Therefore, a small entropy indicates a good clustering.

The *Rand Index* [23] represents the level of agreement between the grouping produced by the evaluated clustering algorithm and the ground truth. It is computed as a fraction where the sum between the number of pairs assigned to the same cluster and the number of pairs assigned to different clusters is the numerator and the total number of pairs is the denominator. The value of *Rand Index* is a number between 0 (the groupings are completely different) and 1 (the groupings are the same).

However, the value of the *Rand Index* converges to 1 as the number of clusters increases. For this reason the *Adjusted Rand Index* quality measure was introduced as the normalized difference of the *Rand Index* and its expected value under the null hypothesis [27].

6.2   Evaluation Results

We evaluate DenLAC's performances on eight synthetic bi-dimensional datasets and two real datasets: Iris [6] and Tetragonula bee species [8] (retrieved from [1]). Each of the eight bi-dimensional datasets holds its particularities:

Algorithm 2. Splitting the density bins into connected components

---

procedure SEPARATECONNECTEDCOMPONENTS($objectsInDensityBin, ef$)
2:    $splitPartitions \leftarrow \emptyset$;
    $clustId \leftarrow 1$;
4:    **for** obj in $objectsInDensityBin$ **do**
      Mark obj as already parsed;
6:      $splitPartitions[clusterId] = splitPartitions[clusterId] \cup obj$;
      $threshold_{kDistance} \leftarrow$ the inflection point of the k-distance plot of $objectsInDensityBin$
8:      $kNeighbours \leftarrow obj$'s nearest neighbors within $threshold_{kDistance} \cdot ef$
      for $kNeighbour$ in $kNeighbours$ **do**
10:        **if** $kNeighbour$ not already parsed **then**
          Mark $kNeigbour$ as already parsed;
12:          $splitPartitions[clustId] = splitPartitions[clustId] \cup kNeighbour$;
          Continue expansion by parsing each of the neighbors neighbor;
14:        **end if**
      end for
16:      Increment $clustId$;   ▷ if we can't expand the current cluster anymore, create a new cluster and start from a new not already parsed point;
    end for
18:    **return** $splitPartitions$;
end procedure

Aggregation [11] is comprised of spherical clusters with different dimensions, Compound [30] contains clusters with varying densities, Pathbased [3], Spiral[3], Jain [15] and Flame [9] consist of arbitrary shaped clusters, Flame also incorporating outliers, R15 [28] and D31 [28] encompass spherical clusters of approximately the same size.

We compare DenLAC's results with the ones obtained by eight popular algorithms: K-Means [18], Clarans [22], Hierarchical agglomerative clustering [19], Birch [31], CURE [12], Gaussian Mixture [2], DBSCAN [7] and Spectral Clustering [21]. The output for the Pathbased synthetic datatset is graphically displayed in Fig. 3.

The evaluation results shown in Table 1a, 1b, 1c indicate that DenLAC obtains the highest overall accuracy: the ARI values are greater than 0.9 for all datasets except Tetragonula and the uncertainty regarding the cluster of a randomly extracted an object given its ground-truth class is smaller than 0.01 for 7 datasets and smaller than 0.3 for all of the datasets.

As expected, the classic algorithms are correct only on particular datasets.

K-Means obtains optimal results solely for the datasets which contain spherical clusters (Iris, Aggregation, D31 and R15 datasets). However, it computes inaccurate results for the Jain, Flame, Pathbased and Spiral datasets.

Because it relies on minimizing the sum of pairwise distances within the clusters formed around medoids, Clarans also favors non-convex clusters, performing best on the R15, D31 and Iris datasets.

Table 1. Evaluation measures values

(a) Entropy evaluation results								
Dataset	DenLAC	Birch	Clarans	Cure	Gaussian mixture	Hierarchical	K-means	DBSCAN	Spectral clustering
Aggregation	0.0111	0.099	0.183	0.0	0.081	0.037	0.062	0.174	0.011
Compound	0.114	0.185	0.315	0.186	0.187	0.279	0.21	0.282	0.18
D31	0.011	0.062	0.184	0.166	0.056	0.05	0.033	0.136	0.035
Flame	0.166	0.682	0.842	0.932	0.606	0.932	0.557	0.94	0.932
Jain	0.0	0.373	0.616	0.726	0.651	0.373	0.49	0.0	0.0
Pathbased	0.098	0.475	0.579	0.459	0.522	0.534	0.487	0.292	0.243
R15	0.008	0.019	0.139	0.046	0.006	0.014	0.006	0.084	0.006
Spiral	0.0	0.98	0.985	0.918	0.999	0.0	0.999	0.0	0.0
Iris	0.0	0.174	0.460	0.115	0.0	0.333	0.0	0.0	0.246
Tetragonula	0.189	0.71	0.71	0.73	0.71	0.71	0.71	0.1	0.8
(b) Rand Index evaluation results								
Dataset	DenLAC	Birch	Clarans	Cure	Gaussian mixture	Hierarchical	K-means	DBSCAN	Spectral custering
Aggregation	0.997	0.917	0.868	1.0	0.944	0.938	0.927	0.927	0.997
Compound	0.966	0.924	0.816	0.922	0.858	0.89	0.843	0.89	0.834
D31	0.997	0.993	0.97	0.971	0.994	0.995	0.997	0.974	0.997
Flame	0.95	0.694	0.603	0.541	0.65	0.541	0.727	0.539	0.541
Jain	1.0	0.759	0.726	0.501	0.512	0.759	0.662	0.975	1.0
Pathbased	0.968	0.757	0.709	0.762	0.729	0.723	0.747	0.819	0.857
R15	0.99	0.997	0.96	0.989	0.999	0.998	0.999	0.971	0.999
Spiral	1.0	0.558	0.536	0.453	0.554	1.0	0.554	1.0	1.0
Iris	1.0	0.912	0.773	0.949	1.0	0.835	1.0	0.92	0.695
Tetragonula	0.888	0.4	0.4	0.36	0.4	0.4	0.4	0.8	0.08
(c) Adjusted Rand Index evaluation results								
Dataset	DenLAC	Birch	Clarans	Cure	Gaussian mixture	Hierarchical	K-means	DBSCAN	Spectral clustering
Aggregation	0.991	0.733	0.586	1.0	0.827	0.8	0.762	0.808	0.99
Compound	0.911	0.809	0.481	0.805	0.596	0.742	0.538	0.743	0.531
D31	0.991	0.884	0.6	0.664	0.897	0.92	0.953	0.682	0.95
Flame	0.9	0.385	0.183	0.013	0.299	0.013	0.453	0.0	0.013
Jain	1.0	0.515	0.356	-0.084	-0.004	0.515	0.324	0.948	1.0
Pathbased	0.929	0.477	0.395	0.487	0.432	0.423	0.46	0.605	0.683
R15	0.989	0.972	0.713	0.914	0.993	0.982	0.993	0.802	0.993
Spiral	1.0	0.017	0.003	0.037	-0.005	1.0	-0.006	1.0	1.0
Iris	1.0	0.803	0.515	0.885	1.0	0.631	1.0	0.81	0.695
Tetragonula	0.658	-0.01	-0.01	0.36	0.004	-0.01	-0.01	0.2	0.001

Hierarchical agglomerative clustering with average linkage splits the elongated clusters and merges sections within neighbouring clusters for the Jain and Pathbased datasets. For the Flame dataset, the results are affected by the presence of the two outliers.

Birch obtains substandard results for non-spherical, elongated clusters such as Spiral, Jain, Flame and Pathbased.

Cure performs well only on the Aggregation, R15 and Iris datasets, since their structure is adequately described by the computed set of representatives.

The Gaussian Mixture algorithm is imprecise on datasets containing non-normally distributed clusters such as Jain, Spiral and Pathbased.

DBSCAN is sensitive to the clusters density discrepancies. For instance, it splits the Compound dataset in three different clusters ignoring the fine details and it overpartitions the Jain dataset into three clusters.

Although Spectral clustering is applicable on arbitrary shaped datasets, its ARI values for the Compound and the Pathbased datasets, 0.531 and 0.683 respectively, are significantly lower than the ones achieved by DenLAC (0.967 and 0.987 respectively).

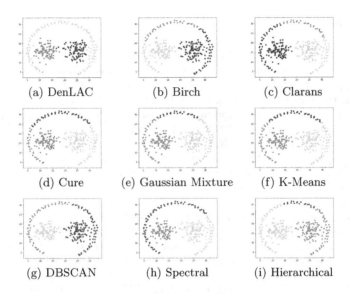

(a) DenLAC   (b) Birch   (c) Clarans

(d) Cure   (e) Gaussian Mixture   (f) K-Means

(g) DBSCAN   (h) Spectral   (i) Hierarchical

Fig. 3. Graphical representation of the results obtained by DenLAC versus other clustering algorithms for the Pathbased synthetic dataset

7   Conclusions

In this paper we present DenLAC, a flexible clustering algorithm which combines the benefits of density based and hierarchical clustering to produce accurate results regardless the input's shape and distribution. As we show in Sect. 6, our method surpasses the established clustering algorithms used for comparison in terms of accuracy, on both synthetic and real datasets. DenLac operates best on elongated, well separated clusters, obtaining perfect scores for the Jain and Spiral synthetic datasets. Aditionally, its accurracy is also maximum for the Iris real dataset and is significantly higher for the Tetragonula dataset as compared with the other methods. Our method requires choosing a number of hyperparameters: the number of clusters, the number of density bins, the expansion factor and the agglomerative clustering linkage method. We propose several heuristics for choosing these values: i) the elbow and silhouette methods are confirmed for determining a dataset's optimal number of clusters, ii) the number of density bins is normally equal to the optimal number of density levels of the cluster tree but should be larger for noisy datasets, iii) the expansion factor is normally equal to 1 but should be smaller for clusters with chains and iv) the preferred linkage method is single linkage; however, for elongated or noisy datasets centroid linkage must be used. We detail these approaches in Sect. 5, Subsect. 5.1. We further intend to incorporate a procedure which computes the number of clusters automatically and improve the nearest neighbour expansion algorithm such that the computation of the expanding factor ef is automatic.

References

1. BWorld Robot Control Software (2020). Accessed 17-Feb-2020
2. Basford, K.E., McLachlan, G.J.: Likelihood estimation with normal mixture models. Appl. Stat. **34**(3), 282 (1985). https://doi.org/10.2307/2347474
3. Chang, H., Yeung, D.Y.: Robust path-based spectral clustering with application to image segmentation. In: International Conference on Computer Vision. IEEE (2005). https://doi.org/10.1109/ICCV.2005.210
4. Chaudhuri, K., Dasgupta, S.: Rates of convergence for the cluster tree. In: Advances in Neural Information Processing Systems, pp. 343–351 (2010)
5. Chen, Y.C.: A tutorial on kernel density estimation and recent advances. Biostatistics Epidemiol. **1**(1), 161–187 (2017). https://doi.org/10.1080/24709360.2017.1396742
6. Dua, D., Graff, C.: UCI machine learning repository (2017)
7. Ester, M., Kriegel, H.P., Sander, J., Xu, X.: A density-based algorithm for discovering clusters a density-based algorithm for discovering clusters in large spatial databases with noise. In: International Conference on Knowledge Discovery and Data Mining, pp. 226–231 (1996)
8. Franck, P., Cameron, E., Good, G., Rasplus, J.Y., Oldroyd, B.: Nest architecture and genetic differentiation in a species complex of Australian stingless bees. Mol. Ecol. **13**(8), 2317–2331 (2004)
9. Fu, L., Medico, E.: Flame, a novel fuzzy clustering method for the analysis of DNA microarray data. BMC Bioinform. **8**(1), 3 (2007). https://doi.org/10.1186/1471-2105-8-3
10. Gaonkar, M.N., Sawant, K.: AutoEpsDBSCAN: DBSCAN with Eps automatic for large dataset. Int. J. Adv. Comput. Theor. Eng. **2**(2), 11–16 (2013)
11. Gionis, A., Mannila, H., Tsaparas, P.: Clustering aggregation. ACM Trans. Knowl. Discov. Data **1**(1), 4-es (2007). https://doi.org/10.1145/1217299.1217303
12. Guha, S., Rastogi, R., Shim, K.: CURE: an efficient clustering algorithm for large databases. ACM SIGMOD Rec. **27**(2), 73–84 (1998). https://doi.org/10.1145/276305.276312
13. Hartigan, J.A.: Consistency of single linkage for high-density clusters. J. Am. Stat. Assoc. **76**(374), 388–394 (1981)
14. Hinneburg, A., Keim, D.A., et al.: An efficient approach to clustering in large multimedia databases with noise. KDD **98**, 58–65 (1998)
15. Jain, A.K., Law, M.H.C.: Data clustering: a user's dilemma. In: Pattern Recognition and Machine Intelligence, pp. 1–10. Springer, Berlin Heidelberg (2005). https://doi.org/10.1007/11590316_1
16. Karypis, G., Han, E.H., Kumar, V.: Chameleon: hierarchical clustering using dynamic modeling. Computer **32**(8), 68–75 (1999). https://doi.org/10.1109/2.781637
17. Liu, P., Zhou, D., Wu, N.: VDBSCAN: varied density based spatial clustering of applications with noise. In: 2007 International Conference on Service Systems and Service Management, pp. 1–4. IEEE (2007)
18. MacQueen, J.: Some methods for classification and analysis of multivariate observations. In: Symposium on Mathematical Statistics and Probability, pp. 281–297. University of California Press, Berkeley, California (1967)
19. Murtagh, F.: A survey of recent advances in hierarchical clustering algorithms. Comput. J. **26**(4), 354–359 (1983). https://doi.org/10.1093/comjnl/26.4.354

20. Nadler, B., Galun, M.: Fundamental limitations of spectral clustering. In: Proceedings of the 19th International Conference on Neural Information Processing Systems, pp. 1017–1024. NIPS 2006. MIT Press, Cambridge, MA, USA (2006)
21. Ng, A.Y., Jordan, M.I., Weiss, Y.: On spectral clustering: analysis and an algorithm. In: Advances in Neural Information Processing Systems 14, pp. 849–856. MIT Press, Cambridge (2002)
22. Ng, R.T., Han, J.: CLARANS: a method for clustering objects for spatial data mining. IEEE Trans. Knowl. Data Eng. **14**(5), 1003–1016 (2002). https://doi.org/10.1109/TKDE.2002.1033770
23. Rand, W.M.: Objective criteria for the evaluation of clustering methods. J. Am. Stat. Assoc. **66**(336), 846–850 (1971). https://doi.org/10.2307/2284239
24. Schubert, E., Sander, J., Ester, M., Kriegel, H.P., Xu, X.: DBSCAN revisited, revisited: why and how you should (still) use DBSCAN. ACM Transactions on Database Systems (TODS) **42**(3), 1–21 (2017)
25. Scott, D.W.: Multivariate Density Estimation: Theory, Practice, and Visualization. John Wiley and Sons, Hoboken (2015)
26. Seo, S.: A review and comparison of methods for detecting outliers in univariate data sets. Ph.D. thesis, University of Pittsburgh (2006)
27. Truică, C.O., Rădulescu, F., Boicea, A.: Comparing different term weighting schemas for topic modeling. In: 2016 18th International Symposium on Symbolic and Numeric Algorithms for Scientific Computing (SYNASC). IEEE, September 2016. https://doi.org/10.1109/SYNASC.2016.055
28. Veenman, C., Reinders, M., Backer, E.: A maximum variance cluster algorithm. IEEE Transactions on Pattern Analysis and Machine Intelligence 24(9), 1273–1280 (sep 2002). https://doi.org/10.1109/TPAMI.2002.1033218
29. Wagner, S., Wagner, D.: Comparing clusterings: an overview. Tech. rep, ETH Zurich (2007)
30. Zahn, C.: Graph-theoretical methods for detecting and describing gestalt clusters. IEEE Trans. Comput. **C-20**(1), 68–86 (1971). https://doi.org/10.1109/t-c.1971.223083
31. Zhang, T., Ramakrishnan, R., Livny, M.: BIRCH: an efficient data clustering method for very large databases. ACM SIGMOD Rec. **25**(2), 103–114 (1996). https://doi.org/10.1145/235968.233324

Acceleration of the Robust Newton Method by the Use of the S-iteration

Krzysztof Gdawiec[1]([✉]) [ID], Wiesław Kotarski[2] [ID], and Agnieszka Lisowska[1] [ID]

[1] Institute of Computer Science, University of Silesia, Będzińska 39,
41–200 Sosnowiec, Poland
{krzysztof.gdawiec,agnieszka.lisowska}@us.edu.pl
[2] Katowice, Poland

Abstract. In this paper, we propose an improvement of the Robust Newton's Method (RNM). The RNM is a generalisation of the known Newton's root finding method restricted to polynomials. Unfortunately, the RNM is slow. Thus, in this paper, we propose the acceleration of this method by replacing the standard Picard iteration in the RNM by the S-iteration. This leads to an essential acceleration of the modified method. We present the advantages of the proposed algorithm over the RNM using polynomiagraphs and some numerical measures. Moreover, we present its possible application to the generation of artistic patterns.

Keywords: Robust Newton · S-iteration · Polynomiography

1 Introduction

Newton's method is used for finding solutions of nonlinear equations [3] and in optimisation techniques [11], but it is not robust. An improvement of this method with respect to its robustness can be achieved, for instance, by the use of a damping factor in Newton's algorithm. This method called the Damped Newton Method, decreases or even eliminates the fractal boundaries between basins [4].

Recently, the Robust Newton Method (RNM) was proposed [10], which radically smooths the boundaries between basins of attraction. That result was obtained by precise controlling of the step's length in Newton's method. Unfortunately, the RNM usually needs a huge number of iterations [10]. Thus, even for not so accurate computations for some points, we need to perform a large number of iterations. This is a drawback of the RNM.

In this paper, we propose a modification of the RNM by introducing the S-iteration instead of the Picard one. This modification allows decreasing the average number of iterations needed to find the solution. We will also analyse, numerically, different aspects of the proposed modification. Moreover, by using

---

W. Kotarski—Independent Researcher.

M. Paszynski et al. (Eds.): ICCS 2021, LNCS 12742, pp. 330–337, 2021.
https://doi.org/10.1007/978-3-030-77961-0_28

various sequences of parameters in the S-iteration, we will generate very complex and intriguing patterns from the dynamics of the modified method.

The paper is organised as follows. In Sect. 2 the RNM is described. In Sect. 3 the proposed modification of the RNM is presented. Section 4 presents the experimental results. And Sect. 5 concludes this paper.

2   The Robust Newton's Method

Newton's method has one drawback – the lack of definition at critical points. To overcome this, the Robust Newton's Method (RNM) was proposed [10]. This method is defined by applying the Geometric Modulus Principle [9] and the Modulus Reduction Theorem [10] to the Newton's method. The RNM guarantees the reduction of the polynomial's modulus in successive iterations. Additionally, the RNM has a few differences in comparison to the Newton's method, i.e., it converges globally, and it finds both the roots and the critical points of polynomials, unlike the Newton's method. Let us follow the definitions from [10].

Let us consider a complex polynomial p of degree $n \in \mathbb{N}$. Assume that $p(z) \neq 0$, $z \in \mathbb{C}$ and let us define [10]: $k := k(z) = \min\{j \geq 1 : p^{(j)}(z) \neq 0\}$, $A(z) = \max\left\{\frac{|p^{(j)}(z)|}{j!} : j = 0, \ldots, n\right\}$, and $u_k := u_k(z) = \frac{1}{k!}p(z)\overline{p^{(k)}(z)}$.

Moreover, let us define $\gamma_k = 2 \cdot Re(u_k^{k-1})$, $\delta_k = -2 \cdot Im(u_k^{k-1})$, and $c_k = \max\{|\gamma_k|, |\delta_k|\}$. Additionally, let θ_k be the angle given by the following formula:

$$\theta_k = \begin{cases} 0, & \text{if } c_k = |\gamma_k| \wedge \gamma_k < 0, \\ \pi/k, & \text{if } c_k = |\gamma_k| \wedge \gamma_k > 0, \\ \pi/(2k), & \text{if } c_k = |\delta_k| \wedge \delta_k < 0, \\ 3\pi/(2k), & \text{if } c_k = |\delta_k| \wedge \delta_k > 0. \end{cases} \tag{1}$$

Then, the RNM for the starting point $z_0 \in \mathbb{C}$ is defined as

$$z_{i+1} = N_p(z_i) := z_i + \frac{C_k(z_i)}{3} \frac{u_k}{|u_k|} e^{\mathbf{i}\theta_k}, \quad i = 0, 1, 2, \ldots, \tag{2}$$

where $\mathbf{i}$ denotes the imaginary unit, i.e., $\mathbf{i} = \sqrt{-1}$, and $C_k(z_i) = \frac{c_k|u_k|^{2-k}}{6A^2(z_i)}$. The term $(u_k/|u_k|)e^{\mathbf{i}\theta_k}$ is called the normalised robust Newton direction at z_i and $C_k(z_i)/3$ is called the step-size [10].

The stopping criterion is given by the following condition: $|p(z_i)| < \varepsilon \vee |p'(z_i)| < \varepsilon$, where $\varepsilon > 0$ is the accuracy. This is a different criterion than in the classical method since the RNM finds both the roots and the critical points.

3   The Robust Newton's Method with the S-iteration

The RNM presented in the previous section uses the Picard iteration for finding fixed points (see Eq. (2)). However, we can use any method from the fixed point theory. One of them is the S-iteration [1] defined in the following way:

$$\begin{cases} z_{i+1} = (1 - \alpha_i)T(z_i) + \alpha_i T(v_i), \\ v_i = (1 - \beta_i)z_i + \beta_i T(z_i), \quad i = 0, 1, 2, \ldots, \end{cases} \tag{3}$$

where $\alpha_i, \beta_i \in [0, 1]$ and T is a mapping. Depending on the type of the mapping and space, the conditions on the sequences α_i and β_i may be expanded to assure the convergence to a fixed point. Let us note that when $\alpha_i = 0$ or $\beta_i = 0$ for all i, the S-iteration reduces to the Picard iteration.

Now, we combine the RNM with the S-iteration, obtaining a new method, called shortly as S-RNM. This combination is done by using the N_p as T in (3) and extending the $[0, 1]$ interval for the parameters to $\mathbb{R}$, obtaining:

$$\begin{cases} z_{i+1} = (1 - \alpha_i)N_p(z_i) + \alpha_i N_p(v_i), \\ v_i = (1 - \beta_k)z_i + \beta_i N_p(z_i), \quad i = 0, 1, 2, \ldots, \end{cases} \tag{4}$$

where $\alpha_i, \beta_i \in \mathbb{R}$.

Let us see the first step of (4) after including the formula for N_p:

$$v_i = (1 - \beta_i)z_i + \beta_i \left(z_i + \frac{C_k(z_i)}{3} \frac{u_k}{|u_k|} e^{i\theta_k} \right) = z_i + \beta_i \frac{C_k(z_i)}{3} \frac{u_k}{|u_k|} e^{i\theta_k}. \tag{5}$$

We can see that if we join β_i with the term representing the step size, i.e., $C_k(z_i)/3$, then we can change the step size in the original RNM. We can look at this step as a predictor step. In the second step of the S-iteration, we combine the values of the original point and the predictor one to obtain the final point in a given iteration. The second step can be treated as a corrector step. Therefore, the S-RNM uses the predictor–corrector strategy – a well-known strategy in numerical methods [11].

4   Experiments

In this section, we present two categories of numerical results: poly-nomiographs [8] of two types (basins of attraction and polynomiographs showing the speed of convergence and dynamics) and the plots presenting various numerical measures: average number of iterations (ANI) [2], convergence area index (CAI) [2] and polynomiograph's generation time [5]. The experiments were performed for $\alpha_i = const = \alpha$ and $\beta_i = const = \beta$.

We have performed the experiments for different polynomials, but due to the lack of space, we present the complete results only for the polynomial $p_3(z) = z^3 - 1$. This polynomial has three roots: $-\frac{1}{2} - \frac{\sqrt{3}}{2}i$, $-\frac{1}{2} + \frac{\sqrt{3}}{2}i$, 1, and only one critical point, namely 0.

To generate the polynomiographs we used the following parameters: area $[-3, 3]^2$, the maximal number of iterations equals to 250, accuracy $\varepsilon = 0.001$, and image resolution 800×800 pixels. In the basins of attraction, we used red, green and blue colours for the roots, yellow one for the critical points, and black for non-convergent points. The values of α and β, used in the S-iteration, were taken from $[0, 20]$ with the step equal to 0.1, which gives 40 401 polynomiographs.

The experiments were performed on the computer with: Intel i5-9600K (@ 3.70 GHz) processor, 32 GB DDR4 RAM, NVIDIA GeForce GTX 1660 Ti with 6 GB GDDR6 SDRAM and Windows 10 (64-bit). The software was implemented in C++, OpenGL and GLSL. The computations were performed on the graphics card with GLSL shaders.

We analyse, first, the basins of attraction of the proposed method for p_3, presented in Fig. 1 for the S-RNM. From these images one can observe that when we increase the values of α or/and β, the borders of the basins are changing. Indeed, a third basin appears between the two ones (Fig. 1(c)). Going further causes that more points are not converging (Fig. 1(d)). This tendency continues up to nearly all non-convergent points. Only some convergence to critical points appears (yellow dots, Figs. 1(e)–(f)). Finally, we can observe something like a chaos with the play of critical points (Figs. 1(g)–(h)). When we look at Fig. 1(b) and Fig. 1(a) (the Picard iteration case), we can observe that the basins of attraction have similar shapes.

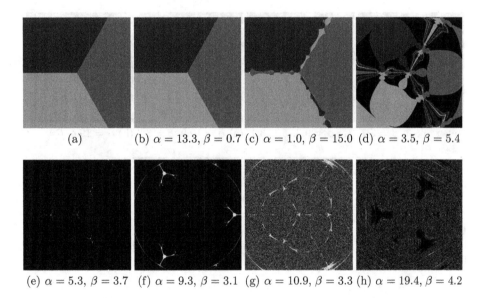

(a)   (b) $\alpha = 13.3$, $\beta = 0.7$ (c) $\alpha = 1.0$, $\beta = 15.0$ (d) $\alpha = 3.5$, $\beta = 5.4$

(e) $\alpha = 5.3$, $\beta = 3.7$ (f) $\alpha = 9.3$, $\beta = 3.1$ (g) $\alpha = 10.9$, $\beta = 3.3$ (h) $\alpha = 19.4$, $\beta = 4.2$

Fig. 1. The examples of basins of attraction for $p_3(z) = z^3 - 1$ for (a) Picard iteration, (b)–(h) S-iteration with various values of α and β. (Color figure online)

Then, we analyse the examples of dynamics polynomiographs of p_3, presented in Fig. 2, for the S-RNM. From these images, we can observe that by increasing the values of α or/and β, the dynamics changes similarly as in the case of the basins of attraction but in a more subtle way. However, the same idea remains – these changes go to the vanishing of convergent points and then to making chaos with the critical points. By comparing the polynomiographs from Figs. 1(b)–(c) and Fig. 1(a) (the Picard case) we see that the use of the S-iteration has lowered the number of iterations needed to find the solution.

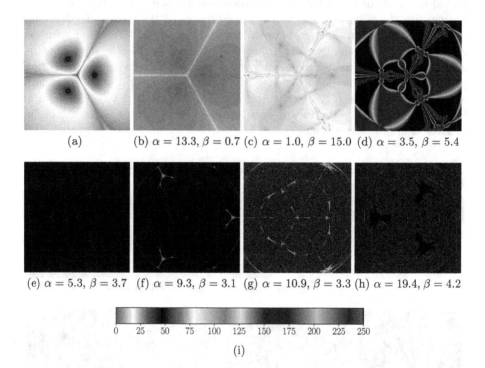

(a)   (b) $\alpha = 13.3, \beta = 0.7$ (c) $\alpha = 1.0, \beta = 15.0$ (d) $\alpha = 3.5, \beta = 5.4$

(e) $\alpha = 5.3, \beta = 3.7$ (f) $\alpha = 9.3, \beta = 3.1$ (g) $\alpha = 10.9, \beta = 3.3$ (h) $\alpha = 19.4, \beta = 4.2$

(i)

Fig. 2. The examples of dynamics polynomiographs for $p_3(z) = z^3 - 1$ for (a) Picard iteration, (b)–(h) S-iteration with various values of α and β. (i) the colour map. (Color figure online)

Next, in Fig. 3 the heatmaps on (α, β)-plane for p_3 are shown, presenting the values of ANI, CAI and time generation, coded in colour. From all these plots one can see that the results are nearly symmetrical along $\beta = \alpha$ line. This tendency is also present in the resulting polynomiographs. In these plots it is possible to point out the values of the parameters α and β for which the S-RNM behaves well, i.e. the method is convergent to the roots or the critical point and has low ANI and time generation parameters (blue areas in Fig. 3(a) and (c), respectively) and CAI parameter is close to 1.0 (the dark red area in Fig. 3(b)). Outside these areas, the algorithm needs much more iterations to find the solution or might be non-convergent. In the mentioned areas there are located α and β parameters for which the S-RNM is essentially better than the RNM. The following values were obtained for the S-RNM in the case of $z^3 - 1$: simultaneously minimal ANI = 5.18 and time = 0.047 s with CAI = 1.0, for $\alpha = 13.3$ and $\beta = 0.7$ (for which the basins of attraction and dynamics polynomiograph are presented in Figs. 1(b) and 2(b), respectively), whereas for the RNM: ANI = 84.89, CAI = 0.99, and time = 0.667 s (the basins of attraction and dynamics polynomiograph are presented in Fig. 1(a) and 2(a)). It means that in this case, the S-RNM needs nearly 17 times iterations less in comparison to the RNM to achieve the same goal in considerably lower computation time.

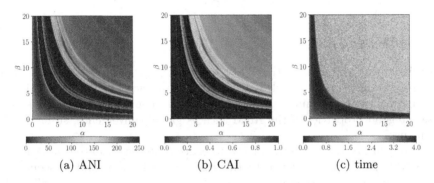

(a) ANI   (b) CAI   (c) time

Fig. 3. The plots of: (a) ANI, (b) CAI, and (c) time of generation for $p_3(z) = z^3 - 1$. (Color figure online)

Dynamics of a discrete dynamical system can give rise to fascinating patterns, see for instance [5,7,12]. The introduction of the S-iteration in the RNM can give birth to very complicated patterns of possible artistic applications. In Fig. 4 we present three patterns obtained with the S-RNM for p_3. The other parameters used to generate these patterns are the following:

(a) $\varepsilon = 0.001$, the maximal number of iterations equal to 250, the area $[-3.2, 3.8] \times [-3.5, 3.5]$ and $\alpha_i = 0.5 \sin(i) \tan(7i) + 0.5$, $\beta_i = 20 \sin(0.25i) \cos(10i) - \cos(0.5i) \tan(7.4i) + 1.25$,

(b) $\varepsilon = 0.01$, the maximal number of iterations equal to 200, the area $[-3, 3]^2$ and $\alpha_i = 25 \sin(0.25i) \cos(0.3i) - \cos(0.5i) \tan(7.4i) + 1.25$, $\beta_i = 0.5$,

(c) $\varepsilon = 0.01$, the maximal number of iterations equal to 50, the area $[-3, 3]^2$ and $\alpha_i = 1.1 \sin(1.1i) \tan(2.5i) + 0.9$, $\beta_i = 1.1 \sin(2i) \tan(1.8i) + 10$.

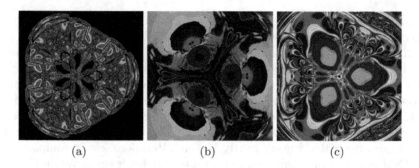

(a)   (b)   (c)

Fig. 4. The examples of artistic patterns obtained by the use of the S-RNM.

5   Conclusions

In this paper, we modified the RNM by replacing the Picard iteration with the S-iteration. The proposed S-RNM significantly accelerates the RNM, but the optimal α and β values are polynomial dependent. The S-RNM does not lose the good properties of the RNM such as global convergence, stability and robustness.

In our future study, we would like to concentrate on two aspects regarding the RNM. The first aspect will be the study on variable α_i and β_i sequences, which might further accelerate the S-RNM. In the second aspect, we will try to replace the S-iteration with other iterations known in literature and collected, for instance, in [6]. This could also lead to the acceleration of the RNM. Moreover, we will investigate the artistic applications of the patterns generated with the new methods.

References

1. Agarwal, R., O'Regan, D., Sahu, D.: Iterative construction of fixed points of nearly asymptotically nonexpansive mappings. J. Nonlinear Convex Anal. **8**(1), 61–79 (2007)
2. Ardelean, G., Balog, L.: A qualitative study of Agarwal et al. iteration procedure for fixed points approximation. Creative Math. Inf. **25**(2), 135–139 (2016)
3. Deuflhard, P.: Newton Methods for Nonlinear Problems. Springer, Heidelberg (2011). https://doi.org/10.1007/978-3-642-23899-4
4. Epureanu, B., Greenside, H.: Fractal basins of attraction associated with a damped Newton's method. SIAM Rev. **40**(1), 102–109 (1998). https://doi.org/10.1137/S0036144596310033
5. Gdawiec, K.: Fractal patterns from the dynamics of combined polynomial root finding methods. Nonlinear Dyn. **90**(4), 2457–2479 (2017). https://doi.org/10.1007/s11071-017-3813-6
6. Gdawiec, K., Kotarski, W.: Polynomiography for the polynomial infinity norm via Kalantari's formula and nonstandard iterations. Appl. Math. Comput. **307**, 17–30 (2017). https://doi.org/10.1016/j.amc.2017.02.038

7. Gdawiec, K., Kotarski, W., Lisowska, A.: On the robust Newton's method with the Mann iteration and the artistic patterns from its dynamics. Nonlinear Dyn. **104**(1), 297–331 (2021). https://doi.org/10.1007/s11071-021-06306-5

8. Kalantari, B.: Polynomial Root-Finding and Polynomiography. World Scientific, Singapore (2009). https://doi.org/10.1142/9789812811837

9. Kalantari, B.: A geometric modulus principle for polynomials. Am. Math. Mon. **118**(10), 931–935 (2011). https://doi.org/10.4169/amer.math.monthly.118.10.931

10. Kalantari, B.: A globally convergent Newton method for polynomials. https://arxiv.org/abs/2003.00372 (2020)

11. Nocedal, J., Wright, S.: Numerical Optimization, 2nd edn. Springer, New York (2006). https://doi.org/10.1007/978-0-387-40065-5

12. Ouyang, P., Tang, X., Chung, K., Yu, T.: Spiral patterns of color symmetry from dynamics. Nonlinear Dyn. **94**(1), 261–272 (2018). https://doi.org/10.1007/s11071-018-4357-0

A New Approach to Eliminate Rank Reversal in the MCDA Problems

Bartłomiej Kizielewicz[ID], Andrii Shekhovtsov[ID], and Wojciech Sałabun[✉][ID]

Research Team on Intelligent Decision Support Systems, Department of Artificial Intelligence and Applied Mathematics, Faculty of Computer Science and Information Technology, West Pomeranian University of Technology in Szczecin, ul. Żołnierska 49, 71-210 Szczecin, Poland
wojciech.salabun@zut.edu.pl

Abstract. In the multi-criteria decision analysis (MCDA) domain, one of the most important challenges of today is Rank Reversal. In short, it is a paradox that the order of alternatives belonging to a certain set is changed when a new alternative is added to that set or one of the current ones is removed. It may undermine the credibility of ratings and rankings, which are returned by methods exposed to the Rank Reversal phenomenon.

In this paper, we propose to use the Characteristic Objects method (COMET), which is resistant to the Rank Reversal phenomenon and combining it with the Technique for Order of Preference by Similarity to Ideal Solution (TOPSIS) and Preference Ranking Organization Method for Enrichment Evaluations II (PROMETHEE II) methods. The COMET method requires a very large number of pair comparisons, which depends exponentially on the number of criteria used. Therefore, the task of pair comparisons will be performed using the PROMETHEE II and TOPSIS methods. In order to compare the quality of both proposed approaches, simple comparative experiments will be presented. Both proposed methods have high accuracy and are resistant to the Rank Reversal phenomenon.

Keywords: Decision analysis · MCDA · Rank Reversal · TOPSIS · PROMETHEE II · COMET

1 Introduction

Multi-Criteria Decision Analysis (MCDA) is an important branch of operational research, where the most crucial challenge is to correctly identify the values of preferences and rankings for a defined set of alternatives with commonly contradictory criteria. Many interesting studies were devoted to the selection of appropriate MCDA methods to the specified problem class [7,21,24]. However, a significant challenge for the MCDA methods is still the Rank Reversal (RR) paradox. There are also many studies conducted on the problem of ranking reversal for MCDA methods, e.g. TOPSIS [9,26], AHP [2], ELECTRE [11] or

© Springer Nature Switzerland AG 2021
M. Paszynski et al. (Eds.): ICCS 2021, LNCS 12742, pp. 338–351, 2021.
https://doi.org/10.1007/978-3-030-77961-0_29

PROMETHEE [23]. The RR phenomenon can define as a change in the ranking of alternatives when the alternative is added or removed in a given set. In this case, the order of priority of decision-makers between the two alternatives changes, which is incompatible with the principle of independence of irrelevant alternatives [1].

Recently, research directions aimed at creating reliable methods fully resistant to the RR phenomenon have become increasingly visible. The most important methods of recent years include Ranking of Alternatives through Functional mapping of criterion sub-intervals into a Single Interval (RAFSI) [27], Stable Preference Ordering Towards Ideal Solution (SPOTIS) [8], and the Characteristic Objects Method (COMET) [17]. Apart from resistance to the RR phenomenon, the COMET method's main advantages are high accuracy and not requiring arbitrary weights. However, the number of required pair comparisons is exponentially dependent on the number of analyzed criteria and the number of characteristic values. Thus, despite its advantages, it is very laborious. That is why the idea was put forward to create a hybrid approach, which would have most of the COMET method's positive features but would not require the expert to make pair comparisons.

In order to create a hybrid approach, we want to test two popular methods such as the Technique for Order of Preference by Similarity to Ideal Solution (TOPSIS) and Preference Ranking Organization Method for Enrichment Evaluations II (PROMETHEE II) methods. These methods have been successfully applied in many scientific works despite the burden of the RR paradox for them [3,4]. They represent two different approaches to decision making [25].

In this paper, we propose a new hybrid approach to combine the advantages of COMET and two other MCDA methods. In both cases, MCDA methods replace the expert when comparing pairs of characteristic objects. As a result, we have obtained two similar approaches, which will be compared with each other, where one is based on TOPSIS method and the second on the PROMETHE II method. At the same time, the use of these classical MCDA methods, together with the COMET method, will also be immune to the rank reversal phenomenon.

The rest of the paper is organized as follows: In Sect. 2, some basic preliminary concepts on MCDA are presented. The PROMETHEE II, TOPSIS and COMET algorithms are presented in Sects. 2.1, 2.2, and 2.3, respectively. The ranking similarity coefficients used in this paper are described in Sect. 2.4. The method for determining criterion weights based on entropy is discussed in Sect. 2.5. In Sect. 3, the proposed approach is presented with numerical examples. In Sect. 4, we present the summary and conclusions.

2   Methods

In this section, the methods used in our work are presented in order to make the newly proposed approach easier to understand.

2.1   The PROMETHEE II Method

The Preference Ranking Organization Method for Enrichment Evaluations II (PROMETHEE II) is a method whose main logic is to compare alternative solutions in pairs. Decision options in this method evaluated according to different types of criteria, which, depending on the type, must be minimized or maximized. This method comes from the PROMETHEE family of methods developed by Brans [6]. It is used in many decision-making problems [14,15,21].

Step 1. The first step is to define a decision-making space at the aid of a decision matrix with n alternatives and criteria in m. The type of criteria (benefit or cost) and the weighting of the criteria must be defined. The sum of criteria weights should be equal to 1.

Step 2. The next stage in the PROMETHEE II method is the normalization of the decision matrix. In this case, linear normalization was used, which is represented by Eqs. (1) (for the benefit type criterion) and (2) (for the cost type criterion).

For beneficial type criteria:

$$r_{ij} = \frac{x_{ij} - \min(x_{ij})}{\max(x_{ij}) - \min(x_{ij})} \tag{1}$$

For cost type criteria:

$$r_{ij} = \frac{\max(x_{ij}) - x_{ij}}{\max(x_{ij}) - \min(x_{ij})} \tag{2}$$

where:

- x_{ij} - value of the decision matrix for column j and row i
- r_{ij} standardized value of column j and row i

Step 3. Calculating the differences d_j of alternatives i^{th} with respect to the other alternatives for each criterion. The pairwise comparison for all alternatives is calculated, where $g_j(a)$ is the value of alternative a in criterion j (3).

$$d_j(a, b) = g_j(a) - g_j(b) \tag{3}$$

Step 4. In the PROMETHEE II method, we use usual preference function to calculate preference values [16]:

$$P(d) = \begin{cases} 0, d \leq 0 \\ 1, d > 0 \end{cases} \tag{4}$$

The most common choice for calculating preferences in PROMETHEE II is the preference function (4) because it has no additional parameters.

Step 5. The aggregated preference values are then determined based on the formula (5) if the sum of weights is equal to 1 (6).

$$\begin{cases} \pi(a,b) = \sum_{j=1}^{k} P_j(a,b)w_j \\ \pi(b,a) = \sum_{j=1}^{k} P_j(b,a)w_j \end{cases} \tag{5}$$

$$\sum_{j=1}^{k} w_j = 1 \tag{6}$$

Step 6. *Leaving and entering outranking flows.* Based on the aggregated preference values, the positive values of (7) and negative values of (8) preference flow ϕ^- and ϕ^+ are calculated.

$$\phi^+(a) = \frac{1}{n-1} \sum_{x \in A} \pi(a,x) \tag{7}$$

$$\phi^-(a) = \frac{1}{n-1} \sum_{x \in A} \pi(x,a) \tag{8}$$

Step 7. *Net outranking flows.* The last step in the PROMETHEE II method is to calculate the net preference flow using Eq. (9). The highest calculated value gets the first position in the ranking.

$$\phi(a) = \phi^+(a) - \phi^-(a) \tag{9}$$

2.2   The TOPSIS Method

The concept of the TOPSIS method is to specify the distance of the considered objects from the ideal and anti-ideal solution [21]. The final effect of the study is a synthetic coefficient which forms a ranking of the studied objects. The best object is defined as the one with the shortest distance from the ideal solution and, at the same time, the most considerable distance from the anti-ideal solution [13,21]. The formal description of the TOPSIS method should be shortly mentioned [4]:

Step 1. Create a decision matrix consisting of n alternatives with the values of criteria k. Then normalize the decision matrix according to the formula (10).

$$r_{ij} = \frac{x_{ij}}{\sqrt{(\sum |x_{ij}|^2)}} \tag{10}$$

where x_{ij} and r_{ij} are the initial and normalized value of the decision matrix.

Step 2. Then create a weighted decision matrix that has previously been normalized according to the Eq. (11).

$$v_{ij} = w_j r_{ij} \tag{11}$$

where v_{ij} is the value of the weighted normalized decision matrix and w_j is the weight for j criterion.

Step 3. Determine the best and worst alternative according to the following formula (12):

$$A_* = \{v_1^*, \cdots, v_n^*\}$$
$$A_/ = \{v_1', \cdots, v_n'\}$$

(12)

where:

$$v_j^* = \{\max(v_{ij}) \text{ if } j \in J; \min(v_{ij}) \text{ if } j \in J'\}$$
$$v_j' = \{\min(v_{ij}) \text{ if } j \in J; \max(v_{ij}) \text{ if } j \in J'\}$$

Step 4. Calculate the separation measure from the best and worst alternative for each decision variant according to the Eq. (13).

$$S_i^* = \sqrt{\sum (v_j^* - v_{ij})^2}$$
$$S_i' = \sqrt{\sum (v_j^* - v_{ij})^2}$$

(13)

Step 5. Calculate the similarity to the worst condition by equation:

$$C_i^* = \frac{S_i'}{(S_i^* + S_i')}$$

(14)

Step 6. Rank the alternatives by their similarity to the worst state.

2.3   The COMET Method

The COMET algorithm the formal notation of the COMET method should be briefly recalled [10, 19]:

Step 1. Definition of the space of the problem - the expert determines the dimensionality of the problem by selecting r criteria, $C_1, C_2, \ldots, C_r$. Then, a set of fuzzy numbers is selected for each criterion C_i, e.g. $\{\tilde{C}_{i1}, \tilde{C}_{i2}, \ldots, \tilde{C}_{ic_i}\}$ (15):

$$C_1 = \left\{\tilde{C}_{11}, \tilde{C}_{12}, \ldots, \tilde{C}_{1c_1}\right\}$$
$$C_2 = \left\{\tilde{C}_{21}, \tilde{C}_{22}, \ldots, \tilde{C}_{2c_1}\right\}$$

(15)

$$\ldots$$

$$C_r = \left\{\tilde{C}_{r1}, \tilde{C}_{r2}, \ldots, \tilde{C}_{rc_r}\right\}$$

where $c_1, c_2, \ldots, c_r$ are the ordinals of the fuzzy numbers for all criteria.

Step 2. Generation of the characteristic objects - the characteristic objects (CO) are obtained with the usage of the Cartesian product of the fuzzy numbers' cores of all the criteria (16):

$$CO = \langle C(C_1) \times C(C_2) \times \cdots \times C(C_r) \rangle$$

(16)

As a result, an ordered set of all CO is obtained (17):

$$CO_1 = \langle C(\tilde{C}_{11}), C(\tilde{C}_{21}), \ldots, C(\tilde{C}_{r1}) \rangle$$
$$CO_2 = \langle C(\tilde{C}_{11}), C(\tilde{C}_{21}), \ldots, C(\tilde{C}_{r1}) \rangle$$
$$\ldots$$
$$CO_t = \langle C(\tilde{C}_{1c_1}), C(\tilde{C}_{2c_2}), \ldots, C(\tilde{C}_{rc_r}) \rangle$$

(17)

where t is the count of COs and is equal to (18):

$$t = \prod_{i=1}^{r} c_i \tag{18}$$

Step 3. Evaluation of the characteristic objects - the expert determines the Matrix of Expert Judgment (MEJ) by comparing the COs pairwise. The matrix is presented below (19):

$$MEJ = \begin{pmatrix} \alpha_{11} & \alpha_{12} & \cdots & \alpha_{1t} \\ \alpha_{21} & \alpha_{22} & \cdots & \alpha_{2t} \\ \cdots & \cdots & \cdots & \cdots \\ \alpha_{t1} & \alpha_{t2} & \cdots & \alpha_{tt} \end{pmatrix} \tag{19}$$

where α_{ij} is the result of comparing CO_i and CO_j by the expert. The function C_{exp} denotes the mental judgement function of the expert. It depends solely on the knowledge of the expert. The expert's preferences can be presented as (20):

$$\alpha_{ij} = \begin{cases} 0.0, f_{exp}(CO_i) < f_{exp}(CO_j) \\ 0.5, f_{exp}(CO_i) = f_{exp}(CO_j) \\ 1.0, f_{exp}(CO_i) > f_{exp}(CO_j) \end{cases} \tag{20}$$

After the MEJ matrix is prepared, a vertical vector of the Summed Judgments (SJ) is obtained as follows (21):

$$SJ_i = \sum_{j=1}^{t} \alpha_{ij} \tag{21}$$

The number of query is equal $p = \frac{t(t-1)}{2}$ because for each element α_{ij} we can observe that $\alpha_{ji} = 1 - \alpha_{ij}$. The last step assigns to each characteristic object an approximate value of preference P_i by using the following Matlab pseudo-code:

```
1: k = length(unique(SJ));
2: P = zeros(t, 1);
3: for i = 1:k
4:     ind = find(SJ == max(SJ));
5:     p(ind) = (k - i)/(k - 1);
6:     SJ(ind) = 0;
7: end
```

In the result, the vector P is obtained, where i-th row contains the approximate value of preference for CO_i.

Step 4. The rule base – each characteristic object and its value of preference is converted to a fuzzy rule as (22):

$$IF \ \ C\left(\tilde{C}_{1i}\right) \ \ AND \ \ C\left(\tilde{C}_{2i}\right) \ \ AND \ \ \dots \ \ THEN \ \ P_i \tag{22}$$

In this way, a complete fuzzy rule base is obtained.

Step 5. Inference and the final ranking - each alternative is presented as a set of crisp numbers, e.g. $A_i = \{a_{i1}, a_{i2}, a_{ri}\}$. This set corresponds to the criteria $C_1, C_2, \dots, C_r$. Mamdani's fuzzy inference method is used to compute the preference of the $i - th$ alternative. The rule base guarantees that the obtained results are unequivocal. The whole process of the COMET method is presented on the Fig. 1.

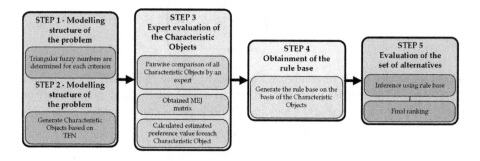

Fig. 1. The flow chart of the COMET procedure [18]

2.4   Similarity Coefficients

Similarity coefficients are measures by which one can compare to what extent the results obtained are similar. In this study, we compared rankings of multi-criteria decision-making methods using the Spearman correlation coefficient (23), Spearman weight correlation coefficient (24), Kendall correlation coefficient (26) and the WS rank similarity coefficient (25) [20].

$$r_s = 1 - \frac{6 \cdot \sum_{i=1}^{n} d_i^2}{n \cdot (n^2 - 1)} \tag{23}$$

$$r_w = 1 - \frac{6 \cdot \sum_{i=1}^{n} (x_i - y_i)^2 ((N - x_i + 1) + (N - y_i + 1))}{n \cdot (n^3 + n^2 - n - 1)} \tag{24}$$

$$WS = 1 - \sum_{i=1}^{n} \left(2^{-x_i} \frac{|x_i - y_i|}{max\{|x_i - 1|, |x_i - N|\}} \right) \tag{25}$$

$$\tau_b = \frac{\sum_{i=1}^{n} \sum_{j=1}^{n} a_{ij} b_{ij}}{\sqrt{\sum_{i=1}^{n} \sum_{j=1}^{n} a_{ij}^2 \sum_{i=1}^{n} \sum_{j=1}^{n} b_{ij}^2}} \tag{26}$$

2.5 Entropy Weighting Method

Shannon's Entropy is a measure used in many areas of informational research [22]. It is a measure of uncertainty and the degree of disorderly elements and states in a specific set. In the case of multi-criteria decision-making methods, it is used to calculate the weighting of criteria [5].

Step 1. Assuming that there is a decision matrix A with dimensions of $n \times m$, where n is the number of criteria and m is the number of alternatives, it should be normalized for the benefit type criterion and the profit type according to formulas 27 and 28 respectively.

$$P_{ij} = \frac{A_{ij}}{\sum_{i=1}^{m} A_{ij}} \tag{27}$$

$$P_{ij} = \frac{\frac{1}{A_{ij}}}{\sum_{i=1}^{m} \frac{1}{A_{ij}}} \tag{28}$$

Step 2. The entropy for each criterion is calculated for a normalized matrix, where Ec is the entropy of criterion C. The entropy of a criterion is defined by the 29 equation.

$$H(X) = E_c = \sum_{j=1}^{n} (P_{ij}) \log\left(\frac{1}{P_{ij}}\right) \tag{29}$$

Step 3. The degree of unreliability of dc for entropy criterion Ec can be calculated using the formula 30.

$$d_c = 1 - E_c, \forall_c \tag{30}$$

Step 4. After calculating the degree of divergence for each criterion, the weights should be calculated according to the formula:

$$w_c = \frac{d_c}{\sum_{j=1}^{n} d_c}, \forall_c \tag{31}$$

3 The Proposed Approaches

The proposed approach is based on the COMET method algorithm. The narrow throat of this method is the necessary number of pair comparisons, which exponentially depends on the number of criteria and characteristic values. Therefore, in step 3 of the algorithm, we propose to use TOPSIS or PROMETHEE

II instead of expert comparisons. Both approaches will be compared with each other in order to determine effectiveness. Critical weighting is determined by the entropy method.

This means that this method will not require the expert to create the MEJ matrix. The mentioned MCDA methods would be generated instead of the expert function f_{exp}. It is worth mentioning that thanks to this hybrid approach the rank reversal phenomenon is not possible, because all characteristic objects are evaluated together, and not the set of alternatives as in the original versions of these methods. The TOPSIS or PROMETHEE II method is used only for the automatic configuration of the COMET method.

3.1   Rank Reversal - Exemplary Study Case

In the purpose of presenting the significance of the rank reversal phenomenon, a simple theoretical example consisting of only two criteria will be presented. It will analyze a set consisting of 10 alternatives, and then another alternative will be added to present the reversal ranking paradox. The range of both criteria is set from zero to one. The weights in both cases were determined using the entropy method (w1 = 0.4874, w2 = 0.5126).

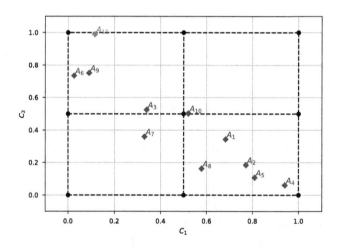

Fig. 2. The decision grid with decision alternatives for the ranking reversal paradox study.

After obtaining the rankings using the classical TOPSIS and PROMETHEE II methods, an investigation for the proposed approach was conducted. For this purpose, the same sets of alternatives were used in the study for PROMETHEE II and TOPSIS itself. In the first step, a decision grid was defined for criteria C_1 and C_2 with values of [0, 0.5, 1]. The characteristic objects from the grid were then evaluated using the PROMETHEE II (and in second approach TOPSIS) method, where the weights were selected by entropy. The evaluated characteristic objects were used in creating the COMET rule base. The COMET model then evaluated two sets of alternatives, a reference set and a set with an additional alternative. As in the case of the PROMETHEE II and TOPSIS methods itself, they were ordered, but in the second set of alternatives, the additional alternative's evaluation was not taken into account. In this way, rankings were obtained for the proposed approach of combining the PROMETHEE II and COMET methods.

Figure 2 shows a defined decision grid for studying the proposed approach combining the PROMETHEE II (or TOPSIS) and COMET methods. Black dots on the grid indicate characteristic objects. A reference set of 10 decision alternatives is shown as a green diamond on the grid, while the additional alternative is shown as a red diamond. The Table 1 shows alternatives with values for each criterion and their rankings. In the case of the assessments of both sets of alternatives, the rankings for the PROMETHEE II method and the COMET method combined approach are different. In case of ranking for the PROMETHEE II method itself, rankings differ significantly, and both are presented in the table. The values of similarity coefficients between PROMETHEE II in both cases (for ten and eleven alternatives) are significant lower, i.e. $r_w = 0.8236$, $r_s = 0.8787$, $WS = 0.8599$ and $\tau_b. = 0.8222$. It is caused by the paradox of reversal rankings, where when adding another alternative to the reference set of alternatives, their rating changes dramatically [12,23]. However, the proposed approach eliminates the ranking reversal paradox using the COMET method, which is resistant to the paradox.

Table 1. Alternatives from the reference set with positions in each ranking.

Alternatives	Criteria		Rankings					
A_i	C_1	C_2	PROMETHEE II	PROMETHEE II (added alternative)	PROMETHEE II + COMET	TOPSIS	TOPSIS (added alternative)	TOPSIS + COMET
A_1	0.6824	0.3438	4.0	3.0	1.0	2.0	2.0	3.0
A_2	0.7716	0.1845	5.0	4.0	4.0	7.0	4.0	6.0
A_3	0.3404	0.5253	2.0	1.0	6.0	4.0	7.0	5.0
A_4	0.9407	0.0608	7.0	7.0	3.0	6.0	3.0	7.0
A_5	0.8096	0.1075	6.0	6.0	5.0	8.0	5.0	8.0
A_6	0.0270	0.7346	8.0	8.0	8.0	5.0	8.0	4.0
A_7	0.3308	0.3600	9.0	9.0	10.0	10.0	10.0	9.0
A_8	0.5780	0.1626	10.0	10.0	9.0	9.0	9.0	10.0
A_9	0.0926	0.7514	1.0	5.0	7.0	3.0	6.0	2.0
A_{10}	0.5206	0.5024	3.0	2.0	2.0	1.0	1.0	1.0

It is worth noting that this example clearly shows that both the PROMETHEE II and TOPSIS methods themselves have shown their lack of resistance to the rank reversal phenomenon. In the two-hybrid methods, there is no rank reversal because PROMETHEE II and TOPSIS do not serve to evaluate alternatives directly but only indirectly by evaluating the characteristic objects. Thus, the proposed approach is systematically free of the rank reversal phenomenon, which has been illustrated in the presented example and directly results from the COMET method's properties.

3.2 Effectiveness

In the purpose of presenting the effectiveness of the approaches considered, the following study was performed. The first step was to evaluate the set generated in each iteration consisting of 3 to 10 alternatives to the three criteria. The first criterion was the patient's heart rate, which was in the range [60, 100], the second criterion was the patient's age, which was in the range [40, 60], and the third criterion was systolic blood pressure, which was in the range [90, 180]. This set was evaluated by an expert function defined as TRI, and this evaluation was a reference preference. The expert function TRI can be defined as follows:

$$TRI = \frac{HR \cdot A^2}{100 \cdot SBP} \tag{32}$$

where, HR is heart rate (bits per minute bpm), A is baseline age (years), and SBP is systolic blood pressure (mmHg).

Algorithm 1: Numerical experiment

1: $N \leftarrow 10000$
2: $criteria \leftarrow 3$
3: **for** num_alt in $[3, 5, 7, 9, 11]$ **do**
4:    **for** $i = 1$ to N **do**
5:       $alternatives \leftarrow generate\_alternatives(num\_alt, criteria)$
6:       $weights \leftarrow entropy\_weights(alternatives)$
7:       $reference\_result \leftarrow tri(alternatives)$
8:       $coefficients \leftarrow Coefficients()$
9:       **for** $method$ in $methods$ **do**
10:          $result \leftarrow method(alternatives, weights)$
11:          $coefficients.add(WS(result, reference\_result))$
12:          $coefficients.add(rw(result, reference\_result))$
13:          $coefficients.add(rs(result, reference\_result))$
14:          $coefficients.add(tau(result, reference\_result))$
15:       **end**
16:       $save\_coefficients(coefficients)$
17:    **end**
18: **end**

In the next step, the same set of alternatives was evaluated using the PROMETHEE II method (and analogical TOPSIS) itself and a combination of the PROMETHEE II and COMET methods. All alternatives were then ranked according to the preferences obtained from the respective approaches. Correlation coefficients were calculated for the obtained rankings with reference rankings, specified in Sect. 2.4. Ten thousand iterations repeated these actions. Algorithm 1 provides details of the next steps of the study.

The Fig. 3 shows a box graph for the value of the rs correlation coefficient to the number of alternatives in the set for the two approaches considered. It shows that proposed approaches have higher level of rankings similarity to reference ranking than for classical TOPSIS and PROMETHEE II methods. Moreover, the rs correlation coefficient values for the two considered cases are not significantly different.

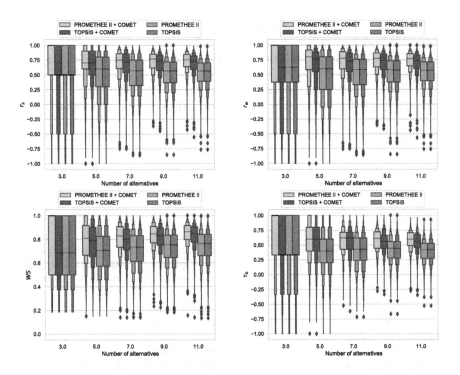

Fig. 3. Experimental results comparing the similarity of the rankings.

Similar results were also obtained for the other coefficients. This clearly shows that the use of the hybrid approach produces better results than classical methods. Additionally, it should be remembered that the proposed approaches are free from the rank reversal phenomenon.

4   Conclusions

The following paper proposes the use of a new approach to decision-making to avoid rank reversal. The proposed hybrid approach combines the COMET method and TOPSIS or PROMETHEE II methods. Thanks to this combination, it is possible to avoid the curse of dimensionality. The COMET method's main disadvantage is that the number of queries to the expert increases exponentially as the number of criteria and characteristic values increases. Therefore, the expert evaluation's mental function is proposed to be replaced by calculations obtained with the TOPSIS or PROMETHEE method. Such an action still guarantees the non-appearance of the rank reversal phenomenon. As shown in the simulation example, the results obtained are better than when using PROMETHEE II and TOPSIS methods alone.

This study poses many further research challenges, where as the main direction for further research may be enumerate:

- testing the influence of the number of characteristic values on the accuracy of the hybrid approach;
- investigating the impact of selecting an optimal grid of characteristic object points on accuracy;
- investigating other methods of preference for PROMETHEE II;
- investigation of other MCDA methods and their possible combination with the COMET method;
- further work on accuracy assessment of the proposed solution.

Acknowledgments. The work was supported by the National Science Centre, Decision number UMO-2018/29/B/HS4/02725.

References

1. de Farias Aires, R.F., Ferreira, L.: The rank reversal problem in multi-criteria decision making: aliterature review. Pesquisa Operacional **38**(2), 331–362 (2018)
2. Barzilai, J., Golany, B.: Ahp rank reversal, normalization and aggregation rules. INFOR: Inf. Syst. Oper. Res. **32**(2), 57–64 (1994)
3. Behzadian, M., Kazemzadeh, R.B., Albadvi, A., Aghdasi, M.: Promethee: a comprehensive literature review on methodologies and applications. Eur. J. Oper. Res. **200**(1), 198–215 (2010)
4. Behzadian, M., Otaghsara, S.K., Yazdani, M., Ignatius, J.: A state-of the-art survey of topsis applications. Expert Syst. Appl. **39**(17), 13051–13069 (2012)
5. Bhowmik, C., Bhowmik, S., Ray, A.: The effect of normalization tools on green energy sources selection using multi-criteria decision-making approach: a case study in India. J. Renew. Sustain. Energy **10**(6), 065901 (2018)
6. Brans, J.P., Vincke, P., Mareschal, B.: How to select and how to rank projects: the promethee method. Eur. J. Oper. Res. **24**(2), 228–238 (1986)
7. Cinelli, M., Kadziński, M., Gonzalez, M., Słowiński, R.: How to support the application of multiple criteria decision analysis? let us start with a comprehensive taxonomy. Omega, p. 102261 (2020)
8. Dezert, J., Tchamova, A., Han, D., Tacnet, J.M.: The spotis rank reversal free method for multi-criteria decision-making support. In: 2020 IEEE 23rd International Conference on Information Fusion (FUSION), pp. 1–8. IEEE (2020)
9. García-Cascales, M.S., Lamata, M.T.: On rank reversal and topsis method. Math. Comput. Model. **56**(5–6), 123–132 (2012)
10. Jankowski, J., Sałabun, W., Wątróbski, J.: Identification of a multi-criteria assessment model of relation between editorial and commercial content in web systems. In: Zgrzywa, A., Choroś, K., Siemiński, A. (eds.) Multimedia and Network Information Systems. AISC, vol. 506, pp. 295–305. Springer, Cham (2017). https://doi.org/10.1007/978-3-319-43982-2_26
11. Liu, X., Ma, Y.: A method to analyze the rank reversal problem in the electre ii method. Omega, p. 102317 (2020)

12. Mareschal, B., De Smet, Y., Nemery, P.: Rank reversal in the promethee ii method: some new results. In: 2008 IEEE International Conference on Industrial Engineering and Engineering Management, pp. 959–963. IEEE (2008)

13. Menouer, T.: KCSS: Kubernetes container scheduling strategy. J. Supercomput. **77**, 4267–4293 (2020)

14. Menouer, T., Cérin, C., Darmon, P.: Accelerated promethee algorithm based on dimensionality reduction. In: Hsu, C.-H., Kallel, S., Lan, K.-C., Zheng, Z. (eds.) IOV 2019. LNCS, vol. 11894, pp. 190–203. Springer, Cham (2020). https://doi.org/10.1007/978-3-030-38651-1_17

15. Palczewski, K., Sałabun, W.: Influence of various normalization methods in promethee ii: an empirical study on the selection of the airport location. Procedia Comput. Sci. **159**, 2051–2060 (2019)

16. Podvezko, V., Podviezko, A.: Dependence of multi-criteria evaluation result on choice of preference functions and their parameters. Technol. Econ. Dev. Econ. **16**(1), 143–158 (2010)

17. Sałabun, W.: The characteristic objects method: a new distance-based approach to multicriteria decision-making problems. J. Multi-Criteria Decis. Anal. **22**(1–2), 37–50 (2015)

18. Sałabun, W., Palczewski, K., Wątróbski, J.: Multicriteria approach to sustainable transport evaluation under incomplete knowledge: electric bikes case study. Sustainability **11**(12), 3314 (2019)

19. Sałabun, W., Piegat, A.: Comparative analysis of MCDM methods for the assessment of mortality in patients with acute coronary syndrome. Artif. Intell. Rev. **48**(4), 557–571 (2017)

20. Sałabun, W., Urbaniak, K.: A new coefficient of rankings similarity in decision-making problems. In: Krzhizhanovskaya, V., et al. (eds.) ICCS 2020. LNCS, vol. 12138, pp. 632–645. Springer, Cham (2020). https://doi.org/10.1007/978-3-030-50417-5_47

21. Sałabun, W., Wątróbski, J., Shekhovtsov, A.: Are mcda methods benchmarkable? a comparative study of topsis, vikor, copras, and promethee ii methods. Symmetry **12**(9), 1549 (2020)

22. Shannon, C.E.: A mathematical theory of communication. Bell Syst. Tech. J. **27**(3), 379–423 (1948)

23. Verly, C., De Smet, Y.: Some results about rank reversal instances in the promethee methods. Int. J. Multicriteria Decision Making 71 **3**(4), 325–345 (2013)

24. Wątróbski, J., Jankowski, J., Ziemba, P., Karczmarczyk, A., Zioło, M.: Generalised framework for multi-criteria method selection. Omega **86**, 107–124 (2019)

25. Wątróbski, J., Jankowski, J., Ziemba, P., Karczmarczyk, A., Zioło, M.: Generalised framework for multi-criteria method selection: rule set database and exemplary decision support system implementation blueprints. Data Brief **22**, 639 (2019)

26. Yang, W.: Ingenious solution for the rank reversal problem of topsis method. Math. Problems Eng. **2020**, 1–12 (2020)

27. Žižović, M., Pamučar, D., Albijanić, M., Chatterjee, P., Pribićević, I.: Eliminating rank reversal problem using a new multi-attribute model—the rafsi method. Mathematics **8**(6), 1015 (2020)

Validating Optimal COVID-19 Vaccine Distribution Models

Mahzabeen Emu©, Dhivya Chandrasekaran$^{(\boxtimes)}$©, Vijay Mago©, and Salimur Choudhury©

Department of Computer Science, Lakehead University,
Thunder Bay, ON P7B 5E1, Canada
{memu,dchandra,vmago,schoudh1}@lakeheadu.ca

Abstract. With the approval of vaccines for the coronavirus disease by many countries worldwide, most developed nations have begun, and developing nations are gearing up for the vaccination process. This has created an urgent need to provide a solution to optimally distribute the available vaccines once they are received by the authorities. In this paper, we propose a clustering-based solution to select optimal distribution centers and a Constraint Satisfaction Problem framework to optimally distribute the vaccines taking into consideration two factors namely priority and distance. We demonstrate the efficiency of the proposed models using real-world data obtained from the district of Chennai, India. The model provides the decision making authorities with optimal distribution centers across the district and the optimal allocation of individuals across these distribution centers with the flexibility to accommodate a wide range of demographics.

Keywords: COVID-19 · Constraint satisfaction problem · Vaccine distribution · Operational research · Policy making

1 Introduction

The ongoing pandemic caused by the Severe Acute Respiratory Syndrome Coronavirus 2 (SARS-CoV-2) called the coronavirus disease (COVID-19), has not only caused a public health crisis but also has significant social, political, and economic implications throughout the world [2,13,19]. More than 92.1 million cases and 1.9 million deaths have been reported worldwide as of January 2021 [6]. One of the widely used solutions, in preventing the spread of infectious diseases is vaccination. Vaccination is defined by WHO as "*A simple, safe, and effective way of protecting people against harmful diseases before they come into*

M. Emu and D. Chandrasekaran—Equal contribution.

contact with them. It uses the body's natural defenses to build resistance to specific infections and makes the immune system stronger[1]. Various researchers and pharmaceutical corporations began research work on identifying potential vaccines to combat the spread of COVID-19. Of the number of vaccines under trial, Tozinameran (BNT162b2) by Pfizer and BioNTech, and mRNA-1273 by Moderna have achieved an efficacy of more than 90% [12,15]. Various nations have begun the process of approval of these vaccines for mass distribution and have placed orders to facilitate a continuous flow of supply of vaccines to meet the demands.

In this paper we propose a Constraint Satisfaction Programming (CSP) framework based model to optimize the distribution of vaccine in a given geographical region. We aim to maximize the distribution of vaccine among the group of the population with higher priority while minimizing the average distance travelled by any individual to obtain the vaccine. In order to justify the efficiency of the proposed model we compare the performance of the following four optimization models namely,

- Basic Vaccine Distribution Model (B-VDM)
- Priority-based Vaccine Distribution Model (P-VDM)
- Distance-based Vaccine Distribution Model (D-VDM)
- Priority in conjunction with Distance-based Vaccine Distribution Model (PD-VDM).

and present how the model PD-VDM provides the most optimal solution for the distribution of vaccines. We perform a Case Study using the demographic data obtained from Chennai - a well-renowned city in Southern India. This case study highlights how the model can be used by decision making authorities of a city with an population of 5.7 million. The model aids the decision making authorities to choose an optimal number of vaccine distribution centers (DCs), and to optimally assign an individual to a hospital such that the individuals in the priority groups are vaccinated first while minimizing the distance they travel to the vaccine DCs. In Sect. 2 we discuss the various steps and challenges involved in the process of vaccine distribution, in Sect. 3 we describe in detail the procedure followed to build the proposed models. The case-study is discussed in Sect. 4 followed by the discussion of results in Sect. 5. We conclude with future research directions in Sect. 6.

2   Related Work

In order to effectively distribute vaccines, it is necessary to understand the supply chain of vaccines. The supply chain of vaccines is divided into four major components namely the product, production, allocation, and finally the distribution [7]. The first concern of the decision making authorities is to decide on

---

[1] https://www.who.int/news-room/q-a-detail/vaccines-and-immunization-what-is-vaccination.

which vaccine to choose for distribution in their country, province, or region. For example, for countries in tropical regions, the storage temperature of the vaccine is an important factor. Similarly, while developed countries are able to afford vaccines at a higher price, most developing countries prefer the vaccine which has an affordable price. In the present scenario, three of the prominent vaccines in play are the BNT162b2, mRNA-1273, and AZD1222 [10]. They have storage temperatures of $-70\,°C$, $-20\,°C$, and $0\,°C$ and cost USD 20, USD 50, and USD 4 respectively[2]. Based on the storage it is safe to assume that while countries in temperate regions like the United States, Canada, and Russia would have the option of purchasing any one of these vaccines while tropical countries like India, Bangladesh, Pakistan would prefer AZD1222. The cost of the said vaccines also has a significant impact on the decision-making process of developing nations which mostly cater to a greater number of people. Once the product is chosen, the production of vaccines has to be scheduled according to the demand. Factors that are taken into consideration at this stage include the production time, capacity for manufactures, supply-demand analysis, and so on [3]. Depending on the stage of the epidemic and the severity, the demand for the vaccine and the optimal timeline for the supply of vaccines may vary. Allocation and distribution stages go hand in hand in the vaccine supply chain. Depending on the distribution strategy, the allocation of vaccines at any level is tuned to achieve the best result. Allocation at a global level may depend on priority established through contracts and agreements among pharmaceuticals and governments. However, once a country receives the vaccines, further allocating the vaccines to provinces, states, or subgroups of the population is a critical decision that in turn has an impact on the distribution strategy. The distribution stage of the supply chain addresses the challenges of establishing an effective routing procedure, infrastructure of the vaccine distribution centers, inventory control, workforce, etc.

Operation Research (OR) involves the development and use of various statistical and mathematical problem-solving strategies to aid in the process of decision making. Various OR models are proposed over time to optimize the distribution of vaccines from the distribution centers. Ramirez-Nafarrate et al. [17] proposes a genetic algorithm to optimize the distribution of vaccines by minimizing the travel and waiting time. Huang et al. [9] formulated a vaccine deployment model for the influenza virus that ensures geographical priority based equity in Texas. However, their mathematical model might have generalization issues when applied to smaller or larger than state-level regions. Lee et al. [11] developed the RealOpt© a tool to aid in identifying optimal location for vaccine distribution centers, resource allocation and so on. Researchers have also provided models to accommodate specific locations, for example Aaby et al. [1] proposes a simulation model to optimize the allocation of vaccine distribution centers at Montgomery county, Maryland. This model aims to minimize the vaccination time and increase the number of vaccinations. While the above models

---

[2] https://www.nytimes.com/interactive/2020/science/coronavirus-vaccine-tracker.html.

consider the distribution centers to be stationary, Halper et al. [8] and Rachaniotis et al. [16] consider the vaccine distribution centers to be mobile and address this as a routing problem. While the later model proposes that various mobile units serve different nodes in a network, the latter considers that a single mobile units serves various areas with a goal to minimize the spread of infection. Some of the OR models used in epidemics' control include non-linear optimization, Quadratic Programming (QP), Integer Linear Programming (ILP) and Mixed Integer Linear Programming (MILP) [5]. ILP, MILP, and QP models are not suitable for many practical use cases due to its time expensive nature and infeasibility issues prevailing with irrational model designs. The mathematical design process and selection of pre-defined numerical bounds can lead to several technical glitches in the models. Despite the apparent similarities with ILP and MILP, CSP can eliminate all the aforementioned drawbacks and ensure sub-optimal solution in non-deterministic polynomial time by applying boolean satisfiability problem formulation. In the field of computer science, CSP is considered as a powerful mechanism to address research challenges regarding scheduling, logistics, resource allocation, and temporal reasoning [4]. Hence in this article we employ CSP to propose four models to optimize the distribution minimizing the traveling distance and maximizing the vaccination of high priority population.

3 Methodology

In this section, we initially determine the optimal number and location of vaccine DCs using K-medoids clustering algorithm. Provided with the various locations of possible vaccine DCs, the algorithm determines the optimal number of clusters into which the region can be divided into, in order to effectively distribute the vaccines across the chosen region. On selecting the number of clusters and the cluster heads, we further propose four different vaccine distribution simulation models to optimize vaccine distribution based on two factors namely distance and priority. The clustering algorithm and the simulation models are explained in the subsections below.

3.1   Selection of Optimal Vaccine Distribution Center

Our proposed Algorithm 1 imitates the core logic of K-medoids clustering technique [14]. Firstly, the algorithm determines the optimal number of vaccine DCs to be selected from a set of hospitals $\tilde{\mathcal{H}}$ based on silhouette score analysis [18], as mentioned in line number 1. We determine the silhouette score for 2 to $\tilde{h}_{\tilde{H}-1}$ number of potential vaccine DCs, where $\tilde{\mathcal{H}} = \{\tilde{h}_1, \tilde{h}_2, \tilde{h}_3, ..., \tilde{h}_{\tilde{\mathcal{H}}}\}$. Then, we select the optimal number of vaccine DCs as η that retains the highest silhouette score. As per the line number 2, we randomly select η hospitals as vaccine DCs into $\mathcal{H}$. Later, we initiate the clustering process. Each hospital is assigned to its closest vaccine DC to form η clusters, according to line numbers 5–8. The cluster information indicating which hospital is associated to which vaccine DC are recorded in $\mathcal{C}$. Next, the algorithm reassigns the vaccine DCs $\mathcal{H}$ to the ones with the

---

Algorithm 1: K-medoids algorithm to choose vaccine DCs from a set of hospitals

---

Input: $\tilde{\mathcal{H}}$: A set of hospitals, *dist*: Squared matrix representing the distances of one hospital to every other hospital

Result: $\mathcal{H}$: A set of COVID-19 vaccine DCs

1   $\eta \leftarrow$ Determine the number of optimal vaccine distribution centers using silhouette score

2   $\mathcal{H} \leftarrow$ Randomly select η hospitals from $\tilde{\mathcal{H}}$

3   $\mathcal{C} \leftarrow \varnothing$

4   **while** *there is no change in $\mathcal{H}$* **do**

5   **foreach** $\tilde{h}_a \in \tilde{\mathcal{H}}$ **do**

6   **foreach** $h_b \in \mathcal{H}$ **do**

7            $h_b \leftarrow$ Find the closest h_b to $\tilde{h}_a$ using *dist* matrix

8         $\mathcal{C} \leftarrow \mathcal{C} \cup (\tilde{h}_a, h_b)$

9   **foreach** $h_b \in \mathcal{H}$ **do**

10         $temp \leftarrow \varnothing$

11   **foreach** $\tilde{h}_a \in \tilde{\mathcal{H}}$ **do**

12   **if** $(\tilde{h}_a, h_b) \in \mathcal{C}$ **then**

13               $temp \leftarrow temp \cup \tilde{h}_a \cup h_b$

14         $q \leftarrow \underset{\hat{h}_a \in temp}{\operatorname{argmin}} \sum_{h_a^* \in temp} dist(\hat{h}_a, h_a^*)$

15   Swap h_b with $\tilde{h}_q$ in $\mathcal{C}$

16   Update h_b by $\tilde{h}_q$ in $\mathcal{H}$

---

minimum total distance to all other hospitals under the same cluster, executed in line numbers 9–16. We let the algorithm repeat the entire clustering process until vaccine DC assignments do not change. Hence, the termination criteria depends on the stability of the clustering process. To summarize, we employ this algorithm with inputs of a set of hospitals/potential vaccine DCs, $\tilde{\mathcal{H}}$ and a 2D *dist* matrix defining the distances of one hospital to every other hospital. The output of the algorithm is the set of optimally selected vaccine DCs $\mathcal{H}$ based on the distance metric. The primary idea is to choose vaccine DCs optimally in a sparse manner to facilitate reachability for people living in any part of the considered region.

3.2   System Model for Vaccine Distribution

In this subsection, we proceed by explaining the system model for vaccine distribution. We denote $\mathcal{H} = \{h_1, h_2, h_3, ..., h_{\mathcal{H}}\}$ to be the set of COVID−19 vaccine distribution centers selected by Algorithm 1, where $\mathcal{H} \subseteq \tilde{\mathcal{H}}$ and h_i is the ith vaccine DC in $\mathcal{H}$. Moreover, we define $\mathcal{U} = \{u_1, u_2, u_3, ..., u_{\mathcal{U}}\}$ as the set of available staff, where every $u_j \in \mathcal{U}_i$. We denote $\mathcal{U}_i$ as the available set of staff in vaccine DC h_i and $\mathcal{U}_i \subseteq \mathcal{U}$. Subsequently, we can infer that $\mathcal{U} = \cup_{i=1}^{|\mathcal{H}|} \mathcal{U}_i$. We further assume that $\mathcal{E} = \{e_1, e_2, e_3, ..., e_{\mathcal{E}}\}$ represents the set of people required to be vaccinated, and the kth person is e_k. In order to specify the priority of people for vaccination purpose, we use the set $\mathcal{P} = \{p_1, p_2, p_3, ..., p_{\mathcal{P}}\}$. Hence, p_k defines the priority level of a person $e_k \in \mathcal{E}$, and $|\mathcal{E}| = |\mathcal{P}|$. It is noteworthy that, the higher the priority level, the faster the vaccination service deployment is desired. The distance between a distribution center $h_i \in \mathcal{H}$ and a specific person $e_k \in \mathcal{E}$ is represented using $\mathcal{D}_{i,k}$. In this research, we consider the solution binary decision variable as $x_{i,j,k} \in \{0, 1\}$. The value of the decision variable $x_{i,j,k}$ is 1, in case a distribution center $h_i \in \mathcal{H}$ allocates a staff $u_j \in \mathcal{U}_i$ to vaccinate a person $e_k \in \mathcal{E}$, otherwise remains 0.

3.3   Problem Formulation

In this paper, we formulate the opted vaccine distribution research enigma as a CSP model. The CSP framework includes a set of aforementioned decision variables that should be assigned values in such a way that a set of hard constraints are satisfied. Hard constraints are essential to be satisfied for any model to reach a feasible solution. Suppose, our proposed model iterates over $T = \{t_1, t_2, ..., t_T\}$ times to complete vaccination, where each time instance $t_n \in T$ refers to per time frame for vaccine deployment decision making. Finally, $\mathcal{N}$ represents the total amount of available vaccine throughout the entire time periods T. All of our proposed CSP based models are subject to the following hard constraints that have been translated into integer inequalities, for any time frame $t_n \in T$:

$$\mathcal{C}1 : \sum_{e_k \in \mathcal{E}} x_{i,j,k} \le 1, \ \forall_{h_i \in \mathcal{H}}, \ \forall_{u_j \in \mathcal{U}_i} \tag{1}$$

$$\mathcal{C}2 : \sum_{h_i \in \mathcal{H}} \sum_{u_j \in \mathcal{U}_i} x_{i,j,k} \le 1, \ \forall_{e_k \in \mathcal{E}} \tag{2}$$

$$\mathcal{C}3 : x_{i,j,k} \in \{0, 1\}, \ \forall_{h_i \in \mathcal{H}}, \ \forall_{u_j \in \mathcal{U}_i}, \forall_{e_k \in \mathcal{E}} \tag{3}$$

The constraint $\mathcal{C}1$ verifies that every staff from any distribution center can vaccinate at most one person at a time. Thus, for every staff $u_j \in \mathcal{U}_i$ of distribution center $h_i \in \mathcal{H}$, either there is a unique person $e_k \in \mathcal{E}$ assigned for vaccination, or the staff remains unassigned. Then, constraint $\mathcal{C}2$ ensures that every person $e_k \in \mathcal{E}$ is allocated at most one vaccine through a single staff $u_j \in \mathcal{U}_i$ from a unique distribution center $h_i \in \mathcal{H}$. Finally, $\mathcal{C}3$ is a binary constraint representing the value of decision variable to be 1, in case a staff $u_j \in \mathcal{U}_i$ of distribution center $h_i \in \mathcal{H}$ is assigned to vaccinate a person $e_k \in \mathcal{E}$, otherwise 0, as mentioned previously.

$$\mathcal{C}4 : \sum_{t_n \in T} \sum_{h_i \in \mathcal{H}} \sum_{u_j \in \mathcal{U}_i} \sum_{e_k \in \mathcal{E}} x_{i,j,k} \le \mathcal{N} \tag{4}$$

The constraint $\mathcal{C}4$ confirms that the total vaccine distribution should not be more than the available vaccine by any means throughout the entire periods T considered for vaccine deployment. Now, let us assume Ω be the set of all the feasible solutions that satisfy all hard constraints.

$$\Omega = \{x_{i,j,k} \mid \mathcal{C}1, \mathcal{C}2, \mathcal{C}3, \mathcal{C}4\} \tag{5}$$

Apart from hard constraints, our proposed CSP formulation incorporates a set of soft constraints as well. Whilst hard constraints are modeled as inequalities, soft constraints are outlined through expressions intended to be eventually minimized or maximized. Soft constraints are not mandatory for finding a solution, rather highly desirable to improvise the quality of the solutions based on the application domain. The soft constraint $C5$ strives to maximize the number of overall vaccinated people. The focus of another soft constraint $C6$ remains to vaccinate the people with higher priority levels beforehand. In other words, this constraint leads to maximize the summation of priorities of all vaccinated people. Subsequently, the soft constraint $C7$ refers that every people should be vaccinated by staff from the nearest vaccine distribution center.

$$C5 : \max_{\Omega} \sum_{h_i \in \mathcal{H}} \sum_{u_j \in \mathcal{U}_i} \sum_{e_k \in \mathcal{E}} x_{i,j,k} \tag{6}$$

$$C6 : \max_{\Omega} \sum_{h_i \in \mathcal{H}} \sum_{u_j \in \mathcal{U}_i} \sum_{e_k \in \mathcal{E}} x_{i,j,k} \times p_k \tag{7}$$

$$C7 : \min_{\Omega} \sum_{h_i \in \mathcal{H}} \sum_{u_j \in \mathcal{U}_i} \sum_{e_k \in \mathcal{E}} x_{i,j,k} \times \mathcal{D}_{i,k} \tag{8}$$

3.4   Variations of Vaccine Distribution Models

By leveraging different combinations of soft constraints, we propose four different variations of vaccine distribution models: **a)** Basic - Vaccine Distribution Model (B-VDM), **b)** Priority based - Vaccine Distribution Model (P-VDM), **c)** Distance based - Vaccine Distribution Model (D-VDM), and **d)** Priority in conjunction with Distance based - Vaccine Distribution Model (PD-VDM). B-VDM is a rudimentary vaccine distribution model that solely concentrates on the soft constraint $C5$ to maximize the overall vaccine distribution, irrespective of any other factors. A gain co-efficient α has been introduced to the ultimate objective function of the model in Eq. 9.

$$C5 \iff \max_{\Omega} \sum_{h_i \in \mathcal{H}} \sum_{u_j \in \mathcal{U}_i} \sum_{e_k \in \mathcal{E}} \alpha \times x_{i,j,k} \tag{9}$$

P-VDM ensures maximum vaccine distribution among the higher priority groups of people, by reducing soft constraints $C5$ and $C6$ into one objective function in Eq. 10. We denote β as the gain factor associated to soft constraint $C6$.

$$C5 \wedge C6 \iff \max_{\Omega} \sum_{h_i \in \mathcal{H}} \sum_{u_j \in \mathcal{U}_i} \sum_{e_k \in \mathcal{E}} x_{i,j,k} \times (\alpha + \beta \times p_k) \tag{10}$$

Contrarily, D-VDM encourages vaccination of the people located closer to distribution centers. This model can be specifically useful for rural regions, where distribution centers and people are sparsely located, including higher travelling expenses. For this model, we incorporate soft constraints $C5$ and $C7$, by multiplying gain coefficients α and γ, respectively.

$$C5 \wedge C7 \iff \max_{\Omega} \sum_{h_i \in \mathcal{H}} \sum_{u_j \in \mathcal{U}_i} \sum_{e_k \in \mathcal{E}} x_{i,j,k} \times (\alpha - \gamma \times \mathcal{D}_{i,k}) \tag{11}$$

Finally, the PD-VDM merges all the soft constraints simultaneously. Our proposed PD-VDM considers maximization of vaccine distribution in priority groups and minimization of distance factored in transportation expenditure, collaboratively. Furthermore, α, β, and γ have been introduced as gain factors to equilibrate the combination of soft constraints and then presented as a multi-objective function in the Eq. 12. Hence, this model can optimize priority and distance concerns conjointly based on the adapted values of gain factors.

$$\mathcal{C}5 \wedge \mathcal{C}6 \wedge \mathcal{C}7 \iff \max_{\Omega} \sum_{h_i \in \mathcal{H}} \sum_{u_j \in \mathcal{U}_i} \sum_{e_k \in \mathcal{E}} x_{i,j,k} \times (\alpha + \beta \times p_k - \gamma \times \mathcal{D}_{i,k}) \quad (12)$$

The gain parameters of all the proposed models can be tuned to balance or incline towards more focused convergent vaccine distribution solutions. For instance, α, β, and γ are individually responsible for maximum distribution, maximization of priorities, and minimization of distance focused vaccine distribution solutions, respectively. Moreover, the policymakers can exploit these models and adjust gain parameters according to the region specifics and domain knowledge of vaccine distribution centers and population density. The regulation of these gain factors can aid to analyze and figure out the applicability of our various proposed models relying on different contextual targets, environment settings, and demand-supply gaps.

4 Case Study - Chennai, India

We demonstrate the proposed models using real-world data obtained from one of the popular cities in the southern part of India - Chennai. As listed by Rachaniotis et al. [16] most of the articles in literature make various assumptions to demonstrate the performance of their models, Similarly, we make a few reasonable assumptions to accommodate the lack of crucial data required to implement the model. Various input parameters of the models and their method of estimation or assumptions made to reach the decisions are described below.

4.1 Clustering Phase

- The entire city is divided into 15 zones for administrative purposes and we assume the distribution of vaccines is carried out based on these 15 zones as well.
- To determine the vaccine distribution centers, we assume that the vaccines will be distributed from hospitals or primary health centers. While there are approximately 800 hospitals in Chennai, based on 'on-the-ground' knowledge, we select 45 hospitals (3 hospitals per zone) to enable us to determine the distance between the hospitals. The selected hospitals are classified as private (PVT) and public funded (PUB) based on their administration. At least one publicly funded hospital or primary health center is chosen per zone.

– A 45×45 distance matrix is constructed with each row and column representing the hospitals. The cells are populated with the geographic distance obtained from Google Maps[3]. Using these distance measures, the optimal vaccine DCs are chosen by implementing k-medoids clustering algorithm described in Sect. 3.1.

4.2 Vaccine-Distribution Phase

– **Total population to be vaccinated ($\mathcal{E}$):** As per the census records collected in 2011, Chennai has a population of 4.6 million distributed across an area of $175 \, \text{km}^2$ with a population density of 26,553 persons/km^2. Based on the growth in the overall population of India, we estimate the current population of Chennai to be 5,128,728 with a population density of 29,307/km^2.
– **Set of Vacccine Distribution centers ($\mathcal{H}$):** The optimal vaccine DCs are chosen from using the clustering phase of the model using silhouette width such that the chosen DCs are evenly spread across the entire district.
– **Staff for vaccination per DC ($\mathcal{U}$):** Based on the capacity of the hospital in terms of facilities, workforce, etc. the hospitals are classified as 'SMALL', 'MED', and 'LARGE'. We assume that small, medium, and large hospitals allocate 5, 20, and 40 health-care workers for vaccination purposes respectively.
– **Priority levels ($\mathcal{P}$):** While the priority levels can be decided by the authorities based on wide variety of parameters, for our simulation we assume that the priority depends on the age of the individual such that the elderly people are vaccinated earlier. The distribution of population across various age groups is calculated based on the age-wise distribution calculated during the census 2011. Table 1 shows the distribution of the population across six priority groups.

Table 1. Distribution of the population of Chennai across 6 age groups

Age group	0–9	10–19	20–54	55–64	65–74	75+
% of distribution	14.02	15.34	56.38	7.87	4.09	2.31

– **Time per vaccination (t):** Based on the data provided during the recent trial dry run carried out in India, the government estimates to carry out 25 vaccinations in 2 h span, we approximate that the time taken for the vaccination of one person to be 5 min.
– **Time per vaccination ($\mathcal{N}$):** Keeping in mind the deficit in the supply of vaccines in the early stages of vaccination we assume that only vaccine doses for 50% of the total population is available currently. However, the number can be increased based on increase in production[4].

---

[3] https://www.google.com/maps/.
[4] https://www.cnn.com/2020/12/18/asia/india-coronavirus-vaccine-intl-hnk/index.html.

These described parameters can be tuned by the decision making authorities to accommodate the distribution at the area under consideration. For all the experimental settings, we consider the gain factors α, β, and γ as $\frac{1}{4} \times |\epsilon|$, $\frac{1}{4} \times \frac{1}{|\mathcal{P}|} \times |\epsilon|$, and 1 respectively. Yet, as mentioned earlier, these gain factors can be explored and set according to the solutions desired by policy makers and decision making authorities.

5   Results and Discussion

We demonstrate the proposed models in two scenarios with randomly generated data and two scenarios with real world data from Chennai. We discuss in detail the inferences from each scenario in this section. In each scenario we compare the four models and highlight how PD-VDM optimizes the two parameters taken into consideration - priority and distance.

5.1   Random-Simulation

For the two random scenarios we consider that there are 12 vaccine DCs in total and based on the silhouette width measures as shown in Fig. 1a we select three DCs optimally distributed across the area. We assume the total population size ($\mathcal{E}$) to be 200 and the total number of vaccines available ($\mathcal{N}$) to be 85. Of the three chosen DCs we assume that each has a capacity ($\mathcal{U}$) of 15, 30, 45. The model considers five different priority levels ($\mathcal{P}$) with 1 being the least and 5 being the most. The population is distributed among these priority levels as shown in Table 2. To demonstrate the impact of the population distribution parameter, we run the simulation model under two different distributions namely,

Table 2. Distribution of population across priority groups in random simulation.

Priority group	1	2	3	4	5
Raw count	43	35	50	45	27

- Random-case - 1 (RC-1): Uniform random distribution
- Random-case - 2 (RC-2): Poisson like distribution where the population is dense at some regions and spares at others.

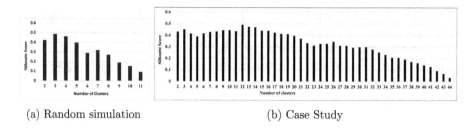

(a) Random simulation   (b) Case Study

Fig. 1. Selection of optimum number of DCs based on Silhouette width

The distribution of vaccines at any time instance t_n by all four models, for both random cases is depicted in Fig. 2. The percentage of individuals vaccinated in each priority group in random case 1 and random case 2 are depicted in Fig. 3a and b respectively. The B-VDM vaccinates 41.86%, 37.14%, 42%, 46.67%, and 44.44% of individuals across priorities 1 to 5 respectively, in both RC1 and RC2. We can see that 55.56% of the individuals from the highest priority group are left out. The D-VDM optimizes only the distance parameter and vaccinates 37.21%, 51.43%, 38%, 51.11% and 33.33% of individuals across priorities 1 to 5 respectively in RC1 and almost the same results in RC2. Again we can notice, that a greater percentage of the individuals in the highest priority group are not vaccinated. While both P-VDM and PD-VDM attempt to vaccinate 100% of the high priority individuals in both RC1 and RC2, on studying the average distance travelled by each individual of the population as depicted in Fig. 4a,

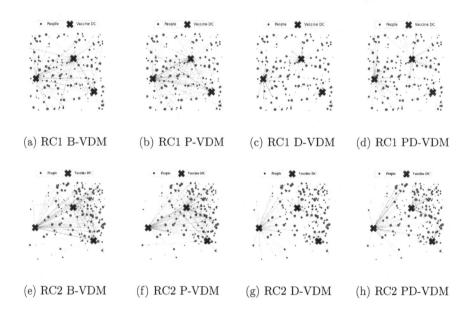

(a) RC1 B-VDM   (b) RC1 P-VDM   (c) RC1 D-VDM   (d) RC1 PD-VDM

(e) RC2 B-VDM   (f) RC2 P-VDM   (g) RC2 D-VDM   (h) RC2 PD-VDM

Fig. 2. Snapshot of the simulation of the models in the random case studies

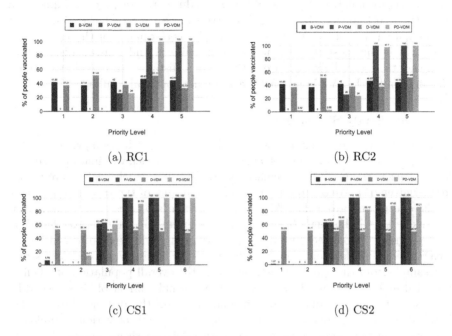

Fig. 3. Distribution of vaccine across various priority groups

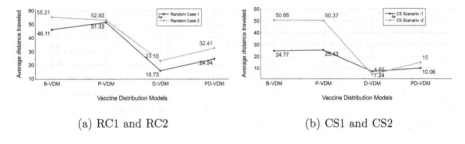

Fig. 4. Average distance travelled by the proposed models

Table 3. Distribution of population across six priority groups for Case study scenarios

Priority group	1	2	3	4	5	6
Case-study scenario 1	546	598	2199	307	160	90
Case-study scenario 2	2817	3082	11331	1583	823	464

we can clearly identify that the PD-VDM reduces the distance parameter by more than 40% in both RC1 and RC2. We can also see that the 'distribution of population' parameter does not impact the performance of the models and PD-VDM is the most efficient across both Uniform and Poisson distribution. Although there is a slight increase in the average distance travelled in the second distribution the PD-VDM still achieves the most optimal results.

5.2   Case-Study Simulation

In both the case-study scenarios we consider that the vaccination process continues for five hours each day and as mentioned in Sect. 4 each vaccination takes (t_n) 5 mins. Hence there are 60 vaccinations carried out by each health care worker for any given day. We compile and present the results of the vaccination process at the end of any given day. Although in the initial stages of the vaccination it is likely that individuals from one priority group will only be vaccinated, as time progresses people across various priority groups will need to be considered. Hence the population considered by the model for each day is taken as a stratified sample from the age-wise distribution of the overall population of Chennai provided in Table 1. Initially we identify the optimal number of DCs needed for effectively serving the district of Chennai based on the distance between the hospitals considered. On analyzing the silhouette width of various 'number of clusters' as depicted in Fig. 1b we present two different scenarios namely,

- Case study scenario - 1 (CS1): Three optimal DCs
- Case study scenario - 2 (CS2): Twelve optimal DCs

Case Study Scenario - 1: For this scenario, we consider 3 optimal vaccine DCs $(\mathcal{H})$ with capacities $(\mathcal{U})$ of 5, 20, 40 since the chosen DCs fall under 'SMALL', 'MED' and 'LARGE' respectively. Based on these factors the total population to be vaccinated $(\mathcal{E})$ in any given data is estimated to be 3,900. As mentioned in Sect. 4 we assume, that the total vaccines $(\mathcal{N})$ available to be half the total population which in this case sums upto 1,950. The entire population falls under six priority groups $(\mathcal{P})$ based on their age. The actual number of people in each priority group is provided in Table 3.

Case Study Scenario - 2: For this scenario, we consider 12 optimal vaccine DCs $(\mathcal{H})$ and among these 12 chosen DCs there are 3, 2 and 7 DCs with a staff capacity $(\mathcal{U})$ of 5 ('SMALL'), 20 ('MED'), and 40 ('LARGE') respectively. Hence the total population $(\mathcal{E})$ that can be vaccinated at any given 5 h day is 20100. Similar to scenario 1 we assume that the total vaccines $(\mathcal{N})$ available to be half the total population which amounts to 10,050 and that the population is distributed across six different priority groups $(\mathcal{P})$ based on their age as shown in Table 3.

The vaccination percentage across the priority groups for both case study scenarios are provided in Fig. 3c and d respectively. Unlike the random scenarios, it is interesting to note that in both CS1 and CS2, the B-VDM though not optimized to satisfy a specific hard constraint, it produces results that are almost identical to P-VDM vaccinating 100% of the highest priority group individuals. This can be attributed to the similarity in the effect of α and β gain parameters in both these models in CS1 and CS2. Though the PD-VDM model vaccinates less than 90% of the individuals in the three highest priority groups we can clearly see that it significantly reduces the 'average distance travelled' by an individual in the population by more than 70%, which makes it more efficient than all the other models. While D-VDM achieves the least 'average distance travelled' value it sacrifices vaccinating almost 50% of the high priority groups. Thus, we demonstrate how PD-VDM efficiently distributes the available vaccines by modifying various parameters like the distribution of population, total population, total number of available vaccines, the number of vaccine DCs and the capacity of each DCs. The model provides flexibility for the decision-making authorities of any given demographic to optimize the distribution of vaccine in the desired region. Moreover, the models developed using CSP are easier to design, adapt, and control compared to other existing combinatorial optimization techniques (e.g., ILP, MILP, and QP) that are often very computationally expensive in terms of memory and time with very rigid search space navigation restrictions.

6 Conclusion and Future Work

In this paper, we propose an optimization model (PD-VDM) based on Constraint Satisfaction Programming framework to find the most effective way to distribute vaccines in a given demographic region, in terms of distance and a priority (age, exposure, vulnerability, etc.). We compare the efficiency of the model with three other models which take into consideration either one or none of the two optimization constraints. While this model can be adapted across a wide variety of scenarios as demonstrated in our case studies, it is essential to understand that due to resource constraints we have demonstrated only two of the many available constraints. Expanding the scope of the model to allow optimizing a wide variety of parameters can be carried out in future research, along with an attempt to replace some of the assumptions made in our model with real world data.

References

1. Aaby, K., Herrmann, J.W., Jordan, C.S., Treadwell, M., Wood, K.: Montgomery county's public health service uses operations research to plan emergency mass dispensing and vaccination clinics. Interfaces **36**(6), 569–579 (2006)
2. Ali, I., Alharbi, O.M.: COVID-19: disease, management, treatment, and social impact. Sci. Total Environ. **728**, 138861 (2020)
3. Begen, M.A., Pun, H., Yan, X.: Supply and demand uncertainty reduction efforts and cost comparison. Int. J. Prod. Econ. **100**(180), 125–134 (2016)

4. Bordeaux, L., Hamadi, Y., Zhang, L.: Propositional satisfiability and constraint programming: a comparative survey. ACM Comput. Surv. (CSUR) **38**(4), 12-es (2006)
5. Brandeau, M.L.: Allocating resources to control infectious diseases. Operations Research and Health Care. ISORMS, vol. 70, pp. 443–464. Springer, Boston, MA (2005). https://doi.org/10.1007/1-4020-8066-2_17
6. Dong, E., Du, H., Gardner, L.: An interactive web-based dashboard to track COVID-19 in real time. Lancet. Infect. Dis **20**(5), 533–534 (2020)
7. Duijzer, L.E., van Jaarsveld, W., Dekker, R.: Literature review: the vaccine supply chain. Eur. J. Oper. Res. **268**(1), 174–192 (2018)
8. Halper, R., Raghavan, S.: The mobile facility routing problem. Transp. Sci. **45**(3), 413–434 (2011)
9. Huang, H.C., Singh, B., Morton, D., Johnson, G., Clements, B., Meyers, L.: Equalizing access to pandemic influenza vaccines through optimal allocation to public health distribution points. PLOS ONE **12**, e0182720 (2017). https://doi.org/10.1371/journal.pone.0182720
10. Kaur, S.P., Gupta, V.: COVID-19 vaccine: a comprehensive status report. Virus Res. **288**, 198114 (2020)
11. Lee, E.K., Pietz, F., Benecke, B., Mason, J., Burel, G.: Advancing public health and medical preparedness with operations research. Interfaces **43**(1), 79–98 (2013)
12. Mahase, E.: COVID-19: Moderna vaccine is nearly 95% effective, trial involving high risk and elderly people shows. Br. Med. J. (BMJ) **371**, 4471 (2020)
13. Nicola, M., et al.: The socio-economic implications of the coronavirus and COVID-19 pandemic: a review. Int. J. Surg. **78**, 185–193 (2020)
14. Park, H.S., Jun, C.H.: A simple and fast algorithm for K-medoids clustering. Exp. Syst. Appl. **36**(2), 3336–3341 (2009)
15. Polack, F.P., et al.: Safety and efficacy of the BNT162b2 mRNA COVID-19 vaccine. N. Engl. J. Med. **383**, 2603–2615 (2020)
16. Rachaniotis, N.P., Dasaklis, T.K., Pappis, C.P.: A deterministic resource scheduling model in epidemic control: a case study. Eur. J. Oper. Res. **216**(1), 225–231 (2012)
17. Ramirez-Nafarrate, A., Lyon, J.D., Fowler, J.W., Araz, O.M.: Point-of-dispensing location and capacity optimization via a decision support system. Prod. Oper. Manag. **24**(8), 1311–1328 (2015)
18. Rousseeuw, P.J.: Silhouettes: a graphical aid to the interpretation and validation of cluster analysis. J. Comput. Appl. Math. **20**, 53–65 (1987)
19. Xiong, J., et al.: Impact of COVID-19 pandemic on mental health in the general population: a systematic review. J. Affect. Disord. **277**, 55–64 (2020)

RNACache: Fast Mapping of RNA-Seq Reads to Transcriptomes Using MinHashing

Julian Cascitti$^{(\boxtimes)}$, Stefan Niebler, André Müller, and Bertil Schmidt

Johannes Gutenberg-Universität Mainz Institute of Computer Science,
55128 Mainz, Germany
jcascitt@students.uni-mainz.de,
{stnieble,muellan,bertil.schmidt}@uni-mainz.de

Abstract. The alignment of reads to a transcriptome is an important initial step in a variety of bioinformatics RNA-seq pipelines. As traditional alignment-based tools suffer from high runtimes, alternative, alignment-free methods have recently gained increasing importance. We present a novel approach to the detection of local similarities between transcriptomes and RNA-seq reads based on context-aware minhashing. We introduce RNACache, a three-step processing pipeline consisting of minhashing of k-mers, match-based (online) filtering, and coverage-based filtering in order to identify truly expressed transcript isoforms. Our performance evaluation shows that RNACache produces transcriptomic mappings of high accuracy that include significantly fewer erroneous matches compared to the state-of-the-art tools RapMap, Salmon, and Kallisto. Furthermore, it offers scalable and highly competitive runtime performance at low memory consumption on common multi-core workstations. RNACache is publicly available at: https://github.com/jcasc/rnacache.

Keywords: Bioinformatics · Next-generation sequencing · RNA-seq · Transcriptomics · Read mapping · Hashing · Parallelism · Big Data

1 Introduction

Obtaining data from the sequencing of RNA (RNA-seq) is a major advancement in medical and biological sciences that allows for the measurement of gene expression. It has thus become an important technique for gaining knowledge in a wide range of applications including drug development and understanding of diseases [23]. The increasing availability and large size of next generation sequencing (NGS) data has also established the need for highly optimized big data methods to align (or map) the produced reads to reference sequences [20]. Since it is estimated that over hundreds of millions of human genomes and transcriptomes will be sequenced by 2025, finding fast RNA-seq mapping algorithms is of high importance to research [22].

© Springer Nature Switzerland AG 2021
M. Paszynski et al. (Eds.): ICCS 2021, LNCS 12742, pp. 367–381, 2021.
https://doi.org/10.1007/978-3-030-77961-0_31

Classical *read mapping* aims to identify the best alignment(s) of each read to a given reference genome. In contrast, RNA-seq often requires the mapping of each produced read to a reference transcriptome, consisting of a collection of genes, where each gene is represented by several alternative transcripts (known as *isoforms*). Due to alternative splicing, many isoforms of the same gene can be transcribed, which often contain highly similar subsequences. Thus, unlike mapping of genomic NGS data, where redundancies are usually less common, highly similar regions between isoforms and homologs have to be taken into account when mapping RNA-seq data. Because the number of potential origins tends to be high, the process of determining the true origin of each sequencing read becomes complex. This, in turn, can result in inaccurate expression estimation of transcripts if not accounted for [6].

From an algorithmic perspective, established short-read aligners such as BWA-MEM [11] and Bowtie2 [9] are based on *seed-and-extend* approaches which map sequencing reads to a reference genome by first identifying seeds using FM-index based data structures. Seeds are extended (e.g., by using fast versions of dynamic programming based alignment algorithms) in order to verify whether a seed is actually contained in a full alignment. Unfortunately, the computation of traditional alignments for a large number of reads typically exhibits long run times. The problem is further exacerbated by the multiplicity of similar potential origins in transcriptome data, which can entail high numbers of seeds.

Consequently, a number of methods have recently been introduced for fast mapping of RNA-seq reads to reference transcriptomes. These approaches are based on the concepts of *lightweight-alignment* or *quasi-mapping* which rely on the observation that the task of identifying the most likely isoform(s) each read may originate from does not require precise and expensive alignment computation. RapMap [21] works by locating consecutive k-mers within a suffix array of concatenated target transcripts and finding the longest possible substring shared between read and target. Sailfish [18] and Selective Alignment (RapMapSA) [19] both extend RapMap by adding a number of criteria and filters. Kallisto [2] generates pseudoalignments by determining the set of transcripts containing all of the reads k-mers using a de-Bruijn-graph. While these approaches clearly outperform traditional aligners such as STAR [5] in terms of runtime, the number of reported potential read origins (called *hits-per-read*) can be high. Furthermore, scalability (with respect to taking advantage of an increasing number of cores of modern CPUs) is often limited while memory consumption can be high.

In this paper, we present RNACache – a new algorithm for fast and memory-efficient mapping of RNA-seq reads to a given reference transcriptome. Our method utilizes *minhashing*, a technique [3] known from processing big data to map reads based on an approximate Jaccard-Index by analyzing k-mer subsamples [10]. Subsequently, a selection of task-specific filters is applied utilizing statistical information gathered during the mapping of all reads in order to significantly reduce the number of reported hits-per-read compared to previous approaches while still maintaining high recall. Furthermore, we take advantage of multi-threading to design a high-speed yet memory-efficient implementation that can exploit large number of cores available in modern workstations. Our performance evaluation shows that RNACache produces a lower number of hits-per-read than RapMap, RapMapSA, Salmon, and Kallisto while achieving higher recall at the same time. In addition, RNACache is able to outperform all other tested mapping-based tools in terms of runtime and memory consumption.

2   Method

We adopt minhashing – a locality sensitive hashing (LSH) based data subsampling technique. It has been successfully employed by search engines to detect near duplicate web pages [4] but has recently gained popularity in NGS analysis with example applications including genome assembly [1], sequence clustering [16], metagenomics [13], food sequencing [8], and single-cell sequencing [15]. To our knowledge, we are the first to apply this concept in the context of RNA-seq mapping.

We apply minhashing for memory-efficient transcriptome database (hashmap) construction and querying. Query outputs are used for initial read assignments. These are further refined by two filters (online and coverage filter) in order to derive final read classifications. The workflow of RNACache is illustrated in Fig. 1.

2.1   Database Construction

In order to construct an index, reference transcriptome sequences (also called *transcripts* or *targets*) are covered in slightly overlapping (default: $k - 1$ nucleotides) *windows*. From each window a preset maximum number of k-mers (substrings of length k) possessing the lowest hash values (referred to as *features*) are selected to form what is called a *sketch*. A hash-map is constructed that maps every *feature* to its corresponding locations, i.e., a list of transcript and window identifiers in which it occurs. Strandedness is handled by substituting every k-mer with the lexicographical minimum of itself and its reverse-complement, referred to as its *canonical* k-mer, prior to hashing.

2.2   Querying and Initial Read Assignment

When querying a read, it is *sketched* similarly to a reference sequence and a lookup of each of its features is performed on the index yielding a list of reference transcript locations which share features with the read, also referred to as *hits* or *matches*. In this way, we determine which windows of which targets share features, respectively k-mers, with the queried read. The number of features shared by the read and a single target window can be interpreted as an approximation of the Jaccard-Index of their respective k-mer sets, i.e., as a measure of their similarity and hence of a local similarity between corresponding target transcriptome sequence and read.

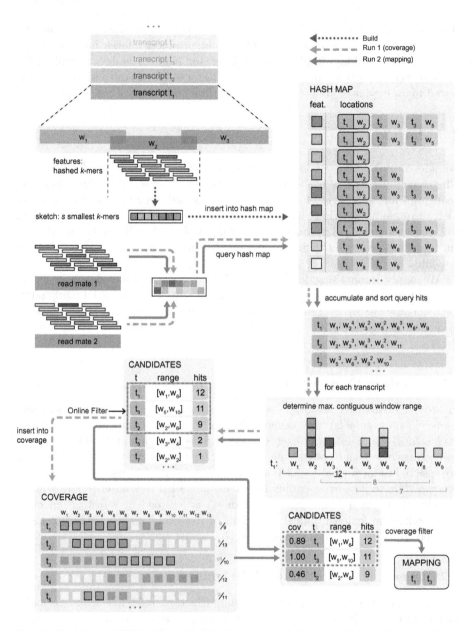

Fig. 1. Workflow of RNA-Cache. Build: Transcripts (t_j) are partitioned into windows (w_i). The s smallest features (hashes of canonical k-mers) of each window are computed and inserted into the hash map (database). Run 1 and Run 2: The hash map is queried with the s smallest features of each read. The returned hits are accumulated per transcript and candidates identified by determining the contiguous window range spanning the most hits on each transcript. After online-filtering, candidates are inserted into the coverage list, which is used to filter candidates in Run 2 (mapping) in order to determine the final mapping.

In order to determine larger regions of similarity, a *candidate* region is subsequently selected from each reference transcriptome sequence by determining the *contiguous-window-range* of a maximum preset length spanning the most feature hits. This maximum window range corresponds to the maximum expected *insert size*, i.e. fragment length for paired-end reads. Thus, *candidates* can be described as tuples of the form $(tgt, (b, e), hits)$, where tgt is an ID of the target sequence, (b, e) is a tuple of window IDs, spanning the closed interval $[b, e]$ of windows on the target sequence, and $hits$ is the number of features shared by the target window interval and the read.

Note that, features which occur in a very high number of locations (default: more than 254) are considered uninformative and are therefore deleted from the index. As these features correspond to k-mers appearing in many different reference sequences, they offer little discriminatory information regarding the origin of a queried read.

After querying, RNACache applies an online filter (see Sect. 2.3) and a subsequent post-processing (coverage) filter (see Sect. 2.4) to select the most plausible transcript(s) among them, creating the final mapping of the read set. The mapping generated in this way seeks to minimize the total number of candidates per read while maximizing the ratio of reads to which the correct original transcript is assigned (*recall*).

2.3 Online Filtering

Candidate targets which a considered read might originate from are refined by an online filtering process based on absolute and relative hit counts. Consider a read r's set of candidates $C(r) = \left\{ c_1 = (tgt_1, (b_1, e_1), hits_1), \ c_2, \ldots \right\}$ (see Sect. 2.2). Online filtering will permit the candidate set

$$F_o(C(r)) = \left\{ \left(tgt, (b, e), hits \right) \in C(r) \ \middle| \ hits > t^{min}, \ \frac{hits}{hits^{max}} > t^{cutoff} \right\} \quad (1)$$

where $hits^{max} = \max_{c_k \in C(r)} hits_k$.

Informally, this filter permits only those candidates whose number of hits exceeds t^{min} in absolute terms and is greater than a fraction of t^{cutoff} relative to the highest number of hits of any candidate of the same read. The second relative filter has the advantage of innately scaling with other parameters and properties of the input and reference data which affect the number of hits produced by well-fitting reads.

2.4 Post-processing (Coverage) Filter

After online filtering, some reads can still map equally well to different targets that share highly similar subsequences. However, it is unlikely that all of these potential transcripts of origin are actually expressed in a transcriptomic sample. Thus, a post processing filter is applied in order to distinguish between expressed and non-expressed isoforms containing highly similar subsequence regions. For this filter it is necessary to consider the complete set of reads as a whole, rather than filter candidates solely on a per-read basis (as done by the online filter).

RNA-seq data sets typically have a large coverage [14]. Thus, we can assume with high confidence that every window of every transcript actually expressed in the sample will appear as a match of at least some reads during the querying process. By keeping track of all target windows *covered* in this way, we can identify target regions with no matching reads, and discard the target as unlikely to be expressed in the sample. Thus, we can use the information gained by reads mapping to a sequence's unique regions, or alternatively the lack thereof, to reason about the correct mapping of reads to non-unique target regions. For this purpose, we use the concept of *coverage* as follows.

Coverage. We define the *coverage* of a target as the ratio of its number of *covered* windows to its number of total windows. A window is said to be *covered* if the following two conditions hold:

1. It is an element of the *contiguous window range* of a *candidate* of at least one read.
2. It shares at least one feature with that read.

Formally, we define $\mathfrak{R}$ as the set of all input reads and $W(tgt)$ as the set of windows of target tgt. Then, the *covered window set* of tgt is defined as:

$$covered(tgt) = \left\{ w_{tgt,j} \in W(tgt) \,\Big|\, \exists r \in \mathfrak{R} : \exists (tgt, (b, e), hits) \in F_o(C(r)) : j \in [b, e] \right.$$

$$\left. \wedge \underbrace{S(w_{tgt,j}) \cap S(r) \neq \emptyset}_{Condition\ 2} \right\}, \tag{2}$$

where $w_{tgt,j} \in W(tgt)$ is the j-th window of target tgt while $S(w_{tgt,j})$ and $S(r)$ denote the *sketch* of $w_{tgt,j}$ and r respectively.

Furthermore, we define the *coverage statistic* of a target tgt as:

$$cov(tgt) = \frac{|covered(tgt)|}{|W(tgt)|}. \tag{3}$$

In the following, the term *coverage* refers to either a target's *covered window set* or *coverage statistic*, unless the distinction is relevant.

Condition 2 ensures that only those windows are counted towards a target's coverage which contribute to the candidate's match count, i.e., share features with the read. This is primarily relevant in the case of paired-end reads where it reduces false positives in the coverage caused by the inclusion of regions that differ from the sequenced fragment in the region located between the read mates. Condition 2 can also be omitted by using a runtime parameter to handle weakly expressed input data sets or transcriptomes that contain transcripts that are too short for their center regions to appear in paired-end reads.

Coverage-Based Filtering. The post-processing coverage filter admits those candidates whose target's coverage statistic exceeds a fraction t^{cov} of the highest coverage statistic of any candidate of the same read.

Formally:

$$F_{cov}(F_o(C(r))) = \left\{ c = (tgt, (b, e), hits) \mid c \in F_o(C(r)), \right.$$
$$\left. \frac{cov(tgt)}{\max\limits_{c_k \in F_o(C(r))} cov(tgt_k)} \geq t^{cov} \right\} \qquad (4)$$

Per-read maximum-normalization allows the filter to automatically scale with properties of the read and reference transcriptome as well as other parameters that affect overall coverage of the dataset and has empirically proven to increase mapping accuracy compared to a simple coverage threshold. Additionally, it has the property of never rejecting the most covered candidate of any read, meaning that it cannot by itself cause a read to remain entirely unmapped. However, the user can disable this normalization in favor of a simple coverage threshold, if desired.

Coverage Data Structure. In order to keep track of coverage, an additional data structure in form of a hash map is constructed while reads are queried, in which targets are mapped to a list of their covered windows, based on the aforementioned definition and the *candidates* resulting from online filtering. Subsequent application of the post-processing filter would require all candidates of each read to be stored. For typical RNA-seq read set sizes, this can become infeasible; e.g. the required memory for a medium-sized set of $\approx 50M$ reads and ≈ 5 candidates per read can already be expected to be on the order of 2.5 GB. Different parameters regarding online filtering and very large read sets can quickly increase this requirement by at least an order-of-magnitude. Thus, the default behavior of RNACache is not to store candidates. Instead, the sketching and querying process, including online-filtering, is repeated after the coverage data structure has been completely assembled.

2.5   Parallelization

Input sequences are processed by multiple CPU threads in parallel that communicate work items with the help of a concurrent queue.

The database build phase uses three threads. One thread reads transcripts from the input files, one sketches sequences into hash signatures and one inserts sketches into the hash table. Using more than three threads does not improve performance since our hash table does not allow for concurrent insertion.

The query phase uses one producer thread for reading the input RNA-seq reads from file and multiple consumer threads to accelerate the time-consuming classification of reads. The producer thread places batches of input reads (default is 4096 per batch) in the queue from which consumer threads then extract them for processing. The individual results of each consumer thread are successively merged into global data structures and/or written to a results file.

3   Evaluation

We assessed the performance of RNACache in terms of mapping accuracy and speed using simulated and real RNA-seq data, and compared it to a number of existing state-of-the-art tools performing transcriptomic mappings without full alignments:

- RapMap v.0.6.0 [21],
- RapMapSA v.0.6.0 [19],
- Kallisto v.0.46.2 [2],
- Salmon v.1.2.1 [17],

as well as to Bowtie2 v.2.4.1 [9], a commonly used full alignment tool.

Simulated datasets were generated using Flux simulator (v.1.2.1) [7] by applying appropriate parameters for human transcriptome RNA-seq experiments. The utilized reference transcriptome consists of 100,566 transcripts of protein-coding genes, taken from GENCODE Human Release 34 (GRCh38.p13). Reads were generated from this transcriptome using corresponding annotations from the same release.

Our tests were conducted on a workstation with a dual 22-core Intel(R) Xeon(R) Gold 6238 CPU (i.e. 88 logical cores total), 187 GiB DDR4 RAM (NUMA), two PC601 NVMe SK Hynix 1TB in RAID0 running Ubuntu 18.04.4.

With the exception of Salmon, whose binary release was used directly, all tools were built using g++ v.7.5.0. As all considered tools other than RNACache and Kallisto use SAM format output, Samtools [12] was utilized to convert their output into BAM format on-the-fly.

3.1   Accuracy Evaluation Using Simulated Reads

To assess accuracy, datasets of ≈ 48 million paired-end reads of length 2×76 base-pairs were generated. Mapping quality is assessed in terms of three measures:

- *Recall*, defined as the fraction of total input reads whose mapping result includes the true origin,
- *Hits-per-read* or *HPR*, defined as the mean number of returned candidates of a mapped read, i.e. disregarding unmapped reads,
- *True-hit-rate* or *THR*, which is the ratio of the number of true origins found and the total number of candidates assigned.

This form of evaluation is based on the evaluation of RapMap [21], in which a read is considered a "True Positive" if its mapping includes its true origin, regardless of the number of other candidates. Furthermore, *HPR* and *THR* can be considered a conceptual counterpart to the *precision* quality in binary classification tasks.

Ground truth mappings were extracted from read headers generated by the simulator. Bowtie2 has been included to allow for a comparison to an established full alignment-based tool. Note that it uses a "k" parameter specifying the maximum number of distinct, valid alignments it will search for. As this setting has a direct effect on the HPR statistic, Bowtie2 was executed with three different settings: $k = 1$ (def.), $k = 20$, $k = 200$.

Table 1 shows the results whereby the experiments were repeated five times with different read data and the reported values are averages. In addition to recall, HPR, and THR, the percentage of reads that have been mapped to at least one transcript is provided in the first row ("aligned").

Table 1. Mapping results for simulated reads (bold denotes best value).

	RNACache	RapMap	RapMapSA	Salmon	Kallisto	Bowtie2		
						$k = 1$	$k = 20$	$k = 200$
Aligned (%)	**98.2**	97.9	83.7	86.5	91.8	96.4	96.9	97.5
Recall (%)	**96.4**	96.0	83.4	86.1	89.4	41.2	95.0	96.0
HPR	3.9	5.0	5.0	5.0	4.6	**1.0**	4.8	5.8
THR	25.2	19.7	20.0	19.9	21.3	**42.7**	20.5	17.1

RNACache achieves the highest percentage of aligned reads and highest recall, while it is second best for HPR and THR. It is able to assign the correct original transcript to 96.4% of all reads and produces fewer than 4 mappings per read on average. While RapMap comes close in terms of recall, it produces more than 25% more false transcripts assignments ($HPR = 5.0$). RapMapSA and Salmon fail to map large portions of reads. Kallisto has similarly low recall. Bowtie2 with $k = 1$ achieves the best HPR and THR at the expense of the worst recall (only $\approx 40\%$). The results of Bowtie2 with medium setting ($k = 20$) are in the vicinity of the other tools while being balanced more towards lower HPR/recall. Even Bowtie2 with highest k-value ($k = 200$) is not capable of beating RNACache in terms of recall, while accumulating the largest amount of false matches (HPR ≈ 5.8).

Impact of Filtering. In order to analyze the effectiveness of our introduced filters, we evaluated their impact to the RNACache processing pipeline in terms of the percentage of aligned reads, recall, HPR, and THR. The results in Table 2 show that the application of the online filter refines the initial read-to-reference mappings by improving both HPR and THR by over an order-of-magnitude while only slightly decreasing recall. The additional coverage filter can improve HPR and THR even further at only a minor decrease of recall.

3.2   Runtime and Parallel Scalability

In addition to accuracy, runtimes were measured in a standard usage scenario. All tools were executed with 88 threads in total on our test system. Query runtimes

Table 2. Impact of filtering steps on the mapping results of RNACache.

	No filter	Online	Online + coverage
Aligned (%)	99.0	98.2	98.2
Recall (%)	97.6	96.9	96.4
HPR	60.2	5.3	3.9
THR	1.6	18.7	25.2

are shown in Fig. 2. Breakdown of absolute runtimes in build (index) and query phase (including disk I/O) as well as peak memory consumption are provided in Table 3. With the exceptions of Kallisto and RNACache, outputs were piped into Samtools for live SAM-to-BAM conversion, for which 8 threads (4 in the case of Bowtie2) were deducted and assigned to Samtools [12]. This represents a common toolchain in bioinformatics when working with the involved tools, as handling large SAM output files is impractical.

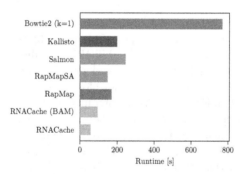

Fig. 2. Query runtimes on synthetic data including SAM-to-BAM conversion and hard-disk I/O.

RNACache outperforms all other tools in terms of runtime, in both its default and BAM output modes. The minhash-based hash map construction (build phase) of RNACache is faster than the corresponding index data structure construction of the other methods; i.e. it outperforms suffix array construction used by RapMap, RapMapSA, and Salmon by at least 1.9×, deBruijn Graph construction used by Kallisto by 2.1×, and FM-index construction used by Bowtie2 by 16.9×. In its default output mode, the query phase of RNACache achieves speedups of 2.9, 2.5, 4.2, 3.4, and 13.2 compared to RapMap, RapMapSA, Salmon, Kallisto, and Bowtie2 (best case, $k = 1$), respectively. Furthermore, RNACache exhibits a low memory footprint.

An advantage of RNACache is that it provides its own significantly more compact output format in which only the identifiers of the query read and of its

Table 3. Build (index creation)/Query runtimes and peak memory (RAM) consumption in a common SAM/BAM conversion chain, including conversion and disk I/O (best values in bold).

	RNACache		RapMap		Salmon	Kallisto	Bowtie2		
	def.	BAM	def.	SA			$k = 1$	$k = 20$	$k = 200$
build [s]	**54.8**	**54.8**	102.5	102.5	117.9	188.2	931.0	931.0	931.0
query [s]	**58.4**	96.1	170.5	148.3	244.7	199.6	769.2	1968.9	5976.2
memory [GB]	2.6	8.5	107.0	107.0	18.3	16.3	**1.5**	3.7	21.4

mapped targets, as well as the number of shared minhashing features, are stored. Optionally, the mapping can be output in the BAM format, which is more suited for alignment-based methods. While this format offers compatibility with many downstream tools expecting traditional alignments, none of the tested alignment-free methods actually produce sufficient information to utilize all of the format's columns to their full extent in the way alignment-based methods can.

To evaluate parallel thread scalability of the compared mapping algorithms, we benchmarked all tools using various thread counts with minimal disk I/O and without subsequent format conversions. Note that for Bowtie2 thread counts lower than 8 were not used for the setting of $k = 200$, as corresponding runtimes would be exceedingly long. Disk I/O was minimized by ensuring that all data was preloaded into page cache beforehand and output was suppressed to the extent possible. The benchmark results are shown in Fig. 3. Data sets of ≈ 20 mio. reads were used, benchmarks were repeated three times and results averaged.

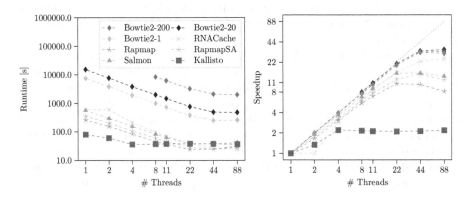

Fig. 3. Query runtimes and speedups of all tested tools for varying numbers of CPU threads using synthetic data. (Suppressed I/O, no format conversion.)

Bowtie2 exhibits the highest parallel efficiency (measured in terms of speedup divided by the number of utilized threads) for large numbers of threads but still has the longest overall runtime. The mapping-based methods (RNACache, RapMap, RapMapSA, Salmon, Kallisto) generally offer superior runtime performance compared to Bowtie2 but exhibit lower parallel efficiency. RNACache outperforms all other mapping-based tools when executed with high thread counts and exhibits superior parallel scalability when more than 22 threads are used.

The noticeable drop in parallel efficiency for more than 22 threads is most likely related to the 2-socket NUMA architecture of our benchmarking system. Additionally, we can observe that hyper-threading (which is used when going from 44 to 88 threads) is not always beneficial when physical cores are exhausted, increasing runtimes due to higher scheduling and inter-thread communication overhead in some cases.

Our results suggest that a minhashing-based method can offer better scalability and lower memory consumption than other state-of-the-art mapping algorithms, and thus might lend itself to an effective port to many-core architectures such as CUDA-enabled GPUs.

3.3 Evaluation Using Experimental Data

As ground truth data is usually not available for experimental data, our evaluation on real RNA-seq data was performed by assessing the concordance of RNACache to other tools. We have used the NCBI GEO accession SRR1293902 sample consisting of 26M 75bp paired-end reads (Illumina HiSeq) and the same reference transcriptome as in Sect. 3.1. For each read r_i, we have compared the set of transcripts $M_{RNACache}(r_i) = \{t_{j_1}, t_{j_2}, ...\}$ assigned to it by RNACache, to the sets $M_{kallisto}(r_i), ..., M_{RapMap}(r_i)$ returned by other tools by accumulating the occurrences of various set relations $(\subset, \supset, =, ...)$ per tool. The results are shown in Fig. 4.

We can observe that RNACache shares over half of all mappings with each tool, with the exception of Bowtie2's default setting, which limits the size of paired alignments to one. In comparison with most tools, RNACache assigns an equal or subset mapping to a large majority of reads ($\approx 80\%$), in line with the already observed hits-per-read values, while leaving fewer reads unmapped. Kallisto constitutes an exception among mapping-based tools in that its stricter rejection of sub-optimal alignments leads to RNACache assigning a larger number of reads to superset mappings. Furthermore, only a vanishingly small number of reads left unmapped by RNACache were mapped by any other tool.

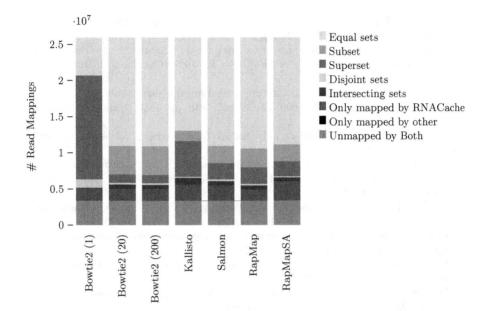

Fig. 4. Number of reads whose mapping by RNACache is either: equal ☐ , a subset ▦, a superset ▪ , a disjoint set ☐, an intersecting set ◾ in comparison to the other tool's mapping, a non-empty set where the other tool's mapping is empty ▪ , or an empty set whereas the other tools' mapping is non-empty ■, or also empty ▨. (Color figure online)

4 Conclusion

We have presented RNACache, a minhashing-based method for determining local sequence similarities by analyzing intersections of local pseudo-random k-mer subsets and applied it to the important task of transcriptomic mapping of RNA-seq data. Using a multi-step filtering process utilizing read-global transcript coverage information, our approach is able to refine initial read-to-target mappings in order to produce transcriptomic mappings of high accuracy. Our performance evaluation shows that RNACache is able to outperform the state-of-the-art mapping tools RapMap, RapMapSA, Salmon, and Kallisto in terms of recall, hits-per-read, and true-hit-rate while having the lowest memory footprint. Furthermore, it offers the fastest runtimes among the tested tools on common multi-core workstations. RNACache and the scripts and parameters used in its evaluation are publicly available at https://github.com/jcasc/rnacache.

References

1. Berlin, K., Koren, S., Chin, C.S., et al.: Assembling large genomes with single-molecule sequencing and locality-sensitive hashing. Nat. Biotech. **33**, 623–630 (2015)

2. Bray, N.L., Pimentel, H., Melsted, P., Pachter, L.: Near-optimal probabilistic RNA-seq quantification. Nat. Biotech. **34**(5), 525–527 (2016)
3. Broder, A.Z.: On the resemblance and containment of documents. In: Proceedings. Compression and Complexity of SEQUENCES 1997 (Cat. No.97TB100171), pp. 21–29 (1997)
4. Broder, A.Z.: Identifying and filtering near-duplicate documents. In: Giancarlo, R., Sankoff, D. (eds.) CPM 2000. LNCS, vol. 1848, pp. 1–10. Springer, Heidelberg (2000)
5. Dobin, A., et al.: Star: ultrafast universal RNA-seq aligner. Bioinformatics **29**(1), 15–21 (2013)
6. Garber, M., Grabherr, M.G., Guttman, M., Trapnell, C.: Computational methods for transcriptome annotation and quantification using RNA-seq. Nat. Methods **8**(6), 469–477 (2011)
7. Griebel, T., et al.: Modelling and simulating generic RNA-seq experiments with the flux simulator. Nucleic Acids Res. **40**(20), 10073–10083 (2012)
8. Kobus, R., et al.: A big data approach to metagenomics for all-food-sequencing. BMC Bioinformatics **21**(1), 1–15 (2020)
9. Langmead, B., Salzberg, S.L.: Fast gapped-read alignment with bowtie 2. Nat. Methods **9**(4), 357–359 (2012)
10. Leskovec, J., Rajaraman, A., Ullman, J.D.: Mining of Massive Data Sets. Cambridge University Press, Cambridge (2020)
11. Li, H.: Aligning sequence reads, clone sequences and assembly contigs with BWA-MEM (2013)
12. Li, H., et al.: The sequence alignment/map format and SAMtools. Bioinformatics **25**(16), 2078–2079 (2009)
13. Müller, A., Hundt, C., Hildebrandt, A., et al.: Metacache: context-aware classification of metagenomic reads using minhashing. Bioinformatics **33**(23), 3740–3748 (2017)
14. Nellore, A., et al.: Rail-RNA: scalable analysis of RNA-seq splicing and coverage. Bioinformatics **33**(24), 4033–4040 (2017)
15. Niebler, S., Müller, A., Hankeln, T., Schmidt, B.: Raindrop: rapid activation matrix computation for droplet-based single-cell RNA-seq reads. BMC Bioinformatics **21**(1), 1–14 (2020)
16. Ondov, B.D., Treangen, T.J., Melsted, P., et al.: Mash: fast genome and metagenome distance estimation using minhash. Genome Biol. **17**, 132 (2016)
17. Patro, R., Duggal, G., Love, M.I., Irizarry, R.A., Kingsford, C.: Salmon provides fast and bias-aware quantification of transcript expression. Nat. Methods **14**(4), 417–419 (2017)
18. Patro, R., Mount, S.M., Kingsford, C.: Sailfish enables alignment-free isoform quantification from RNA-seq reads using lightweight algorithms. Nat. Biotechnol. **32**(5), 462–464 (2014)
19. Sarkar, H., Zakeri, M., Malik, L., Patro, R.: Towards selective-alignment: bridging the accuracy gap between alignment-based and alignment-free transcript quantification. In: Proceedings of the 2018 ACM International Conference on Bioinformatics, Computational Biology, and Health Informatics, pp. 27–36. BCB 2018. ACM (2018)
20. Schmidt, B., Hildebrandt, A.: Next-generation sequencing: big data meets high performance computing. Drug Discovery Today **22**(4), 712–717 (2017)
21. Srivastava, A., Sarkar, H., Gupta, N., Patro, R.: RapMap: a rapid, sensitive and accurate tool for mapping RNA-seq reads to transcriptomes. Bioinformatics **32**(12), i192–i200 (2016)

22. Stephens, Z.D., et al.: Big data: astronomical or genomical? PLoS Biol. **13**(7), e1002195 (2015)
23. Wang, Z., Gerstein, M., Snyder, M.: RNA-seq: a revolutionary tool for transcriptomics. Nat. Rev. Genet. **10**(1), 57–63 (2009)

Digital Image Reduction for Analysis of Topological Changes in Pore Space During Chemical Dissolution

Dmitriy Prokhorov$^{(\boxtimes)}$ [ID], Vadim Lisitsa [ID], and Yaroslav Bazaikin [ID]

Sobolev Institute of Mathematics SB RAS, 4 Koptug Avenue,
Novosibirsk 630090, Russia

Abstract. The paper presents an original algorithm for reducing three-dimensional digital images to improve persistence diagrams computing performance. These diagrams represent topology changes in digital rocks pore space. The algorithm has linear complexity because removing the voxel is based on the structure of its neighborhood. We illustrate that the algorithm's efficiency depends heavily on the pore space's complexity and the size of the filtration steps.

Keywords: Persistence homology · Digital image reduction · Chemical dissolution

1 Introduction

Green energy and environmental geotechnologies such as CO_2 sequestration [15], and geothermal exploration [3,12,17] rises new challenges in reservoir studies. In particular, reactive fluid transport and changes in the pore space geometry and topology due to chemical fluid-solid interaction become the dominant factor of the macroscopic properties (elastic stiffness, electric conductivity, seismic velocities, hydraulic permeability) changes of the aquifers [2,13,14,25]. However, theoretical investigations and numerical simulations of the reactive transport are based on the reservoir-scale models of poromechanics, transport in porous media and coupled modems [4,22,26,29]. These models use empirical relations between porosity, pore space geometry and topology, permeability, tortuosity, elastic stiffness, and others [26].

In recent years, a number of papers were published on experimental [2,15], and numerical study [11,23,27] of carbonates dissolution due to the reactive fluid injection. These experiments show that changes in the pore space geometry due to the rock matrix's chemical dissolution strongly depend on the reaction rate, flow rate, and mineral heterogeneity, resulting in the various scenarios of

The research is supported by the Russian Science Foundation grant no. 21-71-20003. Numerical simulations were performed using "Polytechnic RSC Tornado" (SPBSTU, Russia).

M. Paszynski et al. (Eds.): ICCS 2021, LNCS 12742, pp. 382–393, 2021.
https://doi.org/10.1007/978-3-030-77961-0_32

macroscopic properties changes [2,16]. Recently we presented the research on estimation of the pore space topology changes in 2D case, where the topology is characterized by three Betti numbers representing the number of the connected components of pore space (isolated pores), number of the connected components of the matrix, and the Euler number [16]. We also showed that the evolution of the pore space topology is related to the changes in the physical properties of rock samples. Thus, it can be used to measure pore space changes. In the 3D case, the Betti 1 number, representing the number of channels in the pore space, has the main effect on the rocks' transport properties. However, calculation of the cycles and their evolution requires numerically intense algorithms.

The dynamics of rock matrix dissolution can be expressed as a set of sequential digital images of rock. The sequence corresponds to discrete time, and each digital image represents a spatial sampling of the rock, for example, a tomographic image. In computational topology, such sequence (if it is monotonous) is called filtration, and the natural thought is to count topological filtration invariants called persistent Betty numbers. One of the advantages of persistent Betty numbers is that they evaluate the filtration's topological complexity (i.e., the number of relative homological cycles taken relative to the filtration level). Another property that is very important for applications is their stability with respect to filtration perturbation. It means that a small error in the data leads to a small error in the persistent diagram (the persistent diagram contains all the necessary information about persistent Betty numbers).

There are 0, 1, and 2-dimensional non-zero Betti numbers in three-dimensional space. Moreover, the calculation of 0 and 2-dimensional Betti numbers (number of isolated pores and number of isolated matrix components) are based on disjoint-set-union data structure [8], and the duality of digital spaces does not require high computational resources. A completely different situation is with one-dimensional ones; they can be calculated by the Edelsbrunner-Lettsher-Zomorodyan algorithm, which has cubic complexity from the image size [8].

The paper's main idea is to use the image reduction algorithm compatible with the Edelsbrunner-Letscher-Zomorodyan algorithm for calculating persistent diagrams. We carried out a comparative test on rock samples obtained by statistical modeling methods, in particular by truncated Gaussian field method [10], and on real dissolved samples [1,2]. It is shown that the reduction algorithm makes it possible to accelerate the calculation of persistent Betty numbers. However, the acceleration depends on the porosity, correlation length of the samples, and the discrete time step's size.

2   Digital Images

Let us define a regular spatial grid with points $p_I = (x_{i_1+\frac{1}{2}}, y_{i_2+\frac{1}{2}}, z_{i_3+\frac{1}{2}})$, where $I \in N^3$, and $x_{i_1+\frac{1}{2}} = h_x(i_1 + \frac{1}{2})$, $y_{i_2+\frac{1}{2}} = h_y(i_2 + \frac{1}{2})$, $z_{i_3+\frac{1}{2}} = h_z(i_3 + \frac{1}{2})$, with h_x, h_y, and h_z are the grid steps. Note, that we are interested in the topology of the digital images, thus, we may state $h_x = h_y = h_z = 1$. Now, we can introduce the grid cell or voxel as $C_I = \{(x,y,z) \in R^3 | i_1 \le x \le i_1+1, i_2 \le y \le i_2+1, i_3 \le$

$z \leq i_3 + 1\}$. Using these notations the segmented digital image can be defined as a piece-wise constant function $F(x, y, z)$ mapping rectangular spatial domain $D \subseteq R^3$ to a finite subset of integer numbers $A = \{0, 1, ..., M\}$. A natural choice for the digital rock physics applications is a binary image; i.e., $A = \{0, 1\}$, where 0 corresponds to the *background* and 1 represents the *foreground*. In particular, if fluid flow [5,6] or electric current in porous space is studied, 1 represents pore space and 0 corresponds to the rock matrix.

The set of voxels $X = \cup_{I=1}^{N} C_I$, so that $F(\boldsymbol{x} \in X) = const$, forms a topological space $X(X, T)$ if the topology T is introduced. The topology T on X can be defined in several ways [8]. In particular, we deal with *6-neighborhood rule*; that is the voxels are neighbors if they share a face, and *26-neighborhood rule*; that is the voxels are neighbors if they share either a face, or an edge, or a vertex. $X = (X, T)$ is called a digital image's topological space or a three-dimensional digital image (Fig. 1).

Fig. 1. A fragment of a digital image and its topological implementation for cases of 6- and 26-neighborhood rule

In numerical modeling, space's topology is indirectly defined by choice of the numerical method. In particular, if the fluid flow is simulated using the finite-difference or finite-volume method, the topology of the pore space corresponds to the 6-neighborhood rule because flows are determined through the cells' faces [9,16]. Therefore, the voxels included in the complement (rock matrix) has 26-neighborhood rule; thus, advanced numerical approximations such as rotated grids [24], or finite-elements on hexagonal grids should be utilized if coupled problems are solved [18,21].

Let us consider sequence of digital images with the following topological spaces $\{X_i\}_0^m$ such as $\emptyset = X_0 \subseteq X_1 \subseteq X_2 \subseteq ... \subseteq X_m = X$, and call this sequence *filtration* (Fig. 2). Filtration of a binary image can also be presented as an artificial multi-component digital image:

$$F(\boldsymbol{x}) = \begin{cases} 0, & \boldsymbol{x} \in \text{background of } X \\ k, & \boldsymbol{x} \in X_k \backslash X_{k-1} \end{cases} \qquad (1)$$

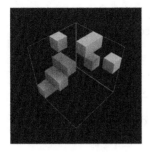 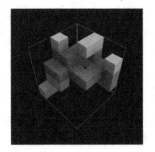 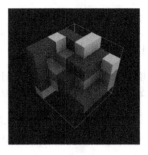

Fig. 2. Filtration of the binary image $X_0 \subseteq X_1 \subseteq X_2$, where voxels from

For each pair of indices $0 \leq i < j \leq m$, the embedding $X_i \subseteq X_j$ induces a homomorphism of p-dimensional homology groups (we consider cellular homology with coefficients in Z_2):

$$f_p^{i,j} : H_p(X_i) \to H_p(X_j). \tag{2}$$

The p-dimensional persistent homology group is the image of the homomorphism considered above: $H_p^{i,j} = Im(f_p^{i,j})$. The rank of this group $\beta_p^{i,j}$ is called *p-dimensional persistent Betty number.*

3   Edelsbrunner Algorithm

The Edelsbrunner-Lettsher-Zomorodyan algorithm [8] is a common way to calculate persistent Betty numbers. The original implementation is described for simplicial complexes. A *k-simplex* σ is the convex hull of $k + 1$ affinely independent points $S = \{v_0, v_1, ..., v_k\}$. A simplex τ defined by $T \subseteq S$ is a face of σ. And a *simplicial complex* is finite set of simplices such that

1. $\sigma \in K$, τ is a face of $\sigma \Rightarrow \tau \in K$
2. $\sigma, \tau \in K \Rightarrow \sigma \cap \tau$ is a face of σ and face of τ

The nested sequence of simplicial complexes $\emptyset = K_0 \subseteq K_1 \subseteq K_2 \subseteq ... \subseteq K_m = K$ is *filtration*.

Therefore, the first task that arises in calculating persistent homologies of a digital image is the triangulation of the image and the definition of filtration on the resulting complex. It is worth noting that triangulation is not required, because cubic homology is equivalent to simplicial. So the image can be converted to *unit cubic complex*, which is more natural. It is defined in the same way as simplicial, but instead of k-simplexes cubical complex consists of k-dimensional unit cubes.

Thus, in the case of the 26-neighborhood rule, each voxel can be considered a three-dimensional unit cube. Moreover, in the case of the 6-neighborhood rule, it is convenient to use the more efficient approach described in [28].

Here is a description of the Edelsbrunner-Lettsher-Zomorodyan algorithm. Let all the simplexes in the complex are numbered according to the filtration [8]. It means that

1. If σ_{i_1} is a face of σ_{i_2} then $i_1 \leq i_2$.
2. If $\sigma_{i_1} \in K_i$ and $\sigma_{i_2} \in K_{i+j} \setminus K_i$ for $j > 0$ then $i_1 \leq i_2$

We have a sequence of simplices $\sigma_1, \sigma_2, ..., \sigma_n$. The data is stored in a linear array $R[1..n]$, whose elements are lists of simplexes.

Algorithm 1. Edelsbrunner-Lettsher-Zomorodyan algorithm

1: **for** $j \leftarrow 1$ **to** m **do**
2:     $L \leftarrow$ list of faces of the σ_j
3:     $R[j] \leftarrow NULL$
4:     **while** $L \neq NULL$ **and** $R[i] \neq NULL$, where i is the largest number of simplices in L **do**
5:         $L \leftarrow L \triangle R[i]$
6:     **end while**
7:     **if** $L \neq NULL$ **then**
8:         $R[i] \leftarrow L$
9:     **end if**
10: **end for**

The fulfillment of the last condition gives the following information - a cycle born at time i is destroyed at time j. Moreover, the failure corresponds to the birth of the cycle at the time of j. Iterations of the inner while loop are called *collisions*. They take most of the running time of the algorithm.

Consider an example in which σ_k and σ_{k+1} are triangles and $\sigma_{k-5}, ..., \sigma_{k-1}$ are their edges, where σ_{k-1} is common. And let these simplices belong to $K_l \setminus K_{l-1}$. Collisions occur when σ_{k-2} and σ_{k-1} are added. It means the birth of one-dimensional cycles. These cycles die when σ_k and σ_{k+1} are added. Since all these simplices belong to one filtration step, they are not considered in the final result. There can be many collisions in large complexes, including cubic ones that do not carry information about persistent homology groups. Therefore, the question arises. Is it possible to reduce their number? The answer to this question is positive; one way is to remove the corresponding simplexes.

4   Reduction Algorithm

The reduction algorithms for simplicial and cubic complexes based on retraction are quite simple. Their main idea is the sequential removal of free faces, which means a retraction of simplexes (cubes). More efficient co-reduction algorithms are also known now [19]. Among other things, they are used for calculating persistent homologies [20].

This part of the paper describes an algorithm based on retraction. The version of the algorithm described in [7] adapted for 3-D digital images with the 6-neighborhood rule.

The algorithm takes the digital image representing filtration $\{X_i\}_0^n$ as an input. After that, it sequentially removes voxels according to rules described below, beginning with voxels that have value n and down to 1.

1. All conditions are checked in voxel neighborhood of size $3 \times 3 \times 3$.
2. Number of connected components of the foreground of X_n with and without current voxel is the same.
3. Number of connected components of the background of X_n is equal to 1.
4. Euler characteristic of the foreground of X_n with and without current voxel is the same.
5. Current voxel is not in any parallelepiped consisted of foreground of X_n with size $1 \times 1 \times 2$ or $1 \times 2 \times 2$ or $2 \times 2 \times 2$ that contains voxels with greater value.

If the voxel is removed, its value becomes equal to 0, and the algorithm rechecks his neighbors.

It is easy to show that the reduction algorithm has linear complexity of the number of voxels in the image. Checking the rule has a constant running time. Moreover, each voxel is checked no more than seven times because it has only six neighbors.

5   Numerical Experiments

5.1   Statistical Models

In the first series of experiments, the algorithm was applied to filtration obtained by "uniform" dissolution of the rock described in [16]. Here was assumed that the reagent concentration is constant, which leads to the same speed of movement of the pore space – rock matrix interface. The original images were obtained by the truncated Gaussian field method [10]. The main parameters in image generation were porosity and correlation length. The size of individual pores in the image depends on the correlation length.

We generated 160 samples of the size 250^3 voxels. We considered the porosity varied from 0.05 to 0.2 with the increment of 0.05, and the correlation length varied from 5 to 20 voxels with the step of 5. Thus, for each pair of parameters (porosity, correlation length), ten statistical realizations were generated. The simulation process had 100 time steps for all samples.

For all of these images, 1-dimensional barcodes were calculated before and after reduction. In each experiment, they were equal up to permutation. This fact confirms the correctness of the implementation of the algorithm.

The time of 1-dimensional barcodes computation was also measured. The Fig. 3 shows the acceleration coefficient averaged by the samples' parameters. It is important to note that the acceleration coefficient was calculated without taking into account the time spent on reduction. It is because the Edelsbrunner-Lettsher-Zomorodyan algorithm has cubic complexity, and the reduction algorithm is linear. Therefore the time spent on reduction becomes insignificant when image size is increasing.

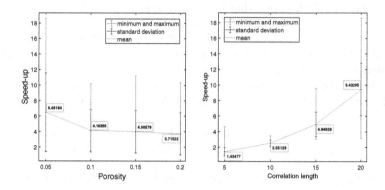

Fig. 3. The average value of the acceleration of the calculation of persistent Betty numbers depending on porosity (left) and correlation length (right)

The graphs show that the highest acceleration was achieved for samples with low porosity and high correlation length. It is because the samples with larger surface area dissolve faster and form more complex topological structure. Averaging within each group of samples also confirms this assumption (Tables 1, 2, 3, 4).

Table 1. The average acceleration of the algorithm depending on the porosity and correlation length

$l;\rho$	0.05	0.1	0.15	0.2
5	1.9	1.4	1.3	1.2
10	3.1	2.5	2.4	2.2
15	6.9	5.1	4.3	3.5
20	14.0	7.7	8.0	8.0

Table 2. The standard deviation of the algorithm acceleration depending on the porosity and correlation length

$l;\rho$	0.05	0.1	0.15	0.2
5	0.98	0.06	0.06	0.09
10	0.31	0.18	0.09	0.10
15	1.58	0.88	0.38	0.22
20	2.83	2.08	1.66	1.63

Table 3. Maximum algorithm acceleration depending on porosity and correlation length

$l;\rho$	0.05	0.1	0.15	0.2
5	4.6	1.5	1.4	1.3
10	3.5	2.8	2.5	2.3
15	9.6	6.5	4.9	3.8
20	18.6	10.2	11.2	10.4

Table 4. Minimum algorithm acceleration depending on porosity and correlation length

$l;\rho$	0.05	0.1	0.15	0.2
5	1.5	1.3	1.2	1.1
10	2.5	2.2	2.2	2.0
15	4.7	3.9	3.8	3.0
20	10.8	3.1	6.0	5.9

5.2   Pore-Scale Dissolution by CO_2 Saturated Brine in a Multi-mineral Carbonate at Reservoir Conditions

The significant difference between these tests is that they were done on real images obtained as a result of the dissolution described in [1]. Here we present only the characteristics of the images. The original images had a size of 1000^3 with a voxel length of 5.2 μm and corresponded to heterogeneous rock samples consisting of 86.6% of dolomite and 11.1% of calcite. Ten images were showing the dynamics of dissolution for each experiment.

Filtration was obtained from these images assuming that pore space voxels do not become rock voxels at the later steps. Fragment of size $400 \times 340 \times 400$ was cut out from sample AH, and of size $260 \times 320 \times 400$ from AL because the main part of the pore space is located where fluid forms the channel.

Table 5. The results of the algorithm for calculating persistent Betty numbers with and without reduction

Sample	AH	AL
ELZ time. sec	2485	445
Reduction time. sec	240	101
ELZ time after reduction. sec	35	15
Acceleration	71	29.67
Total acceleration	9.04	3.84

We measured CPU time of the computation of one-dimensional Betti numbers by Edelsbrunner-Lettsher-Zomorodyan algorithm (ELZ in the table) before and after reduction for both fragments. The time of the reduction was also measured. The results are shown in the Table 5. The acceleration here is much higher than for the first series of tests. It is because the reduction algorithm does not remove voxels with neighbors with a higher filtration step, and these samples have only ten filtration steps, and they are quite large.

Figures 4, 5, 6 shows that the reduction does not preserve the geometry of the image. Preservation of geometry is not necessary, and it is enough to preserve topology.

 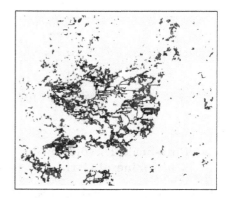

Fig. 4. Horizontal slices of the AH fragment before and after reduction. (Increase in the warmth of color corresponds to an increase in the filtration step.) (Color figure online)

 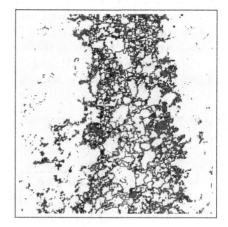

Fig. 5. Vertical slices of the AH fragment before and after reduction. (Color figure online)

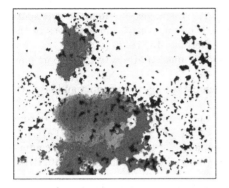 

Fig. 6. Horizontal slices of the AL fragment before and after reduction. (Color figure online)

6   Conclusion

The paper presents an algorithm for the reduction of a digital image of a porous medium. The algorithm is applicable to speed up the calculation of persistent Betty numbers, which are used to characterize the changes in the pore space's structure during the chemical dissolution of rock. It is shown that the use of the algorithm makes it possible to accelerate the calculation of persistent Betty numbers up to 70 times that leads to possibility of processing samples of sizes up to 500^3 voxels by using a single computational node within acceptable wall-clock time, that, less than 15 min (on the machine with Intel(R) Core(TM) i7-3770K CPU 3.5 GHz processor and 32 GB RAM installed, which was used for performed tests). Acceleration depends on the complexity of the pore space structure and the dynamics of the rock dissolution process.

The reduction algorithm allows parallelization. Thus, the first part of the future work is implementation and testing of parallel reduction. The second part is to study another methods of reducing input data for Edelsbrunner algorithm, such as co–reduction [20] and acyclic complex [7]. Then, we will be able to find the most efficient combination of reduction and these two methods for digital images of rock. It will allows us to quickly solve topological optimization problems in material design.

References

1. Al-Khulaifi, Y., Lin, Q., Blunt, M., Bijeljic, B.: Pore-scale dissolution by CO2 saturated brine in a multi-mineral carbonate at reservoir conditions: impact of physical and chemical heterogeneity (2019). https://doi.org/10.5285/52b08e7f-9fba-40a1-b0b5-dda9a3c83be2
2. Al-Khulaifi, Y., Lin, Q., Blunt, M.J., Bijeljic, B.: Pore-scale dissolution by CO2 saturated brine in a multimineral carbonate at reservoir conditions: impact of physical and chemical heterogeneity. Water Resour. Res. **55**(4), 3171–3193 (2019)

3. Alt-Epping, P., Waber, H.N., Diamond, L.W., Eichinger, L.: Reactive transport modeling of the geothermal system at Bad Blumau, Austria: implications of the combined extraction of heat and CO2. Geothermics **45**, 18–30 (2013)

4. Amikiya, A.E., Banda, M.K.: Modelling and simulation of reactive transport phenomena. J. Comput. Sci. **28**, 155–167 (2018)

5. Andra, H., et al.: Digital rock Physics benchmarks - part i: imaging and segmentation. Comput. Geosci. **50**, 25–32 (2013)

6. Bazaikin, Y., et al.: Effect of CT image size and resolution on the accuracy of rock property estimates. J. Geophys. Res. Solid Earth **122**(5), 3635–3647 (2017)

7. Dlotko, P., Wagner, H.: Simplification of complexes for persistent homology computations. Homol. Homotopy Appl. **16**, 49–63 (2014). https://doi.org/10.4310/HHA.2014.v16.n1.a3

8. Edelsbrunner, H., Harer, J.: Computational Topology. An Introduction. American Mathematical Society, Providence (2010)

9. Gerke, K.M., Karsanina, M.V., Katsman, R.: Calculation of tensorial flow properties on pore level: exploring the influence of boundary conditions on the permeability of three-dimensional stochastic reconstructions. Phys. Rev. E **100**(5), 053312 (2019)

10. Hyman, J.D., Winter, C.L.: Stochastic generation of explicit pore structures by thresholding Gaussian random fields. J. Comput. Phys. **277**, 16–31 (2014)

11. Jones, T.A., Detwiler, R.L.: Mineral precipitation in fractures: using the level-set method to quantify the role of mineral heterogeneity on transport properties. Water Resour. Res. **55**(5), 4186–4206 (2019)

12. Kaya, E., Zarrouk, S.J.: Reinjection of greenhouse gases into geothermal reservoirs. Int. J. Greenhouse Gas Control **67**, 111–129 (2017)

13. Krahenbuhl, R.A., Martinez, C., Li, Y., Flanagan, G.: Time-lapse monitoring of CO2 sequestration: a site investigation through integration of reservoir properties, seismic imaging, and borehole and surface gravity data. Geophysics **80**(2), WA15–WA24 (2015)

14. Lamy-Chappuis, B., Angus, D., Fisher, Q.J., Yardley, B.W.D.: The effect of CO2-enriched brine injection on the mechanical properties of calcite-bearing sandstone. Int. J. Greenhouse Gas Control **52**(Supplement C), 84–95 (2016)

15. Lebedev, M., Zhang, Y., Sarmadivaleh, M., Barifcani, A., Al-Khdheeawi, E., Iglauer, S.: Carbon geosequestration in limestone: pore-scale dissolution and geomechanical weakening. Int. J. Greenhouse Gas Control **66**, 106–119 (2017)

16. Lisitsa, V., Bazaikin, Y., Khachkova, T.: Computational topology-based characterization of pore space changes due to chemical dissolution of rocks. Appl. Math. Model. **88**, 21–37 (2020). https://doi.org/10.1016/j.apm.2020.06.037

17. Lucas, Y., Ngo, V.V., Clément, A., Fritz, B., Schäfer, G.: Modelling acid stimulation in the enhanced geothermal system of soultz-sous-forêts (Alsace, France). Geothermics **85**, 101772 (2020)

18. Mokbel, D., Abels, H., Aland, S.: A phase-field model for fluid-structure interaction. J. Comput. Phys. **372**, 823–840 (2018)

19. Mrozek, M., Batko, B.: Coreduction homology algorithm. Discrete Comput. Geom. **41**(1), 96–118 (2009)

20. Mrozek, M., Wanner, T.: Coreduction homology algorithm for inclusions and persistent homology. Comput. Math. Appl. **60**(10), 2812–2833 (2010)

21. Nestola, M.G.C., et al.: An immersed boundary method for fluid-structure interaction based on variational transfer. J. Comput. Phys. **398**, 108884 (2019)

22. Nooraiepour, M., Bohloli, B., Park, J., Sauvin, G., Skurtveit, E., Mondol, N.H.: Effect of brine-CO2 fracture flow on velocity and electrical resistivity of naturally fractured tight sandstones. Geophysics **83**(1), WA37–WA48 (2018)

23. Pereira Nunes, J.P., Blunt, M.J., Bijeljic, B.: Pore-scale simulation of carbonate dissolution in micro-CT images. J. Geophys. Res. Solid Earth **121**(2), 558–576 (2016)

24. Saenger, E.H., Gold, N., Shapiro, S.A.: Modeling the propagation of the elastic waves using a modified finite-difference grid. Wave Motion **31**, 77–92 (2000)

25. Sim, C.Y., Adam, L.: Are changes in time-lapse seismic data due to fluid substitution or rock dissolution? a CO2 sequestration feasibility study at the Pohokura field, New Zealand. Geophys. Prospect. **64**(4), 967–986 (2016)

26. Steefel, C.I., et al.: Reactive transport codes for subsurface environmental simulation. Comput. Geosci. **19**(3), 445–478 (2015)

27. Trebotich, D., Adams, M.F., Molins, S., Steefel, C.I., Shen, C.: High-resolution simulation of pore-scale reactive transport processes associated with carbon sequestration. Comput. Sci. Eng. **16**(6), 22–31 (2014)

28. Wagner, H., Chen, C., Vuçini, E.: Efficient computation of persistent homology for cubical data. In: Peikert, R., Hauser, H., Carr, H., Fuchs, R. (eds.) Topological Methods in Data Analysis and Visualization II. Mathematics and Visualization, pp. 91–106. Springer, Berlin (2012). https://doi.org/10.1007/978-3-642-23175-9_7

29. Wojtacki, K., Daridon, L., Monerie, Y.: Computing the elastic properties of sandstone submitted to progressive dissolution. Int. J. Rock Mech. Min. Sci. **95**, 16–25 (2017)

Oil and Gas Reservoirs Parameters Analysis Using Mixed Learning of Bayesian Networks

Irina Deeva[1]([✉]), Anna Bubnova[1], Petr Andriushchenko[1],
Anton Voskresenskiy[1,2], Nikita Bukhanov[1,3], Nikolay O. Nikitin[1],
and Anna V. Kalyuzhnaya[1]

[1] ITMO University, Saint-Petersburg, Russia
[2] Gazpromneft-GEO, Saint-Petersburg, Russia
[3] Gazpromneft Science and Technology Center, Saint-Petersburg, Russia

Abstract. In this paper, a multipurpose Bayesian-based method for data analysis, causal inference and prediction in the sphere of oil and gas reservoir development is considered. This allows analysing parameters of a reservoir, discovery dependencies among parameters (including cause and effects relations), checking for anomalies, prediction of expected values of missing parameters, looking for the closest analogues, and much more. The method is based on extended algorithm MixLearn@BN for structural learning of Bayesian networks. Key ideas of MixLearn@BN are following: (1) learning the network structure on homogeneous data subsets, (2) assigning a part of the structure by an expert, and (3) learning the distribution parameters on mixed data (discrete and continuous). Homogeneous data subsets are identified as various groups of reservoirs with similar features (analogues), where similarity measure may be based on several types of distances. The aim of the described technique of Bayesian network learning is to improve the quality of predictions and causal inference on such networks. Experimental studies prove that the suggested method gives a significant advantage in missing values prediction and anomalies detection accuracy. Moreover, the method was applied to the database of more than a thousand petroleum reservoirs across the globe and allowed to discover novel insights in geological parameters relationships.

Keywords: Bayesian networks · Structural learning · Causal inference · Missing values prediction · Oil and gas reservoirs · Similarity detection

1 Introduction

The problem of choosing oil and gas reservoir development strategy is one of the most crucial decisions made at the early stages of reservoir development by every oil and gas producer company. Almost all decisions related to fluid production

© Springer Nature Switzerland AG 2021
M. Paszynski et al. (Eds.): ICCS 2021, LNCS 12742, pp. 394–407, 2021.
https://doi.org/10.1007/978-3-030-77961-0_33

have to be made at the early stages, characterized by high uncertainty and lack of information about geological and production reservoir parameters. The reservoir could be fully characterized only at mature development stages. The experience of commissioning new reservoirs shows that most of the project decisions made at the early stages of the reservoir development have a crucial impact on the development strategy and the entire project's economic feasibility. A common method of investigating these new reservoirs is to examine a subsample that is close to or similar to them. Most often, the wrong selection of analogues leads to the fact that the reserves of the reservoir are overestimated, which leads to a discrepancy between the forecast and actual production levels, which results in an overestimated net present value (NPV) forecast for the entire project [5]. Incorrect selection of analogues can even lead to the fact that the actual NPV does not fall into the predicted distribution of probable NPV.

In a significant part of geological companies, analogues search is performed by an expert who manually selects a reservoir which properties resemble properties of the target one. The result of this procedure is a list of reservoir names with similar properties. Analogues are also used at mature stages of reservoir development, for example, to find successful cases of increasing oil recovery. There is another way to find reservoir analogues, namely using the similarity function [16]. The difference between the two approaches consists mainly in distributions shapes of reservoir analogues due to more narrow search space made by industry expert [21]. It was also founded that using a manual approach, some experts limit themselves only to local analogues and completely ignore global analogues. A decision of using only local analogues may not be optimal, especially at the early stages of project development [19]. On the other hand, the results of reservoir analogues performed by ranked similarity function are characterized by a broader distribution of reservoir parameters. In order to find reservoir analogues by similarity function more feasible, reconstruction of missing values should be performed. Based on this, it is highly desirable to have a more automated and mathematically proved instrument to determine the most likely reservoir parameters at the early stages of geological exploration. So far, a few works have been done on topics related to inputting missing values using machine learning to reservoir parameters datasets [20]. For this reason, our efforts also addressed the search for an automated solution for the analysis and modelling of reservoir properties with the usage of machine learning techniques.

2   Related Work

In this section, we summarize, in abbreviated form, the pros and cons of methods that are used for modelling and could potentially be the basis for reservoir analysis. They are described in more detail in our previous work [1].

One common way of modelling is Markov Chain Monte Carlo (MCMC) method. The algorithm walks through the space of all possible combinations of values and moves from one state to a new state that differs in the i-th variable with probability estimated from the dataset. However, this approach for high-dimensional problems requires large amounts of memory and time. Unfortunately, this model does not find the relationship between the parameters explicitly and is difficult to interpret. Nevertheless, MCMC is quite common, ready-made libraries exist for it, and this method is used to solve practical problems in oil and gas data. For example, Gallagher et al. [6] showed the applicability of MCMC method to the task of modelling distributions of geochronological ages, sea-level, and sedimentation histories from two-dimensional stratigraphic sections.

Another modelling approach is the Conditional Iterative Proportional Fitting (CIPF). CIPF is designed to work with multi-dimensional contingency tables. The basic idea is to iteratively fit marginal and conditional distributions, gradually approaching the desired joint distribution in terms of Kullback-Leibler distance. However, this requires knowledge of the dependency structure, but even so, the method converges extremely slowly. There is a simplified and faster version called IPF, which assumes a toy model with an independent set of variables. Li and Deutsch [13], for example, use it to estimate and model facies and rock types. On the positive side, this approach is reduced to the sequential solution of linear equations, and there is a ready-made library for it.

A third approach is to use a copula. Formally, this is a multivariate distribution on an n-dimensional unit cube obtained using inverse distribution functions for some fixed family of distributions. Based on Sklar's theorem, modelling any joint distribution is reduced to approximation by copulas. For example, Han et al. [9] use this approach to determine the tectonic settings. And in the work of Hernández-Maldonado et al. [11] they model complex dependencies of petrophysical properties such as porosity, permeability, etc. Unfortunately, expert selection of the copula family is required. The basic method takes into account only the pairwise dependence; more complex models require knowledge of the dependence structure. It must be considered that the main application of copulas is to work with quantitative values.

In short, our work seeks to circumvent the above limitations on the volume and type of data. Our goal is an interpretive approach that will allow us to combine the tools needed for oil and gas reservoir analysis on a single base. To do this, we turn to Bayesian networks, taking into account the specifics of the domain and the techniques adopted for analysis.

3   Problem Statement

Let us temporarily step aside and look at this issue from the side of statistics, not geology. The first thing that is important to us when we analyze data is to present it in the most effective way that allows us to extract as much hidden information as possible. And oil and gas reservoirs analysis impose some limitations: we need to work with mixed data, i.e. continuous, categorical, and discrete data. The objects under study have a large number of features.

The relationships of the parameters are not always linear and not always obvious, even to a specialist, but we need to be able to identify them for further interpretation. We also would like to have a possibility of taking into account the expert's opinion of the cause and effect links. There are additional restrictions on the time of the algorithm, as well as the need for certain functionality such as filling gaps and finding anomalies. So here is an additional challenge of making an optimal choice among the existing modelling approaches. Based on all of the above we focus our efforts on developing a tool that is able to (1) be interpretable and include some portion of expert knowledge (in the way of composite AI), (2) be multipurpose as a core for partial algorithms (analogues search, gaps filling, probabilistic inference, etc.), (3) be efficiently computed.

As the most promising basis, we choose Bayesian networks (BN). From a graphical point of view, it is a directed acyclic graph (DAG), any vertex of which represents some characteristic of the object. This structure also stores information for the vertices about the value and conditional distribution of the corresponding characteristic. Let $Pa_{X_i}^G$ denote the parents of the vertex X_i in the structure G, $NonDescendants_{X_i}$ denote the vertices in the graph that are not descendants of the vertex X_i [12]. Then, for each vertex X_i:

$$(X_i \perp NonDescendants_{X_i} | Pa_{X_i}^G) \tag{1}$$

Then the multivariate distribution of P in the same space is factorized according to the structure of the graph G, if P can be represented as:

$$P(X_1, ..., X_n) = \prod_{i=1}^{n} P(X_i | Pa_{X_i}^G) \tag{2}$$

However, to use Bayesian networks, it is necessary to solve two problems: learning the network structure and distributions parameters in the network nodes. Unfortunately, learning the graph structure is a complex and resource-intensive task. The number of DAGs grows super-exponentially with the number of vertices [17]. However, there are several approaches [18] to solve this problem, which we will discuss below.

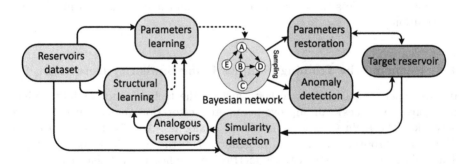

Fig. 1. The pipeline of the proposed Bayesian approach to the oil and gas reservoir analysis. Data preprocessing block is not represented directly to simplify the scheme.

The purpose of this paper is to demonstrate a new flexible approach in the context of a comprehensive analysis of oil and gas reservoirs (see Fig. 1

for details). We chose the Bayesian network model as the core because this probabilistic model generally meets the formulated criteria. There is an extensive theoretical basis for various probabilistic analysis problems, and examples of specialized implementations on small oil and gas datasets [14,15]. However, Bayesian networks are capable of handling larger amounts of data, and the functionality can be represented as a complex structure of complementary elements. In this paper, we propose to implement this idea on the basis of oil and gas reservoirs data, taking into account the specificity of the domain and typical auxiliary techniques such as searching for analogues.

4   Algorithms and Methods

4.1   MixLearn@BN: Algorithm for Mixed Learning of Bayesian Networks

Making the inference of reservoir parameters requires a complex algorithm with several crucial properties. The algorithm MixLearn@BN allows us to combine (1) learning the network structure on homogeneous data subsets, (2) assigning the structure by an expert, and (3) learning the distribution parameters on mixed data (discrete and continuous). Algorithm 1 demonstrates the pseudocode of the proposed complex algorithm. For structure learning, we use the Hill Climbing algorithm [2,7] with the scoring function from K2 algorithm [4]. To learn the parameters, we use a mixed approach, within which the distributions at the nodes can be of three types:

- Conditional probabilities tables (CPT), if the values at the node are discrete and its parents have a discrete distribution;
- Gaussian distribution, if the values at the node are continuous and its parents have a continuous distribution;
- Conditional Gaussian distribution, if the node values are continuous, and among the parents there are discrete nodes and nodes with continuous distribution.

4.2   Reducing the Training Samples Using Similarity Detection

In this paper, in particular, we test whether the technique for finding similar reservoirs (analogues) can improve parameter restoration quality with Bayesian networks. The main task is to find these analogues. And for this, we need to define an appropriate measure of proximity or distance to the target reservoir. The main requirement for distance metrics are defined by task features:

1. The data is represented by a set of values of different types. Metrics on categorical variables rely on whether or not the value matched, whereas, for quantitative ones, it matters how much it does not match. Once quantitative variables are discretized, we can use the general approach for categorical ones, but there is a risk that we miss valuable information.

Algorithm 1. Comprehensive Bayesian Network Learning Algorithm

---

1: **procedure** BAYESIAN NETWORK LEARNING($D, edges, remove\_edges$)
2:     Input: $D = \{ \mathbf{x}_1, ..., \mathbf{x}_n \}$, edges which an expert wants
3:     to add $edges$, boolean $remove\_edges$ that allows removing $edges$
4:     Output: Bayesian network$\{ V, E, DistributionsParameters \}$
5:     $discrete_D = $ Discretization(D)
6:     $\{ V, E \} = $ HillClimbingSearch($discrete\_D$)   ▷ Structure learning
7:     **if** $remove\_edges = false$ **then**
8:         $\{ V, E \} = \{ V, E \} \cup edges$
9:     **end if**
10:     $bn_parameters = $ Parameters learning($D, \{ V, E \}$)   ▷ Parameters learning
11:     **return** Bayesian network$\{ V, E, bn\_parameters \}$
12: **end procedure**
13: **procedure** PARAMETERS LEARNING()$D, \{ V, E \}$
14:     Input: $D = \{ \mathbf{x}_1, ..., \mathbf{x}_n \}$, structure of BN $\{ V, E \}$
15:     Output: dictionary with distributions parameters for each node in BN
16:     $params = $ empty dictionary
17:     **for** $node$ in $BN\_structure$ **do**
18:         **if** $node$ is discrete **and** parents($node$) are discrete **then**
19:             $params[node] = $ CPT($node$, parents($node$), D)
20:         **end if**
21:         **if** $node$ is continuous **and** parents($node$) are continuous **then**
22:             $mean, var = $ parameters from Gaussian($node$,D)
23:             $coef = $ coefficients from BayesianLinearRegression(parents($node$),$node$,
24:             D)
25:             $params[node] = \{ mean, var, coef \}$
26:         **end if**
27:         **if** $node$ is continuous **and** parents($node$) are continuous **and** discrete **then**
28:             $cont_parents = $ parents_continuous($node$)
29:             $disc_parents = $ parents_discrete($node$)
30:             $node\_params = \emptyset$
31:             $combinations = $ all combinations of $disc\_parents$ values
32:             **for** $\{ v_1, ..., v_k \}$ in $combinations$ **do**
33:                 $subsample = \{ \mathbf{x}_i : x_{ij_1} = v_1, ..., x_{ij_k} = v_k \}$
34:                 $mean, var = $ parameters from Gaussian($node$, $subsample$)
35:                 $coef = $ coefficients from BayesianLinearRegression($cont\_parents$,
36:                 $node$, $subsample$)
37:                 $node\_params \cup \{ mean, var, coef \}$
38:             **end for**
39:             $params[node] = node\_params$
40:         **end if**
41:     **end for**
42:     **return** $params$
43: **end procedure**

---

2. Variables are interdependent, and mismatches cannot be considered as unrelated in any way. It is unclear how to account for this.
3. Some variables are more valuable than others. There is a weighted option for most distances and events, but it is not obvious how to select optimal weights.

The following is a description of the distances between the objects u and t involved in the experiments. We begin by determining Gower's general similarity coefficient $S(u, t)$ [8]. A auxiliary coefficient $S_j(u, t)$ is considered for each jth variable. And $S(u, t)$ is their weighted average with weights w_j. In the unweighted version, $w_j = 1$ is assumed. On categorical variables, $S_j(u, t)$ is 0 or 1, depending on whether the categories match. On quantitative variables, it is the modulus of the difference of the normalized values. Note that this is a similarity coefficient, not a distance. However, it is easy to turn it into distance with the following transformation: $dist_G(u, t) = 1 - S(u, t)$.

There are also measures of a different nature than Gower's coefficient. This is cosine distance that has proven itself in the task of ranking search engine results. Applying this distance requires prior preparation of values. For a categorical variable, the value is assumed to be 1 at the target. And 0 or 1 on the object being compared, depending on whether the categories match. For quantitative variables, values are normalized.

We also investigated the performance of the filtering function in the experiments. It depends on only one parameter ε, which for quantitative variables says that a value is close if $|u_j - t_j| \leq \varepsilon \cdot range(j)$. And for categorical, it checks for category matching. First, the set of analogues includes objects that are close to the target in all variables. Then for all but one, and so on.

4.3 Method for Parameters Restoration with Bayesian Networks Learning on Analogues

Using similarity metrics allows you to find the closest reservoirs to a target reservoir. Combining the search for analogous reservoirs and learning Bayesian networks with MixLearn@BN allows formulating a method for solving the problem of parameters restoration (Fig. 1), the structure of which implies the following steps:

1. Select the target reservoir from a dataset;
2. Looking for N nearest reservoirs according to the distance metric;
3. Learn the structure and parameters at the nearest reservoirs with algorithm MixLearn@BN;
4. Initialize the nodes of the Bayesian network with the values of those parameters that are not missing;
5. Sample the missing values from this Bayesian network by forward sampling with evidence [10];
6. For categorical values, the gap is filled with the most frequent category in the sample;
7. For continuous values the gap is filled with the average value in the sample.

5   Experiments and Results

5.1   Exploratory Data Analysis with Bayesian Networks

The dataset used in the study was collected from open sources and presented by plain database contains 1073 carbonate and clastic reservoirs from all over the world. The parameters that define the dataset contain categorical and continuous values: reservoir depth and period, depositional system and environment, tectonic regime, structural setting and trapping mechanism, lithology type, gross and net thicknesses, reservoir porosity and permeability.

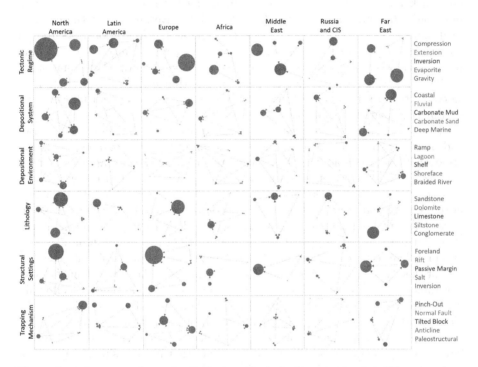

Fig. 2. Bayesian networks are for six main geological parameters across different regions worldwide. Each network has five nodes, which are described on the right margin.

Several Bayesian networks were built to demonstrate that their output could be used to get qualitative data insights. The x-axis on Fig. 2 signifies region division based on geographical closeness (CIS stands for Commonwealth of Independent States) and can be considered as filters of the dataset. The y-axis indicates some categorical parameters which characterize reservoirs in the dataset. The network nodes' color and size signify the value of a categorical parameter (presented on the right portion of the figure) and several such occurrences, respectively. Generally, the figure demonstrates relationships between selected reservoir parameters within regions.

For instance, from Fig. 2 following conclusion could be made: in Europe and Africa, extension tectonic regime and rift structural setting prevail. Extension and rift are closely related in terms of tectonic. Extension tectonic regime causes extension of continental lithosphere which carriers origin of rift structural setting, so the area undergoes extensional deformation (stretching) by the formation and activity of normal faults. The East African Rift System and the North Sea rift could be examples of extensional regimes in the regions. Also, there are arrows from normal faults to tilted blocks that could indicate causal inference between these parameters, and if so, this is in agreement with geological knowledge. A similar approach (in terms of analyzing statistical model with and domain knowledge) was performed by analyzing the causal inference between reservoir parameters and well log data [20]. They have found that those statistical models could produce conclusions that mimic interpretation rules from a domain knowledge point of view, which opens up an opportunity to reveal causal relations between features. In general, such plots could be used for qualitative analysis of prevailing parameters or patterns within regions as an alternative to conventional screening performed by a geologist using manual filtering and spatial visualization techniques.

The x-axis on Fig. 3 signifies the geological period used to filter the dataset and build networks. The figure shows a relation of reservoir parameters within geological timeframes according to International Chronostratigraphic Chart [3]. The Neogene, Cretaceous, Jurassic, and Triassic are characterized by the fact that coastal and fluvial depositional systems and sandstone lithology dominate in these geological periods. It does not contradict domain knowledge that sandstone tends to dominate in coastal and fluvial depositional systems as it the characterized by relatively high flow energy. In Paleozoic, sandstone lithology also prevalent, but carbonate mud depositional system is predominant. The reason for this discrepancy may be due to missing values in the depositional system. It is difficult to draw conclusions from the figure and confirm them with domain knowledge (as was done for the previous one) because different portions of the planet at the same geological time undergo various geological processes.

The possible implementation of this workflow can be easily extended to the internal company database. Apart from the fact that it has a similar number of parameters, it has many instances subdivided in hierarchical order into reservoirs, formations, and wells, respectively. This allows us to work internally with the same approach using an internal database, but here we present some results using the dataset obtained from public sources.

5.2   Bayesian Networks Application Experiments

In this section, we carried out several experiments to study the Bayesian network's ability to restore missing values in parameters and detect abnormal values and compare various metrics for analogues searching in terms of the restoration accuracy. Firstly a Bayesian network was learned on the selected parameters of all reservoirs in the dataset. For structure learning, quantile data discretization

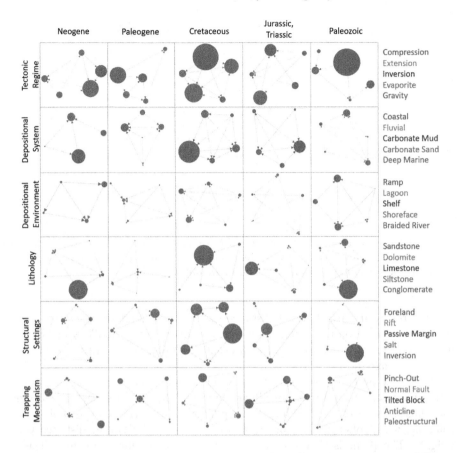

Fig. 3. Bayesian networks are for six main geological parameters across different stratigraphy periods. Each network has five nodes, which are described on the right margin.

was performed. The resulting Bayesian network is shown in Fig. 4. The structure of the experiment is as follows:

1. A reservoir is selected from the dataset;
2. The Bayesian network is learned from all reservoirs except the selected one;
3. In the selected reservoir, the parameters are deleted and restored with the Bayesian network;
4. Recovery results are saved;
5. The steps are repeated for the next reservoirs in the dataset.

To increase parameter restoration accuracy, we can learn Bayesian networks only on similar reservoirs because analogous reservoirs are more homogeneous subsamples. Our study uses several distance metrics to search for similar reservoirs (Sect. 4.2). We apply method (Sect. 4.3) with 40 nearest reservoirs to the

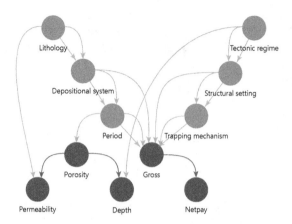

Fig. 4. The result of structural learning of the Bayesian network on the parameters of the reservoir data. The red nodes represent continuous variables and blue nodes represent the categorical variables. (Color figure online)

selected reservoir with different distance metrics. The number of nearby reservoirs was selected on the assumption that it cannot be too small to avoid overfitting, but it cannot be too large to prevent dissimilar reservoirs from entering the list of analogues.

Table 1 shows the average restoration result over leave-one-out cross-validation tests. An approach of filling the most frequent category and average was chosen as baseline. We should pay particular attention to the tables' last columns, which offer results for subsamples built at the Gower distance with weights. The point is that the Gower distance penalty for non-coincidence of categorical is greater than for non-coincidence of continuous ones. Therefore, if you equalize the penalties for all parameters, you can increase the prediction of continuous parameter values while maintaining a good prediction accuracy for categorical values. We first analyzed the average Gower distance penalties for all parameters; the spread of penalties is shown in Fig. 5. Then, to equalize the penalties, we assign the weights of the continuous parameters equal to the average penalty ratio for categorical to the average penalty for continuous ones. This means that categorical parameters are taken without weight (equal to 1), and continuous parameters are taken with a selected weight (5.8). An increase in the continuous ones' accuracy indicates that the selected value of the weights equalizes the penalties and allows you to accurately search for similar reservoirs both in terms of categorical and continuous parameters.

The most challenging case of anomalies in reservoir data is when the reservoirs themselves' continuous parameters are not out of the range of possible values, but in combination with other characteristics, they are impossible. We experimented with testing the Bayesian network's ability to search for such anomalous cases. In each continuous parameter, 10% of the values were randomly changed; such a change simulates the anomaly in the parameter. Then

Table 1. The values of accuracy score for categorical parameters restoration and RMSE for continuous parameters restoration.

Parameter	All dataset		On analogs			
	Baseline	MixLearn@BN	Cosine	Gower	Filtering	Gower with weights
Accuracy for the categorical parameters						
Tectonic regime	0.48	0.48	0.86	0.85	**0.91**	0.78
Period	0.27	0.37	**0.65**	0.63	0.62	0.64
Depositional system	0.35	0.56	**0.81**	0.78	0.78	0.72
Lithology	0.57	0.57	**0.81**	**0.81**	**0.81**	**0.81**
Structural setting	0.48	0.56	0.72	**0.74**	0.73	0.72
Trapping mechanism	0.28	0.54	**0.77**	**0.77**	0.75	**0.77**
RMSE for the continuous						
Gross	436	**297.3**	358.7	375.1	394.1	323.6
Netpay	82.8	72.04	91.02	69.5	75.01	**68.2**
Porosity	7.6	5.8	6.6	5.8	5.3	**4.2**
Permeability	1039	830.07	901.8	873.6	936.4	**703.3**
Depth	1110	962.42	1097	908	1010	**875.3**

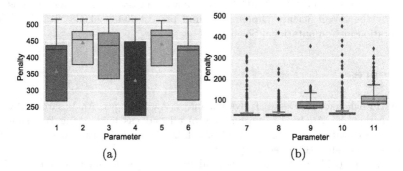

(a)   (b)

Fig. 5. The spread of penalties for categorical parameters (Tectonic regime, Period, Depositional system, Lithology, Structural setting, Trapping mechanism) (a) and continuous parameters (Gross, Netpay, Porosity, Permeability, Depth) (b) for the Gower distance, taken for all reservoirs from the dataset.

the nodes of the Bayesian network were initialized with other parameters, and the checked parameter was sampled. If the current value fell outside the interval of two standard deviations of the sample, it was recognized as an anomaly. ROC-AUC metrics for searching anomalous values of reservoirs parameters [Gross, Netpay, Porosity, Permeability, Depth] are equal to [0.85, 0.97, 0.8, 0.71, 0.7].

6   Conclusion

In this paper, a multipurpose method for analysis of heterogeneous data was presented using data from oil and gas reservoirs. This method consists of constructing Bayesian networks with mixed learning algorithm MixLearn@BN to enhance the accuracy. First, the learning of distribution parameters on mixed data was proposed. Thus, the accuracy of restoring continuous parameters has increased. Secondly, an approach was proposed for training the structure and parameters of a Bayesian network on a subsample of similar reservoirs. This made it possible to make the distributions of most parameters more unimodal, which led to a significant increase in restoration accuracy in all parameters at once. To find similar reservoirs, we used several distance metrics that can work with discrete and continuous data. The highest restoration accuracy for most categorical variables was obtained for the Gower distance. Using the Gower distance with weights allowed us to maintain sufficient accuracy for categorical parameters and significantly improve continuous ones' accuracy. Also, the Bayesian network showed fairly good accuracy in searching for anomalies.

In the future, it would be interesting to compare results for different numbers of analogs. It would also be interesting to describe other practical cases, such as quality control, studying one parameter's effect on another, and others.

Acknowledgement. We would like to thank Gazprom Neft for the provided reservoir dataset. This research is partially financially supported by The Russian Scientific Foundation, Agreement №19-11-00326. Participation in the ICCS conference was supported by the NWO Science Diplomacy Fund project №483.20.038 "Russian-Dutch Collaboration in Computational Science".

References

1. Andriushchenko, P., Deeva, I., Kalyuzhnaya, A., Bubnova, A., Voskresenskiy, A., Bukhanov, N.: Analysis of parameters of oil and gas fields using Bayesian networks. In: Data Science in Oil & Gas, vol. 2020, pp. 1–10. European Association of Geoscientists & Engineers (2020)
2. Chickering, D.M.: Optimal structure identification with greedy search. J. Mach. Learn. Res. **3**(Nov), 507–554 (2002)
3. Cohen, K., et al.: 2013 The ICS International Chronostratigraphic Chart. episodes 36, pp. 199–204. Episodes 36 (Sept 2013). https://doi.org/10.18814/epiiugs/2013/v36i3/002
4. Cooper, G.F., Herskovits, E.: A Bayesian method for the induction of probabilistic networks from data. Mach. Learn. **9**(4), 309–347 (1992)

5. De Jager, G.: The influence of geological data on the reservoir modelling and history matching process (2012). https://doi.org/10.4233/uuid:b06f57c2-24b5-407b-a69c-e68c8bd7ff01

6. Gallagher, K., Charvin, K., Nielsen, S., Sambridge, M., Stephenson, J.: Markov chain monte carlo (MCMC) sampling methods to determine optimal models, model resolution and model choice for earth science problems. Mar. Pet. Geol. **26**(4), 525–535 (2009)

7. Gámez, J.A., Mateo, J.L., Puerta, J.M.: Learning Bayesian networks by hill climbing: efficient methods based on progressive restriction of the neighborhood. Data Min. Knowl. Discovery **22**(1), 106–148 (2011)

8. Gower, J.C.: A general coefficient of similarity and some of its properties. Biometrics **27**, 857–871 (1971)

9. Han, S., Li, M., Zhang, Q., Li, H.: A mathematical model based on Bayesian theory and Gaussian copula for the discrimination of gabbroic rocks from three tectonic settings. J. Geol. **127**(6), 611–626 (2019)

10. Henrion, M.: Propagating uncertainty in Bayesian networks by probabilistic logic sampling. In: Machine Intelligence and Pattern Recognition, vol. 5, pp. 149–163. Elsevier (1988)

11. Hernández-Maldonado, V., Díaz-Viera, M., Erdely, A.: A multivariate Bernstein copula model for permeability stochastic simulation. Geofísica Int. **53**(2), 163–181 (2014)

12. Koller, D., Friedman, N.: Probabilistic Graphical Models: Principles and Techniques. MIT Press, Cambridge (2009)

13. Li, Y., Deutsch, C.V.: Implementation of the iterative proportion fitting algorithm for geostatistical facies modeling. Nat. Resour. Res. **21**(2), 221–231 (2012)

14. Martinelli, G., Eidsvik, J., Sinding-Larsen, R., Rekstad, S., Mukerji, T.: Building Bayesian networks from basin-modelling scenarios for improved geological decision making. Pet. Geosci. **19**(3), 289–304 (2013)

15. Masoudi, P., Asgarinezhad, Y., Tokhmechi, B.: Feature selection for reservoir characterisation by Bayesian network. Arab. J. Geosci. **8**(5), 3031–3043 (2015)

16. Perez-Valiente, M., Rodriguez, H., Santos, C., Vieira, M., Embid, S.: Identification of reservoir analogues in the presence of uncertainty. In: Society of Petroleum Engineers - SPE Intelligent Energy International 2014 (Apr 2014). https://doi.org/10.2118/167811-MS

17. Robinson, R.W.: Counting labeled acyclic digraphs. New directions in the theory of graphs, pp. 239–273 (1973)

18. Scutari, M., Graafland, C.E., Gutiérrez, J.M.: Who learns better Bayesian network structures: accuracy and speed of structure learning algorithms. Int. J. Approximate Reasoning **115**, 235–253 (2019)

19. Sun, S.Q., Wan, J.: Geological analogs usage rates high in global survey. Oil Gas J. **100**, 49–50 (2002)

20. Voskresenskiy, A., Bukhanov, N., Filippova, Z., Brandao, R., Segura, V., Brazil, E.: Feature selection for reservoir analogues similarity ranking as model-based causal inference (Sept 2020). https://doi.org/10.3997/2214-4609.202035170

21. Voskresenskiy, A., et al.: Variations in ranked list of reservoir analogs as an effect of search preferences **2020**(1), 1–5 (2020). https://doi.org/10.3997/2214-4609.202053163. https://www.earthdoc.org/content/papers/10.3997/2214-4609.202053163

Analytic and Numerical Solutions of Space-Time Fractional Diffusion Wave Equations with Different Fractional Order

Abhishek Kumar Singh⬤ and Mani Mehra$^{(\boxtimes)}$⬤

Indian Institute of Technology Delhi, New Delhi 110016, India
mmehra@maths.iitd.ac.in

Abstract. The aim of this manuscript is to investigate analytic and numerical solutions of space–time fractional diffusion wave equations with different fractional order (α and β). After deriving analytic solution, an implicit unconditional stable finite difference method for solving space-time fractional diffusion wave equations is proposed. The Gerschgorin theorem is used to study the stability and convergence of the method. Furthermore, the behavior of the error is examined to verify the order of convergence by numerical example.

Keywords: Implicit finite difference method · Riesz space-fractional derivative · Stability · Convergence

1 Introduction

In fluid mechanics [10], physics [3], biology [4], system control [13], hydrology [12], finance [11,14–16] and various other engineering fields [7], we often encounter complex systems which cannot be modeled by the conventional integer order differential equations. In such scenarios, the use and study of differential equations with non-integer powers of the differentiation order, namely the "fractional differential equations (FDEs)" proves to be extremely beneficial. FDEs describe the memory and hereditary properties of different substances which the conventional models are incapable to take care of.

Several numerical methods have been used to obtained approximate solutions to FDEs including finite difference method, finite element method and the spectral method. The finite difference method is the most efficient numerical method and is a powerful tool owing to its easy implementation. It is used for solving the time and/or space fractional diffusion wave equations. For instance, Khader *et al.* [6] obtained the numerical solutions of time fractional diffusion wave equations by using the Hermite formula. Sun and Wu [17] and Sweilam *et al.* [18] obtained the numerical solution of time and two sided space fractional wave equations,

Supported by University Grants Commission, New Delhi-110002, India.

M. Paszynski et al. (Eds.): ICCS 2021, LNCS 12742, pp. 408–421, 2021.
https://doi.org/10.1007/978-3-030-77961-0_34

respectively, using finite difference scheme. Chen and Li [1] and Zhang *et al.* [19] used a compact finite difference scheme to obtain the approximate solution for a time fractional diffusion wave equations.

This paper is concerned with the space-time fractional diffusion wave equation. We derived the finite difference approximation for the following initial boundary value problem:

$$\begin{cases} CD_{0,t}^{\alpha}u(x,t) = -(-\Delta)^{\beta/2}u(x,t) + g(x,t), & 0 < t < T, 0 < x < 1, \\ u(0,t) = u(1,t) = 0, & 0 < t < T, \\ u(x,0) = v_1(x), \ \frac{\partial u(x,0)}{\partial t} = v_2(x), & 0 < x < 1, \end{cases} \tag{1}$$

where $1 < \alpha < 2$, $1 < \beta < 2$, $CD_{0,t}^{\alpha}$ denote the left-side Caputo fractional derivative of order α with respect to t and $-(-\Delta)^{\beta/2}$ is the Riesz space-fractional derivative of order β.

The article is drafted as follows. Section 2 provides some basic definitions and theorems along with the derivation of the analytic solution of the proposed problem. Section 3 describes the formulation of the implicit finite difference scheme. In Sect. 4, we present the unconditional stability of the implicit finite difference scheme. We also investigate the convergence and error estimate of the scheme. In Sect. 5, a numerical example with known exact solution is presented to verify the convergence.

2 Preliminaries

In this Section, we present some basic concepts required for our work. These have been taken from [2,8,9].

Definition 1. *The left and right Riemann-Liouville derivatives with $\beta > 0$ of the given function $f(t)$, $t \in (a,b)$ are defined as*

$$RLD_{a,t}^{\beta}f(t) = \frac{1}{\Gamma(m-\beta)}\frac{d^m}{dt^m}\int_a^t (t-s)^{m-\beta-1}f(s)ds,$$

and

$$RLD_{t,b}^{\beta}f(t) = \frac{(-1)^m}{\Gamma(m-\beta)}\frac{d^m}{dt^m}\int_t^b (s-t)^{m-\beta-1}f(s)ds,$$

respectively, where m is a positive integer satisfying $m-1 \le \beta < m$.

Definition 2. *The left and right Grünwald $-$ Letnikov derivatives with order $\beta > 0$ of the given function $f(t)$, $t \in (a,b)$ are defined as*

$$GLD_{a,t}^{\beta}f(t) = \lim_{\substack{h \to 0 \\ Nh=t-a}} h^{-\beta}\sum_{j=0}^{N}(-1)^j\binom{\beta}{j}f(t-jh),$$

and

$$GLD_{t,b}^{\beta}f(t) = \lim_{\substack{h \to 0 \\ Nh=b-t}} h^{-\beta}\sum_{j=0}^{N}(-1)^j\binom{\beta}{j}f(t+jh).$$

Definition 3. *The left Caputo derivative with order $\alpha > 0$ of the given function $f(t)$, $t \in (a, b)$ is defined as*

$$CD_{a,t}^{\alpha}f(t) = \frac{1}{\Gamma(m-\alpha)} \int_{a}^{t} (t-s)^{m-\alpha-1} f^{m}(s)ds,$$

where m is a positive integer satisfying $m - 1 < \alpha \leq m$.

Definition 4. *A real or complex-valued function $f(x)$, $x > 0$, is said to be in the space C_{γ}, $\gamma \in \mathbb{R}$, if there exists a real number $p > \gamma$ such that*

$$f(x) = x^{p} f_{1}(x),$$

for a function $f_{1}(x)$ in $C([0, \infty])$.

Definition 5. *A function $f(x)$, $x > 0$, is said to be in the space C_{γ}^{m}, $m \in \mathbb{N}_{0} = \mathbb{N} \cup \{0\}$, if and only if $f^{m} \in C_{\gamma}$.*

Definition 6 (Multivariate Mittag-Leffler fuction). *A multivariate Mittag-Leffler function $E_{(a_{1},...,a_{n}),b}(z_{1}, ..., z_{n})$ of n complex variables $z_{1}, ..., z_{n} \in \mathbb{C}$ with parameters $a_{1}, ..., a_{n}, b$, is defined as*

$$E_{(a_{1},...,a_{n}),b}(z_{1}, ..., z_{n}) = \sum_{k=0}^{\infty} \sum_{\substack{l_{1}+...+l_{n}=k \\ l_{1},...,l_{n} \geq 0}} (k; l_{1}, ..., l_{n}) \frac{\prod_{i=1}^{n} z_{i}^{l_{i}}}{\Gamma(b + \sum_{i=1}^{n} a_{i}l_{i})},$$

in terms of multinomial coefficients

$$(k; l_{1}, ..., l_{n}) = \frac{k!}{l_{1}!...l_{n}!}, \quad k, l_{1}, ..., l_{n} \in \mathbb{N},$$

where $b > 0$, and all $a_{i} > 0$. In particular, if $n = 1$, the multivariate Mittag-Leffler function is reduced to the Mittag-Leffler function

$$E_{a_{1},b}(z_{1}) = \sum_{k=0}^{\infty} \frac{z_{1}^{k}}{\Gamma(b + ka_{1})}, \quad a_{1} > 0, \ b > 0, \ z_{1} \in \mathbb{C}.$$

The Mittag-Leffler function $E_{a_{1},b}(z_{1})$ is a two parameter family of entire functions in z_{1} of order a_{1}^{-1}. It generalizes the exponential function in the sense that $E_{1,1}(z_{1}) = e^{z_{1}}$.

Definition 7. *The Riesz derivative with order $\beta > 0$ of the given function $f(x)$, $x \in (a, b)$ is defined as*

$$_{RZ}D_{x}^{\beta}f(x) = c_{\beta}(_{RL}D_{a,x}^{\beta}f(x) + {}_{RL}D_{x,b}^{\beta}f(x)),$$

where $c_{\beta} = -\frac{1}{2cos(\beta\pi/2)}$, $\beta \neq 2k + 1, k = 0, 1, \cdots$. $_{RZ}D_{x}^{\beta}f(x)$ is sometimes expressed as $\frac{d^{\beta}}{d|x|^{\beta}}$ or $-(-\Delta)^{\beta/2}$.

Using Definition (7) in Eq. (1), we get

$$\begin{cases} {}_cD_{0,t}^{\alpha}u(x,t) = c_{\beta}({}_{RL}D_{0,x}^{\beta}u(x,t) + {}_{RL}D_{x,1}^{\beta}u(x,t)) + g(x,t), & 0 < t < T, 0 < x < 1, \\ u(0,t) = u(1,t) = 0, & 0 < t < T, \\ u(x,0) = v_1(x), \ \frac{\partial u(x,0)}{\partial t} = v_2(x), & 0 < x < 1, \end{cases} \quad (2)$$

where $T > 0$ is a fixed time, ${}_{RL}D_{0,x}^{\beta}$ is the left Riemann-Liouville derivative and ${}_{RL}D_{x,1}^{\beta}$ is the right Riemann-Liouville derivative.

Theorem 1. *Let* $\mu > \mu_1 > \cdots > \mu_n \geq 0$, $m_i - 1 < \mu_i \leq m_i$, $m_i \in \mathbb{N}_0$, $\lambda_i \in \mathbb{R}$, $i = 1, \cdots, n$. *The initial value problem*

$$\begin{cases} ({}_cD_{0,t}^{\mu}y)(t) - \sum_{i=1}^{n} \lambda_i ({}_cD_{0,t}^{\mu_i}y)(t) = g(t), \\ y^{(k)}(0) = c_k \in \mathbb{R}, \ k = 0, \cdots, m-1, \ m-1 < \mu \leq m, \end{cases} \quad (3)$$

where g is assumed to lie in C_{-1} if $\mu \in \mathbb{N}$ or in C_{-1}^1 if $\mu \notin \mathbb{N}$, then (3) has a unique solution in the space C_{-1}^m of the form

$$y(t) = y_g(t) + \sum_{k=0}^{m-1} c_k u_k(t),$$

where

$$y_g(t) = \int_0^t s^{\mu-1} E_{(.),\mu}(s) g(t-s) ds,$$

and

$$u_k(t) = \frac{t^k}{k!} + \sum_{i=l_k+1}^{n} \lambda_i t^{k+\mu-\mu_i} E_{(.),k+1+\mu-\mu_i}(x), k = 0, ..., m-1,$$

fulfills the initial conditions $u_k^l(0) = \delta_{kl}$, $k, l = 0, ..., m-1$. Here,

$$E_{(.),\beta}(t) = E_{\mu-\mu_1,...,\mu-\mu_1,\beta}(\lambda_1 t^{\mu-\mu_1}, ..., \lambda_n t^{\mu-\mu_n}).$$

Proof. See [9].

Remark 21. *In Theorem 1, the natural numbers l_k, $k = 0, ..., m-1$, are determined from the condition $m_{l_k} \geq k+1$ and $m_{l_k+1} \leq k$. In the case $m_i \leq k$, $i = 0, \cdots, m-1$, we set $l_k = 0$ and if $m_i \geq k+1, i = 0, \cdots, m-1$ then $l_k = n$.*

Theorem 2. *Let H be a Hilbert space. If $\{\lambda_n\}_{n\geq 1}$ and $\{\phi_n\}_{n\geq 1}$ are the eigen-values and eigenvectors associated to an operator A in H, then $\{\lambda_n^\alpha\}_{n\geq 1}$ and $\{\phi_n\}_{n\geq 1}$ are the eigenvalues and eigenvectors to the fractional operator A^α, $-1 < \alpha \leq 1$.*

Proof. See [2]

Now we derive one lemma required for the present study using the aforementioned theorems.

Lemma 1. *The solution u of the Problem (1) with $1 < \alpha < 2$ and $1 < \beta \leq 2$ is given by*

$$u(x,t) = \sum_{n=1}^{\infty} u_n(t)\phi_n(x) = \sum_{n=1}^{\infty} \left[\int_0^t s^{\alpha-1} E_{\alpha,\alpha}(-a_n s^\alpha) g_n(t-s) ds + \sum_{k=0}^{1} c_{k,n} y_k(t) \right] \phi_n(x), \quad (4)$$

Proof: In order to solve the non-homogeneous Eq. (1), first we solve the corresponding homogeneous equation (by replacing $g(x,t) = 0$). We first Substitute $u(x,t) = X(x)T(t)$ in the homogeneous equation and obtain a fractional differential equation in $X(x)$:

$$\begin{cases} (-\Delta)^{\beta/2} X(x) = aX(x), & 0 < x < 1, \\ X(0) = 0 = X(1), \end{cases} \quad (5)$$

and a fractional linear differential equation with the Caputo derivative in $T(t)$

$$_C D_{0,t}^\alpha T(t) + aT(t) = 0, \quad (6)$$

where the parameter a is a positive constant.
Applying Theorem (2) with $A = -\Delta$, $\alpha = \beta/2, H = L^2(0,1)$ and $D(A) = H^2(0,1) \cap H_0^1(0,1)$ there exists $\{\lambda_n^{\beta/2}\}_{n\geq 1}$, $\{\phi_n\}_{n\geq 1}$ eigenvalues and eigenvector of Problem (5). Note that for $a = a_n$, $n \geq 1$ we have $X = X_n = \phi_n$ and $a_n = \lambda_n^{\beta/2}$. We now seek a solution of (1) of the form

$$u(x,t) = \sum_{n=1}^{\infty} u_n(t)\phi_n(x), \quad (7)$$

where we assume that the series can be differentiated term by term. In order to determine $u_n(t)$, we expand $g(x,t)$ in the orthonormal complete system $\{\phi_n\}_{n\geq 1}$

$$g(x,t) = \sum_{n=1}^{\infty} g_n(t)\phi_n(x), \quad (8)$$

where

$$g_n(t) = \int_0^1 g(x,t)\phi_n(x)dx.$$

Substituting (7), (8) into (1) yields

$$\sum_{n=1}^{\infty} \phi_n(x)[{_CD_{0,t}^\alpha}u_n(t)] = \sum_{n=1}^{\infty} \phi_n(x)[-a_n u_n(t) + g_n(t)], \tag{9}$$

where we have used the fact that $\phi_n(x)$ is a solution of (5). Because $\{\phi_n\}_{n\geq 1}$ is an orthonormal system, multiplying both members of (9) by ϕ_n and integrate over $(0, 1)$ we get

$$_CD_{0,t}^\alpha u_n(t) + a_n u_n(t) = g_n(t). \tag{10}$$

On the other hand, because $u(x,t)$ satisfies the initial condition in (1), we must have

$$\sum_{n=1}^{\infty} \partial_t^{(k)} u_n(0)\phi_n(x) = v_k(x), \quad k = 0, 1, \; x \in [0, 1],$$

which implies

$$\partial_t^{(k)} u_n(0) = \int_0^1 v_k(x)\phi_n(x)dx := c_{k,n}, \quad n \geq 1, \; k = 0, 1. \tag{11}$$

Finally, for each value of n, (10) and (11) constitute a fractional initial value problem. According to Theorem (1), the fractional initial value problems (10) and (11) has the analytic solution

$$u_n(t) = \int_0^t s^{\alpha-1} E_{\alpha,\alpha}(-a_n s^\alpha g_n(t-s)ds + \sum_{k=0}^{1} c_{k,n} y_k(t), \quad t \geq 0, \tag{12}$$

where

$$y_k(t) = \frac{t^k}{k!} + \sum_{i=l_k+1}^{1} \xi_i t^{k+\alpha} E_{\alpha,k+1+\alpha}(\xi_i t^\alpha), \quad k = 0, 1.$$

According to Remark (21) $l_k = 0$, then

$$y_k(t) = \frac{t^k}{k!} + \xi_1 t^{k+\alpha} E_{\alpha,k+1+\alpha}(\xi_1 t^\alpha), \quad k = 0, 1$$

with $\xi_1 = -a_n$. Hence we get the analytic solution of the initial boundary value problem (1)

$$u(x,t) = \sum_{n=1}^{\infty} u_n(t)\phi_n(x) = \sum_{n=1}^{\infty} \left[\int_0^t s^{\alpha-1} E_{\alpha,\alpha}(-a_n s^\alpha)g_n(t-s)ds + \sum_{k=0}^{1} c_{k,n} y_k(t) \right] \phi_n(x).$$

$\square$

3   Finite Difference Approximation

To establish the numerical approximation scheme, we define the following nota-tions. Let $\tau = T/K$ be the grid size in time direction with $t_n = n\tau$ $(n = 1, \cdots, K)$

and $h = 1/N$ be the grid size in spatial direction with $x_i = ih$ ($i = 0, 1, \cdots, N$). Also, let $u_i^n \approx u(x_i, t_n)$, and $g_i^n = g(x_i, t_n)$. The finite difference approximation for Caputo fractional derivative appeared in problem (2) is derived as follows [8]

$$_C D_{0,t}^{\alpha} u(x_i, t_{n+1}) = \frac{1}{\Gamma(2-\alpha)} \sum_{j=0}^{n} \int_{t_j}^{t_{j+1}} s^{1-\alpha} u''(x_i, t_n - s) ds.$$

On each subinterval $[t_j, t_{j+1}]$, $u''(x_i, t_{n+1} - s)$ is approximated by $\frac{u(x_i, t_{n-j-1}) - 2u(x_i, t_{n-j}) + u(x_i, t_{n-j+1})}{\tau^2}$, then the derived L2 scheme is

$$_C D_{0,t}^{\alpha} u(x_i, t_{n+1}) \approx \frac{\tau^{-\alpha}}{\Gamma(3-\alpha)} \sum_{j=0}^{n} b_j [u(x_i, t_{n-j-1}) - 2u(x_i, t_{n-j}) + u(x_i, t_{n+1-j})],$$

(13)

where $b_j = (j+1)^{2-\alpha} - j^{2-\alpha}$. For the approximation of left and right Riemann-Liouville fractional order derivative, we use the left and right shifted *Grünwald–Letnikov* formula (one shift) defined as [8]

$$_{RL} D_{0,x}^{\beta} u(x_i, t_n) \approx \frac{1}{h^{\beta}} \sum_{j=0}^{i+1} w_j^{(\beta)} u(x_{i-j+1}, t_n),$$

(14)

and

$$_{RL} D_{x,1}^{\beta} u(x_i, t_n) \approx \frac{1}{h^{\beta}} \sum_{j=0}^{N-i+1} w_j^{(\beta)} u(x_{i+j-1}, t_n),$$

(15)

respectively.

Let $\lambda = \frac{\tau^{\alpha} \Gamma(3-\alpha)}{h^{\beta} 2\cos(\beta\pi/2)}$, and $\lambda' = \tau^{\alpha} \Gamma(3-\alpha)$ then using (13), (14) and (15) at node point (x_i, t_{n+1}) in (2) we have the following implicit difference scheme

$$\lambda \left[\sum_{j=0}^{i+1} w_j^{(\beta)} u_{i-j+1}^{n+1} + \sum_{j=0}^{N-i+1} w_j^{(\beta)} u_{i+j-1}^{n+1} \right] + u_i^{n+1} = \lambda' g_i^{n+1} - \sum_{j=1}^{n} b_j (u_i^{n-j-1}$$
$$- 2u_i^{n-j} + u_i^{n-j+1}) - u_i^{n-1} + 2u_i^n$$

(16)

We know that $\frac{\partial u(x,0)}{\partial t} = v_2(x)$, therefore

$$u_i^{-1} = u_i^1 - 2\tau v_2(x_i), \quad i = 0, 1, \cdots, N,$$

hence for $n = 0$ we have

$$\lambda \left[\sum_{j=0, j\neq 1}^{i+1} w_j^{(\beta)} u_{i-j+1}^1 + \sum_{j=0, j\neq 1}^{N-i+1} w_j^{(\beta)} u_{i+j-1}^1 \right] + 2(1 + \lambda w_1^{(\beta)}) u_i^1 = \lambda' g_i^1 + 2\tau v_2(x_i) + 2v_1(x_i),$$

(17)

for $n \neq 0$, Eq. (16) is rewritten as

$$
\lambda[\sum_{j=0,j\neq1}^{i+1} w_j^{(\beta)} u_{i-j+1}^{n+1} + \sum_{j=0,j\neq1}^{N-i+1} w_j^{(\beta)} u_{i+j-1}^{n+1}] + (1 + 2\lambda w_1^{(\beta)}) u_i^{n+1} = \lambda' g_i^{n+1} + \sum_{j=1}^{n-1}(-b_{j-1}
$$

$$
+2b_j - b_{j+1}) u_i^{n-j} + (2 - b_1) u_i^n - b_n u_i^1 + (2b_n - b_{n-1}) u_i^0 + 2b_n \tau v_2(x_i).
$$
$$(18)$$

Equations (17) and (18) can be written as

$$
\begin{cases} A_1 U^1 = F_1, \\ A_2 U^{n+1} = (2 - b_1) U^n + \sum_{j=1}^{n-1}(-b_{j-1} + 2b_j - b_{j+1}) U^{n-j} - b_n U^1 + (2b_n - b_{n-1}) U^0 + F_2, \end{cases}
$$
$$(19)$$

where $U^n = \begin{pmatrix} u_1^n \\ u_2^n \\ \vdots \\ u_{N-1}^n \end{pmatrix}$, $F_1 = \lambda' g^1 + 2\tau v_2(x) + 2v_1(x)$, $F_2 = \lambda' g^{n+1} + b_n 2\tau v_2(x)$,

$$
A_1 = (a_{i,j}^{(1)})_{(N-1),(N-1)} = \begin{cases} 2(1 + \lambda w_1^{(\beta)}), & \text{for } i = j \\ \lambda(w_0^{(\beta)} + w_1^{(\beta)}), & \text{for } j = i - 1 \text{ or } j = i + 1 \\ \lambda w_{i-j+1}^{(\beta)}, & \text{for } j < i - 1 \\ \lambda w_{j-i+1}^{(\beta)}, & \text{for } j > i + 1, \end{cases}
$$

$$
A_2 = (a_{i,j}^{(2)})_{(N-1),(N-1)} = \begin{cases} (1 + 2\lambda w_1^{(\beta)}), & \text{for } i = j \\ \lambda(w_0^{(\beta)} + w_1^{(\beta)}), & \text{for } j = i - 1 \text{ or } j = i + 1 \\ \lambda w_{i-j+1}^{(\beta)}, & \text{for } j < i - 1 \\ \lambda w_{j-i+1}^{(\beta)}, & \text{for } j > i + 1. \end{cases}
$$

4   Stability and Convergence

In this section, a theorem [5] is used to proved the stability of implicit finite difference scheme which is discussed in Sect. 3. Furthermore, the convergence of the scheme is also derived. We denote $\|A_1\| = \|A_1\|_\infty = \max_{1 \leq i \leq N-1}\{\sum_{j=1}^{N-1} |a_{i,j}^{(1)}|\}$ and $\|A_2\| = \|A_2\|_\infty = \max_{1 \leq i \leq N-1}\{\sum_{j=1}^{N-1} |a_{i,j}^{(2)}|\}$.

Lemma 2. $\|A_1^{-1}\| \leq 1$ and $\|A_2^{-1}\| \leq 1$.

Proof. We apply the Gerschgorin theorem (see [5] for details)to conclude that every eigenvalue of the matrices A_1 and A_2 have a magnitude strictly larger than 1.

Note that $w_0^{(\beta)} = 1$, $w_1^{(\beta)} = -\beta$, $w_j^{(\beta)} = (-1)^j \frac{\beta(\beta-1)\cdots(\beta-j+1)}{j!}$, $j = 1, 2, 3, \ldots$,

then for $1 < \beta \leq 2$ and $j \geq 2$, we have $w_j^{(\beta)} \geq 0$.

We also know that for any $\beta > 0$,

$$
(1 + z)^\beta = \sum_{m=0}^{\infty} \binom{\beta}{m} z^m, \quad |z| \leq 1. \tag{20}
$$

Substituting $z = -1$ in (20) yields $\sum_{j=0}^{\infty} w_j^{(\beta)} = 0$, and then $-w_1^{(\beta)} > \sum_{j=0, j\neq 1}^{i} w_j^{(\beta)}$, i.e. $\sum_{j=0}^{i} w_j^{(\beta)} < 0$ for any $i = 1, 2, \cdots m$.

Note that non diagonal elements of matrices A_1 and A_2 are same. According to the Gerschgorin theorem, the eigenvalues of the matrix A_1 lie in the union of $N-1$ circles centered at $a_{i,i}^{(1)}$ with radius $r_i = \sum_{k=1, k\neq i}^{N-1} |a_{i,k}^1|$ and the eigenvalues of the matrix A_2 lie in the union of $N-1$ circles centered at $a_{i,i}^{(2)}$ with the same radius. Using the properties of A_1 and A_2, we have

$$a_{i,i}^{(2)} = 1 + 2\lambda w_1^{(\beta)} = 1 - 2\lambda\beta, \quad a_{i,i}^{(1)} = 2 + 2\lambda w_1^{(\beta)} = 2 - 2\lambda\beta,$$

where $\lambda = \frac{\tau^\alpha \Gamma(3-\alpha)}{h^\beta 2\cos(\beta\pi/2)} < 0$, then we conclude that:

$$r_i = \sum_{k=1, k\neq i}^{N-1} |a_{i,k}^{(1)}| = \sum_{k=1, k\neq i}^{N-1} |a_{i,k}^{(2)}|$$
$$\leq 2|\lambda|\beta.$$

We conclude that eigenvalues of the matrix A_1 and A_2 satisfy $|\rho| \geq 1$ and matrix A_1 and A_2 are diagonally dominant. Then A_1 and A_2 are invertible and eigenvalues are less than or equal to 1 in magnitude. Therefore,

$$\|A_1^{-1}\| \leq \frac{1}{\min_{1\leq i\leq M-1}\{|a_{i,i}^{(1)}| - \sum_{j\neq i, j=1}^{N-1} |a_{i,j}^{(1)}|\}} \leq 1.$$

Similarly, $\|A_2^{-1}\| \leq 1$.   □

Now, the stability of implicit finite difference scheme derived in Sect. 3 is proved in the following theorem with help of Lemma 2.

Theorem 3. *The implicit finite difference scheme defined by (19) to the space-time fractional diffusion wave equation (2) with $1 < \alpha < 2$ and $1 < \beta < 2$ is unconditionally stable.*

Proof. To prove the stability of (19), let $\tilde{u}_i^n$ and u_i^n, $(i = 1, \cdots, N-1; n = 1, \cdots, K)$ be the approximate solution of (19). We denote the corresponding error by $\varepsilon_i^n = \tilde{u}_i^n - u_i^n$ and $\varepsilon^n = (\varepsilon_1^n, \cdots, \varepsilon_{N-1}^n)^t$. Then ε^n satisfies if $n = 0$

$$A_1\varepsilon^1 = \varepsilon^0,$$

if $n \geq 0$

$$A_2\varepsilon^{n+1} = \sum_{j=1}^{n-1}(-b_j + 2b_j - b_{j+1})\varepsilon^{n-j} + (2 - b_1)\varepsilon^n + (2b_n - b_{n+1})\varepsilon^0 + b_n\varepsilon^1.$$

Let us prove $\|\varepsilon^n\| \leq C\|\varepsilon^0\|$, $n = 0, 1, 2, \cdots$ by induction, where C is some positive constant. In fact, if $n = 0$

$$\varepsilon^1 = A_1^{-1}\epsilon^0,$$

from that

$$\|\varepsilon^1\| = \|A_1^{-1}\varepsilon^0\| \le \|A_1^{-1}\|\|\varepsilon^0\|.$$

Since $\|A_1^{-1}\| \le 1$, from the Lemma (2), we have $\|\varepsilon^1\| \le \|\varepsilon^0\|$.

Now assume that $\|\varepsilon^s\| \le C\|\varepsilon^0\|$ for all $s \le n$, we will prove it is also true for $s = n + 1$. For that we use the properties of the function $f(x) = (x+1)^{2-\alpha} - x^{2-\alpha}(x \ge 0)$, we say that $b_j^{-1} \le b_n^{-1}$, $j = 0, 1, \cdots, n$; $(2b_j - b_{j-1} - b_{j+1}) > 0$ and $(2 - b_1) > 0$, we have

$$\|\varepsilon^{n+1}\| = \|\sum_{j=1}^{n-1}(-b_j + 2b_j - b_{j+1})A_2^{-1}\varepsilon^{n-j} + (2 - b_1)A_2^{-1}\varepsilon^n + (2b_n - b_{n+1})A_2^{-1}\varepsilon^0 + b_n A_2^{-1}\varepsilon^1\|$$

$$\le \sum_{j=1}^{n-1}(-b_j + 2b_j - b_{j+1})\|A_2^{-1}\varepsilon^{n-j}\| + (2 - b_1)\|A_2^{-1}\varepsilon^n\| + (2b_n - b_{n+1})\|A_2^{-1}\varepsilon^0\| + b_n\|A_2^{-1}\varepsilon^1\|$$

$$\le \sum_{j=1}^{n-1}(-b_j + 2b_j - b_{j+1})\|\varepsilon^{n-j}\| + (2 - b_1)\|\varepsilon^n\| + (2b_n - b_{n+1})\|\varepsilon^0\| + b_n\|\varepsilon^1\|$$

$$\le \Big(\sum_{j=1}^{n-1}(-b_j + 2b_j - b_{j+1}) + (2 - b_1) + (2b_n - b_{n+1}) + b_n\Big)C\|\varepsilon^0\|$$

$$\le C_1\|\varepsilon^0\|.$$

Hence the scheme is unconditionally stable. $\qquad\square$

Now, for the convergence of implicit finite difference scheme, we proved the following lemma.

Lemma 3. *Suppose that $u(x_i, t_n)$ is the exact solution of (2) at grid point (x_i, t_n), u_i^n is the solution of difference equations (17) and (18), then there exists positive constant M such that*

$$\|e^n\|_\infty \le b_{n-1}^{-1}M(\tau^3 + \tau^\alpha h), \quad n = 1, 2, \cdots, K,$$

where $\|e^n\|_\infty = \max\limits_{1 \le i \le N-1}|e_i^n|$, M is a constant independent of h and τ.

Proof. Define $e_i^n = u(x_i, t_n) - u_i^n$, $i = 1, 2, \cdots, N-1$, $n = 1, 2, \cdots, K$, notice that $e^0 = 0$, we have from (17) and (18) if $n = 0$,

$$\lambda\Big[\sum_{j=0,j\neq1}^{i+1} w_j^{(\beta)}e_{i-j+1}^1 + \sum_{j=0,j\neq1}^{N-i+1} w_j^{(\beta)}e_{i+j-1}^1\Big] + 2(1 + \lambda w_1^{(\beta)})e_i^1 = R_i^1,$$

if $n > 0$

$$\lambda\Big[\sum_{j=0,j\neq1}^{i+1} w_j^{(\beta)}e_{i-j+1}^{n+1} + \sum_{j=0,j\neq1}^{N-i+1} w_j^{(\beta)}e_{i+j-1}^{n+1}\Big] + (1 + 2\lambda w_1^{(\beta)})e_i^1$$

$$= \sum_{j=1}^{n-1}(-b_{j-1} + 2b_j + b_{j+1})e_i^{n-j} + (2 - b_1)e_i^n - b_n e_i^1 + R_i^{n+1},$$

where $|R_i^{n+1}| \leq M_0(\tau^3 + \tau^\alpha h), i = 1, 2, \cdots, N-1; n = 1, 2, \cdots, K-1$, and M_0 is positive constant independent of τ and h.

We use the mathematical induction method to prove the theorem. If $n = 1$, suppose $|e_l^1| = \max_{1 \leq i \leq N-1} |e_i^1|$, and we know that $w_1^{(\beta)} = -\beta$, $\sum_{j=0, j \neq 1}^N w_j^{(\beta)} \leq \beta$ and $\lambda < 0$, hence

$$|e_l^1| \leq 2(1 - \lambda\beta)|e_l^1| + \lambda \sum_{j=0, j \neq 1}^{l+1} w_j^{(\beta)}|e_{l-j+1}^1| + \lambda \sum_{j=0, j \neq 1}^{N-l+1} w_j^{(\beta)}|e_{l+j-1}^1|$$

$$\leq \left| 2(1 - \lambda\beta)e_l^1 + \lambda \sum_{j=0, j \neq 1}^{l+1} w_j^{(\beta)}e_{l-j+1}^1 + \lambda \sum_{j=0, j \neq 1}^{N-l+1} w_j^{(\beta)}e_{l+j-1}^1 \right|$$

$$= |R_l^1| \leq C\tau^\alpha(\tau^{3-\alpha} + h) = C\tau^\alpha b_0^{-1}(\tau^{3-\alpha} + h).$$

Suppose that if $n \leq s$, $\|e^s\|_\infty \leq M\tau^\alpha b_{s-1}^{-1}(\tau^{3-\alpha} + h)$ hold, then when $n = s+1$, let $|e_l^{s+1}| = \max_{1 \leq i \leq N-1} |e_i^{s+1}|$. Using the properties of the function $f(x) = (x+1)^{2-\alpha} - x^{2-\alpha}(x \geq 0)$, we say that $b_j^{-1} \leq b_n^{-1}$, $j = 0, 1, \cdots, n$; $(2b_j - b_{j-1} - b_{j+1}) > 0$ and $(2 - b_1) > 0$, we have

$$|e_l^{s+1}| \leq (1 - 2\lambda\beta)|e_l^{s+1}| + \lambda \sum_{j=0, j \neq 1}^{l+1} w_j^{(\beta)}|e_{l-j+1}^{s+1}| + \lambda \sum_{j=0, j \neq 1}^{N-l+1} w_j^{(\beta)}|e_{l+j-1}^{s+1}|$$

$$\leq \left| (1 - 2\lambda\beta)e_l^{s+1} + \lambda \cdot \sum_{j=0, j \neq 1}^{l+1} w_j^{(\beta)}e_{l-j+1}^{s+1} + \lambda \sum_{j=0, j \neq 1}^{N-l+1} w_j^{(\beta)}e_{l+j-1}^{s+1} \right|$$

$$\leq \sum_{j=1}^{s-1}(2b_j - b_{j-1} - b_{j+1})|e_l^{s-j}| + (2 - b_1)|e_l^s| + b_s|e_l^1| + M_0\tau^\alpha(\tau^3 + h)$$

$$\leq M_0 b_s^{-1}\tau^\alpha(\tau^3 + h)(b_s + \sum_{j=1}^{s-1}(2b_j - b_{j-1} - b_{j+1}) + (2 - b_1) + 1)$$

$$\leq 3M_0 b_s^{-1}\tau^\alpha(\tau^3 + h) = M b_s^{-1}\tau^\alpha(\tau^3 + h).$$

Thus $\|e^{s+1}\|_\infty \leq M b_s^{-1}\tau^\alpha(\tau^{3-\alpha} + h)$.   □

Since

$$\lim_{n \to \infty} \frac{b_n^{-1}}{n^\alpha} = \lim_{n \to \infty} \frac{n^{-\alpha}}{(n+1)^{2-\alpha} - n^{2-\alpha}} = \frac{1}{2 - \alpha}.$$

Hence, there exists constant $C > 0$, such that

$$\|e^n\|_\infty \leq n^\alpha C(\tau^3 + \tau^\alpha h) = (n\tau)^\alpha C(\tau^{3-\alpha} + h), \quad n = 1, 2, \cdots, K.$$

When $n\tau \leq T$, we get the following theorem:

Theorem 4. *Suppose that $u(x_i, t_n)$ is the exact solution of (2) at grid point (x_i, t_n), u_i^n is the solution of difference equations (17) and (18), then there exists positive constant C, such that*

$$|u(x_i, t_n) - u_i^n| \leq C(\tau^{3-\alpha} + h), \; i = 1, 2, \cdots, N - 1; \; n = 1, 2, \cdots, K.$$

5   Numerical Results

In this section, we present the numerical results of space-time fractional wave equation to illustrate the presented scheme that is employed in our study. We take

$$u(x, t) = (t^{2+\alpha} + t + 2)x^4(1 - x)^4,$$

is exact solution of (2) which implies that $v_1(x) = 2x^4(1 - x)^4$, $v_2(x) = x^4(1 - x)^4$, $g(x, t) = B + C + D$, where,

$$B = x^4(1 - x)^4 \left[\frac{\Gamma(3 + \alpha)}{\Gamma(3)} t^2 \right],$$

$$
C = \frac{(t^{2+\alpha} + t + 2)}{2\cos(\beta\pi/2)} \left[\frac{\Gamma(9)}{\Gamma(9 - \beta)} x^{8-\beta} - \frac{4\Gamma(8)}{\Gamma(8 - \beta)} x^{7-\beta} + \frac{6\Gamma(7)}{\Gamma(7 - \beta)} x^{6-\beta} - \frac{4\Gamma(6)}{\Gamma(6 - \beta)} x^{5-\beta} \right.
$$
$$
\left. + \frac{\Gamma(5)}{\Gamma(5 - \beta)} x^{4-\beta} \right],
$$

$$
D = \frac{(t^{2+\alpha} + t + 2)}{2\cos(\beta\pi/2)} \left[\frac{\Gamma(9)}{\Gamma(9 - \beta)} (1 - x)^{8-\beta} - \frac{4\Gamma(8)}{\Gamma(8 - \beta)} (1 - x)^{7-\beta} + \frac{6\Gamma(7)}{\Gamma(7 - \beta)} (1 - x)^{6-\beta} \right.
$$
$$
\left. - \frac{4\Gamma(6)}{\Gamma(6 - \beta)} (1 - x)^{5-\beta} + \frac{\Gamma(5)}{\Gamma(5 - \beta)} (1 - x)^{4-\beta} \right].
$$

We measure the absolute maximum error between exact solution and the finite difference approximation U^K. The numerical results by using implicit finite difference scheme with $\alpha = 1.5$, $\tau = 0.001$ and $t_K = T = 1$ are presented in Table 1.

Table 1. The absolute maximum error and convergence rates of the implicit finite difference approximation (19) with $\alpha = 1.5$, $t_K = 1$ and $\tau = 0.001$

1/h	$\beta = 1.2$	Rate	$\beta = 1.5$	Rate	$\beta = 1.8$	Rate
8	0.005900	-	0.001500	-	0.000352	-
16	0.003800	0.6347	0.000947	0.6634	0.000073	2.2772
32	0.002200	0.7885	0.000529	0.8412	0.000042	0.8101
64	0.001200	0.8745	0.000278	0.9286	0.000029	0.5137
128	0.000617	0.9597	0.000141	0.9761	0.000016	0.8786
256	0.000314	0.9745	0.000070	1.0103	0.000007	1.1098
512	0.000157	0.9940	0.000034	1.0503	0.000003	1.2224

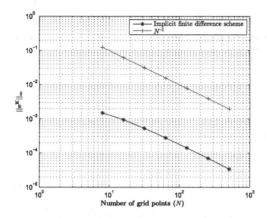

Fig. 1. Absolute maximum error between the numerical solution and exact solution for $\alpha = 1.5$, $\beta = 1.5$, and $T = 1$

We have proved in Sect. 5 that proposed implicit finite difference scheme is first order accurate in spatial variable. In order to show the convergence rate numerically, we take $\tau = 0.001$. In Fig. 1, absolute maximum error between the numerical solution and exact solution for $\alpha = 1.5$, $\beta = 1.5$, and $T = 1$ is plotted with respect to the number of grid points. It can be observed from Fig. 1 that the proposed implicit finite difference scheme exhibits approximately first order convergence rate.

6   Conclusion

In this paper, we provide the implicit finite difference scheme for solving the space-time fractional wave equation as described in (2). The unconditional stability, the rate of convergence and the error estimate of the implicit finite difference scheme are discussed and proved rigorously. The performance of the new scheme is investigated through a numerical example with a known exact solution. From the obtained numerical results in Table 1, we conclude that the numerical solution is in excellent agreement with the exact solution when our scheme is employed to the space-time fractional wave equation as described in (2).

References

1. Chen, A., Li, C.: Numerical solution of fractional diffusion-wave equation. Numer. Funct. Anal. Optim. **37**(1), 19–39 (2016)
2. Fino, A.Z., Ibrahim, H.: Analytical solution for a generalized space-time fractional telegraph equation. Math. Meth. Appl. Sci. **36**(14), 1813–1824 (2013)
3. Hilfer, R.: Fractional calculus and regular variation in thermodynamics. In: World Scientific (2000)
4. Iomin, A., Dorfman, S., Dorfman, L.: On tumor development: fractional transport approach. arXiv preprint arXiv preprint q-bio/0406001 (2004)

5. Isaacson, E., Keller, H.B.: Analysis of Numerical Methods. Wiley, New York (1966)
6. Khader, M.M., Adel, M.H.: Numerical solutions of fractional wave equations using an efficient class of FDM based on the Hermite formula. Adv. Differ. Equ. **2016**(1), 34 (2016)
7. Kilbas, A., Srivastava, H., Trujillo, J.: Theory and Applications of Fractional Differential Equations. Elsevier Science Limited (2006)
8. Li, C., Zeng, F.: Numerical Methods for Fractional Calculus. Chapman and Hall/CRC (2015)
9. Luchko, Y., Gorenflo, R.: An operational method for solving fractional differential equations with the caputo derivatives. Acta Math. Vietnam **24**(2), 207–233 (1999)
10. Mainardi, F., Paradisi, P.: A model of diffusive waves in viscoelasticity based on fractional calculus. In: Proceedings of the 36th IEEE Conference on Decision and Control, vol. 5, pp. 4961–4966 (1997)
11. Meerschaert, M.M., Scalas, E.: Coupled continuous time random walks in finance. Phys. A **370**(1), 114–118 (2006)
12. Meerschaert, M.M., Zhang, Y., Baeumer, B.: Particle tracking for fractional diffusion with two time scales. Comput. Math. Appl. **59**(3), 1078–1086 (2010)
13. Podlubny, I.: Fractional Differential Equations, vol. 198 (1998)
14. Raberto, M., Scalas, E., Mainardi, F.: Waiting-times and returns in high-frequency financial data: an empirical study. Phys. A **314**(1–4), 749–755 (2002)
15. Singh, A.K., Mehra, M.: Uncertainty quantification in fractional stochastic integro-differential equations using Legendre wavelet collocation method. In: Krzhizhanovskaya, V.V., et al. (eds.) ICCS 2020. LNCS, vol. 12138, pp. 58–71. Springer, Cham (2020). https://doi.org/10.1007/978-3-030-50417-5_5
16. Singh, A.K., Mehra, M.: Wavelet collocation method based on Legendre polynomials and its application in solving the stochastic fractional integro-differential equations. J. Comput. Sci. **51**, 101342 (2021). https://doi.org/10.1016/j.jocs.2021.101342
17. Sun, Z., Wu, X.: A fully discrete difference scheme for a diffusion-wave system. Appl. Numer. Math. **56**(2), 193–209 (2006)
18. Sweilam, N.H., Khader, M.M., Nagy, A.: Numerical solution of two-sided space-fractional wave equation using finite difference method. J. Comput. Appl. Math. **235**(8), 2832–2841 (2011)
19. Zhang, Y., Sun, Z., Zhao, X.: Compact alternating direction implicit scheme for the two-dimensional fractional diffusion-wave equation. SIAM J. Numer. Anal. **50**(3), 1535–1555 (2012)

Chebyshev-Type Rational Approximations of the One-Way Helmholtz Equation for Solving a Class of Wave Propagation Problems

Mikhail S. Lytaev[✉][ID]

St. Petersburg Federal Research Center of the Russian Academy of Sciences,
14-th Linia, V.I., No. 39, Saint Petersburg 199178, Russia

Abstract. This study is devoted to improving the efficiency of the numerical methods for solving the pseudo-differential parabolic equation of diffraction theory. A rational approximation on an interval is used instead of the Padé approximation in a vicinity of a point. The relationship between the pseudo-differential propagation operator, variations of the refractive index, and the maximum propagation angle is established. It is shown that using the approximation on an interval is more natural for this problem and allows using a more sparse computational grid than when using the local Padé approximation. The proposed method differs from the existing ones only in the coefficients of the numerical scheme and does not require significant changes in the implementations of the existing numerical schemas. The application of the proposed approach to the tropospheric radio-wave propagation and underwater acoustics is provided. Numerical examples quantitatively demonstrate the advantages of the proposed approach.

Keywords: Wave propagation · Helmholtz equation · Parabolic equation · Diffraction · Rational approximation

1 Introduction

A wide class of wave propagation problems can be effectively tackled by the parabolic equation (PE) method [6,13] and its wide-angle higher-order approximations. Initially, the PE method was proposed by Leontovich and Fock [12] for tropospheric radio-wave propagation problems. PE method allows handling variations of the tropospheric refractive index, irregular terrain, rough sea surface [13], vegetation [23] and backscattering [20,30]. There are works on the application of the PE method in a substantially three-dimensional urban environment [15]. Numerical methods for solving PE have been particularly developed in computational underwater acoustics studies [5,7,26]. Wide-angle approximations and higher-order finite-difference numerical schemas were developed. In modern

© Springer Nature Switzerland AG 2021
M. Paszynski et al. (Eds.): ICCS 2021, LNCS 12742, pp. 422–435, 2021.
https://doi.org/10.1007/978-3-030-77961-0_35

studies, the PE method is usually considered as the one-way Helmholtz equation, which is the generalization of the standard Leontovich-Fock PE [10]. This method is also widely used in geophysics [19], optics [3] and quantum mechanics [28]. The use of the principle of universality of mathematical models [25] promotes the mutual exchange of numerical methods for the PE between different subject areas.

The wide popularity of the PE method in the wave propagation studies is due to its strict deterministic nature and, at the same time, its high computational efficiency. Several computational programs for various purposes have been developed on the basis of the PE method: AREPS [4], PETOOL [21], RAM [24], CARPET [11]. At the same time, the problem of developing fast and reliable numerical schemas for solving PE remains relevant. Most of the works on the numerical solution of the PE are purely theoretical in nature and do not take into account the features of using these numerical algorithms in complex software systems. It is important not only to develop an efficient numerical algorithm but also to determine the limits of its applicability depending on the input data. At the same time, these algorithms should work autonomously, without an expert's intervention. Suitable artificial parameters of numerical schemas, such as the grid steps and approximation order, should be selected automatically based on the input data and the required accuracy [16].

To solve the above mentioned problems, a deep theoretical analysis of the numerical scheme is required. In this paper, we analyze the structure of the pseudo-differential propagation operator, which enables us to establish the relationship between the input parameters of the algorithm and the required approximation accuracy. This analysis allowed us to choose a more suitable approximation of the propagation operator than the existing ones.

The paper is organized as follows. The next section briefly describes the mathematical formulation of the problem. Section 3 is devoted to the development and analysis of various rational approximations of the propagation operator. Section 4 analyzes the results of numerical simulations for various propagation scenarios.

2   Problem Statement

Complex wave field component $\psi(x, z)$ follows the two-dimensional Helmholtz equation

$$\frac{\partial^2 \psi}{\partial x^2} + \frac{\partial^2 \psi}{\partial z^2} + k^2 n^2(x, z)\psi = 0, \qquad (1)$$

where $n(x, z)$ is the refractive index of the medium, $k = 2\pi/\lambda$ is the wavenumber, λ is the wavelength. The schematic description of the problem under consideration is shown in Fig. 1. Depending on the specifics of a particular task, function ψ is subject to the impedance boundary condition [13] or transparent boundary condition [8, 18, 27] on the lower and upper boundaries of the computational domain. The wave process is generated by the initial Dirichlet condition of the form

$$\psi(0, z) = \psi_0(z)$$

with known function $\psi_0(z)$, which corresponds to the radiation pattern of the source. Depending on the specific task, function ψ can respond to the electromagnetic field [13] or the acoustic pressure field [6].

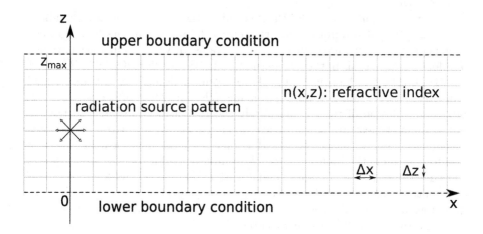

Fig. 1. Schematic description of the considered problem.

Step-by-step solution with longitudinal step Δx for the outgoing waves can be written using the pseudo-differential propagation operator as follows [13]

$$u^{n+1} = \exp\left(ik\Delta x\left(\sqrt{1+L} - 1\right)\right)u^n, \qquad (2)$$

where

$$u^n(z) = u(n\Delta x, z),$$

$$Lu = \frac{1}{k^2}\frac{\partial^2 u}{\partial z^2} + \left(n^2(x, z) - 1\right)u,$$

$$u(x, z) = e^{-ikx}\psi(x, z). \qquad (3)$$

3 Approximation of the Propagator

Using the definition of a pseudo-differential operator [29], we can rewrite propagation operator (2) using the Fourier transform as follows

$$\tilde{u}^{n+1}(k_z) = \exp\left(ik\Delta x\left(\sqrt{1 - \frac{k_z^2}{k^2} + (n^2(x, z) - 1)} - 1\right)\right)\tilde{u}^n(k_z), \qquad (4)$$

where

$$\tilde{u}^n(k_z) = \frac{1}{\sqrt{2\pi}} \int\limits_{-\infty}^{+\infty} u^n(z) e^{-ik_z z} dk_z, \tag{5}$$

$$u^n(z) = \frac{1}{\sqrt{2\pi}} \int\limits_{-\infty}^{+\infty} \tilde{u}^n(k_z) e^{ik_z z} dz. \tag{6}$$

The physical meaning of the Fourier transform (5)–(6) is the decomposition of a vertical wavefront u^n into plane waves. Vertical wavenumber k_z is related to the angle between the plane wave direction and the positive x-axis direction θ (propagation angle) as follows

$$k_z = k \sin \theta.$$

Next, we use a rational approximation of order $[n/m]$

$$\exp\left(ik\Delta x\left(\sqrt{1+\xi}-1\right)\right) \approx \frac{1 + \sum_{l=1}^{m} \tilde{a}_l L^l}{1 + \sum_{l=1}^{n} \tilde{b}_l L^l} = \prod_{l=1}^{p} \frac{1 + a_l \xi}{1 + b_l \xi}, \tag{7}$$

where

$$\xi = -\frac{k_z^2}{k^2} + \left(n^2(x, z) - 1\right). \tag{8}$$

The selection of coefficients and the properties of this approximation will be clarified in the next subsections.

Taking into consideration p-1 new temporary functions $\tilde{v}_l(z)$, expression (4) can be approximately rewritten as follows

$$\begin{cases} \left(1 + b_1\left(-\frac{k_z^2}{k^2} + n^2 - 1\right)\right)\tilde{v}_1^n = \left(1 + a_1\left(-\frac{k_z^2}{k^2} + n^2 - 1\right)\right)\tilde{u}^{n-1} \\ \left(1 + b_l\left(-\frac{k_z^2}{k^2} + n^2 - 1\right)\right)\tilde{v}_l^n = \left(1 + a_l\left(-\frac{k_z^2}{k^2} + n^2 - 1\right)\right)\tilde{v}_{l-1}^n \quad l = 2 \ldots p-1 \\ \cdots \\ \left(1 + b_p\left(-\frac{k_z^2}{k^2} + n^2 - 1\right)\right)\tilde{u}^n = \left(1 + a_p\left(-\frac{k_z^2}{k^2} + n^2 - 1\right)\right)\tilde{v}_{p-1}^n. \end{cases} \tag{9}$$

Applying the inverse Fourier transform to each line of the system (9), we obtain the following system of one-dimensional second-order differential equations

$$\begin{cases} (1 + b_1 L) v_1^n = (1 + a_1 L) u^{n-1} \\ (1 + b_l L) v_l^n = (1 + a_l L) v_{l-1}^n \quad l = 2, \ldots, p-1 \\ \cdots \\ (1 + b_p L) u^n = (1 + a_p L) v_{p-1}^n. \end{cases} \tag{10}$$

System (10) can be solved sequentially from top to bottom. Next, we use the Numerov method [18] with transverse grid step Δz to approximate the second derivative. Then, each line of system (10) can be numerically solved by the tridiagonal matrix method in linear time.

Note that the overall complexity of the propagation algorithm is

$$O\left(\frac{x_{max}}{\Delta x} \cdot \frac{z_{max}}{\Delta z} \cdot p\right),$$

where x_{max} and z_{max} are longitudinal and transverse sizes of the computational domain respectively.

3.1   Padé approximation

Lets now return to the features of approximation (7). In [5] it is proposed to use the Padé approximations [2] to calculate the coefficients a_l and b_l of expansion (7). This approach is called the split-step Padé method. The basis of the Padé approximation is the expansion of the function in the vicinity of the point $\xi = 0$ according to the Taylor series. Then the coefficients of the Taylor expansion are recalculated into a_l and b_l using specially elaborated numerical methods [2]. It is important that the approximation obtained in this way is localized at a specific point ($\xi = 0$).

We introduce the absolute error of approximation of the propagation operator at each longitudinal step

$$R(\xi, a_1 \dots a_p, b_1 \dots b_p, \Delta x) = |\exp\left(ik\Delta x\left(\sqrt{1+\xi}-1\right)\right) - \frac{\prod_{l=1}^p 1 + a_l\xi}{\prod_{l=1}^p 1 + b_l\xi}|.$$

In all further examples, we consider the absolute approximation error equal to 10^{-3} to be acceptable. Note that this error refers to a single step along the longitudinal coordinate, and it accumulates during the step-by-step propagation.

To assess the quality of Padé approximation of the propagator, we consider the dependence of the absolute error of the approximation R on the value ξ for various approximation parameters shown in Fig. 2. As expected, the maximum precision of the Padé approximation is achieved near the point $\xi = 0$, monotonically decreasing as we move away from it. It is clearly seen that an increase in the order of approximation and a decrease in the step Δx lead to a more precise approximation on a larger interval.

3.2   Rational Approximation on an Interval

One can clearly see from expression (8) that the value of ξ is affected by the propagation angle θ and the variations of the refractive index $n(x, z)$. In the case of a homogeneous medium $\xi \in [-\sin^2\theta_{max}, 0]$, where θ_{max} is the maximum propagation angle, which can be estimated based on the geometry of a particular problem. Table 1 shows how the interval size increases depending on the maximum propagation angle.

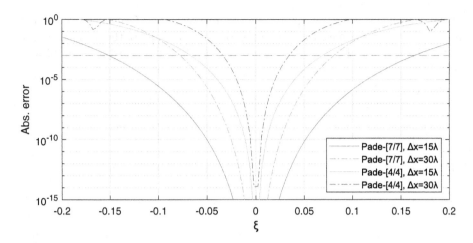

Fig. 2. Dependence of the Padé approximation error on the value ξ for various approximation orders ([4/4] and [7/7]) and longitudinal grid steps Δx (15λ and 30λ).

Table 1. The maximum propagation angle θ_{max} and the value of the interval, on which the approximation should be performed.

θ_{max},°	1	2	5	10	20	45	90
Interval $[-\sin^2 \theta_{max}, 0]$	$[-0.0003,0]$	$[-0.001,0]$	$[-0.008,0]$	$[-0.03, 0]$	$[-0.1, 0]$	$[-0.5, 0]$	$[-1.0, 0]$

Let's assume that the values of the function $n^2(x, z) - 1$ belong to the set $[t_{min}, t_{max}]$. Then $\xi \in [-\sin^2 \theta_{max} + t_{min}, t_{max}]$.

Thus, we naturally come to the necessity of constructing an approximation of function (7) on the interval instead of a local approximation in the vicinity of point $\xi = 0$. Next, we consider the following two rational approximations on an interval, implemented in the Chebfun library [1]: Clenshaw-Lord method [2] (*chebpade* function in Chebfun) and the rational interpolation in the Chebyshev roots [22] (*ratinterp* function in Chebfun). Clenshaw-Lord method is based on the Chebyshev series expansion instead of the Taylor series, thus it is also known as Chebyshev-Padé method.

Figure 3 demonstrates the dependence of the absolute approximation error R on the propagation angle θ for the Padé approximation, Clenshaw-Lord approximation, and interpolation in Chebyshev roots. In all three cases, the approximation order is [7/7], $\Delta x = 200\lambda$, so all three considered configurations are computationally equivalent. It is clearly seen that with the selected parameters, the Padé approximation can provide the required accuracy only for propagation angles up to 6°. Approximation on the interval, in turn, allows one to take into account the propagation angles up to 10° with the same approximation order and the value of Δx. The local nature of the Padé approximation is clearly observable. Namely, it has excessive accuracy at small propagation angles, while the accuracy monotonically decreases with increasing propagation angle. The error of the Clenshaw-Lord approximation and interpolation in Chebyshev nodes does not exceed the threshold for the entire interval $[0; 10°]$. At the same time, the latter gives slightly better accuracy, so we will use it in all further examples.

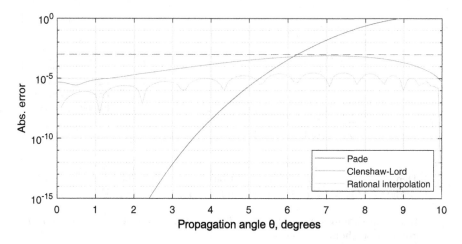

Fig. 3. Dependence of the approximation error R on the propagation angle for the Padé approximation, Clenshaw-Lord approximation, and rational interpolation in the Chebyshev roots. In all cases $\Delta x = 200\lambda$, rational approximation order is $[7/7]$.

4   Numerical Results and Discussion

This section presents the results of numerical simulation obtained by the proposed method of rational interpolation and Padé approximation. An implementation of the rational interpolation method from the Chebfun library [1] was used. The rest of the functionality, including the step-by-step numerical scheme and transparent boundary conditions, is implemented in the Python 3 library [14] developed by the author.

4.1   Radio-Wave Propagation

In the first example, we consider the diffraction of the electromagnetic waves on an impenetrable wedge located on a perfectly conductive surface. A transparent boundary condition is established at the upper boundary of the computational domain. Problems with such geometry arise when computing the tropospheric radio-wave propagation over irregular terrain [13]. The Gaussian horizontally polarized antenna [13] is located at an altitude of 150 m above the surface and emits a harmonic signal at a frequency of 3 GHz. A 150 m high wedge is located at a distance of 3 km from the source of radiation. The wedge is modeled by a staircase approximation. The results obtained by the Padé approximation method and the proposed method are shown in Fig. 4 and 5. It is clearly seen that for the selected longitudinal grid step $\Delta x = 25\lambda$, the proposed method yields a correct result with an approximation order of [4/5]. At the same time, to achieve the same accuracy using the Padé approximation with the same grid size, the order of [10/11] is required. Using the order of [4/5] for the Padé approximation leads to incorrect results in the diffraction zone behind the obstacle at large propagation angles. Thus, in this example, the proposed method is more than twice as fast as the Padé approximation. The dependence of the approximation error on the propagation angle, depicted in Fig. 6, confirms these conclusions.

It should be noted that the transparent boundary condition [9,17] originally developed for Padé approximations also works correctly for the proposed numerical scheme. However, this experimental observation requires further mathematical analysis.

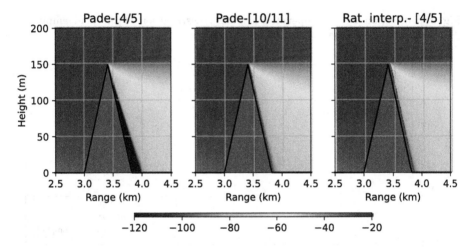

Fig. 4. Diffraction on the impenetrable wedge. Spatial distribution of the field amplitude ($20 \log |\psi(x,z)|$). In all examples $\Delta x = 25\lambda$, $\Delta z = 0.25\lambda$.

4.2   Propagation in a Shallow Water

In the following example, we will consider diffraction on a permeable wedge located on a permeable surface. Similar problems arise when calculating the acoustic pressure field in an inhomogeneous underwater environment. The source 50 Hz acoustic waves is located at a depth of 50 m. The upper surface between water and air is considered smooth. The wedge and the surface on which it is located are modeled by spatial variations of the refractive index n. The field is calculated simultaneously in water and sediment, and a transparent boundary condition is set at the lower boundary. The refractive index in this case is expressed as follows

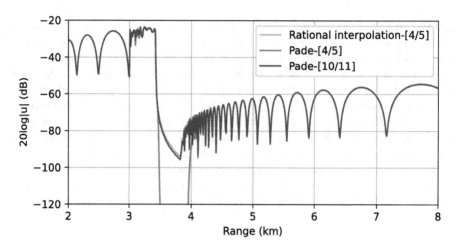

Fig. 5. Diffraction on the impenetrable wedge. Distribution of the field amplitude ($20 \log |\psi(x,z)|$) at the height of $5\,\mathrm{m}$.

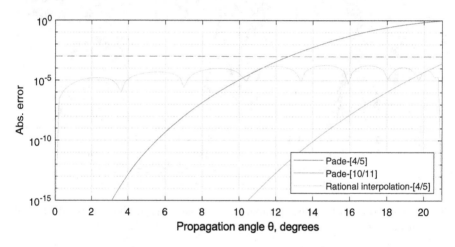

Fig. 6. Diffraction on the impenetrable wedge. Dependence of the approximation error on the propagation angle for the Padé approximation and rational interpolation, $\Delta x = 25\lambda$.

$$n^2(x,z) = \left(\frac{c_0}{c(x,z)} \right)^2,$$

where $c(x,z)$ is the sound speed, $c_0 = 1500$ m/s is the reference sound speed. Constant c_0 can be chosen arbitrary, based on the fact that the wave number in this case is expressed as $k_0 = 2\pi f/c_0$. In this example, the sound speed in the water is $1500\,\mathrm{m/s}$, in the sediment and inside the wedge is $1700\,\mathrm{m/s}$. Additional damping of 0.5 dB per wavelength is also posed inside the sediment and wedge.

For the sake of simplicity, we do not consider variations of the density in this paper, although this should be done in the future works. For the peculiarities of the mathematical formulation of computational hydroacoustics problems, we refer the reader to [6].

The results of the numerical simulation are shown in Fig. 7 and 8. The proposed scheme requires the use of the approximation order of [2/3] when $\Delta x = 6\lambda$, $\Delta z = 0.03\lambda$, while the split-step Padé method for the same computational grid requires the order of [5/6], that is, twice the computational cost. The approximation was based on the segment $\xi \in [-0.23, 0]$. The dependence of the approximation error on ξ is shown in Fig. 9.

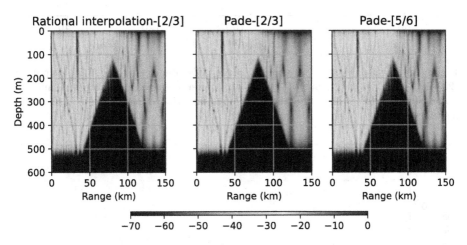

Fig. 7. Diffraction by the penetrable wedge. Spatial distribution of the field amplitude $(20 \log |\psi(x, z)|)$. In all examples $\Delta x = 6\lambda$, $\Delta z = 0.03\lambda$.

4.3   Analysis of the Computational Grid Density

In the last example, we demonstrate how the required density of the computational grid changes with increasing the maximum propagation angle θ_{max} for the Padé approximation and the proposed method. For both methods, an approximation order of [7/8] is used. We require that at a distance of $x_{max} = 10^5\lambda$ from the source, an absolute error should not exceed $tol = 10^{-3}$. Than, following [16], we can compute the optimal value of Δx by solving the following minimization problem

$$n_x = \frac{x_{max}}{\Delta x} \to \min$$

under condition

$$\max_{\xi \in [-\sin^2 \theta_{max}, 0]} R(\xi, a_1 \dots a_p, b_1 \dots b_p, \Delta x) \cdot n_x < tol.$$

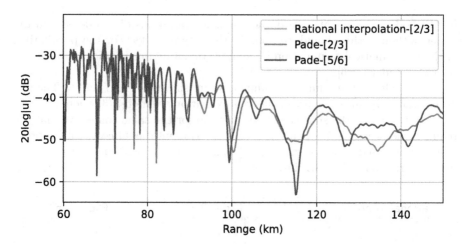

Fig. 8. Diffraction by the penetrable wedge. Distribution of the field amplitude $(20 \log |\psi(x,z)|)$ at the depth of $50\,\mathrm{m}$. The difference between the rational interpolation of the order of $[2/3]$ and the Padé approximation of the order of $[5/6]$ is not distinguishable.

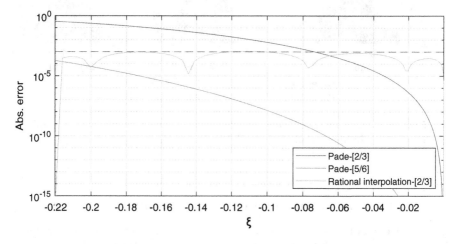

Fig. 9. Diffraction by the penetrable wedge. Dependence of the approximation error on ξ for Padé approximation and rational interpolation, $\Delta x = 6\lambda$.

The last expression is obtained based on the assertion that the error R accumulates at each step, and the total number of steps is n_x. Table 2 shows the minimization results for the maximum propagation angles θ_{max}, equal to $3°$, $10°$, $20°$, $45°$, $60°$ and $80°$. It can be seen that the proposed method makes it possible to use a much more sparse computational grid. Interestingly, that for propagation angles of $60°$ and $80°$ the specified algorithm could not find reasonable values of Δx for the Padé approximation method at all.

We are to keep in mind that a more rigorous mathematical error analysis of the proposed numerical scheme should be carried out in the future.

Table 2. Optimal values of the longitudinal grid step Δx for Padé approximation and rational interpolation (larger is better). The order of approximation in both cases is equal to [7/8]. '-' indicates that the target accuracy was not achieved at reasonable values of Δx.

Propagation angle θ	3°	10°	20°	45°	60°	80°
Optimal Δx for the Padé approximation	600λ	50λ	11λ	1.5λ	-	-
Optimal Δx for the proposed method	2600λ	200λ	40λ	8λ	4λ	1.4λ

5 Conclusion

It was shown that the use of rational interpolation on the interval gives an opportunity to decrease the computational time of the numerical scheme for solving the unidirectional Helmholtz equation by 2–5 times. It was shown that the approximation of the propagation operator on the segment is more natural for the considered problem. The proposed approach does not require significant changes to the existing numerical schemas and its implementations. A significant increase in the performance of a numerical scheme can be achieved only by changing its coefficients, without increasing the density of the computational grid or the order of approximation. The relationship between the input data of the algorithm (variations of the refractive index and the maximum propagation angle) and the required approximation was established. This allows us to implement more accurate and reliable numerical algorithms adapted to a particular situation. It is shown that the proposed method can be effectively applied in computational hydroacoustics and tropospheric radio wave propagation.

Other rational approximations should be studied in the future works. It is also necessary to study the stability and limitations of the proposed method, although the conducted computational experiments give reason to believe that the proposed numerical scheme, as well as the split-step Padé method, is absolutely stable. In this paper, the two-dimensional propagation was considered, but the proposed method can be generalized to a more general three-dimensional case in the future works.

References

1. Chebfun (2020). https://www.chebfun.org/
2. Baker, G.A., Graves-Morris, P.: Pade Approximants, vol. 59. Cambridge University Press (1996)

3. Bekker, E.V., Sewell, P., Benson, T.M., Vukovic, A.: Wide-angle alternating-direction implicit finite-difference beam propagation method. J. Light. Technol. **27**(14), 2595–2604 (2009)
4. Brookner, E., Cornely, P.R., Lok, Y.F.: Areps and temper-getting familiar with these powerful propagation software tools. In: IEEE Radar Conference, pp. 1034–1043. IEEE (2007)
5. Collins, M.D.: A split-step pade solution for the parabolic equation method. J. Acoust. Soc. Am. **93**(4), 1736–1742 (1993)
6. Collins, M.D., Siegmann, W.: Parabolic Wave Equations with Applications. Springer, New York (2019). https://doi.org/10.1007/978-1-4939-9934-7_4
7. Collins, M.D., Westwood, E.K.: A higher-order energy-conserving parabolic equation for range-dependent ocean depth, sound speed, and density. J. Acoust. Soc. Am. **89**(3), 1068–1075 (1991)
8. Ehrhardt, M.: Discrete transparent boundary conditions for Schrodinger-type equations for non-compactly supported initial data. Appl. Numer. Math. **58**(5), 660–673 (2008)
9. Ehrhardt, M., Zisowsky, A.: Discrete non-local boundary conditions for split-step padé approximations of the one-way Helmholtz equation. J. Comput. Appl. Math. **200**(2), 471–490 (2007)
10. Fishman, L., McCoy, J.J.: Derivation and application of extended parabolic wave theories. i. The factorized Helmholtz equation. J. Math. Phys. **25**(2), 285–296 (1984)
11. Huizing, A.G., Theil, A.: CARPET Version 3. Computer-Aided Radar Performance Evaluation Tool, User's manual (2019)
12. Leontovich, M.A., Fock, V.A.: Solution of the problem of propagation of electromagnetic waves along the earth's surface by the method of parabolic equation. J. Phys. USSR **10**(1), 13–23 (1946)
13. Levy, M.F.: Parabolic Equation Methods for Electromagnetic Wave Propagation. The Institution of Electrical Engineers, UK (2000)
14. Lytaev, M.S.: Python wave proragation library (2020). https://github.com/mikelytaev/wave-propagation
15. Lytaev, M., Borisov, E., Vladyko, A.: V2i propagation loss predictions in simplified urban environment: a two-way parabolic equation approach. Electronics **9**(12), 2011 (2020)
16. Lytaev, M.S.: Automated selection of the computational parameters for the higher-order parabolic equation numerical methods. In: Gervasi, O., et al. (eds.) ICCSA 2020. LNCS, vol. 12249, pp. 296–311. Springer, Cham (2020). https://doi.org/10.1007/978-3-030-58799-4_22
17. Lytaev, M.S.: Nonlocal boundary conditions for split-step padé approximations of the Helmholtz equation with modified refractive index. IEEE Antennas Wirel. Propag. Lett. **17**(8), 1561–1565 (2018)
18. Lytaev, M.S.: Numerov-pade scheme for the one-way Helmholtz equation in tropospheric radio-wave propagation. IEEE Antennas Wirel. Propag. Lett. **19**, 2167–2171 (2020)
19. Meng, X., Sun, J., Wei, P., Xu, Y., Xu, Z.: Superwide-angle seismic migration by including the back-scattered wave through domain decomposition and biaxial parabolic approximation. J. Petrol. Sci. Eng. **198**, 108194 (2021)
20. Ozgun, O.: Recursive two-way parabolic equation approach for modeling terrain effects in tropospheric propagation. IEEE Trans. Antennas Propag. **57**(9), 2706–2714 (2009)

21. Ozgun, O., et al.: Petool v2.0: parabolic equation toolbox with evaporation duct models and real environment data. Comput. Phys. Commun. **256**, 107454 (2020)
22. Pachón, R., Gonnet, P., Van Deun, J.: Fast and stable rational interpolation in roots of unity and Chebyshev points. SIAM J. Numer. Anal. **50**(3), 1713–1734 (2012)
23. Permyakov, V.A., Mikhailov, M.S., Malevich, E.S.: Analysis of propagation of electromagnetic waves in difficult conditions by the parabolic equation method. IEEE Trans. Antennas Propag. **67**(4), 2167–2175 (2019)
24. Porter, M.: Ocean acoustics library (2020). https://oalib-acoustics.org/
25. Samarskii, A.A., Mikhailov, A.P.: Principles of Mathematical Modelling: Ideas, Methods, Examples. Taylor and Francis (2002)
26. Sanders, W.M., Collins, M.D.: Nonuniform depth grids in parabolic equation solutions. J. Acoust. Soc. Am. **133**(4), 1953–1958 (2013)
27. Schmidt, F., Friese, T., Yevick, D.: Transparent boundary conditions for split-step pade approximations of the one-way Helmholtz equation. J. Comput. Phys. **170**(2), 696–719 (2001)
28. Schwendt, M., Pötz, W.: Transparent boundary conditions for higher-order finite-difference schemes of the schrödinger equation in (1+ 1)d. Comput. Phys. Commun. **250**, 107048 (2020)
29. Taylor, M.: Pseudodifferential Operators and Nonlinear PDE, vol. 100. Springer, Heidelberg (2012)
30. Vavilov, S.A., Lytaev, M.S.: Modeling equation for multiple knife-edge diffraction. IEEE Trans. Antennas Propag. **68**(5), 3869–3877 (2020)

Investigating In Situ Reduction via Lagrangian Representations for Cosmology and Seismology Applications

Sudhanshu Sane[1]([✉]), Chris R. Johnson[1], and Hank Childs[2]

[1] SCI Institute at University of Utah, Salt Lake City, USA
ssane@sci.utah.edu
[2] University of Oregon, Eugene, USA

Abstract. Although many types of computational simulations produce time-varying vector fields, subsequent analysis is often limited to single time slices due to excessive costs. Fortunately, a new approach using a Lagrangian representation can enable time-varying vector field analysis while mitigating these costs. With this approach, a Lagrangian representation is calculated while the simulation code is running, and the result is explored after the simulation. Importantly, the effectiveness of this approach varies based on the nature of the vector field, requiring in-depth investigation for each application area. With this study, we evaluate the effectiveness for previously unexplored cosmology and seismology applications. We do this by considering encumbrance (on the simulation) and accuracy (of the reconstructed result). To inform encumbrance, we integrated in situ infrastructure with two simulation codes, and evaluated on representative HPC environments, performing Lagrangian in situ reduction using GPUs as well as CPUs. To inform accuracy, our study conducted a statistical analysis across a range of spatiotemporal configurations as well as a qualitative evaluation. In all, we demonstrate effectiveness for both cosmology and seismology—time-varying vector fields from these domains can be reduced to less than 1% of the total data via Lagrangian representations, while maintaining accurate reconstruction and requiring under 10% of total execution time in over 80% of our experiments.

Keywords: Lagrangian analysis · In situ processing · Vector data

1 Introduction

High-performance computing resources play a key role in advancing computational science by enabling modeling of scientific phenomena at high spatiotemporal resolutions. A challenge with regard to studying the output of a simulation is the prohibitively large size of the total data generated. Compromise in the form of storing a subset of the data can impact the extent and accuracy of subsequent post hoc exploratory analysis and visualization. In particular, for accurate time-varying vector field analysis and visualization, access to the full spatiotemporal resolution is required. Since storing the entire simulation output is expensive, scientists resort to temporal subsampling or lossy compression, and often limit analysis

© Springer Nature Switzerland AG 2021
M. Paszynski et al. (Eds.): ICCS 2021, LNCS 12742, pp. 436–450, 2021.
https://doi.org/10.1007/978-3-030-77961-0_36

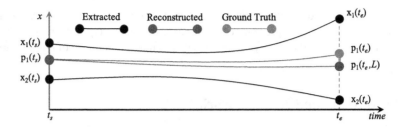

Fig. 1. Notional space-time visualization of Lagrangian representations for a time-varying 1D flow. The black trajectories are computed in situ and encode the behavior of the vector field between start time t_s and end time t_e. In a post hoc setting, a Lagrangian-based advection scheme L, i.e., a technique to interpolate the extracted data, is used to calculate the trajectory of a new particle p_1. The red trajectory is the trajectory reconstructed post hoc and the blue trajectory is the ground truth. The end location of the red trajectory deviates by a margin of error from the ground truth. (Color figure online)

to individual time slices. An emerging paradigm to address large data challenges is the use of in situ processing to perform runtime analysis/visualization or data reduction to support exploratory post hoc analysis.

Lagrangian analysis is a powerful tool to study time-varying vector fields and is widely employed for ocean modeling applications [28]. The notion of calculating a Lagrangian representation or *flow map*, i.e., sets of particle trajectories, "online" (in situ) for "offline" (post hoc) exploration was first proposed by Vries et al. [29] for an ocean modeling simulation. Figure 1 illustrates the approach. More recently, multiple works have advanced Lagrangian research along axes such as strategies for in situ extraction of reduced Lagrangian representations [1,19,22], post hoc reconstruction [6,10,21], and theoretical error analysis [4,5,9].

An open challenge for time-varying vector field exploration is predicting the uncertainty and variability in accuracy for different analysis techniques. Although the effectiveness of Lagrangian representations for any possible time-varying vector field that can be produced by a scientific simulation remains an open question, prior theoretical demonstration of Lagrangian techniques [1,4–6,9,10,19–21] on analytical, SPH, climate and ocean modeling data, and practical application in ocean activity analysis [23], has provided encouraging results. Using Lagrangian representations, the quality of post hoc reconstruction depends on the vector field itself, as well as configuration specifics such as sampling strategy and frequency of storage. Thus, to investigate the potential benefits of Lagrangian representations for a broader range of applications and to gauge its viability in practice, we leverage the recent developments of runtime in situ infrastructure that enable the straightforward extraction via APIs to study Lagrangian representations for cosmology and seisomology applications.

In this paper, our unique contribution is an investigation of Lagrangian representations to encode self-gravitating gas dynamics of a cosmology simulation and seismic wave propagation of a seisomology simulation. We measure the effectiveness of the technique by considering in situ encumbrance and post hoc accuracy. For both applications, our experiments show that Lagrangian representations offer high data reduction, in many cases requiring less than 1% storage of the complete time-varying vector fields, for a small loss of accuracy. Further, our study shows Lagrangian representations are viable to compute in representative HPC environments, requiring under 10% of total execution time for data analysis and visualization in the majority of configurations tested.

2 Background and Related Work

2.1 Frames of Reference

In fluid dynamics, there are two frames of reference to observe fluid motion: Eulerian and Lagrangian. With the Eulerian frame of reference, the observer is in a fixed position. With the Lagrangian frame of reference, the observer is attached to a fluid parcel and is moving through space and time.

Storage of a flow field in an Eulerian representation is typically done by means of its velocity field. A velocity field v is a time-dependent vector field that maps each point $x \in \mathbb{R}^d$ in space to the velocity of the flow field for a given time $t \in \mathbb{R}$

$$v : \mathbb{R}^d \times \mathbb{R} \to \mathbb{R}^d, \, x, t \mapsto v(x,t) \tag{1}$$

In a practical setting, a flow field at a specific time/cycle is defined as vector data on a fixed, discrete mesh. Time-varying flow is represented as a collection of such data over a variety times/cycles.

Storage of a flow field in a Lagrangian representation is done by means of its flow map $F_{t_0}^t$. The flow map is comprised of the starting positions of massless particles x_0 at time t_0 and their respective trajectories that are interpolated using the time-dependent vector field. Mathematically, a flow map is defined as the mapping

$$F_{t_0}^t(x_0) : \mathbb{R} \times \mathbb{R} \times \mathbb{R}^d \to \mathbb{R}^d, \, t \times t_0 \times x_0 \mapsto F_{t_0}^t(x_0) = x(t) \tag{2}$$

of initial values x_0 to the solutions of the ordinary differential equation

$$\frac{d}{dt} x(t) = v(x(t), t) \tag{3}$$

In a practical setting, the flow map is represented as sets of particle trajectories calculated in the time interval $[t_0, t] \subset \mathbb{R}$. The stored information, encoded in the form of known particle trajectories (i.e., a Lagrangian representation), encodes the behavior of the time-dependent vector field over an interval of time.

2.2 Lagrangian Analysis

Within the vector field analysis and visualization community, Lagrangian methods have been increasingly researched in the past decade. In this paper, we focus on the use of Lagrangian methods to store time-varying vector fields in situ and enable subsequent post hoc analysis. In sparse temporal settings, Lagrangian representations are expected to perform better than their Eulerian counterparts. The key intuition behind this expectation is that Lagrangian representations capture the behavior of the flow field over an interval of time, as opposed to the state at a single time slice. However, in addition to the frequency of temporal sampling, the nature of the vector field and spatial sampling resolution impacts the quality of reconstruction.

Agranovsky et al. [1] conducted the seminal work to evaluate the efficacy of reduced Lagrangian representations. To maintain domain coverage, the study proposed the use of uniform spatial sampling to extract sets of temporally non-overlapping basis trajectories. Sane et al. [20] studied performance across a range of spatiotemporal configurations when operating using a fixed storage budget. The experiments in these works were conducted in a theoretical in situ setting, i.e., files were loaded from disk. Most recently, Jakob et al. [10] trained a DNN to upsample FTLE visualizations derived from reduced Lagrangian representations. To generate training data, they first computed Lagrangian representations of a 2D flow field using a tightly-coupled integration with an open-source CFD solver on HPC resources and reported computation costs. However, the grid size of 4×4 per rank used in the study is not representative of real-world applications. Thus, the current literature lacks in situ encumbrance measurements in representative settings.

Lagrangian representations of a time-varying vector field can be extracted by adopting various strategies. Sane et al. [21] explored computing trajectories of variable duration and placement. Rapp et al. [19] applied their void-and-cluster sampling technique to identify a representative set of scattered samples. Although these strategies improved accuracy, they increased computation costs and are presently limited to single node settings. To address scalability challenges of extracting a Lagrangian representation in distributed memory, Sane et al. [22] explored an accuracy-performance tradeoff and demonstrated the use of a communication-free model that stored only trajectories that remain within the rank domain during the interval of computation.

Prior works have presented research pertaining to post hoc reconstruction using Lagrangian-based interpolation schemes. Hlawatsch et al. [8] proposed a hierarchical reconstruction scheme that can improve accuracy, but relies on access to data across multiple time intervals. Chandler et al. [6] proposed a modified k-d tree as a search structure for Lagrangian data extracted from an SPH simulation. Further, Chandler et al. [5] identified correlations between Lagrangian-based interpolation error and divergence in the flow field. Bujack et al. [4] evaluated the use of parameter curves to fit interpolated pathline points to improve the aesthetic of trajectories calculated using Lagrangian data. Lastly, Hummel et al. [9] provided theoretical error bounds for error propagation that can occur when calculating trajectories using Lagrangian representations.

2.3 Time-Varying Vector Field Reduction

Although Eulerian representations have been shown to be susceptible to temporal sparsity [1, 18, 20, 27], temporal subsampling remains the widely used solution to limit data storage. Our study adds to this body of work by using temporal subsampling for comparison. Multiple works have proposed single time step vector field reduction strategies while maintaining an Eulerian representation [13, 25, 26]. Although these techniques could be used to reduce and store data more frequently, they do not inherently address the challenge of increasing temporal sparsity.

In a recent large-scale tornadic supercell thunderstorm study [15], Leigh Orf modi-fied the I/O code to use a hierarchical data format and lossy floating-point compression via ZFP [12]. ZFP provides dynamic accuracy control by allowing the user to specify a maximum amount of deviation. Orf stated that although ZFP is effective for scalar fields that do not require differentiation during post hoc analysis, only a very small value of deviation can be chosen for each component of velocity to maintain accu-rate time-varying vector field reconstruction. Thus, ZFP allowed a limited amount of compression to vector field data without introducing significant uncertainty to post hoc analysis. The technique provided an average reduction of 30% of total uncompressed vector field data, with regions of high gradient resulting in less compression. Overall, Orf referred to the use of lossy compression as unfortunate but necessary.

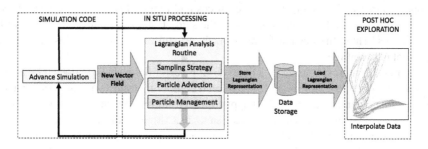

Fig. 2. Schematic of the Lagrangian in situ reduction and post hoc exploration workflow.

3   In Situ Reduction via Lagrangian Representations

This section describes the instantiation we consider for our study. Figure 2 shows a high-level description of the Lagrangian in situ reduction post hoc exploration (L-ISR-PHE) workflow. For our study, we focused on the current best practices in this space. To describe our instantiation, the remainder of this section is divided based on the two phases: in situ reduction and post hoc exploration.

In Situ Reduction. Both simulations we considered partitioned space amongst ranks, with each rank owning one portion of the vector field. Our in situ routines followed this pattern, with an instance of our Lagrangian analysis routine executing for each rank, accessing its portion of the vector field. Further, for both simulations we were interested in capturing time-varying vector field behavior across the entire domain. Thus, for our in situ data reduction strategy, we prioritized domain coverage. Similar to Agranovsky et al. [1], we used uniform spatial sampling and a predetermined interval to store/reset particles. Thus, we computed sets of temporally non-overlapping basis trajectories over the duration of the simulation. Each set of basis trajectories encodes the behavior of the time-varying vector field over a specific interval of time. Our particle termination followed the local Lagrangian flow map model from Sane et al. [22], where particles are terminated once they reach the end of the interval or exit the block. Our implementation had two main knobs that control the total data storage and quality of reconstruction:

number of basis trajectories, i.e., spatial sampling resolution, and frequency of storing information to disk, i.e., storage interval. The effect of these settings varies depending on the underlying vector field.

We used the Ascent [11] in situ infrastructure and VTK-m [14] library to implement L-ISR. The Ascent API can be used to perform tightly-coupled integration with an application code and access various in situ analytics capabilities. The VTK-m Lagrangian filter on each rank operated independently and maintained its own list of particles. We used the existing particle advection infrastructure available in VTK-m [17]. RK4 particle advection is implemented using VTK-m worklets (kernels) that offer performance portability by utilizing the underlying hardware accelerators. In our implementation, each Lagrangian filter stored the displacement of each particle (three double), as well as its validity (one Boolean), i.e., whether the particle remained within the domain during the interval of calculation. Overall, computing a Lagrangian representation increased the runtime memory cost on the simulation by approximately by four one-dimensional simulation "fields". Simulations often have tens to hundreds of fields defined on the simulation grid, and thus, this cost would likely be acceptable for most simulations.

To compute a Lagrangian representation, the simulation invoked Ascent after every cycle it advanced. Ascent accessed the simulation vector field data and consequently invoked the Lagrangian filter. The Lagrangian filter used the vector field to advance particles, and triggered the storage of trajectories at the end of an interval. For integration, all the steps involved—creating an instance of Ascent, specifying parameters, and invoking the VTK-m Lagrangian filter—required only 23 lines of code (C++).

Post Hoc Exploration. For post hoc analysis, new particle trajectories are computed to explore the time-varying vector field. To construct new particle trajectories, we first identified which basis trajectories to follow and then performed interpolation. Based on the study of accuracy of various Lagrangian-based advection schemes in [2], our study employed a Delaunay triangulation to identify the neighborhood of valid basis trajectories and second-order barycentric coordinates for interpolation. We used the CGAL [24] library to construct and search the Delaunay triangulation. After constructing new pathlines or deriving new scalar fields from the basis trajectories, we used VisIt [7] to generate visualizations.

4   Study Overview

This section provides an overview of our study. It is organized as follows: runtime environment (4.1), simulation codes (4.2), experiments (4.3), and evaluation metrics (4.4).

4.1   Runtime Environment

Our study used the Summit supercomputer at ORNL. A Summit compute node has two IBM Power9 CPUs, each with 21 cores running at 3.8 GHz and 512 GBytes of DDR4 memory. Nodes on Summit also have enhanced on-chip acceleration with each CPU connected via NVLink to 3 GPUs, for a total of 6 GPUs per node. Each GPU is an NVIDIA Tesla V100 with 5120 CUDA cores, 6.1 TeraFLOPS of double precision performance, and 16 GBytes of HBM2 memory. Lastly, it has a Mellanox EDR 100G InfiniBand, Non-blocking Fat Tree as its interconnect topology.

4.2 Simulation Codes

Nyx: In this cosmological simulation [3], baryonic matter is evolved by solving the equations of self-gravitating gas dynamics. We derived the velocity field using the fields of momentum and density of the baryonic gas. The simulation involves particles gravitating toward high-density regions to form multiple clusters across the domain. The distribution of high-density clusters and their formation is of interest to scientists. To study the distribution, scientists currently perform statistical analysis of gas particle density at a single time slice. We investigated the potential of reduced Lagrangian representation to accurately visualize the particle evolution and the distribution of high-density clusters using pathlines. The Nyx simulation we built executed as a single rank using two CPUs on a single Summit compute node.

SW4: In this seismology simulation [16], seismic wave propagation is studied using a fourth-order method. The application simulates waves radiating from the epicenter through viscoelastic media. We used the 3D time-varying displacement vector defined at each grid point as input. We investigated how accurately we can derive and visualize the field encoding displacement over time in two regions: at the epicenter and away from the epicenter. The SW4 simulation we built executed using six ranks per compute node on Summit. Each rank was allocated a GPU for execution.

4.3 Experiments

For each application in this study, we organized our experiments to inform in situ encumbrance and post hoc accuracy. We considered four evaluation criteria (EC). To inform in situ encumbrance, we measured the execution time (EC1) and runtime memory usage (EC2) for in situ processing. To inform post hoc accuracy, we measured the size of data artifacts (EC3) and the reconstruction quality of time-varying vector field data (EC4). Next, we identified four factors that when varied produce the workloads we want to evaluate for our study:

- **Number of particles:** the spatial sampling resolution denoted using **1:X**, where X is the reduction factor. For example, a 1:8 configuration states that one basis particle is used for every 8 grid points ($\approx$12.5% of the original data size).
- **Storage interval:** the number of cycles between saves and denoted by **I**.
- **Grid size:** the spatial resolution of the mesh.
- **Concurrency:** the scale of the execution and underlying parallelization hardware.

Rather than consider a complete cross-product of options for every workload factor, we sampled the space of possible options. Our goal was to provide coverage and allow us to see the impact of certain workload factors, all while staying within our compute budget. For Nyx, we ran 18 experiments, with 6 informing in situ encumbrance (varying **1:X**, grid size) and 12 informing post hoc accuracy (varying **1:X**, **I**). For SW4, we ran 11 experiments, with 7 informing in situ encumbrance (varying **1:X**, grid size, concurrency) and 4 informing post hoc accuracy (varying **1:X**). The specific options selected are presented along with the results in Sect. 5.

4.4 Evaluation Metrics

We selected our evaluation metrics based on the evaluation criteria listed in Sect. 4.3.

For EC1, we measured the average cost of invoking the Lagrangian VTK-m filter through Ascent every cycle, **Step**, in seconds. Additionally, we presented the percentage of simulation time spent on data analysis and visualization, or **DAV%**. We used $\mathbf{Sim}_{cycle}$ to denote the average time required for a single simulation cycle in seconds.

For EC2, we measured **InSituMem**, the runtime memory cost incurred by every compute node to maintain the state (current position) of particles at runtime in Bytes.

For EC3, we measured the total data storage (**DS**) required on the file system and report it in Bytes stored. In addition to I/O being infrequently performed, we observed that for the scale of study we conducted, Summit provided fast write times. In comparison to performing in situ processing every cycle, we found the I/O write cost to be negligible.

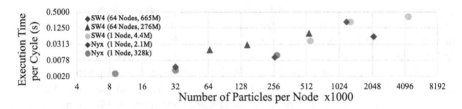

Fig. 3. Lagrangian in situ reduction cost per cycle for all in situ encumbrance experiments. The SW4 simulation executes with six ranks (each allocated one GPU) sharing memory on every node. The Nyx simulation executes on a single rank using all the cores of two CPUs on a single node. The legend includes concurrency and number of simulation grid points in parenthesis and both axes use logarithmic scales.

For EC4, we considered both a statistical and qualitative analysis. For Nyx, we derived pathlines from the basis trajectories and measured the reconstruction error by calculating the average Euclidean distance of interpolated points from the ground truth (precomputed using the complete simulation data) for each trajectory. We visualized the distribution of pathline reconstruction error for every configuration using violin plots, and for a subset of configurations, the pathline clustering directly. For SW4, we derived a field encoding magnitude of displacement over time from the basis trajectories. In this case, we visualized and compared the derived field to the ground truth time-varying displacement field using violin plots and isosurfaces.

5 Results

Our results are organized as follows. Sections 5.1 and 5.2 present findings from our study investigating reduced Lagrangian representations for cosmology and seismology applications, respectively. Tables 1 and 2 provide information pertaining to in situ encumbrance experiments, such as concurrency information, spatial dimensions,

Sim_{cycle}, number of particles per compute node, **InSituMem** per compute node, **Step**, and **DAV%**, for each application. Figure 3 shows the execution time per cycle for all the in situ encumbrance experiments. Figures 4, 5, 6, and 7 show the results of our post hoc accuracy evaluation. For each application, the figures are annotated with configuration specifics such as the **DS**, **1:X**, and **I**. Further, Lagrangian and Eulerian tests are distinguished explicitly in the captions or are labeled LT and ET, respectively, where T is the test number.

5.1 Nyx Cosmology Simulation

In Situ Encumbrance. Using all the cores of two CPUs on a single compute node, we used OpenMP to parallelize the Nyx simulation and Lagrangian VTK-m filter. We tested two options for grid size - 69^3 and 129^3 - on a single rank, and three particle advection workloads (1:1, 1:8, 1:27) each. In a single compute node hour, the simulation performed approximately 300 and 38 cycles when using 69^3 and 129^3 grid sizes, respectively. An 8X increase in grid size resulted in a proportional increase in Sim_{cycle} but only a small increase in particle advection costs for the same number of particles. In practice, we would expect a single rank to operate on between 32^3 to 256^3 grid points, and thus our workloads provided a representative estimate of **DAV%**.

An encouraging finding was the low in situ encumbrance when performing L-ISR on the CPUs. Depending on the setup of various simulations and the form of integration for in situ processing, future work can consider offloading L-ISR computation to CPUs. Overall, considering the longer Sim_{cycle} times for the Nyx simulation, and parallel computation coupled with low memory latency when using CPUs, the highest in situ encumbrance to extract a Lagrangian represenation was 0.1% of the simulation time or under 0.06s to compute 2.1M basis trajectories per cycle.

Table 1. In situ encumbrance evaluation and experiment configurations for the Nyx simulation executing on CPUs.

Nodes	Ranks	Dimensions	Sim_{cycle}	Particles	InSituMem	Step	DAV%
1	1	$65 \times 65 \times 65$	10.9s	9k	0.2MB	0.0025s	0.023%
				32k	0.8MB	0.0033s	0.030%
				274k	6.8MB	0.0122s	0.0112%
		$129 \times 129 \times 129$	88.3s	78k	1.9MB	0.0044s	0.005%
				262k	6.5MB	0.0101s	0.011%
				2.1M	53.6MB	0.0596s	0.067%

Post Hoc Accuracy. To evaluate the usefulness of Lagrangian representations to encode time-varying self-gravitating gas dynamics, we considered a 69^3 grid over 400 cycles, three options for data reduction (1:1, 1:8, 1:27) and four options for **I** (25, 50, 100,

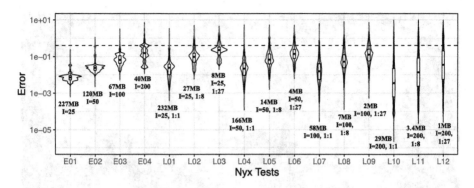

Fig. 4. Accuracy results for the Nyx experiments. Each violin plot shows the distribution of the particle reconstruction error for a specific configuration and the horizontal blue dashed line in the chart represents an error equivalent to a single grid cell side. The error axis uses a logarithmic scale. While Eulerian configurations contain greater uncertainty as the value of storage interval **I** increases, the Lagrangian representations offer the opportunity for improvements in accuracy. Additionally, we find high reconstruction accuracy relies on a high spatial sampling resolution as well.

200). We constructed pathlines for 50,000 randomly placed particles during post hoc analysis. We visualize the distribution of reconstruction error for all tests in Fig. 4.

The self-gravitating gas dynamics of this simulation produce a vector field that captures the transport of randomly distributed particles to multiple high-density clusters. Particles travel with increasing velocity as clusters increase in density. For this data, we found that Eulerian temporal subsampling performs better for small values of **I**. This result can be expected given reconstruction using an Eulerian representation and fourth-order Runge Kutta interpolation remain more accurate than second-order barycentric coordinates interpolation employed to interpolate Lagrangian representations [4,9]. However, as the value of **I** increases, the distribution of error for the Lagrangian tests indicates that a larger percentage of samples are reconstructed more accurately. In particular, this is true when a high spatial sampling resolution is used. Thus, particle evolution in this cosmology simulation can be tracked more accurately when a dense set of basis trajectories integrated for a long duration are interpolated. In contrast, Eulerian representations become less effective at reconstructing the vector field due to increased numerical approximation.

We used pathlines with manually set transfer functions to visualize the evolution and clustering of particles in regions of high density. The total size of the simulation vector field data used to compute the ground truth is 5.3 GB. We visualized a random subset of 10,000 pathlines in Fig. 5 for configurations with **I** set to 25. The Lagrangian representations demonstrate the ability to closely reconstruct regions where dense clusters are formed while requiring a fraction of the total simulation data size. For example, the 1:8 Lagrangian configuration enables the visualization of transport to dense clusters while requiring only 27MB, i.e., a 200X data reduction of the uncompressed vector field.

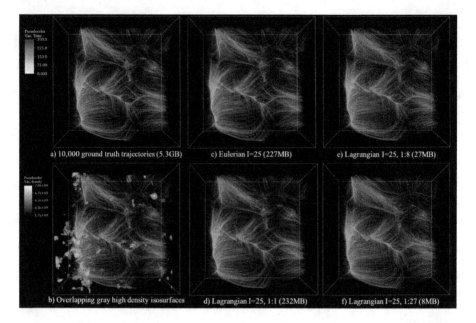

Fig. 5. Pathline visualization of baryonic particles evolution in self-gravitating gas dynamics of Nyx simulation. Using 10,000 randomly seeded particles, we visualize pathlines over 300 cycles. To focus on regions where particles cluster to form dense regions, we set opacity of the pathline to be directly proportional to time. Thus, we are able to focus on clustering as well as provide context of transport toward these regions. Lagrangian representations are able to reconstruct the ground truth trajectories and capture clustering accurately when high spatial sampling is used (1:1, 1:8). However, when using a 1:27 data reduction factor, some clusters are visualized less clearly.

5.2   SW4 Seismology Simulation

In Situ Encumbrance. For the SW4 simulation, we considered five grid sizes at varying concurrencies. In each case, we used all six GPUs available on a compute node to execute the simulation and L-ISR. For all L-ISR workloads tested, the execution time required per cycle remained under 0.5 s on average, and the maximum in situ memory required by a node was 112 MB to compute the trajectories for 4.4M particles. The cost for performing L-ISR was most dependent on the number of particles and only slightly impacted by increasing grid sizes. Referencing Fig. 3, although the SW4 experiments used six GPUs, we found execution time to be slower than Nyx experiments due to the use of shared memory by multiple ranks (each has its own data block) and the high cost of launching kernels on the GPU for limited amounts of computation (each basis particle advances by only a single step/cycle each invocation).

Post Hoc Accuracy. We studied the reconstruction of the time-varying displacement vector field encoding wave propagation by considering four options for data reduction (1:1, 1:8, 1:27, 1:64) and one option for I (250). The ground truth was computed using data defined on a regular mesh containing 4.5M grid points over 2000 simulation cycles and required 245 GB. The displacement was highest near the epicenter

Table 2. In situ encumbrance evaluation and experiment configurations for the SW4 simulation executing on GPUs. Particles and **InSituMem** are per compute node.

Nodes	Ranks	Dimensions	Sim_{cycle}	Particles	InSituMem	Step	DAV%
1	6	$251 \times 251 \times 70$	0.35s	555k	13.89MB	0.0412s	11.67%
		$335 \times 335 \times 93$	2.02s	1.3M	33.16MB	0.2125s	10.48%
		$501 \times 501 \times 139$	7.58s	4.4M	111.13MB	0.3309s	4.365%
64	384	$1001 \times 1001 \times 276$	1.6s	66k	1.6MB	0.0194s	1.201%
			1.5s	146k	3.6MB	0.0295s	1.944%
			1.3s	540k	13.5MB	0.0798s	6.175%
		$1335 \times 1335 \times 368$	2.9 s	1.2M	31.9 MB	0.2095s	7.074%

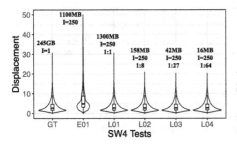

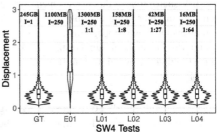

(a) High displacement near the epicenter.   (b) Low displacement away from the epicenter.

Fig. 6. Violin plots of the distribution of particle displacement for the ground truth (GT), one Eulerian configuration and four Lagrangian configurations. The Eulerian configuration, with access to a limited number of time slices, overestimates the displacement. The Lagrangian representation captures displacement in both settings, in regions near and away from the epicenter, accurately.

and reduced as waves propagate further away. For each simulation run, we measured the displacement of 200,000 samples reconstructed near the epicenter (Fig. 6a) and 90,000 samples reconstructed in six regions away from the epicenter (Fig. 6b). Here, we directly compared against the distribution of ground truth (GT) displacement. In both cases, Lagrangian representations offered significant data reduction while maintaining high accuracy. We found that as the number of basis trajectories extracted reduces, the displacement for some samples near the epicenter can be underestimated. In contrast, using a temporally subsampled Eulerian representation (E01) results in significant overestimation of displacement. This result can be expected since temporal subsampling fails to capture the transient nature of wave propagation, whereas Lagrangian representations encoding behavior over an interval of time remain accurate. Compared to Fig. 6a, the ground truth in Fig. 6b has smaller displacement and a multimodal distribution, which is the result of samples collected from six regions of the domain away from the epicenter.

Figure 7 visualizes field encoding displacement over time near the epicenter using multiple semi-opaque isosurfaces. Although regions of highest displacement can be underestimated as the data reduction factor increases, the overall structure is well

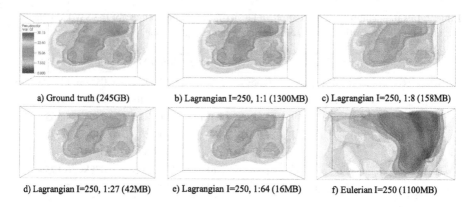

a) Ground truth (245GB)   b) Lagrangian I=250, 1:1 (1300MB)   c) Lagrangian I=250, 1:8 (158MB)

d) Lagrangian I=250, 1:27 (42MB)   e) Lagrangian I=250, 1:64 (16MB)   f) Eulerian I=250 (1100MB)

Fig. 7. Visualization of the displacement field derived from reduced Lagrangian representations near the epicenter using multiple isosurfaces. The ground truth is computed using 2000 cycles of the seismic wave propagation simulation. Although at higher data reduction factors regions of high displacement are underestimated, Lagrangian representations are capable of accurately reconstructing the overall feature structure.

preserved using highly compressed Lagrangian representations. In all cases, Lagrangian representations required less than 1% of the storage of the complete vector data.

6   Conclusion

Accurate exploratory analysis and visualization of time-varying vector fields is challenging. On the one hand, it can be performed accurately if the entire spatiotemporal resolution is available. However, storing all the data to disk for exploratory post hoc analysis is expensive. On the other hand, if subsets of the data are stored, predicting uncertainty and variability of accuracy for analysis techniques post hoc is difficult. In this context, Lagrangian representations computed using the full spatiotemporal resolution via in situ processing demonstrate the potential to enable accurate exploratory time-varying vector field analysis for reduced data storage costs.

For wider adoption and consideration of Lagrangian representations, an important step is characterization of effectiveness across a broad range of real-world applications. In this paper, we investigated in situ reduction via Lagrangian representations for vector fields from Nyx cosmology and SW4 seismology simulations. For the Nyx cosmology simulation, we found that Lagrangian representations are sensitive to both the spatial and temporal sampling rate, notably providing higher reconstruction accuracy when basis trajectories are computed using a high spatial and low temporal resolution. For the SW4 seismology simulation, we found Lagrangian representations are well suited to capture the transient seismic waves and offer high data reduction options for a small loss of accuracy. For both simulations, the percentage of execution time spent on computing the Lagrangian representation in situ was under 10% in most cases. Overall, we believe the findings of this study demonstrates that two computational science simulations considered benefit from Lagrangian representations for time-varying vector field exploration. This finding also provides confidence that more computational areas can benefit, and we encourage future work in this direction.

Acknowledgment. This research was supported by the Exascale Computing Project (17-SC-20-SC), a collaborative effort of the U.S. Department of Energy Office of Science and the National Nuclear Security Administration. This research used resources of the Oak Ridge Leadership Computing Facility at the Oak Ridge National Laboratory, which is supported by the Office of Science of the U.S. Department of Energy under Contract No. DE-AC05-00OR22725.

References

1. Agranovsky, A., Camp, D., Garth, C., Bethel, E.W., Joy, K.I., Childs, H.: Improved post hoc flow analysis via lagrangian representations. In: 4th IEEE Symposium on Large Data Analysis and Visualization, LDAV, pp. 67–75 (2014)
2. Agranovsky, A., Camp, D., Joy, K.I., Childs, H.: Subsampling-based compression and flow visualization. In: Visualization and Data Analysis 2015, vol. 9397, pp. 207–220. International Society for Optics and Photonics, SPIE (2015)
3. Almgren, A.S., Bell, J.B., Lijewski, M.J., Lukić, Z., Van Andel, E.: Nyx: a massively parallel AMR code for computational cosmology. Astrophysical. J. **765**(1), 39 (2013)
4. Bujack, R., Joy, K.I.: Lagrangian representations of flow fields with parameter curves. In: IEEE Symposium on Large Data Analysis and Visualization (LDAV), pp. 41–48 (2015)
5. Chandler, J., Bujack, R., Joy, K.I.: Analysis of error in interpolation-based pathline tracing. In: Proceedings of the Eurographics/IEEE VGTC Conference on Visualization: Short Papers, pp. 1–5. Euro graphics Association (2016)
6. Chandler, J., Obermaier, H., Joy, K.I.: Interpolation-based pathline tracing in particle-based flow visualization. IEEE Trans. Visual. Comput. Graphics **21**(1), 68–80 (2015)
7. Childs, H.: Visit: An end-user tool for visualizing and analyzing very large data (2012)
8. Hlawatsch, M., Sadlo, F., Weiskopf, D.: Hierarchical line integration. IEEE Trans. Visual. Comput. Graphics **17**(8), 1148–1163 (2011)
9. Hummel, M., Bujack, R., Joy, K.I., Garth, C.: Error estimates for lagrangian flow field representations. In: Proceedings of the Eurographics/IEEE VGTC Conference on Visualization: Short Papers, pp. 7–11. Euro graphics Association (2016)
10. Jakob, J., Gross, M., Günther, T.: A fluid flow data set for machine learning and its application to neural flow map interpolation. IEEE Trans. Visual. Comput. Graphics (Proc. IEEE Scientific Visualization), **27**(2), 1279–1289 (2020)
11. Larsen, M., et al.: The alpine in situ infrastructure. In: Proceedings of the In Situ Infrastructures on Enabling Extreme-Scale Analysis and Visualization, pp. 42–46. ACM (2017)
12. Lindstrom, P., Isenburg, M.: Fast and efficient compression of floating-point data. IEEE Trans. Visual. Comput. Graphics **12**(5), 1245–1250 (2006)
13. Lodha, S.K., Faaland, N.M., Renteria, J.C.: Topology preserving top-down compression of 2d vector fields using bintree and triangular quadtrees. IEEE Trans. Visual. Comput. Graphics **9**(4), 433–442 (2003)
14. Moreland, K., et al.: Vtk-m: Accelerating the visualization toolkit for massively threaded architectures. IEEE Comput. Graphics Appl. **36**(3), 48–58 (2016)
15. Orf, L.: A violently tornadic supercell thunderstorm simulation spanning a quarter-trillion grid volumes: Computational challenges, i/o framework, and visualizations of tornadogenesis. Atmosphere **10**(10), 578 (2019)
16. Petersson, N.A., Sjögreen, B.: Wave propagation in anisotropic elastic materials and curvilinear coordinates using a summation-by-parts finite difference method. J. Comput. Phys. **299**, 820–841 (2015)

17. Pugmire, D., et al.: Performance-Portable Particle Advection with VTK-m. In: Eurographics Symposium on Parallel Graphics and Visualization. The Eurographics Association (2018)
18. Qin, X., van Sebille, E., Sen Gupta, A.: Quantification of errors induced by temporal resolution on lagrangian particles in an eddy-resolving model. Ocean Model. **76**, 20–30 (2014)
19. Rapp, T., Peters, C., Dachsbacher, C.: Void-and-cluster sampling of large scattered data and trajectories. IEEE Trans. Visual. Comput. Graphics **26**(1), 780–789 (2019)
20. Sane, S., Bujack, R., Childs, H.: Revisiting the evaluation of in situ lagrangian analysis. In: Eurographics Symposium on Parallel Graphics and Visualization. The Eurographics Association (2018)
21. Sane, S., Childs, H., Bujack, R.: An interpolation scheme for VDVP lagrangian basis flows. In: Eurographics Symposium on Parallel Graphics and Visualization. The Eurographics Association (2019)
22. Sane, S., et al.: Scalable in situ lagrangian flow map extraction: demonstrating the viability of a communication-free model. arXiv preprint arXiv:2004.02003 (2020)
23. Siegfried, L., et al.: The tropical-subtropical coupling in the southeast atlantic from the perspective of the northern benguela upwelling system. PloS one **14**(1), e0210083 (2019)
24. The CGAL Project: CGAL User and Reference Manual. CGAL Editorial Board, 5.2.1 edn. (2021). https://doc.cgal.org/5.2.1/Manual/packages.html
25. Theisel, H., Rossl, C., Seidel, H.: Combining topological simplification and topology preserving compression for 2d vector fields. In: Proceedings of 11th Pacific Conference on Computer Graphics and Applications 2003, pp. 419–423 (2003)
26. Tong, X., Lee, T.Y., Shen, H.W.: Salient time steps selection from large scale time-varying data sets with dynamic time warping. In: IEEE Symposium on Large Data Analysis and Visualization (LDAV), pp. 49–56. IEEE (2012)
27. Valdivieso Da Costa, M., Blanke, B.: Lagrangian methods for flow climatologies and trajectory error assessment. Ocean Model. **6**(3), 335–358 (2004)
28. van Sebille, E., et al.: Lagrangian ocean analysis: fundamentals and practices. Ocean Model. **121**, 49–75 (2018)
29. Vries, P., Döös, K.: Calculating lagrangian trajectories using time-dependent velocity fields. J. Atmos. Oceanic Technol. **18**(6), 1092–1101 (2001)

Revolve-Based Adjoint Checkpointing for Multistage Time Integration

Hong Zhang$^{(\boxtimes)}$ and Emil Constantinescu

Mathematics and Computer Science Division, Argonne National Laboratory,
Lemont, IL 60439, USA
{hongzhang,emconsta}@anl.gov

Abstract. We consider adjoint checkpointing strategies that minimize the number of recomputations needed when using multistage timestepping. We demonstrate that we can improve on the seminal work based on the REVOLVE algorithm. The new approach provides better performance for a small number of time steps or checkpointing storage. Numerical results illustrate that the proposed algorithm can deliver up to two times speedup compared with that of REVOLVE and avoid recomputation completely when there is sufficient memory for checkpointing. Moreover, we discuss a tailored implementation that is arguably better suited for mature scientific computing libraries by avoiding central control assumed in the original checkpointing strategy. The proposed algorithm has been included in the PETSc library.

Keywords: Adjoint checkpointing · Multistage timestepping · Revolve

1 Introduction

Adjoint computation is commonly needed in a wide range of scientific problems such as optimization, uncertainty quantification, and inverse problems. It is also a core technique for training artificial neural networks in machine learning via backward propagation. The adjoint method offers an efficient way to calculate the derivatives of a scalar-valued function at a cost independent of the number of the independent variables. In the derivative computation, the chain rule of differentiation is applied starting with the dependent variables and propagating back to the independent variables; therefore, the computational flow of the function evaluation is reversed. But the intermediate information needed in the reverse computation may not be available and must be either saved beforehand or recomputed. When the storage is insufficient for all the intermediate information, one can checkpoint some selected values and recompute the missing information as needed. This approach gives rise to the adjoint checkpointing

This work was supported in part by the U.S. Department of Energy, Office of Science, Office of Advanced Scientific Computing Research, Scientific Discovery through Advanced Computing (SciDAC) program through the FASTMath Institute under Contract DE-AC02-06CH11357 at Argonne National Laboratory.

© Springer Nature Switzerland AG 2021
M. Paszynski et al. (Eds.): ICCS 2021, LNCS 12742, pp. 451–464, 2021.
https://doi.org/10.1007/978-3-030-77961-0_37

problem, which aims to minimize the recomputation cost, usually in terms of number of recomputed time steps, given limited storage capacity. Griewank and Walther proposed the first offline optimal checkpointing strategy [4] that minimizes the number of recomputations when the number of computation steps is known a priori. The algorithm was implemented in a software called REVOLVE and has been widely used in automatic differentiation tools such as ADtool, ADOL-C, and Tapenade. Many follow-up studies have addressed online checkpointing strategies for cases where the number of computation steps is unknown [5,9,10], and multistage or multilevel checkpointing strategies [1,2,7] have been developed for heterogeneous storage systems (e.g., devices with memory and disk). A common assumption used in developing these algorithms is that memory is considered to be limited and the cost of storing/restoring checkpoints is negligible.

While the problem has been well studied in the context of reversing a sequence of computing operations explicitly, time-dependent problems are often used to model the general stepwise evaluation procedures [4] because of their common sequential nature. In addition, time-dependent differential equations are ubiquitous in scientific simulations, and in their adjoint computation, a time step can be considered as the primitive operation in the sequence to be reversed.

In this paper, we show that the classical REVOLVE algorithm can be improved when multistage time integration methods such as Runge–Kutta methods are used to solve differential equations. A simple modification to the algorithm and the checkpointing settings can lead to fewer recomputations. Performance of the proposed algorithm is demonstrated and compared with that of REVOLVE.

2   Revisiting Optimality of Classic Checkpointing Strategy for Multistage Methods

Here we discuss the adjoint method applied to functionals with ordinary differential equation (ODE) constraints such as $\dot{u} = F(u)$. Previous studies assume implicitly or explicitly that only the solution is saved if it is marked as a checkpoint, as opposed to saving the intermediate stages for the corresponding time step. However, the optimal scheduling based on this assumption will not always work best for all timestepping methods, especially multistage methods.

Multistage schemes, for example, Runge–Kutta methods, have been popular in a wide range of applications; their adjoint counterparts are implemented by FATODE [12] and have recently been used by the open source library PETSc [3,11]. An s-stage explicit Runge–Kutta (ERK) method is

$$
U_i = u_n + \sum_{j=1}^{i-1} h_n\, a_{ij}\, F(U_j), \quad i = 1, \cdots, s,
$$

$$
u_{n+1} = u_n + \sum_{i=1}^{s} h_n\, b_i\, F(U_i).
$$

$$(1)$$

The discrete adjoint of ERK is

$$\lambda_{s,i} = h_n \boldsymbol{f}_{\boldsymbol{u}}^T(\boldsymbol{U}_i) \left(b_i \lambda_{n+1} + \sum_{j=i+1}^{s} a_{ji} \lambda_{s,j} \right), \quad i = s, \cdots, 1$$

$$\lambda_n = \lambda_{n+1} + \sum_{j=1}^{s} \lambda_{s,j},$$

(2)

where λ is the adjoint variable that carries the sensitivity information.

To perform an adjoint step of an ERK scheme, one needs all the stage values from the corresponding forward time step according to the sensitivity equation (2). When using REVOLVE, this is achieved by implementing an action named *youturn*, which takes a forward step followed immediately by an adjoint step. Figure 1a shows that an adjoint step in the reverse run is always preceded by a forward step. Ideally if one checkpointed the solution at every time step, $m - 1$ recomputations would still be required in order to adjoin m time steps. If one checkpointed the stage values instead of the solution for all the time steps, however, no recomputation would be needed in the ideal case, and fewer recomputations may be expected for other cases.

Based on this observation, we extend the existing optimal offline checkpointing scheme to the case where both the solution and stage values are saved. Although saving more information at each time step yields fewer allowed checkpoints, we show that the extended schemes may still outperform the original schemes in certain circumstances, depending on the total number of time steps to be adjoined.

3   Modified Offline Checkpointing Scheme

For convenience of notation, we associate the system solution at each time step with an index. The index of the system solution starts with 0 corresponding to the initial condition and increases by 1 for each successful time step. In the context of adaptive timestepping, a successful time step refers to the last actual time step taken after several attempted steps to determine a suitable step size. Therefore, the failed attempts are not indexed. The adjoint integration starts from the final time step and decreases the index by 1 after each reverse step until reaching 0.

An optimal reversal schedule that is generated by Revolve yields a minimal number of recomputations for a given number of 10 time steps and 3 checkpoints is shown in Fig. 1a. During the forward integration, the solutions at time index 0, 4 and 7 are copied into checkpoints 0, 1, and 2, respectively. After the final step $9 \rightarrow 10$ is finished, the adjoint sensitivity variables are initialized, and the adjoint calculation starts to proceed in the backward direction. The solution and stage values at the last time step are accessible at this point, so the first adjoint step can be taken directly. To compute the adjoint step $9 \rightarrow 8$, one can obtain the forward solution at 9 and the stage values by restoring the checkpoint 2

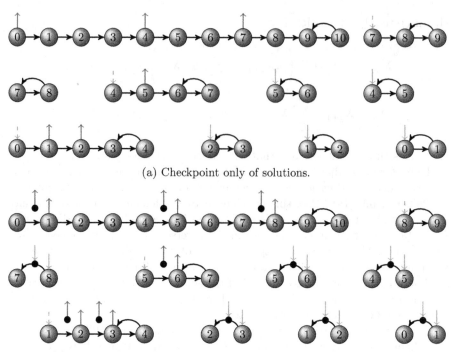

(a) Checkpoint only of solutions.

(b) Checkpoint of both solutions and stage values (denoted by dots).

Fig. 1. From left to right, top to bottom: the processes controlled by (a) REVOLVE and (b) modified REVOLVE. The up arrow and down arrow stand for the "store" operation and "restore" operation, respectively. When a stack is used for holding the checkpoints, the arrows with solid lines correspond to push and pop operations. The down arrow with a dashed line indicates to read the top element on the stack without removing it. Adapted from [11].

and recomputing two forward steps. The checkpoint 2 can be discarded after the adjoint step $8 \rightarrow 7$ so that its storage can be reused in the following calculation.

This schedule results from calling the REVOLVE routine repeatedly and implementing the actions determined. The return value of REVOLVE indicates the calling program to perform one of the actions including *advance*, *takeshot* (*store*), *restore*, *firsturn*, and *youturn*, which are explained in [4] and briefly summarized in Table 1.

The main modification we made is to change every checkpoint position by adding 1 to the index and save the stage values, which are used to compute the solution. For example, if the original `revolve` algorithm determines that the solution at time index i should be checkpointed, we will save the solution at time index $i + 1$ and the stage values for the time step $i \rightarrow i + 1$ as a combined checkpoint with index $i + 1$. The actions prescribed by REVOLVE are essentially mapped to a series of new but similar actions to guarantee the optimality for

Table 1. REVOLVE nomenclature.

REVOLVE actions	
Restore	Copy the content of the checkpoint back to the solution
Takeshot(store)	Copy the solution into a specified checkpoint
Advance	Propagate the solution forward for a number of time steps
Youturn	Take one forward step and then one reverse step for adjoint
Firsturn	Take one reverse step directly for adjoint
REVOLVE parameters	
Check	Number of checkpoint being stored
Capo	Beginning index of the time step range currently being processed
Fine	Ending index of the time step range currently being processed
Snaps	Upper bound on number of checkpoints taken

Table 2. Mapping the REVOLVE output to new actions.

REVOLVE action	New action in modified REVOLVE
Restore to solution i	Copy checkpoint to solution $i + 1$ and the stages
Store solution i	Copy solution $i + 1$ and the stages into a specified checkpoint
Advance from i to j	Propagate the solution from $i + 1$ to $j + 1$
Youturn	Take one reverse step directly

the new checkpointing settings. Table 2 enumerates the mapping we conduct in the modified scheme.

The adjoint of every time step except the last one always starts from a "restore" operation, followed by recomputations from the solution restored. Since the positions of all checkpoints are shifted one time step forward, one fewer recomputation is taken in the recomputation stage before computing an adjoint step. This observation leads to the following proposition for this modified REVOLVE algorithm.

Proposition 1. *Assume a checkpoint is composed of stage values and the resulting solution. Given s allowed checkpoints in memory, the minimal number of extra forward steps (recomputations) needed for the adjoint computation of m time steps is*

$$\tilde{p}(m, s) = (t - 1)\, m - \binom{s + t}{t - 1} + 1, \tag{3}$$

where t is the unique integer (also known as repetition number [4]) satisfying $\binom{s+t-1}{t-1} < m \le \binom{s+t}{t}$.

Proof. Griewank and Walther proved in [4] that the original REVOLVE algorithm would take

$$p(m, s) = t\, m - \binom{s + t}{t - 1} \tag{4}$$

extra forward steps. According to the observation mentioned above, one can save $m - 1$ extra forward steps by using the modified scheme. We can prove by contradiction that no further savings are possible.

If a schedule that satisfies the assumption takes fewer recomputations than (3), one can move all the checkpoints backward by one step and change the content of the checkpoints to be solutions only. The resulting scheme will cause $m - 1$ extra recomputations by construction, thus the total number of recomputations will be less than $t\,m - \binom{s+t}{t-1}$. This contradicts with the optimality result proved in [4].

We remark that the modification can also be applied to the online algorithms in [5,9,10] and to the multistage algorithms in [8]. The saving in recomputations is always equal to the total number of steps minus one, given the same amount of allowable checkpoints.

The choice of using modified or original algorithms clearly results from the tradeoff between recomputation and storage. Given a fixed amount of storage capacity, the choice depends solely on the total number of time steps since the recomputations needed can be determined by these two factors according to Proposition 1. For example, we suppose there are 12 allowed checkpoints if we save only the solution, which means 6 checkpoints are allowed if we add one stage to the checkpoint data and 4 checkpoints for adding two stages. Figure 2 shows the relationship between the recomputations and the total number of steps for these different options. In this example, the crossover point at which saving the stages together with the solution leads to fewer recomputations than saving only the solution occurs when 224 and 41 steps are taken for the two illustrated scenarios (saving one additional stage and two additional stages, respectively). Furthermore, the number of stages saved, determined by the timestepping method, has a significant impact on the range of number of time steps in which the former option is more favorable; for methods with fewer stages, the range is generally larger (compare 224 with 41).

Consequently, one can choose whether or not to save stage values in order to minimize the recomputations for discrete adjonit calculation. The optimal choice depends on the number of steps, the number of stages of the timestepping algorithm, and the memory capacity.

Moreover, when using the modified REVOLVE algorithm, we can save fewer stages than a timestepping method actually has, if the last stage of the method is equal to the solution at the end of a time step. For instance, the Crank–Nicolson method, which can be cast as a two-stage Runge–Kutta method, requires storing only one stage with the solution corresponding to the solutions at the beginning

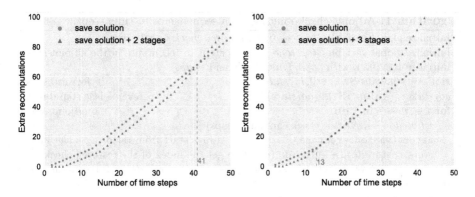

Fig. 2. Comparison in terms of recomputation effort for including different numbers of stages (0,1,and 2) in the checkpoint data. It is assumed that up to 12 solution vectors or stage vectors can be stored in both forward integration and adjoint integration. Adapted from [11].

and the end of a time step. Many classic implicit Runge–Kutta methods have such a feature that enables further improvement of the performance.

4 Using REVOLVE in the discrete adjoint calculation

The REVOLVE library is designed to be an explicit controller for conducting forward integration and adjoint integration in time-dependent applications. The interface requires the user to provide procedures such as performing a forward and backward step and saving/restoring a checkpoint. Thus, incorporating REVOLVE in other simulation software such as PETSc can be intrusive, or even infeasible, especially when the software has its own adaptive time step control and an established framework for time integration. To mitigate intrusion, we use REVOLVE in a different way in the sense that its role becomes more of a "consultant" rather than a "controller." Algorithm 1 describes the workflow for the adjoint calculation with checkpointing. The parameters `capo`, `fine`, and `check` are updated internally by REVOLVE. See Table 1 for descriptions of these parameters. A counter `stepstogo` for the number of steps to advance is computed from these variables. We insert calls to REVOLVE only before a forward step is taken when `stepstogo` is not zero, so as to preserve the original integration process, which is represented by `forwardSweep`. This trick is also applied to the adjoint sweep. One can verify that the resulting schedule is equivalent to the one generated by calling REVOLVE repeatedly (the "controller" mode) based on the following observations:

- If REVOLVE asks to store a checkpoint (`takeshot`), it will return either *advance* or *youturn* (*firsturn*) in the next call.
- In the adjoint sweep, it is always required to restore a checkpoint and recompute one or more steps from this point.

Algorithm 1. Adjoint checkpointing for a sequence of m time steps.

Initialize global variables $capo \leftarrow 0$, $fine \leftarrow m$, $check \leftarrow -1$, $snaps \leftarrow s$

Initialize global variable $stepstogo \leftarrow 0$ ▷ steps to recompute in the adjoint sweep

Initialize REVOLVE with $capo, fine, check,$ and $snaps$

$state \leftarrow$ FORWARDSWEEP$(0, m, state)$ ▷ forward sweep

$adjstate \leftarrow$ ADJOINTSTEP$(adjstate)$ ▷ reverse last step directly

for $i \leftarrow m - 1$ **to** 1 **do**   ▷ adjoint sweep

    $whatodo \leftarrow$ REVOLVE(check,capo,fine,snaps)

    `Assert(whatodo=restore)`   ▷ always start from restoring a checkpoint

    $state, restoredind \leftarrow$ RESTORE$(check)$ ▷ get the index of the restored checkpoint

    $state \leftarrow$ FORWARDSWEEP$(restoredind, i - restoredind, state)$ ▷ recompute from the restored solution

    $adjstate \leftarrow$ ADJOINTSTEP$(adjstate)$

end for

function FORWARDSWEEP$(ind, n, state)$ ▷ advance n steps from the ind-th point

    for $i \leftarrow ind$ **to** $ind + n - 1$ **do** ▷ REVOLVE returns `youturn`/`firsturn` at the end

        REVOLVEFORWARD$(state)$

        $state \leftarrow$ FORWARDSTEP$(state)$

    end for

    return state

end function

function REVOLVEFORWARD$(state)$ ▷ query REVOLVE and take actions

    if $stepstogo = 0$ **then**

        $oldcapo \leftarrow capo$

        $whatodo \leftarrow$ REVOLVE(check,capo,fine,snaps)

        if $whatodo = takeshot$ **then**

            STORE$(state, check)$ ▷ store a checkpoint

            $oldcapo \leftarrow capo$

            $whatodo \leftarrow$ REVOLVE(check,capo,fine,snaps)

        end if

        if $whatodo = restore$ **then**

            RESTORE$(state, check)$ ▷ restore a checkpoint

            $oldcapo \leftarrow capo$

            $whatodo \leftarrow$ REVOLVE(check,capo,fine,snaps)

        end if

        `Assert(whatodo=advance || whatodo=youturn || whatodo=firsturn)`

        $stepstogo \leftarrow capo - oldcapo$

    else

        $stepstogo \leftarrow stepstogo - 1$

    end if

end function

The checkpointing scheme using the modified REVOLVE algorithm is depicted in Algorithm 2. A main difference from Algorithm 1 is that the call to REVOLVE is lagged because the positions of all the checkpoints have been shifted. We note that reducing the counter `stepstogo` by one in the loop for adjoint sweep reflects the fact that one fewer recomputation is needed for each adjoint step.

Algorithm 2. Adjoint checkpointing using the modified REVOLVE algorithm.

Initialize global variables $capo \leftarrow 0$, $fine \leftarrow m$, $check \leftarrow -1$, $snaps \leftarrow s$
Initialize global variable $stepstogo \leftarrow 0$
Initialize REVOLVE with $capo, fine, check$, and $snaps$
$state \leftarrow$ FORWARDSWEEP$(0, m, state)$
$adjstate \leftarrow$ ADJOINTSTEP$(adjstate)$
for $i \leftarrow M - 1$ to 1 **do**
    $restoredind \leftarrow$ REVOLVEBACKWARD$(state)$ ▷ always restore a checkpoint
    $state \leftarrow$ FORWARDSWEEP$(restoredind, i - restoredind, state)$
    $adjstate \leftarrow$ ADJOINTSTEP$(adjstate)$
end for
function FORWARDSWEEP$(ind, n, state)$
    for $i \leftarrow ind$ to $ind + n - 1$ **do**
        $state \leftarrow$ FORWARDSTEP$(state)$
        REVOLVEFORWARD$(state)$
    end for
    return $state$
end function
function REVOLVEBACKWARD$(state)$
    $whatodo \leftarrow$ REVOLVE(check,capo,fine,snaps)
    Assert$(whatodo = restore)$
    $state, restoredind \leftarrow$ RESTORE$(check)$
    $stepstogo \leftarrow \max(capo - oldcapo - 1, 0)$ ▷ need one less extra step since stage
values are saved
    return $restoredind$
end function

Both Algorithms 1 and 2 are implemented under the `TSTrajectory` class in PETSC, which provides two critical callback functions `TSTrajectorySet()` and `TSTrajectoryGet()`. The former function wraps REVOLVEFORWARD in FORWARDSWEEP. The latter function wraps all the statements before ADJOINTSTEP in the **for** loop. This design is beneficial for preserving the established workflow of the timestepping solvers so that the impact to other PETSC components such as TSADAPT (adaptor class) and TSMONITOR (monitor class) is minimized.

PETSC uses a redistributed package[1] that contains a C wrapper of the original C++ implementation of REVOLVE. The parameters needed by REVOLVE can be passed through command line options at runtime. Additional options are provided to facilitate users monitoring the checkpointing process. Listing 1.1 exhibits an exemplar output for `-ts_trajectory_monitor` `-ts_trajectory_view` when using modified REVOLVE to reverse 5 time steps given 3 allowable checkpoints.

By design, PETSC is responsible for the manipulation of checkpoints. A stack data structure with push and pop operations is used to conduct the actions decided by the checkpointing scheduler. The `PetscViewer` class manages deep

---

[1] https://bitbucket.org/caidao22/pkg-revolve.git.

copy between the working data and the checkpoint, which can be encapsulated in either sequential or parallel vectors that are distributed over different processes.

In addition to the offline checkpointing scheme, PETSc supports online checkpointing and multistage checkpointing schemes provided by the REVOLVE package, and the modification proposed in this paper has been applied to these schemes as well. Therefore, checkpoints can be placed on other storage media such as disk. For disk checkpoints, the `PetscViewer` class offers a variety of formats such as binary and HDF5 and can use MPI-IO on parallel file systems for efficiency.

```
TSTrajectorySet: stepnum 0, time 0. (stages 1)
TSTrajectorySet: stepnum 1, time 0.001 (stages 1)
Store in checkpoint number 0 (located in RAM)
Advance from 0 to 2
TSTrajectorySet: stepnum 2, time 0.002 (stages 1)
TSTrajectorySet: stepnum 3, time 0.003 (stages 1)
Store in checkpoint number 1 (located in RAM)
Advance from 2 to 4
TSTrajectorySet: stepnum 4, time 0.004 (stages 1)
TSTrajectorySet: stepnum 5, time 0.005 (stages 1)
First turn: Initialize adjoints and reverse first step
TSTrajectoryGet: stepnum 5, stages 1
TSTrajectoryGet: stepnum 4, stages 1
Restore in checkpoint number 1 (located in RAM)
Advance from 2 to 3
Skip the step from 2 to 3 (stage values already checkpointed)
Forward and reverse one step
TSTrajectoryGet: stepnum 3, stages 1
Restore in checkpoint number 1 (located in RAM)
Forward and reverse one step
Skip the step from 2 to 3 (stage values already checkpointed)
TSTrajectoryGet: stepnum 2, stages 1
Restore in checkpoint number 0 (located in RAM)
Advance from 0 to 1
Skip the step from 0 to 1 (stage values already checkpointed)
Forward and reverse one step
TSTrajectoryGet: stepnum 1, stages 1
Restore in checkpoint number 0 (located in RAM)
Forward and reverse one step
Skip the step from 0 to 1 (stage values already checkpointed)
TSTrajectoryGet: stepnum 0, stages 1
TSTrajectory Object: 1 MPI processes
   type: memory
   total number of recomputations for adjoint calculation = 2
```

Listing 1.1. Monitoring the checkpointing process in PETSc

5   Algorithm Performance

To study the performance of our algorithm, we plot in Fig. 3 the actual number of recomputations against the number of time steps and compare our algorithm with the classic REVOLVE algorithm. For a fair comparison, the same number of checkpointing units is considered.

Figure 3 shows that our modified algorithm outperforms REVOLVE. For 30 checkpointing units and 300 time steps, modified REVOLVE clearly takes fewer recomputations than does REVOLVE, with the maximum gap being 40 recomputations. For 60 checkpointing units and 300 time steps, the savings with modified REVOLVE are 186 recomputations. If the recomputation cost of a time step is fixed, the result implies a speedup of approximately $1.5X$ in runtime for the adjoint calculation. As the number of time steps further increases, REVOLVE is expected to catch up with modified REVOLVE and eventually surpass modified REVOLVE. Furthermore, when the number of time steps is small, which means there is sufficient memory, no recomputation is needed with modified REVOLVE; however, REVOLVE requires the number of recomputations to be at least as many as the number of time steps minus one.

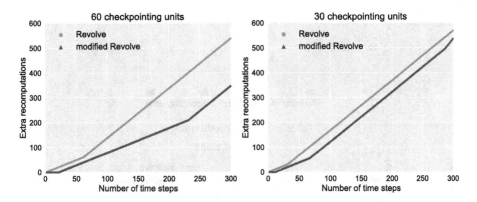

Fig. 3. Performance comparison between REVOLVE and modified REVOLVE. The plotted data is computed for time integration methods with two stages.

Now we demonstrate the performance with a real example. In the experiment, we consider the gradient calculation using an adjoint method to solve a PDE-constrained optimization problem. The objective is to minimize the discrepancy between the simulated result and observation data (reference solution):

$$\underset{U_0}{\text{minimize}} \, \|U(t_f) - U^{ob}(t_f)\|^2 \tag{5}$$

subject to the Gray-Scott equations [6]

$$\dot{u} = D_1 \nabla^2 u - uv^2 + \gamma(1 - u)$$
$$\dot{v} = D_2 \nabla^2 v + uv^2 - (\gamma + \kappa)v, \tag{6}$$

where $U = [u \ v]^T$ is the PDE solution vector. The settings of this example follow [11]. The PDE is solved with the method of lines. The resulting ODE is solved by using the Crank–Nicolson method with a fixed step size 0.5. A centered finite-difference scheme is used for spatial discretization on a uniform grid of size 128×128. The computational domain is $\Omega \in [0, 2]^2$. The time interval is $[0, 25]$. The nonlinear system that arises at each time step is solved by using a Newton-based method with line search, and the linear systems are solved by using GMRES with a block Jacobi preconditioner.

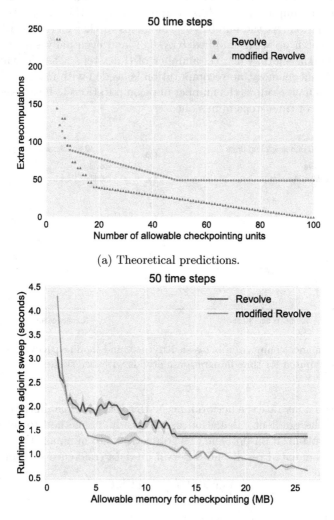

(a) Theoretical predictions.

Fig. 4. Comparison of the timing results of the adjoint calculation with the theoretical predictions for REVOLVE and modified REVOLVE. The Crank–Nicolson method is used for time integration. It consists of two stages, but only the first stage needs to be saved since the second stage is the same as the solution.

Figure 4 shows that the runtime decreases as the allowable memory for checkpointing increases for both schemes and that the best performance achieved by modified REVOLVE is approximately $2.2X$ better than that by REVOLVE. When memory is sufficient, modified REVOLVE requires no recomputation. Thus the estimated speedup of modified REVOLVE over REVOLVE would be approximately $2X$ provided the cost of the forward sweep is comparable to the cost of the adjoint sweep. Another important observation is that the experimental results, despite being a bit noisy, roughly match with the theoretical predictions. The observable oscillation in timing can be attributed mostly to the fact that the cost of solving the implicit system varies across the steps, which is not uncommon for nonlinear dynamical systems.

6 Conclusion

With the abstraction of a sequence of time steps, the classic algorithm REVOLVE provides optimal checkpointing schedules for efficient adjoint computation in many scientific computations. However, it may yield suboptimal performance when directly applied to multistage time integration methods.

In this paper we have considered checkpointing strategies that minimize the number of recomputations under two assumptions: (1) the stage values of a multistage method can be saved as part of a checkpoint, and (2) a stage vector has the same size (therefore the same memory cost) with a solution vector. By extending REVOLVE and redefining the content of a checkpoint, we derived a modified REVOLVE algorithm that provides better performance for a small number of time steps. The performance has been studied numerically. The results on some representative test cases show that our algorithm can deliver up to $2X$ speedup compared with REVOLVE and avoid recomputation completely when there is sufficient memory for checkpointing.

In addition, the implementation of our algorithm is tailored to fit into the workflow of mature scientific computing libraries. Integrating our algorithm into existing frameworks avoids using it as a centralized controller over the entire workflow; thus, it becomes less intrusive to the application codes. The proposed algorithm has been successfully employed in the PETSc library.

References

1. Aupy, G., Herrmann, J.: Periodicity in optimal hierarchical checkpointing schemes for adjoint computations. Optim. Methods Softw. **32**(3), 594–624 (2017). https://doi.org/10.1080/10556788.2016.1230612
2. Aupy, G., Herrmann, J., Hovland, P., Robert, Y.: Optimal multistage algorithm for adjoint computation. SIAM J. Sci. Comput. **38**(3), C232–C255 (2016)
3. Balay, S., et al.: PETSc Users Manual. Technical report ANL-95/11 - Revision 3.7, Argonne National Laboratory (2016)
4. Griewank, A., Walther, A.: Algorithm 799: revolve: an implementation of checkpointing for the reverse or adjoint mode of computational differentiation. ACM Trans. Math. Softw. **26**(1), 19–45 (2000). https://doi.org/10.1145/347837.347846

5. Nagel, W.E., Walter, W.V., Lehner, W. (eds.): Euro-Par 2006. LNCS, vol. 4128. Springer, Heidelberg (2006). https://doi.org/10.1007/11823285

6. Hundsdorfer, W., Ruuth, S.J.: IMEX extensions of linear multistep methods with general monotonicity and boundedness properties. J. Comput. Phys. **225**(2007), 2016–2042 (2007). https://doi.org/10.1016/j.jcp.2007.03.003

7. Schanen, M., Marin, O., Zhang, H., Anitescu, M.: Asynchronous two-level check-pointing scheme for large-scale adjoints in the spectral-element solver Nek5000. Procedia Comput. Sci. **80**, 1147–1158 (2016). https://doi.org/10.1016/j.procs.2016.05.444

8. Stumm, P., Walther, A.: MultiStage approaches for optimal offline checkpoint-ing. SIAM J. Sci. Comput. **31**(3), 1946–1967 (2009). https://doi.org/10.1137/080718036

9. Stumm, P., Walther, A.: New algorithms for optimal online checkpointing. SIAM J. Sci. Comput. **32**(2), 836–854 (2010). https://doi.org/10.1137/080742439

10. Wang, Q., Moin, P., Iaccarino, G.: Minimal repetition dynamic checkpointing algo-rithm for unsteady adjoint calculation. SIAM J. Sci. Comput. **31**(4), 2549–2567 (2009). https://doi.org/10.1137/080727890

11. Zhang, H., Constantinescu, E.M., Smith, B.F.: PETSc TSAdjoint: a discrete adjoint ODE solver for first-order and second-order sensitivity analysis. CoRR abs/1912.07696 (2020), http://arxiv.org/abs/1912.07696

12. Zhang, H., Sandu, A.: FATODE: a library for forward, adjoint, and tangent linear integration of ODEs. SIAM J. Sci. Comput. **36**(5), C504–C523 (2014). https://doi.org/10.1137/130912335

Comprehensive Regularization of PIES for Problems Modeled by 2D Laplace's Equation

Krzysztof Szerszeń$^{(\boxtimes)}$, Eugeniusz Zieniuk , Agnieszka Bołtuć ,
and Andrzej Kużelewski

Institute of Computer Science, University of Bialystok, Konstantego Ciołkowskiego 1M,
15-245 Białystok, Poland
{kszerszen,ezieniuk,aboltuc,akuzel}@ii.uwb.edu.pl

Abstract. The paper proposes the concept of eliminating the explicit computation of singular integrals appearing in the parametric integral equation system (PIES) used to simulate the steady-state temperature field distribution. These singularities can be eliminated by regularizing the PIES formula with the auxiliary regularization function. Contrary to existing regularization methods that only eliminate strong singularities, the proposed approach is definitely more comprehensive due to the fact that it eliminates all strong and weak singularities. As a result, all singularities associated with PIES's integral functions can be removed. A practical aspect of the proposed regularization is the fact that all integrals appearing in the resulting formula can be evaluated numerically with a standard Gauss-Legendre quadrature rule. Simulation results indicate the high accuracy of the proposed algorithm.

Keywords: Computational methods · Regularized PIES · Singular integrals · Potential problems 2D · Bézier curves

1 Introduction

One of the most significant problems to be faced during computer implementation of parametric integral equation system (PIES) is the evaluation of singular integrals. Their presence is related to the fact that PIES is based on the analytical modification of the conventional boundary integral equation (BIE). This modification, previously presented for various types of differential equations [1–3], is aimed to include analytically the shape of the boundary problem directly in the obtained PIES formula. As a result, in opposite to finite and boundary element methods (FEM, BEM), PIES's solutions of boundary value problems are obtained without domain or boundary discretization. Finally, we can introduce some alternative representations of the boundary shape. In the case of 2D problems, it is particularly promising to define the boundary with parametric curves, e.g. Bézier curves. Hence, instead of a mesh of boundary elements with their nodes, we can use parametric curves of different degrees defined by a relatively small set of control points. Moreover, the obtained formal separation between the declared boundary and

© Springer Nature Switzerland AG 2021
M. Paszynski et al. (Eds.): ICCS 2021, LNCS 12742, pp. 465–479, 2021.
https://doi.org/10.1007/978-3-030-77961-0_38

boundary functions in PIES allows to approximate the boundary functions by effective Chebyshev series.

The presence of singular integrals is a common problem both for BIE and PIES methods and is related to their integral kernels dependent on the distance between the so-called source and field points. In the case when these points are close to each other, this distance tends to zero and the kernels are singular. The degree of singularity depends on the form of the kernel functions. Accurate evaluation of singular integrals plays a crucial role in the overall accuracy of solutions of boundary value problems. This is even more difficult because the direct evaluation of singular integrals by the popular Gauss-Legendre (G-L) quadrature may result in an unacceptable accuracy degradation. Due to the importance of the problem, there is an extremely rich literature on this subject with many algorithms for evaluation of the singular integrals. They are mainly dedicated to BEM, among which we can mention nonlinear transformations [4–6], adaptive subdivision [7], variable transformation [8], semi-analytical methods [9, 10], as well as quadrature methods [11]. One of the most promising are regularization methods [12–15].

This paper proposes a new algorithm to eliminate weakly and strongly singular integrals in PIES. The algorithm is based on the regularization of the PIES formula using the auxiliary regularizing function with regularization coefficients. As a result, all obtained regularized integrals are no longer singular and can be evaluated by the standard G-L quadrature. To demonstrate the capability and accuracy of the proposed scheme we present two simulation examples.

2 The Singular Formulation of the PIES

The paper deals with the prediction of the steady-state temperature field distribution inside the 2D domain Ω and on the boundary Γ. The model formulation is based on the linear boundary value problem for the Laplace equation

$$\frac{\partial^2 u}{\partial x_1^2} + \frac{\partial^2 u}{\partial x_2^2} = 0, \tag{1}$$

with Dirichlet u_Γ and Neumann p_Γ boundary conditions, as shown in Fig. 1a.

In the case of practical problems defined for more complex geometries and boundary conditions, we need to use numerical computational methods, for example finite element method (FEM) or boundary element method (BEM). Figure 1b shows modeling of the domain Ω by FEM with finite elements. A similar discretization strategy is related to BEM, but refers only to the boundary Γ as shown in Fig. 1c. Such modeling has gained wide popularity, but in practice it requires to generate a significant number of elements as well as algebraic equations.

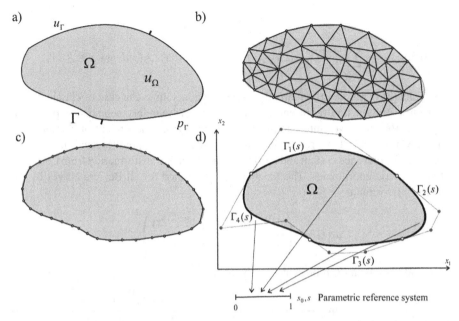

a)

b)

c)

d)

Fig. 1. Problem's domain, boundary shape and boundary conditions (a), modeling of a 2D domain with finite elements (b) and its boundary with boundary elements (c), definition of the boundary with Bézier curves in relation to the parametric reference system in PIES (d).

Here, to overcome some of the limitations observed in the case of FEM and BEM, an alternative approach for solving boundary value problems, called as PIES, is used. The PIES has several advantages and we can identify them as:

- The prediction of the temperature field in the interior of the domain, similar to BIE, is obtained via the analysis of the problem only on the boundary of that domain. This reduces the mathematical dimension of the problem under analysis by one.
- Only the boundary of the domain needs to be defined. The boundary in PIES is parameterized and described in a very general way as a closed parametric curve (e.g. Bézier and NURBS curves). Figure 1d shows practical definition of the boundary from Fig. 1a by joining 4 Bézier curves of degree 3. Bézier curves allow for intuitive description of the boundary using only control points and it is more effective than the classical discretization with boundary elements in BEM.
- The boundary description with Bézier curves is analytically included into the PIES formula used to find solutions on the boundary of the problem. For the Laplace equation this formula is written as [2]

$$0.5u_l(s_0) = \sum_{i=1}^{n} \int_0^1 \left\{ \overline{U}_{lj}^*(s_0, s)p_j(s) - \overline{P}_{lj}^*(s_0, s)u_j(s) \right\} J_j(s)ds, \, s_0, s \in \langle 0, 1 \rangle. \quad (2)$$

Formula (2) is not specified directly on the boundary, as is the case of classical BIE, but on a straight line representing a projection of boundary segments represented by by n Bézier curves into parametric reference system dependent on parameter s, as shown in Fig. 1d. Moreover s_0 is the co-called source point located in the same parametric reference system and $J_j(s)$ is the Jacobian of transformation from Cartesian to parametric coordinates. The Jacobian is determined for all Béziera curves by the following formula

$$J_j(s) = \left[\left(\frac{\partial \Gamma_j^{(1)}(s)}{\partial s} \right)^2 + \left(\frac{\partial \Gamma_j^{(2)}(s)}{\partial s} \right)^2 \right]^{0.5}. \quad (2a)$$

- Integral kernels in (2) represented in the form

$$\overline{U}_{lj}^*(s_0, s) = \frac{1}{2\pi} \ln \frac{1}{[\eta_1^2 + \eta_2^2]^{0.5}}, \quad \overline{P}_{lj}^*(s_0, s) = \frac{1}{2\pi} \frac{\eta_1 n_j^{(1)}(s) + \eta_2 n_j^{(2)}(s)}{\eta_1^2 + \eta_2^2}, \quad (3)$$

include analytically in their mathematical formalism the boundary shape generated by Bézier curves by following relations

$$\eta_1 = \Gamma_l^{(1)}(s_0) - \Gamma_j^{(1)}(s), \, \eta_2 = \Gamma_l^{(2)}(s_0) - \Gamma_j^{(2)}(s), \quad (4)$$

where functions $\boldsymbol{\Gamma}_l(s_0)$, $\boldsymbol{\Gamma}_j(s)$ describe Bézier curves that contain the co-called source point denoted as s_0 and the field point denoted as s, while $n_j^{(1)}(s), n_j^{(2)}(s)$ are the normals to the curves.
- The declaration of Bézier curves in formula (2) is separated from the boundary functions $u_j(s)$ and $p_j(s)$ describing in the physical interpretation temperature and flux on the boundary. In this paper they are approximated by Chebyshev series

$$u_j(s) = \sum_{k=0}^{K-1} u_j^{(k)} T_j^{(k)}(s), \quad p_j(s) = \sum_{k=0}^{K-1} p_j^{(k)} T_j^{(k)}(s), \quad (5)$$

where $u_j^{(k)}, p_j^{(k)}$ denote the coefficients of the series, K is the adopted number of terms of the series, whereas $T_j^{(k)}(s)$ represent the Chebyshev polynomials defined in the parametric reference system in PIES.
- It should be noted, that the form of boundary functions (5) is completely independent from the Bézier curves used to describe the boundary. This is the essence of the formal separation of the boundary shape from the representation of the problem on the boundary in PIES. The direct incorporation of the boundary shape in functions (2) through kernels (3) is the main advantage of PIES compared to traditional BIE. As a result, the proposed PIES is characterized by the fact that its numerical solution does not require, in contrast to the BEM, discretization at the level of the aforementioned boundary declaration as well as the discretization of the boundary functions.

- Having the solutions on the boundary, in the second step we can obtain from them the solutions inside the domain using an additional integral identity represented by formula (10). There is no need to define a representation of the domain, so we have the possibility to find such solutions at any point and at the same time we can freely refine their resolution.

3 Elimination of Singularities from PIES Through the Regularization

We want to eliminate the singularities from the integral functions $\overline{U}^*_{lj}(s_0, s)$, $\overline{P}^*_{lj}(s_0, s)$ through the regularization applied to (2). The presence of the singularities in (2) depends on the distance between the source s_0 and the field s points. If this distance is significant then the kernels $\overline{U}^*_{lj}(s_0, s)$ and $\overline{P}^*_{lj}(s_0, s)$ are easily numerically integrable with standard numerical methods, e.g. with the G-L quadrature rule. However, when this distance goes to zero $s \to s_0$, the integral functions become singular and the values of such integrals tend to infinity. We can identify weak (logarithmic) singularity in $\overline{U}^*_{lj}(s_0, s)$ and strong singularity in $\overline{P}^*_{lj}(s_0, s)$. The direct application of the G-L quadrature to evaluate the singular integrals produces large errors. In previous studies, we evaluated the strongly singular integrals analytically, whereas the weakly singular ones with singular points isolation. Finally, these integrals have been evaluated with satisfactory accuracy, but at the expense of additional complexity.

In order to regularize PIES represented by (2), we introduce below an auxiliary formula (2) but with other boundary functions marked as $\hat{u}_j(s)$, $\hat{p}_j(s)$

$$0.5\hat{u}_l(s_0) = \sum_{j=1}^{n} \int_0^1 \left\{ \overline{U}^*_{lj}(s_0, s)\hat{p}_j(s) - \overline{P}^*_{lj}(s_0, s)\hat{u}_j(s) \right\} J_j(s)ds, \quad l = 1, 2, 3 \ldots n. \quad (6)$$

Next, we assume that $\hat{u}_j(s)$ takes the following form

$$\hat{u}_j(s) = A_l(s_0)\big(\Gamma_j(s) - \Gamma_l(s_0)\big) + u_l(s_0), \quad (7)$$

together with its directional derivative along to the normal vector to the boundary

$$\hat{p}_j(s) = A_l(s_0)n_j(s), \, n_j(s) = n_j^{(1)}(s) + n_j^{(2)}(s), \quad (8)$$

where $A_l(s_0) = \frac{p_l(s_0)}{n_l(s_0)}$, $n_l(s_0) = n_l^{(1)}(s_0) + n_l^{(2)}(s_0)$.

Function (7) is arbitrarily chosen to satisfy Laplace's equation. After subtracting (6) from (2), we get the final formula for the regularized PIES

$$\sum_{j=1}^{n} \left\{ \int_0^1 \overline{U}_{lj}^*(s_0, s)\left[p_j(s) - d_{lj}(s, s_0)p_l(s_0)\right] - \right.$$
$$\left. \int_0^1 \overline{P}_{lj}^*(s_0, s)\left[u_j(s) - u_l(s_0) - g_{lj}(s_0, s)p_l(s_0)\right] \right\} J_j(s)ds = 0, \tag{9}$$

where $d_{lj}(s, s_0) = \frac{n_j(s)}{n_l(s_0)}$, $g_{lj}(s_0, s) = \frac{\mathbf{\Gamma}_j(s) - \mathbf{\Gamma}_l(s_0)}{n_l(s_0)}$.

After solving (9), we obtain solutions $u_j(s)$ and $p_j(s)$ on the boundary in the form of the Chebyshev series (5). Having these solutions on the boundary, in the second step we can obtain solutions inside the domain at point $x \in \Omega$ using the following integral identity

$$u_\Omega(x) = \sum_{j=1}^{n} \int_0^1 \left\{ \overline{U}_j^*(x, s)p_j(s) - \overline{P}_j^*(x, s)u_j(s) \right\} J_j(s)ds, \quad x \in \Omega \tag{10}$$

Formula (10), similarly as (9), includes analytically the boundary generated by Bézier curves via following kernels

$$\overline{U}_j^*(x, s) = \frac{1}{2\pi} \ln \frac{1}{\left[n_1^2 + n_2^2\right]^{0.5}}, \quad \overline{P}_j^*(x, s) = \frac{1}{2\pi} \frac{n_1 n_1^{(j)}(s) + n_2 n_2^{(j)}(s)}{n_1^2 + n_2^2}, \tag{11}$$

where

$$n_1 = x_1 - \Gamma_j^{(1)}(s), \quad n_2 = x_2 - \Gamma_j^{(2)}(s).$$

To determine the solution in the domain, only coefficients $u_j^{(k)}$ and $p_j^{(k)}$ for the Chebyshev series (5) for every Bézier curve which model the boundary have to be taken into account in (10).

4 Numerical Implementation

In order to use (9) for simulating stationary temperature field, the collocation method [16] is applied. The collocation points are placed in the parametric domain of Bézier curves and represent by points s_0. Writing (9) at the collocation points, we obtain a system of algebraic equations approximating PIES with the size determined by the number of parametric curves modeling the boundary and the number of terms in the approximating series (5) on individual curves.

$$
\begin{bmatrix}
\bar{h}_{11}^{(1,0)} & \cdots & h_{1j}^{(1,k)} & \cdots & h_{1n}^{(1,K-1)} \\
\vdots & & \vdots & & \vdots \\
h_{l1}^{(c,0)} & \cdots & \bar{h}_{lj}^{(c,k)} & \cdots & h_{ln}^{(c,K-1)} \\
\vdots & & \vdots & & \vdots \\
h_{n1}^{(K-1,0)} & \cdots & h_{nj}^{(K-1,k)} & \cdots & \bar{h}_{nn}^{(K-1,K-1)}
\end{bmatrix}
\begin{bmatrix}
p_1^{(0)} \\
\vdots \\
p_j^{(k)} \\
\vdots \\
p_n^{(K-1)}
\end{bmatrix}
$$

$$
=
\begin{bmatrix}
\bar{g}_{11}^{(1,0)} & \cdots & g_{1j}^{(1,k)} & \cdots & g_{1n}^{(1,K-1)} \\
\vdots & & \vdots & & \vdots \\
g_{l1}^{(c,0)} & \cdots & \bar{g}_{lj}^{(c,k)} & \cdots & g_{ln}^{(c,K-1)} \\
\vdots & & \vdots & & \vdots \\
g_{n1}^{(K-1,0)} & \cdots & g_{nj}^{(K-1,k)} & \cdots & \bar{g}_{nn}^{(K-1,K-1)}
\end{bmatrix}
\begin{bmatrix}
u_1^{(0)} \\
\vdots \\
u_j^{(k)} \\
\vdots \\
u_n^{(K-1)}
\end{bmatrix}.
\tag{12}
$$

In the absence of regularization and direct application of formula (2) for the solution of the problem on the boundary, all integrals on the main diagonals of matrices $\mathbf{H}$ and $\mathbf{G}$ are singular. The proposed regularization eliminates these singularities and the new formulas for non-singular integrals on the main diagonal in (12) on the basis of (9) are as follows

$$
\begin{aligned}
\left[\bar{h}_{ll}^{(c,k)}\right] = &\int_0^1 \overline{U}^* u\left(s_0^{(l,c)}, s\right) T_l^{(k)}(s) J_l(s) ds \\
&- \sum_{j=0}^n \int_0^1 \frac{n_j^{(1)}(s) + n_j^{(2)}(s)}{n_l^{(1)}\left(s_0^{(c)}\right) + n_l^{(2)}\left(s_0^{(c)}\right)} \overline{U}_{lj}^*\left(s_0^{(c)}, s\right) T_j^{(k)}(s) J_j(s) ds \\
&+ \sum_{j=0}^n \int_0^1 \frac{\Gamma_j^{(1)}(s) - \Gamma_l^{(1)}\left(s_0^{(c)}\right) + \Gamma_j^{(2)}(s) - \Gamma_l^{(2)}\left(s_0^{(c)}\right)}{n_l^{(1)}\left(s_0^{(c)}\right) + n_l^{(2)}\left(s_0^{(c)}\right)} \overline{P}_{lj}^*\left(s_0^{(c)}, s\right) T_j^{(k)}(s) J_j(s) ds,
\end{aligned}
\tag{13}
$$

$$
\left[\bar{g}_{ll}^{(c,k)}\right] = \int_0^1 \overline{P}_{ll}^*\left(s_0^{(c)}, s\right) T_l^{(k)}(s) J_l(s) ds - \sum_{j=0}^n \int_0^1 \overline{P}_{lj}^*\left(s_0^{(c)}, s\right) T_j^{(k)}(s) J_j(s) ds.
\tag{14}
$$

Non-diagonal elements in (12) are calculated on the basis of the following integrals

$$
\left[g_{lj}^{(c,k)}\right] = \int_0^1 \overline{P}_{lj}^*\left(s_0^{(c)}, s\right) T_j^{(k)}(s) J_j(s) ds,
\tag{15}
$$

$$
\left[h_{lj}^{(c,k)}\right] = \int_0^1 \overline{U}_{lj}^*\left(s_0^{(c)}, s\right) T_j^{(k)}(s) J_j(s) ds
\tag{16}
$$

Integrals (15) and (16) are non-singular and have the same form both for (2) and (9). The complete algorithm for solving the regularized PIES is listed below.

Regularized PIES algorithm

Read boundary input data (control points of n Bézier segments)
Read boundary conditions
1: **for** $l \leftarrow 1, n$ **do** //loop over Bézier segments
2:     **for** $j \leftarrow 1, n$ **do** //loop over Bézier segments
3:         **if** $i = j$ **then**
4:             **for** $k \leftarrow 0, K-1$ **do** //loop over Chebyshev series
5:                 **for** $c \leftarrow 1, K$ **do** //loop over collocation points
6:                     **for** $e \leftarrow 1, n$ **do** //loop over Bézier segments
7:

$$[g_{ll}^{(kc)}] \leftarrow G\_L\_integration\left(\overline{P}^*{}_{ll}(s_0^{(c)}, s^{(q)})T_l^{(k)}(s^{(q)})J_l(s^{(q)}) - \overline{P}^*{}_{le}(s_0^{(c)}, s^{(q)})T_e^{(k)}(s^{(q)})J_e(s^{(q)})\right)$$

8:

$$[h_{ll}^{(kc)}] \leftarrow G\_L\_integration\big(\overline{U}^*{}_{ll}(s_0^{(c)}, s^{(q)})T_l^{(k)}(s^{(q)})J_l(s^{(q)})$$
$$-(n_e^{(1)}(s^{(q)}) + n_e^{(2)}(s^{(q)}))/(n_l^{(1)}(s_0^{(c)}) + n_l^{(2)}(s_0^{(c)}))\overline{U}^*{}_{le}(s_0^{(c)}, s^{(q)})T_e^{(k)}(s^{(q)})J_e(s^{(q)})$$
$$-(\Gamma_e^{(1)}(s^{(q)}) - \Gamma_l^{(1)}(s_0^{(c)}) + \Gamma_e^{(2)}(s^{(q)}) - \Gamma_l^{(2)}(s_0^{(c)}))/(n_l^{(1)}(s_0^{(c)}) + n_l^{(2)}(s_0^{(c)}))\overline{P}^*{}_{le}(s_0^{(c)}, s^{(q)})T_e^{(k)}(s^{(q)})J_e(s^{(q)})\big)$$

9:                 **end for**
10:                 **end for**
11:             **end for**
12:             add submatrix $[g_{ll}^{(kc)}]$ to $[g_{ll}]$ and $[h_{ll}^{(kc)}]$ to $[h_{ll}]$
13:         **else**
14:             **for** $k \leftarrow 0, K-1$ **do** //loop over Chebyshev series
15:                 **for** $c \leftarrow 1, K$ **do** //loop over collocation points
16:                     $[g_{lj}^{(kc)}] \leftarrow G\_L\_integration\left(\overline{P}^*{}_{lj}(s_0^{(c)}, s^{(q)})T_j^{(k)}(s^{(q)})J_j(s^{(q)})\right)$
17:                     $[h_{lj}^{(kc)}] \leftarrow G\_L\_integration\left(\overline{U}^*{}_{lj}(s_0^{(c)}, s^{(q)})T_j^{(k)}(s^{(q)})J_j(s^{(q)})\right)$
18:                 **end for**
19:             **end for**
20:             add submatrix $[g_{lj}^{(kc)}]$ to $[g_{lj}]$ and $[h_{lj}^{(kc)}]$ to $[h_{lj}]$
21:         **end if**
22:         add submatrix $[g_{lj}]$ to **G** and $[h_{lj}]$ to **H**
23:     **end for**
24: **end for**
25: applying boundary conditions
26: transform $[\mathbf{H}]\{\mathbf{u}\} = [\mathbf{G}]\{\mathbf{p}\}$ into $[\mathbf{A}]\{\mathbf{x}\} = \{\mathbf{b}\}$
27: solve system of equations $[\mathbf{A}]\{\mathbf{x}\} = \{\mathbf{b}\}$

---

5 Verification of the Approach

The regularization is validated on two examples having analytical solutions. Below, we show how to generate the boundary by Bézier curves and investigate the influence of the minimal distance between the collocation point and the quadrature node on the stability of diagonal integrals and overall accuracy of PIES.

5.1 Example 1

We consider a stationary temperature distribution governed by the Laplace equation in a wrench. As shown in Fig. 2a, the boundary is generated by 14 Bézier curves. Among them, 9 are linear being simply straight lines between two end points. The remaining 5 are the cubic ones each defined by 4 control points and used to define curvilinear parts of the boundary.

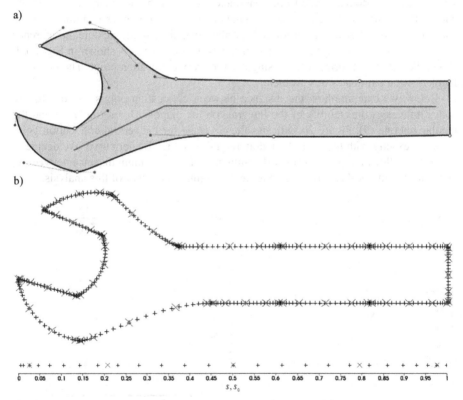

Fig. 2. The boundary of the problem for example 1 defined by Bézier curves together with the analyzed cross-section of the domain solutions (a), distribution of 5 collocation points (red x) and 27 nodes of the G-L quadrature (black +) in the parametric reference system and after mapping to the boundary (b). (Color figure online)

We assume that the expected distribution of temperature on the boundary and inside the domain is described by the following function that depends on the Cartesian coordinates

$$u(x_1, x_2) = -x_1^3 - -x_2^3 + 3x_1^2 x_2 + 3x_1 x_2^2. \tag{17}$$

The value of function (17) is specified on the boundary in the form of Dirichlet boundary conditions, while its normal derivative

$$\frac{du(x_1, x_2)}{dn} = \left(-3x_1^2 + 6x_1 x_2 + 3x_2^2\right)n_1 + \left(-3x_2^2 + 6x_1 x_2 + 3x_1^2\right)n_2, \tag{18}$$

represents the expected analytical solutions on the boundary.

In order to solve the problem on the boundary by (9), we place 5 collocation points at roots of the Chebyshev polynomials of the second kind within the parametric domain of each Bézier cuve. The nodes of the G-L quadrature of degree 27 are defined in the same parametric domain. Due to full parameterization of the boundary, we can freely choose the positions of the collocation points and quadrature nodes identified with s_0 and s in (9) and also in (2). Their mutual distribution in the parametric domain reference system s, s_0 and after mapping to each of the 14 Bézier curves is shown in Fig. 2b. It should be noted that formula (2) is singular for every collocation point. The proposed regularization eliminates this problem.

Below, we examine how the distance between collocation points and quadrature nodes influences the stability of the integrals on the diagonal for formula (2) and the regularized one (9). Figure 2b indicates the coverage for the central collocation point $s_0 = 0.5$ exactly with the central quadrature node for each Bézier curve. We decide to move this collocation point to study the influence of the minimum distance between s_0 and s on stability of these integrals. Figure 3 presents a summary of this analysis.

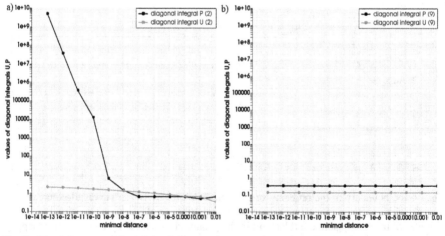

Fig. 3. The influence of the minimal distance between the collocation point and the quadrature node on the stability of diagonal integrals for (2) (a) and (9) (b).

Figure 3a shows that the diagonal integrals in (2) with the direct application of G-L quadrature are unstable. It is especially noticeable for the strongly singular one. In turn, Fig. 3b shows that the diagonal integrals in (9) are stable for the full range of distances between the investigated collocation point and quadrature node. The presented results refer to one selected collocation point from the total number of 70 specified in Fig. 2b. Moreover, these behaviors and dependencies are analogous for all other points. It should be noted that the values of diagonal integrals in (2) and (9) are different due to the regularization. But at this point we are interested in forecasting overall computational stability rather than individual values.

The regularization also allows for obtaining excellent accuracy of the problem under study. Figure 4 shows the solution on the boundary obtained by (9) and in the domain by identity (10) for the case when the minimal distance between the collocation point and the quadrature node is $1e-13$.

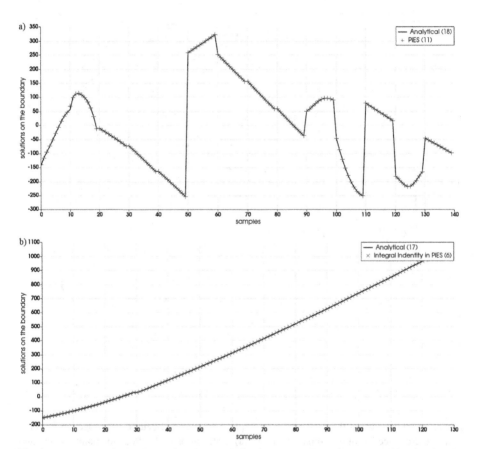

Fig. 4. The obtained solutions with the regularization on the boundary (a) and in the domain (b) when the minimal distance between the collocation point and the quadrature node is $1e-13$. (Color figure online)

The results show excellent agreement with exact solutions (17–18) and confirm the strategy, which is independent from the representation of the boundary shape and the type of applied boundary conditions.

5.2 Example 2

We repeat the analysis given in example 1, but for more complicated shape of the boundary with another boundary conditions. We consider a stationary temperature distribution in a multiply connected domain shown in Fig. 5a. The inner and outer boundaries are described by linear Bézier curves. The geometry of the boundary is thus completely defined by a set of 31 control points.

a)

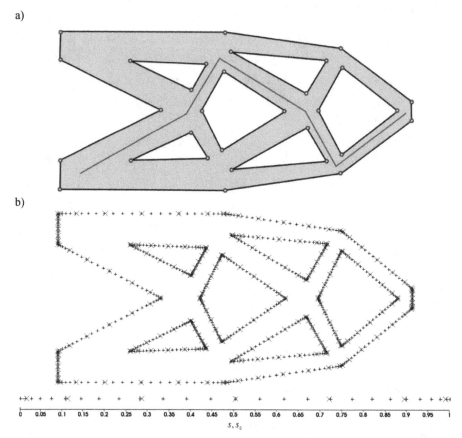

b)

Fig. 5. The boundary of the problem for example 2 defined by Bézier curves together with the analyzed cross-section of the domain solutions (a), distribution of 7 collocation points (red x) and 27 nodes in the G-L quadrature (black +) in the parametric reference system and after mapping to the boundary (b). (Color figure online)

We assume that Dirichlet boundary conditions are posed on the whole boundary. They are calculated on the basis of the following function

$$u(x_1, x_2) = \cos(x_1)\exp(x_2). \tag{19}$$

Function (19) satisfies the Laplace equation and represents the expected analytical temperature distribution inside the multiply connected domain. In turn, the normal derivative of (19) gives the reference analytical solutions of the problem on the boundary

$$\frac{du(x_1, x_2)}{dn} = -\sin(x_1)\exp(x_2)n_1 + \cos(x_1)\exp(x_2)n_2. \tag{20}$$

Table 1 shows the influence of the minimal distance between collocation point and the quadrature node for accuracy of solutions on the boundary obtained by (9) and in the domain by identity (10). The results are computed for 7 collocation points and 27 nodes of the G-L quadrature per Bézier curve. When analyzing the placement of these points and nodes shown in Fig. 5b, we can identify the cases with $s \to s_0$ and $s_0 = s$. Moreover, as in example 1, the central collocation point coincides with the central quadrature node. Therefore, we again decide to move this point to determine the minimum distance, for which we observe the existence and stability of the solutions.

Table 1. The influence of the minimal distance between collocation point and G-L quadrature nod for accuracy of solutions on the boundary and domain.

Minimal distance between collocation and quadrature points	L_2 error norm [%] for solutions on the boundary	L_2 error norm [%] for solutions in the domain
1e−2	0.396046	0.00730305
1e−3	0.397675	0.00739621
1e−4	0.398794	0.00741943
1e−5	0.398975	0.00742293
1e−6	0.398999	0.00742338
1e−7	0.399002	0.00742344
1e−8	0.399002	0.00742345
1e−9	0.399002	0.00742345
1e−10	0.399002	0.00742419
1e−11	0.398997	0.00742788
1e−12	0. 398991	0.00742783
1e−13	0.400376	0.00770156
0.0	0.399002	0.00742345

The results again confirm the stability of solutions for the multi connected domain. We obtained the excellent accuracy of the regularized PIES for a very close distance between collocation points and quadrature nodes.

6 Conclusions

The results indicate the effectiveness of the proposed regularization. It avoids the use of complicated explicit methods for the evaluation of singular integrals and, on the other hand, provides a unified scheme for eliminating these singularities. The approach is also independent from the ways of declaring the boundary with the help of various curves that have already been used in PIES. In the paper, Bézier curves are chosen, but we can apply other ones, e.g. NURBS. Moreover, the separation of the boundary declaration from the approximation of the boundary functions, as in the original PIES's formula is preserved. Thanks to this, in the current paper the boundary functions could be approximated with an effective Chebyshev series.

References

1. Zieniuk, E.: Bézier curves in the modification of boundary integral equations (BIE) for potential boundary-values problems. Int. J. Solids Struct. **40**(9), 2301–2320 (2003)
2. Zieniuk, E., Szerszeń, K.: The PIES for solving 3D potential problems with domains bounded by rectangular Bézier patches. Eng. Comput. **31**(4), 791–809 (2014)
3. Zieniuk, E.: Szerszeń, K: A separation of the boundary geometry from the boundary functions in PIES for 3D problems modeled by the Navier-Lamé equation. Comput. Math. Appl. **75**(4), 1067–1094 (2018)
4. Telles, J.C.F.: A self-adaptive co-ordinate transformation for efficient numerical evaluation of general boundary element integrals. Int. J. Numer. Meth. Eng. **24**(5), 959–973 (1987)
5. Cerrolaza, M., Alarcon, E.: A bi-cubic transformation for the numerical evaluation of the Cauchy principal value integrals in boundary methods. Int. J. Numer. Meth. Eng. **28**(5), 987–999 (1989)
6. Li, X., Su, Y.: Three-dimensional stress analysis of thin structures using a boundary element method with sinh transformation for nearly singular integrals. Comput. Math. Appl. **72**(11), 2773–2787 (2016)
7. Gao, X.W., Yang, K., Wang, J.: An adaptive element subdivision technique for evaluation of various 2D singular boundary integrals. Eng. Anal. Boundary Elem. **32**(8), 692–696 (2008)
8. Xie, G., Zhang, J., Qin, X., Li, G.: New variable transformations for evaluating nearly singular integrals in 2D boundary element method. Eng. Anal. Boundary Elem. **35**(6), 811–817 (2011)
9. Fata, S.N.: Semi-analytic treatment of nearly-singular Galerkin surface integrals. Appl. Numer. Math. **60**(10), 974–993 (2010)
10. Niu, Z., Wendland, W.L., Wang, X., Zhou, H.: A semi-analytical algorithm for the evaluation of the nearly singular integrals in three-dimensional boundary element methods. Comput. Methods Appl. Mech. Eng. **194**(9–11), 1057–1074 (2005)
11. Hayami, K., Matsumoto, H.: A numerical quadrature for nearly singular boundary element integrals. Eng. Anal. Boundary Elem. **13**(2), 143–154 (1994)
12. Chen, H.B., Lu, P., Schnack, E.: Regularized algorithms for the calculation of values on and near boundaries in 2D elastic BEM. Eng. Anal. Boundary Elem. **25**(10), 851–876 (2001)
13. Granados, J.J., Gallego, R.: Regularization of nearly hypersingular integrals in the boundary element method. Eng. Anal. Boundary Elem. **25**(3), 165–184 (2001)
14. Sladek, V., Sladek, J., Tanaka, M.: Regularization of hypersingular and nearly singular integrals in the potential theory and elasticity. Int. J. Numer. Meth. Eng. **36**(10), 1609–1628 (1993)

15. Huang, Q., Cruse, T.A.: Some notes on singular integral techniques in boundary element analysis. Int. J. Numer. Meth. Eng. **36**(15), 2643–2659 (1993)
16. Gottlieb, D., Orszag, S.A.: Numerical analysis of spectral methods: theory and applications. SIAM, Philadelphia (1977)

High Resolution TVD Scheme Based on Fuzzy Modifiers for Shallow-Water Equations

Ruchika Lochab⬤ and Vivek Kumar$^{(\boxtimes)}$⬤

Department of Applied Mathematics, Delhi Technological University,
Delhi 110042, India
{ruchika_phd2k17,vivekaggarwal}@dtu.ac.in

Abstract. This work proposes a new fuzzy logic based high resolution (HR), total variation diminishing (TVD) scheme in finite volume frameworks to compute an approximate solution of the shallow water equations (SWEs). Fuzzy logic enhances the execution of classical numerical algorithms. To test the effectiveness and accuracy of the proposed scheme, the dam-break problem is considered. A comparison of the numerical results by implementing some classical flux limiting methods is provided. The proposed scheme is able to capture both smooth and discontinuous profiles, leading to better oscillation-free results.

Keywords: Shallow water equation · Fuzzy modifier · Limiter

1 Introduction

Shallow-water equations are frequently employed in the situations which involve the modelling of water flow corresponding to various water bodies such as lakes, rivers, reservoirs, and other such variants in which the fluid depth metric is significantly smaller than the horizontal length metric [14,15,21]. The standard SWEs (also known as the Saint Venant equations), were initially introduced about one and a half century ago and still these equations are used in various applications [11–13]. For many practical, real-life models, such as dam-break problems, flood problems, etc., these equations are frequently used. The solutions to such systems are generally non-smooth and produce discontinuities also, so it is essential to have a robust, efficient, and accurate numerical strategy for the Shallow water system and related models. Finite-volume schemes are among the most popular tools to tackle such situations.

The particular case in which SWEs are inviscid in nature, lead to an hyperbolic system of equations, and all the robust numerical strategies [9,10] that have been constructed for hyperbolic conservation laws can be implemented to such equations. From a mathematical standpoint, the hyperbolic equations are well known to admit discontinuous solutions, and their numerical integration is expected to compute such discontinuities sharply and without oscillations. Based on the classical HR-TVD flux limiting schemes, this work addresses a new hybrid

© Springer Nature Switzerland AG 2021
M. Paszynski et al. (Eds.): ICCS 2021, LNCS 12742, pp. 480–492, 2021.
https://doi.org/10.1007/978-3-030-77961-0_39

flux limiting method [7,17,22], which is used in this work to approximate the flow components at the midpoints of cell edges inside the control volumes of a computational domain. Hyperbolic conservation laws govern SWEs. By defining the SWEs with this broader class of equations, the opportunity to exploit the set of techniques and mathematical tools previously established for these computationally complicated situations opens up [12–14].

Motivation behind fuzzy-logic based approach: The fuzzy logic theory has evolved in a number of ways since Zadeh's introduction of fuzzy set theory. Fuzzy logic theory is now commonly used in fluid mechanics, control engineering, information processing, artificial intelligence, strategic planning, and other fields[5,6]. Fuzzy control problems have made significant progress in recent decades being one of the most popular frameworks of fuzzy sets and fuzzy logic. The fuzzy-logic-based control has been commonly used in machine engineering, intelligence control, system recognition, image classification, neural networks, and other areas. In contrast to traditional crisp control, fuzzy logic controller will more accurately model physical reality in a linguistic format, allowing for more efficient method of achieving intelligent management in engineering settings. In Fuzzy mathematics, the concept of fuzzy logic is quite unique as compared to the classical logic, as fuzzy logic works more like the human way of reasoning. In other words, fuzzy logic approach is more easy and understandable. Fuzzy logic has many applications in almost are the industries related to various commercial and practical purposes. In artificial intelligence, Fuzzy Logic helps in simulating the human oriented cognitive processes.

The hybrid method's main merit is its optimized construction using an entirely different concept from fuzzy logic [5], which makes it better than the classical limiters. The optimized fuzzy flux limited scheme is implemented into a one-dimensional structured finite volume model to approximate the shallow water flows. This work concentrates on the optimization of classical numerical methods for observing the behavior of Dam-Break Problem governed by one-dimensional shallow water equation in which discontinuities are present and are important to model. The computational results of the SWEs with the proposed scheme is assessed by experimenting with the dam-break problem [4,8]. The proposed scheme results are compared with those obtained from the classical minmod scheme and the monotonized-central (MC) scheme for validation.

The work is further structured as follows. The numerical flow model is explained in the Sect. 2. In the Sect. 3, the hyperbolic numerical approach is explained. In the frame of uniform mesh and finite volume methods, the proposed new scheme with a brief discussion on fuzzy logic cocepts is introduced in the Sect. 4. After that, a numerical assessment is shown in the Sect. 5. Further, the work is concluded with some remarks and the scope of future work in the Sect. 6.

2   Numerical Model: One-Dimensional Shallow Water Flows

In one space dimension, the SWEs [1] can be described in mathematical form as:

$$q_t + f(q)_x = s, \tag{1}$$

where

$$q = \begin{pmatrix} h \\ hu \end{pmatrix}, \ f(q) = \begin{pmatrix} hu \\ hu^2 + \frac{1}{2}gh^2\alpha \end{pmatrix}, \ s(q) = \begin{pmatrix} 0 \\ -ghZ_x \end{pmatrix}.$$

Here, the height of water is denoted as $h(x,t)$, the fluid velocity as $u(x,t)$, the notation for acceleration due to gravity is g and the bottom surface function is denoted by $Z(x)$. As, the present work is concerned towards hyperbolic conservation laws, so this function $Z(x)$ is taken to be zero. So, the Eq. 1 becomes:

$$\partial_t \begin{pmatrix} h \\ hu \end{pmatrix} + \partial_x \begin{pmatrix} hu \\ hu^2 + \frac{1}{2}gh^2 \end{pmatrix} = \begin{pmatrix} 0 \\ 0 \end{pmatrix}. \tag{2}$$

In this work, the finite-volume framework is used as it prevents any global transformation in the conserved variables, so the overall scheme remains conservative in nature. Here as illustrated in the upcoming sections, both space and temporal discretizations are done in a higher-order accurate manner. A spatial reconstruction, called the MUSCL (Monotone Up-stream Centered Scheme for Conservation Laws) technique, has been considered to obtain higher order accuracy in space.

3   Numerical Approach: Flux Limiting High Resolution Schemes

To formulate the finite-volume framework, the primary task is to discretize the space domain in forms of smaller cells $[x_{i-1/2}, x_{i+1/2}]$ (refer to the Fig. 1), which have a uniform spatial step of length Δx, such that $x_i = \Delta x(i + 1/2)$. Similarly, the temporal domain is discretized into sub intervals $[t^n, t^{n+1}]$ with uniform step size Δt, such that $t^n = \Delta t(n)$. $[x_{i-1/2}, x_{i+1/2}]$ denotes the ith control volume, where $x_i = (i + 1/2)\Delta x$ is the mid-point of this control volume. A numerical integration of the nonlinear conservation laws, discussed in the Sect. 2 requires a finite volume Godunov method of upwind-type. A conservative form related to the homogeneous Eq. 1 is written as:

$$q_i^{n+1} = q_i^n - \frac{\Delta t}{\Delta x} \left(F_{i+\frac{1}{2}} - F_{i-\frac{1}{2}} \right), \tag{3}$$

where

$$q_i^n \approx \frac{1}{\Delta x} \int_{x_{i-\frac{1}{2}}}^{x_{i+\frac{1}{2}}} q(x, t^n) dx \tag{4}$$

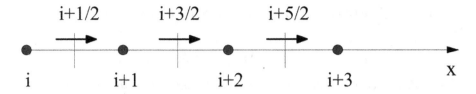

Fig. 1. A spatial representation of the computational grid.

is the cell average corresponding to the spatial components, and $F_{i\pm 1/2}$ denote the numerical flux functions (temporal cell-averages), defined in the following manner:

$$F_{i-\frac{1}{2}} \approx \frac{1}{\Delta t} \int_{t^n}^{t^{n+1}} f(q(x_{i-\frac{1}{2}}, t)) dt. \tag{5}$$

The important task in approximating such conservation laws is the proper selection of the flux presented in the Eq. 5. In general, this construction demands the solutions of various Riemann problems at the cell interfaces.

In a finite-volume HR technique, the numerical flux is calculated by mixing a lower and a higher order flux altogether. For $f(q) = a(q)$ with positive speed a, the mathematical form of the Lax-Wendroff technique is:

$$q_i^{n+1} = q_i^n - \nu(q_i^n - q_{i-1}^n) - \frac{1}{2}\nu(1 - \nu)(q_{i+1}^n - 2q_i^n + q_{i-1}^n), \tag{6}$$

where $\nu = a.\Delta t/\Delta x$ is termed as the Courant number. It is an upwind scheme of first-order along with an extra anti-diffusive term of second-order. The scheme in the Eq. 6 is second-order accurate, but it still does not follows the TVD property. Therefore, the Eq. 6 is further modified by introducing a limiting function, say, ϕ to the term of second order in the following manner:

$$q_i^{n+1} = q_i^n - (q_i^n - q_{i-1}^n)\left[\nu + \frac{1}{2}\nu(1 - \nu)\left(\frac{\phi(r_{i+\frac{1}{2}})}{r_{i+\frac{1}{2}}} - \phi(r_{i-\frac{1}{2}})\right)\right], \tag{7}$$

where the function $r_{i+1/2}$ is defined as:

$$r_{i+\frac{1}{2}} = \frac{q_i^n - q_{i-1}^n}{q_{i+1}^n - q_i^n}. \tag{8}$$

A HR scheme is developed when the limiting function given in the Eq. 8 is positive [3]. To advance the solution in time, the numerical fluxes are calculated as follows:

$$F(q_{i+\frac{1}{2}}) = f_{i+\frac{1}{2}}^l - \phi(r_i)\left(f_{i+\frac{1}{2}}^l - f_{i+\frac{1}{2}}^h\right), \tag{9}$$

here, f^l resembles the low-resolution and f^h resembles the high-resolution [17] numerical flux functions.

Theorem 1 (Harten's Lemma). *A numerical method can be formulated in the incremental form as:*

$$q_i^{n+1} = q_i^n - C_{i-1/2}^n \Delta q_{i-1/2}^n + D_{i+1/2}^n \Delta q_{i+1/2}^n. \tag{10}$$

If $\forall n \in Z$, *and each integral value* i, *the coefficients follow the constraints presented as follows:*

$$C_{i+1/2}^n \geq 0, \tag{11}$$

$$D_{i+1/2}^n \geq 0, \tag{12}$$

$$C_{i+1/2}^n + D_{i+1/2}^n \leq 1, \tag{13}$$

then such a numerical scheme is TVD.

For the implementation of flux limiters [11] in the numerical scheme, the reconstruction step should obey an additional TVD property given as:

$$\phi_{sweby}(r) = \max\{0, \min\{2r, 2\}\}. \tag{14}$$

Table 1. Some TVD flux limiting functions.

Limiter	Representation	Remarks
Minmod	$max(0, min(r, 1))$	Roe, 1986 [20]
Superbee	$max(0, min(2r, 1), min(r, 2))$	Roe, 1986 [20]
Van albada	$\frac{r(r+1)}{r^2+1}$	Van Albada, et al., 1982 [3]
Monotonized central	$max(0, min(min(2r, (1+r)/2, 2)))$	Van Leer, 1977 [11]
Van leer	$\frac{r+|r|}{1+|r|}$	Van Leer, 1974 [21]

Table 1 gives a quick introduction to some of the commonly used flux limiters. For a detailed theory refer to the citations provided with each limiter in the Table 1. Further, a graphical representation of these flux limiters is shown in the Sweby's TVD region [19], as seen in the Fig. 2. The present work highlights a fundamental idea to modify and optimize the classical limiters to enhance the overall numerical outputs.

4   Development of the New Flux Limiter Scheme

The present algorithm consists of optimizing a specific classical flux limiter to form a better hybrid alternative. Many flux limiters are available in the literature to prevent discontinuities [7]. To optimize the classical schemes, some important concepts from the literature of Fuzzy mathematics are also required.

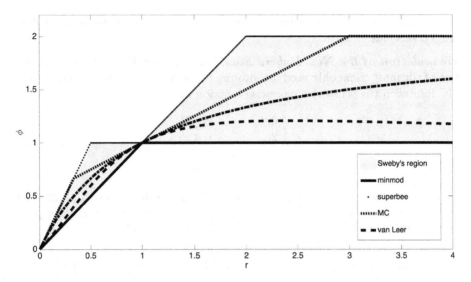

Fig. 2. Graphical representation of the classical Flux Limiters mentioned in the Table 1.

Fuzzy Sets: A fuzzy set is basically a classical (crisp) set having a special property, which allows each member of the considered universal set to get connected with this crisp set by a suitable intensity (called membership value). The membership intensity depends on the degree of compatibility of a particular element with the crisp set. The most commonly used set for membership degrees in fuzzy sets is $[0, 1]$, however this set restricts to the discrete values $\{0, 1\}$ for a crisp set. Mathematically, for a classical set T in the universe of discourse U, a fuzzy set A could be presented as follows: $A = \{(x, \mu(x)) \mid x \in T\}$, where the membership function μ sends the members of the classical set T to the closed interval $[0, 1]$.

Fuzzy Linguistics: These are known as the fuzzy variables originated through a special domain consisting of words, it's members are also known as the linguistic entities in the context of Fuzzy mathematics. These variables help to associate the elements of the universal set with a suitable membership value, using which a relationship of that element could be defined with the concerned fuzzy set [5]. As fuzzy values capture measurement uncertainties as a consequence of initial data sets, these are much more adaptive than the crisp variables to real-life models.

Fuzzy Modifiers: Fuzzy modifiers are an important ingredient in the construction of the new limiter. Fuzzy hedges fine tune the interpretation of the given data by modifying the membership units for the related fuzzy sets. Corresponding to the fuzzy set A defined above, several commonly used fuzzy hedges are: the Dilation modifier ($\{(y, \sqrt[p]{(\mu(y))}) \mid y \in U\}$), the Concentration modifier ($\{(y, (\mu(y))^p)) \mid y \in U\}$) [12], here p is some arbitrary real value.

Further this section comprises of enhancing the classical limiters to establish a new limiter. The procedure in this section is centered on how to use suitable fuzzy modifier operations to fine-tune parameter settings in the classical flux-limited schemes.

Formulation of the New Hybrid Flux Limiter: For this work, an optimization of the most commonly used monotonized central (MC) limiter is shown. The MC limiter in the sense of a piecewise function is:

$$\phi(r) = \begin{cases} 0, & r \leq 0 \\ 2r, & 0 < r \leq 1/3 \\ \frac{1}{2}(1+r), & 1/3 < r \leq 3 \\ 2, & \text{else.} \end{cases} \tag{15}$$

This is improvised by assigning the concentration modifier function of intensity $p = 6$ and $p = 8$ to the smooth and extrema regions respectively, and other parts are kept the same [17]. So, the hybrid limiter turns out to be:

$$\phi(r) = \max\left(0, \min\left(\min\left(\frac{\frac{2}{3}(\frac{9-3r}{8})^6 + 2\frac{3r-1}{8}}{(\frac{9-3r}{8})^6 + \frac{3r-1}{8}}, 2, \frac{\frac{2}{3}(3r)^6}{(1-3r) + (3r)^6}\right)\right)\right). \tag{16}$$

The hybrid flux limiter is shown in the Fig. 3. This procedure opens up infinitely many choices for the flux limiter functions in the context of Fuzzy Mathematics [18,22]. To show the performance of the hybrid fuzzy limiter presented in the Eq. 16, the Shallow water problem, written as the Eq. 2 is approximated in the upcoming section.

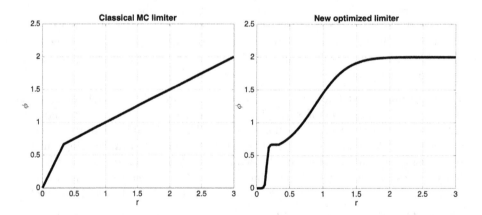

Fig. 3. The hybrid fuzzy flux limiter.

5   Numerical Validation: Using Dam Break Problem

For the numerical computations in this section, the classical limiter functions: the Min-Mod and the MC limiters have been utilised to compare the approximate results obtained from the proposed hybrid limiter. The flux limiting techniques are based on a deterministic finite volume solver. For computational puposes, MATLAB 2015b version has been used with macOS Mojave, RAM 8 GB and 2.3 GHz Intel Core i5.

Throughout this section of numerical validation, the acceleration due to gravity is set to $g = 9.81$ and the standard SI measuring units corresponding to the physical quantities (like s (seconds), kg (kilograms), m (meters), etc.) are omitted in the discussion. For the space discretization, we have used uniform cartesian grids.

To analyse the performance of the hybrid method discussed in this paper, let q_i^n be the numerically computed solution and $q(x_i, t^n)$ be the exact solution corresponding to the ith control volume at the nth time stamp, thus the L_1 error norm, presented here by $\|e_n\|_1$ is written as:

$$\|e_n\|_1 = \sum_{i=1}^{N} |q(x_i, t^n) - q_i^n| \Delta x. \tag{17}$$

and the L_∞ error norm, denoted by $\|e_n\|_\infty$ is given as follows:

$$\|e_n\|_\infty = \max_{1 \le i \le N} |q(x_i, t^n) - q_i^n|. \tag{18}$$

where N represents the computational points.

5.1   Dam Break

In this section, the shallow problem (2) is computed to assess the proposed solution scheme by using various test cases corresponding to the dam-break scenario in a rectangle shaped domain having flat topography (i.e., $Z(x) = 0$, refer 1). The computational domain is $[-1, 1]$, and the step size is $\Delta x = 0.005$. In the next subsection, three test cases have been considered. The test case 1 corresponds to the Riemann problem in height profile, the test case 2 is basically opposite of the test case 1, and the third test case represents a vacuum Riemann problem.

Test Case 1: The initial profile used for approximating this test situation of the dam break problem is:

$$u(x, t = 0) = 0; \quad h(x, t = 0) = \begin{cases} 0.1, & -1 < x \le 0 \\ 2, & 0 < x \le 1. \end{cases} \tag{19}$$

Tables 2, 3 provide the details of point-wise error analysis for the standard flux limiting functions and the proposed method for the L_1 and the L_∞ norms, and the Figs. 4, 5 present the computational results corresponding to the classical MC limiter and the proposed limiting function for 400 computational points.

It is visible from the Fig. 4 that the MC limiting function is able to grasp the solution profile, although slight oscillations can still be seen. However, the computational output appears to be improved for the hybrid limiter, as seen in the Fig. 5.

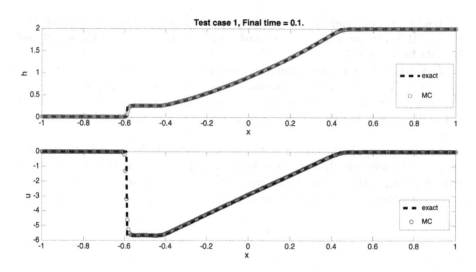

Fig. 4. Approximation results corresponding to the test case 1 obtained by the classical limiter for $N = 400$ control volumes at final time $t = 0.1$.

Table 2. Error analysis based on L_1 norm for the test case 1

Final Time	Minmod (MM)	Monotonized-Central (MC)	Proposed (New)
0.1	1.40e-03	1.12e-03	8.36e-04
0.05	1.36e-03	1.09e-03	1.02e-03
0.02	1.63e-03	1.45e-03	1.37e-03

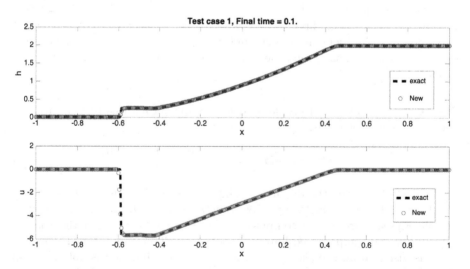

Fig. 5. Approximation results corresponding to the test case 1 obtained by the proposed limiter for $N = 400$ control volumes at final time $t = 0.1$.

Table 3. Error analysis based on L_∞ norm for the test case 1

Final time	Minmod (MM)	Monotonized-Central (MC)	Proposed (New)
0.1	2.13e-01	1.86e-01	1.11e-01
0.05	1.95e-01	1.66e-01	1.05e-01
0.02	2.01e-01	1.97e-01	1.00e-01

Test Case 2. The following is the initial data profile for simulating this dam-break test case:

$$u(x, t = 0) = 0; \; h(x, t = 0) = \begin{cases} 2, & -1 < x \leq 0 \\ 0.1, & 0 < x \leq 1. \end{cases} \tag{20}$$

The MC limiter is clearly able to capture the solution structure, as shown in the Fig. 6, though minor perturbations can still be seen. Nevertheless, as shown in the Fig. 7, the numerical result for the hybrid limiter appears to be improved.

Test Case 3: The initial data for approximating this dam-break test case is:

$$h(x, t = 0) = 0.1; \; u(x, t = 0) = \begin{cases} -2, & -1 < x \leq 0 \\ 2, & 0 < x \leq 1. \end{cases} \tag{21}$$

Although slight disturbances can still be seen in the solution pattern, but the MC limiter is clearly able to capture it, as shown in the Fig. 8. However, as can be seen in the Fig. 9, the numerical result for the hybrid limiter appears to be better.

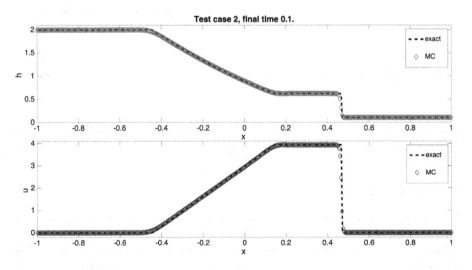

Fig. 6. Approximation results corresponding to the test case 2 obtained by the classical limiter for $N = 400$ grid points at final time $t = 0.1$.

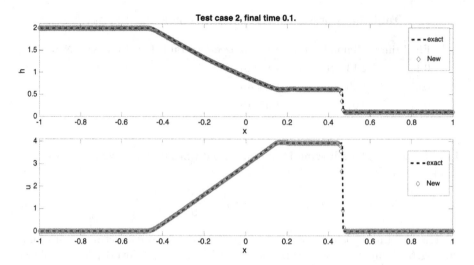

Fig. 7. Approximation results corresponding to the test case 2 obtained by the proposed limiter for $N = 400$ grid points at final time $t = 0.1$.

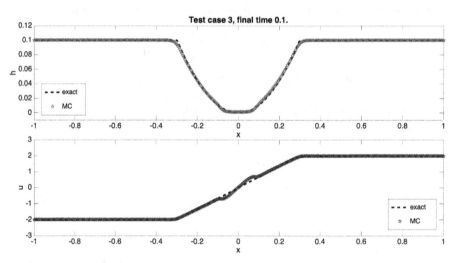

Fig. 8. Approximation results corresponding to the test case 3 obtained by the classical limiter for $N = 400$ grid points at final time $t = 0.1$.

5.2   CPU Time

The comparison of results obtained by various test cases has been the primary focus in the Subsect. 5.1. In terms of the CPU time taken by the various numerical integrations, the proposed scheme requires relatively more CPU time in all of the test cases since the number of operations per time step involved in calculating the fluxes across neighboring cells is greater than the standard flux limiting

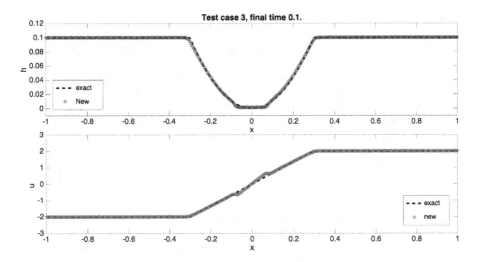

Fig. 9. Approximation results corresponding to the test case 3 obtained by the proposed limiter for $N = 400$ grid points at final time $t = 0.1$.

Table 4. CPU time data in seconds (s)

CPU time	Test case 1	Test case 2	Test case 3
Monotonized-Central (MC)	2.8644 s	2.1922 s	1.40411 s
Proposed (New)	2.0097 s	2.7627 s	1.71645 s

schemes. Refer to the Table 4 for the comparison of CPU times for various test cases considered in the Subsect. 5.1.

6   Conclusion

In summary, we have presented a numerical formulation of SWEs using a fuzzy logic based flux limiting scheme. The approach is based on the physical principles and balance laws of classical fluid mechanics. The distinction between the classical and the proposed method lies in its main new ingredient: Fuzzy modifiers. As future work, this strategy could be further extended to higher dimensional SWEs. The proposed HR method is non-oscillatory, conservative, well balanced, and suitable for shallow water models. The proposed flux limited method is verified against the benchmark dam-break problem with flat bottom topography. Final results are comparable with the classical scheme and show good agreement with analytical solutions also.

Acknowledgement. The authors would like to express their gratitude to the anonymous reviewers for their keen observations and insightful comments.

References

1. LeVeque, R.: Finite-Volume Methods for Hyperbolic Problems. Cambridge University Press, Cambridge (2002)
2. Xu, J., Liao, Z., Hu, Z.: A class of linear differential dynamical systems with fuzzy initial condition. Fuzzy Sets Syst. **158**(21), 2339–2358 (2007)
3. Belotserkovskii, O.M.: Methods of computational gasdynamics. Computational Gasdynamics. ICMS, vol. 40, pp. 219–407. Springer, Vienna (1975). https://doi.org/10.1007/978-3-7091-2732-2_2
4. Hirsch, C.: Numerical Computation of Internal and External Flows. Elsevier, Amsterdam (2007)
5. Klir, G., Yuan, B.: Fuzzy Sets and Fuzzy Logic, Theory and Applications, Prentice Hall PTR, Upper Saddle River (1995)
6. Zadeh, L.A.: Fuzzy sets. Inf. Control **8**, 338–353 (1965)
7. Breuss, M., Dietrich, D.: On the optimization of flux limiters for hyperbolic conservation laws. Numer. Methods Partial. Differ. Equ. **29** 884–896 (2013)
8. Chin, T., Qi, X.: Genetic Algorithms for learning the rule base of fuzzy logic controller. Fuzzy Sets Syst. **97**, 1–7 (1998)
9. Kumar, V., Srinivasan, B.: An adaptive mesh strategy for singularly perturbed convection diffusion problem. Appl. Math. Model. **39**, 2081–2091 (2015)
10. Kumar, V., Rao, R.: Composite scheme using localized relaxation non-standard finite difference method for hyperbolic conservation laws. J. Sound Vib. **311**, 786–801 (2008)
11. Toro, E.F.: Riemann Solvers and Numerical Methods for Fluid Dynamics, 3rd edn. Springer, Berlin (2009)
12. Vázquez-Cendón, M.E.: Improved treatment of source terms in upwind schemes for the shallow water equations in channels with irregular geometry. J. Comput. Phys. **148**(2), 497–526 (1999)
13. Vreugdenhil, C.B.: Numerical Methods for Shallow-Water Flow. Springer, Netherlands (1994)
14. Kurganov, A.: Finite-volume schemes for shallow-water equations. Acta Numer. **27**, 289–351 (2018)
15. Versteeg, H.K., Malalasekera, W.: An Introduction to Computational Fluid Dynamics: the Finite Volume Method. 2nd ed. Pearson Education Ltd, Gosport, Hants (2007)
16. Li, J., Du, Z.: A two-stage fourth order time-accurate discretization for Lax-Wendroff type flow solvers I. Hyperbolic conservation laws, SIAM J. Sci. Comput. **38**(5), A3046–A3069 (2016)
17. Sergeyev, Y.D., Kvasov, D.E. (eds.): NUMTA 2019. LNCS, vol. 11973. Springer, Cham (2020). https://doi.org/10.1007/978-3-030-39081-5
18. Lochab, R., Kumar, V.: An improved flux limiter using fuzzy modifiers for hyperbolic conservation laws. Math. Comput. Simulation **181**, 16–37 (2021)
19. Sweby, P.K.: High resolution schemes using flux limiters for hyperbolic conservation laws. SIAM J. Numer. Anal. **21**(5), 995–1011 (1984)
20. Roe, P.L.: Characteristic-based schemes for the Euler equations. Ann. Rev. fluid Mech. **18**, 337–365 (1986)
21. Van Leer, B.: Towards the ultimate conservative difference scheme. II. Monotonicity and conservation combined in a second-order scheme. J. Comput. Phys. **14**(4), 361–370 (1974)
22. Lochab, R., Kumar, V.: A new reconstruction of numerical fluxes for conservation laws using fuzzy operators. Int. J. Numer. Meth. Fluids. **93**, 1690–1711 (2021). https://doi.org/10.1002/fld.4948

PIES for Viscoelastic Analysis

Agnieszka Bołtuć$^{(\boxtimes)}$ (ID) and Eugeniusz Zieniuk$^{(\boxtimes)}$ (ID)

Institute of Computer Science, University of Bialystok, Bialystok, Poland
{aboltuc,ezieniuk}@ii.uwb.edu.pl

Abstract. The paper presents the approach for solving 2D viscoelastic problems using the parametric integral equation system (PIES). On the basis of Kelvin model the PIES formula in time differential form is obtained. As solving procedure the time marching is adopted, by introducing a linear approximation of displacements. The proposed approach, unlike other numerical methods, does not require discretization even the boundary. It uses curves as a tool for global modeling of boundary segments: curves of the first degree for linear segments and of the third degree for curvilinear segments. The accuracy is steered by the approximation series with Lagrange basis functions. Some test are made and shown in order to validate the proposed approach.

Keywords: PIES · Viscoelasticity · Time marching · Parametric curves

1 Introduction

Many materials together with their elastic properties also show viscous characteristics. They require special treatment, because exhibit time-dependent strain. The two best known methods for solving elastic problems are the finite element method (FEM) [1] and the boundary element method (BEM) [2]. In the literature are also available various procedures for viscoelastic analysis, i.a. the one which transforms a viscoelastic equation into a pseudo-elastic [3] or uses an incremental scheme, where viscous behavior is added to the elastic responses [4]. An alternative to them is time marching process applied in both FEM and BEM [5,6]. It bases on the differential constitutive relation for well-known viscoelastic models and allows for quick reaching steady states by speeding up time integration. Moreover, such approach makes easier changes in boundary conditions and viscous parameters along time. On the other hand it can influence the accuracy especially at small time scales. Referring to the FEM and BEM themselves, also in this approach they require spatial approximation by elements. In FEM the whole domain is divided into finite elements regardless of the problem solved. In BEM it depends on the formulation: one is defined in the domain and thus divides it into cells [5], while another transforms the problem to the boundary modeled by boundary elements [6].

The authors in their own research develop the approach that is an alternative to the above-mentioned element methods. The parametric integral equation

© Springer Nature Switzerland AG 2021
M. Paszynski et al. (Eds.): ICCS 2021, LNCS 12742, pp. 493–499, 2021.
https://doi.org/10.1007/978-3-030-77961-0_40

system (PIES) [8–10] does not require traditional discretization, because the shape is analytically incorporated into its mathematical formalism. For this reason, various curves can be used to model the boundary [11], and their number is determined only by the shape, not by the accuracy expectations. It comes from the fact that in PIES the approximation of the boundary and the solutions are separated. The latter are approximated by special series, and the accuracy is controlled by changing the number of its terms. The efficiency of PIES has been confirmed on various examples from many fields like elasticity [8], acoustics [9] or plasticity [10]. In order to distinguish PIES from the recently popular isogeometric analysis [7] which also bases on CAD design tools, it should be emphasized that the curves used in PIES are integrated into it at the analytical level. As a result, any modification of the shape causes an automatic modification of the PIES formalism. Moreover, the isogeometric analysis very often requires elements, not for modeling, but for the integration. Finally, the approximation of the solutions and the shape in PIES are independent, which allows various approaches to be used for both.

The main aim of the paper is to obtain the PIES formula and the algorithm of its numerical solving for 2D viscoelastic problems. For the sake of simplicity, PIES is developed using the Kelvin viscoelastic model, but other models could be added to the approach following similar steps. The problem is defined only on the boundary, using Bezier curves for its modeling. The time marching methodology is applied using the linear approximation for the displacement time derivative. As a result, accurate solutions obtained by fewer computational resources are expected. The numerical examples included confirm assumptions in comparison to analytical solutions.

2   PIES for Viscoelastic Problems

2.1   Basic Relations for Kelvin Model

The Kelvin viscoelastic model is represented by a purely viscous damper and elastic spring connected in parallel [12]. Therefore, it can be stated that the strains in each component are the same, while the total stress is the sum of stresses in each component

$$\varepsilon_{ij} = \varepsilon_{ij}^e = \varepsilon_{ij}^v, \quad \sigma_{ij} = \sigma_{ij}^e + \sigma_{ij}^v, \tag{1}$$

where σ_{ij}, ε_{ij} indicate the stress and strain tensors, while superscripts e and v stands for elastic and viscous parts.

Stress tensors in (1) can be written in terms of strain components as follows

$$\sigma_{ij}^e = C_{ijlm}\varepsilon_{lm}, \quad \sigma_{ij}^v = \eta_{ijlm}\dot{\varepsilon}_{lm}, \quad \eta_{ijlm} = \gamma C_{ijlm}, \tag{2}$$

where C_{ijlm}, η_{ijlm} are elastic and viscoelastic tensors, which for isotropic materials can be written as a function of one another using a viscosity constant γ. Using (1) and (2), the general Kelvin constitutive relation is obtained

$$\sigma_{ij} = C_{ijlm}(\gamma\dot{\varepsilon}_{lm} + \varepsilon_{lm}). \tag{3}$$

Introducing viscous effects to the global equilibrium equation of the body and neglecting the dynamic terms and the body forces results in

$$\sigma^e_{ij,i} + \sigma^v_{ij,i} = 0. \tag{4}$$

2.2   PIES Formula

PIES for 2D viscoelastic analysis is obtained on the basis of the differential equilibrium Eq. (4) and takes the following form

$$0.5\boldsymbol{u}_l(\bar{s}) + 0.5\gamma\dot{\boldsymbol{u}}_l(\bar{s})$$
$$= \sum_{j=1}^{n} \int_{s_{j-1}}^{s_j} \left\{ \mathbf{U}^*_{lj}(\bar{s}, s)\boldsymbol{p}_j(s) - \mathbf{P}^*_{lj}(\bar{s}, s)\boldsymbol{u}_j(s) - \gamma\mathbf{P}^*_{lj}(\bar{s}, s)\dot{\boldsymbol{u}}_j(s) \right\} J_j(s)ds, \tag{5}$$

where $\boldsymbol{U}^*_{lj}(\bar{s}, s)$, $\boldsymbol{P}^*_{lj}(\bar{s}, s)$ are the displacement and traction boundary fundamental solutions, while $\boldsymbol{p}_j(s)$, $\boldsymbol{u}_j(s)$ and $\dot{\boldsymbol{u}}_j(s)$ are the displacement, traction and displacement time derivative components. The boundary in PIES is defined in 1D parametric reference system, where $s, \bar{s}$ are parameters and they are limited by $s_{l-1} \le \bar{s} \le s_l$, $s_{j-1} \le s \le s_j$ (s_{l-1} and s_{j-1} mark the start of lth and jth segments, and s_l and s_j their ends). $J_j(s)$ is the Jacobian, n is the number of boundary segments and $l, j = 1..n$.

$\mathbf{U}^*_{lj}(\bar{s}, s)$ and $\mathbf{P}^*_{lj}(\bar{s}, s)$ are presented explicitly in [8, 10]. They include analytically integrated shape of the boundary, which can be defined by any parametric curves e.g. Bezier curves of various degrees [11]. The choice of the appropriate representation depends on the complexity of the shape, and the modeling itself is reduced to defining an appropriate number of control points.

3   Shape Modeling

In viscoelastic problems solved by BEM the geometry is defined in two different ways depending on the formulation. One approach bases on the idea of FEM and requires defining the domain, and hence the cells [5]. The process is similar to dividing the area into finite elements (Fig. 1a). The second formulation transforms the domain integral into the boundary by certain algebraic operations [6]. This makes the use of cells unnecessary (Fig. 1b).

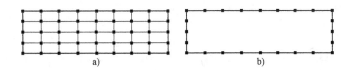

Fig. 1. The geometry modeled by: a) cells, b) boundary elements in BEM

Modeling the geometry in PIES is reduced to posing only control points. Their number and type depends on used curves. When linear segments of the boundary are defined, then only end points of each curve are specified (curves of the first degree). A rectangular shape created by corner points is presented in Fig. 2a. For curvilinear shapes, curves of the third degree (cubic) should be used. They require defining of four control points (two ends and two extra points responsible for the shape). An example of curvilinear geometry created by control points is shown in Fig. 2b.

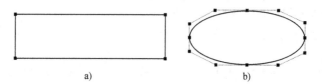

a)   b)

Fig. 2. The geometry modeled by: a) corner points, b) control points in PIES

In both mentioned cases in PIES the classical discretization is not required. The shape is modeled using the smallest number of data (e.g. corner points of the polygons), while the accuracy is provided by means of the appropriately selected number of coefficients in the approximating series (see the next section).

4   Solving Procedure

Solving PIES (5) is reduced to finding unknown functions $u_j(s)$, $p_j(s)$ and $\dot{u}_j(s)$. They are approximated by the following expressions

$$u_j(s) = \sum_{r=0}^{R-1} u_j^r L_j^r(s), \quad p_j(s) = \sum_{r=0}^{R-1} p_j^r L_j^r(s), \quad \dot{u}_j(s) = \sum_{r=0}^{R-1} \dot{u}_j^r L_j^r(s), \quad (6)$$

where $L_j^r(s) = \prod_{w=0, w \neq r}^{R} \frac{s - s_w}{s_r - s_w}$, u_j^r, p_j^r, $\dot{u}_j^r$ are values of boundary functions at collocation points and R is the number of collocation points on jth segment.

After substituting (6) in (5), the PIES approximating form for the viscoelastic problems is obtained

$$0.5u_l(\bar{s}) + 0.5\gamma\dot{u}_l(\bar{s}) = \sum_{j=1}^{n} \sum_{r=0}^{R-1}$$

$$\left\{ p_j^r \int_{s_{j-1}}^{s_j} \mathbf{U}_{lj}^*(\bar{s}, s) - u_j^r \int_{s_{j-1}}^{s_j} \mathbf{P}_{lj}^*(\bar{s}, s) - \gamma\dot{u}_j^r \int_{s_{j-1}}^{s_j} \mathbf{P}_{lj}^*(\bar{s}, s) \right\} L_j^r(s) J_j(s) ds.$$

$$(7)$$

Equation (7) after calculating all integrals and writing it at all collocation points takes the following shortened form

$$\boldsymbol{H}\boldsymbol{u} + \gamma\boldsymbol{H}\dot{\boldsymbol{u}} = \boldsymbol{G}\boldsymbol{p}, \qquad (8)$$

where $\boldsymbol{H} = [H]_{TT}$ and $\boldsymbol{G} = [G]_{TT}$ $(T = 2nR)$ are square matrices of elements expressed by integrals from (7), while $\boldsymbol{u}$, $\dot{\boldsymbol{u}}$, $\boldsymbol{p}$ are vectors containing coefficients of approximation series for displacements, displacement time derivatives and tractions.

In order to solve time differential Eq. (8) it is necessary to approximate velocity in time. It can be done by a simple linear time approximation as follows

$$\dot{\boldsymbol{u}}_{z+1} = \frac{\boldsymbol{u}_{z+1} - \boldsymbol{u}_z}{\Delta t}, \tag{9}$$

where Δt is the assumed time step. Applying (9) into (8) the following system of equations is obtained

$$(1 + \frac{\gamma}{\Delta t})\boldsymbol{H}\boldsymbol{u}_{z+1} = \boldsymbol{G}\boldsymbol{p}_{z+1} + \gamma\boldsymbol{H}\frac{\boldsymbol{u}_z}{\Delta t}. \tag{10}$$

The time marching process is obtained by solving (10) for each time step until the total time interval is reached, assuming that past values $(\boldsymbol{u}_z)$ are known. The boundary conditions along time are prescribed by interchanging columns of $\boldsymbol{H}$ and $\boldsymbol{G}$.

5 Examples

The first example concerns a rectangular bar loaded at its free end. The material is under plane stress with the properties and other analysis parameters presented in Fig. 3a.

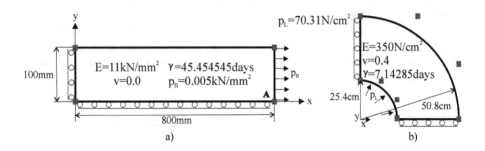

Fig. 3. The considered: a) stressed bar, b) thick cylinder

In PIES the boundary of the bar is modeled using 4 linear Bezier curves (4 corner points). It was also modeled by BEM in two ways. In [5] the domain of the bar is defined by 32 cells. Assuming they are linear, 45 nodes are used (not counting overlapping nodes). In the second approach [6], only the boundary is modeled using 48 boundary elements (48 nodes). In both cases, this is much more data to be modeled than in PIES.

The horizontal displacements at the bottom right corner of the bar (point A in Fig. 3a) are obtained at each time step ($\Delta t = 1$ day). Their comparison with analytical solutions is presented in Fig. 4. As can be seen obtained displacements are very similar to analytical results. Figure 4 also presents dependence of the solutions on the time step length ($\Delta t = 1, 2, 5$ days). As mentioned in the introduction, the time marching can be less accurate at early times, which is visible for longer time steps.

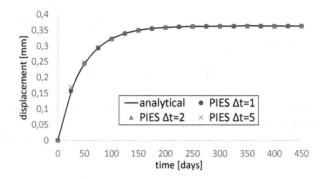

Fig. 4. The displacement at the bottom right corner of the bar (point A in Fig. 3a)

In the second example the behavior of a thick cylinder subjected to an internal pressure is analyzed. The geometry and physical properties are shown in Fig. 3b. The quarter of the cylinder in PIES is modeled by 2 cubic and 2 linear curves (8 control points). For comparison, in [5] BEM requires 12 cells (20 nodes).

The radial displacements of the outer boundary are obtained by PIES and compared with analytical solutions in Fig. 5. As can be seen, the results generated by PIES are very similar to analytical solutions, except some small discrepancies at early times.

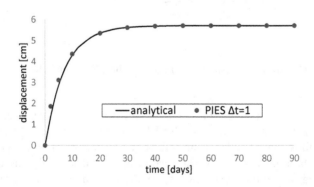

Fig. 5. The radial displacement on the outer boundary

Computational resources necessary to implement the proposed approach are also examined. The bar solutions are obtained within 0.137 s and the solved system of equations contains 32 equations. The quarter of the cylinder requires 0.174 s to be solved, with 48 equations in the system.

6 Conclusions

The paper presents the form of PIES for viscoelastic problems and the approach for its numerical solving. The shape is defined by parametric curves globally, i.e. without traditional discretization. Unknown displacements and tractions are approximated by series with Lagrange basis functions, while the displacement time derivative is obtained by linear approximation. The system of algebraic equations is solved for each time step until the predetermined period of time has elapsed.

The proposed approach is validated on two examples. A high agreement with the analytical results is obtained with much lower effort related to modeling of the geometry. Moreover, fewer time and memory resources are required to obtain such solutions.

The presented technique can be generalized to 3D viscoelastic problems using parametric surfaces for boundary representation. Moreover, it can be helpful for solving viscoplastic problems with surfaces used for modeling the domain. However, this requires additional research and testing.

References

1. Zienkiewicz, O.C.: The Finite Element Methods. McGraw-Hill, London (1977)
2. Aliabadi, M.H.: The Boundary Element Method, vol. 2. Applications in Solids and Structures. John Wiley and Sons Ltd, Chichester (2002)
3. Lemaitre, J., Chaboche, J.L.: Mechanics of Solids. Cambridge University Press, Cambridge (1990)
4. Yadagiri, S., Papi, R.C.: Viscoelastic analysis of near incompressible solids. Comput. Struct. **20**, 817–825 (1985)
5. Mesquita, A.D., Coda, H.B.: A boundary element methodology for viscoelastic analysis: part I with cells. Appl. Math. Model. **31**, 1149–1170 (2007)
6. Mesquita, A.D., Coda, H.B.: A boundary element methodology for viscoelastic analysis: part II without cells. Appl. Math. Model. **31**, 1171–1185 (2007)
7. Cottrell, J.A., Hughes, T.J.R., Bazilevs, Y.: Isogeometric Analysis: Toward Integration of CAD and FEA. John Wiley & Sons, UK (2009)
8. Zieniuk, E., Bołtuć, A.: Non-element method of solving 2D boundary problems defned on polygonal domains modeled by Navier equation. Int. J. Solids. Struct. **43**(25–26), 7939–7958 (2007)
9. Zieniuk, E., Bołtuć, A.: Bézier curves in the modeling of boundary geometry for 2D boundary problems defined by Helmholtz equation. J. Comput. Acoust **14**(3), 353–367 (2006)
10. Bołtuć, A.: Parametric integral equation system (PIES) for 2D elastoplastic analysis. Eng. Anal. Bound Elem. **69**, 21–31 (2016)
11. Salomon, D.: Curves and Surfaces for Computer Graphics. Springer, USA (2006). https://doi.org/10.1007/0-387-28452-4
12. Fung, Y.C.: Foundations of Solid Mechanics. Prentice Hall, Englewood Cliffs (1965)

Fast and Accurate Determination of Graph Node Connectivity Leveraging Approximate Methods

Robert S. Sinkovits[✉]

San Diego Supercomputer Center, University of California San Diego, La Jolla, CA 92093, USA
sinkovit@sdsc.edu

Abstract. For an undirected graph G, the node connectivity K is defined as the minimum number of nodes that must be removed to make the graph disconnected. The determination of K is a computationally demanding task for large graphs since even the most efficient algorithms require many evaluations of an expensive *max flow* function. Approximation methods for determining K replace the *max flow* function with a much faster algorithm that gives a lower bound on the number of node independent paths, but this frequently leads to an underestimate of K. We show here that with minor changes, the approximate method can be adapted to retain most of the performance benefits while still guaranteeing an accurate result.

Keywords: Graph algorithm · Node connectivity · Approximation methods · k-components

1 Introduction

Given an undirected graph, what is the smallest number of nodes that need to be removed so that the graph is broken into two or more disjoint components (i.e. there no longer exist paths connecting all possible pairs of nodes)? Like many problems in computer science, the question is easy to pose, but can be difficult to solve. Although the minimum node degree provides an upper bound, even highly connected graphs can become fragmented by the removal of just a few nodes (Fig. 1). The crux of our contribution is recognizing that the expensive vertex disjoint paths calculations underlying the solution of the node connectivity problem can first be estimated using a much faster approximate algorithm. We can easily determine when the approximation is invalid and revert to the more expensive calculation where necessary, thereby guaranteeing that we get the correct result.

The node connectivity problem is not just of theoretical interest, but is highly relevant to a number of fields. It has been applied to the impact of peer groups on juvenile delinquency [1], economic network models [2], clustering in social networks [3], political polarization [4], community structure [5, 6] and neural connectivity [7].

In the remainder section we define the conventions and nomenclature that will be used throughout the paper and describe the current state of the art in the calculation of the graph node connectivity. This is followed by descriptions of our new faster algorithm (Sect. 2), the implementation (Sect. 3), results of computational experiments (Sect. 4)

© Springer Nature Switzerland AG 2021
M. Paszynski et al. (Eds.): ICCS 2021, LNCS 12742, pp. 500–513, 2021.
https://doi.org/10.1007/978-3-030-77961-0_41

and finally a discussion and future work (Sect. 5). Notation and symbols used repeatedly in the manuscript are summarized in Table 1.

A graph $G = (V, E)$ is specified by a collection of nodes $V(G)$ and the edges $E(G)$ that connect pairs of nodes. An undirected graph is a graph for which the edges do not have a directionality. The numbers of nodes and edges in the graph are denoted by $|V(G)|$ and $|E(G)|$, respectively. Since we are working with a single graph, we simply denote the node and edge counts as $|V|$ and $|E|$. The degree of a node, $d(v)$, is the number of edges that connect to node v and δ is the minimum vertex degree across all nodes in the graph. A set of nodes S whose removal makes the graph disconnected is a cut set and the node connectivity K of a graph is defined as the size of the minimal cut set.

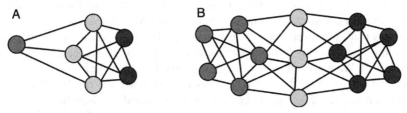

Fig. 1. Removal of gray nodes separates graphs into two disjoint graphs (blue and red nodes). In panel A, the degree of the blue node is the same as the number of nodes that need to be removed. In panel B, although no node has degree less than four, the removal of three nodes is sufficient. (Color figure online)

Table 1. Notation and symbols used in manuscript

Symbol	Definition
v, w	Nodes
v_s	Starting node used in node connectivity algorithms
$d(v)$	Degree of node v
δ	Minimum node degree
$E, |E|$	Edges, number of edges
κ	Local node connectivity
κ_{approx}	Approximate local node connectivity based on shortest paths
K	Graph node connectivity
K_{approx}	Graph node connectivity based on approximate algorithm
n_{target}	Number of target nodes tested in search for optimal v_s
n_{trial}	Number of starting nodes tested in search for optimal v_s
t_{approx}	Run time for approximate node connectivity algorithm based on shortest paths
t_{fast}	Run time for fast node connectivity algorithm (presented in this paper)
t_{orig}	Run time for original node connectivity algorithm
$t_{worst-case}$	Worst-case run time for fast node connectivity algorithm
$V, |V|$	Nodes, number of nodes in graph

2 Related Work

The determination of K is a computationally challenging problem (for a good overview of both node and edge connectivity algorithms, see [8]). Several approaches have been designed for this purpose that rely on the underlying evaluation of a *max flow* function. This in turn is based on Menger's theorem, which states that for a pair of non-adjacent nodes in an undirected graph, the number of node independent paths between the pair (i.e. paths that have no nodes in common other than the starting and ending points) is equal to the number of nodes that must be removed from the graph so that no paths remain between the pair [9]. Let $\kappa(G, v, w)$ be the local node connectivity function that returns the number of independent paths between vertices v and w in graph G. A brute force approach to finding K would be to simply calculate $\kappa(G, v, w)$ for all non-adjacent pairs of nodes in G and take the minimum value. It should be noted that there are several algorithms that do not rely on *max flow*, either addressing specific problems such as confirming 3-connectivity [10] or 4-connectivity [11] or the general case using random methods [12]. We will limit our discussions to only those that use *max flow*.

Several authors have developed schemes that minimize the number of calls to κ, leading to much better scaling. Even and Tarjan presented an algorithm that required $(K + 1)(|V| - \delta - 1) - \frac{1}{2} K(K + 1)$ calls to the *max flow* function [13]. Esfahanian and Hakimi further reduced the number of calls to $(|V| - \delta - 1) + \frac{1}{2} K(2\delta - K - 3)$ [14]. They noted that for each minimum node cut set in G, there exists at least one node that does not belong to the cut set. A randomly chosen node therefore either belongs to every minimum cut set or is outside of at least one minimum cut set. In the first case, it is sufficient to calculate the number of node independent paths between all pairs of non-adjacent neighbors of the selected node (i.e. between neighbors of the selected node that are not themselves neighbors). In the second case, we need to calculate the node connectivity between the selected node and all non-neighboring nodes. This algorithm, which accounts for both cases, can be expressed as follows.

$$G = (V, E)$$
$$K \leftarrow \delta$$
select arbitrary node v_s such that $d(v) == K$
for w in $V \setminus \{\{\text{neighbors}(v_s)\} \cup \{v_s\}\}$
    $K \leftarrow \min(\kappa(G, v_s, w), K)$
for x, y in $\{\text{neighbors}(v_s)\}$ such that x, y non-adjacent
    $K \leftarrow \min(\kappa(G, x, y), K)$

Improvements to the performance of this graph node connectivity algorithm would require either a further reduction in the number of evaluations of κ or the implementation of a faster method for calculating κ. Esfahanian and Hakimi had already considered the former and described a procedure for limiting the number of target nodes in the case where v_s is not a member of the minimum cut set. Although not considered in their paper, identifying a starting node v_s with the absolute smallest number of non-adjacent neighbors from among the minimal degree nodes could lead to a reduction in number of calls to κ, but for large graphs this would make a negligible difference. Regarding the latter option, although it might be possible to find more efficient ways to calculate κ, multiple algorithms already exist and additional progress would be extremely difficult.

Our approach is described in the next section and, while related to the efficient evaluation of κ, does not require the development of a new or refinement of an existing *max flow* algorithm. Instead, we use a combination of the approximate and exact algorithms for κ to boost performance while still guaranteeing an accurate result for K.

3 Faster Computation of Graph Node Connectivity

An approximate algorithm for calculating the local node connectivity, denoted κ_{approx}, is based on the repeated identification of the shortest path between the pair of nodes [15]. The shortest path is first identified and nodes on this path are marked as visited. The next shortest path involving only unvisited nodes is then determined and the nodes on that path are also marked as visited. The process is repeated until no paths remain between the node pair that involve only unvisited nodes. This approximate algorithm is extremely fast since the shortest path calculations are much more efficient than any known *max flow* algorithm. An approximate method for finding K simply replaces the calls to κ with κ_{approx}. The use of this approximation has also been applied to the problem of finding the k-components of a graph, where a k-component is the maximal subgraph with node connectivity equal to at least k [16]. The obvious downside of using κ_{approx} instead of κ is that we trade accuracy for speed. While the approximate local node connectivity algorithm often agrees with the result of the *max flow* function, a single discrepancy can lead to an incorrect result for K.

Fortunately, there is a way that we can exploit the efficiency of the approximation while still obtaining an exact result. We note that κ_{approx} always returns a lower bound on the true local node connectivity since the *max flow* algorithm can identify node independent paths that are not necessarily shortest paths. The condition $\kappa_{approx}(G, v, w)$ $\leq \kappa(G, v, w)$ holds for all (v, w) in G.

This means that we can first perform the approximate calculation and then repeat the exact calculation only if κ_{approx} is less than the value currently assigned to K, resulting in a modified algorithm that can leverage the better performance of the approximation while still guaranteed to yield the correct result. This is illustrated below.

$K \leftarrow \delta$
select arbitrary node v_s such that $d(v) == K$
for w in $V \setminus \{\{neighbors(v_s)\} \cup \{v_s\}\}$
    if $\kappa_{approx}(G, v_s, w) < K$
        $K \leftarrow min(\kappa(G, v_s, w), K)$
for x, y in $\{neighbors(v_s)\}$ such that x, y non-adjacent
    if $\kappa_{approx}(G, x, y) < K$
        $K \leftarrow min(\kappa(G, x, y), K)$

We define the following conventions used in the remainder of the paper. The *original* algorithm refers to Esfahanian and Hakimi, which exclusively uses κ for the local node connectivity calculations. The *fast* algorithm, described above, uses a combination of κ

and κ_{approx} while ensuring the correct result. The *approximate* algorithm (introduced in [15]) relies entirely on κ_{approx} and is not guaranteed to give the correct answer.

The performance benefits depend on the relative costs of evaluating κ_{approx} and κ, along with the number of calls to κ that can be avoided. Let N be the number of calls to κ in the original algorithm. The computational expense is then $N <t(\kappa)>$, where $<t(\kappa)>$ is the average time needed to evaluate κ. In the best case, where $\kappa_{approx}(G, v, w)$ is greater than or equal to the initial value assigned to K for all node pairs (a, b) that are tested, no calls need to be made to κ and the run time of the new algorithm will be $N <t(\kappa_{approx})>$. In the worst case, where $\kappa_{approx}(G, v, w)$ is always less than the running value for K, the run time will be $N <t(\kappa_{approx})> + N <t(\kappa)>$. In general, $\kappa_{approx}(G, v, w)$ will be less than K for a fraction p of the calls and the computational expense is $N <t(\kappa_{approx})> + pN <t(\kappa)>$. Note that in order to avoid having to calculate κ, it is not necessary that κ_{approx} return the correct result. Rather, we only require that $\kappa_{approx}(G, v, w)$ be greater than or equal to the running value for K.

Assuming that $<t(\kappa_{approx})>$ is small compared to $<t(\kappa)>$, the scaling of the fast algorithm with $|V|$, K and δ will be the same as that for the original algorithm but with overall run time multiplied by a factor of p.

There can be significant variation in the value of p for different choices of the starting node v_s. Finding with absolute certainty the best node would incur considerable expense since it entails determining K for each node of minimal degree. Our solution is to consider a subset of the minimal degree nodes and evaluate κ_{approx} between the trial starting node and a random set of non-neighboring nodes. We then choose as v_s the node that satisfies the condition $\kappa_{approx} < \delta$ with the lowest frequency.

$$G = (V, E)$$
$$\text{for } v \text{ in } X \subseteq \{x : d(x) == \delta\}; \text{ where } |X| == n_{trial}$$
$$\quad c(v) \leftarrow 0$$
$$\quad \text{for } w \text{ in } Y \subseteq V \setminus \{neighbors(v)\} \setminus \{v\}; \text{ where } |Y| == n_{target}$$
$$\quad\quad \text{if } \kappa_{approx}(G, v, w) < \delta$$
$$\quad\quad\quad c(v) \leftarrow c(v) + 1$$
$$v_s \leftarrow \{v : c(v) == \min(c(v)\}$$

We want to choose the numbers of trial nodes (n_{trial}) and test evaluations of κ_{approx} for each trial node (n_{target}) to be large enough to obtain a sufficiently small value for p, but not so large that we expend significant computational effort in the search for v_s.

4   Implementation

As our starting point, we use the Esfahanian and Hakimi algorithm as deployed in the *node_connectivity* function of the NetworkX Python package. This enables us to measure the benefits of our modified algorithm in the context of a state-of-the-art implementation and to take advantage of the full NetworkX framework for our benchmarks. The NetworkX version also builds an auxiliary digraph and residual network that are passed as additional arguments to the local node connectivity function, leading to improved performance. The *node_connectivity* function can accept one of several *max flow* functions and we use the current default (Edmonds-Karp in NetworkX 2.1). For the evaluation of κ_{approx} we use the approximate *node_connectivity* function, which can be called after importing the NetworkX approximation module.

The deployment of our new algorithm involves minimal changes to the original version. In the most basic implementation, this only requires the addition of several statements to evaluate $\kappa_{approx}(G, v, w)$ and test whether the result is less than the running value of K. A small amount of extra code allows one to choose between using a particular starting node v_s that is passed as an input argument or automatically selecting v_s from among candidate nodes of minimal degree. We also provide an option that forces the algorithm to simulate the worst case by evaluating κ_{approx} and ignoring the result regardless of how it compares to the running value for K, leading to both κ and κ_{approx} being called for each node pair. Although this option would not be used in practice, it is useful for testing purposes and providing a bound on the worst case.

It should be noted that the NetworkX package specifically or a Python implementation in general will likely have lower performance than an implementation in a compiled language such as C++. We are aware of these limitations, but as described earlier, using NetworkX provides the convenience of being able to work in a complete graph analytics package. The simplicity of our modifications to the original algorithm make them easy enough to employ in an implementation in any language where the user has access to the necessary graph libraries. The performance gains should be comparable to what we see using NetworkX, but the exact speedup will depend on the relative computational expense of the *max flow* and shortest path calculations.

5 Results

Here we compare the performance of our algorithm against the original and approximate algorithms for several types of graphs. We start with a collection of random graphs of varying sizes and densities generated using three different well-known models. We then apply our method to a set of graphs generated from a real-life anonymized social network, starting with the largest embedded 3-component and ending with the largest embedded 7-component. This social network example includes several intermediate graphs generated during the iterative process used to find these embedded components.

In all these benchmarks, the reported times for our method include the overhead associated with choosing an appropriate starting node v_s. We used up to 10 trial nodes (n_{trial}) and 100 target nodes (n_{target}). If there are fewer than 10 nodes with $d(v)$ equal to δ, all minimum-degree nodes are used.

We also address two issues raised earlier. The first regards the dependence of the run time on the choice of the source node v_s. The second concerns the overhead associated with the worst case where κ_{approx} is always less than the running value for K and the evaluation of the exact algorithm for the local node connectivity can never be avoided. All timings are obtained running the benchmarks with Python 3.6.4 and NetworkX 2.1 on a 1.8 GHz Intel Core i5 with 8 GB 1600 MHz DDR3.

5.1 Random Graphs

We present benchmark results for graphs generated using the Barabasi-Albert [17], Erdös-Renyi [18] and Watts-Strogatz [19] models for a range of graph sizes and edge counts in Tables 2, 3, 4, respectively. The Barabasi-Albert model generates scale-free networks in which the degree distribution of the nodes follows a power-law of approximately k^{-3}. This is accomplished using a preferential attachment model whereby nodes of higher degree are more likely to be assigned new edges. The Erdös-Renyi model produces random graphs where each potential edge in the graph is independently created with a fixed probability. The Watts-Strogatz model creates graphs having small-world properties by starting with a ring lattice where each node is connected to a given number of neighbors and then rewiring the edges with a given probability. Keep in mind that rerunning the examples will produce different results since new graphs will be randomly generated and there is a variation in run time even when working with the same graphs.

Table 2. Performance of original and fast algorithms for random Barabasi-Albert graphs using seed equal to $1456789356 + n + m$. The model parameters n and m are the number of nodes and number of edges from the newly added node to existing nodes as the graph is grown, respectively. $|E|$ is the number of edges in graph, K is the node connectivity, t_{approx}, t_{fast} and t_{orig} are the times taken by approximate, fast and original algorithms, respectively, to find the node connectivity and speedup is the ratio t_{orig}/t_{fast}. All times reported in seconds.

n	m	$|E|$	K	t_{approx}	t_{fast}	t_{orig}	Speedup
1,000	4	3,984	4	0.10	0.36	15.31	42.53
1,000	6	5,964	6	0.17	0.45	21.06	46.80
1,000	8	7,936	8	0.24	0.75	28.07	37.43
1,000	10	9,900	10	0.32	1.30	31.14	23.95
2,000	4	7.984	4	0.31	0.75	52.31	69.75
2,000	6	11,964	6	0.46	1.08	69.72	64.56
2,000	8	15,936	8	0.63	1.43	98.15	68.64
2,000	10	19,900	10	1.30	1.99	120.26	60.43
4,000	4	15,984	4	0.91	1.67	238.77	142.98
4,000	6	23.964	6	1.30	2.68	332.70	124.14
4,000	8	31,936	8	1.91	3.15	373.17	118.47
4,000	10	39,900	10	2.68	4.86	446.49	91.87
8,000	4	31.984	4	2.88	3.97	851.88	214.58
8,000	6	47,964	6	2.65	4.48	1,170.44	261.26
8,000	8	63,936	8	4.86	6.92	1,470.57	212.51
8,000	10	79,900	10	5.81	8.95	1,747.80	195.28
16,000	4	63,984	4	5.90	8.22	3,496.00	425.30
16,000	6	95,964	6	7.97	11.31	4,692.47	414.90
16,000	8	127,936	8	15.83	20.85	5,941.14	284.95
16,000	10	159,900	10	14.05	19.52	7,250.68	371.45

Table 3. Performance of original and fast algorithms for random Erdös-Renyi graphs using seed equal to 1456789356 + n + 1000*p. The model parameters n and p are the number of nodes and probability of edge creation, respectively. $|E|$ is the number of edges in graph, K is the node connectivity, t_{approx}, t_{fast} and t_{orig} are the times taken by approximate, fast and original algorithms, respectively, to find the node connectivity and speedup is the ratio t_{orig}/t_{fast}. All times reported in seconds.

n	p	$|E|$	K	t_{approx}	t_{fast}	t_{orig}	Speedup
1,000	0.01	4,965	3	0.10	0.32	16.86	52.69
1,000	0.02	9,988	8	0.31	0.52	34.84	67.00
1,000	0.04	15,004	13	0.59	1.24	58.35	47.06
2,000	0.01	20,019	6	0.61	1.30	113.60	87.38
2,000	0.02	40,007	21	3.60	5.55	311.78	56.18
4,000	0.03	59,761	35	8.58	13.07	648.37	49.61
4,000	0.01	80,443	18	7.24	10.33	986.54	95.50
4,000	0.02	159,306	51	58.99	64.49	3,648.42	56.57
4,000	0.03	240,767	89	111.60	216.73	11,576.66	53.42

Table 4. Performance of original and fast algorithms for random Watts-Strogatz graphs using seed equal to 1456789356 + n + k. The model parameters n and k are the number of nodes and number of neighbors to join in ring topology, respectively. Rewiring probability was set to 0.10 for all graphs. $|E|$ is the number of edges in graph, K is the node connectivity, t_{approx}, t_{fast} and t_{orig} are the times taken by approximate, fast and original algorithms, respectively, to find the node connectivity and speedup is the ratio t_{orig}/t_{fast}. All times reported in seconds. Incorrect result returned by approximate algorithm marked with *.

n	k	$|E|$	K	t_{approx}	t_{fast}	t_{orig}	Speedup
1,000	5	2,000	2	0.23	0.38	11.00	28.95
1,000	7	3,000	3	0.35	0.42	13.45	32.02
1,000	9	4,000	5	0.50	0.75	18.72	24.96
2,000	5	4,000	2	0.54	1.03	38.35	37.23
2,000	7	6,000	3	0.81	0.84	50.69	60.35
2,000	9	8,000	5	1.20	2.19	67.96	31.03
4,000	5	8,000	2	1.23	1.95	129.42	66.37
4,000	7	12,000	3	2.25	2.48	177.92	71.74
4,000	9	16,000	5	3.73	5.71	215.24	37.70
8,000	5	16,000	2	6.22	7.70	592.69	76.97
8,000	7	24,000	3	6.09*	7.80	699.72	89.71
8,000	9	32,000	4	9.34	8.27	841.16	101.71
16,000	5	32,000	2	10.77	13.24	2,167.33	163.70
16,000	7	48,000	3	15.98	16.77	2,488.67	148.40
16,000	9	64,000	4	23.78	21.84	3,259.02	149.22

For the Barabasi-Albert graphs, we observe speedups ranging from $24\times$ to $425\times$, with the speedup increasing with the number of nodes and showing a weak dependence on the parameter m, which specifies the number of edges from the newly added node to existing nodes during the construction of the graph. For the Erdös-Renyi graphs, we obtained speedups ranging from $47\times$ to $95\times$, with weak dependence on the graph size and reduced benefits at larger values of the edge creation probability. For the Watts-Strogatz graphs, speedups ranged from $25\times$ to $149\times$. We obtained noticeably weaker gains for smaller graphs, but this might reflect the overhead in searching for v_s. The performance gains also decrease as the parameter k, which sets the number of neighbors to join in the ring topology, is increased.

Although there is considerable variation in the speedup when applying our algorithm to graphs created using different models and parameters, we consistently observe significant gains of at least one order of magnitude reduction in run time.

For the random graphs, we found that the approximate algorithm nearly always returned the correct result while only taking 30–80% as long as our fast algorithm. While it is tempting to simply rely on the approximate algorithm, there is no guarantee of accuracy as seen in Table 4 and demonstrated in the next section.

5.2 Social Network Example

Table 5 contains the results of benchmarks for subgraphs extracted from a large social network. The performance gains, while still significant, are less dramatic than those obtained when applying our fast algorithm to the random graphs. They range from a $5\times$ speedup for the largest 7-component to $31\times$ for the largest pre-5-component, the penultimate graph generated during the iterative search for the largest 5-component [20]. The fast algorithm did extremely well for the most challenging problem, reducing the

Table 5. Performance of original and fast algorithms on subgraphs of large social network. Naming convention for graphs: X-comp is the largest k-component in graph and pre-X is the largest graph found in last step of iterative process during search for k-component. K is the true node connectivity and K_{approx} is the connectivity reported by the approximate algorithm; $|V|$ (equivalent to n in the random graph models in Tables 2, 3, 4) and $|E|$ are the number of nodes and edges; t_{approx}, t_{fast} and t_{orig} are the times taken by approximate, fast and original algorithms; speedup is the ratio t_{orig} / t_{fast}. All times reported in seconds.

Graph	K	K_{approx}	$|V|$	$|E|$	t_{approx}	t_{fast}	t_{orig}	Speedup
7-comp	7	5	542	3,704	0.37	2.75	14.20	5.16
pre-7	6	6	573	3,889	0.77	1.02	15.82	15.51
6-comp	6	4	3,089	19,002	13.97	69.59	468.29	6.73
pre-6	5	3	3,636	22,166	12.50	45.51	520.58	11.44
5-comp	5	3	9,864	53,732	38.39	210.85	2653.61	12.59
pre-5	4	3	9,923	53,996	37.02	79.87	2508.79	31.41
4-comp	4	2	19,948	94,630	41.15	518.87	9540.75	18.39
3-comp	3	1	37,938	150,453	109.03	1017.04	28351.48	27.88

time to find the connectivity for the largest 3-component from nearly eight hours to about 17 min. Most importantly though, our algorithm produced the correct result whereas the approximate algorithm underestimated the connectivity for seven out of the eight graphs tested.

5.3 Dependence on Choice of Starting Node

The performance of the fast algorithm depends critically on the frequency with which the exact local node connectivity calculation can be avoided. We found, particularly for the subgraphs extracted from the social network, that this frequency in turn depends on the choice of starting node v_s.

Let $n_\kappa(v_s)$ be the number of evaluations of κ given v_s. Figure 2 shows the run time as a function of $n_\kappa(v_s)$ for the fast algorithm applied to the largest 6-component using all 451 minimal degree nodes ($d = 6$) as starting points. The time varies from 51 s to 552 s, with the corresponding values of $n_\kappa(v_s)$ equal to 199 and 2,996, respectively. Although the run time tracks closely with $n_\kappa(v_s)$, there is still some variation. For example, using the starting node that maximized $n_\kappa(v_s)$ with a value of 3,079, the run time was only 479 s. While long compared to the best cases, this is still considerably less than the longest run time. At the other end of the spectrum, there can be a 30% variation in run time for starting nodes with the same or nearly the same value for $n_\kappa(v_s)$. This spread reflects the fact that the overall run time depends not only on $n_\kappa(v_s)$, but also in the variation in the time needed to execute κ for different node pairs.

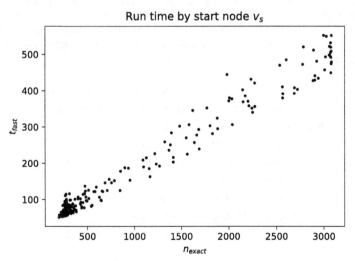

Fig. 2. Dependence of the fast algorithm performance on choice of starting node (v_s) when applied to the largest 6-component in large social network. Each marker represents one of the 451 minimal degree ($d = 6$) nodes. n_{exact} is the number of calls to κ and t_{fast} is the time required by the fast algorithm.

These results reaffirm the importance of carefully selecting v_s. In this particular case, nearly 80% of the starting nodes lead to performance that is at least $4\times$ better than the original algorithm. Evaluating even a small number of choices for v_s before launching the full calculation drastically reduces the likelihood of a poor outcome.

5.4   Worst-Case Performance

Here we compare the worst-case performance of our algorithm against the standard algorithm for a collection of random graphs. As noted in the previous section, the choice for v_s affects the run time in two ways – the number of calls that must be made to κ and the variation in the computational expense of evaluating κ for different node pairs. To allow for a fair comparison, each pair of runs using the original and fast algorithms is done using the same choice for v_s.

The results are shown in Table 6. To simulate the worst case, we evaluate κ_{approx} as usual, but then ignore the result and require that κ still be executed. We find that the fast algorithm takes at most a few percent longer than the original algorithm, although in several instances we measured it to be slightly faster. Since the worst case basically involves running the original algorithm plus the overhead associated with the execution of κ_{approx}, this finding can be attributed to the natural variability in runtime.

Table 6. Performance of original and worst-case fast algorithm. K is the node connectivity; $t_{worst\text{-}case}$ and t_{orig} are the times of the worst case for the fast algorithm and original algorithm, respectively. For the worst-case, the fast algorithm was deliberately modified so that the exact local node connectivity was calculated for every node pair that was considered, even if the approximate node connectivity was greater than or equal to the running value for graph node connectivity. For each choice of model and parameters, test was run on five different random graphs. All times reported in seconds.

Model	K	$t_{worst\text{-}case}$	t_{orig}	Speedup
Barabasi-Albert ($n = 2000$, $m = 4$)	4	67.25	68.10	1.01
Barabasi-Albert ($n = 2000$, $m = 4$)	4	67.48	66.44	0.98
Barabasi-Albert ($n = 2000$, $m = 4$)	4	67.65	67.19	0.99
Barabasi-Albert ($n = 2000$, $m = 4$)	4	68.08	66.78	0.98
Barabasi-Albert ($n = 2000$, $m = 4$)	4	68.82	67.96	0.99
Erdos-Renyi ($n = 1000$, $p = 0.03$)	13	57.87	57.45	0.99
Erdos-Renyi ($n = 1000$, $p = 0.03$)	15	62.67	60.97	0.97
Erdos-Renyi ($n = 1000$, $p = 0.03$)	12	55.81	55.29	0.99

6   Discussion

The main challenge going forward is improving the selection of the starting node v_s. As we described earlier, finding the optimal choice would generally be prohibitively expensive since it requires running the full algorithm for every minimal degree node. Furthermore, it's not entirely obvious that a minimal degree node is the best option. Esfahanian and Hakimi made this decision in order to minimize the number of calls to κ. Choosing a starting node of higher degree might minimize the frequency with which κ_{approx} is less than the running value for K, but these benefits could be offset by a larger number of calculations. Even within our current scheme, additional progress is possible. Our default is to select 10 trial starting nodes and evaluate κ_{approx} for 100 different non-neighboring nodes. Increasing either n_{trial} or n_{target} improves the odds of finding a good choice for v_s, but also increases the overhead costs. To further complicate the situation, the values that strike the right balance between the extra overhead and likelihood of finding an optimal, or close to optimal, choice are probably dependent on the properties of the graph.

Our work focuses on improving the performance of the general algorithm, but it should be kept in mind that there are several simple steps that should be taken before undertaking the more expensive calculations. Graphs with δ equal to two have a maximum K of two and testing for articulation points – nodes whose removal results in a disconnected graph – quickly identifies graphs with K equal to one.

There are several additional avenues for future work. Our proposed improvements are not limited to the Esfahanian and Hakimi algorithm and can easily be incorporated into other schemes that rely on the *max flow* calculation for determining node connectivity [13, 21] or deciding if the node connectivity is at least k [21, 22]. We are also considering parallelization of the algorithm since the local node connectivity calculations can be done independently. One complication is that determining the validity of the shortest paths-based approximation depends on the running value for the overall graph node connectivity. This could be addressed by periodic synchronization between threads and backtracking when required.

Another possible application is to the more difficult problem of identifying the k-components of a graph. The method described by Moody and White [23] relies on the repeated execution of two computationally demanding steps: determining the connectivity of a graph or subgraph and finding all minimum sized cut sets using Kanevsky's algorithm [24]. Replacing the former with our faster implementation can yield immediate performance gains, although the magnitude will depend on the relative amounts of time spent in the two steps. Our preliminary benchmarks indicate that the improvements are modest, generally around 10–15%, but additional work remains to be done. Nonetheless, given the interest in k-components, especially from researchers in the social sciences [1–6], even small improvements will be welcome.

In conclusion, our new graph node connectivity algorithm effectively leverages the performance of the fast approximation to the local node connectivity (κ_{approx}) while still being guaranteed to give the correct result. Although there is significant variation in performance relative to the original algorithm, depending on the size and complexity of the graph, we find that in every instance there are unambiguous benefits. The worst-case scenario, which we have not encountered in any test and that we can simulate only by forcing κ_{approx} and κ to be calculated for every node pair considered, only results in at most a few percent degradation in performance.

The improved algorithm, test data and Jupyter notebook for running the benchmarks can be downloaded at https://github.com/sinkovit/node-connectivity-fast.

Acknowledgments. The author thanks Doug White for many useful discussions and for introducing him to the challenging problems in node connectivity and the identification of k-components in large graphs. Early stages of this work were supported in part by National Science Foundation grants OCI #0910847 Gordon: A Data Intensive Supercomputer and ACI#1053575 XSEDE: eXtreme Science and Engineering Discovery Environment (XSEDE) through the ECSS program.

References

1. Kreager, D.A., Rulison, K., Moody, J.: Delinquency and the structure of adolescent peer groups. Criminol. **49**(1), 95–127 (2011)
2. Mani, D., Moody, J.: Moving beyond stylized economic network models: the hybrid world of the Indian firm ownership network. AJS Am. J. Sociol. **119**(8), 1629 (2014)
3. Moody, J., Coleman, J.: Clustering and cohesion in networks: concepts and measures. International Encyclopedia of Social and Behavioral Sciences (2014)
4. Moody, J., Mucha, P.J.: Portrait of political party polarization. Netw. Sci. **1**(01), 119–121 (2013)
5. Newman, M.E.: Modularity and community structure in networks. Proc. Natl. Acad. Sci. **103**(23), 8577–8582 (2006)
6. Porter, M.A., Onnela, J.-P., Mucha, P.J.: Communities in networks. Notices of the AMS **56**(9), 1082–1097 (2009)
7. Sporns, O.: Graph theory methods for the analysis of neural connectivity patterns. In: Neuroscience databases, pp. 171–185. Springer (2003). https://doi.org/10.1007/978-1-4615-1079-6_12
8. Esfahanian, A.H.: Connectivity algorithms. http://www.cse.msu.edu/~cse835/Papers/Graph_connectivity_revised.pdf. (2013)
9. Menger, K.: Zur allgemeinen kurventheorie. Fundam. Math. **10**(1), 96–115 (1927)
10. Hopcroft, J.E., Tarjan, R.E.: Dividing a graph into triconnected components. SIAM J. Comput. **2**(3), 135–158 (1973)
11. Kanevsky, A., Ramachandran, V.: Improved algorithms for graph four-connectivity. In: Foundations of Computer Science, 1987, 28th Annual Symposium on IEEE (1987)
12. Henzinger, M.R., Rao, S., Gabow, H.N.: Computing vertex connectivity: new bounds from old techniques. J. Algorithms **34**(2), 222–250 (2000)
13. Even, S., Tarjan, R.E.: Network flow and testing graph connectivity. SIAM J. Comput. **4**(4), 507–518 (1975)
14. Esfahanian, A.H., Louis Hakimi, S.: On computing the connectivities of graphs and digraphs. Networks **14**(2), 355–366 (1984)
15. White, D.R., Newman, M.: Fast approximation algorithms for finding node-independent paths in networks. Santa Fe Institute Working Papers Series. Available at SSRN: ssrn.com/abstract_id=1831790. 29 June 2001
16. Torrents, J., Ferraro, F.: Structural cohesion: visualization and heuristics for fast computation. J. Soc. Struct. **16**(8), 1–35 (2015)
17. Barabási, A.-L., Albert, R.: Emergence of scaling in random networks. Sci. **286**(5439), 509–512 (1999)
18. Erdős, P., Rényi, A.: On the strength of connectedness of a random graph. Acta Mathematica Academiae Scientiarum Hungarica **12**(1–2), 261–267 (1961). https://doi.org/10.1007/BF02066689

19. Watts, D.J., Strogatz, S.H.: Collective dynamics of 'small-world' networks. Nature **393**(6684), 440 (1998)
20. Sinkovits, R.S., Moody, J., Oztan, B.T., White, D.R.: Fast determination of structurally cohesive subgroups in large networks. J. Comput. Sci. **17**, 62–72 (2016)
21. Galil, Z., Italiano, G.F.: Fully dynamic algorithms for edge connectivity problems. In: Proceedings of the Twenty-Third Annual ACM Symposium on Theory of Computing. ACM (1991)
22. Even, S.: An algorithm for determining whether the connectivity of a graph is at least k. SIAM J. Comput. **4**(3), 393–396 (1975)
23. Moody, J., White, D.R.: Structural cohesion and embeddedness: a hierarchical concept of social groups. Am. Sociol. Rev. **68**(1), 103–127 (2003)
24. Kanevsky, A.: Finding all minimum-size separating vertex sets in a graph. Networks **23**(6), 533–541 (1993)

An Exact Algorithm for Finite Metric Space Embedding into a Euclidean Space When the Dimension of the Space Is Not Known

Ewa Skubalska-Rafajłowicz[1]($\boxtimes$) and Wojciech Rafajłowicz[2]

[1] Department of Computer Engineering, Wroclaw University of Science and Technology, Wyb. Wyspianskiego 27, 50 370 Wrocław, Poland
ewa.rafajlowicz@pwr.edu.pl
[2] Department of Control Systems and Mechatronics, Wroclaw University of Science and Technology, Wyb. Wyspianskiego 27, 50 370 Wrocław, Poland
wojciech.rafajlowicz@pwr.edu.pl

Abstract. We present a $O(n^3)$ algorithm for solving the Distance Geometry Problem for a complete graph (a simple undirected graph in which every pair of distinct vertices is connected by a unique edge) consisting of $n + 1$ vertices and non-negatively weighted edges. It is known that when the solution of the problem exists, the dimension of the Euclidean embedding is at most n. The algorithm provides the smallest possible dimension of the Euclidean space for which the exact embedding of the graph exists. Alternatively, when the distance matrix under consideration is non-Euclidean, the algorithm determines a subset of graph vertices whose mutual distances form the Euclidean matrix. The proposed algorithm is an exact algorithm. If the distance matrix is a Euclidean matrix, the algorithm provides a geometrically unambiguous solution for the location of the graph vertices.

The presented embedding method was illustrated using examples of the metric traveling salesman problem that allowed us in some cases to obtain high dimensional partial immersions.

Keywords: Isometric embedding · Euclidean distance matrix · Euclidean distance geometry problem · Rigidity of graphs

1 Introduction

In this paper, we concentrate on providing a fast algorithm for solving the Euclidean geometry problem which aims at deciding whether it is possible to find the configuration of points in the Euclidean space, such that the Euclidean distances between each pair of points match a given distance matrix. The dimension of the Euclidean space is not known in advance. Our goal, if any solution exists, is to find the point configuration in the Euclidean space of the smallest dimensionality.

© Springer Nature Switzerland AG 2021
M. Paszynski et al. (Eds.): ICCS 2021, LNCS 12742, pp. 514–524, 2021.
https://doi.org/10.1007/978-3-030-77961-0_42

The existence of a solution is directly related to the Euclidean distance matrix problem (EDM) [5,13]. In this paper, we assume that only distance matrix $D = [d_{ij}]$, $i,j \in V$ is given. Our goal is to check if D is EDM and if the answer is positive, to determine the location of vertices V in the Euclidean space of the smallest dimensionality $m \leq |V| - 1$.

The problem can also be formulated equivalently in the terms of graphs. Namely, given a non-negatively weighted complete graph (defined by the distance matrix), decide whether the smallest dimension m exists and the configuration of points in R^m corresponding to the graph vertices such that Euclidean distances between all pairs of the points are equal to the edge weights.

There are known conditions that guarantee that the real solution of the Euclidean embedding problem exists [5,15,18,19,23], but verifying these conditions has a similar computational complexity as the direct solving of the problem proposed here.

For many years, other problems related to the Euclidean geometry problem have also been considered, such as Euclidean matrix completion problems [1], molecular reconstruction of chemical structures [12,14–17], sensor localization in sensor networks [4,6,10,26], machine learning and dimensionality reduction [7,25], and signal processing [11] among others.

These problems are outside the scope of this paper, but we are convinced that the algorithm proposed in this paper can also be adapted to solve many of the problems mentioned. However, this will be the subject of further research and experimentation.

The outline of the paper is as follows. Section 2 introduces the main ideas, definitions, and properties used throughout this paper. Section 3 describes the proposed embedding algorithm and discuss its computational complexity. Section 4 provides the experimental framework used to illustrate the performance of the algorithm and its possible results when the distance matrix is a non-Euclidean one. Finally, Sect. 5 presents our conclusions as well as some propositions for further research.

2   The Euclidean Distance Matrix Problem and Euclidean Distance Geometry Problem.

The Euclidean Distance Matrix (EDM) problem is formulated as follows. Determine whether a given matrix is Euclidean and a specific Distance Geometry Problem (DGP): given a nonnegatively weighted complete graph defined by the distance matrix, decide whether the smallest dimension m exists such that Euclidean distances between pairs of points in R^m are equal to the edge weights.

Let us denote d_{ij}^2 - a squared distance between nodes $i \in V$ and $j \in V$, by a_{ij} Thus, $A = [a_{ij}]$ is the matrix of squared distances $[d_{ij}^2]_{ij,0:n}$.

Matrix A is a squared distance matrix if and only if all elements on the diagonal of A are zero, the matrix is symmetric, i.e., $a_{ij} = a_{ji}$, $a_{ij} \geq 0$ and (by the triangle inequality)

$$\sqrt{a_{ij}} \leq \sqrt{a_{ik}} + \sqrt{a_{kj}}.$$

Any function $\psi : V \to R^m$ is an embedding of V in R^m.

Theorem 1 ([15],[22]) *A necessary and sufficient condition for the isometric embeddability of a finite metric set (V, d) of $n + 1$ elements in an Euclidean space R^n is that the following statement be true:*
The matrix $[\frac{1}{2}(a_{0i} + a_{0j} - a_{ij})]_{i,j=1:n}$ is positive definite.

There are also known conditions of embeddability based on the Cayley-Menger determinants [15, 23], but discussion of that approach is outside the scope of our paper.

An embedding is locally unique [21], i.e., we say that $p : V \to R^m$ and $q : V \to R^m$ are congruent, if

$$||p_i - p_j|| = ||q_i - q_j||$$

for all pairs $i, j \in V$.

As a consequence, if Euclidean distance matrix $D =$ is given, then any solution $\psi : V \to R^m$ such that $||\psi_i - \psi_j|| = d_{ij}$, $i, j \in V$ is locally unique.

Obviously, any rigid transformation such as a translation, a rotation or a reflection or their composition does not destroy the distance structure of (V, D). More precisely, two complete graphs in the same Euclidean space are congruent if they are related by an isometry which is either a rigid motion (translation and/or rotation), or a composition of a rigid motion and a reflection [8].

A rigid graph is an embedding of a graph in a Euclidean space which is structurally rigid [9, 18]. A graph in Euclidean space R^m is said to be rigid if and only if a continuous motion of the vertices in R^m with keeping the distances between adjacent vertices unchanged preserves the distances between all pairs of graph vertices.

It is well known that every complete graph embedded in R^m is rigid in R^m [2, 3, 9].

3   An Exact Incremental Algorithm for Embedding a Complete Weighted Graph in the Euclidean Space

In this section, we will deal with the algorithmic side of the problem.

A set of $n + 1$ points in the Euclidean space spans a subspace of at most n dimensions, hence the embedding dimension considered should not exceed this number. We assume that metric distance matrix $D = [d_{ij}]$, $ij, = 0 : n$ is given. The naive approach leads to the need to solve the system of $n(n+1)/2$ non-linear equations with a very large number of variables (at least $2n$, at most n^2):

$$||x(i) - x(j)|| = d_{ij}, \ i, j = 0 : n$$

$x(i) \in R^n$, $i = 0 : n$.

It is clear that without loss of generality we can set $x(0) = (0, \ldots, 0)$.

If the system of equations is contradictory, the problem is a non-Euclidean one and an exact embedding in the Euclidean space is not possible.

Analyzing the equivalent form of the previous system of equations, i.e.,

$$\sum_{l=1}^{n}(x_l(i) - x_l(j))^2 = a_{ij}, \ i, j = 0 : n$$

(recall that $a_{ij} = d_{ij}^2$) we can provide a computationally efficient algorithm for embedding a complete graph (a set of objects) in the Euclidean space when such exact immersion exists. Otherwise, the algorithm will stop indicating that it is impossible to preserve the currently considered set of equality constraints. The proposed approach consists of joining another vertex to an already existing partial immersion (as a new point in the Euclidean space). Next, it is necessary to check the possibility of maintaining the distance of the vertex in question, let us say k_{th}, from other $k - 1$ vertices already located in the Euclidean space, so as not to violate the values contained in matrix D. It may possibly require increasing the current dimension of the Euclidean space by one.

In the end, the algorithm provides the dimension number of the embedding and the locally unique embedding of V , or stops when exact embedding is impossible. The exact embedding does not depend on the vertex chosen as a starting point.

We begin analysis with a set of three vertices, let say, vertices labeled by $0, 1$ and 2. Due to the triangle inequality, it is obvious that these vertices can be embedded in R^2 and this embedding is locally unique. Vertex-representing points form a triangle and the shape of this triangle is unique. So, without loss of the generality we can locate vertex v_0 in $x(0) = (0,0)$, v_1 in $x(1) = (d_{01}, 0)$, and v_2 in $x(2) = (x_1(2), x_2(2))$, where $x(2)$ coordinates are obtained by solving the following system of equations:

$$(x_1(2) - x_1(0))^2 + (x_2(2) - x_2(0))^2 = a_{02}, \tag{1}$$

$$(x_1(2) - x_1(1))^2 + (x_2(2) - x_2(1))^2 = a_{12}. \tag{2}$$

Since $x(0) = (0,0)$, (1) simplifies to

$$x_2^2(1) + x_2^2(2) = a_{02} \tag{3}$$

and (2) takes the form

$$(x_2(1) - d_{01})^2 + x_2^2(2) = a_{12}.$$

Substracting the first equation from the second one, we obtain a linear equation:

$$-2d_{01}x_2(1) + a_{01} = a_{12}.$$

Thus, $x_{21} = -0.5(a_{12} - a_{01})/d_{01}$, and consecutively

$$x_2^2(2) = a_{02} - \frac{(a_{12} - a_{01})^2}{4a_{01}}.$$

Due to the triangle inequality,

$$a_{02} - \frac{(a_{12} - a_{01})^2}{4a_{01}} \leq 0.$$

There exist at most two solutions of the system with $x_2(2)$ being a positive or a negative real number. When points $x(0)$, $x(1)$, $x(2)$ are collinear, $x_2(2) = 0$ and the embedding dimension is $m = 1$. As a consequence, all coordinates $x_2(\cdot)$ can be neglected and may be removed.

3.1 Algorithm of the Euclidean Embedding of a 3-vertex Structure

Algorithm 1.

1. Step B1. Set $x(0) = 0.$, $x(1) = d_{01}$ and dimension number $m(1) = 1$.
2. Step B2. Compute $\Delta = a_{02} - \frac{(a_{12} - a_{01})^2}{4a_{01}}$ If $\Delta = 0.$ set $x(2) = (0.5(a_{12} - a_{01})/d_{01})$ and $m(2) = 1$. Otherwise set $m(2) = 2$, $x(0) = (0., 0.)$, $x(1) = (d_{01}, 0.)$, and $x(2) = (-0.5(a_{12} - a_{01})/d_{01}, \sqrt{\Delta})$.

The proposed approach can generalized for larger graphs and larger dimensions. Expanding the set of the immersed vertices one by one leads to the incremental embedding algorithm.

3.2 General Location Algorithm

Locating a new vertex from V in the Euclidean space leads to the following problem:

Problem 1. *We assume that we have given coordinates of $r + 1$ points representing an exact embedding of $V_r \subset V$, $|V_r| = r + 1$ in a Euclidean space, i.e., $(x(0), x(1), \ldots x(r)) \in R^{m(r)}$, where $m(r)$ is a dimension of the Euclidean space, and $m(r) <= r$. Thus, that we have:*

$$||x(i) - x(j)||^2 = a_{ij}, \ i, j \in 0 : r.$$

Find a vector of dimensionality at most $m(r) + 1$ representing a new selected vertex, let us say, $v_{r+1} \in V - V_r$ that does not violate distances from D between all vertices in $V_r \cup v_{r+1}$.

Thus, our goal is to find

$$x(r + 1) = (x_1(r + 1), x_2(r + 1), \ldots, x_{m(r)+1}(r + 1))$$

such that

$$||x(i) - x(r + 1)||^2 = a_{i,r+1}, \ i = 0, 1, \ldots, r, \tag{4}$$

where the dimension of all points in $X_r = [x(i)]_{i=0:r}$ is expanded to $m(r) + 1$, and the $(m(r)+1)$-th coordinates of all vectors in X_r are set to zero. If $x(r+1) \in R^{m(r)+1}$ exists and $x_{m(r)+1}(r + 1) = 0$ the system of $r + 2$ points lays in the m

dimensional quotient space, i.e., $x(0), x(2), \ldots x(r), x(r+1) \in R^s$. All $m(r)+1$ zero coordinates should be removed and $m(r+1) = m(r)$. Otherwise, when $x_{m(r)+1}(r+1) \neq 0$, the new dimension of the embedding Euclidean space is $m(r+1) = m(r)+1$. The lack of any real solution means that the metric d is not a Euclidean metric and an exact immersion in the Euclidean space does not exist.

Taking into account the fact that the number of non-zero coordinates successively considered vertices v_i gradually increase and is equal to the previously determined values of $m(i)$, we can rewrite (4) as a system of linear equations:

$$x(r+1)x^T(i) = \sum_{j=1}^{m(i)} x_j(i)x_j(r+1) = \tag{5}$$

$$\frac{1}{2}[a_{0,i} + a_{0,r+1} - a_{i,r+1}], \; i = 1:r$$

supplemented by a nonlinear equation of the form $||x(r+1)||^2 = a_{0,r+1}$. Notice, that in (5) we have replaced $||x(r+1)||^2$ by $a_{0,r+1}$ and $||x(i)||^2$ by $a_{0,i}$.

If the solution of the linear system (2) exists, but it is such that $||x(r+1)||^2 \neq a_{0,r+1}$ there is no real solution of (4) and the exact embedding of $V(r+1)$ in the Euclidean space is not possible.

To better illustrate the essence of the proposed algorithm, let us first show a simple example.

Example Let $r = 2$, $m(1) = 1$, and $m(2) = 2$ then (2) is of the form

$$x_1(3)x_1(1) = [a_{0,1} + a_{0,3} - a_{1,3}]/2,$$

$$x_1(3)x_1(2) + x_2(3)x_2(2) = [a_{0,2} + a_{0,3} - a_{2,3}]/2.$$

Using results of Step B2, i.e., $x(1) = (d_{01}, 0.)$, and $x(2) = (-0.5(a_{12} - a_{01})/d_{01}, \sqrt{\Delta})$, we can obtain values of $x_1(3)$, $x_2(3)$ and check if $x_1^2(3) + x_2^2(3) = a_{0,3}$. Alternatively, after computing $x_1(3)$ from the first equation, one can obtain $x_2^2(3) = a_{0,3} - x_1^2(3)$. It is obvious that when $a_{0,3} - x_1^2(3) < 0$ the real solution of the problem does not exist.

The system of equations (5) can be written as:

$$\mathcal{X}x^T(r+1) = b, \tag{6}$$

where $x(r+1) = (x_1(r+1), \ldots, x_{m(r)}(r+1))$, $b = \frac{1}{2}[a_{0,i} + a_{0,r+1} - a_{i,r+1}]_{i=1:r}$ and $\mathcal{X} = [x(i)]_{i=1:r} \in R^{r \times m(r)}$ is a triangle-like matrix.

Algorithm

1. Start with Algorithm 1. We assume that previously computed values of $m(1), \ldots, m(r)$ are known, and available. Additionally $m(0) = 0$.
2. **Step 1. Solving (5):**

3. **for** $i \leftarrow 1$ **to** r **do**
   set $m = m(i)$.
4. **if** $m = m(i-1)+1$ compute

$$x_m(r+1) = \frac{(a_{0,i} + a_{0,r+1} - a_{i,r+1})}{2x_{m(i)}(i)} x - \frac{\sum_{j=1}^{m(i)-1} x_j(i)x_j(r+1)}{x_{m(i)}(i)}$$

and go to Step 2.
5. **else** (i.e., when $m = m(i-1)$) recalculate $x_m(r+1)$ as

$$x_m^{new}(r+1) = \frac{(a_{0,i} + a_{0,r+1} - a_{i,r+1})}{2x_{m(i)}(i)} - \frac{\sum_{j=1}^{m(i)-1} x_j(i)x_j(r+1)}{x_{m(i)}(i)}.$$

   if $|x_m(r+1) - x_m^{new}(r+1)| > 0$
   then system of equations (5) is not consistent. Stop and provide adequate information.
6. **else** (i.e., when $|x_m(r+1) - x_m^{new}(r+1)| = 0$) **end**
7. **Step 2. Checking feasibility:**
8. Compute

$$a = \sum_{j=1}^{m(r)} x_j^2(r+1).$$

9. **if** $a > a_{0,r+1}$,
   then   full embedding does not exist. Stop and give adequate information.
   else
   if $a < a_{0,r+1}$
   then set $m(r+1) = m(r)+1$ and $x_{m(r+1)}(r+1) = \sqrt{a_{0,r+1} - a}$,
   else (when $a = a_{0,r+1}$) **then** set $m(r+1) = m(r)$.
10. **end**

System (5) can be over-determined (when $m(r) < r$) but it is consistent when exact embedding exists.

Notice that if two subsequent, previously embedded vertices, let us say v_i and v_{i+1}, have the same embedding dimension (i.e., $m(i) = m(i+1)$) then two subsequent equations from the system (5) are linearly dependent or inconsistent. In the second case the solution does not exist.

The presented algorithm exploits the triangle-like structure of $\mathcal{X}$.

Thus, it allows us to solve (5) performing at most $O(r^2)$ arithmetic operations. Using the ordinary least square method to solve (5) leads to $O(r^3)$ local complexity.

Summarizing, the algorithm of embedding V is as follows:

Given a $n+1 \times n+1$ matrix of squared distances A the algorithm provides the dimension of the embedding $m(n)$ and vector's coordinates $\mathcal{X} = [x(i)]_{i=1:n}$ (recall that $x(0)$ is the zero vector) or information that exact embedding does not exist. Additionally, a partial embedding of the first k vertices ($3 < k < n+1$) is given.

It should be emphasized that the computational complexity of embedding of the whole set of $n + 1$ vertices is in the worst case $O(n^3)$. Using the least square method to solve a local system of equations (5) increases the total complexity to $O(n^4)$ [24].

4   Computational Experiments

The proposed algorithm was extensively tested with problems from the well-know TSPLIB. The TSPLIB proposed by [20] (http://elib.zib.de/pub/mp-testdata/tsp/) is a typical set of benchmarks containing 111 different symmetric traveling salesman problems (TSP) problems.

We are investigating two groups of problems:

- symmetric problems with Euclidean 2D distances (EUC_2D),
- symmetric problems with explicit weights in the form of distance matrices (MATRIX).

As a method of verification of the concept and its accuracy some of the 61 problems of type EUC_2D (two-dimensional problems where coordinates of the cities are known and the distance is calculated as a simple Euclidean norm) were converted to distance matrices. We know that the exact dimension is 2. By this reversed method we can easily verify the accuracy of the result.

Annotation MATRIX indicates that the distances are supplied in the form of matrices. No other information is available. As a rule, such matrices are not Euclidean distance matrices. Such data allow us to obtain some partial Euclidean immersions of the TSP vertices.

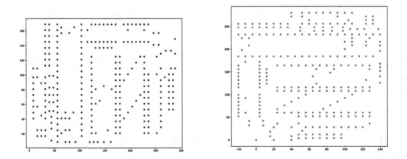

Fig. 1. Original (first) and reconstructed (second) layout of the a280 problem from tsplib.

In Fig. 1 we can see the results of the reconstruction in the comparison to the original data a280 problem from tsplib. It is easy to observe that the resulting image is accurate with respect to rotation and mirroring. In Fig. 2 the reconstruction of the bier127 problem is visualized. In both pictures the vertices

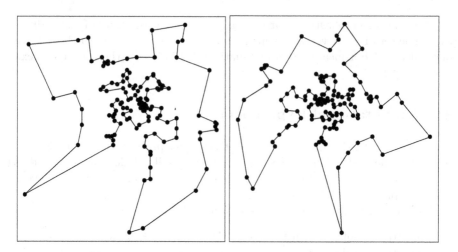

Fig. 2. Original (first) and reconstructed (second) layout of the `bier127` problem from tsplib.

(cities) are connected according to the optimal tour. Further, some metric problems, namely `si175`, `si535` and `si1032`, were examined. Distance matrices are not EDM, as shown by the proposed algorithm. Nevertheless, large parts of the corresponding graphs can be precisely (without any distortion) embedded in the separate Euclidean spaces of large dimensionality.

The experiments were performed in the following way. After the information about the inability to immerse the next vertex in the lastly indicated Euclidean space appeared, the procedure was interrupted and the next subset of vertices was generated. The last rejected vertex was taken as the starting point of the algorithm. A randomly generated vertex was selected as the starting node, and then the order of considering other vertices was established according to the nearest neighbor rule.

In the `si175` problem, 5 subsets of nodes immersed in separate Euclidean spaces of dimension 48, 58, 33, 2, and 29, respectively, and composed of 49, 59, 34, 3, and 30 vertices, was obtained. The `si535` problem has been divided into 8 subgraphs immersed in separate Euclidean spaces of dimension 14, 97, 116, 2, 133, 126, 2 and 37, respectively. Finally, the `si1032` problem was decomposed by the algorithm into 16 separate EDM cliques consisting of 109, 80, 80, 80, 69, 83, 39, 54, 50, 45, 127, 54, 53, 46, 54, and 9 nodes.

5   Concluding Remarks

However, the algorithm presented in this paper can also be used as a method of computing an approximate solution of the embedding problem, the global optimization approach based on minimization of the sum of squared distance distortions (see (7)), examined in [14], provides better approximations from the point of view of the mean squared error criterion minimization.

Unfortunately, the global distortion error defined as

$$\sum_{i=0}^{n-1} \sum_{j=i+1}^{n} \left(||x(i) - x(j)||^2 - a_{ij} \right)^2,$$

or as

$$\sum_{i=0}^{n-1} \sum_{j=i+1}^{n} \left| ||x(i) - x(j)||^2 - a_{ij} \right|^2, \tag{7}$$

is not easy to minimize, when exact embedding does not exist.

The algorithm presented in the paper allows us to easily impose upper bounds on subsequent distortion errors of individual distances (both absolute and relative errors). However, this will be the subject of further research.

Summarizing, the proposed algorithm provides the smallest dimension number of the embedding and the configuration of points in the Euclidean space, such that the Euclidean distances between each pair of points match a given distance matrix. Alternatively, if the matrix is a non-Euclidean one, only a partial solution is generated.

In such a case the partial embedding depends on the vertex selected as a starting point. The problem of partial embedding of the finite metric set is combinatorial in nature and can be formulated in many different ways. In our opinion, it is a new class of open problems that is worthy of further research.

References

1. Alfakih, A., Khandani, A., Wolkowicz, H.: Solving Euclidean distance matrix completion problems via semidefinite programming. Comput. Optim. Appl. **12**, 13–30 (1999)
2. Asimow, L., Roth, B.: The rigidity of graphs. Trans. Amer. Math. Soc. **245**, 279–289 (1978)
3. Barvinok, A.I.: Problems of distance geometry and convex properties of quadratic maps. Discrete Comput. Geom. **13**(2), 189–202 (1995). https://doi.org/10.1007/BF02574037
4. Biswas, P., Lian, T., Wang, T., Ye, Y.: Semidefinite programming based algorithms for sensor network localization. ACM Trans. Sens. Netw. **2**(2), 188–220 (2006)
5. Blumenthal, L.: Theory and Applications of Distance Geometry. Oxford University Press, Oxford (1953)
6. Cao, M., Anderson, B.D.O., Morse, S.: Sensor network localization with imprecise distances. Syst. Control Lett. **55**, 887–893 (2006)
7. Courrieu, P.: Straight monotonic embedding of data sets in Euclidean spaces. Neural Netw. **15**, 1185–1196 (2002)
8. Coxeter, H.S.M.: Introduction to Geometry, 2nd edn. Wiley, Hoboken (1969)
9. Crapo, H.: Structural rigidity. Struct. Topol. **73**(1), 26–45 (1979)
10. Ding, Y., Krislock, N., Qian, J., Wolkowicz, H.: Sensor network localization, Euclidean distance matrix completions, and graph realization. Optim. Eng. **11**(1), 45–66 (2010)

11. Dokmanic, I., Parhizkar, R., Ranieri, J., Vetterli, M.: Euclidean distance matrices: essential theory, algorithms, and applications. IEEE Signal Process. Mag. **32**(6), 12–30 (2015). https://doi.org/10.1109/MSP.2015.2398954
12. Dong, Q., Wu, Z.: A linear-time algorithm for solving the molecular distance geometry problem with exact inter-atomic distances. J. Glob. Optim. **22**, 365–375 (2002)
13. Gower, J.: Euclidean distance geometry. Math. Sci. **7**, 1–14 (1982)
14. Lavor C.: On generating instances for the molecular distance geometry problem. In: Liberti, L., Maculan, N. (eds.) Global Optimization: From Theory to Implementation. Nonconvex Optimization and Its Applications, vol. 84, pp. 405–414. Springer, Boston (2006). https://doi.org/10.1007/0-387-30528-9_14
15. Liberti, L., Lavor, C., Nelson, M.N., Mucherino, A.: Euclidean distance geometry and applications. SIAM Rev. **56**(1), 3–69 (2014)
16. Liberti, L., Lavor, C.: Euclidean Distance Geometry. An Introduction. Springer International Publishing Switzerland, Cham (2017)
17. Liberti, L.: Distance geometry and data science. TOP **28**(2), 271–339 (2020). https://doi.org/10.1007/s11750-020-00563-0
18. Maehara, H.: Euclidean embeddings of finite metric spaces. Discrete Math. **313**, 2848–2856 (2013)
19. Menger, K.: New foundation of Euclidean geometry. Amer. J. Math. **53**, 721–745 (1931)
20. Reinelt, G.: TSPLIB - a traveling salesman problem library. ORSA J. Comput. **3**(4), 376–384 (1991)
21. Saxe, J.B.: Embeddability of weighted graphs in k-space is strongly NP-hard. In: Proceedings of the 17th Allerton Conference in Communications, Control, and Computing, pp. 480–489 (1979)
22. Schoenberg, I.J.: Metric spaces and positive definite functions. Trans. Am. Math. Soc. **44**, 522–536 (1938)
23. Sippl, M., Scheraga, H.: Cayley-menger coordinates. Proc. Natl. Acad. Sci. USA **83**, 2283–2287 (1986)
24. Stroeker, R.J.: On the sum of consecutive cubes being a perfect square. Compositio Mathematica **97**(1–2), 295–307 (1995)
25. Tenenbaum, J.B., De Silva, V., Langford, J.C.: A global geometric framework for nonlinear dimensionality reduction. Science **290**(5500), 2319–2323 (2000)
26. Tasissa, A., Lai, R.: Exact reconstruction of Euclidean distance geometry problem using low-rank matrix completion. IEEE Trans. Inf. Theor. **65**(5), 3124–3144 (2019)

Resolving Policy Conflicts
for Cross-Domain Access Control:
A Double Auction Approach

Yunchuan Guo[1,2], Xiyang Sun[1,2], Mingjie Yu[1,3], Fenghua Li[1,2], Kui Geng[1],
and Zifu Li[1(✉)]

[1] Institute of Information Engineering, Chinese Academy of Sciences, Beijing, China
`lizifu@iie.ac.cn`
[2] School of Cyber Security, University of Chinese Academy of Sciences,
Beijing, China
[3] School of Cyber Security, University of Science and Technology of China,
Hefei, China

Abstract. Policy-mapping mechanisms can efficiently help to realize
the exchange and the sharing of cross-domain information at low cost.
However, due to concerns over policy conflicts, if not sufficient incentives, most selfish domains are often disinterested in helping others to
implement policy mapping cooperatively. Thus an appropriate incentive
mechanism is required. In this paper, we propose an incentive mechanism
to encourage selfish domains to take part in policy mapping and resolve
policy conflicts. Formulating conflict resolution as a double auction and
solving Bayesian Nash equilibrium, we design the optimal asking/bidding
price scheme to maximize the benefits of the domains involved. Simulations demonstrate that our approach can efficiently incentivize selfish
domains to take part in cooperation.

Keywords: Cross-domain collaboration · Conflict resolution ·
Incentive mechanism · Double auction

1 Introduction

Cross-domain collaboration, which enables multiple organizations or systems in
domains via the networks (e.g., Internet, mobile Internet, and Internet of things)
to work together to achieve their own or common goals, has been widely used in
various applications, such as E-government, healthcare [12] and remote offices [2].
Through cross-domain collaboration, one can directly access the resource of the
other domain without the time-consuming manual authorization, thus increasing
work efficiency. To securely realize cross-domain collaboration at low cost, the

Supported by the National Key Research and Development Program of China
(No.2019YFB2101702), the National Natural Science Foundation of China (No.
U1836203) and the Youth Innovation Promotion Association CAS (2019160).

M. Paszynski et al. (Eds.): ICCS 2021, LNCS 12742, pp. 525–539, 2021.
https://doi.org/10.1007/978-3-030-77961-0_43

mapping of access control policies has been recently proposed to logically connect the involved domains without rebuilding a new collaboration system. In this approach, one autonomous domain's roles are mapped to the roles of the other domain. Through this approach, the authorized user of the first domain is automatically allowed to access the resource of the second domain, thus improving interoperability. However, because policy mapping breaks the security boundaries of inter-domain and causes a large amount of policy conflicts [14], domains that carry out policy mapping are suffering from an increasing number of security events, e.g., data breach, data corruption, and privacy leakage. To prevent and mitigate these events, one important thing that should be done is to design an efficient scheme of conflict resolution to achieve a tradeoff between security and interoperability.

Motivation: However, the existing schemes of conflict resolution for cross-domain policy mapping cannot efficiently work without sufficient incentives, because most autonomous domains are selfish and they are often uninterested in resolving policy conflicts (as a result, they do not participate in cross-domain collaboration) for the following reasons: (1) *Autonomy losses.* In most cases, resolving policy conflicts causes autonomy losses. For example, as shown in Fig. 1(a), role-based access control (RBAC) is used to assign permissions for users in domains A and B, where domain A has two roles (i.e., r_1 and r_2) and r_1 inherits all permissions of r_2. Domain B has one role (i.e., r_3). To achieve interoperability, we assume that r_1 of domain A and r_3 of domain B are mapped to r_3 of domain B and r_2 of domain A, respectively, as shown in Fig. 1(b). Under this assumption, a conflict called cyclic inheritance between r_1 and r_3 will be caused, as shown in Fig. 1(c). To resolve this conflict, one possible scheme is to delete the inheritance relationship between r_1 and r_2 (i.e., role r_1 no longer inherits the permissions of role r_2), as shown in Fig. 1(d). From this example, we can see that conflict resolution will decrease autonomy.

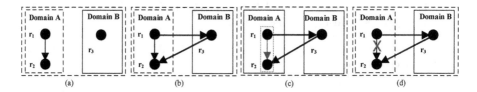

(a)   (b)   (c)   (d)

Fig. 1. (a) RBAC policy graph of domain A and B used in Autonomy losses. (b) RBAC policy after policy mapping. (c) RBAC policy conflicts between domain A and B. (d) RBAC policy after conflict resolution.

(2) *Privacy issues.* In existing schemes, the third party is often required to be responsible for mapping policies between domains and resolving their conflict. To complete its task, the third party has to obtain access control policies of both domains involved (i.e., the accessing domain and the accessed domain). However, because these policies often contain a lot of private or confidential

information, both the involved domains are unwilling to disclose their policies to the third party. For example, if a policy specifies that George can access top-secret files, then it can be inferred that George is a sensitive person. Due to the above reasons, a selfish domain is unwilling to participate in the cross-domain collaboration without appropriate incentives. Therefore, an incentive mechanism is required to incentivize the involved domain to participate in policy mapping and conflict resolution.

Existing incentive mechanism can be roughly divided into two categories: non-game-theoretical approaches [3,19] and game-theoretical approaches [5,6]. In the first approaches, contract mechanisms (e.g., reward contract [3] and judge contract [19]) are often adopted to predefine a set of reward rules to motivate individuals to fulfil their promise agreed in the contract and maximize their utility. For instance, Cruz et al. [3] adopted Ethereum's smart contract to encourage nodes to verify trans-organizational utilization of roles. Zhang et al. [19] designed a judge contract to analyze and penalize subjects' misbehavior and incentive the accessed objects to facilitate the validation of subjects' identity in the Internet of Things. However, in these approaches, a participant often ignores the competitive behavior of other participants and cannot make reasonable decisions. Considering rational individuals, a large amount of efforts are spent on designing the game-theoretical approaches to evaluate the competitive behavior of all participants and select the rational action to maximize its utility. Considering the community-structured behavior, Ren et al. [5] designed an evolutionary game-theoretic framework to incentive users to protect privacy information in OSNs, and Fang et al. [6] proposed an auction-based incentive mechanism to encourage co-owners to carry out a privacy agreement. However, existing incentive mechanism are not suitable for cross-domain collaboration because they ignore interoperation and autonomy.

Contribution: To address the above problem, we investigate conflict resolution from the aspects of the game. Our main contributions are as follows.

(1) Considering the selfishness of domains and inter-domain competition, we formulate an incentive mechanism for conflict resolution of cross-domain policy mapping as a double auction game and propose a conflict resolution framework of cross-domain policy mapping.
(2) After analyzing factors that impact the cost of the accessed domain and value of the accessing domain, we solve Bayesian Nash equilibrium under the assumption the incomplete information and design the optimal asking/bidding price scheme to maximize their interests.
(3) A series of experiments are carried out on a simulated dataset and the experimental results show that our proposed algorithm can efficiently incentivize domains to participate in conflict resolution.

The rest of this paper is organized as follows. In Sect. 2, we discuss the related work. In Sect. 3, we introduce the problem statement and our basic idea. Section 4 formulates the value and the cost in conflict resolution and analyzes its influencing factors. We conduct a double auction game in Sect. 5. Simulations and their analysis are given in Sect. 6. We draw a conclusion in Sect. 7.

2   Related Work

From the aspect of the number of resolution parties, the existing conflict resolution for access control can be divided into two categories: resolution within a single party and resolution among multiple parties.

2.1   Conflict Resolution Within a Singleparty

In this approach, conflict resolution is often formulated into a single-objective or multi-objective optimization problem to maximize the overall goal (e.g., highest precedence [8], resource consumption [18], and policy consistency [16]).

Considering the applicable laws, Huynh et al. [8] minimized the rule graph to resolve the conflicts between regional regulations and patient wishes, and selected the policy with the "highest" precedence (with regard to priority, specificity and modality) as the final policy decision. After translating access behaviors into the canonical representation of query spaces, Yahiaoui et al. [16] used fine-grained algebra expressions to infer and resolve the conflicts of policies and maximize the consistency of attribute-based access control policies. Rather than using inference to seek the solution [16], Omar et al. [11] adopted an answer set programming to search for the candidate resolution and calculate their priority. Extending Petri nets with both time and resource factors, Zeng et al. [18] designed three efficient strategies (i.e., start-early priority strategy, waiting-short priority strategy, and key-activity priority strategy) to resolve the conflict of emergency response processes and minimize resource consumption. To minimize the worst-case performance of services conflicts in smart cities, Ma et al. [9] designed an integer linear programming to generate serval resolution options and adopted a signal temporal logic to evaluate the performance of these options. Although the above schemes can efficiently resolve policy conflicts caused by individual mistake, they ignored individual interest and cannot deal with the policy conflicts caused by the conflict of interest of stakeholders.

2.2   Conflict Resolution Among Multiparty

From the aspect of interests of stakeholders, the existing schemes can be roughly divided into two categories: conflict resolution with optimizing collective interests and conflict resolution with optimizing individual interests.

Conflict Resolution with Optimizing Collective Interests. In this approach, the centralized platform regarded multiple domains as a whole and selected the scheme that maximizes their whole benefits as the final policy. Along this line, Shafiq et al. [14] adopted an integer programming (IP) to approximately avoid conflicts and maximize the global interoperation in the case of an acceptable autonomy loss. To assign permissions without conflicts, Zhu [20] et al. formulated the problem of maximizing the resolution performance into a linear programming problem and adopted the CPLEX optimization package to solve this problem and obtain the approximate optimal solution. Similarly, Samadian et al. [13]

developed a dynamic programming algorithm with a polynomial-time complexity to resolve the contingent conflicts. Although the above schemes can maximize the collective interests, they suffer from two problems: (1) A centralized platform is required to collect the privacy-sensitive policies of the involved domains. As a result, the privacy-sensitive domains are reluctant to provide their policies to the centralized platform. (2) The selfish domain often hopes to maximize not the collective interests but its own interests. Thus, the selfish domain is unwilling to resolve conflicts cooperatively.

Conflict Resolution with Optimizing Individual Interests. A series of mechanisms (e.g., negotiation [10,17] and game [4,7]) have been proposed to express the individual interests and maximize individual interests. For instance, considering privacy preferences and the sharing requirement, Mehregan *et al.* [10] designed a negotiation mechanism to interactively adjust and revise the access permission of the shared data of online social networks (OSNs). Yang *et al.* [17] designed a local supervisory controller to observe the states (i.e., activated or inactivated) of roles of each domain and maximize resolution efficiency through selecting the prevention scheme of conflicts caused by role change. Formulating the multiparty privacy conflict in OSNs into a multi-player noncooperative game, Ding *et al.* [4] established the multiple equilibria to achieve the trade-off between privacy preference and the social influence in OSNs. Using the multiparty control game, Hu *et al.* [7] proposed an optimal conflict resolution algorithm to adjust the privacy setting in OSNs and maximize the benefits of the user who shares data. As shown above, although a large amount of efforts have been spent on resolving policy conflicts of access control, these efforts ignore the factors that affect the cost and value of the domains involved. As a result, they cannot effectively incentivize selfish domains to resolve conflicts cooperatively.

3 Problem Statement and Basic Idea

3.1 Cross-Domain Policy Mapping in RBAC

In our work, cross-domain policy mapping in RBAC is formally specified by hierarchical role graphs, where nodes of a graph can be divided into three categories: user nodes, role nodes, and permission nodes. Figure 2(a) gives an example of hierarchical RBAC roles. In this example, there are two domains (i.e., domain A and domain B), where domain A has 3 roles (i.e., r_1, r_2 and r_3), 3 users (i.e., u_1, u_2, and u_3), and 3 permissions (i.e., p_1, p_2 and p_3); Domain B has 4 roles (i.e., r_4, r_5, r_6 and r_7), 2 users (i.e., u_4 and u_5), and 3 permissions (i.e., p_4, p_5 and p_6).

As shown in Fig. 2, edges between nodes can be divided into 7 categories: user-role assignment($\mapsto$), role-permission assignments($\twoheadrightarrow$), inheritance hierarchy($\rightarrow$, I-hierarchy), Activation hierarchy($\dashrightarrow$, A-hierarchy), role-specific Separation of duty(SOD) constraints ($\overleftrightarrow{RSOD}$), user-specific SoD constraints ($\overleftrightarrow{USOD}$), role mapping($\Longrightarrow$) and the induced role SOD constraint($\overleftrightarrow{InducedRSOD}$), where the edge $u \mapsto r$ represents that user u is assigned role r, edge $r \twoheadrightarrow p$ indicates that role r is assigned permission p, edges $r_1 \rightarrow r_2$ and $r_1 \dashrightarrow r_2$ represent that r_1 can inherit all permissions of r_2 without activation operation and with activation operation, respectively. Edge $r_1 \overleftrightarrow{RSOD} r_2$ (called role-specific SOD constraints, RSoD)denotes no user can be allowed to simultaneously access r_1 and r_2 in the same session, and edge $u_1 \xleftrightarrow[USOD]{r} u_2$ (called the user-specific SOD constraint) indicates that users u_1 and u_2 are not be allowed to access role r in the same session. Edge $r_1:A$ $\implies r_2:B$ indicates role r_1 in domain A is mapped to r_2 in domain B, that is, r_1 is assigned the permissions owned by role r_2. Edge $r_1 \overleftrightarrow{InducedRSOD} r_2$ indicates that the new RSOD constraint between roles role1 and role2 is induced.

3.2 Conflicts Induced by Policy Mapping

For two conflict-free policies, after executing policy mapping, three types of conflicts may be induced, defined as follows.`

Definition 1. (induced cyclic-inheritance conflict, iCIC.) An iCIC happens in one domain (says domain A) if after cross-domain policy mapping, at least one role (says r) in domain A inherits the permissions of roles that are senior to r.

Definition 2. (induced role-specific SoD conflict, iRSODC.) An iRSODC happens in one domain (says domain A) if after cross-domain policy mapping, a user in domain A can simultaneously can be assigned to two conflicting roles of domain A.

Definition 3. (induced user-specific SoD conflict, iUSODC). An iUSODC happens in one domain (says domain A) if after cross-domain policy mapping, two users in domain A can simultaneously access two conflicting roles.

Next, we give an example to illustrate the above conflicts induced by cross-domain mapping.

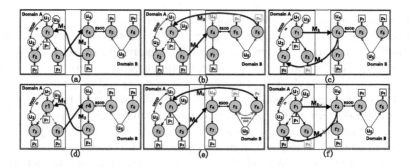

Fig. 2. (a) iCIC, (b) iRSODC, (c) iUSODC. (d) iCIC resolution, (e) iRSODC resolution, and (f) iUSODC resolution.

Example 1. As shown in Fig. 2, we assume that: (1) in domain A, r_1 inherits all permissions of r_3, role r_2 is junior to r_1, and a user-specific SoD constraint is specified between user u_1 assigned to r_2. (2) In domain B, r_4 inherits the permissions of r_7 and a role-specific SoD constraint is specified for r_4 and r_5. Next, we give three conflicts.

iCIC. In Fig. 2(a), if roles r_7 in domain B and r_3 in domain A are mapped to roles r_1 and r_4 respectively, then the junior role r_3 will inherit the permissions of the senior role r_1 through inheritance path $r_3 \rightarrow r_4 \rightarrow r_7 \rightarrow r_1$. Therefore, an iCIC will be induced, as shown in Fig. 2(d).

iRSODC. In Fig. 2(b), if roles r_3 in domain A and r_6 in domain B are mapped to roles r_4 and r_1 respectively, then role r_6 will inherit the permissions of role r_4 along inheritance path $r_6 \rightarrow r_1 \rightarrow r_3 \rightarrow r_4$. Before the two mappings are executed, u_5 can be assigned to r_5 and r_6, simultaneously. However, after these mappings are executed, u_5 will own the permissions of role r_4. Because iRSOD conflict exists between r_4 and r_5, a new RSoD between r_5 and r_6 will be induced, as shown in Fig. 2(e).

iUSODC. In Fig. 2(c), if roles r_1 in domain A and r_4 in domain B are mapped to roles r_4 and r_2 respectively, thens role r_1 will inherit the permissions of role r_2 through inheritance path $r_1 \rightarrow r_4 \rightarrow r_2$. Before the policy mappings are executed, u_1 and u_2 cannot be assigned to r_2 (because there is a user-specific SoD constraints between user u_1 assigned to r_2 and u_2 assigned to r_2), simultaneously. However, after these mappings are executed, user u_1 indirectly obtains the permissions of r_2 along inheritance path $r_1 \rightarrow r_4 \rightarrow r_2$. As a result, an iUSODC between u_1 and u_2 about r_2 will be induced, as shown in Fig. 2(f).

3.3 Basic Idea for Conflict Resolution

Undoubtedly, conflict resolution may cause autonomy loss. For example, to resolve the iCIC in Fig. 2(a), one approach is to revoke the permissions owned by r_3 from the permissions of role r_1, i.e., modify the inherance relationship between r_1 and r_3, as shown in Fig. 2(d) and the accessed domain has to revoke its internal permissions and loses its autonomy. As a result, if domain A is selfish, it doesn't cooperatively take participate in policy mapping.

To encourage domains to cooperate, we regard policy mapping as service and propose an auction-based incentive mechanism to improve the interoperation, where the accessed domain sells its service and wins virtual credits to make up for its autonomy loss or privacy leakage. The accessing domain acting as a buyer pays for the service to the seller. As shown in Fig. 3, our auction can be divided into four steps: Policy Mapping Request, Policy Mapping Response, Asking Price & Bidding Price, and Auction, discussed as follows.

Step 1 (Policy Mapping Request): The accessing domain (says domain A) requests the permissions reqPSET required by domain A to the accessed domain (says domain B).

Step 2 (Policy Mapping Response): Once the policy mapping request is received, domain B matches the requested permission reqPSET and searches the required roles in its own domain. If the match succeeds, domain B sends the matched roles to domain A; Otherwise, policy mapping fails.

Step 3 (Asking Price & Bidding Price): During asking price and bidding price, (1) Domain B evaluates the cost c of policy mapping and seals[1] the asking price p_s. (2) Domain A evaluates the value v of this policy mapping and seals the bidding price p_b. (3) Then, domain A and domain B exchange the sealed bidding price and sealed asking price, respectively.

Step 4 (Auction): Domain A and domain B involved unseals the bidding price p_s and the asking price p_b. If $p_s \leq p_b$ holds, the transaction is concluded and domain A pays $\frac{p_s+p_b}{2}$ to domain B to resolve the conflicts. Otherwise, the transaction fails.

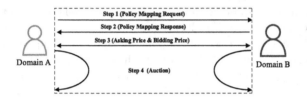

Fig. 3. Framework of conflict resolution.

4   Cost and Value in Conflict Resolution

In this section, we discuss the interoperation value of the accessing domain and the autonomy loss of the accessed domain. Next, we analyze the factors that affect value and cost.

4.1   Influencing Factors

Autonomy Loss and False Gain. For the accessed domain, policy mapping will cause not only autonomy losses but also false gains. Given domain A, we define the autonomy of managing its role at time t as the weighted sum of permissions assigned to the role, and autonomy of domain A at time t as the weighted sum of the autonomy of its all roles at the current time, formally described as follows.

$$Autonomy(A \mid t) = \sum_{r \in role(A|t)} \left(w_r \sum_{p \in perm(r|t)} w_p \right) \tag{1}$$

---

[1] Temporal attribute-based encryption (TABE) can be used to seal the price. Only when the pre-negotiated time is reached, the domain involved can decrypt the sealed price. TABE is out of the scope of our paper, please refer to [1] for more details.

Where functions $role(A \mid t)$ and $perm(r \mid t)$ returns all roles in domain A and all permissions (including the original permissions and the inherited permissions) of role r at time t, respectively. Parameters w_r and w_p represent autonomy weights of role r and permission p, respectively. Given an autonomy function of domain A, we use function $Autochange(A) = Autochange(A \mid before\ mapping) - Autochange(A \mid after\ mapping)$ to denote autonomy change. Autonomy losses $autoloss(A)$ and false gain $falsegain(A)$ caused by cross-domain mapping can be defined as follows, respectively.

$$autoloss(A) = \begin{cases} Autochange(A), if\ Autochange(A) > 0 \\ 0, otherwise \end{cases} \tag{2}$$

$$falsegain(A) = \begin{cases} \mid Autochange(A) \mid, if\ Autochange(A) < 0 \\ 0, otherwise \end{cases} \tag{3}$$

Interoperation. As shown in Sect. 3, if a role (say r_A) in domain A is mapped to a role (say r_B) in domain B, then interoperation of domain A will be increased. Generally, for the accessing domain, greater interoperation means a high value/benefit. In our work, the interoperation $Interoperation(A \rightarrow B)$ brought by mapping roles in domain A to roles in domain B is defined as the weighted sum of B's permissions assigned to roles of domain A, formally described as follows.

$$Interoperation(A \rightarrow B) = \sum_{r \in map-role(A)} \left(w_I(r) \sum_{p \in mapped-perm(r,B)} w_I(p) \right) \tag{4}$$

Where functions $map-role(A)$ and $mapped-perm(r, B)$ return the map roles in domain A and the obtained B's permissions of role r, respectively. Parameters $w_I(r)$ and $w_I(p)$ represent interoperation weights of role r and permission p, respectively.

Privacy Leakage. If a role (say r_A) in domain A is mapped to a role (say r_B) in domain B, then the privacy of domain B will be leaked. For the accessed domain, a greater privacy leakage means a high risk to suffer the malicious attack. In our work, we define the amount $PrivacyLoss(A \rightarrow B)$ of privacy leakage brought by mapping roles in domain A to roles in domain B as the weighted sum of the privacy of its all mapped permissions, formally described as follows.

$$PrivacyLoss(A \rightarrow B) = \sum_{r \in map-role(A)} \left(w_p(r) \sum_{p \in mapped-perm(r,B)} w_p(p) \right) \tag{5}$$

Parameters $w_p(r)$ and $w_p(p)$ represent privacy weights of role r and permission p, respectively.

4.2 Seller's Cost

Generally, more than one candidate schemes (say $S_1...S_n$) can be used to resolve conflicts[2] and cost rely on candidate schemes. As shown in Sect. 3, the accessed domain acts a seller. Given a candidate scheme, its cost depends on the loss caused by policy mapping, including three parts: autonomy loss, false gain, and privacy leakage.

Given the accessing domain A and the accessed domain B, we define B's service cost $Cost(A \rightarrow B)$ caused by a policy mapping from domain A to domain B as follows.

$$Cost(A \rightarrow B) = \min\{cost\,(A \rightarrow B \mid S_1), \ldots, cost\,(A \rightarrow B \mid S_n)\} \qquad (6)$$

where $cost\,(A \rightarrow B \mid S_i) = autoloss\,(A \rightarrow B \mid S_i) + PrivacyLoss\,(A \rightarrow B \mid S_i) + falsegain\,(A \rightarrow B \mid S_i)$ represents cost domain B's service cost if resolution scheme S_i is adopted. $autoloss\,(A \rightarrow B \mid S_i)$ denotes autonomy losses of domain B in candidate scheme S_i, $PrivacyLoss\,(A \rightarrow B \mid S_i)$ and $falsegain\,(A \rightarrow B \mid S_i)$ are similar.

4.3 Buyer's Value

The accessing domain acts a buyer and its value depends on benefits caused by policy mapping, including two parts: interoperation improvements and false gain.

Given the accessing domain A and the accessed domain B, value of A $Value(A \rightarrow B)$ induced by a policy mapping as follows.

$$Value(A \rightarrow B) = Interoperation(A \rightarrow B) - falsegain(A \rightarrow B) \qquad (7)$$

5 Double Auction Game and Its Analysis

In our work, we formulate conflict resolution in cross-domain collaboration as an auction game, where cooperation with resolving conflicts is regarded as service, the accessing domain plays as a buyer, and the accessed domain plays as a seller. In this game, the key aspect is to ask price p_s and the bid price p_b to maximize the interests of the involved domains. In our game, the accessing domain and the accessed domain bid at the same time. If the asking price p_s is less than or equals the bidding price p_b, then the auction concludes at the trading price $\frac{p_s+p_b}{2}$; otherwise, the auction fails, where (1) the asking price p_s depends on cost c of a seller and buyer's value v predicted by the seller, (2) the bidding price p_b relies on value v to the buyer and seller's cost c predicted by the buyer.

---

[2] For example, as shown in Fig. 2(d), there are two candidate schemes $S_1 = $ {remove the mapping $r_7 : B \implies r_1 : A$ or $r_3 : A \implies r_4 : B$} and $S_2 = $ {modify the inheritance relationship between r_1 and r_3} of resolving conflicts: if we delete mapping $r_7:B \implies r_1:A$ or $r_3:A \implies r_4:B$, then no conflict can be found.

5.1   Double Auction with Incomplete Information

In the above auction, if the seller knows the service value to the buyer and the buyer also knows the seller's service cost, then this auction is complete. However, in practice, service cost and service value are often private information of a seller and a buyer, respectively. Thus, the assumption of incomplete information is more reasonable. That, although the seller does not accurately know the value to the buyer and the buyer does not know the seller's cost, the probability distributions of cost and value are assumed to be their common knowledge. In our work, both cost and value are assumed to obey the two-parameters Burr XII distribution, which is widely adopted in the economic field [15].

The probability density function of two-parameter Burr XII function with shape parameter α and τ is as follows.

$$burr(c \mid \alpha, \tau) = \frac{\alpha \tau c^{\tau-1}}{(1 + c^{\tau})^{\alpha+1}}, c \geq 0, \alpha \geq 1, \tau \geq 1 \tag{8}$$

Now if $\alpha > 1$, then $burr(c \mid \alpha, \tau)$ is a unimodal function which peak at $c = \left(\frac{\alpha-1}{(\alpha\tau+1)^{\tau+1}}\right)^{\frac{1}{\alpha}}$. If $\alpha = 1$, $burr(c \mid \alpha, \tau)$ is an L-shape function. The culmulative distribution function of two-parameters Burr XII is as follows:

$$Burr(c \mid \alpha, \tau) = 1 - \frac{1}{(1 + c^{\tau})^{\alpha}} \tag{9}$$

In this paper, the cost c and the value v have the truncated Burr XII distribution over an interval $[0, U]$, that is, their probability distribution function is given by:

$$truncated - burr(c \mid \alpha, \tau) = \frac{burr(c|\alpha,\tau)}{Burr(U|\alpha,\tau)-Burr(0,|\alpha,\tau)}, 0 \leq c \leq U, \alpha \geq 1, \tau \geq 1 \tag{10}$$

We assume that the asking price p_s of a seller and the bidding price p_b of a buyer are proportional to cost c and value v, that $p_s(c) = \beta_s c$ and $p_b(v) = \beta_b v$ and where $\beta_b > 1$ and $0 < \beta_s < 1$ are linear parameters. Given the above assumption, we can calculate the expected benefits $E_{accessed}(p_s, p_b(v))$ of the accessed domain and the expected benefits $E_{access}(p_s(c), p_b)$ of the accessing domain, as follows.

$$E_{accessed}(p_s, p_b(v)) = E\left[\frac{p_s + p_b(v)}{2} - c \mid p_b(v) \geq p_s\right] \tag{11}$$

$$E_{access}(p_s(c), p_b) = E\left[v - \frac{p_s(c) + p_b}{2} \mid p_s(c) \leq p_b\right] \tag{12}$$

5.2   Bayesian Nash Equilibrium

Because the above game is incomplete, we should solve Bayesian Nash equilibrium, where Bayesian Nash equilibrium is defined as follows: the strategic combination $(p_s^*(c), p_b^*(v))$ is a Bayesian Nash Equilibrium if $p_s^*(c) = \max_{p_s} E_{accessed}(p_s, p_b(v))$ and $p_b^*(v) = \max_{p_b} E_{access}(p_s(c), p_b)$. Next, we solve it.

According to statistical probability, we have the following formula:

$$p_s^*(c) = \max_{p_s} E_{accessed}(p_s, p_b(v)) = \max_{p_s} \left[\frac{p_s + E[p_b(v)|p_b(v) \geq p_s]}{2} - c \right] Prob\{p_b(v) \geq p_s\}$$
$$= \max_{p_s} \left(\frac{p_s}{2} - c \right) \left(1 - \frac{1}{\left(1 + \frac{p_s}{\beta_b}^\tau\right)^\alpha} \right) + \frac{1}{2} \int_{\frac{p_s}{\beta_b}}^U \beta_b \frac{\alpha \tau x^\tau}{(1 + x^\tau)^{\alpha+1}} dx \tag{13}$$

$$p_b^*(v) = \max_{p_b} E_{access}(p_s(c), p_b) = \max_{p_b} \left[v - \frac{p_b + E[p_s(c)|p_s(c) \leq p_b]}{2} \right] Prob\{p_s(c) \leq p_b\}$$
$$= \max_{p_b} \left(v - \frac{p_b}{2} \right) \left(1 - \frac{1}{\left(1 + \frac{p_b}{\beta_s}^\tau\right)^\alpha} \right) + \frac{1}{2} \int_0^{\frac{p_b}{\beta_s}} \beta_s \frac{\alpha \tau x^\tau}{(1 + x^\tau)^{\alpha+1}} dx \tag{14}$$

Since the integral part of Formula (13) and Formula (14) cannot be calculated, we convert it power series (the center of the series is equal to zero) as the approximate solution, gives.

For the accessed domain's benefit function:

$$p_s^*(c) = \max_{p_s} \frac{p_s}{2} - c - \frac{p_s}{2\left(1 + \frac{p_s}{\beta_b}^\tau\right)^\alpha} + \frac{c}{\left(1 + \frac{p_s}{\beta_b}^\tau\right)^\alpha}$$
$$+ \frac{\tau \alpha \beta_b}{2} \left[\sum_{n=0}^\infty \frac{(-1)^n[(\alpha+1)(\alpha+2)\ldots\ldots(\alpha+n)]}{n!(\tau n + \tau + 1)} \left(U^{\tau n + \tau + 1} - \beta_b^{\tau n + \tau + 1} \right) \right] \tag{15}$$

For the accessing domain's benefit function:

$$p_b^*(v) = \max_{p_b} v - \frac{p_b}{2} + \frac{p_b}{2\left(1 + \frac{p_b}{\beta_s}^\tau\right)^\alpha} - \frac{v}{\left(1 + \frac{p_b}{\beta_s}^\tau\right)^\alpha}$$
$$+ \frac{\tau \alpha \beta_s}{2} \left[\sum_{n=0}^\infty \frac{(-1)^n[(\alpha+1)(\alpha+2)\ldots\ldots(\alpha+n)]}{n!(\tau n + \tau + 1)} \left(\frac{p_b^{\tau n + \tau + 1}}{\beta_s} - 0 \right) \right] \tag{16}$$

We can easily obtain its solutions of Formula (15) Formula (16) and use either bisection or Newton's method to get the equilibrium strategy of the accessed domain and the accessing domain.

6   Experiment Evaluation

In our experiment, three scales (i.e., large-scale, medium-scale and little-scale) of RBAC policies [21] are adopted to simulate the double auction in cross-domain collaboration. In the large-scale dataset, there are 1008 users, 314 roles and 34 role-specific SOD constraints; In the medium-scale dataset, there are 503 users, 137 roles and 17 role-specific SOD constraints; In the little-scale dataset, there are 10 users, 10 roles and 1 role-specific SOD constraint. To simulate our auction, we set the parameter of the Burr XII distribution to be $\alpha = 2$, $\tau = 5$ (the two parameters fit the actual value and cost distribution), $U = 0.8$ (that is upper bound of cost and value), $\beta_s = 0.95$, and $\beta_b = 1.05$ (i.e., about 5% gain).

Auction Benefits: For each double auction, the benefits of the accessing domain and the accessed domain are defined as $\frac{p_s+p_b}{2} - c$ and $v - \frac{p_s+p_b}{2}$, respectively. Fig. 4 shows the total benefits of the involved domains over the number of auctions. From this figure, we can see that the benefits of the domain increase with the number of auctions. Thus, a rational and selfish domain actively takes part in conflict resolution.

Autonomy Loss v.s. Interoperation: In this experiment, we compare autonomy loss and interoperation of our approach with Shafiq's approach [14]. In Shafiq's approach, to guarantee security, an autonomous domain has to set the upper bounds of the acceptable autonomy loss. In our experiment, the upper bound is set to 10% and 50%, respectively. For simplicity, we write the two schemes corresponding to the two parameters as 10%-autonomy-loss and 50%-autonomy-loss, respectively. Figure 5 shows the autonomy loss and interoperation with the number of role mapping. From Fig. 5, we can see that the interoperation of our approach approximates or exceeds the interoperation of the 50%-autonomy-loss scheme, while the autonomy loss of our approach approximates or is lower than the autonomy loss of the 10%-autonomy-loss scheme. In other words, our approach can approximately achieve trade-off between maximizing interoperation and minimizing autonomy loss.

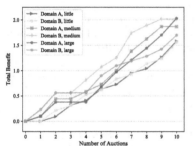

Fig. 4. Auction benefits

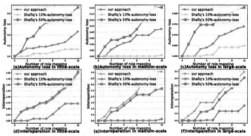

Fig. 5. Autonomy loss *v.s.* Interoperation

7 Conclusion

In this paper, we design an incentive mechanism to motivate domains to participate in conflict resolution. In detail, considering the selfishness and rationality of the involved domain, a game-theoretic approach is proposed to maximize their utility while achieving a tradeoff between security and interoperability. The simulation demonstrates the effectiveness of the conflict resolution approach.

References

1. Calistabebe, P., Akila, D.: Quantitative sørensen-dice indexed damgård-jurik cryptosystem for secured data access control in cloud. In: IOP Conference Series: Materials Science and Engineering. vol. 993, p. 012093. IOP Publishing (2020)

2. Chen, H.C.: Collaboration IoT-based RBAC with trust evaluation algorithm model for massive IoT integrated application. Mob. Netw. Appl. **24**(3), 839–852 (2019). https://doi.org/10.1007/s11036-018-1085-010.1007/s11036-018-1085-0

3. Cruz, J.P., Kaji, Y., Yanai, N.: Rbac-sc: role-based access control using smart contract. IEEE Access **6**, 12240–12251 (2018)

4. Ding, K., Zhang, J.: Multi-party privacy conflict management in online social networks: a network game perspective. IEEE/ACM Trans. Networking **28**(6), 2685–2698 (2020)

5. Du, J., Jiang, C., Chen, K., Ren, Y., Poor, H.V.: Community-structured evolutionary game for privacy protection in social networks. IEEE Trans. Inf. Forensics Secur. **13**(3), 574–589 (2018)

6. Fang, L., Yin, L., Guo, Y., Wang, Z., Li, F.: Resolving access conflicts: an auction-based incentive approach. In: MILCOM 2018–2018 IEEE Military Communications Conference (MILCOM), pp. 1–6. IEEE (2018)

7. Hu, H., Ahn, G.J., Zhao, Z., Yang, D.: Game theoretic analysis of multiparty access control in online social networks. In: Proceedings of the 19th ACM Symposium on Access Control Models and Technologies, pp. 93–102 (2014)

8. Huynh, N., Frappier, M., Pooda, H., Mammar, A., Laleau, R.: Sgac: a multi-layered access control model with conflict resolution strategy. Comput. J. **62**(12), 1707–1733 (2019)

9. Ma, M., Stankovic, J.A., Feng, L.: Cityresolver: a decision support system for conflict resolution in smart cities. In: 2018 ACM/IEEE 9th International Conference on Cyber-Physical Systems (ICCPS), pp. 55–64. IEEE (2018)

10. Mehregan, P., Fong, P.W.: Policy negotiation for co-owned resources in relationship-based access control. In: Proceedings of the 21st ACM on Symposium on Access Control Models and Technologies, pp. 125–136 (2016)

11. Omar, I.Y., Laborde, R., Wazan, A.S., Barrère, F., Benzekri, A.: Egovernment service security policy: obligation conflict resolution in xacmlv3. In: Proceedings of the International Conference on Security and Management (SAM), p. 89. The Steering Committee of The World Congress in Computer Science (2016)

12. Salehi, A., Rudolph, C., Grobler, M.: A dynamic cross-domain access control model for collaborative healthcare application. In: 2019 IFIP/IEEE Symposium on Integrated Network and Service Management (IM), pp. 643–648. IEEE (2019)

13. Samadian, H., Tuiyot, D., Valera, J.: Dynamic programming approach in conflict resolution algorithm of access control module in medical information systems. In: Arai, K., Kapoor, S., Bhatia, R. (eds.) FICC 2020. AISC, vol. 1129, pp. 672–681. Springer, Cham (2020). https://doi.org/10.1007/978-3-030-39445-5_49

14. Shafiq, B., Joshi, J.B., Bertino, E., Ghafoor, A.: Secure interoperation in a multidomain environment employing RBAC policies. IEEE Trans. Knowl. Data Eng. **17**(11), 1557–1577 (2005)

15. Tadikamalla, P.R.: A look at the burr and related distributions. International Statistical Review/Revue Internationale de Statistique, pp. 337–344 (1980)

16. Yahiaoui, M., Zinedine, A., Harti, M.: Deconflicting policies in attribute-based access control systems. In: 2018 IEEE 5th International Congress on Information Science and Technology (CiSt), pp. 130–136. IEEE (2018)

17. Yang, B., Hu, H.: Secure conflicts avoidance in multidomain environments: a distributed approach. IEEE Trans. Syst. Man Cybern. Syst. (99), 1–12 (2019)

18. Zeng, Q., Liu, C., Duan, H., Zhou, M.: Resource conflict checking and resolution controller design for cross-organization emergency response processes. IEEE Trans. Syst. Man Cybern. Syst. **50**(10), 3685–3700 (2019)

19. Zhang, Y., Kasahara, S., Shen, Y., Jiang, X., Wan, J.: Smart contract-based access control for the internet of things. IEEE Internet Things J. **6**(2), 1594–1605 (2018)
20. Zhu, H., Sheng, Y., Zhou, X., Zhu, Y.: Group role assignment with cooperation and conflict factors. IEEE Trans. Syst. Man Cybern. Syst. **48**(6), 851–863 (2016)
21. Zhu, T., Li, F., Jin, W., Guo, Y., Fang, L., Cheng, L.: Cross-domain access control policy mapping mechanism for balancing interoperability and autonomy. J. Commun. **41**(9), 29–48 (2020)

An Adaptive Network Model for Procrastination Behaviour Including Self-regulation and Emotion Regulation

Hildebert Moulie[1], Robin van den Berg[1], and Jan Treur[2(✉)]

[1] Computational Science, University of Amsterdam, Amsterdam, Netherlands
{hildebert.moulie,robin.vanden.berg}@student.uva.nl
[2] Social AI Group, Vrije Universiteit Amsterdam, Amsterdam, Netherlands
j.treur@vu.nl

Abstract. In this paper, the goal is to model both the self-control and the emotion regulation dynamics involved in the process of procrastination. This is done by means of a temporal-causal network, incorporating learning and control of the learning. Additionally, the effect of stress regulation-therapy on the process of procrastination was investigated. The model's base level implementation was verified by making sure the aggregated impact matches the node values for certain stationary points and the model's Hebbian learning behaviour was also mathematically shown to be correctly implemented. The results proved this model's ability to model different types of individuals, all with different stress sensitivities. Therapy was also shown to be greatly beneficial.

Keywords: Procrastination · Adaptive · Emotion regulation · Self-regulation

1 Introduction

Procrastination is defined as the act of delaying or postponing something. The problem has been increasing in size over the years [19]. It was estimated by Steel that approximately 80–95% of college students procrastinate [19]. Furthermore, Harriott and Ferrari [7] found that an estimated 20% of adults are self-proclaimed chronic procrastinators. Apart from the self-destructive consequences of procrastination, [17] also has shown that persistent procrastination can lead to mental and physical health problems such as depression, anxiety and even cardiovascular diseases.

For a long time, procrastination was regarded as a problem of self-control and time management. However, in current academia, there has been a growing amount of research that has focused on the emotional backdrop of procrastination [5]. It is often found that emotional thresholds, such as stress or fear of the result of an action, are what stimulate procrastination.

In this paper, an attempt was made at modelling both the self-control and the emotion-regulation dynamics involved in the process of procrastination. This is done by means of a temporal-causal network, incorporating first- and second-order adaptation for controlled learning. We set out to unveil the dynamics of the system. Additionally, the effect of

M. Paszynski et al. (Eds.): ICCS 2021, LNCS 12742, pp. 540–554, 2021.
https://doi.org/10.1007/978-3-030-77961-0_44

stress regulation-therapy on the process of procrastination was investigated. As a starting point, research articles within psychology were analysed and used as a basis for the implementation. Subsequently, in the Methodology, we elaborate on the implementation of the model and translate the psychological connections to a computational model. Here, the experiments carried out are also presented. Thereafter, the results from the example scenarios are discussed and the report is finalised with a conclusion and recommendations for further research.

2 Background Knowledge

In order to create a model representing the process of procrastination, it is important to first look at existing literature in order to create a model compliant with past studies. Our findings include the following statements based on published papers and articles.

According to an article by Onwuegbuzie [12], using a regression method, it was found that 25% of academic procrastination was a direct result of self-regulation while 14% were linked to anxiety, depression and self-esteem. Additionally, the importance of self-regulation amongst other self-variables was found to be the highest for predicting procrastination tendencies in another paper [8]. Correlation results indicated that students with intrinsic reasons for pursuing academic tasks procrastinated less than those with less autonomous reasons, this once more confirms the importance of self-regulation [14]. Failure of self-control is often the result of conflicting goals. In this instance, the conflicting actions of instant gratification and pursuit of long-term goals [16, 20]. Additionally, procrastination was found to stem from the anxiety linked to possible failure as was reported by students [13]. Moreover, it was found by Steel that there exist significant relations between procrastination and task aversiveness, task delay, self-efficacy and impulsiveness [19].

However, it is important to note that there is a difference between active and passive procrastination [2, 3]. The former is where people postpone doing a task but are able to meet a deadline and are satisfied with the outcome in the end while the latter is where people are unable to perform the task on time. The passive procrastinators are often troubled by their ability to achieve, subsequently provoking feelings of guilt and depression leading to more procrastination and thus to failure of the task [3]. Therefore, passive procrastination can be linked primarily to the emotional regulation.

Procrastination can lead to small boosts in enjoyment, this is why students often check social media when procrastinating [11]. Furthermore, Tice et al. [20] describe that the desire for evasion of emotional distress increases the inclination towards choices that render immediate pleasure. In this paper, we assume that the activities with which one procrastinates induce direct enjoyment. Therefore, we state a two-way relation between procrastination and anxiety/stress [9, 20]. Next to anxiety and stress, guilt and shame can also result from procrastination, both contributing to adverse mental health issues [4].

Furthermore, another method that was found to be very efficient against procrastination is therapy [22]. Therapy is not only great to deal with procrastination, it is also a great method used to fight against stress. Indeed, in this study by Gammon and Morgan-Samuel, it was shown that helping students with a tutorial support made these students end up with significantly less stress than students from a control group that did not benefit from therapy [6, 18]. Next to stress-control therapy, therapy can also be focused on self-compassion since low self-compassion has been found to be one of the linking factors between procrastination and stress [15].

3 The Modeling Approach Used

In this section, the network-oriented modeling approach to causal modeling adopted from [21] is briefly introduced. Following these, a *temporal-causal network model* is characterised by; here X and Y denote nodes (also called states) of the network with network connections for how they causally affect each other:

- *Connectivity characteristics*
  Connections from a state X to a state Y and their weights $\omega_{X,Y}$
- *Aggregation characteristics*
  For any state Y, some combination function $\mathbf{c}_Y(..)$ defines the aggregation that is applied to the causal impacts $\omega_{X,Y}X(t)$ on Y from its incoming connections from states X
- *Timing characteristics*
  Each state Y has a speed factor η_Y defining how fast it changes for given impact.

The following difference (or differential) equations that are used for simulation purposes and also for analysis of temporal-causal networks incorporate these network characteristics $\omega_{X,Y}$, $\mathbf{c}_Y(..)$, η_Y in a standard numerical format:

$$Y(t + \Delta t) = Y(t) + \eta_Y\big[\mathbf{c}_Y(\omega_{X_1,Y}X_1(t), \ldots, \omega_{X_k,Y}X_k(t)\big) - Y(t)]\Delta t \qquad (1)$$

for any state Y and where X_1 to X_k are the states from which Y gets its incoming connections. The generic Eq. (1) is hidden in the dedicated software environment; see [21], Ch 9. Within the software environment described there, around 40 useful basic combination functions are included in a combination function library; see Table 1 for the ones used in this paper. The selected ones for a model are assigned to states Y by specifying combination function weights $\gamma_{i,Y}$ and their parameters used by $\pi_{i,j,Y}$.

The above concepts allow for the design of network models and their dynamics in a declarative manner, based on mathematically defined functions and relations. The idea is that the network characteristics that define the design of the network model, are used as input for the dedicated software environment. Within this environment the generic difference Eq. (1) is executed for all states, thus generating simulation graphs as output. Note that 'network characteristics' and 'network states' are two distinct concepts for a network. Self-modeling is a way to relate these distinct concepts to each other in an interesting and useful way:

- A *self-model* is making the implicit network characteristics (such as connection weights ω or excitability thresholds τ) explicit by adding states for these characteristics; thus the network gets a self-model of part of the network structure; as self-models can change over time, this can easily be used to obtain an *adaptive network*.

Table 1. Basic combination functions from the library used in the presented model

	Notation	Formula	Parameters
Identity	$\mathbf{id}(V)$	V	–
Advanced logistic sum	$\mathbf{alogistic}_{\sigma,\tau}(V_1, ...,V_k)$	$\left[\dfrac{1}{1+e^{-\sigma(V_1+...+V_k-\tau)}} - \dfrac{1}{1+e^{\sigma\tau}}\right](1 + e^{-\sigma\tau})$	Steepness $\sigma > 0$ Threshold τ
Hebbian learning	$\mathbf{hebb}_\mu(V_1, V_2, W)$	$V_1 V_2(1 - W) + \mu W$	Persistence factor $\mu > 0$
Scaled sum	$\mathbf{ssum}_\lambda(V_1, ...,V_k)$	$\dfrac{V_1+...+V_k}{\lambda}$	Scaling factor λ

- In this way, multiple self-modeling levels can be created where network characteristics from one level relate to states at a next level. This can cover *second-order* or *higher-order adaptive networks*; see [21], Ch 4.

Adding a self-model for a temporal-causal network is done in the way that additional network states $\mathbf{W}_{X,Y}$, $\mathbf{C}_{i,Y}$, $\mathbf{P}_{i,j,Y}$, $\mathbf{H}_Y$ (*self-model states*) are added as nodes to the network for some of the states Y of the base network and some of their related network structure characteristics for connectivity, aggregation and timing (in particular, some from $\omega_{X,Y}$, $\gamma_{i,Y}$, $\pi_{i,j,Y}$, η_Y):

(a) **Connectivity self-model**

- Self-model states $\mathbf{W}_{X_i,Y}$ are added to the network representing connectivity characteristics, in particular connection weights $\omega_{X_i,Y}$

(b) **Aggregation self-model**

- Self-model states $\mathbf{C}_{j,Y}$ are added to the network representing aggregation characteristics, in particular combination function weights $\gamma_{i,Y}$
- Self-model states $\mathbf{P}_{i,j,Y}$ are added representing aggregation characteristics, in particular combination function parameters $\pi_{i,j,Y}$

(c) **Timing self-model**

- Self-model states $\mathbf{H}_Y$ are added to the network representing timing characteristics, in particular speed factors η_Y

The notations $\mathbf{W}_{X,Y}$, $\mathbf{C}_{i,Y}$, $\mathbf{P}_{i,j,Y}$, $\mathbf{H}_Y$ for the self-model states indicate the referencing relation with respect to the characteristics $\omega_{X,Y}$, $\gamma_{i,Y}$, $\pi_{i,j,Y}$, η_Y: here $\mathbf{W}$ refers to ω, $\mathbf{C}$ refers to γ, $\mathbf{P}$ refers to π, and $\mathbf{H}$ refers to η, respectively. For the processing, these self-model states define the dynamics of state Y in a canonical manner according to Eqs. (1) whereby $\omega_{X,Y}$, $\gamma_{i,Y}$, $\pi_{i,j,Y}$, η_Y are replaced by the state values of $\mathbf{W}_{X,Y}$, $\mathbf{C}_{i,Y}$, $\mathbf{P}_{i,j,Y}$, $\mathbf{H}_Y$ at time t, respectively. The dynamics of the self-model states themselves are defined in the standard manner based on the generic difference Eq. (1) by their incoming connections and other network characteristics (such as combination functions and speed factors) used to fully embed them in the created *self-modeling network*. As the self-modeling network that is the outcome of the addition of a self-model is also a temporal-causal network model itself, as has been shown in detail in [21], Ch 10, this construction can easily be applied iteratively to obtain multiple levels of self-models.

4   The Designed Adaptive Network Model

In order to describe the behaviour and emotional dynamics involved in procrastination, a computational model was developed in the form of a temporal-causal network. To this extent, we utilised a dedicated modeling environment implemented in MATLAB [21], Ch. 9. A graphical representation of the model can be found in Fig. 1 and an overview of the states in Table 2. The model is a multilevel self-modeling network model consisting of three levels. The base level addresses the interactions between the different emotional and behavioural states. Level 1 addresses the first-order adaptivity by a first-order self-model of base level connections, which allows for evolving connection weights within the base level. In addition, level 2 influences the speed by which the states on the first level change (adaptive learning rate). A more detailed description of the different levels can be found in Table 2. The role matrices for the network characteristics defining the model can be found in the Appendix at https://www.researchgate.net/publication/350 108642.

The base level was designed using the psychological research described in Sect. 2. State X_1 represents the importance of the task at hand and X_2 is the stimulus to make progress on that same task. This pressure is associated with the importance of the task and portion of the task that remains. State X_3 is the central node of this network and constitutes procrastination. The activation value of the state denotes the amount of procrastination. Connected to this node, three main feedback loops can be distinguished, i.e., $\{X_2; X_3; X_6\}$, $\{X_3; X_5; X_6; X_7; X_8\}$ and $\{X_3; X_4; X_{10}\}$, henceforth named L_1, L_2 and L_3 respectively. Loop L_1 embodies the effects of the stimulus X_2, this stimulus increasing the amount of experienced stress X_6, while also decreasing the amount of procrastination X_3. On its turn, procrastination increases stress and vice versa. Loop L_2 delineates a part of the balance between instant gratification and long term satisfaction. Here, we see a mutual exclusion of anxiety/stress X_6 with general happiness X_8. Furthermore, procrastination X_3 induces shame X_5, which subsequently reduces the procrastination-induced joy X_7. Lastly, loop L_3 contains part of the behavioural system involved in limiting procrastination. It features self-control X_4 limiting the amount of procrastination X_3 as well as the total work done X_{10}, which positively influences the amount of self-control X_4. The strength of the connection between the latter is determined by past experience X_{14}. Furthermore, we see that procrastination X_3 logically decreases the amount of work done X_{10}.

Outside of these three loops, two individual states influence the overall dynamics. Firstly, completed work state X_{10} increases the amount of self-control X_4 based on the perspective that one is often more inclined to continue working on a task after starting

Table 2. States in the model

Number	State name	Description	Level
X_1	Task importance	The importance of the task at hand	Base level
X_2	Stimulus	The stimulus to do work coming from a certain task	
X_3	Procrastination	The act of procrastinating a task	
X_4	Self-control	The ability to force oneself to tackle the task at hand	
X_5	Shame	A task that was supposed to be done	
X_6	Anxiety/stress	Emotion induced by fear of the result of one's actions	
X_7	Joy	Procrastination-induced relief	
X_8	General happiness	The happiness about life in general	
X_9	Doing work	The rate of progress on work	
X_{10}	Work done	The amount of work done	
X_{11}	Stress control state	Control state for the stress/anxiety	
X_{12}	Therapy	Therapy to increase control over the stress/anxiety	
X_{13}	$\mathbf{W}_{X_6,X_{11}}$	Self-model state for stress-induced learning representing connection weight $\omega_{X_6,X_{11}}$	First-order self-model
X_{14}	$\mathbf{W}_{X_4,X_9}$	Self-model state for learning based on past experiences representing connection weight ω_{X_4,X_9}	
X_{15}	$\mathbf{H}_{\mathbf{W}_{X_6 X_{11}}}$	Self-model state for speed factor (adaptive learning rate) $\eta_{\mathbf{W}_{X_6 X_{11}}}$ of self-model state $\mathbf{W}_{X_6,X_{11}}$	Second-order self-model
X_{16}	$\mathbf{H}_{\mathbf{W}_{X_4 X_9}}$	Self-model state for speed factor (adaptive learning rate) $\eta_{\mathbf{W}_{X_4 X_9}}$ of self-model state $\mathbf{W}_{X_4,X_9}$	

it. It also positively feeds back into general happiness X_8 through pride or an obtained reward. Furthermore, a higher completed work X_{10} means that less of the task remains, therefore lowering the stimulus X_2. Secondly, we have the anxiety/stress control state X_{11} and the therapy state X_{12}. The therapy state X_{12} increases the amount of control one has over anxiety/stress X_6, while the anxiety/stress control state X_{11} itself allows for lowering the anxiety/stress X_6 [22].

On top of the base level, there is the first-order self-model level for learning. This learning is materialised by therapy-induced learning through **W**-state $\mathbf{W}_{X_6,X_{11}}$, also named X_{13}, representing an adaptive connection weight from anxiety/stress X_6 to the anxiety/stress control state X_{11}. By therapy stimulating the stress-control, a quickened negative feedback loop is expected in the case of heightened stress. Moreover, **W**-state $\mathbf{W}_{X_4,X_9}$, also called X_{14}, models the learning from past experiences and represents the adaptive connection weight of the connection from work rate X_4 to self-control X_9. Here, it is assumed that doing work increases self-control over time. Both connections are learned by the Hebbian learning adaptation principle, which describes an often used form of plasticity [21]. The combination function for this can be found in Table 1.

The second-order self-model level controls the learning by influencing the speed factors (learning rate) of the **W**-states in the first-order self-model level. To this end, **H**-state $\mathbf{H}_{\mathbf{W}_{X_6 X_{11}}}$, also called X_{15}, represents the speed factor of therapy-induced learning and, therefore, changes how fast the modelled individual can learn from therapy. Similarly, **H**-state $\mathbf{H}_{\mathbf{W}_{X_4 X_9}}$, also called X_{16}, represents the speed factor of X_{14} and thus changes how fast we learn from past experiences. Interesting dynamics are found when looking at X_{16}'s incoming connections. Since emotions change our perception of the work we are doing, connections were added from the negative emotions towards X_{16}. These negative emotions stimulate the rate at which one learns from past work, resulting from the underlying idea that one would want to avoid such feeling in subsequent work.

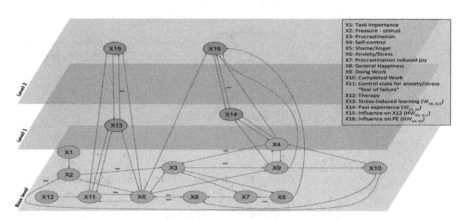

Fig. 1. The connectivity of the adaptive network model.

5   Experimental Setup

The goal of this research is to computationally explore the dynamics of the procrastination by taking into account both the behavioural and the emotional aspects. To this extent, we modeled a variety of situations, three of which will be discussed here (for another one, Experiment 3, see the Appendix at https://www.researchgate.net/publication/350 108642).

Experiment 1 Stress Sensitivity. In the primary experiment, three individuals are modeled exhibiting different susceptibilities to stress. Where, for instance, the case of high susceptibility could represent an individual close to a burn-out, while the individual with low susceptibility could model a person with differently placed priorities, therefore not so much influenced by the task at hand. To model this effect, we change the speed with which the stress/anxiety X_6 changes. The used speed factors for η_{X_6} are values from $\{0.05, 0.15, 0.25\}$.

Experiment 2 Stress-Control Therapy. In a second experiment, we examine the effect of stress-control therapy. Here, it was chosen to keep the standard value for the stress speed factor $\eta_{X_6} = 0.15$. Since the value of the therapy state X_{12} remains constant throughout the simulation, the initial value was altered. The initial values for the different simulations of this experiment were chosen from $\{0, 0.1, 0.2\}$.

Experiment 4 The Effect of a Stress Control Therapy. In Experiment 4, the individuals modelled in the previous experiment were taken as a basis. Using these same setups, therapy was added to varying levels in an attempt to combat stress. For therapy to be added, the initial value needs to be increased from its value of 0 of the baseline values as shown in Table 10 in the Appendix. Here the goal was to obtain a peak stress level as close as possible between the three individuals to examine the additional behaviour. To do so, the values 0.15 (anxious), 0.1 (average), and 0.02 (confident) were used.

6 Results of the Simulation Experiments

In this section the results of the main simulation experiments are discussed.

Base Scenario. To look into the results of the experiments, a baseline simulation first was established. To do so, the values specified in Tables 5, 6, 7, 8, 9 and 10 of the Appendix at Linked Data at https://www.researchgate.net/publication/350108642 were used. The results of this simulation can be observed in Fig. 2 top row. Inspecting this, it is clear that there are positive correlations between the procrastination and the stress, a relation that was previously demonstrated by empirical research [17, 18]. Furthermore, the model shows positive correlations between shame and procrastination as well as shame and stress, which reflects the results found in [15]. Moreover, these graph show a negative correlation between procrastination and general happiness as well as between stress and general happiness, which is also supported by literature [18]. Following the base situation, the effect of stress sensitivity was evaluated as well as the results from a stress control therapy. Subsequently, three different types of individuals were modeled on which the effect of therapy was tested.

Stress Sensitivity Experiment. In order to test the stress sensitivity, the speed factor of the stress/anxiety state X_6 was altered. It was first lowered to 0.05 from 0.15 used in the base scenario, thus yielding the simulation shown in the middle row of Fig. 2. Here, lowering the sensitivity to stress results in a much slower stress increase than in the base scenario. For a stress speed factor of 0.15, the stress peak is reached at $t = 40.3$ with a stress value of 0.7944 while with a stress speed factor of 0.05, the stress peak is reached at $t = 61.61$ with a much lower value of 0.6701. This change is to be expected and it also affects other nodes as a result. Procrastination, and in turn also shame, since they are very closely related, sees its evolution being much slower. Indeed, the peak is reached at $t = 37.86$ with a value of 0.2826 when a stress speed factor of 0.15 is used while it is reached at $t = 54.27$ with a value of 0.2112 with a stress speed factor of 0.05. This more stable procrastination over time could be the result of the stress being less intense and therefore causing less abrupt psychological changes. Finally, the speed factor adaptation of therapy state X_{15} is strongly affected by stress which explains why in the middle row of Fig. 2 a much more sudden original increase can be observed in comparison to the base scenario. The peak is also higher and reached earlier with a higher stress speed factor. In the scenario shown in the bottom row of Fig. 2, the stress speed factor was increased to a value of 0.25.

While the difference between this simulation and the baseline one is not as significant as the one between the simulations shown in the middle row in Fig. 2 and the baseline, the impact of a higher stress remains very clear. Indeed, this simulation shows a faster rate of increase for stress at first with a peak at 0.8174 reached at $t = 31.27$ but also a higher procrastination, and shame, with a peak of 0.2972 reached at $t = 30.3$. The speed factor adaptation of therapy state X_{15} is also impacted in the same way but to a lesser extent with a peak of 0.88 at $t = 35$. Overall, these impacts make a lot of sense as a person more sensitive to stress is expected to have their stress peak faster and higher when given an important task. The impacts on procrastination, while not as large, remain present. These can be asserted to what was described in Tice et al. [20]'s paper, which is that, as explained in Sect. 2, procrastination is often used to combat emotional distress. Furthermore, it can be concluded that the correlations noted in the base simulation also apply across variations in different individuals, therefore confirming the agreement between the model results and empirical psychological research.

Stress-Control Therapy Experiment. Now that different values for stress speed factors were analysed, the effects of therapy regarding stress are taken into consideration, the level of procrastination and the time to fully finish the task at hand. In order to analyse those results, the same baseline as for the previous analysis was used. The top row in Fig. 2 shows the results using an initial value for therapy of 0, this means that therapy was absent from the model. For the current experiment, first the therapy initial value was set to 0.1, the results of which can be observed in the top row of Fig. 3. Note that a speed factor of 0.15 was used for the stress state (X_6) as this was the baseline value.

Here, a few things can be observed. First, the level of stress was decreased a lot from a peak of 0.7944 in the base scenario simulation compared to a peak of 0.6969 here. Secondly, the procrastination level and therefore also the shame/anger are much lower throughout with a peak of around 0.2329 compared to the 0.2826 of the baseline simulation. This makes sense, given that the subject is helped to deal with his procrastination through therapy. Overall and most importantly, this simulation makes the individual complete his task much quicker than he did in the previous one finishing it at $t = 83.22$ compared to the original $t = 97.95$. Furthermore, the general happiness at the end of the

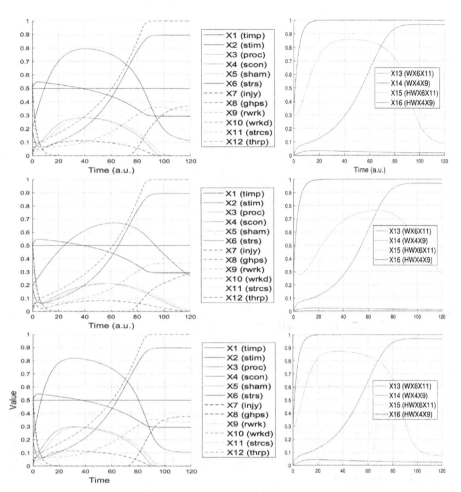

Fig. 2. The state values of the base level (left) and adaptation levels (right) with varying speed factors of the stress/anxiety state: 0.15 (top row) 0.05 (middle row), and 0.25 (bottom row)

simulation is increased in comparison to the simulation without the influence of therapy. In order to further test the effects of therapy, the initial value of therapy was changed to 0.2. This change yielded the results shown in the middle row of Fig. 3. In this second variation of the initial value for therapy, the same positive impacts can be observed but to a greater extent when compared to the baseline simulation. While therapy helps even more than it did, doubling the therapy's initial value isn't causing as great of an impact as introducing therapy into the model. Here the stress peaks slightly earlier at $t = 28.78$ with a value of 0.55 while procrastination, closely followed by shame, peaking at $t = 26.54$ with a value of 0.1685. The speed at which the task is completed is also further improved with a completion time of $t = 80.67$. Overall, this shows that while therapy helps get work done quicker while also lowering stress and procrastination, it does not scale linearly. Therapy being an efficient method to circumvent procrastination was also

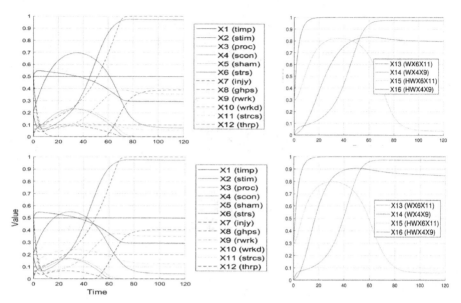

Fig. 3. The state values of the base level (left) and adaptation levels (right) with varying initial values for therapy: 0.1 (top row) and 0.2 (bottom row)

shown in Sect. 2 [22]. It was also shown in Sect. 2 that therapy greatly helps in dealing with stress [6], thus also matching our simulations of this experiment.

The Effect of a Stress-Control Therapy. Continuing from the previous experiment, the three modeled persons were used to test the effect of stress-control therapy. First, we modeled P_1 with an initial value for therapy changed to 0.15. The results can be see in the graph of the upper row in Fig. 4. Then, we look at the averagely-stressed individual P_2; see Fig. 4 bottom row left.

Here the initial value for therapy (X_{12}) was changed to 0.1. Finally, we simulate the confident person P_3 using an initial value for X_{12} of 0.02; see Fig. 4 bottom row right. Here, the inclusion of therapy changed the results in several ways. Since the level of therapy was adjusted to obtain a similar peak value for stress across the three individuals, analysing the behaviour of X_6 is a good place to start. P_1's stress peaks at $t = 17.39$ with a value of 0.7027 while P_2 peaks with a value of 0.6969 at $t = 34.40$ and P_3 with 0.6979 at $t = 54.66$. While these peak times are very different in the same way they were in the previous experiment, we can say that the values are very close with a maximum deviation across them which can be rounded to 0.05%. The experimental setup here had as a first goal to get the modeled individuals with a very close stress peak value to compare what the other dynamics would show. This experimental setup attempt can therefore be considered successful.

Stress also keeps the same trend throughout the simulation as in the previous experiment, all, however, with lower amplitudes. The peaks are also reached quicker than they were. This can be attributed to the therapy helping to deal with the stress a lot quicker than would have otherwise been possible. Secondly, we look into procrastination. Here,

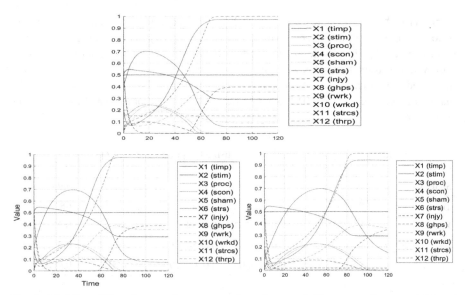

Fig. 4. Modeling different types of persons: anxious (top row left), average (top row right) and confident (middle row left). Modeling the effect of therapy: (middle row right) (bottom row left) (bottom row right)

X_3 peaks with 0.2458 at $t = 17.99$, 0.2329 at $t = 31.77$ and 0.2259 at $t = 48.94$ for P_1, P_2 and P_3, respectively. Just like for stress, therapy has a very positive effect on procrastination, not only are the peaks lower than the ones observed in the previous experiment, they also come significantly earlier. Finally, the times at which the work is completed are $t = 79.81$, $t = 83.22$ and $t = 94.58$ for P_1, P_2 and P_3, respectively, while they were all very close in the previous experiment. In comparison to the previous experiment, the work was completed 17.12%, 15.04% and 4.13% faster for P_1, P_2 and P_3, respectively. P_1 received more therapy than P_2 and P_2 more than P_3. Overall, these results demonstrate a very positive impact for therapy and while a more anxious individual could potentially benefit more from it, any individual subject to therapy seems to see significant improvement in their emotional states and task efficiency. Moreover, an increase in general happiness is observed for all individuals.

7   Verification by Analysis of Stationary Points

To verify the behaviour of the implemented network model against the conceptual specification, analysis of stationary points was performed. As a stationary point for a state Y is a point where $dY(t)/dt = 0$, from (1) the following general criterion for it can be derived:

$$\eta_Y = 0 \text{ or}$$
$$c_Y(\omega_{X_1,Y}X_1(t), \ldots, \omega_{X_k,Y}X_k(t)) = Y(t) \tag{2}$$

where X_1 to X_k are the states from which Y gets its incoming connections and the left hand side of the last line is denoted by aggimpact. We verify that the aggregated impact defined by the left hand side of (2) matches the state value for some stationary points observed in a simulation. This was done for base states X_3, X_5, X_7, X_8 and X_9, all using the scaled sum function as a combination function (see Table 3). As seen in that table, the maximum deviation is 0.000013, which provides evidence that the base level functions as intended.

Table 3. Verification of the model using temporary stationary points.

State X_i	X3	X5	X7	$X9_1$	$X9_2$
Time point t	27.26	30.70	25.44	10.09	65.37
$X_i(t)$	0.1657438	0.163772	0.0654626	0.061891	0.3715222
Aggimpact$_{X_i}(t)$	0.1657392	0.163767	0.0654619	0.061904	0.3715170
Deviation	– 0.0000046	0.000005	0.0000007	– 0.000013	0.0000052

To verify the Hebbian learning behaviour of the model, its behaviour was also analysed by checking X_{13} and X_{14}. For this analysis, we used the simulation shown in Fig. 3 bottom row right, as the learning is most pronounced there. Locating the stationary point in the graph, we find that this occurs for X_{13} at $t = 4.961$ with a value of $W = 0.90435484$. The incoming connections are from X_6 and X_{11}, which at the time of the stationary point have values of 0.43416206 and 0.21773694, respectively. Based on (2), the relation derived in [21], Ch 3, Sect. 3.6.1, is as follows:

$$W = \frac{V_1 V_2}{1 - \mu + V_1 V_2} \tag{3}$$

Filling in the above values in (3), right hand side, yields $W = 0.904336$ which is a deviation of 0.000018 from the observed value 0.9043548 for W. Similarly, for X_{14}, a stationary point at $t = 7.349$ was found with a value of $W = 0.972595236$. Based on the incoming states X_4 and X_9 for X_{14}, which have values of 0.974656883 and 0.364125516 at that time point, it was found $W = 0.972595039$. Again, the analysis result matches well with a deviation of 0.000000197. This provides evidence that also the learning behaves as expected.

8   Conclusion and Future Research

In this paper, it was endeavoured to create a model describing procrastination including both the behavioural and emotional components. To this extent, an adaptive network model was created featuring both first- and second-order adaptation by using self-models. The simulations created with the model show the dynamics and correlations found in psychology research. This leads us to believe the main dynamics of the model are valid. To test the model more extensively, it will be required to obtain empirical data that demonstrates the evolution over time; unfortunately, this is currently not available.

In the current state of the model, therapy is included as a constant level starting at the beginning of the simulation. For future research, this can be modeled in a more detailed manner. Furthermore, one may address adaptive variation of the threshold of the stress node, through which one could also regulate an individual's sensitivity to stress. Lastly, it is found in the literature that self-compassion has been shown to have a significant correlation with the level of stress experienced by the procrastinator [15]. It has also been observed that therapy can be of help in this aspect and as such it could also be included in the model in future research [1, 10]. The developed computational model may be used as a basis to advise therapists about timing and duration of certain therapies for their clients.

References

1. Birnie, K., Speca, M., Carlson, L.E.: Exploring self-compassion and empathy in the context of mindfulness-based stress reduction (MBSR). Stress. Health 26(5), 359–371 (2010)
2. Choi, J.N., Moran, S.V.: Why not procrastinate? Development and validation of a new active procrastination scale. J. Soc. Psychol. 149(2), 195–212 (2009)
3. Chu, A., Choi, J.: Rethinking procrastination: effects of active procrastination behavior on positive attitudes and performance. J. Soc. Psychol. 145, 254–264 (2005)
4. Fee, R.L., Tangney, J.P.: Procrastination: a means of avoiding shame or guilt? J. Soc. Behav. Pers. 15(5), 167–184 (2000)
5. Flett, A.L., Haghbin, M., Pychyl, T.A.: Procrastination and depression from a cognitive perspective: an exploration of the associations among procrastinatory automatic thoughts, rumination, and mindfulness. J. Rational-Emot. Cognitive-Behav. Ther. 34(3), 169–186 (2016). https://doi.org/10.1007/s10942-016-0235-1
6. Gammon, J., Morgan-Samuel, H.: A study to ascertain the effect of structured student tutorial support on student stress, self-esteem and coping. Nurse Educ. Pract. 5(3), 161–171 (2005)
7. Harriott, J., Ferrari, J.R.: Prevalence of procrastination among samples of adults. Psychol. Rep. 78(2), 611–616 (1996). https://doi.org/10.2466/pr0.1996.78.2.611
8. Klassen, R.M., Krawchuk, L.L., Rajani, S.: Academic procrastination of undergraduates: low self-efficacy to self-regulate predicts higher levels of procrastination. Contemp. Educ. Psychol. 33(4), 915–931 (2008). https://doi.org/10.1016/j.cedpsych.2007.07.001
9. Lay, C.H.: Trait procrastination and affective experiences: describing past study behavior and its relation to agitation and dejection. Motiv. Emot. 18(3), 269–284 (1994)
10. Lee, W.K., Bang, H.J.: The effects of mindfulness-based group intervention on the mental health of middle-aged Korean women in community. Stress. Health 26(4), 341–348 (2010)
11. Myrick, J.G.: Emotion regulation, procrastination, and watching cat videos online: who watches internet cats, why, and to what effect? Comput. Hum. Behav. 52, 168–176 (2015)
12. Onwuegbuzie, A.J.: Academic procrastination and statistics anxiety. Assess. Eval. High. Educ. 29(1), 3–19 (2004)
13. Schouwenburg, H.C.: Academic procrastination. In: Joseph, R.F., Judith, L.J., William, G.M. (eds.) Procrastination and Task Avoidance, pp. 71–96. Springer, Boston (1995)
14. Senecal, C., Koestner, R., Vallerand, R.J.: Self-regulation and academic procrastination. J. Soc. Psychol. 135(5), 607–619 (1995)
15. Sirois, F.: Procrastination and stress: exploring the role of self-compassion. Self Identity 13, 128–145 (2014). https://doi.org/10.1080/15298868.2013.763404
16. Sirois, F., Pychyl, T.: Procrastination and the priority of short-term mood regulation: consequences for future self. Soc. Pers. Psychol. Compass 7(2), 115–127 (2013)

17. Sirois, F.M., Melia-Gordon, M.L., Pychyl, T.A.: 'I'll look after my health, later': an investigation of procrastination and health. Pers. Individ. Differ. **35**(5), 1167–1184 (2003)

18. Stead, R., Shanahan, M.J., Neufeld, R.W.: 'I'll go to therapy, eventually': procrastination, stress and mental health. Pers. Individ. Differ. **49**(3), 175–180 (2010)

19. Steel, P.: The nature of procrastination: a meta-analytic and theoretical review of quintessential self-regulatory failure. Psychol. Bull. **133**(1), 65–94 (2007)

20. Tice, D., Bratslavsky, E., Baumeister, R.: Emotional distress regulation takes precedence over impulse control: if you feel bad, do it! J. Pers. Soc. Psychol. **80**, 53–67 (2001). https://doi.org/10.1037/0022-3514.80.1.53

21. Treur, J.: Network-oriented modeling for Adaptive Networks: Designing Higher-Order Adaptive Biological, Mental and Social Network Models. SSDC, vol. 251. Springer, Cham (2020). https://doi.org/10.1007/978-3-030-31445-3

22. van Eerde, W., Klingsieck, K.B.: Overcoming procrastination? A meta-analysis of intervention studies. Educ. Res. Rev. **25**, 73–85 (2018)

Improved Lower Bounds for the Cyclic Bandwidth Problem

Hugues Déprés[1], Guillaume Fertin[2($\boxtimes$)] [ID], and Eric Monfroy[3] [ID]

[1] ENS de Lyon, Computer Science Department, Lyon, France
hugues.depres@ens-lyon.fr
[2] Université de Nantes, LS2N (UMR 6004), CNRS, Nantes, France
guillaume.fertin@univ-nantes.fr
[3] LERIA, Université d'Angers, Angers, France
eric.monfroy@univ-angers.fr

Abstract. We study the classical CYCLIC BANDWIDTH problem, an optimization problem which takes as input an undirected graph $G = (V, E)$ with $|V| = n$, and asks for a labeling φ of V in which every vertex v takes a unique value $\varphi(v) \in [1; n]$, in such a way that $B_c(G, \varphi) = \max\{\min_{uv \in E(G)}\{|\varphi(u) - \varphi(v)|, n - |\varphi(u) - \varphi(v)|\}\}$, called the *cyclic bandwidth of G*, is minimized.

We provide three new and improved lower bounds for the CYCLIC BANDWIDTH problem, applicable to any graph G: two are based on the neighborhood vertex density of G, the other one on the length of a longest cycle in a cycle basis of G. We also show that our results improve the best known lower bounds for a large proportion of a set of instances taken from a frequently used benchmark, the Harwell-Boeing sparse matrix collection. Our third proof provides additional elements: first, an improved sufficient condition yielding $B_c(G) = B(G)$ (where $B(G) = \min_\varphi\{\max_{uv \in E(G)}\{|\varphi(u) - \varphi(v)|\}\}$ denotes the *bandwidth of G*); second, an algorithm that, under some conditions, computes a labeling reaching $B(G)$ from a labeling reaching $B_c(G)$.

Keywords: Cyclic bandwidth problem · Graph labeling · NP-hard problem · Optimization · Lower bounds

1 Introduction

Let $G = (V, E)$ be an undirected graph without loops and multiple edges. Let $|V| = n$, and let φ be a bijection from V to $[n]$, where $[n]$ denotes the set $\{1, 2, 3 \ldots n\}$. In the rest of the paper, φ will be called a *cb-labeling* of G (where "*cb*" stands for "cyclic bandwidth"). For any value $p \in \mathbb{Z}$, we let $|p|_n = \min\{|p|, n - |p|\}$. The Cyclic Bandwidth of G induced by φ, which we denote $B_c(G, \varphi)$, is defined as follows: $B_c(G, \varphi) = \max_{uv \in E}\{|\varphi(u) - \varphi(v)|_n\}$. The *Cyclic Bandwidth* of a graph G, denoted $B_c(G)$, is the minimum value $B_c(G, \varphi)$ among all possible labelings φ, and such a labeling will be called an

© Springer Nature Switzerland AG 2021
M. Paszynski et al. (Eds.): ICCS 2021, LNCS 12742, pp. 555–569, 2021.
https://doi.org/10.1007/978-3-030-77961-0_45

optimal cb-labeling. We are now ready to define the CYCLIC BANDWIDTH problem, which is the following optimization problem: given a graph $G = (V, E)$ with n vertices, find $B_c(G)$. In the following, we shall denote by φ^* any optimal cb-labeling of G, i.e., any labeling of G satisfying $B_c(G, \varphi^*) = B_c(G)$.

The CYCLIC BANDWIDTH problem has been introduced in [6], in the context of designing a ring interconnection network between computers. It can be seen as a variant of the well-known BANDWIDTH MINIMIZATION problem, whose formal introduction is due to Harper [3] in 1964 – see also the survey from Chinn et al. [1] for a more detailed historical account. The BANDWIDTH MINIMIZATION problem also asks for a *bm-labeling* φ (where "*bm*" stands for "bandwidth minimization"), i.e., a bijection from V to $[n]$. The value computed from φ is in that case $B(G, \varphi) = \max_{uv \in E}\{|\varphi(u) - \varphi(v)|\}$. The BANDWIDTH MINIMIZATION problem asks for a labeling φ^* such that $B(G, \varphi^*)$ is minimized, and the latter value is denoted $B(G)$. Any optimal labeling will be called an *optimal bm-labeling* of G. Note that there is a strong connection between $B(G)$ and $B_c(G)$, starting with the trivial fact that for any graph G, $B_c(G) \leq B(G)$.

The CYCLIC BANDWIDTH has been extensively studied. It has been shown to be NP-hard, even in the case of trees with maximum degree 3 [8]. The value of $B_c(G)$ has also been determined when G belongs to a specific class of graphs, e.g., paths, cycles, Cartesian products of paths (resp. of cycles, of paths and cycles), full k-ary trees, complete graphs, complete bipartite graphs, and unit interval graphs [2,4,7,8]. Some other works studied the relationship between $B_c(G)$ and $B(G)$, and notably aimed at determining sufficient conditions under which the equality $B(G) = B_c(G)$ holds [5,9]. In particular, it has been shown that $B(T) = B_c(T)$ for any tree T [4]. Another set of results is concerned with determining bounds for $B_c(G)$, and notably lower bounds, for general graphs [15]. First, it is easy to see that for any graph G, $B_c(G) \geq \frac{\Delta(G)}{2}$, where $\Delta(G)$ denotes the maximum degree of G. Besides, for any graph G, we have $\frac{B(G)}{2} \leq B_c(G) \leq B(G)$, the leftmost bound being from [9]. Other lower bounds have been obtained, most of these results being based on the density or on a relevant cycle basis of the studied graph (see, e.g., [4,15] which are in connection with our results).

Finally, more recent papers are concerned with designing efficient heuristics for the CYCLIC BANDWIDTH problem (see, e.g., [12–14]) where both execution time and upper bounds for $B_c(G)$ are examined – the latter being tested on a subset of the classical Harwell-Boeing sparse matrix collection (see, e.g., https://math.nist.gov/MatrixMarket/collections/hb.html).

Our goal in this paper is to provide lower bounds for $B_c(G)$ that apply to *any* graph G. We essentially prove three lower bounds: two of them rely on the graph density (Theorems 1 and 3), and both improve the best known results; the third one relies on a relevant cycle basis for G (Theorem 9), and also improves the best known related result. The latter result also presents improved sufficient conditions under which $B_c(G) = B(G)$, and provides a relabeling algorithm which, under these conditions, and given a cb-labeling φ of G, computes a *bm-*

labeling φ' of G such that $B_c(G, \varphi) = B(G, \varphi')$. We also successfully applied our results to a benchmark of 28 Harwell-Boeing graphs.

This paper is organized as follows: in Sect. 2 we discuss our graph density lower bounds, and in Sect. 3 our lower bound based on a relevant cycle basis of G. Section 4 describes our results on the abovementioned benchmark.

2   Graph Density Lower Bounds

In this section, we present, in Theorems 1 and 3, two new lower bounds on $B_c(G)$ that apply to any graph G. As discussed later, they can be seen as generalizations of previously existing lower bounds.

Theorem 1. *Let G be a graph, v any vertex of G, and $i \geq 1$ an integer. Let $N_i(v)$ denote the set of vertices within distance i from v in G. Then we have $B_c(G) \geq \left\lceil \frac{|N_i(v)| - 1}{2i} \right\rceil$.*

In order to prove Theorem 1, the following lemma is needed.

Lemma 2. *Let $d \geq 0$ be an integer, and let $u_0, u_1, \ldots, u_d$ be a path of length d in G. Let φ^* be an optimal cb-labeling of G. Then, $|\varphi^*(u_d) - \varphi^*(u_0)|_n \leq d \cdot B_c(G)$.*

Proof. First, note that $|x - y|_n$ satisfies the triangle inequality. Since φ^* is an optimal cb-labeling of G, by definition we have $|\varphi^*(u_{i+1}) - \varphi^*(u_i)|_n \leq B_c(G)$ for any $0 \leq i < d$. Consequently, summing the above expression for all $0 \leq i < d$ yields the required result.   □

Proof. (of Theorem 1). Let $i \geq 1$ be an integer, and recall that $N_i(v)$ denotes the set of vertices lying at distance less than or equal to i from v in G. Let φ^* be an optimal cb-labeling of G, and by extension let $\varphi^*(N_i(v))$ be the set of labels used by φ^* to label the vertices of $N_i(v)$. From Lemma 2 we conclude that $\varphi^*(N_i(v)) \subseteq [\varphi^*(v) - i \cdot B_c(G); \varphi^*(v) + i \cdot B_c(G)] \bmod n$, since for any vertex $u \in N_i(v)$ there exists a path of length at most i between v and u.

The above interval $[\varphi^*(v) - i \cdot B_c(G); \varphi^*(v) + i \cdot B_c(G)] \bmod n$ contains $2i \cdot B_c(G) + 1$ distinct values. Since φ^* is injective, we have that $|N_i(v)| \leq 2i \cdot B_c(G) + 1$, which concludes the proof.   □

Note that Theorem 1 generalizes the trivial lower bound $B_c(G) \geq \left\lceil \frac{\Delta(G)}{2} \right\rceil$ (where $\Delta(G)$ denotes the maximum degree of G): take for this $i = 1$ and any vertex v of degree $\Delta(G)$. Our second result has the same flavor as Theorem 1 above. However, instead of relying on $N_i(v)$, it considers the number of vertices that lie at distance at most i from *either one of two vertices u or v* of G, where $uv \in E(G)$.

Theorem 3. *Let $G = (V, E)$ be a graph, u and v two distinct vertices of G, $i \geq 1$ an integer, and $N_i(u, v)$ the set of vertices within distance i from either u or v in G. Then, for every edge $uv \in E$, we have $B_c(G) \geq \left\lceil \frac{|N_i(u,v)| - 1}{2i + 1} \right\rceil$.*

Proof. The arguments are somewhat similar to those used in proof of Theorem 1. Take any edge $uv \in E(G)$, and observe an optimal cb-labeling φ^* of G. First, note that $|\varphi^*(u) - \varphi^*(v)|_n \leq B_c(G)$ by definition. Moreover, by Lemma 2, we know that the set of labels used by φ^* to label the vertices of $N_i(u, v)$ is included in a closed interval of size $(2i + 1) \cdot B_c(G)$, since any two vertices of $N_i(u, v)$ are connected by a path of length at most $2i + 1$. Since φ^* is injective, we conclude from the above that $|N_i(u, v)| \leq (2i + 1) \cdot B_c(G) + 1$, which proves the result. □

We note that, although Theorems 1 and 3 look similar, none of these two results strictly contains the other. Note also that Theorem 3 generalizes a result for lower bounds on trees proved in [8]. Due to lack of space, the above two claims are not proved here. We finally note that the lower bounds of Theorems 1 and 3 are tight for several well-known classes of graphs, such as paths, cycles, complete graphs, and caterpillars.

3 Cycle Basis Lower Bound

The whole current section is devoted to proving Theorem 9, which provides a lower bound for $B_c(G)$ based on both $B(G)$ and the length ℓ of the longest cycle in a cycle basis of G. Theorem 9 also gives an improved sufficient condition under which $B_c(G) = B(G)$, also based on ℓ. Theorem 9 can be considered as somewhat similar to Theorem 3.1 from [4], but is actually different as it provides a lower bound for $B_c(G)$.

Another interesting point is that, on the way to proving Theorem 9, we also prove the following result, of independent interest: there exists a labeling algorithm (namely, Algorithm 1) that, when $\ell < \frac{n}{B_c(G)}$, computes a bm-labeling φ' of G starting from a cb-labeling φ of G, such that $B(G, \varphi') \leq B_c(G, \varphi)$. In particular, if φ is optimal, so is φ', and thus $B_c(G) = B(G)$. The interest of Algorithm 1 lies in the fact that, when $\ell < \frac{n}{B_c(G)}$, a labeling φ satisfying $B_c(G, \varphi) = k$ does not necessarily satisfy $B(G, \varphi) = k$ – and in that case another labeling φ' is needed in order to reach $B(G, \varphi') = k$.

Theorem 9 relies on Lemma 7, which we prove first. Before that, we start with some definitions.

Definition 4. *Let G_1 and G_2 two graphs such that $V(G_1) = V(G_2) = V$. The pairwise symmetric difference of G_1 and G_2, denoted $psd(G_1, G_2)$, is the graph G' whose set of vertices is V, and whose edges are those belonging to exactly one of the two graphs G_1 and G_2. By extension, the symmetric difference $sd(\mathcal{G})$ of a set $\mathcal{G} = \{G_1, G_2 \ldots G_p\}$ of graphs, each built on the same set V of vertices, is $psd(G_1, psd(G_2(\ldots psd(G_{p-1}, G_p))))$.*

Definition 5. *A cycle basis of a graph G is a set of cycles $\mathcal{C}$ of G such that any cycle of G either belongs to $\mathcal{C}$ or can be obtained by symmetric difference of a subset of $\mathcal{C}$.*

Finally, we use a definition from [4], which will prove useful in the following.

Definition 6. *Given a graph $G = (V, E)$, let $S \subset V \times V$ be such that $(u, v) \in S$ iff $uv \in E(G)$. Let φ be a cb-labeling of V, and let $\alpha_\varphi : S \to \{-1, 0, 1\}$ denote the following function:*

$$\alpha_\varphi(u, v) = \begin{cases} 0 & \text{if } |\varphi(u) - \varphi(v)| \leq B_c(G, \varphi) \\ 1 & \text{if } \varphi(u) - \varphi(v) \geq n - B_c(G, \varphi) \\ -1 & \text{if } \varphi(u) - \varphi(v) \leq B_c(G, \varphi) - n \end{cases}$$

In other words, $\alpha_\varphi(u, v) = 0$ whenever u and v have "close enough labels". Otherwise, $\alpha_\varphi(u, v) = \pm 1$, depending on which vertex among u and v has the greatest label. Hence, α_φ is not symmetric, since for any edge uv in $E(G)$, $\alpha_\varphi(u, v) = -\alpha_\varphi(v, u)$. In the following, we will say that an edge $uv \in E(G)$ has a *zero* α_φ when $\alpha_\varphi(u, v) = 0$, and a *non-zero* α_φ otherwise.

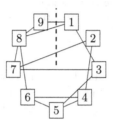

Fig. 1. Illustration of Definition 6 on a graph G with $n = 9$ with a cb-labeling φ such that $B_c(G, \varphi) = 4$. Each edge uv crossing the dashed vertical line has a non-zero $\alpha_\varphi(u, v)$.

Fig. 1 illustrates the above definition. This figure represents a graph G with $n = 9$ vertices, together with a cb-labeling φ satisfying $B_c(G, \varphi) = 4$ (note that φ is not optimal, since $B(G) = 2$ in this case). Vertices of G are displayed along a cycle, ordered (clockwise) by their labels in φ. For readability, we assume that labels and vertex identifiers are the same.

It can be seen that all edges having a non-zero α_φ are the ones that cross the dashed vertical line – because that line separates edges whose end vertices carry labels that are strictly more than $B_c(G, \varphi) = 4$ apart. The important thing to notice here is that, for any edge uv of G, the sign of the corresponding α_φ depends in which direction the dashed line is crossed, for example $\alpha_\varphi(7, 2) = \alpha_\varphi(8, 1) = \alpha_\varphi(9, 1) = 1$ (the dashed line is crossed from left to right), while $\alpha_\varphi(2, 7) = \alpha_\varphi(1, 8) = \alpha_\varphi(1, 9) = -1$ (the dashed line is crossed from right to left). In other words, when computing $\alpha_\varphi(u, v)$ for an edge $uv \in E(G)$, an orientation is given to that edge.

Let us define $s_\alpha(C) = \alpha_\varphi(x_p, x_0) + \sum_{i=0}^{p-1} \alpha_\varphi(x_i, x_{i+1})$ for any cycle $C = (x_0, x_1, \ldots x_p, x_0)$ of G, and let us call $s_\alpha(C)$ the *α-sum of C*.

It can then be seen that, if C_1 and C_2 are two cycles in G, and if $C = psd(C_1, C_2)$ is also a cycle, not all edges of C necessarily have the same orientation (induced by α_φ) as in C_1 or C_2. Such a case is illustrated in Fig. 2, where, e.g., the edge connecting vertices 2 and 5 in C_1 (left) is oriented from 2 to 5 in the α-sum of C_1, while in $C = spd(C_1, C_2)$ (right), the same edge is oriented from 5 to 2 in the α-sum of C. As a consequence, in this case, we have $s_\alpha(C_1) = s_\alpha(C_2) = 0$ and $s_\alpha(C) = 2 \neq 0$. The example shown in Fig. 2 shows

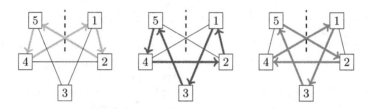

Fig. 2. A graph G with $n = 5$ vertices, such that $B_c(G) = 2$, in which the labels of an optimal cb-labeling φ are also the vertices identifiers. Edges in C_1 (left), C_2 (middle), and $C = spd(C_1, C_2)$ (right) are arbitrarily oriented to compute their respective α-sum. It can be seen that, in any case, $s_\alpha(C_1) = s_\alpha(C_2) = 0$, while $s_\alpha(C) = 2$.

that s_α is not necessarily conserved by symmetric difference. However, in the following lemma, which is a stronger version of Lemma 3.2 from [4], we are able to show that all cycles have a 0 α-sum under certain conditions.

Lemma 7. *Let G be a graph, and φ an optimal cb-labeling of G. If the set of cycles of G of length at most $\ell < \frac{n}{B_c(G)}$ contains a cycle basis, then for any cycle C of G, $s_\alpha(C) = 0$.*

Proof. Let φ be an optimal cb-labeling of G. Assume, by contradiction, that there is a cycle C of G such that $s_\alpha(C) \neq 0$. Remove from G all edges that have a non-zero α_φ, and let $S_1, S_2, .., S_p$ denote the connected components of the resulting graph. For each connected component S_i, $1 \leq i \leq p$ and any subgraph $G' = (V', E')$ of G, we define the following function $\gamma_i(G')$: $\gamma_i(G') = \sum_{e \in E(G')} \gamma_i(e)$, where for any edge $uv \in E(G')$, $\gamma_i(uv) = 1$ if $\alpha(u, v) \neq 0$ and at least one vertex (say x) among u, v satisfies (i) $\varphi(x) > \frac{n}{2}$ and (ii) $x \in V(S_i)$; and $\gamma_i(uv) = 0$ otherwise.

One of the interests of γ_i is that, unlike α_φ, this funtion is symmetric: any edge uv will contribute the same way to γ_i, whether it is considered from u to v or from v to u. We will now prove Lemma 7 through a series of three claims.

Claim 1. Let $1 \leq i \leq p$, and let G_1 and G_2 be two subgraphs of G such that $\gamma_i(G_1)$ and $\gamma_i(G_2)$ are even. Then, $G' = psd(G_1, G_2)$ is such that $\gamma_i(G')$ is even.

Proof. (of Claim 1). Take any $1 \leq i \leq p$, and let G_1 and G_2 be two subgraphs of G such that both $\gamma_i(G_1)$ and $\gamma_i(G_2)$ are even. Then $\gamma_i(psd(G_1, G_2)) = \gamma_i(G_1) + \gamma_i(G_2) - 2 \cdot \sum_{e \in E(G_1) \cap E(G_2)} \gamma_i(e)$, since all edges e present in both G_1 and G_2 are removed from $psd(G_1, G_2)$. Since all numbers in the above equation are even, $\gamma_i(sd(G_1, G_2))$ is also even. $\qquad\square$

Claim 2. Any cycle C' of G whose length ℓ satisfies $\ell < \frac{n}{B_c(G)}$ is such that $\gamma_i(C')$ is even for any $1 \leq i \leq p$.

Proof. (of Claim 2). Observe a cycle $C' = (x_0, x_1, ..., x_\ell)$ (with $x_0 = x_\ell$) of G, of length $\ell < \frac{n}{B_c(G)}$, and consider the set $\mathcal{L}$ of labels that are used to label $V(C')$. We claim that $\mathcal{L}$ belongs to a (cyclic) interval I of cyclic length strictly less than $\frac{n}{2}$. In order to show this, let u (resp. v) be the vertex of C' whose label is minimum (resp. maximum), and let $\varphi_{\min} = \varphi(u)$ (resp. $\varphi_{\max} = \varphi(v)$). Now let us consider the set $\mathcal{L}$ of labels that are used to label $V(C')$. There are three possible cases for $\mathcal{L}$: first, suppose $\mathcal{L} \subseteq [\varphi_{\min}; \varphi_{\max}]$. Then, $I = [\varphi_{\min}; \varphi_{\max}]$, and I is of length strictly less than $\frac{n}{2}$ by definition of the cyclic bandwidth and since $B_c(G) < \frac{n}{\ell}$. Second, suppose $\mathcal{L} \subseteq [1; \varphi_{\min}] \cup [\varphi_{\max}; n]$. For the same reasons as in the previous case, we conclude that the interval $I = [1; \varphi_{\min}] \cup [\varphi_{\max}; n]$ is of cyclic length strictly less than $\frac{n}{2}$. Third, suppose we are not in one of the two above cases, and let us show that this case cannot happen: indeed, this means that at least one vertex $x \neq u, v$ (resp. $y \neq u, v$) of $V(C')$ satisfies $\varphi(x) \in [1; \varphi_{\min}[\cup]\varphi_{\max}; n]$ (resp. $\varphi(y) \in]\varphi_{\min}; \varphi_{\max}[$. We will say that x (resp. y) has a *small* (resp. *high*) label. Now let P_x denote the path from u to v, that contains x, and that follows the edges of C'. Let S_x be the sum of $|\varphi(a) - \varphi(b)|_n$ over all edges ab in P_x. Then, we know that $S_x \geq \varphi_{\max} - \varphi_{\min}$. Now consider the path P_y going from u to v that contains y, and that follows the edges of C'. Let S_y be the sum of $|\varphi(c) - \varphi(d)|_n$ over all edges cd in P_y. Then we know that $S_y \geq n - \varphi_{\max} + \varphi_{\min}$. Hence, $S_x + S_y \geq n$. We will now show that, since we supposed $\ell < \frac{n}{B_c(G)}$, this cannot happen. Indeed, we have

$$S_x + S_y = \sum_{i=0}^{\ell-1} |\varphi(x_{i+1}) - \varphi(x_i)|_n \leq \ell \cdot B_c(G),$$ because for any two neighbors x_{i+1} and x_i in C' we have $|\varphi(x_{i+1}) - \varphi(x_i)|_n \leq B_c(G)$, by definition of $B_c(G)$. Moreover, we know by hypothesis that $\ell \cdot B_c(G) < n$, thus we conclude that $S_x + S_y < n$, a contradiction.

We just proved that $\mathcal{L}$ belongs to a (cyclic) interval I of cyclic length strictly less than $\frac{n}{2}$. Let us now observe interval I. There are two cases: first, if either 1 or n is not in I, then all edges uv of C' satisfy $\alpha_\varphi(u, v) = 0$, by definition of α_φ and because $B_c(G) \leq \frac{n}{2}$ for any graph G. Thus, every edge of C' contributes to zero in $\gamma_i(C')$, and consequently $\gamma_i(C') = 0$ for any $1 \leq i \leq p$.

Second, if both 1 and n are present in I, take any integer $1 \leq i \leq p$ and partition the vertices of C' in two sets: the first set V_1 contains every vertex $u \in V(C')$ such that $\varphi(u) > \frac{n}{2}$ and $u \in V(S_i)$; the second set is $V_2 = V(C') - V_1$.

Let uv be an edge of C'. We will say that uv is a *crossing* edge if $u \in V_1$ and $v \in V_2$, or vice-versa – this term refers to the fact that, as we will prove it, a crossing edge uv necessarily satisfies $\alpha(u, v) \neq 0$, and thus, crosses the dashed vertical as in Fig. 1. Take any edge uv in C'. By definition, $\gamma_i(uv) \in \{0, 1\}$. Our goal here is to show that $\gamma_i(uv) = 1$ iff uv is a crossing edge.

($\Leftarrow$) Suppose uv is a crossing edge, and suppose wlog that $u \in V_1$ and $v \in V_2$. In that case, $\varphi(v) < \frac{n}{2}$. Otherwise, we would have $\alpha_\varphi(u, v) = 0$ (by definition of α_φ, because the labels of u and v are separated by less than $B_c(G)$), which implies that both vertices u and v belong to S_i, since $u \in V(S_i)$. But in that case v would belong to V_1, since $v \in V(S_i)$ and $\varphi(v) > \frac{n}{2}$. This is a contradiction since v cannot simultaneously belong to V_1 and V_2. Thus, every crossing edge uv with $u \in V_1$ and $v \in V_2$ is such that $\varphi(v) < \frac{n}{2}$, and consequently $\alpha_\varphi(u, v) \neq 0$. This implies that for any such edge uv, $\gamma_i(uv) = 1$.

($\Rightarrow$) Now suppose $\gamma_i(uv) = 1$, which implies $\alpha_\varphi(u, v) \neq 0$. Thus, there are two cases: $\alpha_\varphi(u, v) > 0$ or $\alpha_\varphi(u, v) < 0$. First, if $\alpha_\varphi(u, v) > 0$, and since $\gamma_i(uv) = 1$, then $u \in V(S_i)$ by definition. Moreover, $\alpha_\varphi(u, v) > 0$ implies (by definition) $\varphi(u) \geq \varphi(v) + n - B_c(G, \varphi)$. Since $B_c(G, \varphi) \leq \frac{n}{2}$ for any graph G and any labeling φ, and since $\varphi(v) \geq 1$, we conclude that $\varphi(u) > \frac{n}{2}$. We then know that $u \in V_1$ by definition. Moreover, $v \notin V_1$, otherwise $uv \in E(S_i)$, a contradiction to the fact that all edges in $E(S_i)$ have a zero α_φ. Thus, we conclude that $v \in V_2$, and that uv is a crossing edge. Now if $\alpha_\varphi(u, v) < 0$, by a symmetrical argument as above, we conclude that $v \in V_1$ and $u \in V_2$, i.e., uv is a crossing edge as well.

We thus know that each edge $uv \in E(C')$ contributes to 1 to γ_i iff uv is a crossing edge. This allows us to prove the claim: if we follow the edges of C' to compute $\gamma_i(C')$, the number of crossing edges encountered is necessarily even, since C' starts and ends in the same vertex. Since only a crossing edge increases γ_i (by exactly 1), we conclude that $\gamma_i(C')$ is even.   □

Claim 3. There exists an integer $1 \leq j \leq p$ and a cycle C' of G such that $\gamma_j(C') = 1$.

Proof. (of Claim 3). Recall that we suppose there exists a cycle C in G such that $s_\alpha(C) \neq 0$. For each $1 \leq i \leq p$, let Σ_i be the sum of the $\alpha_\varphi(u, v)$ for each edge uv of C that has an extremity in S_i, counting uv twice in Σ_i if both $u, v \in V(S_i)$. The sum of the Σ_is over every $1 \leq i \leq p$ is thus equal to $2s_\alpha(C)$.

Since we supposed $s_\alpha(C) \neq 0$, we know there exists at least one j, $1 \leq j \leq p$, such that $\Sigma_j \neq 0$. Thus, there exists at least one edge $uv \in E(C)$ such that (i) at least one of $u, v \in V(S_j)$, and (ii) $\alpha_\varphi(u, v) \neq 0$.

Now let us look at every edge along C (following the natural order of the cycle), and let us observe each pair (uv, wx) of edges (with possibly $uv = wx$) satisfying the following: (i) $\alpha_\varphi(u, v) \neq 0$ and $u \in V(S_j)$ and (ii) $\alpha_\varphi(w, x) \neq 0$ and $x \in V(S_j)$. Because $\Sigma_j \neq 0$, such a pair (again with possibly $uv = wx$) must exist. Moreover, among such pairs, and since $\Sigma_j \neq 0$, there necessarily exists a pair, say (ab, cd), such that $\alpha_\varphi(a, b) = \alpha_\varphi(c, d)$.

There are now two cases to consider. First, if $ab = cd$, then we construct the cycle C_1 containing edge ab and edges of a path $P_{a,b}$ joining a to b in S_j (we know there exists such a path, since $a, b \in V(S_j)$, and S_j is connected by definition). By definition also, $P_{a,b}$ only contains zero α_φ edges. Thus $\gamma_j(C_1) = 1$, since only edge ab contributes to 1 to $\gamma_j(C_1)$ – all other edges of C_1 contribute to 0.

Second, if $ab \neq cd$, take the path $P = ab \ldots cd$ in C that connects a to d through ab and cd. Path P has no internal vertex in S_j by construction. Consider a second path $P' = a \ldots d$ connecting a to d within S_j (again, P' exists since $a, d \in V(S_j)$ by definition, and since S_j is connected). Note that P and P' only intersect in a and d. Then, we can construct cycle C_2, which is the concatenation of P and P'. What is important to observe is that only one of the two edges ab or

cd will contribute to 1 in the computation of $\gamma_j(C_2)$ – the other will contribute to 0. Moreover, all other edges from P contribute to 0 to $\gamma_j(C_2)$ (because none contains a vertex of S_j), and the same conclusion is reached with any edge of P': by definition, each such edge belongs to S_j, and thus is a zero α_φ edge.   □

We are now ready to prove Lemma 7. Suppose the set of cycles of G of length at most $\ell < \frac{n}{B_c(G)}$ contains a cycle basis $\mathcal{C}$. Observe an optimal cb-labeling φ of G, and suppose by contradiction that there exists a cycle C in G such that $\alpha_\varphi(G) \neq 0$. By Claim 3, we know that there exists a cycle C' and an integer j such that $\gamma_j(C') = 1$. Besides, we know by definition that C' can be obtained by symmetric difference of elements of $\mathcal{C}$.

However, Claims 2 and 1 respectively yield that any cycle in $\mathcal{C}$ has even γ_i and thus that any symmetric difference of elements of $\mathcal{C}$ also has even γ_i. We thus get the desired contradiction, which in turn proves Lemma 7.   □

Note that Lemma 7 remains correct if φ is not an optimal labeling (but in that case, $B_c(G)$ should be replaced by $B_c(G, \varphi)$ in the lemma's statement).

The next lemma provides a sufficient condition yielding $B(G) = B_c(G)$, improved from [4]. It is the last step towards proving Theorem 9, and needs Lemma 7 to be proved.

Lemma 8. *If the set of cycles of G of length at most $\ell < \frac{n}{B_c(G)}$ contains a cycle basis, then $B_c(G) = B(G)$.*

Proof. Suppose G satisfies the conditions of the lemma, and let φ be a cb-labeling of G. We will show in the following that Algorithm 1 below computes a bm-labeling φ' of G satisfying $B(G, \varphi') \leq B_c(G, \varphi)$. If φ is optimal, then $B_c(G, \varphi) = B_c(G)$, and consequently $B(G, \varphi') = B_c(G) = B(G)$ (i.e., φ' is an optimal bm-labeling) since $B(G) \geq B_c(G)$ for any graph G.

We need Claims 4 and 5 to show that $B(G, \varphi') \leq B_c(G, \varphi)$.

Claim 4. For any adjacent vertices u and v, $\beta(v) = \beta(u) + \alpha_\varphi(u, v)$.

Proof. (of Claim 4). First suppose v has been discovered by u during the Depth-First Search (DFS) in Algorithm 1: then Line 6 of Algorithm 1 yields $\beta(v) = \beta(u) + \alpha_\varphi(u, v)$. Symmetrically, the result is proved if u has been discovered by v during the DFS, since $\alpha_\varphi(u, v) = -\alpha_\varphi(v, u)$ for any labeling φ and any edge $uv \in E(G)$. Now suppose none of u and v have discovered the other during the DFS. Since u and v are in the same connected component, by property of DFS, there exists a vertex s such that $\beta(s) = 0$ and two paths $P_u = u_0u_1u_2...u_p$ (with $u_0 = s$ and $u_p = u$) and $P_v = v_0v_1v_2...v_q$ (with $v_0 = s$ and $v_q = v$), such that $\beta(u_{i+1}) = \beta(u_i) + \alpha_\varphi(u_i, u_{i+1})$ for any $0 \leq i \leq p - 1$ and $\beta(v_{i+1}) = \beta(v_i) + \alpha_\varphi(v_i, v_{i+1})$ for any $0 \leq i \leq q - 1$. Therefore, we have $\beta(u) = \sum_{i=0}^{p-1} \alpha_\varphi(u_i, u_{i+1})$ and $\beta(v) = \sum_{i=0}^{q-1} \alpha_\varphi(v_i, v_{i+1})$.

Suppose first that P_u and P_v only share vertex s. In that case, we obtain a cycle C_{uv} by concatenating paths P_u, P_v and edge uv. Since we are in the conditions of Lemma 7, we conclude that $s_\alpha(C_{uv}) = 0$. Observe that

---

Algorithm 1: Transformation of a cb-labeling φ into a bm-labeling solution φ' such that $B(G, \varphi') \le B_c(G, \varphi)$, under the conditions of Lemma 8

Data: A graph $G = (V, E)$ and a cb-labeling φ of G
Result: A bm-labeling φ' of G

1 **for** *each connected component S of G* **do**
2   Choose a vertex s in S and set $\varphi^*(s) = \varphi(s)$ and $\beta(s) = 0$
3   Apply a Depth-First Search from s
4   **for** *each newly visited vertex y discovered by a vertex x* **do**
5        $\beta(y) = \beta(x) + \alpha_\varphi(x, y)$
6        $\varphi^*(y) = \varphi(y) + n \cdot \beta(y)$
7   **end**
8 **end**
9 **for** *i from 1 to n* **do**
10   Set $\varphi'(i)$ to be the rank of $\varphi^*(i)$
11 **end**
12 Return φ'

$s_\alpha(C_{uv}) = S_{P_u} + \alpha_\varphi(u, v) + S_{P_v}$, where $S_{P_u} = \sum_{i=0}^{p-1} \alpha_\varphi(u_i, u_{i+1}) = \beta(u)$ and $S_{P_v} = \sum_{i=0}^{q-1} \alpha_\varphi(v_{i+1}, v_i)) = -\sum_{i=0}^{q-1} \alpha_\varphi(v_i, v_{i+1}) = -\beta(v)$. Hence we conclude that $\beta(u) + \alpha_\varphi(u, v) - \beta(v) = 0$, which is the desired result.

If P_u and P_v share more than vertex s, consider the common vertex between both paths having the largest index $r > 0$ in $P_u = u_0 u_1 \ldots u_p$, and let t be such that $u_r = v_t$. Thus, we can write $P_u = P_r \cdot P$ (where $P_r = u_0 u_1 \ldots u_r$) and $P_v = P_t \cdot P'$ (where $P_t = v_0 v_1 \ldots v_t$), and in that case $S_{P_u} = S_{P_r} + S$ and $S_{P_v} = S_{P_t} + S'$, where $S_{P_r} = \sum_{i=0}^{r-1} \alpha_\varphi(u_i, u_{i+1})$, $S = \sum_{i=r}^{p-1} \alpha_\varphi(u_i, u_{i+1})$, $S_{P_t} = \sum_{i=0}^{t-1} \alpha_\varphi(v_i, v_{i+1})$ and $S' = \sum_{i=t}^{q-1} \alpha_\varphi(v_i, v_{i+1})$. Again, if we concatenate P, P' and uv, we get a cycle C_{rt} and by Lemma 7 we have $s_\alpha(C_{rt}) = 0$. Thus, it remains to show that $S_{P_r} - S_{P_t} = 0$ to conclude. This is done by induction on r. When $r = 0$, we are in the case where P_u and P_v only share vertex s, which has already been discussed. When $r = 1$, $P_r = u_0 u_1 = s u_1$ is limited to one edge. If $t = 1$, then $v_1 = u_1$ and $P_t = P_r = s u_1$ and by antisymmetry of α_φ we have $S_{P_r} - S_{P_t} = \alpha_\varphi(s, u_1) + \alpha_\varphi(u_1, s) = 0$. If $t > 1$, then concatenating edge P_r and path P_t yields a cycle C', for which we know by Lemma 7 again that $s_\alpha(C') = S_{P_r} - S_{P_t} = 0$. Now if $r > 1$, apply induction hypothesis to r', where r' is the largest index (with $r' < r$) of a vertex of P_u also belonging to P_v. Let t' be the corresponding index in P_v. Then, $P_u = P_{r'} \cdot P_{r'r} \cdot P$ and $P_u = P_{t'} \cdot P_{t't} \cdot P'$, where $P_{r'r}$ (resp. $P_{t't}$) is a path from r' to r (resp. from t' to t). By induction hypothesis, $S_{P_{r'}} - S_{P_{t'}} = 0$. Moreover, concatenating $P_{r'r}$ and $P_{t't}$ yields a cycle C'', which satisfies, by Lemma 7, $s_\alpha(C'') = 0$. Since $s_\alpha(C'') = S_{P_{r'r}} - S_{P_{t't}}$, we obtain altogether that $(S_{P_{r'}} + S_{P_{r'r}}) - (S_{P_{t'}} + S_{P_{t't}}) = 0$. But $S_{P_{r'}} + S_{P_{r'r}} = S_{P_r}$ and $S_{P_{t'}} + S_{P_{t't}} = S_{P_t}$, hence we conclude $S_{P_r} - S_{P_t} = 0$, and the proof by induction holds. Altogether, we have $S_{P_r} - S_{P_t} = 0$ in all cases. This allows us to conclude that $\beta(v) = \beta(u) + \alpha_\varphi(u, v)$. $\qquad\square$

Claim 5. For any two adjacent vertices x and y in G, $|\varphi^*(y) - \varphi^*(x)| = |\varphi(y) - \varphi(x)|_n$.

Proof. (of Claim 5). If $\alpha_\varphi(x, y) = 0$, then $\beta(y) = \beta(x)$ by Claim 4. Thus, we have $|\varphi^*(y) - \varphi^*(x)| = |\varphi(y) - \varphi(x)|$. However, by definition of α_φ, we know that $|\varphi(y) - \varphi(x)| \leq B_c(G, \varphi)$ and thus $|\varphi(y) - \varphi(x)|_n = |\varphi(y) - \varphi(x)| = |\varphi^*(y) - \varphi^*(x)|$. Suppose now $\alpha_\varphi(x, y) = 1$. Then, by Claim 4 again, $\beta(y) = \beta(x) + 1$, and consequently $|\varphi^*(y) - \varphi^*(x)| = |\varphi(y) - \varphi(x) + n| = n - (\varphi(x) - \varphi(y))$ (because $\alpha_\varphi(x, y) = 1$ implies $\varphi(x) > \varphi(y)$). We then conclude that $|\varphi^*(y) - \varphi^*(x)| = |\varphi(y) - \varphi(x)|_n$. Finally, if $\alpha_\varphi(x, y) = -1$, by a symmetrical argument as above, we get that $|\varphi^*(y) - \varphi^*(x)| = |\varphi(y) - \varphi(x) - n| = n - (\varphi(y) - \varphi(x))$ (because $\alpha_\varphi(x, y) = -1$ implies $\varphi(x) < \varphi(y)$). Finally, $|\varphi^*(y) - \varphi^*(x)| = |\varphi(y) - \varphi(x)|_n$. $\square$

We are now ready to finish proof of Lemma 8. First, note that φ^* is injective, since for any vertex v, $\varphi^*(v)$ is congruent to $\varphi(v)$ modulo n. Moreover, since φ^* is injective, φ' is also injective. As $|V| = n$, φ' is a bijection from V to $[n]$. Moreover, $|\varphi'(y) - \varphi'(x)|$ is equal to the number of vertices v for which $\varphi^*(v)$ lies between $\varphi^*(y)$ (included) and $\varphi^*(x)$ (excluded). Since φ^* is injective, this number is less than or equal to $|\varphi^*(y) - \varphi^*(x)|$, so we have $|\varphi'(y) - \varphi'(x)| \leq |\varphi^*(y) - \varphi^*(x)|$. Since we know by Claim 5 that $|\varphi^*(y) - \varphi^*(x)| = |\varphi(y) - \varphi(x)|_n$, we conclude that φ' is a bm-labeling satisfying $B(G, \varphi') \leq B_c(G, \varphi)$. In particular, if φ is an optimal cb-labeling, then $B_c(G, \varphi) = B_c(G)$ and thus $B(G, \varphi') \leq B_c(G)$. But since for any graph G, $B(G) \geq B_c(G)$, we conclude that $B(G, \varphi') = B_c(G)$. Consequently φ' is an optimal bm-labeling, and $B(G) = B_c(G)$. $\square$

Our main result of the section is now stated.

Theorem 9. *Let G be a graph and let ℓ be the length of the longest cycle of a cycle basis of G. Then, we have $B_c(G) \geq \min\{B(G), \lceil \frac{n}{\ell} \rceil\}$. In particular, if $B(G) \leq \lceil \frac{n}{\ell} \rceil$, then $B(G) = B_c(G)$.*

Proof. Suppose first $B_c(G) \geq \lceil \frac{n}{\ell} \rceil$. Since $B(G) \geq B_c(G)$ for any graph, we have $B(G) \geq \lceil \frac{n}{\ell} \rceil$, which proves the result. Now suppose $B_c(G) < \lceil \frac{n}{\ell} \rceil$. In that case we have $B_c(G) < \frac{n}{\ell}$, since $B_c(G)$ is an integer, and by Lemma 8 we conclude that $B(G) = B_c(G)$. $\square$

4 Applying Our Results on 28 Harwell-Boeing Instances

We now illustrate how the results we obtained in Sects. 2 and 3 improve previous knowledge, by applying them on instances taken from the Harwell-Boeing sparse matrix collection. This benchmark contains a large number of matrices (that we interpret as graphs here) derived from several industrial applications. This collection has often been used as a benchmark for both the BANDWIDTH MINIMIZATION and CYCLIC BANDWIDTH problems, see e.g. [10–14]. Before going further, we note that all lower bound results provided in previous sections can be computed in polynomial time. Hence, computational performances are not provided here, as they do not play a major role.

Table 1. Summary of results applied to a set of 28 instances from the Harwell-Boeing repository, sorted by increasing $n = |V(G)|$. The rightmost column presents the best lower bound obtained by our results (i.e., the maximum value obtained by Theorems 1, 3, and 9). In the leftmost and rightmost column, asterisks indicate optimality, while bold values in the rightmost column indicate a strict improvement compared to previous knowledge. Note that the "Best Known Upper Bound" for `bcspwr03` we report here is the only result not imported from [13] (where the reported bound is 10). Indeed, we know by [11] that $B(G) \leq 9$ for that instance; since $B_c(G) \leq B(G)$ for any graph G, we get a(n optimal) upper bound of 9 for `bcspwr03`.

Graph G	n	m	Best known	Lower bound		$B(G) =$	Lower bound	Lower bound from our results
			Upper bound	Density	ℓ	$B_c(G)$?	Cycle basis	
jgl009	9	50	4*	4	3	no	3	4*
rgg010	10	76	5*	5	3	no	4	5*
jgl011	11	76	5*	5	3	no	4	5*
can_24	24	92	5*	5	3	yes	5	5*
pores_1	30	103	7*	6	4	yes	7	7*
ibm32	32	90	9*	8	5	no	7	8
bcspwr01	39	46	4*	4	13	no	3	4*
bcsstk01	48	176	12	9	4	no	12	**12***
bcspwr02	49	59	7	6	8	yes	7	**7***
curtis54	54	124	8*	8	9	no	6	8*
will57	57	127	6	6	4	yes	6	**6***
impcol_b	59	281	17	14	6	no	10	**14**
ash85	85	219	9	8	3	yes	9	**9***
nos4	100	247	10	9	4	yes	10	**10***
dwt_234	117	162	11	10	8	yes	11	**11***
bcspwr03	118	179	9*	9	10	yes	9	9*
bcsstk06	420	3720	45	31	4	yes	37	**37**
bcsstk07	420	3720	45	31	4	yes	37	**37**
impcol_d	425	1267	35	24	21	no	21	**24**
can_445	445	1682	46	36	10	?	45	**45**
494_bus	494	586	28	24	19	?	25	**25**
dwt_503	503	2762	41	29	8	yes	29	**29**
sherman4	546	1341	27	21	4	yes	21	**21**
dwt_592	592	2256	29	22	3	yes	22	**22**
662_bus	662	906	38	36	23	?	29	**36**
nos6	675	1290	16	15	4	yes	15	**15**
685_bus	685	1282	32	26	22	yes	30	**30**
can_715	715	2975	60	52	16	?	45	52

We took a set of 28 Harwell-Boeing instances, that have recently been specifically used for the CYCLIC BANDWIDTH problem [12–14]. Our results are summarized in Table 1. The three leftmost columns describe the instance: respectively its usual name, its number n of vertices, and m of edges. The next column ("Best Know Upper Bounds") provides the best upper bounds results concerning B_c, which are taken from the recent paper by Ren et al. [13]. Optimal values for $B_c(G)$ are marked with an asterisk.

The five remaining columns are computations and informations derived from the present work: in a nutshell, column "Lower Bound Density" contains the best result obtained by Theorems 1 and 3 (each theorem being applied for all possible values of i and any vertex v in the case of Theorem 1, and for all possible values of i and any edge uv in the case of Theorem 3), while column "Lower Bound Cycle Basis" contains the best results obtained by Theorem 9. Note that for computing the values in this column, we need to determine the length ℓ of the longest cycle in a cycle basis for G (column "ℓ") ; then, based on ℓ and on known upper bounds on $B(G)$ [11], we can (most of the time, see column "$B(G) = B_c(G)$?") determine whether $B(G) = B_c(G)$. If $B(G) = B_c(G)$, then we can invoke the best known lower bounds from $B(G)$ taken from [10]. If not, we apply Theorem 9 and take the minimum between $\frac{n}{\ell}$ and (the best lower bound on) $B(G)$. Finally, the righmost column, "Lower Bound from our results", simply takes the maximum value between the one in column "Lower Bound Density" (Theorems 1 and 3) and the one in column "Lower Bound Cycle Basis" (Theorem 9).

In the rightmost column of Table 1, which summarizes our results, values in bold show a strict improvement from previous lower bound results (based on the lower bounds for $B_c(G)$ from [13]), while an asterisk shows that the lower bound we obtain is actually optimal.

We first note that, among the 28 instances, our density lower bound is strictly better than our cycle basis lower bound 10 times, while the opposite happens 11 times (and thus it is a draw for the 7 remaining instances). This shows that both types of lower bounds are useful for improvement.

We also observe that Ren et al. [13] have determined the optimal value for $B_c(G)$ for 9 instances (the 7 first together with curtis54 and bscpwr03), hence our results cannot strictly improve them. It should be noted however, that for 8 out of these 9 instances, our lower bounds results are actually optimal (as shown by the asterisks in the rightmost column). Our results also show that, among the 19 remaining instances, we have been able to determine the optimal values for $B_c(G)$ for 6 of them, since our new lower bounds appear to be matching the upper bounds for B_c provided by [13]. This clearly shows the interest of our work in aiming at increasing the lower bounds.

Concerning the 13 remaining instances, our results have drastically reduced the gap between lower and upper bounds for B_c, which now ranges from 1 (instances can445 and nos6) to 12 (instance dwt_503), with an average gap of 5.54. Previously, based on the results presented in [13], the gap between lower and upper bounds ranged between 8 and 40, with an average of 24.31. Over the 28 studied instances, there are only two cases for which our results have neither reached the optimal value for B_c nor strictly improved previous lower bounds – namely instances ibm32 and can_715.

5 Conclusion and Open Questions

In this paper, we have studied the CYCLIC BANDWIDTH problem, and have particularly focused on determining new lower bounds for $B_c(G)$ for any graph G. We have, on the way, provided a relabeling algorithm (Algorithm 1) of a graph G that, under specific conditions, computes a labeling φ' for the BANDWIDTH MINIMIZATION problem, starting from a labeling φ for CYCLIC BANDWIDTH such that $B(G, \varphi') \leq B_c(G, \varphi)$. We have applied our results to a benchmark of 28 Harwell-Boeing instances already used in [12–14], and showed how our results greatly improve the best current knowledge.

The results we provide here may open the door to further improvements. In particular, it would be worth investigating whether they could help determining $B_c(G)$ for new families of graphs. They could also, in a more general setting, help better characterize properties of graphs satisfying $B_c(G) = B(G)$.

Concerning the benchmark that we used, note that among the 28 Harwell-Boeing instances from Table 1, we have not been able to determine whether $B_c(G) = B(G)$ for 4 instances, and investigating that question could be of interest. Finally, we observe that there exists graphs for which Algorithm 1, starting from a suboptimal labeling φ for CYCLIC BANDWIDTH, is able to compute an optimal labeling φ' for both BANDWIDTH MINIMIZATION and CYCLIC BANDWIDTH. Hence, it would be interesting to apply Algorithm 1 on the instances from Table 1 for which conditions of Lemma 8 are fulfilled, in order to determine whether this technique helps improving the knowledge on B and/or B_c.

References

1. Chinn, P.Z., Chvátalová, J., Dewdney, A.K., Gibbs, N.E.: The bandwidth problem for graphs and matrices - a survey. J. Graph Theor. **6**(3), 223–254 (1982)
2. Chung, F.R.: Labelings of graphs. Sel. Top. Graph Theor. **3**, 151–168 (1988)
3. Harper, L.H.: Optimal assignments of numbers to vertices. J. Soc. Ind. Appl. Math. **12**(1), 131–135 (1964)
4. Hromkovič, J., Müller, V., Sýkora, O., Vrťo, I.: On embedding interconnection networks into rings of processors. In: Etiemble, D., Syre, J.-C. (eds.) PARLE 1992. LNCS, vol. 605, pp. 51–62. Springer, Heidelberg (1992). https://doi.org/10.1007/3-540-55599-4_80
5. Lam, P.C.B., Shiu, W.C., Chan, W.H.: Characterization of graphs with equal bandwidth and cyclic bandwidth. Discrete Math. **242**(1–3), 283–289 (2002)
6. Leung, J.Y., Vornberger, O., Witthoff, J.D.: On some variants of the bandwidth minimization problem. SIAM J. Comput. **13**(3), 650–667 (1984)

7. Lin, Y.: A level structure approach on the bandwidth problem for special graphs. Ann. N. Y. Acad. Sci. **576**(1), 344–357 (1989)
8. Lin, Y.: The cyclic bandwidth problem. Syst. Sci. Math. Sci. **7**, 282–288 (1994)
9. Lin, Y.: Minimum bandwidth problem for embedding graphs in cycles. Networks **29**(3), 135–140 (1997)
10. Martí, R., Campos, V., Piñana, E.: A branch and bound algorithm for the matrix bandwidth minimization. Eur. J. Oper. Res. **186**(2), 513–528 (2008)
11. Pop, P., Matei, O., Comes, C.A.: Reducing the bandwidth of a sparse matrix with a genetic algorithm. Optimization **63**(12), 1851–1876 (2014)
12. Ren, J., Hao, J., Rodriguez-Tello, E.: An iterated three-phase search approach for solving the cyclic bandwidth problem. IEEE Access **7**, 98436–98452 (2019)
13. Ren, J., Hao, J.K., Rodriguez-Tello, E., Li, L., He, K.: A new iterated local search algorithm for the cyclic bandwidth problem. Knowl.-Based Syst. **203**, 106136 (2020)
14. Rodriguez-Tello, E., Romero-Monsivais, H., Ramírez-Torres, G., Lardeux, F.: Tabu search for the cyclic bandwidth problem. Comput. Oper. Res. **57**, 17–32 (2015)
15. Zhou, S.: Bounding the bandwidths for graphs. Theoret. Comput. Sci. **249**(2), 357–368 (2000)

Co-evolution of Knowledge Diffusion and Absorption: A Simulation-Based Analysis

Kashif Zia[1]([⊠]), Umar Farooq[2], Sanad Al-Maskari[1], and Muhammad Shafi[1]

[1] Sohar University, Sohar, Oman
kzia@su.edu.om
[2] Department of Computer Science, University of Science and Technology Bannu,
Bannu, Khyber Pakhtunkhwa, Pakistan

Abstract. The paper utilizes agent-based simulations to study diffusion and absorption of knowledge. The causal relation of diffusion on absorption is established in order. The process of diffusion and absorption of knowledge is governed by network structure and the dynamics of the recurring influence, conceptualized and modeled as legitimacy, credibility, and strategic complementarity; again a causal relation between the three in order. If not stationary, the agents can also move to acquire either random walk or profile-based mobility modes. Therefore, the co-evolution of network structure due to the mobility of the agents and the dynamics of the recurring influence of ever-changing neighborhood is also modeled. The simulation results reveal that – (i) higher thresholds for legitimacy and credibility determine slower, (ii) higher number of early adopters results into faster, and (iii) a scheduled and repeated mobility (the profile-based mobility) results in faster – absorption of knowledge.

Keywords: Knowledge diffusion · Knowledge adoption · Network structure · Recurring influence · Agent-based model

1 Introduction

Humans tend to be similar to their peers and get influence from others [5]. In many contexts, such as innovation adoption [25] and collective action [22], these tendencies become decisive. For example, rapid diffusion of information is often associated with innovation adoption. However, a more thoughtful understanding of innovation adoption, for example, is heavily dependent on its underlying dynamics in terms of (i) the evolution of the underlying network and (ii) the modalities of interaction among peers [4]. These two factors are relevant to any scenario related to *social influence*, in general.

Social influence plays an important role in a range of behavioral phenomena observed around us [30]. However, there are ongoing challenges in quantifying social influence due to a number of reasons. For instance, people tend to engage in the same activities as their peers, thus making it difficult to identify who was influenced by whom temporally. Similarly, two persons may be influenced by the same source but in a spatially incremental way, thus, making it difficult

© Springer Nature Switzerland AG 2021
M. Paszynski et al. (Eds.): ICCS 2021, LNCS 12742, pp. 570–584, 2021.
https://doi.org/10.1007/978-3-030-77961-0_46

to identify the exact source of information. Analysis of temporal and spatial causalities is an ongoing research area in social influence analysis [5].

The research on understanding social influence revolves around the effectiveness of the mechanisms of diffusion and absorption, the impact of social structures and populations, the impact of dynamics of the structures, dynamics of interactions between the agents, and the impact of agents' mobility – all related to underlying information contents. However, not going into the academic discussion of how knowledge can be differentiated from information, in this paper, we have termed this content as knowledge – some novel thing or stream or data introduced into the network [12].

In the presence of many inter-dependent, overlaying, and overlapping aspects (as hinted in the paragraph above) the challenge of understating social influence is real. In addition to that, there are some general considerations stated below:

1. The increased availability of social networking and interaction data has resulted in an exponential growth in related domains of science, however, social network analysis in general and knowledge diffusion/absorption in particular, are comparatively new disciplines.
2. Social networks comprise of two parts, namely, the network structure, and the interaction between peers. The network not only influences the interactions but also gets influenced by the interactions. Social network dynamics, therefore, not only occur from network to interactions but also from the interactions towards the network. However, most of the studies on knowledge diffusion/absorption focus on one way dynamics, mostly, analyzing the impact of network structure on knowledge diffusion.
3. Knowledge absorption – a truer essence of knowledge diffusion – and knowledge diffusion are, most of the time, considered analogous, which is not the case. Therefore, the differentiation between diffusion and absorption of knowledge should be clearly marked when dealing with social influences.

Elaborating the last point further, it is of great importance to differentiate between diffusion and absorption of knowledge. The term diffusion is used to indicate the availability of knowledge. If a peer p comes to know about a new thing from its networked peers, the knowledge is said to be successfully diffused to p. It is, however, irrelevant whether p is affected by the disseminated knowledge or not. When p is influenced by the knowledge, the knowledge is said to be truly absorbed and thus able to lead towards subsequent actions.

With a focus on the argument that "diffusion should not be considered sufficient for the absorption of knowledge", in this paper, we have tackled important modalities of social influence namely (i) the evolution of social structures and populations, (ii) the evolution of friendship structures and interactions, and (iii) agents able to acquire various mobility modes. The underpinning of diffusion and absorption mechanism is based on factors of "strategic complementarity", "credibility" and "legitimacy", motivated from Centola et al.'s model of complex contagion [9].

Using an agent-based model we have studied the conditions leading towards diffusion and absorption of knowledge. The main purpose of research is to see what emerges at a global scale due to localized interactions of the agents, while all these aspects were combined in a realistic and meaningful way. The simulation experiments revealed interesting results, which could not only validate a large body of domain-specific research results scattered all across social networking literature, but also provide an outlook of what to expect when looking at the things crossing across the borders.

The rest of the paper is structured as follows. Next section is about the background and motivation for this work. It not only introduces the diffusion and absorption models but also enlists the contribution of this work. In the following section, the conceptual model of the proposed work is presented, proceeded by a detailed description of the agent-based model. Then the simulation results are provided. The last section concludes the paper.

2 Background and Motivation

Research done on social influence particularly in terms of social networking can easily be divided into research which is performed on social structures and inter-action dynamics and research on the models of diffusion and absorption of the knowledge. We present related work in both domains. In this section, the terms information and innovation should be considered synonym to knowledge. Similarly, the differentiation between dissemination and diffusion should be considered a synonym to the differentiation between adoption and absorption.

2.1 Networking Structure and Interactions

Social networking [15] is considered as one of the best means of knowledge dissemination these days and considered as the best platform to investigate the mechanisms of information and innovation diffusion. One of the most important factor in this regards is networking structure [9]. One important networking features in this context is the nature of ties in the network. The understanding about "the strength of weak ties" is an established fact now, given by Granovetter [14] as: "whatever is to be diffused can reach a larger number of people, and traverse a greater social distance, when passed through weak ties rather than strong". Conversely, the notion of "the weakness of strong ties" is also true resulting in localization of information diffusion, due to propagation of information only in a closely knit network. It means that strong relational ties are structurally weak, and vice versa, where relational ties are individual social ties and a structural tie represents its ability of propagating information. Consequently, information spreads more rapidly in a small-world network structure in which a few long ties augment mostly tightly-knit local communities [30]. Also, people who interact more often have greater tendency to influence each other [16]. On the other hand, people who interact infrequently could have more diverse social networks, resulting in providing novel information [8].

Contrarily, with a more broader focus, authors in [24] defined "knowledge network", and emphasized that social relationships and the network relationships play a vital role in knowledge creation, diffusion, and absorption. Furthermore, authors in [20] argued that strong interpersonal ties are more effective than weak ties to enhance knowledge transfer and learning. Their thesis is that strong ties help to establish trust, which increases awareness to access each other's knowledge. When it comes to knowledge creation, weak ties allows access to disconnected or distinct partners, and results in diverse information and have a positive effect on creativity [23]. Authors in [13] discuss how strong ties relate to job finding on Facebook's social network. Several other examples of research on the influence of tie strength on information dissemination can be listed [6,17, 26,28,29].

Another important network structure is that a network having a power-law structure is most conducive for information dissemination. If the network is following the power-law structure, the authors in [21] have proved that it would result in disseminating information to "a large number of nodes in a small amount of time is possible". Also in such networks, "large scale dissemination can be achieved with simple resend rules (i.e., they do not require sophisticated centralized planning)". Moreover, node centrality has been seen as pivotal to maximize the information dissemination in a social network [18]. But this is typically true for social networks which are generally scale-free networks. Also, there are noted aberrations to these basic principles. For example, authors in [27] discuss that not the centre but the periphery of the network have a decisive role in spontaneous collective action. This example separates purely network-based dissemination from physical activity, thus, evidencing the fact that mere dissemination/diffusion cannot guarantee innovation adoption/absorption.

From the above related work, we conclude that social influence is directly related to contagion diffusion and absorption. In terms of network structure, social influence is mostly about type and modality of connectivity. However, there are contradictory views on that. In one case, a special kind of network such as a scale-free network supports quick diffusion due to central nodes. In the second case, even in a scale-free network, peripheral nodes are more important, indicating the importance of real-life implications [31] when the knowledge is *absorbed*. As we have already mentioned that knowledge diffusion should be differentiated from knowledge absorption. Therefore, we opted to use a regular network (for the diffusion of knowledge) to handle this confusion. Further, in our model, the knowledge absorption mechanism builds on diffused knowledge and considers tie strength as a decision parameter.

2.2 Diffusion and Absorption of Knowledge

Knowledge dissemination happens through diffusion models. At a very basic level, it can be a Susceptible-Infectious (SI) model [3]. SI and extended models use a threshold-based mechanism where a node becomes infected if a designated fraction of the neighborhood (the threshold) is already infected. We have used this basic principle to enrich our diffusion model.

Quantifying the timing of interaction and recurrence [10] is another important aspect of knowledge diffusion. For example, authors [2] provide a general discussion on the impact of social networks on human behavior. According to [4], "the propagation of adoption depends on several factors from the frequency of contacts to burstiness and timing correlations of contact sequences. More specifically, burstiness is seen to suppress cascades sizes when compared to randomised contact timings, while timing correlations between contacts on adjacent links facilitate cascades." Similarly, authors in [19] credit two factors, 1) When did someone in your friends adopted an innovation and 2) The number of exposures, but they discredit personal traits, such as number of friends (followers) and date of joining the network.

However, knowledge diffusion alone is not enough to make people act and bring a change – the situation termed as absorption. Taking knowledge absorption as an extension of diffusion, we enrich our diffusion model based on findings reported in [4]: "what drives the adoption of a node is the number of recent contacts from adopted individuals, such that multiple contacts from the same adopted individual have the same effect as the same number of contacts from multiple adopted sources."

Like the above, the absorption of knowledge is considered as an extension of diffusion in most cases, but it can be considered as a mechanism having its own dynamics. For example, a thesis advocating it and in other words contradicting the notion of the "strength of weak/long ties" differentiates between mere dissemination and potentially a more demanding collective action. The seminal work is from Centola and Macy [9]. The authors postulate that "network structures that are highly efficient for the rapid dissemination of information are often not conducive to the diffusion of collective action based on the information". Authors in [9] also provide a more discrete specification capturing the soul of the argument as: "The "strength of weak ties" applies to the spread of information and disease but not too many types of social diffusion which depend on influence from prior adopters, such as participation in collective action, the use of costly innovations, or compliance with emergent norms. For these contagions, we contend that long ties are not strong in either of Granovetter's meanings, relational or structural."

Information and diseases are simple contagions requiring only one source to spread. Complex contagions require two or more sources of activation. According to Centola and Macy [9], four factors contribute to a complex contagion: (i) late adopters waiting for early adopters, termed as *strategic complementarity*, (ii) *credibility* provided by neighbors who have already adopted an innovation, (iii) *legitimacy* provided by close friends who have already adopted an innovation, and (iv) "expressive and symbolic impulses in human behavior that can be communicated and amplified in spatially and socially concentrated gatherings" [11] termed as *emotional contagion*. Further, they define "a contagion as *uncontested* if activation depends solely on the number of neighbors who are activated, without regard to the number who are not activated." Whereas, a contagion is *contested* if it also depends on persons who are not activated. The implications of this, according to them, are: in case of uncontested contagions, "The larger the number of neighbors, the greater the chance of becoming activated", and in case of contested contagions, "the more neighbors an actor has, the lower the susceptibility to activation". Examples of complex contagion are the spread of participation in collective action and norms and social movements. Naturally, these usually fall into the category of contested contagions. We adopt this model (with some refinements) as our basic absorption model typically in an uncontested environment.

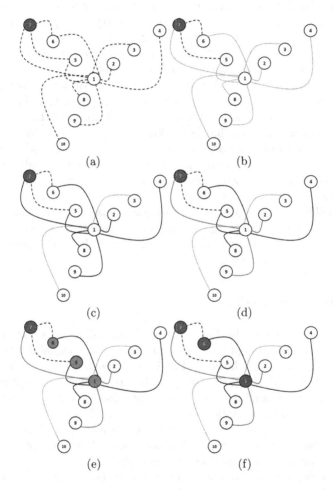

Fig. 1. An overview of the proposed model.

2.3   Contribution of the Paper

Most of the models detailed above are analytical in nature. However, there is a lot of potential in analyzing the models using a bottom-up approach, thus, providing an opportunity to have a behavioral-based implementation at an individual level. Agent-Based Modeling (ABM) [7] provides an approach to model a population at an individual level, with detailed temporal and spatial resolution, including the stochasticity of interactions and mobility. Therefore, through this work, we intend to enrich a relatively thin body of research done in knowledge diffusion and absorption modeling in ABM domain.

In our earlier work [1], we have shown that the late absorbents are affected by the early absorbents, but only when the mobility model is closer to human mobility (a planned, scheduled, and repeated mobility). Early absorbents do not affect late absorbents if all agents are stationary or acquiring random walk mobility. Also, with an increase in the percentage of early absorbents, the number of final absorbents increases. All the other varying factors, such as interaction radius, threshold values, etc. do not have a substantial impact.

But this work is different from our earlier work. In our earlier model, the phases of diffusion and absorption were not distinct. In fact, knowledge diffused to an agent was considered sufficient for comparison. In this paper, we have introduced a formal mechanism of knowledge diffusion, which was lacking before. Hence, we can differentiate between diffusion and absorption quantitatively. More specifically, in this paper, we provide a thorough study about conditions leading to knowledge diffusion and absorption in a proximity-based regular network of agents. The study also intends to quantify the relationship between credibility with legitimacy in time and spatial domains and its relationship to early vs. late adopters. Lastly, and most importantly, knowledge diffusion and absorption in an environment of dynamically evolving tie strengths due to agents' mobility is also studied.

3   Conceptual Model

The model is constituted by four sequential phases, followed by an optional mobility phase. The conceptual model for this work is explained using a regular network of 10 nodes shown in Fig. 1. It explains the interaction, consolidation, diffusion, and absorption phases, thus providing a conceptual outlook of the dynamically changing network structure due to continuous interactions. The interaction and consolidation imitate the strength of ties, where the friends are the strongest, contacts being the weak while mere connectivity means no tie. Then, subsequent processes of diffusion followed by absorption (based on principles of complex contagion) are conceptualized.

We use nodes labelled from one (1) to ten (10). Dotted lines are representing the connections between the nodes. Green color nodes are the nodes that have absorbed the knowledge. The initial setup is presented in Fig. 1(a), in which only node seven (7) has absorbed the knowledge. All the nodes, in parallel, invoke interaction, consolidation, diffusion and absorption processes in order.

Interact: Taking the example of node one (1), Fig. 1(b) illustrates the interaction process, which allows a node to interact with all its neighboring nodes. For example, node one (1) has nine (9) neighbors to interact with (gray color lines represent this).

Consolidate: The consolidate process is split into two parts namely, determining the *contact* nodes among the interacting nodes and then determining the *friend* nodes among those who have become contact nodes. These two steps use 'k' and 'm' thresholds to determine the contact and friend nodes respectively, where 'k' and 'm' relate tie strengths to the frequency of interactions. For example, six (6) out of nine (9) neighbors for node one (1) become contact nodes (represented using dark lines), as illustrated in Fig. 1(c). Finally, Fig. 1(d) shows that three (3) out of six (6) contact nodes became friends (represented as red lines).

Diffuse: Figure 1(e) shows the number of nodes that receive the knowledge diffused. In this example, suppose to three nodes named one(1), five (5), and six (6), the knowledge has been diffused.

Absorb: The knowledge absorption depends on the credibility and legitimacy of the connected nodes. Credibility is concerned with all the connected nodes and legitimacy is concerned with the friends only. Both thresholds should be fulfilled in order for knowledge to be absorbed. In this example, the thresholds of credibility and legitimacy are 1/9 and 1/3 correspondingly. That is why two nodes (namely one (1) and six (6)) absorbed the knowledge, as shown in Fig. 1(f).

This is, however, a static view of the proceeding, not truly capturing the dynamics of the model. In the case of mobility, some connectivity may not transform into a tie. However, some ties may become strong and vice versa. The model runs at each iteration and for all the agents.

4   Agent-Based Model

The ABM of the above conceptual model operates in a 2D space comprising a grid of cells. The agents reside on top of the cells and can move on the grid (if required). The simulation runs in discrete time, each time unit termed as an iteration. At each iteration, each agent in a population of agents acts according to the model specification given below. Agents perform calculations and act in a sequence, where the order of the sequence is randomly shuffled for each iteration, thus, maintaining fairness between agents. We distribute the model into four sequential procedures.

Interact. Within a radius **r**, an agent updates its neighborhood; the *neighbor_list*. In case of no mobility (stationary case), nothing would change in the data associated with a neighbor; its identity, and its corresponding discovery frequency. But, if the agents are moving (random walk or profile-based mobility cases), the discovery frequency serves as a number to distinguish between more frequent and less frequent neighbors. The agents which were neighbors of an agent at time $t - 1$ and are not neighbors anymore (at current time t) remain in the list of neighbors and do not get deleted. In this way, we have complete interaction history of all the agents.

Consolidate. A fraction k of agents which are in *neighbor_list* and *still* neighbors of an agent would be added into a *contact_list* of an agent. Along with contact identity, the associated data stored is contact-making time (current time). Again, the old contacts still remain and we do not delete them. Thus we have a list of contacts, the latest identifiable by time at which they were converted from neighbors to contacts.

A fraction m of agents which are in the *contact_list* and *still* neighbors of an agent would be added to the *friend_list* of an agent. Along with the identity of a friend, the associated data stored is friend-making time (current time). Yet again, the old friends still remain and we do not delete them. Thus we have a list of friends, the latest identifiable by time at which they were converted from contacts into friends.

Diffusion. An agent is considered *wobbling* if it has received new knowledge. If an agent is not wobbling, the new knowledge is not diffused to it. Hence, with probability p, the agent which is not wobbling would start wobbling based on the following equation:

$$p = c \times N \times \tau \times (W_t/N) \times ((N - W_t)/N)) \tag{1}$$

where N is the count of neighbors of an agent and W_t is the count of agents in the neighborhood of an agents who are wobbling. And c and τ are two constants acting as sensitivity parameters (both set to 0.5). With a probability of p (equal to 0.5 again) – if a random float is less than the value of p – the current agent starts wobbling itself. Wobbling equates to diffused influence, which does not guarantee knowledge absorption.

Absorption. This applies to all the agents who did not absorb yet but are able to wobble. The measure of *credibility* is calculated as:

$$credibility = N_A/N \tag{2}$$

where N_A is the count of the agents in the neighborhood who have already absorbed the knowledge and N is the total number of neighborhood agents. Next, the friends who have already absorbed the knowledge are counted, designated by F_A. Then the measure *legitimacy* is calculated as:

$$legitimacy = F_A/N_F \tag{3}$$

where N_F is the total number of friends. Finally, if *credibility* and *legitimacy* is greater than $th1$ and $th2$ respectively, the agent is considered to have absorbed the knowledge.

Mobility. There are three **mobility** modes under which the whole mechanism operates.

1. Mobility 1: No mobility – all agents are stationary.
2. Mobility 2: Random walk – agents choose a direction to move randomly at each iteration.
3. Mobility 3: Profile-based walk – agents build some random locations to move to, and they move from one location to another. This equates to a planned, scheduled, and repeated mobility.

5 Simulation

5.1 Simulation Setup

The simulation world consists of a grid of size 100×100. This equals 10000 cells, each having a unique xy-coordinate. An agent population equal to $x\%$ was generated for a simulation run, for example, if $x = 25\%$, the population equals 2500 agents. An initial population of $y\%$ of these agents is considered to have absorbed the knowledge already – referred to as *early adopters*. If $y = 5\%$, 125 agents out of 2500 would be early adopters.

Towards the analysis of the simulation model, we have focused on three parameters:

1. **threshold 1 (th1):** the threshold (the percentage of agents in the contact list) to measure the credibility of the information being disseminated. Values are 0.1 (10%), 0.2 (20%), and 0.3 (30%).
2. **threshold 2 (th2):** the threshold (the percentage of agents in the friend list) to measure the legitimacy of the information being disseminated. Values are 0.1 (10%), 0.2 (20%) and 0.3 (30%).
3. **mobility mode:** stationary, random walk, and profile-based walk.

All of the above three parameters are permuted to form different cases representing all possible combinations. Each of these cases is executed under two "aggregation" strategies. In the **basic** strategy, the two thresholds (stated above) are used without any change. Whereas, in the **local** strategy, both thresholds are normalized according to their relative difference. The other parameters such as radius (range of influence of an agent), population size and density, the percentage of early adopters, and values like k and m are kept constant.

5.2 Simulation Results

Lower threshold values provide less resistance to adoption as a result of the achievement of credibility and legitimacy easily. For example, if both th1 and th2 are set to 0.1, it only requires 10% of the contacts and 10% of the friends who have adopted for an adoption to occur. It should be possible in all cases and should not take that much time. This is what is apparent in Fig. 2 (a). The adoption occurs almost right after diffusion (wobbling) and it happens very early in the simulation. Further, it is noted that changing mobility mode does not affect this pattern at all. That is also valid for adoption strategies. So, the graph shown in Fig. 2 (a) is for all possible mobility modes and for both "aggregation for adoption" strategies. Further, it only takes a few iterations before the whole population of agents has adopted. A simulation view representing this is shown in Fig. 3 (I) in the case of random walk mobility and basic adoption strategy.

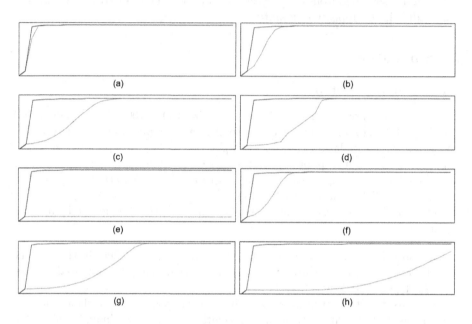

Fig. 2. X-axis of the graph shows time (simulation iteration) and y-axis shows the number of agents. Blue line represents the wobbling agents and green line represents agents who have adopted. (Color figure online)

As the value of th1 increases, the resistance to adoption increases, typically to gain credibility. However, if the value of th1 still supports credibility, the adoption would most certainly happen (in many initial configurations), however, it will be delayed. This is what is apparent in Fig. 2 (b). The adoption occurs but late. However, it is not that late if th1 values are increased further, that is to 0.3 (compare Fig. 2 (b) with Fig. 2 (c)). Again, changing mobility and adoption strategy do not affect this pattern at all. A simulation view representing this is shown in Fig. 3 (II) in the case of random walk mobility and basic adoption strategy (th1 = 0.3, th2 = 0.1).

In case of th1 = 0.3 and th2 = 0.1, and in profile-based mobility, a very different pattern emerges. Analyzing Fig. 2 (d), it is apparent that the adoption happens in phases. A simulation view representing this is shown in Fig. 3 (III). Also, the adoption depends on the relative positioning of the early adopters. In many random simulation setting, the adoption would not happen at all if th1 = 0.2 or more (see graph shown in Fig. 3 (III)). But, instead of applying quantitative analysis, we applied qualitative analysis unless we have an educated

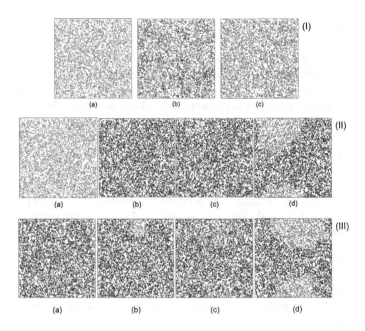

Fig. 3. Blue agents represent the wobbling agents and green agents represent agents who have adopted. (I) at t= 0, (b) at t= 1, and (c) at t= 2. (II) at t= 0, (b) at t= 1, (c) at t= 2, and at t = 10. (III)) at t= 1, (b) at t= 3, (c) at t= 5, and (d) at t = 7. (Color figure online)

knowledge of where to put the early adopters and what should be their relative positioning, which remains as future work.

These results show that just wobbling (diffusion) in fact happens very quickly in all the cases. However, the absorption may happen late or never, based on the values of thresholds for achieving credibility and legitimacy. Hence, there is a clear-cut difference between diffusion and absorption (although related) and these two aspects should not be treated equally. A lower threshold for legitimacy (th2) would let the agents acquire the absorption sooner or later depending on the threshold for credibility (th2) – lower threshold, sooner, and higher threshold, later. Stationary and random walk behave exactly the same, whereas, profile-based mobility (typically with high th1) produces absorption in phases of increasing intensity.

As th2 is increased, the resistance to absorb would increase, and generally, it would take more time to absorb. A comparison between Fig. 2 (b) (th1 = 0.2, th2 = 0.1) and Fig. 2 (f) (th1 = 0.2, th2 = 0.2), and Fig. 2 (e) (th1 = 0.3, th2 = 0.1) and Fig. 2 (g) (th1 = 0.2, th2 = 0.3) is sufficient to establish this fact. Finally, another case is when both th1 and th2 are quite high. Occasionally, it would generate absorption as shown in Fig. 2 (h), but most of the time, there would be no absorption, similar to the graph shown in Fig. 2 (e).

6   Conclusion

This paper presents a framework integrating the models for information dissemination/diffusion and adoption/absorption using an agent-based modeling paradigm. In particular, we provide an application of the Centola and Macy's information/innovation dissemination and adoption model [9] in a realistic setting. Sub-models of discrete spatial configuration (a grid of cells) and of proximity-based networking are integrated with an agent-based specification of the innovation adoption. Consequently, a thorough study about conditions leading to innovation dissemination and adoption is presented. Additionally, we quantify the relationship of late vs. early adopters in different conditions. The study also intends to quantify the relationship between credibility with legitimacy in time and spatial domains. Lastly, and most importantly, innovation dissemination in different mobility modes is studied in a proximity-based regular network.

The study revealed the following qualitative results:

- The proximity and strong ties between people in a proximity-based regular networks play an important role in dissemination and ultimate adoption of information.
- This dictates that the social interactions among individuals are a key factor for the disseminating and adaptation of information in a society.
- It was witnessed that as the number of early adopters was increased, it convinced more late adopters to adopt information.
- To start with less number of adopters, it will take more time to disseminate and ultimately convince people to adopt an innovation.
- Late adopters are, however, influenced by early adopters only when the latter category people had planned, scheduled, and repeated interaction with the former category.

Comparison with real experiments on social group and evaluation with real datasets such as social networks, which could greatly improve the proposed work, will be taken up as future work.

References

1. Al-Maskari, S., Zia, K., Muhammad, A., Saini, D.K.: Impact of mobility mode on innovation dissemination: an agent-based simulation modeling. In: Manoonpong, P., Larsen, J.C., Xiong, X., Hallam, J., Triesch, J. (eds.) SAB 2018. LNCS (LNAI), vol. 10994, pp. 3–14. Springer, Cham (2018). https://doi.org/10.1007/978-3-319-97628-0_1
2. Althoff, T., Jindal, P., Leskovec, J.: Online actions with offline impact: How online social networks influence online and offline user behavior. In: Proceedings of the Tenth ACM International Conference on Web Search and Data Mining, pp. 537–546. ACM (2017)
3. Anderson, R.A., et al.: Transmission dynamics and epidemiology of BSE in british cattle. Nature **382**(6594), 779 (1996)

4. Backlund, V.P., Saramäki, J., Pan, R.K.: Effects of temporal correlations on cascades: threshold models on temporal networks. Phys. Rev. E **89**(6), 062815 (2014)
5. Bakshy, E., Rosenn, I., Marlow, C., Adamic, L.: The role of social networks in information diffusion. In: Proceedings of the 21st International Conference on World Wide Web, pp. 519–528. ACM (2012)
6. Bapna, R., Gupta, A., Rice, S., Sundararajan, A.: Trust and the strength of ties in online social networks: an exploratory field experiment. MIS Q. Manage. Inf. Syst. **41**(1), 115–130 (2017)
7. Bonabeau, E.: Agent-based modeling: methods and techniques for simulating human systems. Proc. Nat. Acad. Sci. **99**(suppl 3), 7280–7287 (2002)
8. Burt, R.S.: Structural holes: the social structure of competition. Harvard University Press (2009)
9. Centola, D., Macy, M.: Complex contagions and the weakness of long ties. Am. J. Soc. **113**(3), 702–734 (2007)
10. Cheng, J., Adamic, L.A., Kleinberg, J.M., Leskovec, J.: Do cascades recur? In: Proceedings of the 25th International Conference on World Wide Web, pp. 671–681. International World Wide Web Conferences Steering Committee (2016)
11. Collins, R.: Emotional energy as the common denominator of rational action. Rationality Soc. **5**(2), 203–230 (1993)
12. Cowan, R., Jonard, N.: Network structure and the diffusion of knowledge. J. Econ. Dyn. Control **28**(8), 1557–1575 (2004)
13. Gee, L.K., Jones, J., Burke, M.: Social networks and labor markets: how strong ties relate to job finding on facebook's social network. J. Labor Econ. **35**(2), 485–518 (2017)
14. Granovetter, M.S.: The strength of weak ties. Am. J. Sociol. **78**(6), 1360–1380 (1973)
15. Greenwood, S., Perrin, A., Duggan, M.: Social media update 2016. Pew Res. Center **11**, 83 (2016)
16. Hill, S., Provost, F., Volinsky, C.: Network-based marketing: Identifying likely adopters via consumer networks. Stat. Sci. **21**(2), 256–276 (2006)
17. Jia, P., MirTabatabaei, A., Friedkin, N.E., Bullo, F.: Opinion dynamics and the evolution of social power in influence networks. SIAM Rev. **57**(3), 367–397 (2015)
18. Kandhway, K., Kuri, J.: Using node centrality and optimal control to maximize information diffusion in social networks. IEEE Trans. Syst. Man Cybern. Syst. **47**(7), 1099–1110 (2017)
19. Kooti, F., Mason, W.A., Gummadi, K.P., Cha, M.: Predicting emerging social conventions in online social networks. In: Proceedings of the 21st ACM International Conference on Information and Knowledge Management, pp. 445–454. ACM (2012)
20. Levin, D.Z., Cross, R.: The strength of weak ties you can trust: the mediating role of trust in effective knowledge transfer. Manage. Sci. **50**(11), 1477–1490 (2004)
21. Mishori, R., Singh, L.O., Levy, B., Newport, C.: Mapping physician twitter networks: describing how they work as a first step in understanding connectivity, information flow, and message diffusion. J. Med. Internet Res. **16**(4), e107 (2014)
22. Olson, M.: The Logic of Collective Action: Public Goods and the Theory of Groups, Second Printing with New Preface and Appendix, vol. 124. Harvard University Press, Cambridge (2009)
23. Perry-Smith, J.E.: Social yet creative: the role of social relationships in facilitating individual creativity. Acad. Manage. J. **49**(1), 85–101 (2006)
24. Phelps, C., Heidl, R., Wadhwa, A.: Knowledge, networks, and knowledge networks: a review and research agenda. J. Manage. **38**(4), 1115–1166 (2012)

25. Rogers, E.M., Shoemaker, F.F.: Communication of innovations; a cross-cultural approach (1971)
26. Shi, Z., Rui, H., Whinston, A.B.: Content sharing in a social broadcasting environment: evidence from twitter. MIS Q. **38**(1), 123–142 (2014)
27. Steinert-Threlkeld, Z.C.: Spontaneous collective action: peripheral mobilization during the Arab spring. Am. Polit. Sci. Rev. **111**(2), 379–403 (2017)
28. Wang, Y., Wu, J.: Social-tie-based information dissemination in mobile opportunistic social networks. In: 2013 IEEE 14th International Symposium and Workshops on a World of Wireless, Mobile and Multimedia Networks (WoWMoM), pp. 1–6. IEEE (2013)
29. Watts, D.J., Dodds, P.S.: Influentials, networks, and public opinion formation. J. Consum. Res. **34**(4), 441–458 (2007)
30. Watts, D.J., Strogatz, S.H.: Collective dynamics of 'small-world' networks. Nature **393**(6684), 440–442 (1998)
31. Yang, Yu., Chen, E., Liu, Q., Xiang, B., Xu, T., Shad, S.A.: On approximation of real-world influence spread. In: Flach, P.A., De Bie, T., Cristianini, N. (eds.) ECML PKDD 2012. LNCS (LNAI), vol. 7524, pp. 548–564. Springer, Heidelberg (2012). https://doi.org/10.1007/978-3-642-33486-3_35

Estimation of Road Lighting Power Efficiency Using Graph-Controlled Spatial Data Interpretation

Sebastian Ernst[✉] and Leszek Kotulski

Department of Applied Computer Science, AGH University of Science
and Technology, Al. Mickiewicza 30, 30 -059 Kraków, Poland
{ernst,kotulski}@agh.edu.pl

Abstract. Estimation of street lighting energy requirements is a task crucial for both investment planning and efficiency evaluation of retrofit projects. However, this task is time-consuming and infeasible when performed by hand. This paper proposes an approach based on analysis of the publicly available map data. To assure the integrity of this process and automate it, a new type of graph transformations (Spatially Triggered Graph Transformations) is defined. The result is a semantic description of each lighting situation. The descriptions, in turn, are used to estimate the power necessary to fulfil the European lighting standard requirements, using pre-computed configurations stored in a 'big data' structure.

Keywords: Spatially-triggered graph transformations · Big data · Lighting system · Spatial data analysis

1 Introduction

Design of street lighting installations is a compromise between the safety of road users and the operational costs. These costs are proportional to the amount of consumed energy, which directly translates into the amount of CO_2 emitted during its production. Recent studies also point out an additional (and unmeasurable) cost: the negative impact of light pollution during the night on humans and other organisms [6].

In Europe, the requirements for road and street lighting are set by the EN 13201 standard [5], and there are several software tools used to verify designs (prepared by a human designer) in that respect, such as Dialux[1] or Relux[2]. Unfortunately, it is not feasible for human designers to consider all possible configurations for a given set of street parameters, which leads to simplifications, such as averaging of values to reduce the number of cases to be considered. For instance, industry-standard designs do not take slight variations in pole spacing

---

[1] https://www.dial.de/en/dialux/.
[2] https://relux.com/en/.

© Springer Nature Switzerland AG 2021
M. Paszynski et al. (Eds.): ICCS 2021, LNCS 12742, pp. 585–598, 2021.
https://doi.org/10.1007/978-3-030-77961-0_47

(e.g.: 33, 34, 32, 35 m) into consideration, but instead use one, averaged value (e.g. 33 m). The result of such an approach is increased energy consumption.

Let us consider the Polish market. In projects supporting lighting retrofit, the National Fund for Environmental Protection and Water Management requires that the energy consumption is reduced by at least 50%. A designer is able to provide such results using the aforementioned traditional design tools. However, there is room for improvement. Automatic photometric optimisation systems [10, 11], based on artificial intelligence solutions, can provide designs with energy consumption reduction of 70% to 80%. This has been practically proven in several lighting retrofit designs for cities of Tbilisi (100,000 lighting points), Washington (54,000 lighting points) and numerous Polish cities.

This additional 30% is very important from the ROI (return on investment) point of view. However, direct inclusion of consumption reduction requirements in public tender requirements is controversial. Since this value is always calculated using the power consumption of the initial installation, it is susceptible to inadequacies of that original configuration: if the road was underlit (and the EN 13201 was not fulfilled, which means new poles may need to be installed), the reduction factor will be lower (or there will be none at all); if it was overlit, this value may be much higher.

Thus, it is essential to provide city officials with means of obtaining a realistic estimate of the power required by a well-designed lighting installation and basing their requirements on that. However, executing a full design cycle during the project preparation phase is infeasible. Moreover, the photometric calculations alone may take days, even for an automated tool.

Therefore, we suggest that this estimation is carried out based on the analysis of publicly available map data. This paper presents a methodology that allows for such analysis and allows for quick estimation of the requirements based on pre-computed designs.

The structure of the paper is as follows. In Sect. 2, we characterise the factors which influence the design of street lighting infrastructures, and, as a result, their energy requirements. Section 3 describes the concept of spatially-triggered graph transformations (STGT), a mechanism used to identify spatial relationships among map objects and express them in a formal graph structure. In the following Sect. 4, we propose a data warehouse-like approach to store the results of photometric calculations and use them for quick estimation of the energy requirements of outdoor lighting. Finally, Sect. 5 provides concludes the presented work and provides an insight into works planned in the future.

2   Determinants of Street Lighting Design and Efficiency

The EN 13201 standard defines the rules assigning so-called *lighting classes* to individual lit areas [4], and specifies the required lighting parameter values (luminance, uniformity, etc.) for each class [5]. To fulfil these requirements in each individual street or sidewalk, many factors must be taken into consideration, including parameters of the area itself (e.g. its width), placement of the lamp-posts (e.g. spacing) and parameters of the infrastructure itself (e.g. pole height, arm length, fixture type, etc.).

Figure 1 provides a visual representation of some of these parameters, and their complete list is presented in Table 1. The rightmost column presents the maximum reasonable number of variants (granulation) for each parameter. Combinatorially, this leads to more than 2.75×10^{22} situations for one design (1.6×10^{21} is the product of the values; each combination shall be analysed for each surface type). For a large city, this may result in a need to prepare several hundred different designs, which is time-consuming. A possible solution would be to pre-calculate these values; however, with this many combinations, both the calculations and the storage of results seem infeasible. Fortunately, the number of possible variants can be reduced by introducing formal relations among these parameters, as presented in Sect. 4.

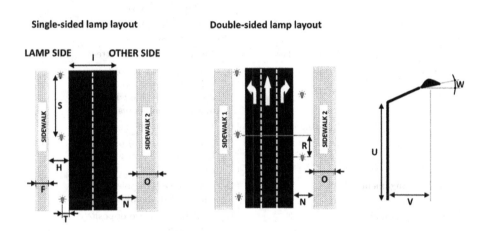

Fig. 1. Street situations

3 Modelling and Interpretation of Spatial Relationships

The first, essential step in estimating the power needs of an area related to street lighting is identification and interpretation of the lighting situations, e.g. the values of parameters regarding the objects illuminated by a given lamp or series of lamps. As stated in the previous section, we aim at extracting as much information as possible using publicly available road maps, such as OpenStreetMap[3].

The street infrastructure (e.g. roads, sidewalks, cycle lanes, etc.) is represented in such sources as lines, described by the following attributes:

- the *spatial* component, e.g. the shape of the line,
- the *attribute* component, describing, among others:

---

[3] https://www.openstreetmap.org.

Table 1. Parameters of the design designation

Variable label			Description	
Number of fixtures		E	Number of fixtures installed on the pole	2
Sidewalks (lamp side)	Width	F	Width (in meters) of the sidewalk on the street side where an analyzed lamp is installed	5
	Lighting class	G	Lighting class of a sidewalk	12
	Distance from street	H	Width (in meters) of the gap between the street and the sidewalk	6
STREET	Width	I	Street width (in meters)	26
	Number of lanes	J	Number of street lanes	5
	Surface type	K	Street surface type: from RTAB1 to RTAB16	16
	Q0	L	Value of the q0 factor for the street surface	10
	Lighting class	M	Lighting class of the street	12
Sidewalks (lamp side)	Width	N	Width (in meters) of the sidewalk on the street side opposite to a side where an analyzed lamp is installed	6
	Lighting class	O	Lighting class of the sidewalk	12
	Distance from street	P	Width (in meters) of a gap between the street and the sidewalk	6
Lamp arrangament		Q	Arrangement of poles/fixtures: single-sided, double-sided	2
Shift in fixture alligment		R	Shift between two opposite lamp rows	30
Distance between pools		S	Average distance (in meters) between poles	60
Pole distance from street		T	Distance (in meters) from the pole to street (setback)	6
Pole hight		U	Pole height (in meters)	10
Arm lenght		V	Arm length (in meters)	6
Tilt angle		W	Fixture tilt angle (in degrees)	10
Fixture type		X	Fixture model	40
LDT file		Y	Name of a file containing a light distribution matrix	10
Fixture power		Z	Fixture power (in watts)	20

- object *type*: road (including road class), walkway (sidewalk, path, steps), cycle lane, etc.; this attribute is always provided,

- *number of lanes*; this is provided optionally, but can be inferred from the road type in most cases,
- *width* in meters; also provided optionally, but can be inferred from the number of lanes and road type.

However, the relationships between individual objects are usually absent in the source data; these include:

– *continuity* – denoting if a certain fragment of a road is a continuation of another one; this defines the *road network* structure, which can usually be inferred from the map (by identifying common nodes), but may also be missing e.g. due to editing mistakes,
– *intersections* – a concept similar to continuity, denoting if two objects intersect; it may represent the intersection of two roads, but also an intersection of a walkway and a road, which can (and should) be interpreted as a pedestrian crossing,
– *distance* – the distance between the closest points of two spatial objects (road to road, road to building, lamppost to road, etc.),
– *parallelism*[4] – identification of objects which run parallel to each other; this relationship is *never* represented in an explicit way.

The last mentioned relationship, parallelism, is crucial in many applications, including street lighting. This is because the same infrastructure is often used to illuminate different objects:

– a single lamp (fixture) often illuminates a road and its sidewalks,
– a single pole may be used to host lamps illuminating dual carriageways, or a road and a walkway separated by a median island.

3.1 Graph Representation and Generation

Identification of spatial relationships between objects can be a time-consuming task, as it involves theanalysis of geometric shapes. Also, due to the huge number of possible combinations and the complexity of the process, such analysis may easily get out of hand when performed using only traditional tools, such as spatial databases (e.g. PostGIS[5]) or toolkits (e.g. GeoPandas[6]).

Because of this, we propose that the identified relationships are stored in a graph called *SRG* (*Spatial Relationship Graph*). After this phase, the 'raw' relationships will be further analysed to infer useful knowledge; in this case, the knowledge pertains to identified lighting situations, as described in Sect. 2.

---

[4] The use of this term may be confusing, at it is often used to describe parallel execution of logic on separate processors in concurrent applications. Please note that in this paper, it is always used to denote *geometric* parallelism of shapes.
[5] http://postgis.org.
[6] https://geopandas.org.

Spatially-Triggered Graph Transformations. For this purpose, we suggest the use of spatially-triggered graph transformations (STGT). The proposed mechanism is based on the formal approach supported by the well-known graph transformation structure [1,8].

In essence, this methodology maintains a graph, where the structure and/or attributes are modified when:

(i) a certain spatial relationship is identified in the raw data, or
(ii) a specialised, semantic relationship is inferred from a more general one.

Its main advantage is that it allows the analysis to be performed automatically at a large scale (e.g. an entire city), while maintaining a form that can be reviewed and, if necessary, altered by the operator (as they are explicitly expressed according to their semantics). One of the major drawbacks of the approach presented in [9] (which doesn't use a graph to coordinate the analytical process) is the complex representation of both the analytic algorithms and their intermediate results. The presented STGT-based approach remedies this, allowing for easy backtracking and improvement.

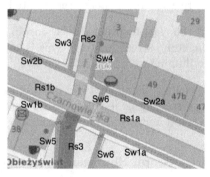

(a) Initial state of map data   (b) Estimated object shapes

Fig. 2. Fragment of a city map used for spatial analysis

A Simple Example. Let us provide a simple, intuitive example. Figure 2a presents the initial state of the map data, as downloaded from OpenStreetMap. To maintain clarity, we shall focus on the part to the east of the junction, e.g. on objects Rs_{1a}, Sw_{1a}, Sw_{1b} and Sw_6. Already at this stage, we may infer the following relationships:

1. Sw_{1a} is parallel to Rs_{1a},
2. Sw_{2a} is parallel to Rs_{1a},
3. Sw_6 intersects with Rs_{1a}.

Furthermore, having estimated the width of the individual objects, we may obtain their estimated shapes, as presented in Fig. 2b[7]. This lets us supplement the identified relationships with the following attributes:

1. The average distance between Sw_{1a} and Rs_{1a} is 0.5 m.
2. The average distance between Sw_{2a} and Rs_{1a} is 6 m.

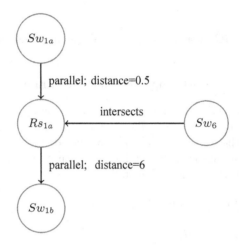

Fig. 3. Initial state of the SRG

Thus, the initial state of the SRG is presented in Fig. 3. Subsequently, the following analyses and transformations may be performed:

1. If two objects are parallel to one another and their distance is less than 1 m, we shall assume they are neighbouring.
2. If the distance is greater than 1 m, we shall assume there is a green area (that does not need to be illuminated) between them.
3. If a walkway intersects with a road, we shall assume there is a pedestrian crossing and derive its offset (in metres) along the road.

As a result, we arrive at the specialized form of the SRG, the SRG_L (spatial relationship graph for lighting purposes), presented in Fig. 4 This structure contains descriptions of all lighting situations in a form that allows them to be subjected to photometric calculations or, as described further in this paper, to look up the power values in a dedicated data structure.

---

[7] Please note that neither the algorithms used to detect the spatial relationships nor those used to estimate the road width are not presented here in detail. They are, however, a subject of individual research tracks; their results will be published in future papers.

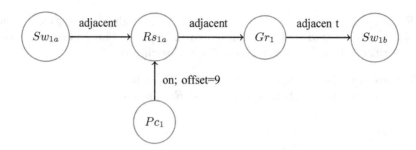

Fig. 4. The SRG_L graph, representing an identified lighting situation

3.2 Graph Generation Using a Formal Grammar

All of the transformations described at the intuitive level in Sect. 3.1 need to be performed in a controlled manner. The correctness of the proper graph generation is ensured by the graph transformation rules defined in a graph grammar. The grammar also allows for a hierarchical structure, as described later in this section. Such a grammar is defined as follows.

Definition 1. *A graph grammar Ω is a tuple:*

$$\Omega = (\Sigma_\Omega, \Gamma_\Omega, \Delta_\Omega, \Phi_\Omega, S_\Omega, \Pi_\Omega)$$

where:

- Σ_Ω *is the set of node labels,*
- $\Delta_\Omega \subset \Sigma_\Omega$, *is the set of terminal node labels,*
- Γ_Ω *is the set of edge labels,*
- Φ_Ω *is the set of transformation rules,*
- S_Ω *is the starting graph,*
- Π_Ω *is the graph grammar validation condition, that verifies the current state of the graph.*

We use Π_Ω as the validation condition of the graph grammar because sometimes it is necessary to execute a sequence of transformations and some of the intermediate states may not form a correct graph. In graph transformations, a common practice is to informally assume such a validation rule by introducing non-terminal nodes, as the final graph may not have any non-terminal nodes.

$\pi \in \Phi_\Omega$ is a transformation rule that transforms one graph into another one. π is denoted as a set of two graphs, lhs and rhs. For a given graph G, application of the π transformation rule is defined as follows:

- the lhs graph is removed from G creating G';
- the lhs graph is added to G' (but at this moment these graphs are separated);
- all edges in G that contain one of the nodes belonging to $V_{lhs} \cap V_{rhs}$ and the second to $V_G \setminus V_{lhs}$ are restored in $G' \cup rhs$;

– all edges in G that contains removed nodes ($V_{lhs} \setminus V_{rhs}$) are removed.

Formally, the language generated by a graph grammar is defined as all validated graphs generated by a sequence of productions applied to the starting graph S_Ω (denoted as $L(\Omega)$).

In practice, we will execute only some sequences of a production – one that can be designated by analysing an independent, separate dataset. In this case, this graph grammar will be triggered by spatial relationships identified in GIS data.

Lighting design is usually based on the concept of a *road segment*, which is a uniform fragment of a street – one which can be described as a single lighting situation. The most complex representation of the segment is presented in Fig. 5.

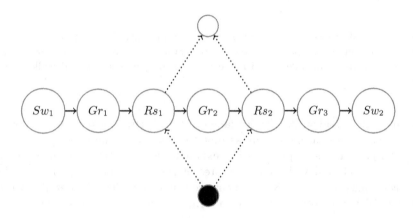

Fig. 5. Segments

We will elaborate on how this extended form is achieved using the following example. Initially, a segment is represented by a single node labelled by Rs. As shown before, by analysing GIS data, we can identify spatial relationships between the carriageways side-walks and green (separation) areas. Later, these relationships are transformed into lighting situation descriptions. Several examples of correct representations are presented in Fig. 6.

The examples presented up to this point only represent a single road segment, which is assumed to be of homogenous structure. As stated in Sect. 3, another spatial relationship detected among objects is *continuity*. Therefore, regardless of the internal structure of these segments, each segment may be linked with other segments with such a relationship. For clarity, this can be used to introduce a hierarchy in the graph structure.

The left-hand side of Fig. 7 the top level of this hierarchy. Nodes labelled as SS (street segments) represent contracted road segments, where their internal structure is not visible. Since lighting of irregular areas (such as junctions or squares) is designed using a calculation methodology different from that used for

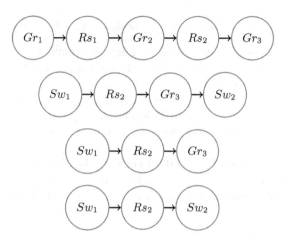

Fig. 6. Example representations of different lighting situations: a dual carriageway with surrounding green islands, a single carriageway with two sidewalks (one of them separated), a single carriageway with one sidewalk and one with two sidewalks on each side.

regular roads or sidewalks. Nodes labelled as FF (free-form shapes) represent such areas (in this case, one may assume that they are junctions).[8]

The right-hand side presents the expanded form of the segment structure. The small black and white circles represent the input/output points to which the edges linked to a top-level segment are connected when the detailed view of the segment is presented (see [7] for a formal description of the concept).

3.3   Relation to Previous Results

Due to the flexibility of possible graph transformations, the proposed SRG structure can be easily used to obtain graph structures previously defined within our research.

For instance, it can easily be used to obtain the Semantic Environment Graph (SEG), described in [3].

Also, because the SRG maintains (and can even enhance) the network aspect of the map data, it can be used to automatically generate a Traffic Flow Graph (see [2]). This type of graph is used for modelling vehicle traffic to increase the feasibility of dynamic lighting control [12,13] in areas with sparse sensor coverage.

---

[8] The estimation of power needed to illuminate them cannot be based on their length (W/m), but rather on their area. Its accuracy may be lower, but illumination of irregular areas constitutes a small percentage of the overall power consumption by city lighting.

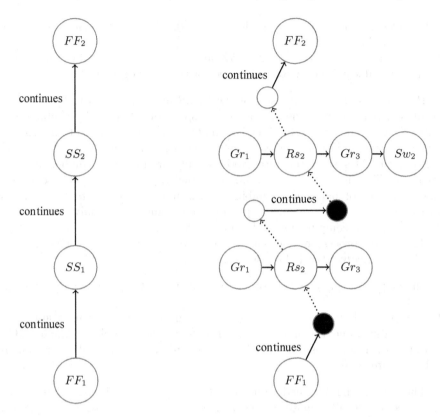

Fig. 7. Hierarchical view of road segments in contracted (left) and expanded form (right)

4 Energy Requirement Estimation of Outdoor Lighting

$STGT$ transformations allow us to generate SRG_L graphs, which describe lit areas in such a way that all factors significant for lighting design are present. To allow for a complete design, the formalism needs to be supplemented with lamp locations; however, here we do not aim at providing a detailed design, but to estimate the amount of energy required by a lighting installation to fulfil the lighting norms.

The concept of *lighting situations* (formally expressed with graphs generated by $STGT_L$) is key for this purpose. Each situation is described by a part of the graph presented in Fig. 7 and additional parameters such as pole height, arm length, distance from the street, distance between the poles, lighting class and other non-structural information described in Table 1.

Therefore, we propose a multi-cube structure, LightCalc, where the main dimension is the lighting situation. For each lighting situation in LightCalc, we calculate and store a set of data, including:

– the parameters identifying the situation,

- the average power [W/km] in the most optimal (energy-efficient) design of this situation,
- the average power [W/km] after 5% and 10% disturbances are introduced into the values of parameters obtained for the most optimal design[9].

However, the issue of the size of the LightCalc cube needs to be resolved. As indicated in Sect. 2, the combinatorial space of all possible combinations makes the computations unfeasible. Fortunately, experience gained from projects involving large-scale photometric designs (Tbilisi, Washington) shows that in practice, the AI system generates and use only around 10–12 million lighting situations. This is a good size estimate of the LightCalc cube; it also means that if the set of parameter combinations is narrowed down to those which actually occur in real life (which is possible due to automatic analysis of lighting situations), both computation and storage is feasible.

The actual process of estimating the power requirements of city lighting installations involves the following steps:

1. The map data is analysed in order to build the SRG_L graph, as described in Sect. 3.
2. The LightCalc multi-cube is scanned for situations that are most similar to the identified one; the aforementioned susceptibility, modelled using 5%/10% value disturbances, is used to select a more 'robust' estimate if several possibilities are present.

The sum of the obtained values can be used as a real-life estimate of the power required to illuminate the set of streets under consideration. In practical situations, if the value is provided as the maximum power of installations offered in a public tender, a certain margin of error (e.g. 20%) should be added to compensate differences occurring from different (less-than-optimal) pole spacings or setbacks.

5   Conclusions

The paper tries to address the problem of estimating the energy required to illuminate a given area of a city according to the applicable standards (e.g. EN 13201). The presented concept is based on automatic analysis of publicly-available map data in order to explicitly identify semantic relationships among the represented objects. These are used to infer the parameters which influence the required lamp parameters (e.g. their power). Using this knowledge, the most probable installation power is estimated by reviewing previously calculated designs in a multi-cube structure.

---

[9] This value is stored to represent the susceptibility of the required power to parameter value fluctuations, which in turn can be used to estimate whether a certain power value is likely to occur in real-life situations, or if a greater margin should be assumed to make the estimate more realistic.

The contributions of this paper are two-fold. On one hand, it proposes a method for consistent extraction of knowledge from spatial relationships identified between objects represented in map (geographic) datasets. A new type of graph transformations (Spatially Triggered Graph Transformation) is defined to coordinate and automate the analytical process and generate the appropriate data structure for lighting design.

The second contribution pertains directly to quick estimation of energy requirements of outdoor lighting. Instead of performing the time-consuming calculations on demand, the semantic relations expressed in the LightCalc multicube structure allow us the estimate the power of an optimally-designed installation using designs prepared for segments with a similar structure.

In practice, this task is very important for city officials. It lets them use a realistic estimate of the required power and compare it to the current, legacy installation in order to mitigate the risk of not meeting the power reduction factors required by funding agencies. It can also be useful to prepare realistic requirements in public tender procedures to enforce the proper quality of the offered configurations, e.g. by specifying their maximum total power.

This methodology could also prove useful for the funding agencies themselves, by letting them reformulate the environmental impact requirements imposed upon the beneficiaries. Requiring a certain reduction of energy usage in relation to the current installation favours cities with over-lighting, but hinders retrofit in areas where the lighting was inadequate (e.g. due to excessive lamp spacing, which translates to too few lamps).

Finally, it should be noted that the application area of the proposed spatially-triggered graph transformations (and the resulting semantic model) is much broader than just for optimisation of street lighting. A semantic description of an area is considerably easier to analyse than a set of shapes. This allows for planning and analysis, both on the micro and the macro level (e.g. to evaluate the availability of public transportation or green areas in relation to residential buildings).

References

1. Ehrig, H., Ehrig, K., Prange, U., Taentzer, G.: Fundamentals of Algebraic Graph Transformation. MTCSAES. Springer, Heidelberg (2006). https://doi.org/10.1007/3-540-31188-2
2. Ernst, S., Komnata, K., Łabuz, M., Środa, K.: Graph-based vehicle traffic modelling for more efficient road lighting. In: Zamojski, W., Mazurkiewicz, J., Sugier, J., Walkowiak, T., Kacprzyk, J. (eds.) DepCoS-RELCOMEX 2019. AISC, vol. 987, pp. 186–194. Springer, Cham (2020). https://doi.org/10.1007/978-3-030-19501-4_18
3. Ernst, S., Łabuz, M., Środa, K., Kotulski, L.: Graph-based spatial data processing and analysis for more efficient road lighting design. Sustainability 10(11), 3850 (2018). https://doi.org/10.3390/su10113850
4. European Committee for Standarization: CEN/TR 13201-1: Road lighting – Part 1: Guidelines on selection of lighting classes. Technical report, European Committee for Standarization (December 2014)

5. European Committee for Standarization: EN 13201-2: Road lighting – Part 2: Performance requirements. Technical report (December 2014)
6. Hölker, F., et al.: The dark side of light: a transdisciplinary research agenda for light pollution policy. Ecol. Soc. **15**(4), (2010)
7. Kotulski, L., Wydawnictwa, A.G.H.I.S.S.K., AGH., W.: Rozproszone Transformacje Grafowe: Teoria i Zastosowania. Redakcja Wydawnictw AGH (2013)
8. Rozenberg, G. (ed.): Handbook of Graph Grammars and Computing by Graph Transformation. WSPC, New Jersey, Singapore (January 1997)
9. Ernst, S., Starczewski, J.: How spatial data analysis can make smart lighting smarter. In: ACIIDS 2021: 13th Asian Conference on Intelligent Information and Database Systems. Accepted for Publication (2021)
10. Sędziwy, A.: A new approach to street lighting design. LEUKOS **12**(3), 151–162 (2016). https://doi.org/10.1080/15502724.2015.1080122
11. Sędziwy, A., Basiura, A.: Energy reduction in roadway lighting achieved with novel design approach and LEDs. LEUKOS **14**(1), 45–51 (2018). https://doi.org/10.1080/15502724.2017.1330155
12. Wojnicki, I., Kotulski, L.: Empirical study of how traffic intensity detector parameters influence dynamic street lighting energy consumption: a case study in Krakow. Poland Sustain. **10**(4), 1221 (2018). https://doi.org/10.3390/su10041221
13. Wojnicki, I., Kotulski, L.: Improving control efficiency of dynamic street lighting by utilizing the dual graph grammar concept. Energies **11**(2), 402 (2018). https://doi.org/10.3390/en11020402

Embedding Alignment Methods in Dynamic Networks

Kamil Tagowski$^{(\boxtimes)}$ ⓘ, Piotr Bielak ⓘ, and Tomasz Kajdanowicz ⓘ

Department of Computational Intelligence, Wroclaw University of Science and Technology, Wrocław, Poland
{kamil.tagowski,piotr.bielak,tomasz.kajdanowicz}@pwr.edu.pl

Abstract. In recent years, dynamic graph embedding has attracted a lot of attention due to its usefulness in real-world scenarios. In this paper, we consider discrete-time dynamic graph representation learning, where embeddings are computed for each time window, and then are aggregated to represent the dynamics of a graph. However, independently computed embeddings in consecutive windows suffer from the stochastic nature of representation learning algorithms and are algebraically incomparable. We underline the need for embedding alignment process and provide nine alignment techniques evaluated on real-world datasets in link prediction and graph reconstruction tasks. Our experiments show that alignment of Node2vec embeddings improves the performance of downstream tasks up to 11 pp compared to the not aligned scenario.

Keywords: Dynamic graphs · Graph embedding · Embedding alignment

1 Introduction

Node representation learning is pervasive across multiple applications, like social networks [13,21], spatial networks [24,25] or citation networks [9,21]. The vast majority of node embedding methods are trained in an unsupervised manner, providing an automated way of discovering node representations for static networks. However, the body of knowledge for dynamic graph node embedding methods is rather unaddressed [4]. There are not many approaches to deal with real-world scenarios, where the structure of the network evolves and node embedding depends on such dynamics.

The embedding of dynamic graphs can be performed according to two scenarios: continuous and discrete-time approaches. The continuous approach allows to handle a single event that triggers updates of node embeddings. The latter setting that is commonly utilized, involves the aggregation of graph data into snapshots and computes embeddings for each one of them. Such snapshot embeddings are further combined into a single node embedding that captures the whole graph evolution. Unfortunately, such decomposition of the embedding process suffers from the stochastic nature of representation learning algorithms.

© Springer Nature Switzerland AG 2021
M. Paszynski et al. (Eds.): ICCS 2021, LNCS 12742, pp. 599–613, 2021.
https://doi.org/10.1007/978-3-030-77961-0_48

Embeddings of consecutive snapshots are algebraically incomparable due to the transformations (artifacts) induced by the embedding methods. Therefore, there exists an research gap of how to deal with these unwanted transformations. The expected outcome is to map embeddings from particular snapshots into a common space. This can be achieved by **embedding alignment methods** that mitigate linear transformations and provide the ability to compare embeddings along with consecutive snapshots. Performing downstream tasks on nonaligned node embedding vectors may provide inconclusive results.

In this paper, we focus on several node embedding alignment methods that allow finding unified representation for nodes in dynamic networks using static network embedding approaches (in our case: node2vec). Based on extensive experiments on several real-world datasets for link prediction and graph reconstruction tasks, we demonstrate that node embedding alignment is crucial and allows to increase performance up to 11 pp compared to not aligned embeddings.

We summarize our contributions as follows. (1) We formulate aligner performance measures (AMPs) for evaluating alignment algorithms, regardless of the downstream tasks. (2) We propose nine embedding alignment methods for graph. (3) We provide a comprehensive evaluation showing that alignment is an indispensable operation in dynamic graph embedding based on a discrete approach, while dealing with node2vec embeddings.

This paper is structured as follows: in Sect. 2 we discuss other work related to our topic. Then, we discuss applications of graph embedding alignment and emphasize its importance (Sect. 3.2). We also formulate aligner performance measures in Sect. 3.3. Next, we propose several methods for dynamic graph embedding alignment (Sect. 3.4) and evaluate them in downstream tasks as well as by means of introduced measures (Sect. 4). We conclude our work and point out future directions in Sect. 5.

2   Related Works

The literature on static node embedding methods is very rich [4]. We can distinguish many approaches that are based on random-walks: DeepWalk [18], Node2vec [13], metapath2vec [9]; graph neural networks: GCN [14], GAT [23]; and matrix factorization: LLE [19], Laplacian Eigenmaps [1], HOPE [17]. Even though all of them are very powerful concepts, their applicability to dynamic graph embeddings is very limited. Embedding alignment is a tool that makes static embedding usable. Indeed, embedding alignment is crucial in many machine learning areas, e.g., in machine translation [12], cross-graph alignment [5,6,8], dynamic graph embedding [3,20,22]. Embedding alignment techniques are often based on solving Orthogonal Procrustes problem to obtain a linear transformation between pairs of embeddings [6]. We can also distinguish approaches that utilize adversarial training [5,8]. Dynamic Graph Embedding methods provide an embedding update mechanism for changes in the graph structure (appearing or disappearing nodes and edges). Embedding update may be performed in an online manner with the arrival of single events [15,16] or with arrival of a new batch (graph snapshot) [3,11,20,22]. In the tNodeEmbed [20] and LCF [22] methods, alignment is achieved by solving the Orthogonal Procrustes problem using all common nodes to obtain the transformation matrix. In FILDNE [3], the authors do not follow this scenario and they provide a mechanism for selecting only a subset of nodes used in the alignment process.

3   Graph Embedding Alignment

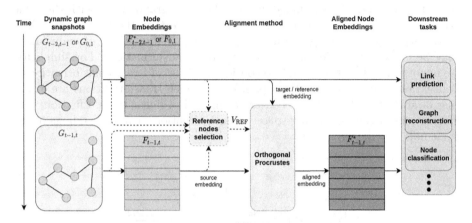

Fig. 1. Graph embedding alignment in the whole graph processing pipeline. For a dynamic graph in the form of snapshots, compute node embeddings, then (optionally) determine the reference nodes $\mathcal{V}_{\mathrm{REF}}$ and align the newest embedding $F_{t-1,t}$ to a given target / reference embedding (*previous* one $F_{t-2,t-1}$, or the *first* one overall $F_{0,1}$). Such aligned embeddings can be used in downstream tasks, improving the performance compared to non-aligned embeddings.

3.1   Notation and Problem Statement

We denote a dynamic graph $G_{0,T}$ as a tuple $(\mathcal{V}_{0,T}, \mathcal{E}_{0,T})$, where $\mathcal{V}_{0,T}$ is the set of all nodes (vertices) observed between timestamp 0 and T, and $\mathcal{E}_{0,T}$ is the set of edges in the same timestamp range. We model such a dynamic graph as a series of snapshots $G_{0,1}, G_{1,2}, \ldots, G_{T-1,T}$.

A node embedding function $f : \mathcal{V} \to \mathbb{R}^{|\mathcal{V}| \times d}$ maps every node $v \in \mathcal{V}$ into a low-dimensional vector representation of size $d, d << |\mathcal{V}|$, resulting in a node embedding matrix F, where each row represents an embedding of a single node.

An embedding alignment is a function $g : \mathbb{R}^{|\mathcal{V}| \times d} \to \mathbb{R}^{|\mathcal{V}| \times d}$ that transforms (aligns) a given node embedding matrix F respective to another one, producing an aligned node embedding matrix F^*. The method is trained on the observed change of embedding for a subset of nodes – called reference nodes – $\mathcal{V}_{\mathrm{REF}} \subseteq \mathcal{V}$.

3.2   The Importance of Embedding Alignment

Node embedding methods capture the structure of graphs and encode it in low-dimensional representation vectors for every node. The final form of the embedding space and the actual positions of node vectors highly depend on the optimized cost function as well as the optimization procedure itself. For instance, in random walk-based methods (like node2vec), there are three sources: (1) the stochastic nature of random walk generation, (2) the random initialization of embedding vector values, and (3) the order of node-context pairs used for training the Skip-gram model influence the final node vectors. It results in the situation that calculating the embedding for the same graph twice, with the same parameters of the embedding method, we observe that the node embeddings in the second run end up in different positions. Deformations of embeddings may be caused by a wide family of geometric transformations. For simplicity, we hypothesize that it is enough to consider a subset of linear transformations – translation, scaling and rotations (see: Fig. 2). Such transformations are examples of affine transformations, i.e., are composed of linear transformations and translations of the embedding space. In such situations, two embeddings of the same node are incomparable.

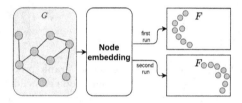

Fig. 2. Two runs of embedding calculation of the same graph.

The problem becomes fundamental in the case of dynamic graph representation learning, where embeddings are computed for every snapshot independently. In downstream tasks, these embeddings are often combined to obtain a representation for the whole dynamic graph, e.g., as in itebielak2020fildne. To obtain a rational combination of snapshots' embeddings, we are forced to align them. All rotations, scaling, and translations must be eliminated (see: Fig. 3). Note that proper alignment requires some transformation anchors (nodes). Sophisticated methods may automatically learn to perform the alignment against the common nodes between consecutive graphs. Notwithstanding, we developed a much simpler and computationally less complex methodology to select a subset of common nodes present in both graph snapshots. It significantly widens the applicability of the alignment to large scale networks. Our intuition is that nodes whose local structure has significantly changed, should not have been used to perform the alignment. The selection of appropriate *reference nodes* influences the performance of downstream tasks.

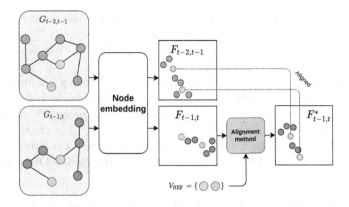

Fig. 3. Embedding of dynamic graph snapshots with alignment method applied.

3.3 Alignment Performance Measures

In this subsection, we introduce a novel set of alignment performance measures (*APM*) that constitute the criteria an alignment algorithm should meet. We also propose measures to evaluate the fit of alignment methods to criteria.

Overall, graph representation learning methods derive low-dimensional vector embeddings for different entities in the graph, i.e., nodes, edges, or subgraphs. From now on, we will present a case of node embedding approach, but it can be easily transfigured to edge or subgraph related problems.

The aim of node embedding methods is, generally speaking, to encode structural information in vector representations by placing embeddings of similar nodes near in the embedding space and keeping dissimilar nodes at a further distance. The definitions of "distance" and "similarity" as well as "structural information" depend on the properties of the representation we want to achieve (e.g., Euclidean distance, cosine similarity, considering first or second-order neighborhoods).

The family of random walk-based embedding approaches (e.g., DeepWalk [18], Node2vec [13], metapath2vec [9]) explicitly preserve the distances between nodes in the graph. Thus, we postulate the first *APM* which is related to distances:

APM 1. *The pairwise distances of vectors in the embedding space should be preserved during alignment.*

Changing the relative distances corrupts the information encoded by the embedding algorithm. To measure the magnitude of changes in relative distances, we propose **Pairwise Embedding Distance (PED)**:

$$\mathbf{PED}(F, F^*) = \frac{1}{|\mathcal{V}| * (|\mathcal{V}| - 1)} \sum_{\substack{(u,v) \in \mathcal{V} \times \mathcal{V} \\ u \neq v}} |D^{(F)}(u, v) - D^{(F^*)}(u, v)|, \quad (1)$$

where F is the initial node embedding matrix, F^* is the matrix after alignment, $|\cdot|$ denotes the absolute value. For the distance measure $D^{(F)}(u,v)$ between embeddings of nodes u and v in the embedding matrix F, we use the L_2 distance, but one can employ other ones, like the cosine distance.

This performance measure quantifies how the embedding alignment method mitigates translation and rotation. High values of this metric indicate that the embedding structure is corrupted during the alignment process. Contrary, if the value is equal to zero $(PED(F, F^*) = 0)$, the embedding is perfectly preserved.

Besides translation and rotation, we also consider the scaling transformation, proposing the following APM. We propose the **Scaling Score Distance (SSD**; see Eq. 2).

APM 2. *The scaling of distances between reference nodes after alignment should be the same for all other nodes.*

$$\mathbf{SSD}(F, F^*) = \left| \frac{1}{|\mathcal{V}_{\text{REF}}| * (|\mathcal{V}_{\text{REF}}| - 1)} \sum_{\substack{(u_R, v_R) \in \mathcal{V}_{\text{REF}} \times \mathcal{V}_{\text{REF}} \\ u_R \neq v_R}} \frac{D^{(F^*)}(u_R, v_R)}{D^{(F)}(u_R, v_R)} \right.$$
$$\left. - \frac{1}{|\mathcal{V} \backslash \mathcal{V}_{\text{REF}}| * (|\mathcal{V} \backslash \mathcal{V}_{\text{REF}}| - 1)} \sum_{\substack{(u, v) \in (\mathcal{V} \backslash \mathcal{V}_{\text{REF}}) \times (\mathcal{V} \backslash \mathcal{V}_{\text{REF}}) \\ u \neq v}} \frac{D^{(F^*)}(u, v)}{D^{(F)}(u, v)} \right|.$$
$$(2)$$

We also address the requirement of preserving the same positions of reference nodes in two snapshots of a dynamic graph. We assume that the alignment is performed according to those nodes. This leads us to our third APM and we propose the **Reference Nodes Distance (RND**; see Eq. 3).

APM 3. *After the alignment of the embedding of the second snapshot, the vectors of reference nodes must be placed in the same positions as in the embedding of the first snapshot.*

$$\mathbf{RND}(F_1, F_2^*) = \frac{1}{|\mathcal{V}_{\text{REF}}|} \sum_{u_R \in \mathcal{V}_{\text{REF}}} D^{(F_1, F_2^*)}(u_R), \qquad (3)$$

$\mathcal{V}_{\text{REF}}$ is the set of reference nodes and $D^{(F_1, F_2^*)}(u_R)$ is the L_2 distance between u_R's embedding vectors in F_1 and F_2^*, respectively. The most desired case assumes a score equal to zero.

Assuming that the embedding algorithm introduces rotations, scaling, and translations, the alignment algorithm is performing perfectly if all the proposed measures are equal to zero.

3.4 Dynamic Graph Embedding Alignment Methods

The embedding alignment problem might be addressed in two major ways: either (i) using post-hoc alignment of already computed embedding matrices, or (ii) adding an auxiliary loss to the embedding method that ensures alignment of node embeddings. The first one assumes the situation where all embeddings of consecutive static networks are already computed and the networks are not available at the time of dynamic embedding. On the other hand, the latter setting benefits from the availability of two network snapshots as it can address embedding deformation artifacts. In this work, we consider the post-hoc setting only leaving the second approach for the future work. To keep our proposed alignment methods computationally trackable, we focus in this work on the methods inspired by matrix alignment using the Orthogonal Procrustes problem [20]. Thus, we omit computationally expensive methods based on neural networks, like [8]. In this section, we will present how the embedding alignment techniques work and how they select reference nodes.

Given two matrices $A \in \mathbb{R}^{n \times d}$ and $B \in \mathbb{R}^{n \times d}$ with matching rows, the Orthogonal Procrustes method find a transformation matrix Q such that:

$$\operatorname*{argmin}_{Q : Q^\intercal Q = I} ||BQ - A||_2^2 \tag{4}$$

It can be solved as $Q_{\text{opt}} = UW^\intercal$ with $U\Sigma W^\intercal$ being the Singular Value Decomposition (SVD) of $B^\intercal A$. In the case of dynamic graph embedding alignment, the A and B matrices are two node embeddings.

As we already pointed out in Sect. 3.2, for dynamic graphs, we will select a subset of nodes, whose characteristics were the same (or at least similar) in both graph snapshots. Although, for completeness, we evaluate also the approach with all common nodes as reference nodes and call it **Procrustes Aligner (PA)**.

A simple yet quite restrictive approach, called **Procrustes Unchanged Aligner (PUA)**, is to select all nodes whose neighborhood does not change between snapshots, i.e., nodes having the same neighbors in both snapshots.

Another method for selecting only a subset of nodes as the reference nodes was already presented in FILDNE [3]. The authors proposed to use a node's activity (**activity function**), in the form of the *multi-degree* centrality, and then compare these activities in both snapshots using the function presented in Eq. 5 to finally obtain a nodes' score s (**scoring function**).

$$s(a_{t-1}^{(v)}, a_t^{(v)}) = |a_{t-1}^{(v)} - a_t^{(v)}| \left(\frac{\pi}{2} - \arctan(\max\{a_{t-1}^{(v)}, a_t^{(v)}\}) \right), \tag{5}$$

where $a_t^{(v)}$ is the v node's activity in snapshot $G_{t-1,t}$. This method is evaluated in our experiments as **FILDNE Aligner (FA)**.

We further explore this idea and propose to apply other activity functions. Due to the fact that the embedding is performed on dynamic graphs, we propose to utilize temporal (dynamic) node centrality measures from a body of knowledge related to complex network analysis, e.g. [26]. We postulate to measure the activity using: temporal betweenness **TB**, temporal closeness **TC**, temporal k-shell score **TK**, and temporal degree deviation **TDD**. See details of the measures in [26]. We choose L1-norm as the scoring function:

$$s(a_{t-1}^{(v)}, a_t^{(v)}) = |a_{t-1}^{(v)} - a_t^{(v)}| \tag{6}$$

Both for the FILDNE and all temporal node-activity-based aligners, we still need to select the actual reference nodes based on the computed nodes' scores. In our experiments, we consider the **top percent** scenario from FILDNE [3] as the other ones can be equivalently applied. We select the top **p** percent of the lowest scores: $\mathrm{select}(S, \mathcal{V}) = \mathcal{V}_{\mathrm{REF}} \subseteq \mathrm{sort}_S(\mathcal{V})$, s.t. $|\mathcal{V}_{\mathrm{REF}}| = \mathbf{p}|\mathcal{V}|$, where S are the node scores.

We also further develop another perspective of the scoring function that deeper exploits the structural graph properties, instead of node activity only. We propose the **Edge Jaccard Aligner (EJA)**, which for every node existing in two snapshots, computes the Jaccard distance of their neighbouring edges:

$$s(E_{t-1}^{(v)}, E_t^{(v)}) = 1 - \frac{|E_{t-1}^{(v)} \cap E_t^{(v)}|}{|E_{t-1}^{(v)} \cup E_t^{(v)}|}, \tag{7}$$

where $E_t^{(v)}$ is the set of edges in snapshot $G_{t-1,t}$ connected to node v.

The last proposed aligner is the **Embedding Neighbor Jaccard Aligner (ENJA)**, which utilizes the computed node embeddings. For all nodes existing in two snapshots, it extracts n percent of the closest neighbors in both embedding spaces, and then computes the Jaccard distance of neighbor sets:

$$s(F_{t-1}^{(v)}, F_t^{(v)}) = 1 - \frac{|\mathcal{CN}_n(F_{t-1}^{(v)}) \cap \mathcal{CN}_n(F_t^{(v)})|}{|\mathcal{CN}_n(F_{t-1}^{(v)}) \cup \mathcal{CN}_n(F_t^{(v)})|}, \tag{8}$$

where $\mathcal{CN}_n(F_t^{(v)})$ is the set of the top n percent of closest neighbors of node v in the the embedding F_t.

4   Experiments

We evaluate all the proposed alignment methods on real-world datasets on two downstream tasks: link prediction and graph reconstruction. Moreover, we provide the analysis of the introduced Alignment Performance Measures. The code, as well as the computational environment configuration (DVC pipeline), is made publicly available at https://gitlab.com/tgem/embedding-alignment to ensure reproducibility.

4.1   Datasets

We perform experiments on nine real-world datasets (see Table 1). Each of them is split based on the timestamp frequency: daily (ia-hypertext), monthly (enron-employees, radoslaw-email, fb-forum, fb-messages) and yearly (bitcoin-alpha, bitcoin-otc, ppi, ogbl-collab). The total number of snapshots varies from 3 to 9. As dataset time characteristics are not ideal and some of the generated snapshots were too small, we performed the following operations: we merged the first snapshot into the second one (bitcoin-alpha, bitcoin-otc, fb-messages); we merged the last snapshot to the second last one, as validation on tiny snapshots would be biased (bitcoin-alpha, bitcoin-otc, employees, ia-radoslaw-email); we also ignored first four snapshots (ppi, ogbl-collab), as merging them would result in a much bigger time-span than other snapshots.

Table 1. Statistics of graph datasets. $|\mathcal{V}|$ - number of nodes, $|\mathcal{E}|$ - number of edges, **Directed** - whether the graph is directed or not

DATASET	$|\mathcal{V}|$	$|\mathcal{E}|$	DIRECTED	TIMESPAN	NUMBER OF SNAP- SHOTS	SNAPSHOT TIMESPAN
HYPERTEXT	113	20 818	×	2.5 DAYS	3	1 DAY
ENRON- EMPLOYEES	151	50 572	×	37.9 MONTHS	6	6 MONTHS
RADOSLAW- EMAIL	167	82 927	√	9 MONTHS	9	1 MONTH
FB-FORUM	899	33 720	×	5.5 MONTHS	5	1 MONTH
FB-MESSAGES	1 899	61 734	×	7.2 MONTHS	7	1 MONTH
BITCOIN- ALPHA	3 783	24 186	√	5.2 YEARS	5	1 YEAR
BITCOIN-OTC	5 881	35 592	√	5.2 YEARS	5	1 YEAR
PPI	16 386	141 836	×	24 YEARS	5	5 YEARS
OGBL- COLLAB	233 513	1 171 947	×	34 YEARS	7	5 YEARS

4.2 Node Embeddings

To compute node embeddings we utilize the Node2vec method implemented in the PyTorch Geometric library [10]. We embed each snapshot separately. Using the Hyperopt optimizer [2] (restricted to 200 iterations), we performed Node2vec hyperparameter search. We recomputed all embeddings 25 times to handle the stochastic nature of Node2vec (random initialization and random walks). The obtained embeddings were evaluated on a link prediction task on the same snapshot, i.e., embedding $F_{t-1,t}$ was evaluated against graph $G_{t-1,t}$ for all snapshots.

4.3 Embedding Aggregation

In our setting, all downstream tasks require a single embedding for a given node that captures its whole history in the dynamic graph. Hence, we need to combine nodes' embeddings from all snapshots. There are several approaches, like the simple averaging, convex combination with Bayesian inference, [3], exponential decaying, linear combination [22], or deep neural networks [11,20]. We decided to choose the most computationally efficient one – averaging node embedding vectors from all snapshots.

608 K. Tagowski et al.

4.4 Embedding Alignment

We evaluate all the proposed aligners (Sect. 3.4) accompanied with not aligned embeddings **N/AL** as a baseline. As shown in Fig. 1, one could either align a given snapshot embedding $F_{t-1,t}$ to its *previous* one $F^*_{t-2,t-1}$ or the **first embedding overall** $F_{0,1}$. We decided to use the latter setting, as it provides a common reference space for all the following snapshots. Moreover, we performed a grid search on the link prediction task using aggregated embeddings. The percent parameter was evaluated for a range $p \in \{0.1, 0.2, \ldots, 0.9\}$ and (for ENJA aligner) the parameter n was evaluated for following values: $n \in \{0.1, 0.2, \ldots 1.0\}$.

4.5 Link Prediction

Setup. Link prediction evaluation predicts the existence of edges in the last snapshot based on previous ones. We combine the snapshot embeddings $F_{0,1}, \ldots, F_{T-2,T-1}$ using average operator. The link prediction dataset is generated from the last snapshot $G_{T-1,T}$, with edges in the graph as existing links (class 1). Additionally, we sample an equal number of non-existing edges (class 0). We split the dataset into train (75%) and test (25%). Using edge representations, obtained from the Hadamard product, we train a Logistic Regression classifier. To measure the performance, we report the mean and standard deviation of the AUC metric over 25 runs.

Results. The results can be found in Table 2 (upper half). Notice that for the vast majority of cases, the alignment of node embeddings improves the overall link prediction performance. We can gain up to 10 and 11pp (bitcoin-alpha with EJA, enron-employees with ENJA) over non-aligned embeddings. In the case of the Procrustes Unchanged aligner (PUA), we observe that five out of nine datasets could not be aligned. This occurs because this aligner relies on nodes, whose neighborhood was precisely the same between snapshots and for those datasets there were snapshot pairs with an empty set of reference nodes. Moreover, such a restrictive selection criterion affects the performance – see bitcoin-alpha and bitcoin-otc, where the aligned embeddings perform worse than not aligned ones $(-0.95\text{pp}$ and -0.44pp, respectively). On the other hand, the biggest loss of -1.03pp was for hypertext using ENJA, but comparing the standard deviations of the results ENJA provides a more robust embedding. Among the best aligners we find TB (fb-forum, fb-messages, ppi), EJA (radoslaw-email, bitcoin-alpha) and ENJA (enron-employees, bitcoin-otc).

4.6 Graph Reconstruction

Setup. Graph reconstruction evaluation aims at reproducing the graph structure based on nodes' embedding. In our case, we expect to reconstruct the whole dynamic graph $G_{0,T}$ from the dynamic embedding of all snapshots $F_{0,T} = \text{Avg}(F_{0,1}, \ldots, F_{T-1,T})$. We compute the mean Average Precision score (mAP) for graphs [7]. This metric captures local graph properties, i.e. for any node and its embedding vector it checks how many of the nearest vectors in the embedding space (in the sense of euclidean norm) are actually first-order neighbors of this node (see FILDNE [3] for details). Similarly to link prediction results, we also report the mean and standard deviation of the mAP metric over 25 runs.

Results. The results can be found in Table 2 (lower half). For all but one case, we observe an improvement in the mAP metric values – up to about 8 pp (for enron-employees with PA and hypertext with TB). The only worse result occurs in hypertext with EJA (-0.53 pp). The best performing aligners are PA (bitcoin-otc, fb-messages, enron-employees, ppi) and TB (fb-forum, hypertext, radoslaw-email).

4.7   Impact of the Node Fraction Taken in the Alignment Process

In Table 2 we only reported the best scores for a particular aligner. We performed also grid search over parameter p, see Sect. 4.4, which describes percent of reference nodes taken from all common nodes. We compared each result with PA aligner (its mean and std result) that used all common nodes. It turned out that for bitcoin-alpha, bitcoin-otc, fb-forum, fb-messages, ia-hypertext, ppi and ogbl-collab datasets it was sufficient to take only 10% of nodes, whereas for enron-employees 20% and for radoslaw-email 30 of common nodes to achieve comparative results. Such feature is especially crucial when dealing with large datasets, as it shortens computation time of Orthogonal Procrustes.

4.8   Embedding Alignment Performance Metrics

We compute the metrics mentioned in Sect. 3.3 for all aligned embeddings across all datasets and alignment algorithms. We observe that for the PED and SSD metrics we receive values close to zero, which proves the alignment process preserves the information encoded in the node embeddings. In the case of the RND metric, we obtain values greater than zero, but after careful investigation, we show that the distance between reference nodes decreases after alignment, respective to the targeted embedding (see: Fig. 4).

Overall, all the results obtained in our study on embedding alignment allow claiming that (1) alignment is essential for embedding algorithms and (2) provide superior results in downstream tasks like link prediction and graph reconstruction.

Table 2. Downstream task results (link prediction and graph reconstruction). **N/AL** denotes evaluation of not aligned embeddings and ✗ denotes a scenario where embeddings could not be aligned due to missing reference nodes. We present values as the mean and standard deviations over 25 embedding retrains with gain (or loss) of aligned embeddings over not aligned ones (difference of mean values) in parenthesis. **Bold** values mark the best results for a single dataset. We performed a Friedman test with Nemenyi post-hoc to confirm that all alignment methods are statistically different from N/AL case. There were no statistically significant differences between alignment methods. The "*" symbol denotes methods that are significantly worse than the best method.

LINK PREDICTION (AUC)

ALIGNER	BITCOIN ALPHA	BITCOIN OTC	FB FORUM	FB MESSAGES	ENRON EMPLOYEES	HYPERTEXT	RADOSLAW EMAIL	OGBL COLLAB	PPI
N/AL	47.72 ± 14.44 *	70.19 ± 5.63 *	83.29 ± 2.80 *	59.50 ± 6.80 *	74.29 ± 3.95 *	87.99 ± 12.83	84.32 ± 1.34 *	81.24 ± 0.45 *	59.08 ± 0.80 *
PA	$50.80 \pm 13.95_{(+3.08)}$	$79.37 \pm 2.83_{(+9.18)}$	$90.36 \pm 1.36_{(+7.07)}$	$64.46 \pm 8.33_{(+4.96)}$	$84.48 \pm 1.37_{(+10.19)}$	$87.56 \pm 6.31_{(-0.43)}$	$92.89 \pm 0.38_{(+8.57)}$ *	$\mathbf{82.48 \pm 0.53}_{(+1.24)}$ *	$60.22 \pm 0.78_{(+1.14)}$
PUA	$46.77 \pm 14.17_{(-0.95)}$	$69.75 \pm 5.41_{(-0.44)}$ *	$83.72 \pm 2.53_{(+0.43)}$ *	✗	✗	✗	$86.55 \pm 1.05_{(+2.23)}$ *	✗	✗
FA	$50.91 \pm 14.78_{(+3.19)}$	$77.48 \pm 4.76_{(+7.29)}$	$90.47 \pm 1.28_{(+7.18)}$	$63.60 \pm 6.61_{(+4.10)}$	$83.09 \pm 1.39_{(+8.80)}$ *	$87.95 \pm 5.59_{(-0.04)}$	$92.96 \pm 0.38_{(+8.64)}$	$82.40 \pm 0.67_{(+1.16)}$	$60.26 \pm 0.74_{(+1.18)}$
TB	$53.42 \pm 13.35_{(+5.70)}$	$78.91 \pm 3.75_{(+8.72)}$	$91.07 \pm 1.37_{(+7.78)}$	$\mathbf{66.79 \pm 6.74}_{(+7.29)}$	$79.83 \pm 2.04_{(+5.54)}$ *	$88.95 \pm 4.68_{(+0.96)}$	$92.70 \pm 0.40_{(+8.38)}$ *	$82.19 \pm 0.53_{(+0.95)}$	$\mathbf{60.53 \pm 0.57}_{(+1.45)}$ *
TC	$55.21 \pm 12.61_{(+7.49)}$	$78.02 \pm 3.75_{(+7.83)}$	$90.03 \pm 1.90_{(+6.74)}$	$64.85 \pm 7.26_{(+5.35)}$	$83.10 \pm 2.08_{(+8.81)}$ *	$90.37 \pm 4.45_{(+2.38)}$	$92.78 \pm 0.42_{(+8.46)}$ *	$82.03 \pm 0.51_{(+0.79)}$ *	$59.80 \pm 0.66_{(-0.72)}$ *
TK	$56.91 \pm 12.77_{(+9.19)}$	$79.28 \pm 2.82_{(+9.09)}$	$90.71 \pm 1.64_{(+7.42)}$	$62.68 \pm 6.31_{(+3.18)}$	$83.50 \pm 0.97_{(+9.21)}$ *	$90.42 \pm 2.02_{(+2.43)}$	$92.93 \pm 0.37_{(+8.61)}$ *	$82.41 \pm 0.55_{(+1.17)}$	$60.11 \pm 0.80_{(+1.03)}$
TDD	$56.91 \pm 12.77_{(+9.19)}$	$79.28 \pm 2.82_{(+9.09)}$	$90.71 \pm 1.64_{(+7.42)}$	$62.68 \pm 6.31_{(+3.18)}$	$83.50 \pm 0.97_{(+9.21)}$ *	$90.42 \pm 2.02_{(+2.43)}$ *	$92.93 \pm 0.37_{(+8.61)}$ *	$82.41 \pm 0.55_{(+1.17)}$	$60.11 \pm 0.80_{(+1.03)}$
EJA	$\mathbf{57.75 \pm 12.10}_{(+10.03)}$	$79.06 \pm 3.95_{(+8.87)}$	$89.64 \pm 1.73_{(+6.35)}$	$63.17 \pm 7.69_{(+3.67)}$	$84.46 \pm 1.25_{(+10.17)}$	$88.55 \pm 5.00_{(+0.56)}$	$\mathbf{93.23 \pm 0.31}_{(+8.91)}$ *	$82.42 \pm 0.65_{(+1.18)}$	$60.14 \pm 0.77_{(+1.06)}$
ENJA	$56.60 \pm 12.06_{(+8.88)}$	$\mathbf{79.67 \pm 2.87}_{(+9.48)}$	$90.56 \pm 1.49_{(+7.27)}$	$64.70 \pm 7.93_{(+5.20)}$	$\mathbf{85.29 \pm 1.29}_{(+11.00)}$	$86.96 \pm 4.80_{(-1.03)}$	$93.02 \pm 0.43_{(+8.70)}$	$82.38 \pm 0.46_{(+1.14)}$	$60.26 \pm 0.75_{(+1.18)}$

GRAPH RECONST. (MAP)

ALIGNER	BITCOIN ALPHA	BITCOIN OTC	FB FORUM	FB MESSAGES	ENRON EMPLOYEES	HYPERTEXT	RADOSLAW EMAIL	OGBL COLLAB	PPI
N/AL	25.18 ± 0.20 *	28.09 ± 0.36 *	11.30 ± 0.29 *	14.43 ± 0.19 *	44.53 ± 1.85 *	52.48 ± 1.15 *	38.85 ± 1.19 *	11.73 ± 0.05 *	8.90 ± 0.07 *
PA	$27.40 \pm 0.25_{(+2.22)}$	$\mathbf{30.61 \pm 0.31}_{(+2.52)}$ *	$14.08 \pm 0.32_{(+2.78)}$ *	$\mathbf{16.05 \pm 0.24}_{(+1.62)}$ *	$\mathbf{52.53 \pm 0.90}_{(+8.00)}$ *	$57.46 \pm 0.74_{(+4.98)}$	$45.27 \pm 0.42_{(+6.42)}$ *	$\mathbf{11.99 \pm 0.04}_{(+0.26)}$ *	$\mathbf{9.43 \pm 0.05}_{(+0.53)}$ *
PUA	$25.21 \pm 0.28_{(+0.03)}$	$28.23 \pm 0.39_{(+0.14)}$ *	$11.56 \pm 0.36_{(+0.26)}$ *	✗	✗	✗	$42.94 \pm 0.59_{(+4.09)}$ *	✗	✗
FA	$26.44 \pm 0.24_{(+1.26)}$	$30.44 \pm 0.35_{(+2.35)}$ *	$14.15 \pm 0.31_{(+2.85)}$	$15.90 \pm 0.23_{(+1.47)}$	$51.46 \pm 1.09_{(+6.93)}$	$59.04 \pm 0.93_{(+6.56)}$	$45.91 \pm 0.43_{(+7.06)}$ *	$11.97 \pm 0.05_{(+0.24)}$ *	$9.06 \pm 0.09_{(+0.16)}$
TB	$27.34 \pm 0.28_{(+2.16)}$	$30.53 \pm 0.34_{(+2.44)}$ *	$\mathbf{14.35 \pm 0.38}_{(+3.05)}$ *	$15.98 \pm 0.25_{(+1.55)}$	$50.93 \pm 1.06_{(+6.40)}$	$\mathbf{60.42 \pm 0.87}_{(+7.94)}$ *	$\mathbf{46.67 \pm 0.41}_{(+7.82)}$ *	$11.88 \pm 0.04_{(+0.15)}$ *	$9.39 \pm 0.05_{(+0.49)}$
TC	$\mathbf{27.48 \pm 0.26}_{(+2.30)}$ *	$30.53 \pm 0.32_{(+2.44)}$	$14.11 \pm 0.31_{(+2.81)}$	$15.76 \pm 0.23_{(+1.33)}$	$52.46 \pm 1.12_{(+7.93)}$	$55.76 \pm 0.88_{(+3.28)}$ *	$46.22 \pm 0.52_{(+7.37)}$	$11.85 \pm 0.04_{(+0.12)}$ *	$9.19 \pm 0.08_{(+0.29)}$
TK	$26.70 \pm 0.31_{(+1.52)}$ *	$30.47 \pm 0.32_{(+2.38)}$ *	$13.97 \pm 0.32_{(+2.67)}$ *	$15.90 \pm 0.24_{(+1.47)}$ *	$52.16 \pm 1.07_{(+7.63)}$	$54.60 \pm 0.70_{(+2.12)}$ *	$44.64 \pm 0.58_{(+5.79)}$ *	$11.98 \pm 0.05_{(+0.25)}$ *	$9.39 \pm 0.05_{(+0.49)}$
TDD	$26.70 \pm 0.31_{(+1.52)}$ *	$30.47 \pm 0.32_{(+2.38)}$ *	$13.97 \pm 0.32_{(+2.67)}$ *	$15.90 \pm 0.24_{(+1.47)}$ *	$52.16 \pm 1.07_{(+7.63)}$	$54.60 \pm 0.70_{(+2.12)}$ *	$44.64 \pm 0.58_{(+5.79)}$ *	$11.98 \pm 0.05_{(+0.25)}$ *	$9.39 \pm 0.05_{(+0.49)}$
EJA	$26.70 \pm 0.30_{(+1.52)}$ *	$30.38 \pm 0.34_{(+2.29)}$ *	$13.84 \pm 0.30_{(+2.54)}$ *	$15.96 \pm 0.22_{(+1.53)}$	$51.10 \pm 1.08_{(+6.57)}$	$51.95 \pm 0.51_{(-0.53)}$ *	$46.04 \pm 0.42_{(+7.19)}$	$\mathbf{11.99 \pm 0.04}_{(+0.26)}$ *	$9.39 \pm 0.05_{(+0.49)}$
ENJA	$26.70 \pm 0.25_{(+1.52)}$ *	$30.54 \pm 0.30_{(+2.45)}$ *	$14.10 \pm 0.29_{(+2.80)}$ *	$15.92 \pm 0.26_{(+1.49)}$ *	$51.70 \pm 0.87_{(+7.17)}$	$57.42 \pm 0.85_{(+4.94)}$ *	$46.16 \pm 0.57_{(+7.31)}$	$11.97 \pm 0.04_{(+0.24)}$ *	$9.42 \pm 0.05_{(+0.52)}$

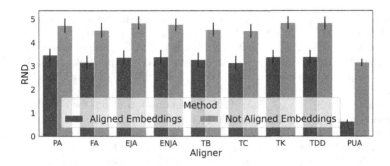

Fig. 4. Visualization of RND measure. We compare RND computed using: (1) non-aligned embedding and (2) aligned embeddings. We aggregate values across all datasets and show that in the RND for aligned embeddings are lower than for not aligned case. This indicates that the alignment is successful.

5   Conclusions and Future Work

In this paper, we emphasize the importance of node embedding alignment in dynamic graph embedding. We formulate three aligner performance measures for the evaluation of alignment algorithms. We propose several embedding alignment methods for dynamic graphs. According to our experimental evaluation, alignment is an indispensable operation in dynamic graph embedding. Furthermore, embedding alignment improves the performance of downstream tasks up to 11 pp compared to the not aligned scenario. We plan to exploit other approaches of embedding alignment that are directed to edges and subgraphs.

Acknowledgment. The project was partially supported by The National Science Centre, Poland, the research project no. 2016/21/D/ST6/02948, 2016/23/B/ST6/01735 and statutory funds of Department of Computational Intelligence.

References

1. Belkin, M., Niyogi, P.: Laplacian eigenmaps for dimensionality reduction and data representation. Neural Comput. **15**(6), 1373–1396 (2003)
2. Bergstra, J., Yamins, D., Cox, D.: Making a science of model search: hyperparameter optimization in hundreds of dimensions for vision architectures. In: International Conference on Machine Learning, pp. 115–123 (2013)
3. Bielak, P., Tagowski, K., Falkiewicz, M., Kajdanowicz, T., Chawla, N.V.: Fildne: a framework for incremental learning of dynamic networks embeddings (2020)
4. Chami, I., Abu-El-Haija, S., Perozzi, B., Ré, C., Murphy, K.: Machine learning on graphs: A model and comprehensive taxonomy. arXiv preprint arXiv:2005.03675 (2020)

5. Chen, C., et al.: Unsupervised adversarial graph alignment with graph embedding. CoRR abs/1907.00544 (2019). http://arxiv.org/abs/1907.00544
6. Chen, X., Heimann, M., Vahedian, F., Koutra, D.: CONE-align: consistent network alignment with proximity-preserving node embedding. pp. 1985–1988. Association for Computing Machinery, New York (2020). https://doi.org/10.1145/3340531.3412136
7. De Sa, C., Gu, A., Ré, C., Sala, F.: Representation tradeoffs for hyperbolic embeddings. Proc. Mach. Learn. Res. **80**, 4460 (2018)
8. Derr, T., Karimi, H., Liu, X., Xu, J., Tang, J.: Deep adversarial network alignment. CoRR abs/1902.10307 (2019), http://arxiv.org/abs/1902.10307
9. Dong, Y., Chawla, N.V., Swami, A.: metapath2vec: scalable representation learning for heterogeneous networks. In: Proceedings of the 23rd ACM SIGKDD International Conference on Knowledge Discovery and Data Mining, pp. 135–144 (2017)
10. Fey, M., Lenssen, J.E.: Fast graph representation learning with PyTorch Geometric. In: ICLR Workshop on Representation Learning on Graphs and Manifolds (2019)
11. Goyal, P., Chhetri, S.R., Canedo, A.: dyngraph2vec: capturing network dynamics using dynamic graph representation learning. Knowl.-Based Syst. **187**, 104816 (2020)
12. Grave, E., Joulin, A., Berthet, Q.: Unsupervised alignment of embeddings with wasserstein procrustes. In: Chaudhuri, K., Sugiyama, M. (eds.) Proceedings of Machine Learning Research. Proceedings of Machine Learning Research, vol. 89, pp. 1880–1890. PMLR (16–18 April 2019). http://proceedings.mlr.press/v89/grave19a.html
13. Grover, A., Leskovec, J.: node2vec: scalable feature learning for networks. In: Proceedings of the 22nd ACM SIGKDD International Conference on Knowledge Discovery and Data Mining, pp. 855–864 (2016)
14. Kipf, T.N., Welling, M.: Semi-supervised classification with graph convolutional networks. arXiv preprint arXiv:1609.02907 (2016)
15. Lee, J.B., Nguyen, G., Rossi, R.A., Ahmed, N.K., Koh, E., Kim, S.: Dynamic node embeddings from edge streams (2019)
16. Ma, Y., Guo, Z., Ren, Z., Tang, J., Yin, D.: Streaming graph neural networks. In: Huang, J., Chang, Y., et al. (eds.) Proceedings of the 43rd International ACM SIGIR Conference on Research and Development in Information Retrieval, SIGIR 2020, Virtual Event, China, 25–30 July 2020. pp. 719–728. ACM (2020) https://doi.org/10.1145/3397271.3401092
17. Ou, M., Cui, P., Pei, J., Zhang, Z., Zhu, W.: Asymmetric transitivity preserving graph embedding. In: Proceedings of the 22nd ACM SIGKDD International Conference on Knowledge Discovery and Data Mining, pp. 1105–1114 (2016)
18. Perozzi, B., Al-Rfou, R., Skiena, S.: Deepwalk: online learning of social representations. In: Proceedings of the 20th ACM SIGKDD International Conference on Knowledge Discovery and Data Mining, pp. 701–710 (2014)
19. Roweis, S.T., Saul, L.K.: Nonlinear dimensionality reduction by locally linear embedding. Science **290**(5500), 2323–2326 (2000)
20. Singer, U., Guy, I., Radinsky, K.: Node embedding over temporal graphs. In: Proceedings of the 28th International Joint Conference on Artificial Intelligence, pp. 4605–4612. AAAI Press (2019)
21. Tang, J., Qu, M., Wang, M., Zhang, M., Yan, J., Mei, Q.: Line: large-scale information network embedding. In: Proceedings of the 24th International Conference on World Wide Web, pp. 1067–1077 (2015)
22. Trivedi, P., Büyükçakır, A., Lin, Y., Qian, Y., Jin, D., Koutra, D.: On structural vs. proximity-based temporal node embeddings (2020)

23. Veličković, P., Cucurull, G., Casanova, A., Romero, A., Lio, P., Bengio, Y.: Graph attention networks. arXiv preprint arXiv:1710.10903 (2017)
24. Wu, Z., Pan, S., Long, G., Jiang, J., Zhang, C.: Graph wavenet for deep spatial-temporal graph modeling. arXiv preprint arXiv:1906.00121 (2019)
25. Xu, D., Wei, C., Peng, P., Xuan, Q., Guo, H.: Ge-gan: a novel deep learning framework for road traffic state estimation. Transp. Res. Part C Emerg. Technol. **117**, 102635 (2020)
26. Yu, E.Y., Fu, Y., Chen, X., Xie, M., Chen, D.B.: Identifying critical nodes in temporal networks by network embedding. Sci. Rep. **10**(1), 1–8 (2020)

The OpenPME Problem Solving Environment for Numerical Simulations

Nesrine Khouzami[1]([✉]), Lars Schütze[1], Pietro Incardona[1,2,3],
Landfried Kraatz[1], Tina Subic[1,2,3], Jeronimo Castrillon[1],
and Ivo F. Sbalzarini[1,2,3,4]

[1] Faculty of Computer Science, Technische Universität Dresden, Dresden, Germany
{nesrine.khouzami,lars.schutze,pietro.incardona,landfried.kraatz,
tina.subic,jeronimo.castrillon,ivo.sbalzarini}@tu-dresden.de
[2] Max Planck Institute of Molecular Cell Biology and Genetics, Dresden, Germany
[3] Center for Systems Biology Dresden, Dresden, Germany
[4] Cluster of Excellence Physics of Life, TU Dresden, Dresden, Germany

Abstract. We introduce OpenPME, the Open Particle-Mesh Environment, a problem solving environment that provides a Domain Specific Language (DSL) for numerical simulations in scientific computing. It is built atop a domain metamodel that is general enough to cover the main types of numerical simulations: simulations using particles, meshes, and hybrid combinations of particles and meshes. Using model-to-model transformations, OpenPME generates code against the state-of-the-art C++ parallel computing library OpenFPM. This effectively lowers the programming barrier and enables users to implement scalable simulation codes for high-performance computing (HPC) systems using high-level abstractions. Plenty of recent research has shown that higher-level abstractions and problem solving environments are well suited to alleviate low-level implementation overhead. We demonstrate this for OpenPME and its compiler on three different test cases—particle-based, mesh-based, and hybrid particle-mesh—showing up to 7-fold reduction in the number of lines of code compared to a direct OpenFPM implementation in C++.

Keywords: Domain specific compiler · Particle-mesh methods

1 Introduction

Computer simulations are the third pillar of science, alongside theory and experiment. Scientists increasingly rely on simulated models to investigate scales that are not experimentally accessible and nonlinearities that are not theoretically treatable. However, the complexity of modern high-performance computing (HPC) systems, and the complexity of HPC programming models increasingly limit implementation efficiency and performance portability. Alleviating this has therefore become a key research focus for HPC [25]. Examples are problem solving environments (PSE) [6], which reduce the programming barrier by

© Springer Nature Switzerland AG 2021
M. Paszynski et al. (Eds.): ICCS 2021, LNCS 12742, pp. 614–627, 2021.
https://doi.org/10.1007/978-3-030-77961-0_49

providing higher-level domain-specific abstractions. Since higher-level abstractions enable more powerful compiler transformations, this often leads to performance improvements over hand-written code [21]. A PSE typically consists of a Domain Specific Language (DSL) and an Integrated Development Environments (IDE) [6]. In scientific computing, such approaches have successfully been proposed for array programming [28,29], finite-element simulations [9,16,18,22,23], and for tensor expressions in numerical simulations [3,24]. Most numerical simulations can be expressed in terms of particles, meshes, or a combination thereof. Meshes are often used to discretize continuous fields, e.g., for finite-differences or finite volumes. Particles can be used to represent either discrete objects, like atoms in a molecule or cars in road traffic, or as arbitrarily discretization points for continuous fields. The classic Particle-In-Cell (PIC) codes in plasma physics [10] are an example of a combination of particles and meshes. Existing DSLs for particle-mesh simulations include PPML [2], PPME [14], and FDPS [12]. Of these, FDPS and PPME are limited to particles, with FDPS focusing on N-body simulations. PPML supports both particles and meshes, but relies on the discontinued Fortran library PPM [27]. To the best of our knowledge, there exists no DSL that supports particles, meshes, and hybrid particle-mesh codes on a modern and actively maintained platform library. This paper presents a DSL and IDE for particle, mesh, and hybrid particle-mesh simulations on parallel computers, the Open Particle-Mesh Environment (OpenPME). OpenPME is based on the Open Framework for Particles and Meshes (OpenFPM) [11], a recent and actively maintained open-source C++ library. Compared to writing C++ code for OpenFPM, OpenPME reduces the implementation effort and lowers the entry barrier for users. It hides the distributed-memory constructs of OpenFPM and provides error messages that are easier to understand than the template-engine errors of the C++ compiler.

Our DSL rests on a metamodel that captures all five types of simulations: particles (continuous and discrete), mesh (continuous), and two types of hybrid simulations. The OpenPME compiler implements a staged compilation process, where high-level DSL constructs are reduced to an intermediate metamodel. The metamodel provides a language-independent representation and enables model-to-model transformations. This allows the OpenPME compiler to automatically inject OpenFPM communication operations for distributed-memory programs, freeing the user of having to write them. In a second step, the metamodel is translated to OpenFPM C++ code. We showcase the expected productivity increase and DSL design of OpenPME using three real-world examples that highlight the seamless support for hybrid particle-mesh simulations and the automatic insertion of data-movement and parallel communication primitives by the compiler.

2   Background and Motivation

We introduce the general structure of particle-mesh simulations, describe the OpenFPM library, and detail our motivation for developing OpenPME.

2.1   Hybrid Particle-Mesh Simulations

Particle-mesh methods provide a universal framework for numerical simulations in scientific computing. They can be used to simulate both discrete and continuous models, either deterministically or stochastically [20, 26]. When simulating discrete models, particles directly represent entities in the model. In continuous models, particles correspond to mathematical discretization points. Particles are point objects that have a position and a list of properties. They can *interact* with other particles through pairwise interactions, and then *evolve* as a consequence of the interactions, i.e., particles change their position and/or their properties. These interactions can be deterministic or probabilistic. Hybrid particle-mesh methods are used to obtain more efficient simulations or to simulate multi-physics models. Particle-to-mesh and mesh-to-particle interpolation allows transferring data between the two representations [10].

2.2   The OpenFPM Library

OpenFPM [11] is an open-source C++ template library for implementing scalable parallel particle-mesh simulations on multi-CPU and multi-GPU computer hardware. It provides multiple layers of abstraction: based on memory allocators and memory-layouting abstractions, OpenFPM implements single-core data structures. Using data-decomposition and network communication abstractions, these are then transformed to multi-core and distributed-memory data structures. Finally, a library of frequently used numerical solvers is implemented using these data structures. A profiling interface and transparent in-situ visualization of simulation results [8] complete the library.

In OpenFPM, particles can carry any composite container of C++ objects as properties, and simulations can be performed in arbitrary-dimensional domains. It guarantees transparent memory layout conversion (e.g., between CPU and GPU) and run-time dynamic load-balancing to distribute data evenly and adapt to changes in local mesh resolution or particle density. OpenFPM includes checkpointing, parallel file I/O, and communications abstractions, including the `ghost_get` operator to transparently communicate boundary data between processors in a domain decomposition.

2.3   Motivation

OpenFPM heavily relies on C++ templates and template meta-programming to achieve its flexibility and performance. This renders the source code more complex, leading to programming mistakes and cryptic error messages from the C++ compiler. A typical example of OpenFPM code is shown in Fig. 1 for a simulation of the three-dimensional Navier-Stokes equation of fluid mechanics (only one dimension shown). The more concise OpenPME three-dimensional representation in Fig. 9, point ④ will be discussed in Sect. 4.3. Not only is the C++ more error-prone, but also makes it impossible for the compiler to detect semantic errors that arise from accessing objects in the wrong place or

```
g_dwp.template get<rhs>(key)[x]=
  fac1*(g_vort.template get<vorticity>(key.move(x,1))[x]+
  g_vort.template get<vorticity>(key.move(x,-1))[x])+
  fac2*(g_vort.template get<vorticity>(key.move(y,1))[x]+
  g_vort.template get<vorticity>(key.move(y,-1))[x])+
  fac3*(g_vort.template get<vorticity>(key.move(z,1))[x]+
  g_vort.template get<vorticity>(key.move(z,-1))[x])-
  2.0f*(fac1+fac2+fac3)*
  g_vort.template get<vorticity>(key)[x]+
  fac4*g_vort.template get<vorticity>(key)[x]*
  (g_vel.template get<velocity>(key.move(x,1))[x]-
  g_vel.template get<velocity>(key.move(x,-1))[x])+
  fac5*g_vort.template get<vorticity>(key)[y]*
  (g_vel.template get<velocity>(key.move(y,1))[x]-
  g_vel.template get<velocity>(key.move(y,-1))[x])+
  fac6*g_vort.template get<vorticity>(key)[z]*
  (g_vel.template get<velocity>(key.move(z,1))[x]-
  g_vel.template get<velocity>(key.move(z,-1))[x]);
```

Fig. 1. OpenFPM C++ code snippet to calculate the x-component of the three-dimensional Navier-Stokes equation.

misplaced operators in the formula. User also must manually place `ghost_get` communication operations to account for data exchange on a parallel computer. These calls must be in the right place and include the correct properties to be communicated, which is not always obvious for users. Extraneous `ghost_get` calls reduce the scalability and performance of the simulation.

OpenPME's goal is to allow computational scientists to write efficient simulations with domain-specific abstractions and error messages. OpenPME's abstractions enable high-level optimizations that would otherwise require complex and brittle analysis in a custom C++ compiler pass. We demonstrate this by automatically placing `ghost_get` operations.

3   OpenPME Design and Implementation

We detail the design and implementation of OpenPME based on a metamodel that captures all (hybrid) particle-mesh methods.

3.1   Design Overview

OpenPME introduces an intermediate layer between the user's simulation application and the OpenFPM C++ library, as illustrated in Fig. 2. It is based on two metamodels representing particle-mesh simulations and the C++ code using OpenFPM. Model-to-model transformations identify syntactic elements and semantic relations that enable, e.g., expression rewriting or automatic insertion of communication statements. The generated code executes on any platform supported by OpenFPM. This includes multi-core CPUs, GPUs, distributed-memory CPU clusters, and multi-GPU clusters. Parallel efficiency and scalability are inherited from OpenFPM [11]. OpenPME is implemented in Jetbrains MPS.[1]

---

[1] Jetbrains Meta Programming System (MPS) https://www.jetbrains.com/mps/.

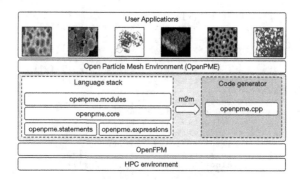

Fig. 2. Modular architecture of OpenPME. An intermediate layer between HPC application domains interfacing OpenFPM for HPC architecture systems.

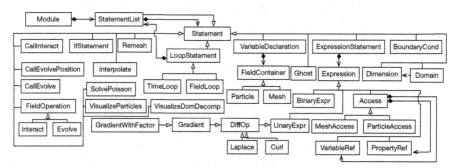

Fig. 3. Excerpt from the metamodel to describe particle-mesh simulations.

The simulation metamodel is implemented as a stack of sub-languages (see Fig. 2). The use of sub-languages enables separation of concerns and keeps the implementation modular, extensible, and maintainable. Lower levels provide foundational elements of the language that are used to build more complex features higher in the stack.

The OpenFPM C++ metamodel captures the domain of imperative, object-oriented programs. It is implemented in the `openpme.cpp` language (see Fig. 2). This model is expressive enough to generate the complete C++ code of a simulation using OpenFPM. We also use this model as an intermediate representation when lowering simulation code using model-to-model transformation.

3.2   Metamodel for Particle-Mesh Simulations

The metamodel for particle-mesh simulations builds on our prior work in [14]. It supports three types of simulations: particle-only, mesh-only, and hybrid particle-mesh of both continuous and discrete models. Figure 3 shows an excerpt of the model as UML diagram. An OpenPME program consists of modules, each with a sequence of statements. A statement can declare variables or a boundary condition, or define the simulation domain and dimensionality (see Fig. 3). A `TimeLoop`

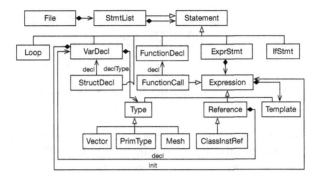

Fig. 4. Excerpt from the metamodel to describe OpenFPM C++ programs.

statement supports different time-discretization methods as first-class concepts in the model. The model supports basic data types found in most programming languages, as well as domain-specific types such as `Particle` or `Mesh`. `Expression` statements are binary or unary arithmetic and logical expressions that access constants or particle and mesh properties. There are first-class concepts to express computations on a complete field, such as differential operators (`DiffOp`), e.g., a gradient. Domain-specific types allow specifying `Interpolate` statement to transform data from particles to meshes and vice-versa.

3.3 OpenPME DSL Syntax

The syntax of the OpenPME DSL allows for both imperative and declarative programs. Its declarative nature allows hiding loops and conditional statements, e.g., when specifying particle interaction and evolution (see Fig. 7, lines 25–26). Another design principle is to remain close to the mathematical notation of the equations to be simulated. This can be seen in Fig. 8, where the Partial Differential Equation (PDE) to be simulated is explicitly expressed. Properties of both particles and meshes can be accessed in bulk which frees the user from having to iterate through a loop (see Fig. 9). This hides OpenFPM's iterator classes, where using templates is mandatory to reach high performance.

3.4 Metamodel for OpenFPM C++

The second metamodel captures features of imperative, object-oriented C++ code that compiles against the OpenFPM library. Model-to-model transformations from the metamodel for simulations allows detecting and avoiding programming mistakes. An excerpt of the model is shown in Fig. 4 as a UML diagram.

This metamodel has a single root concept `File` containing a list of statements. Statements can be, e.g., the declaration of a variable (`VarDecl`) of primitive type or of any OpenFPM class, such as distributed vectors and meshes. Computations are expressed through expression statements (`ExprStmt`), with

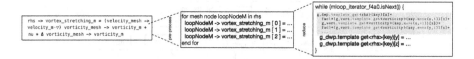

Fig. 5. Code generation for the 3D Navier-Stokes equation

most common ones being arithmetic operations, references to variables, or function calls. The model captures function calls to the OpenFPM API in a generic way, which is only specified during code generation. This allows to adapt to future changes in the OpenFPM API. Template parameters can be inferred by semantic analysis during code generation.

3.5 Model-to-model Transformation and Code Generation

Code generation in OpenPME is implemented by multiple model-to-model transformations refining and optimizing the program, a model-to-model transformation lowering the program to our intermediate representation, and a final text-generation stage that produces C++ code. Models in MPS are directed graphs that have distinct spanning trees, which correspond to abstract syntax trees. A model-to-model transformation maps an input graph to an output graph, where the output graph may consist of elements from different metamodels.

To illustrate how these transformations can be chained, consider ④ in Fig. 9, which computes a complete field stored in a mesh. Because the computation happens on the whole field, the expression is first refined to be contained in a `FieldLoop` (see Fig. 3) to yield a point-wise calculation over all mesh nodes. In a subsequent transformation, the assignment to the property is replaced by an assignment to each dimension of the property, rewriting the expression on the right-hand side to consider only the respective dimension. While lowering the expression into its intermediate representation, the differential operator (here $\nabla \times$) is replaced by finite differences. In the last transformation, the intermediate program is traversed, generating C++ code for each element. For instance, the access to properties of the mesh is replaced by templated C++ code where each property is represented by its template argument (see Fig. 5).

Communication between processors is hidden in the OpenPME DSL and automatically inserted during code generation. For example, particle positions can only change due to particle interactions. Interaction happens, by design, in particle loops. Therefore, whenever the position of a particle is assigned, an OpenFPM communication statement can be inserted after the respective loop.

4 Evaluation

We evaluate OpenPME in three use-cases representative of different types of simulations: First, we consider a molecular dynamics simulation of a Lennard-Jones gas, which is representative of a particles-only simulation. Second, we use

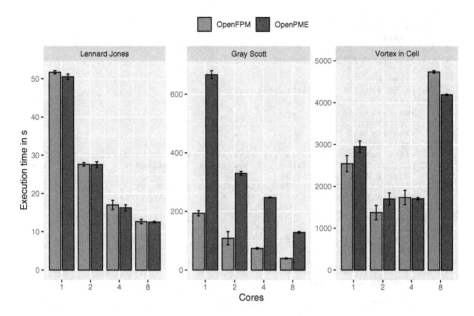

Fig. 6. Execution time of simulations written directly in OpenFPM versus those generated by OpenPME for the three uses-cases with a varying amount of cores.

OpenPME to simulate the Gray-Scott reaction-diffusion model in a representative mesh-only simulation. Third, we simulate incompressible fluid dynamics by solving the Navier-Stokes equations using a Vortex Method representative of hybrid particle-mesh simulations. We compare how many lines of code (LOC) were necessary to implement each use-case. The experiments were repeated 10 times and scaled from 1 to 8 cores on a HPC cluster. The overall result is shown in Fig. 6 where each is discussed in the following sections.

4.1   Lennard-Jones: Particles-Only Simulation of a Discrete Model

We consider a molecular dynamics simulation in three dimensions (3D) where particles represent atoms in a gas that interact according to the pairwise Lennard-Jones [13], exerting a force on both atoms as a function of their distance. Figure 7 shows an excerpt of the corresponding OpenPME program. The simulation process is encapsulated in the methods `interact` and `evolve` defined on the particles sets (lines 14–18). The simulation uses Verlet time stepping, where the particle velocities and positions are advanced alternatively with the force recomputed in-between. Calls to `interact` and `evolve` must specifying the property to use (lines 25–26). The communications required when particles move across processor boundaries, as well as to update the ghost layers with the new velocities from the neighboring processors, are automatically added by the compiler and do not need to be specified in the OpenPME program. Finally, the program computes the potential energy of the system using an imperative

```
 1 Module Lennard Jones
 2  initialization
 3   dimension: 3
 4   domain_size: box((0.0,0.0,0.0),(1.0,1.0,1.0))
 5   boundary_conditions: periodic
 6   ...
 7  simulation
 8   type of simulation: particle-based
 9   Particle sets:
10    name particles
11    properties
12     velocity d:3
13     force d:3
14    Define interact in particles with self as p_force, neighbor as q_force
15     p_force->force += 24.0 * 2.0 * sigma/r^7 - sigma6/r^4 *
16       diff(p_force,q_force)
17    Define evolve in particles with self as p_velocity
18     p_velocity->velocity += 0.5 * dt * p_velocity->force
19    ...
20   Body:
21    ...
22    interact force, particles
23    time loop start:0 stop: 10000
24     ...
25     evolve velocity, particles
26     interact force, particles
27     ...
28     for particle p_energy in particles
29       ...
30        E += 2.0 * sigma12/rn_e^6 - sigma6/rn_e^3 - shift
31     end for
32    end timeloop
33   visualization
34    output file: "particles"
35 end module
```

Fig. 7. OpenPME program for Lennard-Jones molecular dynamics simulation.

iteration over particles (line 29). At the end, particle data is stored in a file and visualized according to the specifications in the visualization phase (line 34). As can be seen in Fig. 6 both approaches perform similar. The implementation with OpenPME requires 57 LOC instead of 151 LOC using OpenFPM.

4.2   Gray-Scott: Mesh-Only Simulation of a Continuous Model

To illustrate the use of OpenPME for a mesh-only simulation of a continuous model, we simulate the Gray-Scott reaction-diffusion model given by the PDEs [7]

$$\frac{\partial u}{\partial t} = D_u \Delta u - uv^2 + F(1-u), \quad \frac{\partial v}{\partial t} = D_v \Delta v + uv^2 - (F+k)v \qquad (1)$$

that govern the space-time evolution of the concentration fields $u(\boldsymbol{x}, t)$ and $v(\boldsymbol{x}, t)$ of two chemicals U and V. D_u and D_v are the diffusion constants of the two chemicals, and F and k are the chemical reaction rates. This model is nonlinear and shows self-organized emergence of sustained spatiotemporal patterns. The OpenPME program in Fig. 8 shows a 3D model on a regular Cartesian mesh

```
1  Module Gray Scott                    17  Body:
2   initialization                      18   ...
3   ...                                 19   Load Old from "init_mesh.hdf5"
4   simulation                          20   time loop start: 0 stop: 5000
5    type of simulation: mesh-based     21    New->u = Old->u + dt * du *
6    Meshes:                            22     △Old->u - Old->u *
7     name Old                          23     Old->v^2 + F *
8     properties                        24     1.0 - Old->u
9      u d:1                            25    New->v = Old->v + dt * dv *
10     v d:1                            26     △Old->v + Old->u *
11     size{128,128,128}                27     Old->v^2 - F + k * Old->v
12     name New                         28    copy from New to Old
13     properties                       29    Resync Ghost Old<u,v>
14      u d:1                           30   end timeloop
15      v d:1                           31   ...
16      size{128,128,128}               32  end module
```

Fig. 8. OpenPME program for Gray-Scott simulation on a mesh.

of $128 \times 128 \times 128$ nodes, discretizing the two scalar ($d = 1$) fields U and V (lines 6–16). Time stepping is done using the explicit Euler method. The Laplace operator Δ over the continuous fields u and v is discretized using central finite differences. The corresponding OpenPME expressions closely mirror the terms and notation in the model PDEs. The DSL compiler can automatically insert OpenFPM communication operators and distributed OpenFPM data structures. In this case, two identically distributed meshes are required in order to read and write simultaneously. We are on average $3.25\times$ slower than OpenFPM original code. The reason is that consecutive bulk access of mesh properties results in independent mesh loops that are merged in the hand-written OpenFPM C++ program which requires more communication and computation. This is one of many optimizations for the code generation we are targeting in future work. For this simulation, OpenPME requires 40 LOC instead of 100 LOC in OpenFPM.

4.3 Vortex-in-Cell: Hybrid Particle-Mesh Simulation of a Continuous Model

We simulate incompressible fluid dynamics as governed by the 3D Navier-Stokes equations in vorticity form using a Vortex Method simulation [4]. This is a hybrid particle-mesh method where particles represent fluid elements that carry vorticity, and a mesh is used to solve the Poisson equation for the flow velocity field, which in turn moves the vorticity particles.

The OpenPME program in Fig. 9 starts by initializing the flow field on a mesh and creating particles at the mesh nodes using the `remesh` statement ①. After initializing the particles, the simulation enters the time loop with a fixed time step `dt = 0.0125`. In each time step, the vorticity is interpolated from particles to mesh in order to solve the Poisson equation for the flow velocity on the mesh ②. Computing and storing the right-hand side of the Poisson equation requires a second mesh `phi` ③. Next, the vortex stretching term is computed on the mesh, where the Laplace operator Δ is automatically discretized by central finite differences ④. As all properties were calculated on meshes, we have to

Fig. 9. OpenPME program for a Vortex Method simulation of incompressible flows.

interpolate back to the particles ⑤. This is used to update particle positions ⑥. The generated code performs on average as OpenFPM. For this simulation the parallel efficiency drops the more cores are used [11]. For the measured problem size 2 cores give the best performance. This implementation of a Vortex Method in OpenPME requires 73 LOC instead of 508 LOC in OpenFPM which is a 7-fold reduction.

5 Related Work

Many frameworks have been proposed to ease the use of HPC systems for scientific computing and simulations.

Among internal DSLs, Blitz++ [29] is a C++ template library and DSL for defining mesh stencils from a high-level mathematical specification library. Liszt [5] is a portable DSL, embedded into Scala, to implement PDE solvers on meshes, supporting different parallel programming models. Freefem++ [9] is a DSL for finite-element methods allowing users to define analytic and finite-element functions using abstractions like meshes and differential operators.

Other approaches use external DSLs. Examples include the Unified Form Language (UFL) [1] to define PDEs and finite-element simulations. There are several optimizing compilers that generate low-level code from UFL, e.g., the FEniCS Form Compiler (FFC) [15]. The Firedrake project [22] is an extension to FEniCS that adds composing abstractions like parallel loop operations. Users express simulations in a high-level specification translated to an abstract syntax tree by the COFFEE [18] compiler, able to apply suitable low-level optimizations.

Many DSLs use domain-specific optimizations. Examples include PyOP2 [23], a DSL for unstructured-mesh simulations executing numerical kernels in parallel, which enhances FFC to generate optimized kernels from finite-element expressions written in UFL. Devito [17] is a DSL that uses symbolic Python expressions to provide optimized stencil computations. Saiph [19] is an optimizing DSL for computational fluid dynamics, implemented in Scala, with a custom compiler to translate code to an intermediate representation that is subsequently translated to C++. In the CFDlang [24] DSL, the locality of operators is exploited to optimize tensor operations for performance in computational fluid dynamics.

6   Conclusions and Future Work

We presented OpenPME, a PSE for particle, mesh, and hybrid particle-mesh simulations on parallel and high-performance computers. It is based on two novel metamodels, covering the domain of particle, mesh, and hybrid particle-mesh simulations, while also covering the imperative object-oriented C++ API of OpenFPM. It supports simulations of all types for both continuous and discrete models. The OpenPME compiler uses a sequence of model-to-model transformations between the two metamodels in order to automatically inject the required communication and synchronization operations for distributed-memory codes, and to translate OpenPME programs to OpenFPM C++ code. This leverages the scalability, GPU support, and performance of OpenFPM to generate efficient HPC applications. The OpenPME language stack is modularly composed of multiple sub-languages, improving maintainability and extensibility. The OpenFPM metamodel is only specified during code generation, hence flexibly adapting to future API changes in OpenFPM.

OpenPME programs can be developed independently at a high level of abstraction, close to the mathematical model. It frees users from having to explicitly deal with the C++ template constructs of OpenFPM, from having to use explicit iterators, and from having to decide on and place communication abstractions in a distributed-memory parallel program. This enables high-level optimizations not otherwise possible, renders simulation programs more compact, easier to read, and more accessible to new users. In the benchmarks presented here the code size of simulations is reduced up to a factor of 7 when implementing them in OpenPME versus directly writing C++ code for OpenFPM. The generated code performs in general as the hand-written C++ code using OpenFPM.

In the future, we plan to further improve OpenPME by adding features to support optimized memory layouts in OpenFPM, like space-filling curves as sub-domain or mesh-node indices in order to improve cache efficiency. The high-level OpenPME syntax cleanly separates the model specification from the choice and parameters of the numerical methods used to simulate the model. This will enable auto-tuning in the compiler to automatically select simulation parameters (e.g., mesh resolution or time step size) and discretization methods to reduce the simulation runtime required to reach a certain target accuracy. A more declarative syntax could allow domain experts to apply manual optimizations and gives us more control over the generated code. The OpenPME PSE – the DSL, IDE, and compiler – are available as open-source software.[2]

Acknowledgement. This work was partly supported by the German Research Foundation (DFG) within the OpenPME project (number 350008342) and by the Federal Ministry of Education and Research (Bundesministerium für Bildung und Forschung, BMBF) under funding code 031L0160 (project "SPlaT-DM – computer simulation platform for topology-driven morphogenesis").

References

1. Alnæs, M.S., Logg, A., Ølgaard, K.B., Rognes, M.E., Wells, G.N.: Unified form language: a domain-specific language for weak formulations of partial differential equations. ACM Trans. Math. Softw. **40**(2), 1–37 (2014)
2. Awile, O., Mitrovic, M., Reboux, S., Sbalzarini, I.F.: A domain-specific programming language for particle simulations on distributed-memory parallel computers. In: Proceedings of III International Conference on Particle-Based Methods (PARTICLES). Stuttgart, Germany (2013)
3. Baumgartner, G., et al.: Synthesis of high-performance parallel programs for a class of ab initio quantum chemistry models. Proc. IEEE **93**(2), 276–292 (2005)
4. Cottet, G.H., Koumoutsakos, P.: Vortex Methods - Theory and Practice. Cambridge University Press, New York (2000)
5. DeVito, Z., et al.: Liszt: a domain specific language for building portable mesh-based PDE solvers. In: Proceedings of SC 2011, p. 1. ACM Press (2011)
6. Gallopoulos, E., Houstis, E., Rice, J.R.: Computer as thinker/doer: problem-solving environments for computational science. IEEE Comput. Sci. Engrg. **1**(2), 11–23 (1994)
7. Gray, P., Scott, S.K.: Sustained oscillations and other exotic patterns of behavior in isothermal reactions. J. Phys. Chem. **89**(1), 22–32 (1985)
8. Gupta, A., Incardona, P., Aydin, A.D., Gumhold, S., Günther, U., Sbalzarini, I.F.: An architecture for interactive in situ visualization and its transparent implementation in OpenFPM. In: Proceedings of Workshop In Situ Infrastructures for Enabling Extreme-scale Analysis and Visualization (ISAV), SC 2020, pp. 20–26. ACM (2020)
9. Hecht, F.: New development in Freefem++. J. Numer. Math. **20**(3–4), 251–266 (2012)
10. Hockney, R.W., Eastwood, J.W.: Computer Simulation using Particles. Institute of Physics Publishing, Bristol (1988)
11. Incardona, P., Leo, A., Zaluzhnyi, Y., Ramaswamy, R., Sbalzarini, I.F.: OpenFPM: a scalable open framework for particle and particle-mesh codes on parallel computers. Comput. Phys. Commun. **241**, 155–177 (2019)
12. Iwasawa, M., Tanikawa, A., Hosono, N., Nitadori, K., Muranushi, T., Makino, J.: Implementation and performance of FDPS: a framework for developing parallel particle simulation codes. Publ. Astron. Soc. Japan **68**(4), 54 (2016)
13. Jones, J.E.: On the determination of molecular fields. II. from the equation of state of a gas. Proc. Roy. Soc. London A **106**(738), 463–477 (1924)

---

[2] OpenPME is available under https://github.com/Nesrinekh/OpenPME.

14. Karol, S., Nett, T., Castrillon, J., Sbalzarini, I.F.: A domain-specific language and editor for parallel particle methods. ACM Trans. Math. Softw. **44**(3), 34:1–34:32 (2018)
15. Logg, A., Ølgaard, K.B., Rognes, M.E., Wells, G.N.: FFC: the FEniCS form compiler. In: Logg, A., Mardal, KA., Wells, G. (eds.) Automated Solution of Differential Equations by the Finite Element Method, vol. 84, pp. 227–238. Springer, Berlin (2012) https://doi.org/10.1007/978-3-642-23099-8_11
16. Luporini, F., Ham, D.A., Kelly, P.H.J.: An algorithm for the optimization of finite element integration loops. ACM Trans. Math. Softw. **44**(1), 1–26 (2017)
17. Luporini, F., et al.: Architecture and performance of Devito, a system for automated stencil computation. ACM Trans. Math. Softw. **46**(1), 1–28 (2020)
18. Luporini, F., et al.: Cross-loop optimization of arithmetic intensity for finite element local assembly. ACM Trans. Arch. Code Opt. **11**(4), 1–25 (2015)
19. Macià, S., Mateo, S., Martínez-Ferrer, P.J., Beltran, V., Mira, D., Ayguadé, E.: Saiph: Towards a DSL for high-performance computational fluid dynamics. In: Proceedings of Real World DSL Workshop 2018, pp. 6:1–6:10. ACM, New York (2018)
20. Oñate, E., Owen, D.R.J.: Particle-based methods: fundamentals and applications. Springer, New York (2013) https://doi.org/10.1007/978-94-007-0735-1
21. Püschel, M., et al.: SPIRAL: code generation for DSP transforms. Proc. IEEE **93**(2), 232–275 (2005)
22. Rathgeber, F., et al.: Firedrake: automating the finite element method by composing abstractions. ACM Trans. Math. Softw. **43**(3), 1–27 (2016)
23. Rathgeber, F., et al.: Pyop2: a high-level framework for performance-portable simulations on unstructured meshes. SC Companion, pp. 1116–1123 (2012)
24. Rink, N.A., et al.: CFDlang: high-level code generation for high-order methods in fluid dynamics. In: Proceedings of 3rd International Workshop on Real World DSLs, pp. 5:1–5:10. ACM, New York (2018)
25. Sbalzarini, I.F.: Abstractions and middleware for petascale computing and beyond. Intl. J. Distr. Syst. Tech. **1**(2), 40–56 (2010)
26. Sbalzarini, I.F.: Modeling and simulation of biological systems from image data. Bioessays **35**(5), 482–490 (2013)
27. Sbalzarini, I.F., Walther, J.H., Bergdorf, M., Hieber, S.E., Kotsalis, E.M., Koumoutsakos, P.: PPM - a highly efficient parallel particle-mesh library for the simulation of continuum systems. J. Comput. Phys. **215**(2), 566–588 (2006)
28. Šinkarovs, A., Bernecky, R., Vießmann, H.N., Scholz, S.B.: A rosetta stone for array languages. In: Proceedings of 5th ACM SIGPLAN International Workshop on Libraries, Languages, and Compilers for Array Programming, pp. 1–10 (2018)
29. Veldhuizen, T.L.: Blitz++: the library that thinks it is a compiler. In: Langtangen, H.P., Bruaset, A.M., Quak, E. (eds.) Advances in Software Tools for Scientific Computing, vol. 10, pp. 57–87. Springer, Berlin (2000) https://doi.org/10.1007/978-3-642-57172-5_2

Building a Prototype for Easy to Use Collaborative Immersive Analytics

Daniel Garrido[1,2]([✉]), João Jacob[1,2], and Daniel Castro Silva[1,2]

[1] Faculty of Engineering, University of Porto, Porto, Portugal
{dlgg,joajac,dcs}@fe.up.pt
[2] Artificial Intelligence and Computer Science Laboratory (LIACC),
Rua Dr. Roberto Frias S/n, 4200-465 Porto, Portugal

Abstract. The increase in the size and complexity of today's datasets creates a need to develop and experiment with novel data visualization methods. One of these innovations is immersive analytics, in which extended reality technologies such as virtual reality headsets are used to present and study data in virtual worlds. But while the use of immersive analytics dates back to the end of the 20th century, it wasn't until recently that collaboration in these data visualization environments was taken in consideration. One of the problems currently surrounding this field is the lack of availability of easy to use cooperative data visualization tools that take advantage of the modern, easily attainable head mounted display virtual reality solutions. This work proposes to create an accessible collaborative immersive analytics framework that users with low virtual reality background can master, and share, regardless of platform. With this in mind, a prototype of a visualization platform was developed in Unity3D that allows users to create their own visualizations and collaborate with other users from around the world. Additional features such as avatars, resizable visualizations and data highlighters were implemented to increase immersion and collaborative thinking. The end result shows promising qualities, as it is platform versatile, simple to setup and use and is capable of rapidly enabling groups to meet and analyse data in an immersive environment, even across the world.

Keywords: Immersive analytics · Data visualization · Virtual reality · Immersive environments · Immersive collaboration

1 Introduction

With the establishment of the importance of data in today's world, which is increasing in size and complexity, it became necessary to create new ways to visualize and analyse it [9]. Immersive analytics is one of the areas that brings forward novel visualization and interaction models to the area of data visualization and analytics.

Immersive Analytics (IA), as defined by Dwyer *et al.*, is "the use of engaging, embodied analysis tools to support data understanding and decision making".

© Springer Nature Switzerland AG 2021
M. Paszynski et al. (Eds.): ICCS 2021, LNCS 12742, pp. 628–641, 2021.
https://doi.org/10.1007/978-3-030-77961-0_50

Its goal is to allow the user to close the distance between himself and the data through the use of virtual environment technologies [15].

The increasing demands of large and multidimensional datasets are also making it unfeasible for a single expert to be able to tackle the analysis of large quantities of data. It has become necessary to create multidisciplinary teams with varying areas of expertise and even using different methodologies to solve problems when using these large datasets. This created new challenges in data visualization, such as determining how to present the data to different users, how to allow simultaneous data manipulation by multiple users, or how to let users socialize, among other challenges [22].

Research has proven that groups of people collaborating perform better in a variety of tasks than when alone, producing more accurate results [21]. Immersion also has a positive impact in user performance when used for collaboration [30]. Unsurprisingly, a collaborative immersive environment brings these benefits to the area of data analytics [3].

While immersive analytics is a research topic that is garnering interest, collaboration has yet to be well studied. Since there is a clear benefit in using collaboration to aid immersive data visualization, it should be easier for everyone to have access to this resource, in an easy-to-use package.

Furthermore, several immersive analytics experts (such as Chandler *et al.* [9], Hackatorn and Margolis [19], and Skarbez *et al.* [35], to cite a few) have pointed out that a huge research opportunity is the integration of collaboration in IA, creating Collaborative Immersive Analytics (CIA). Nowadays, with millions of people required to work from home, being able to "meet" with colleagues in an immersive environment to discuss, analyse and solve problems is welcoming.

To achieve this, an extension of an open-source immersive data visualization tool to include collaboration capabilities is proposed. The goal is to allow two or more users (immersed or not) to share a virtual environment where they can discuss a common data visualization of their choice. In addition, creating this visualization and setting up the virtual environment should be accessible to those not well versed in Virtual Reality (VR)/Data Visualization (DataViz).

This resulted in the implementation of a Unity3D framework that built upon the capabilities of the already existing DXR [34], by integrating it into a client-server room-based network handled by PHOTON PUN. User avatars were also implemented, alongside interaction tools that promote collaboration and data exploration. The expected contributions of this work are manifold. First off, the resulting research gap analysis from the state-of-the-art in CIA. Second, a proposed design for a CIA framework. And last, a prototype for easy to use CIA.

The remainder of this document is structured as follows: Sect. 2 explores some of the most recent related work in this field; Sect. 3 presents the proposed solution, the chosen technologies and the system architecture; Sect. 4 details the design choices and development of the prototype; Sect. 5 demonstrates how the platform can be used for collaborative immersive analytics and outlines the design of the user studies; and finally, Sect. 6 concludes the document and presents the plans for future work.

2   Literature Review

The field of immersive collaboration is extensive, as it can be applied to many areas. This literature review starts with an overview of some of these areas, followed by a study of a collection of immersive analytics frameworks, and ending with a review of works related to this project.

2.1   Collaboration in Immersive Environments

The study of collaboration in immersive environments has been expanding in the past few years, with the number of areas taking advantage of this concept in virtual environments rising.

Education in one of such areas that fits perfectly to incorporate collaboration, as the act of teaching is in itself collaborative. This led to the creation of immersive worlds where students can see and experience things they wouldn't otherwise be able to, like the virtual Neurorobotics Lab created by Matthes *et al.* [29]. Zikky *et al.* were able to take students to a trip through the solar system. Their platform also opens the possibility for teachers and students to meet when far away and can accommodate larger class sizes [39].

Collaborative content creation is an area that faces a lot of challenges, especially not stepping on the work of other users. This is also true in immersive environments and is what Xia *et al.* worked on. They presented 3 new interaction concepts, which together give single users more powerful editing tools and groups of users less friction between them [37]. The work of Pereira *et al.* presents a generalized framework for collaboration in immersive environments [32].

Interestingly, another area gaining traction for immersive collaboration is human-robot interaction. Training robots to collaborate with human beings can be dangerous, and thus immersive simulations were created to allow humans and machines to collaborate safely during testing. The work by Matsas *et al.* [28] and de Giorgio *et al.* [18] are perfect examples of what immersive cooperation can achieve in this area.

2.2   Immersive Analytics Frameworks

Frameworks capable of generating many different types of 2D and 3D data visualizations for flat screens exist for quite some time. Sometimes mentioned as visualization authoring tools, these tools have evolved to become both easy to use and powerful, with Protovis [4], D3 [5] and Vega-Lite [33] being examples of notable ones. However, only recently frameworks made specifically for immersive analytics have emerged.

These frameworks facilitate the creation of this type of visualization by providing users with the tools required to present information efficiently and legibly, but without the necessary knowledge and effort associated with implementing them from the ground up.

Chronologically, the first example found of this kind of frameworks is Glance, from 2016 by Filonik *et al.*. It focused on providing dynamic representations of multidimensional data using geometric primitives and transformations. While pioneering in this area, this framework lacks the capability to easily be integrated in other works, something the more recent ones possess [17].

On the other hand, the first VR framework dedicated to data visualization and analysis was ImAxes in 2017 by Cordeil *et al.* [11]. It is based on the abstraction of data into physical axes that the user can grab and combine to create data visualizations. With this, the user can create 2D and 3D scatter plots as well as parallel coordinate plots, without interacting with any menus. The authors also noted that this level of physicality in the environment and interactions is advantageous for collaborative analysis, as physical cues like "gesticulation, passing, personal and shared spaces" facilitates collaboration, making it more instinctive. ImAxes was later used as the basis for a user study in 2020, in which participants were only asked to explore a dataset, and produced interesting results [2]. Of note, they found that legibility and fatigue weren't a problem, even for 60 min sessions, and that users rarely moved, preferring to stay in place and work around them.

In 2019, three additional frameworks were presented that expanded on the concept of ImAxes, further simplifying the process of creating immersive data visualizations, even for those who are not proficient in the area. Once again, Cordeil *et al.* [10], and Sicat *et al.* [34] created toolkits for the Unity3D game engine called AITK (Immersive Analytics ToolKit) and DXR (Data visualization applications for eXtended Reality), respectively. Both were designed to work in virtual and mixed reality environments and accept common CSV (Comma-Separated Values) files as data input. The major difference between the two is that IATK is made specifically to visualize and analyse graphs, while DXR is more open to the type of data representation, but doesn't include the same interactive tools for data analysis that IATK has.

The last studied framework was developed upon the WebVR specification to run in web browsers. Like the previous two, VRIA utilizes a CSV file for input data and is restricted to 3D visualization of graphs like scatter plots and bar charts. User tests revealed that while subjects took longer to perform the given analysis tasks, they felt more engaged in the task at hand [7].

2.3   Related Work

The presented works focus on the collaborative immersive analytics projects that make use of Head Mounted Displays (HMD), since this is the technology that will be used for this work. The first example, authored by Donalek *et al.*, dates to 2014 and used the then recent Oculus Rift DK1. They started by using OpenSimulator, an open-source platform for sharing immersive environments. A data plot was then added through the platform's scripting language. This concept was then replicated in Unity3D to add more advanced features. This work showed it was possible to create good immersive visualizations that can be used with much cheaper hardware than the previously used CAVE systems [13].

Two year later, in 2016, Cordeil *et al.* presented a couple of immersive solutions for air traffic controllers. The first one situated the user in a virtual remote air traffic controller. The second one allowed air traffic experts located in distant sites to meet in a virtual environment where they can study flight trajectories together. The users' position is shared through the use of avatars, they can point and select paths as well as filter them, features that contribute to the collaboration between users [12].

A different approach to collaborative immersive analytics was presented by Butscher *et al.* in 2018. Instead of relying in a virtual environment, cameras in the front of the HMD capture the real world to create an augmented reality. This was used in conjunction with a touch-sensitive tabletop for user input. A 3D parallel coordinates visualization placed on top of the interactive table allowed the users to analyse multidimensional data [8].

A similar approach was presented by Ens *et al.* in 2020, in the form of the system called Uplift, which also uses a tabletop display in conjunction with augmented reality HMDs [16]. This project focused on accessibility and ease of use and targets scenarios of collaboration in quick bursts. Where this approach sets itself apart is in the interaction methods, provided by "tangible widgets". These are objects that are tracked in space, and which users can manipulate to modify the presented visualizations, thus increasing accessibility, engagement, as well as promoting collaboration through shared visualization controls.

A distinct solution for collaborative immersive analytics was researched by Sun *et al.* in 2019, by using a large, wall-sized display where various users can gather to collaborate while using OST-HMDs (Optical See-Through Head-Mounted Displays) [36]. While this implementation of CIA is similar to the two previous ones, the researchers introduced a new concept: giving different users distinct levels of access to information. Information displayed on the wall is open to all users, but it is supplemented with information each user can only see through his HMD. To promote collaboration, the users must verbally request information controlled by other users. The authors noted that in an initial phase, users tend to work alone, and progressively started collaborating as time went by, when they realised they required additional information.

The remainder of the reviewed literature dates from 2019, showing that a new interest in this topic has again emerged, probably due to the easy access to new generation, inexpensive HMDs. The first of these works was developed by Nguyen *et al.* and explores the visualization of multidimensional data using Star Coordinates and Star Plot graphs. The users can interact with the data simultaneously, and are represented by avatars to show what they are doing and improve social connectivity. Data analytics tools are also provided to the users to help them create decision trees as "a way to visualise the data in a top-down categorical approach", facilitating data analysis [31].

Another interesting project was the work of Bailey *et al.*. They developed a prototype to visualize microscopy data in virtual reality to which multiple users could join. While the main focus of the work is visualizing the microscopy data, they also spent some effort in implementing embodiment features such as full body avatars and personalized user faces. Interaction was handled through "physical" toggles (instead of a 2D Graphical User Interface (GUI)), such as switches and sliders, that the user can interact with his own hands [1].

The final reviewed work is FIESTA, a system for collaborative immersive analytics created by Lee *et al.*. It focuses on giving the users freedom to utilize the virtual space as they see fit, like creating visualizations and position them as they please. Several users can do this at the same time in the same room and then visually and verbally communicate their insights with each other. This system focuses on creating a shared, immersive data analysis environment for teams to present and discuss data through the use of several conventional 2D/3D visualizations like scatter plots [26].

A year later, in 2020, the authors conducted a user test based on the framework to ascertain collaborative performance with different visualization types [25]. One of the conclusions was that users usually start by creating individual work zones and only later join together to share their thoughts and findings. During this initial phase users engaged in "territorial" behaviour, by isolating in their zones and not using visualizations that were not created by them.

In conclusion, the works by Donalek *et al.*, Nguyen *et al.*, Butscher *et al.* and Ens *et al.* present platforms which take advantage of 3D space and immersion to present data in a novel way to groups of people collaborating. This means that with these systems, the user has a limited choice of type and number of visualizations. This is especially true for the systems presented by Cordeil *et al.* and Bailey *et al.*, which were designed to visualize a specific type of data (flight trajectories and microscopy images, respectively). The FIESTA project by Lee *et al.* gives the users freedom to create as many visualization as they wants and arrange them afterwards in a collaborative, shared environment. It also simplifies the data input process, since the visualizations are generated by IATK [10]. However, it is currently limited to 2D and 3D scatterplots and the authors have not made it open-surce, like many of the other similar frameworks such as DXR, IATK and ImAxes.

3   Proposed Solution

As mentioned in Sect. 1, the objective is to create a CIA framework more approachable to users such as researchers and scientists. The analysed state of the art shows that a solution that combines freedom of visualization choice and ease of use is not available. The proposed solution is to create an immersive environment for modern VR HMDs that allows the user to input his data and visualization styling and collaborate with other users in the same virtual space.

3.1   Technology Choices

As the majority of the mentioned projects in Sect. 2, this solution will be build on top of the Unity3D game engine. Other alternatives such as Unreal Engine don't offer the same quantity and variety of community developed tools useful for this project. Unity3D enables the use of existing tools made specifically for VR that suit this implementation, as detailed below. It will also allow authoring the project for a plethora of client devices' hardware and software combinations.

Before implementation, it is necessary to choose the tools that will give shape to the solution. The first decision is what framework to use to create the visualizations. Several options presented in Sect. 2.2 are suitable, although only three of them are open-source and ready to integrate in Unity3D: ImAxes [11], DXR [34] and IATK [10]. ImAxes was first rejected, as it provided poor customization support and visualization type variety. As mentioned previously, IATK and DXR are very similar in approach, but thanks to its superior visualization style diversity, better documentation and additional support for both AR and VR HMDs, DXR was chosen. Additionally, it was designed to strike a good balance between the ease of use for inexperienced users and the flexibility in creating data visualizations, akin to Vega-Lite [33].

To interface with VR in Unity3D the VRTK[1] toolkit was used. It provides seamless integration and runtime switching of the most popular VR Software Development Kits (SDKs), and many premade scripts for locomotion and interaction. It also includes VRSimulator, making it possible for users without a VR headset to also join the immersive environment through the desktop and interact with other users. A potential alternative would be OpenVR, although it's lack of desktop emulator and closed-source nature contradict the ethos of this solution. Additionally, VRTK is activly supported by the community.

Lastly, a network framework capable of making two or more users share a virtual location and interact with each other synchronously was necessary. Two popular networking Unity3D free add-ons are currently widely regarded: Mirror and PHOTON PUN[2]. Mirror is based on UNET, the stock Unity3D networking solution, and is designed to accommodate large scale networks with several clients connected to one host (server). On the other hand, PHOTON PUN is designed around smaller lobbies that the users find through a central server hosted by PHOTON. Subsequent traffic is also handled by the central PHOTON server. While Mirror is regarded to be more scalable, the simplicity of PHOTON PUN lobby centered design and in the cloud central server simplifies between user remote connection and lobby sharing. For these reasons PHOTON was chosen as the network backbone of this solution.

3.2 System Architecture

PHOTON PUN works in a client-server model, with a central server hosted by PHOTON to which different Unity3D instances can connect via an AppID. Users can join different rooms, but for the purpose of this prototype they will all connect to the same one, as seen in Fig. 1. The first user to connect is the one to instantiate the visualization, becoming its master.

---

[1] More information available online at https://vrtoolkit.readme.io/v3.3.0.
[2] More information available online at: https://www.photonengine.com/pun.

When it comes to the individual instances, VRTK is the glue that connects all the other modules together. It is responsible for handling the virtual reality hardware SDKs, which in turn are responsible for presenting the user with the visualization and gathering the required user data such as the headset and controllers' position. This information is then used to update the user's avatar position. VRTK also manages the interactions with the data visualization, which is externally created by DXR. PHOTON's client is responsible for updating the data visualization and user avatars in the server, and receiving the updates from the other users. This architecture can be visualize in Fig. 2.

4   Implementation

Margery *et al.* set the base level of cooperation in virtual environments at the ability of being able to see other users and communicate with them [27]. Subsequently, the first development step was to create a virtual scene that two or more users can share simultaneously and be able to visualize each other's presence.

To achieve this, PHOTON PUN was used to create the necessary network infrastructure, which consists of a client-server architecture hosted in PHOTON's cloud, allowing users to join the visualization. When a user joins, a representation of the headset and hand controllers is created. This simplified avatar follows the movements of the user in the 3D space, which are then shared with the other connected users via the PHOTON connection with the *Photon View* and *Photon Transform View* scripts.

When it came to integrating the data visualization component into the developed system, a couple of design considerations had to be weighted, based on the state-of-the-art work presented before, in Sect. 2. The first one was related to the amount of visualizations available for the users to share. All but one of the referenced contributions focus on presenting only one. FIESTA, the collaborative immersive analytics framework developed by Lee *et al.*, is the only one

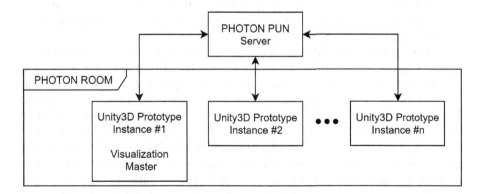

Fig. 1. Integration of PHOTON PUN client server architecture to enable collaboration between users.

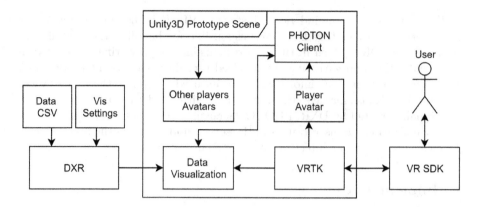

Fig. 2. Architecture of the proposed visualization scene.

that enables multiple visualizations, with specific features for tracking visualization ownership. While a departure from previous contributions, it encourages the users to seclude from each other and work alone, defeating the purpose of enabling collaboration.

The second consideration was the size of the visualization. All the presented papers use visualizations with small sizes, that can be seen as a whole by the user without moving his head. As explored by Yang *et al.* [38] and Kraus *et al.* [24], this size of visualization is more often advantageous when compared to room scale ones. However, both authors agree that this style of visualization is regarded by the users as the most immersive and engaging. Since both visualization sizes have different strong points, it is up to user to decide which one to use.

One of the example visualization in DXR was then imported to the scene as a prefab. The visualization is instantiated by the first user to join the room, which at this point becomes the visualization master. In the current prototype iteration only the visualization master of an object can manipulate it, to avoid having to deal with synchronization problems. Despite this limitation, the master can grab the visualization and position it in any orientation and scale it up or down by grabbing with both hands.

In the context of collaborative analytics, simply sharing the visualizations and presenting them to other users is not enough. One of the most important aspects of visualization collaboration is guaranteeing that all users share a common ground, i.e., the shared information is perceived the same way by all the participants, eliminating any possible ambiguities [20]. Through PHOTON, it is already guaranteed that the visualization is viewed equally by all, but when discussing individual data points it might become hard to convey which one is being referenced. To prevent this and help the users indicate which part of the visualization is being discussed, a pointer and highlighter interaction was developed and added to the prototype. The user can use the HMD hand controller to point a laser that selects the intersected data point on the visualization, highlighting it and showing its information to that user. To the other users sharing the

(a)   (b)

(c)   (d)

Fig. 3. Two users sharing an immersive environment and analysing data. Figure (a) shows one user grabbing the visualization and orienting it towards the other user; In Fig. (b) the users share the same visualization, but scaled up; In Fig. (c) the user is highlighting a specific data point, which the other user in Fig. (d) can also see as being highlighted.

visualization, this selected data point will be highlighted, preventing uncertainties when discussing specific data points.

5   Prototype Testing

To test the developed functionalities, a virtual environment was created with a simplified version of an example visualization created with DXR. For this test, two instances of the prototype were executed side by side in the same machine. The first instance (the master of the visualization) used an Oculus Rift CV1 with Oculus controllers while the second instance used the keyboard and mouse to control a virtual VR system through the built-in VR Simulator in VRTK, with similar functionality to the real HMD.

As can be seen in Fig. 3a, both users can see each other's avatar, allowing them to know where they are and where their hands are pointing to. The participants can then gather around the visualization and the master can manipulate it to provide a better view. To obtain a more immersive data visualization, it can be scaled up, turning the play area into the visualization itself, as shown in Fig. 3b. Using the controllers, the users can point to a specific data element, highlighting it to the other viewer, as depicted in Figs. 3c and 3d.

Being in a immersive environment with nothing but the data and other users increases attention and engagement. Additionally, a big difference in immersion is felt when using an enlarged visualization rather than a small one. Having the data surrounding the users creates a feeling of being inside or part of the visualization. It is also much easier for users to have a meaningful discussion about the data, thanks to the ability to "travel" to the area of interest. Aiming the pointer to highlight specific data points also becomes easier due to the increased size of the data points.

The following test scenarios have been projected to evaluate the implemented features in this prototype:

a) *Immersive Collaboration with Interactions*: Immerse a team of people in the visualization and task them with finding specific insight from it. The users have access to all the developed tools, such as avatars and highlighter.
b) *Immersive Collaboration without Interactions*: The same as the previous one, but the users don't have access to the developed tools.
c) *Regular Collaboration*: The same as scenario a), but using conventional data analytics visualization tools.
d) *Single User*: The same as scenario a), but with a team of only one person.

For all these tests the teams will be evaluated by the same performance metrics. These include time to complete tasks, correctness of the answers and number of interactions by type. To evaluate the amount of collaboration both verbal communication time and amount of shared focus time can be used.

In the end, questionnaires will be handed out to the participants to gather their opinions on the prototype, their preferences and perceived immersion and level of collaboration. This includes the Simulator Sickness Questionnaire (SSQ) [23] and System Usability Scale (SUS) [6], which are commonly used in this field. A less acknowledged questionnaire, developed by Dupont *et al.*, will also be used. It was especially designed to evaluate immersive and collaborative performance in applications using Immersive Collaborative Environments [14].

6   Conclusion and Future Work

An immersive data analytic environment was created that allows users to collaborate in analysing information using modern virtual reality HMD technology. By showing a representation of the position of the other participants a greater sense of immersion and cooperation is given to the users. The inclusion of a data highlighting system also aids the collaborative discussion process. Furthermore, the use of DXR and PHOTON PUN to create the visualization and shared virtual space, respectively, contributes to the main objective of creating an immersive collaborative workspace in which it is simple to input data, change visualization styles and setup the necessary networking, even for remote scenarios. This work also helps further extend the state-of-the-art of this growing field.

The developed work stands as a prototype that can now be expanded into a fully fledged visualization tool. Before implementing other features, the user tests detailed above will be carried out. Afterwards, it is crucial to add a verbal communication system to complement the already implemented positional communication through the avatars. Subsequently, more data interaction tools must be added, such as filtering, creation of clusters and providing statistical analysis tools on those clusters. Furthermore, a new networking architecture could be integrated to bypass the restrictions of ownership and manipulation of the visualization, to further increase cooperation. Finally, implementing an asynchronous collaboration system is worthy of consideration, as it is currently even more under-researched than synchronous CIA.

References

1. Bailey, B.J., et al.: Multi-user immersive virtual reality prototype for collaborative visualization of microscopy image data. In: The 17th International Conference on Virtual-Reality Continuum and Its Applications in Industry. VRCAI 2019, Association for Computing Machinery, New York (2019)
2. Batch, A., et al.: There is no spoon: evaluating performance, space use, and presence with expert domain users in immersive analytics. IEEE Trans. Visual Comput. Graphics **26**(1), 536–546 (2020)
3. Billinghurst, M., Cordeil, M., Bezerianos, A., Margolis, T.: Collaborative immersive analytics. Immersive Analytics. LNCS, vol. 11190, pp. 221–257. Springer, Cham (2018). https://doi.org/10.1007/978-3-030-01388-2_8
4. Bostock, M., Heer, J.: Protovis: a graphical toolkit for visualization. IEEE Trans. Visual Comput. Graphics **15**(6), 1121–1128 (2009)
5. Bostock, M., Ogievetsky, V., Heer, J.: D^3 data-driven documents. IEEE Trans. Visual Comput. Graphics **17**(12), 2301–2309 (2011)
6. Brooke, J.: Sus: A 'quick and dirty' usability scale. In: Usability Evaluation In Industry, 1 edn, pp. 189–194. Taylor & Francis (1996)
7. Butcher, P.W.S., John, N.W., Ritsos, P.D.: Vria - a framework for immersive analytics on the web. In: Extended Abstracts of the 2019 CHI Conference on Human Factors in Computing Systems, CHI EA 2019, Association for Computing Machinery, New York (2019)
8. Butscher, S., Hubenschmid, S., Müller, J., Fuchs, J., Reiterer, H.: Clusters, trends, and outliers: how immersive technologies can facilitate the collaborative analysis of multidimensional data. In: Proceedings of the 2018 CHI Conference on Human Factors in Computing Systems, CHI 2018, Association for Computing Machinery, New York (2018)
9. Chandler, T., et al.: Immersive analytics. In: 2015 Big Data Visual Analytics (BDVA), pp. 1–8 (September 2015)
10. Cordeil, M., et al.: Iatk: an immersive analytics toolkit. In: 2019 IEEE Conference on Virtual Reality and 3D User Interfaces (VR), pp. 200–209 (March 2019)

11. Cordeil, M., Cunningham, A., Dwyer, T., Thomas, B.H., Marriott, K.: Imaxes: immersive axes as embodied affordances for interactive multivariate data visualisation. In: Proceedings of the 30th Annual ACM Symposium on User Interface Software and Technology, UIST 2017, pp. 71–83. ACM, New York (2017)

12. Cordeil, M., Dwyer, T., Hurter, C.: Immersive solutions for future air traffic control and management. In: Proceedings of the 2016 ACM Companion on Interactive Surfaces and Spaces, ISS 2016, pp. 25–31. Companion, Association for Computing Machinery, New York (2016)

13. Donalek, C., et al.: Immersive and collaborative data visualization using virtual reality platforms. In: 2014 IEEE International Conference on Big Data (Big Data) (October 2014)

14. Dupont, L., Pallot, M., Christmann, O., Richir, S.: A universal framework for systemizing the evaluation of immersive and collaborative performance. In: ACM (ed.) VRIC 2018, Virtual Reality International Conference - Laval Virtual. Laval, France (April 2018)

15. Dwyer, T., et al.: Immersive analytics: an introduction. Immersive Analytics. LNCS, vol. 11190, pp. 1–23. Springer, Cham (2018). https://doi.org/10.1007/978-3-030-01388-2_1

16. Ens, B., et al.: Uplift: a tangible and immersive tabletop system for casual collaborative visual analytics. IEEE Trans. Visual Comput. Graphics **27**, 1193–1203 (2020)

17. Filonik, D., Bednarz, T., Rittenbruch, M., Foth, M.: Glance: Generalized geometric primitives and transformations for information visualization in ar/vr environments. In: Proceedings of the 15th ACM SIGGRAPH Conference on Virtual-Reality Continuum and Its Applications in Industry, VRCAI 2016, vol. 1. pp. 461–468. Association for Computing Machinery, New York (2016)

18. de Giorgio, A., Romero, M., Onori, M., Wang, L.: Human-machine collaboration in virtual reality for adaptive production engineering. procedia manufacturing 11, 1279–1287, In: 27th International Conference on Flexible Automation and Intelligent Manufacturing, FAIM2017, 27–30 June 2017. Modena, Italy (2017)

19. Hackathorn, R., Margolis, T.: Immersive analytics: Building virtual data worlds for collaborative decision support. In: 2016 Workshop on Immersive Analytics (IA), pp. 44–47 (March 2016)

20. Heer, J., Agrawala, M.: Design considerations for collaborative visual analytics. Inf. Vis. **7**(1), 49–62 (2008)

21. Hill, G.W.: Group versus individual performance: are n+1 heads better than one? Psychol. Bull. **91**(3), 517–539 (1982)

22. Isenberg, P., Elmqvist, N., Scholtz, J., Cernea, D., Ma, K.L., Hagen, H.: Collaborative visualization: definition, challenges, and research agenda. Inf. Vis. **10**(4), 310–326 (2011)

23. Kennedy, R.S., Lane, N.E., Berbaum, K.S., Lilienthal, M.G.: Simulator sickness questionnaire: an enhanced method for quantifying simulator sickness. Int. J. Aviat. Psychol. **3**(3), 203–220 (1993)

24. Kraus, M., Weiler, N., Oelke, D., Kehrer, J., Keim, D.A., Fuchs, J.: The impact of immersion on cluster identification tasks. IEEE Trans. Visual Comput. Graphics **26**(1), 525–535 (2020)

25. Lee, B., Hu, X., Cordeil, M., Prouzeau, A., Jenny, B., Dwyer, T.: Shared surfaces and spaces: Collaborative data visualisation in a co-located immersive environment. IEEE Trans. Visual Comput. Graphics **27**(2), 1171–1181 (2020)

26. Lee, B., Cordeil, M., Prouzeau, A., Dwyer, T.: Fiesta: a free roaming collaborative immersive analytics system. In: Proceedings of the 2019 ACM International Conference on Interactive Surfaces and Spaces, ISS 2019, pp. 335–338. Association for Computing Machinery, New York (2019)

27. Margery, D., Arnaldi, B., Plouzeau, N.: A general framework for cooperative manipulation in virtual environments. In: Gervautz, M., Schmalstieg, D., Hildebrand, A. (eds.) Virtual Environments '99, pp. 169–178. Springer, Vienna (1999). https://doi.org/10.1007/978-3-7091-6805-9_17

28. Matsas, E., Vosniakos, G.C., Batras, D.: Prototyping proactive and adaptive techniques for human-robot collaboration in manufacturing using virtual reality. Robot. Comput.-Integr. Manuf. **50**(C), 168–180 (2018)

29. Matthes, C., et al.: The collaborative virtual reality neurorobotics lab. In: 2019 IEEE Conference on Virtual Reality and 3D User Interfaces (VR), pp. 1671–1674 (March 2019)

30. Narayan, M., Waugh, L., Zhang, X., Bafna, P., Bowman, D.: Quantifying the benefits of immersion for collaboration in virtual environments. In: Proceedings of the ACM Symposium on Virtual Reality Software and Technology, VRST 2005, pp. 78–81. Association for Computing Machinery, New York (2005)

31. Nguyen, H., Ward, B., Engelke, U., Thomas, B., Bednarz, T.: Collaborative data analytics using virtual reality. In: 2019 IEEE Conference on Virtual Reality and 3D User Interfaces (VR), pp. 1098–1099 (March 2019)

32. Pereira, V., Matos, T., Rodrigues, R., Nóbrega, R., Jacob, J.: Extended reality framework for remote collaborative interactions in virtual environments. In: 2019 International Conference on Graphics and Interaction (ICGI), pp. 17–24 (2019)

33. Satyanarayan, A., Moritz, D., Wongsuphasawat, K., Heer, J.: Vega-lite: a grammar of interactive graphics. IEEE Trans. Visual Comput. Graphics **23**(1), 341–350 (2017)

34. Sicat, R., et al.: DXR: a toolkit for building immersive data visualizations. IEEE Trans. Visual Comput. Graphics **25**(1), 715–725 (2019)

35. Skarbez, R., Polys, N.F., Ogle, J.T., North, C., Bowman, D.A.: Immersive analytics: theory and research agenda. Front. Robot. AI **6**, 82 (2019)

36. Sun, T., Ye, Y., Fujishiro, I., Ma, K.: Collaborative visual analysis with multi-level information sharing using a wall-size display and see-through hmds. In: 2019 IEEE Pacific Visualization Symposium (PacificVis), pp. 11–20 (2019)

37. Xia, H., Herscher, S., Perlin, K., Wigdor, D.: Spacetime: enabling fluid individual and collaborative editing in virtual reality. In: Proceedings of the 31st Annual ACM Symposium on User Interface Software and Technology, UIST 2018, pp. 853–866. Association for Computing Machinery, New York (2018)

38. Yang, Y., Cordeil, M., Beyer, J., Dwyer, T., Marriott, K., Pfister, H.: Embodied navigation in immersive abstract data visualization: Is overview+detail or zooming better for 3d scatterplots? IEEE Trans. Visual Comput. Graphics **27**(2), 1214–1224 (2020)

39. Zikky, M., Fathoni, K., Firdaus, M.: Interactive distance media learning collaborative based on virtual reality with solar system subject. In: 2018 19th IEEE/ACIS International Conference on Software Engineering, Artificial Intelligence, Networking and Parallel/Distributed Computing (SNPD), pp. 4–9 (June 2018)

Implementation of Auditable Blockchain Voting System with Hyperledger Fabric

Michał Pawlak and Aneta Poniszewska-Marańda[✉][iD]

Institute of Information Technology, Lodz University of Technology, Lodz, Poland
{michal.pawlak,aneta.poniszewska-maranda}@p.lodz.pl

Abstract. An efficient democratic process requires a quick, fair and fraud-free election process. Many electronic-based voting systems have been developed to fulfil these requirements but there are still unsolved issues with transparency, privacy and data integrity. The development of distributed ledger technology called blockchain creates the potential to solve these issues. This technology's rapid advancement resulted in numerous implementations, one of which is Hyperledger Fabric, a secure enterprise permissioned blockchain platform. In this paper, the implementation of an Auditable Blockchain Voting System in Hyperledger Fabric is presented to showcase how various platform components can be used to facilitate electronic voting and improve the election process.

Keywords: E-voting · Blockchain · Hyperledger Fabric · Auditable Blockchain Voting System

1 Introduction

An efficient and honest democratic process requires a fair and fraud-free election process [1]. For that reason, the voting process is secured by complex measures. However, that is not enough and elections are still vulnerable to various attacks. What is more, the most common traditional paper-based systems are slow and manipulation-prone. All this undermines trust in such systems and reduces participation in the main democratic process.

There is ongoing research aiming to solve this problem, which resulted in many different solutions, including electronic-based ones [2]. These electronic voting systems provide many advantages, such as quick result calculation, improved ballot presentation, reduced costs and convenient usage due to the possibility of unsupervised voting through a network. However, these systems face many challenges and issues [3]. The most prominent of which is lack of transparency, difficulties with voters' authorization and authentication, and enforcement of data integrity and privacy [4].

Many of these problems may be solved with the application of blockchain technology, which is distributed in a peer-to-peer network system of ledgers. Peers cooperate on validation and management of stored data. This technology enables the creation of anonymous, transparent, manipulation-resistant systems.

M. Paszynski et al. (Eds.): ICCS 2021, LNCS 12742, pp. 642–655, 2021.
https://doi.org/10.1007/978-3-030-77961-0_51

Since its introduction in 2008, there have been many new and more advanced implementations of blockchain technology. One of them is Hyperledger Fabric, which is an open-source customizable blockchain solution designed for enterprise use. For that reason, it provides complex permission and policy management. Furthermore, Hyperledger Fabric supports smart contracts in popular programming languages like Java, Go and Node.js, which allow for quick development of advanced business logic on top of the blockchain [5].

This paper intends to present an implementation of Auditable Blockchain Voting System (ABVS) concepts in Hyperledger Fabric. The main focus is on how the e-voting blockchain network can be organized in Hyperledger Fabric and how various Hyperledger Fabric components can be used to facilitate and secure electronic voting.

The remaining parts of this paper are organized as follows. Section 2 presents the theoretical background of electronic voting and presents a technical description of Hyperledger Fabric. Section 3 provides an overview of works related to the field of electronic voting and blockchain. In Sect. 4, the implementation of the Auditable Blockchain Voting System in Hyperledger Fabric is detailed. Section 5 analyses voting properties concerning the presented system while Sect. 6 presents conclusions.

2 Electronic Voting and Blockchain in Hyperledger Fabric

The Council of Europe defines an electronic voting system or e-voting systems as any type of election or referendum, which utilizes electronic means to at least facilitate vote casting [6]. Thus, the term covers a wide variety of different solutions and implementations. In general, electronic voting systems can be classified concerning two characteristics, namely: *supervision* and *remoteness* [7–9]. The classification is presented in Fig. 1.

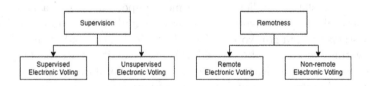

Fig. 1. Classification of electronic voting systems

Supervision divides electronic voting systems depending on the circumstances of voting. In *supervised* systems, elections and referendums are conducted from polling stations under official supervision. This type of e-voting is secure against coercion and vote selling, but it is inconvenient, time-consuming and expensive. On the other hand, *unsupervised* systems allow voting to be conducted from any location as long as voters can access the required facilities to transfer their votes. This type of e-voting is fast and convenient but vulnerable to vote selling, coercion and manipulation.

Remoteness categorizes electronic voting systems concerning the method of transporting cast ballots for counting. *Remote* systems instantly transfer cast ballots to a remote location via a chosen communication channel, e.g. the Internet. These systems provide fast results and reduced overhead but are vulnerable to attacks on the communication channel and data manipulation during transfer. Non-remote systems store cast ballots locally and provide local results combined to provide a final result after an election ends. These systems are secure against attacks during transfer but are prone to local manipulations and errors during counting.

Electronic voting systems must aim to satisfy the following criteria to be considered safe and secure [7,10,11]: eligibility, privacy, correctness/completeness, fairness, transparency, verifiability, availability.

Eligibility describes a requirement of allowing only authenticated and authorized voters to cast their votes. *Privacy* ensures that voters cannot be linked to their votes, so only the voters themselves know how they voted. *Correctness/Completeness* requires that only valid votes are counted and a given electoral law is enforced. *Fairness* ensures that e-voting systems do not in any way influence election results. *Transparency* is a requirement that all procedures and components are clear and understandable to voters. *Verifiability* requires that e-voting systems can be verified against their requirements and safety criteria. Finally, *availability* describes a requirement that e-voting systems should allow all eligible voters to vote and not prevent anyone from participating in elections.

There exist many different electronic voting systems and each has its own advantages and disadvantages. However, electronic voting systems aim to provide the following advantages: (i) reduction and prevention of frauds by minimising human involvement; (ii) acceleration of result processing; (iii) improvement of ballot readability to reduce the number of spilled ballots; (iv) reduction of costs by minimising overhead.

Unfortunately, electronic voting systems must also face many technical, procedural and legislative challenges. The most important one is a lack of trust, which is a result of: (i) inadequate transparency and understanding of electronic voting systems by common voters; (ii) lack of widely accepted standards against which e-voting systems can be verified; (iii) vulnerability to attacks by privileged insiders and system providers; (iv) costs of implementation and infrastructure.

There is still ongoing research on securing voting systems or solving some (or all) challenges of electronic voting systems. This paper aims to provide a solution to verifiability and transparency issues with the use of blockchain technology implemented in Hyperledger Fabric, which is described in the following section.

Blockchain technology consists of a distributed system of ledgers stored in a chain-like data structure of connected blocks and a consensus algorithm that collectively validates the contents of the blocks in a peer-to-peer network. In general, blockchains can be divided into [8,12]:

- public and private,
- permissionless and permissioned.

Public blockchains allow everyone to join a blockchain network and read data contained within, for example, Bitcoin and Ethereum. In contrast, *private* blockchains allow only a selected group of entities to access blockchain data, for instance, MultiChain and Hyperledger Fabric. On the other hand, *permissionless* blockchains allow anyone to join and participate in a consensus algorithm, while *permissioned* blockchains divide participants concerning their permissions, for example, MultiChain and Hyperledger Fabric.

There are numerous blockchain implementations, one of which is already mentioned Hyperledger Fabric developed as an open-source project by the Linux Foundation, which is an esteemed developer community. The Hyperledger Fabric is a part of a whole family of solutions and tools, all designed with the following principles in mind [13]: modularity and extensibility, interoperability, security, token agnosticism, rich and easy-to-use APIs.

Hyperledger Fabric fulfils these principles by the implementation of the following design features, which are its main components [5,13]: assets, ledger, smart contracts and chaincode, consensus, privacy components, security and membership services.

Assets are represented by a collection of key-value pairs and enable to exchange of valuables over a Hyperledger Fabric blockchain network. Assets form the main business objects that are stored in an immutable *ledger* made of connected blocks. In essence, assets represent facts about a system and a ledger stores current and historical states of them.

However, every system needs some kind of business logic. In the case of Hyperledger Fabric-based systems, it is provided by *smart contracts*. They are executable programs, which define common terms, data, rules, concept definitions and processes involved in the interaction between involved parties. In order to utilize these programs, they must be packed and deployed on a blockchain network. In the context of Hyperledger Fabric, packed and deployed smart contracts are called a *chaincode*. Chaincode can interacts with a ledger primarily with *put*, *get*, *delete* and query operations. Furthermore, chaincode provides a rich API that provides methods for interaction between chaincodes, chaincode events and client-related requests. Every such transaction is recorded in a ledger. It is important to note that chaincode is stored on blockchain nodes, called peers, and can only be executed by them when explicitly installed.

However, before a transaction can be committed to a block and a ledger, it must first be endorsed. How this process is conducted depends on an endorsement policy, which describes which members (*organisations*) of a network must approve generated transactions. A model of a smart contract with a connected application is presented in Fig. 2. The model presents a smart contract for car exchange, which allows querying for a car, updating car properties and transferring car ownership between organizations. Furthermore, it has an associated policy, requiring both organizations to approve a transaction before it can be committed.

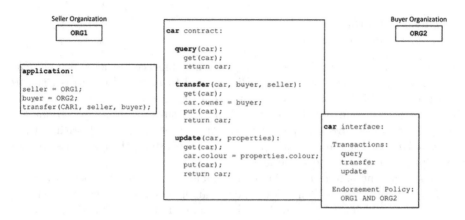

Fig. 2. Model of a smart contract, adapted from [5]

Because of the distributed nature of a blockchain network, participating nodes must agree on a common state of the network and contained within data to achieve a *consensus*. To achieve that, Hyperledger Fabric's network is made of two types of nodes. The first type, already mentioned, consists of *peers*, which store copies of ledgers and validate transactions by creating *transaction proposals* (executed transaction without updating the network's state). Valid transactions are then passed to *orderer nodes*, which elect from themselves a leader, which is allowed to modify the network's state by creating and committing new blocks. The leader is selected via the Raft algorithm [14]. It is important to note that these roles are not exclusive and a single node can fulfil multiple roles.

To provide privacy and separation in a blockchain network, Hyperledger Farbic allows creating *consortiums* and *channel*. A *consortium* is a set of organizations which intend to cooperate with each other. A *channel* is a separated communication mechanism that allows members of a consortium to freely communicate with each other in separation from other involved organizations, which can have their own consortiums and channels in the same network.

Security is provided by issuing, via Public Key cryptography, cryptographic certificates to organizations, network components and client applications. As a result, each entity can be identified, and its rights and privileges within the system can be managed to a significant degree. When combined with separation provided by consortiums and channels, Hyperledger Fabric delivers an environment suitable for private and confidential operations. In the context of security, the *Membership Service Provider* (MSP) must be mentioned. MSP is a component abstracting membership operations and provides mechanisms for issuing certificates, validating certificates and user authentication.

Figure 3 presents a model of the described components. As can be seen, there are two peer nodes and one orderer node connected to a single channel. Each node stores its own copy of the channel's ledger and a copy of the chaincode. The whole network interacts with an application that triggers smart contracts stored in the chaincode.

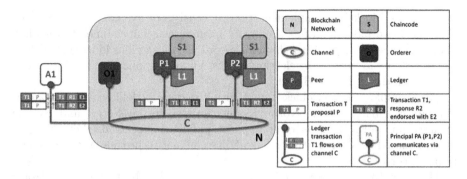

Fig. 3. A model of a single Hyperledger Fabric channel, adapted form [5]

Hyperledger Fabric components are highly configurable and modular, which allows a high degree of customisation [5,13]. This makes this solution ideal for many applications, which require security and privacy. Moreover, for these reasons, it was chosen as a platform for an e-voting solution.

3   Related Works

There are numerous publications regarding blockchain-based electronic voting. In [15] presents an overview of electronic voting systems created with blockchain technology. The authors review multiple research articles and describe presented in the systems, along with their advantages and disadvantages.

The authors of [16] describe an implementation of the electronic voting system in Hyperledger Fabric-based on an approach presented in [17]. The described approach allows conducting voting traditionally with paper and electronically with software. In addition to Hyperledger Fabric, the system utilizes blind signatures, secret sharing schemes and identity mixer to provide security and privacy.

[8] presents results of investigation of scalability and performance scalability constraints of e-voting systems based on the blockchain technology. The authors conducted experiments concerning voting population, block size, block generation rate and transaction speed on both permissionless and permissioned blockchain, namely, Bitcoin and MultiChain.

The author of [7] demonstrates a model of the blockchain-based electronic voting system that intends to provide coercion resistance, receipt-freeness and universal verifiability using zero-knowledge Succinct Non-iterative Arguments of Knowledge and Bitcoin blockchain implementation.

Chaintegrity introduced in [18] is intended to achieve scalability, verifiability and robustness in large scale elections. The main used components are smart contracts, homomorphic and Paillier threshold encryptions, and a counting Bloom filter and Merkle hash tree. Furthermore, the authors provide extensive documentation of conducted testing, including performance evaluation.

There exist some commercial blockchain-based electronic voting systems. One of the most prominent is Agora [19]. It is a Swiss-made e-voting system based on a custom blockchain implementation. It consists of five components/layers: Bulletin Board blockchain functioning as a communication channel, Cotena logging mechanism, Bitcoin blockchain for recording transactions in ledgers, Valeda global network for validating election results, Votapps application layer for interaction with the Bulleting Board.

An example of a non-blockchain e-voting system is Hellios, an open audit, remote and unsupervised e-voting system written in JavaScript and Python Django framework. Helios is a web-application following centralized client-server architecture. The system utilizes Sako-Kilian mixnet to provide anonymity and to prove correctness.

However, the woks presented above possess the drawbacks for an electronic voting system targeting low-coercion risk, Internet based elections. Not only the system has to be auditable, safe and transparent, but also minimize the work voters have to perform to keep the system integral. The traits can be achieved by decentralizing the system through blockchain usage.

The proposed approach presents an implementation of Auditable Blockchain Voting System concepts in Hyperledger Fabric together with e-voting blockchain network organized in Hyperledger Fabric and various Hyperledger Fabric components to facilitate and secure electronic voting.

4   Auditable Blockchain Voting System with Hyperledger Fabric

This section presents the Auditable Blockchain Voting system implementation in Hyperledger Fabric. The system was designed as a remote and supervised voting system. However, it is possible to use it in an unsupervised environment. ABVS is intended to improve the existing voting system in Poland. The following subsections will present various aspects of ABVS in separate subsections.

4.1   Auditable Blockchain Voting System Overview

Details and an initial idea behind Auditable Blockchain Voting System can be found in [4], but in general a process utilized in ABVS can be divided into three phases (Fig. 4).

In the *election setup* phase, a set of trusted public and private organizations is defined by the National Electoral Commission (NEC). The selected organizations will provide the blockchain infrastructure required to run a Hyperledger Fabric network. In the next step, NEC sets up an ABVS hyperledger-based blockchain network. This consists of creating a Certificate Authority, which provides complying with X.509 standard certificates and distributes them between accepted organizations, which made up the ABVS network. Furthermore, NEC generates Vote Authentication Tokens (VITs), which are one-time numerical codes similar to Indexed Transaction Authentication Numbers (iTANs) that allow citizens to cast their votes. VITs are then distributed over the country, and the citizens can retrieve them from local offices after authorization and authentication. All

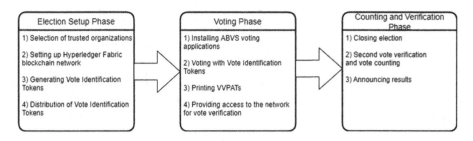

Fig. 4. Auditable Blockchain Voting System overview

retrievals are recorded in a blockchain, so it is possible to verify if only authorized people obtained the VITs.

In the *voting* phase, ABVS certified voting apps are installed in polling stations. In order to vote, the voters provide one of the obtained VITs and select their chosen candidate. The ABVS applications transfer votes to the blockchain network, where they are validated. The voters receive a validated and accepted transaction from which they generate VVPATs to leave a physical trail. The voters can also use their VITs to verify the presence of their votes in the blockchain. Furthermore, as long as the voters have their VITs, they can keep casting votes because only the newest one will be counted. It is also important to note that all votes are encrypted and can only be decrypted after the election time is over, so elections remain fair.

In the *counting and verification* phase, the election is closed and second verification is conducted, ensuring that all votes were cast by distributed VITs. After validation is over, the votes are calculated and announced.

4.2 Auditable Blockchain Voting System Network

The organisation of the Auditable Blockchain Voting System network is presented in Fig. 5. The network consists of a single consortium made of an NEC node, which is an initial organisation that stated initiated the network, and a set of trusted organisations $Orgs = \{Org_1, Org_2, ..., Org_n\}$, which are selected before the formation of the network. Each organisation adds to the network a number of orderer-peer nodes $Nodes = \{Org_1Node_1, Org_2Node_2, ..., Org_nNode_n\}$. The network is organised into two channels: VITs-Distribution Channel and Voting Channel.

VITsDistribution Channel is used connected to VITsDistribution Applications, which allows voters to obtain their voting tokens. After the ABVS network is set up, NEC generates a predefined number of VITs, which are recorded as blockchain transactions in the channels' ledger. To obtain the code, each willing voter must report to a local office, authorize and authenticate to ensure eligibility and use the application to send a request for VIT. The ABVS network will return one of the generated VITs and record its retrieval to ensure that only retrieved VITs are used in given voting.

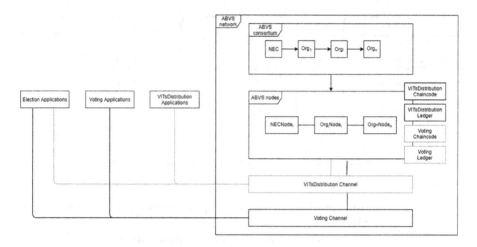

Fig. 5. Auditable Blockchain Voting System network organization

Voting Channel is used to manage the election itself, and it contains the main business logic of ABVS. Through this channel, elections are generated, and votes are cast and recorded for further counting. This is done via dedicated client applications designated as *Voting Applications*.

Furthermore, both channels are connected to *Election Applications*, which are designed for election process administrators to manage the election process. The applications allow creating an election, generating VITs, validate elections and produce results.

4.3   Auditable Blockchain Voting System Smart Contracts

ABVS utilizes three main assets presented in Fig. 6. The *election* represents the current election and contains fields describing what the election is about (*electionGoal*), a list of candidates, and the start and end dates of the given election. A second asset consists of *VITs* and each consists of a given Election asset and a map of index-value pairs used for the given election. The final assets are *Votes*, each consisting of a given Election, VIT used for casting a vote, value of a vote and a vote location, which describes an electoral district or a university department in case of more local elections.

Each node in the ABVS system has installed chaincode, which allows its execution. There are three main smart contracts included in the ABVS chaincode, one for each asset.

Election contract and election applications provide an interface, as shown in Fig. 7. The election contract allows administrators to:

1. query for elections,
2. create elections for specific goal, candidates and dates,
3. get an election for use in other smart contracts,

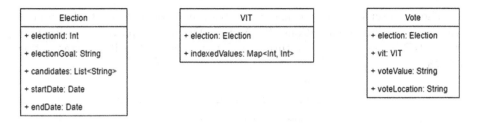

Fig. 6. Auditable Blockchain Voting System assets

Fig. 7. ABVS election application and contract interfaces

4. generate VITs for creating a specific number of VITs codes for a given election,
5. validate an election by comparing used and distributed VITs,
6. calculate results by decrypting and counting votes.

The endorsement policy for this contract requires endorsement from NEC and at least $\frac{n}{2} + 1$ of other organisations. The election applications' interface utilizes the methods from the contract interface except for the *getElection()* method, which is for internal use only.

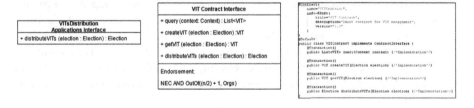

Fig. 8. ABVS VITsDistribution applications and VIT contract interfaces

VIT contract interface provides access to methods for:

1. querying the blockchain for specific VITs,
2. creating a new VIT for a given election,
3. getting a specific VIT for use in other methods and smart contracts,
4. distributing a VIT to voters and saving this operation.

The endorsement policy consists of a required endorsement of NEC and at least $\frac{n}{2}+1$ of other organisations. The VITsDistribution applications utilize only the *getVIT()* method, which returns a VIT to voters and logs it as a transaction (Fig. 8).

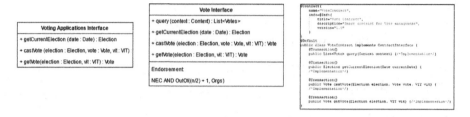

Fig. 9. ABVS Voting applications and Vote contract interfaces

Figure 9 presents interfaces provided by voting applications and Vote contract. The contract allows to:

1. query the blockchain for a specific Vote,
2. get the current election via smart contract communication,
3. cast a vote for a given election by providing a VIT,
4. get a vote by providing a VIT, which is a method for verification by voters.

The endorsement policy consists of a required endorsement of NEC and at least $\frac{n}{2}+1$ of other organisations. The voting applications provide methods for getting a the current election to facilitate the process, casting votes by providing a VIT, and getting a vote for verification by providing a VIT.

4.4   Auditable Blockchain Voting System Security Components

ABVS utilizes standard Certificate Authorities (CAs) distributing X.509 certificates to the involved entities. Thus it uses out-of-the-box distribution via Root CAs, Intermediate CAs and MSPs. However, ABVS also requires role functionality and role-based access control to the provided interfaces. To accomplish that, Attribute-Based Access Control was implemented utilizing an additional type attribute, which is added to each certificate to restrict access to various functionalities.

The methods that can be accessed by a *VoterApp* role, which represents client applications for retrieving VIT and casting votes, are: *getVIT* and *castVote*. *OrgMember* role is designed for organizations, which do not initiate elections and should not have any additional options to affect the election. Its methods are: *getElection, validateElection, calculateResults* and all query methods. Finally, *NECAdmin* roles is designed with the same privileges as OrgMember, but can create an election and generate VITs for a given election.

Furthermore, it is important to note that all transactions remain encrypted until the end of the election, and no involved entity can perform any operation besides adding new transactions. Only voters can check their votes before the election ends.

5   Property Analysis

Eligibility is achieved in two ways. Firstly, all organizations participating in the network use public-key cryptography and are certified by the CA created by NEC, making unauthorized access to the network difficult and easily detectable. Secondly, authentication and authorization for the voters are achieved by the fact that the system was designed as a supervised system, which means that the voting takes place under official control.

$$Eligibility = \{publicKey \cup certifcationCA\} \vee \{supervisedSystem\}$$

Privacy is achieved by allowing supervision during the election process. The voters authenticate and authorize themselves before an election committee and then proceed to cast their vote using their VITs, which are in no way linked to specific voters besides physical ownership. The voters cast their votes anonymously from the privacy of polling stations. This has two major disadvantages. Firstly, it requires voters to travel to the polling stations, which nullifies one of the greatest advantages of electronic voting. Secondly, voters can be forced to provide their VITs to show how they voted. However, this is mitigated by the fact that ABVS allows multiple voting and counts only the last vote cast by a given voter.

$$Privacy =$$
$$\{supervisedSystem \cup authenticatation \cup authorization \cup voterVIT\}$$

Correctness/completeness is achieved because only authenticated and authorized voters are allowed to participate in an election, so only eligible votes are cast. Furthermore, the voting applications will ensure that only valid votes are transferred to the ABVS network. Finally, all transactions are validated by the ABVS network itself, so no invalid votes should be counted.

$$Correctness =$$
$$\{authenticatation \cup authorization \cup ABVSnetworkValidation\}$$

Fairness is achieved by encryption of data stored on the ABVS network through the use of private-collections functionality of Hyperledger Fabric. Moreover, access to this data is restricted to the specific role of the OrgMember, which restricts who can view the stored data. Finally, every operation in the network is saved as a transaction, so the property is further reinforced because it is relatively easy to identify eventual leaks of the data, which may affect election results.

$$Fairness =$$
$$\{dataEncryption \cup HyperledgerFabricPrivate\_collections$$
$$\cup OrgMember\_role \cup transaction\_operation\}$$

Transparency and verifiability are achieved because of blockchain inherent properties and because each operation on the Hyperledger network is saved. Furthermore, all available operations are based on smart contracts, which are public and thus can be validated. Lastly, voters can use their VITs to view their cast votes to see if their vote was saved correctly.

$Transparency =$
$\{blockchainProperty \cup operationSaving \cup smartContracts \cup voterVIT\}$

Availability is achieved because, in essence, the ABVS is an evolution of the traditional paper-based methods and uses similar procedures, so it is no less available than its predecessors.

$Availability = \{ABVS\_properties\}$

6   Conclusions

An honest and fair democracy requires a quick and efficient election system. In order to improve the traditional paper-based voting, numerous electronic systems were designed. However, many suffer from issues with transparency, verifiability and privacy. Rapidly developed blockchain technologies may offer a solution to these still not solved problems.

Hyperledger Fabric is one of the blockchain-based platforms for developing applications on top of it. It is an open-source project by The Linux Foundation created to provide a blockchain-based solution for enterprise private and permissioned network. It is characterized by high customizability and modularity in every aspect, from identity management to smart contract validation and consensus algorithm. These reasons made this platform ideal for implementing the electronic voting system because such a system requires a specific set of settings, which may not be available in more popular platforms like Bitcoin or Ethereum.

In this paper, high-level implementation details of the Auditable Blockchain Voting System were presented. The most important components of ABVS were presented concerning various Hyperledger Fabric components and how these components interact. Models of ABVS network, smart contracts and privacy settings were shown and described. In the future, the implementation will be tested to identify the most optimal parameters and settings for a quick and scalable e-voting system. Furthermore, implementation in the MultiChain platform will be developed to verify the portability of the ABVS model.

References

1. Lehoucq, F.: Electoral fraud: causes, types, and consequences. Ann. Rev. Polit. Sci. **6**(1), 233–256 (2003)
2. Willemson, J.: Bits or paper: which should get to carry your vote? J. Inf. Secur. Appl. **38**, 124–131 (2018)
3. De Faveri, C., Moreira, A., Araújo, J.: Towards security modeling of e-voting systems. In: Proceedings of IEEE 24th International Requirements Engineering Conference Workshops (REW), Beijing, China (2016)

4. Pawlak, M., Guziur, J., Poniszewska-Marańda, A.: Voting process with blockchain technology: auditable blockchain voting system. In: Xhafa, F., Barolli, L., Greguš, M. (eds.) INCoS 2018. LNDECT, vol. 23, pp. 233–244. Springer, Cham (2019). https://doi.org/10.1007/978-3-319-98557-2_21
5. Hyperledger: A Blockchain Platform for the Enterprise: Hyperledger Fabric. Linux Foundation, 4 September 2019. https://hyperledger-fabric.readthedocs.io/en/latest/. Accessed 15 Sep 2019
6. Caarls, S.: E-voting handbook: Key steps in the implementation of e-enabled elections. Council of Europe, November 2010. https://www.coe.int/t/dgap/goodgovernance/Activities/E-voting/E-voting%202010/Biennial_Nov_meeting/ID10322%20GBR%206948%20Evoting%20handbook%20A5%20HD.pdf
7. Dimitriou, T.: Efficient, coercion-free and universally verifiable blockchain-based voting. Comput. Netw. **174**, 107234 (2020)
8. Khan, K.M., Arshad, J., Khan, M.M.: Investigating performance constraints for blockchain based secure e-voting system. Future Gener. Comput. Syst. **105**, 13–26 (2019)
9. National Democratic Institute. Common Electronic Voting and Counting Technologies. https://www.ndi.org/e-voting-guide/common-electronic-voting-and-counting-technologies. Accessed 22 Jan 2018
10. Council of Europe - Committee of Ministers: Recommendation CM/Rec (2017) 51 of the Committee of Ministers to member States on standards for e-voting, 14 June 2017. https://search.coe.int/cm/Pages/result_details.aspx?ObjectID=0900001680726f6f
11. Council of Europe - Committee of Ministers: Recommendation Rec (2004) 11 of the Committee of Ministers to member states on legal, operational and technical standards for e-voting, Strasbourg: Council of Europe Publishing (2005)
12. Drescher, D.: Blockchain Basics: A Non-Technical Introduction in 25 Steps, 1st edn. Apress, Frankfurt (2017)
13. Gaur, N., Desrosiers, L., Ramakrishna, V., Novotny, P., Baset, S.A., O'Dowd, A.: Hands-On Blockchain with Hyperledger. Packt Publishing Ltd, Building decentralized applications with Hyperledger Fabric and Composer, Birmingham (2018)
14. Ongaro, D., Ousterhout, J.: In Search of an Understandable Consensus Algorithm (Extended Version), 20 May 2014. https://raft.github.io/raft.pdf. Accessed 10 Aug 2020
15. Xiao, S., Wang, X.A., Wang, W., Wang, H.: Survey on blockchain-based electronic voting. In: Barolli, L., Nishino, H., Miwa, H. (eds.) INCoS 2019. AISC, vol. 1035, pp. 559–567. Springer, Cham (2020). https://doi.org/10.1007/978-3-030-29035-1_54
16. Kirillov, D., Korkhov, V., Petrunin, V., Makarov, M., Khamitov, I.M., Dostov, V.: Implementation of an e-voting scheme using hyperledger fabric permissioned blockchain. In: Misra, S. (ed.) ICCSA 2019. LNCS, vol. 11620, pp. 509–521. Springer, Cham (2019). https://doi.org/10.1007/978-3-030-24296-1_40
17. He, Q., Su, Z.: A new practical secure e-voting scheme. Information Security Conference (SEC 1998) (1993)
18. Zhang, S., Wang, L., Xiong, H.: Chaintegrity: blockchain-enabled large-scale e-voting system with robustness and universal verifiability. Int. J. Inf. Secur. https://doi.org/10.1007/s10207-019-00465-8 (2019)
19. Agora Technologies: Agora_Whitepaper_v0.2.pdf 2015. https://agora.vote/Agora_Whitepaper_v0.2.pdf. Accessed 20 Apr 2018
20. Adida, B.: Helios: Web-based Open-Audit Voting. In: Proceedings of 17th USENIX Security Symposium (2008)

Quantum Data Hub: A Collaborative Data and Analysis Platform for Quantum Material Science

Shweta Purawat[1], Subhasis Dasgupta[1], Luke Burbidge[1], Julia L. Zuo[2], Stephen D. Wilson[2], Amarnath Gupta[1], and Ilkay Altintas[1(✉)]

[1] University of California San Diego, La Jolla, CA, USA
{shpurawat,sudasgupta,lburbidge,a1gupta,ialtintas}@ucsd.edu
[2] University of California Santa Barbara, Santa Barbara, CA, USA
{jlzu,stephendwilson}@ucsb.edu

Abstract. Quantum materials research is a rapidly growing domain of materials research, seeking novel compounds whose electronic properties are born from the uniquely quantum aspects of their constituent electrons. The data from this rapidly evolving area of quantum materials requires a new community-driven approach for collaboration and sharing the data from the end-to-end quantum material process. This paper describes the quantum material science process in the NSF Quantum Foundry with an overarching example, and introduces the Quantum Data Hub, a platform to amplify the value of the Foundry data through data science and facilitation of: (i) storing and parsing the metadata that exposes programmatic access to the quantum material research lifecycle; (ii) FAIR data search and access interfaces; (iii) collaborative analysis using Jupyter Hub on top of scalable cyberinfrastructure resources; and (iv) web-based workflow management to log the metadata for the material synthesis and experimentation process.

Keywords: Quantum material science · FAIR · Data management · Collaboration platform · JupyterHub

1 Introduction

Quantum materials research is a rapidly growing domain of materials research, seeking for novel compounds whose electronic properties are born from the uniquely quantum aspects of their constituent electrons. Electronic states whose order can be defined locally, such as superconductivity and collective magnetism, emerge in quantum materials as well as electronic states forming non-local order, such as topologically nontrivial band structures and many-body entangled states. These and other states are sought to form the basis of the coming revolution in quantum-based electronics and can allow quantum information to be harnessed for next-generation computing and sensing applications.

ⓒ Springer Nature Switzerland AG 2021
M. Paszynski et al. (Eds.): ICCS 2021, LNCS 12742, pp. 656–670, 2021.
https://doi.org/10.1007/978-3-030-77961-0_52

Although there are material research data facilities built around generation, ingestion and sharing of data, the data from this rapidly evolving area of quantum materials require a community-driven approach for collaboration and sharing the data from the end-to-end quantum material process. It is critical to establish a community network and a collaborative data management and analytical ecosystem to couple data to theory to materials development and to complement the growing number of theory-forward materials prediction databases in the field.

UC Santa Barbara's NSF Quantum Foundry[1], funded by the National Science Foundation, is a next generation materials foundry that develops materials and interfaces hosting the coherent quantum states needed to power the coming age of quantum-based electronics. Its mission is to develop materials hosting unprecedented quantum coherence, train the next generation quantum workforce, and to partner with industry to accelerate the development of quantum technologies.

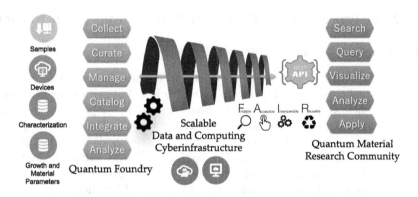

Fig. 1. The conceptual data pipeline for the Quantum Data Hub.

This paper describes the quantum material science process with an overarching example from the Quantum Foundry, and introduces the Quantum Data Hub (QDH), a platform to amplify the value of the Foundry data through data science. As depicted in Fig. 1, the QDH collects, curates and manages the experimental and theoretical scientific data, often not in a searchable and queriable form until now. The data, which includes large amounts of physical objects (e.g. samples/devices), characterization data, and growth and material parameters serves as a backbone of our efforts, to aid the data-driven discovery and development of new materials with engineered functionalities. The data is made readily searchable for analysis using advanced cyberinfrastructure tools for automated workflows, machine learning, and statistical analysis. The QDH enables data cleaning to provide users with FAIR [19] views over it. The data collected is served through RESTful APIs that can serve the raw, cleaned and versions of

---

[1] Quantum Foundry Website: https://quantumfoundry.ucsb.edu/.

analytical data products in a scalable fashion. This scalable approach allows for the simultaneous availability of such data to many processing modules.

Contributions. In this paper, we present the data and analysis components of the Quantum Data Hub to facilitate: (*i*) storing and parsing the metadata that exposes programmatic access to the quantum material research lifecycle involving experimentation and synthesis; (*ii*) FAIR data search and access interfaces with access control; (*iii*) collaborative analysis using Jupyter Hub on top of dynamic cyberinfrastructure resources; and (*iv*) web-based workflow management to log the metadata for the material synthesis and experimentation process. We also present a case study for powder synthesis and measurement process.

Outline. The rest of this paper is organized as follows. In Sect. 2, we describe the quantum material research process and a case study for powder synthesis. Section 3 introduces the Quantum Data Hub architecture and its main components. We review related work in Sect. 4 and conclude in Sect. 5.

2   Quantum Material Research Process and Data Model

2.1   Quantum Material Research Process

The past 15 years have witnessed a revolution in the computational modeling and theoretical prediction of quantum materials with tailored electronic properties. Experimental assessment of these predictions however proceeds at a much slower pace due to the bottleneck of the laborious and often iterative process of experimentally synthesizing newly predicted materials. Due to the difficulty in predicting and modeling inorganic reaction pathways, diffusion, and grain growth at elevated high temperatures, the materials growth synthesis process in the quantum materials domain remains dominated by chemically informed starting points followed by onerous trial and error iteration.

The research process itself starts with a prediction of a new material with desired functional properties. This is followed by developing a plan to synthesize the new compound, and starting points are typically chosen based off of a researcher's prior experience synthesizing related materials or via reported synthesis conditions of similar compounds. A starting point for the reaction conditions/processing space is chosen which involves the choice of the starting reagents, a thermal profile for reacting the reagents, and the choice of the correct processing space for the reaction to occur (e.g., what gas environment should be used; what type of furnace/heating source; what type of reaction vessel; etc.).

Once the initial conditions are chosen, the experiment is executed and then the product is analyzed via a number of experimental probes to ascertain what material was created. This typically involves x-ray structural analysis, various forms of chemical fingerprinting such as energy dispersive spectroscopy, and composition analysis via electron microscopy. Once the composition of the created material is ascertained, then the original conditions of the reaction are modified to push toward the reaction toward the desired result. In quantum materials research, the most common goal is to create a high purity, single phase sample of the desired compound for follow on study.

Once the desired compound is created with the requisite purity, then the electronic properties of the compound are explored via a number of complementary probes. This can include bulk measurements of electrical resistivity, the magnetic susceptibility, heat capacity, more advanced characterization with optical spectroscopy, angle-resolved photoemission, and scanning tunneling microscopy. Depending on the hypothesis being tested and the experiments needed, the form factor for a sample may need to be a macroscopic, single crystal of the new compound, rather than a multigrain powder (a collection of microscopic crystallites). Developing the necessary parameters for achieving crystal growth of a given compound then requires further experiment design and iterative growth/testing steps. We also note here that quantum materials are also heavily explored in thin film form, which entails additional complexities (substrate type, growth orientation, etc.) beyond the broad overview of "bulk" materials synthesis detailed above.

Once a high enough purity sample is created in the appropriate form factor (e.g., power, single crystal, thin film), the lifecycle of experimental exploration can be long. Measurement by multiple complementary probes is common, and many materials are tuned chemically following their initial measurement in order to test new hypothesis formed from the characterization data. Changing the composition of the starting material to address these hypothesis begins the iterative synthesis process again, which feeds into the *theory, synthesis, characterization* loop. The synthesis step in this loop is a major bottleneck for the field.

2.2   Example: Powder Synthesis and Measurement Process

An example use case of the synthesis and measurement of the chemical $GaNb_4Se_8$ is illustrated in Fig. 2. The preparation of $GaNb_4Se_8$ involves several steps involving multiple instruments and processes before the figure of merit of the sample, its chemical purity, is confirmed using X-ray powder diffraction. This measurement step is common to many solid-state chemical reactions and serves as a node where the user then decides whether to proceed with other measurement processes, represented by magnetometry and synchrotron diffraction in Fig. 2. The Quantum Data Hub (QDH) provides a number of features to make such an analysis simpler and streamlined within the quantum material research process.

Ease of Integrated Data Access and Analysis. Throughout the synthesis and measurement workflow, all metadata and measurement results can be uploaded from any laboratory with a computer and an internet connection. The QDH provides a highly portable analysis environment for measurement data such as magnetometry. Typically, such magnetic data would be transferred from the instrument to the user's computer, where they perform analysis with a variety of methods including user-written Python scripts. On the QDH, the data and user-written Python analysis libraries can be uploaded and run anywhere, greatly streamlining the synthesis and measurement workflow. The QDH also

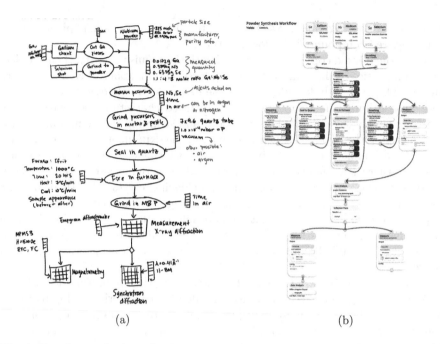

Fig. 2. Powder synthesis and measurement process for the chemical GaNb$_4$Se$_8$: (a) An excerpt from a lab notebook outlining the process. (b) The Quantum Data Hub representation of the workflow for the process outlined in the lab notebook.

allows the user to access information about their samples and data analysis online.

Capturing Reusable Synthesis Conditions. Quantum materials, as well as materials science, has historically relied on published journal articles for reported materials synthesis conditions. There have been a number of efforts to use published synthesis conditions to datamine and machine learn material synthesis to remove the synthesis bottleneck in materials research. However, researchers lose vital information in the publication process. Failed synthesis conditions provide invaluable data points for exploring a new chemical phase space but typically only successful synthesis conditions are published. The Quantum Data Hub retains failed synthesis conditions with query-able metadata associated with the material and its synthesis processes. By creating a database of materials and their synthesis conditions, the QDH can serve as a highly organized and useful dataset for data science efforts in quantum materials science.

Extracting Physical Properties. Other limitations in the field of quantum materials research include compilations of measured physical properties of candidate quantum materials. In fact, compiling data exhibiting the physical properties of novel materials presents many additional challenges including the same challenges of compiling published synthesis conditions. Measured properties are seldom published in raw data form. Instead, researchers present measured data in a variety of formats, including plots, tables, and other graphics. Additionally, data published in journals are often processed to exhibit certain features of the magnetic properties of a material, making it difficult for automated data extraction. The QDH automatically creates a searchable database of physical properties that are important to quantum materials as raw data, retaining a maximal amount of information and unpublished materials property measurements.

One such use case involves the magnetic properties of materials. Magnetic properties are very important to many quantum materials. Magnetometry measurements are common as an initial characterization method of a novel material. There are a variety of different descriptors of magnetic properties, including ordering temperatures, but in quantum materials research, often the most important magnetic features are qualitative, such as the general shape of the magnetization as a function of temperature or applied magnetic field. This is particularly true for quantum materials that are of interest to leverage exotic magnetic properties. Existing databases of experimental magnetic properties focus on limited descriptors such as ordering temperatures and common magnetic properties such as ferromagnetism and antiferromagnetism. By design, a lot of experimental information is lost in these databases. However, if researchers had access to raw magnetic data, initial assessments of the magnetic properties of a material are incredibly quick to a trained eye. The QDH will enable users to search through and quickly evaluate as-measured magnetic data in the form that they choose.

The QDH allows searching through materials and their properties, enabling users to quickly assess the magnetic properties of materials with all the associated metadata of the measurement and material. This will accelerate the initial bottleneck of selecting materials candidates and synthesizing them as well as enabling data science initiatives in quantum materials research to connect materials descriptors such as chemistry or atomic structure to novel physical properties.

2.3   The Quantum Foundry Data Model

The Quantum Foundry Data Model (QFDM) builds on the premise that the data activities of the Foundry is centered around scientific processes and their products. The processes include the synthesis of new materials, taking a newly synthesized material through a series of instruments and computations to measure complex properties, recording these measurements and computational results, evaluation of these results, publishing the results in scientific venues, and possibly using the products of one synthesis process as the raw ingredients of another synthesis process. The data model captures the essential descriptions and order of these processes, as well as all artifacts produced at different stages of these processes.

Formally, the QFDM is a *federated heterogeneous data model*, which is a multi-part model, each part expressed with a different modeling language and implemented in a different store, yet schematically connected through explicit

```
"processSteps": [
 {"@id": "_:b1",
 "processName": "http://sweetontology.net/procPhysical/Shorten",
 "processParameters": [
    {   "cuttingSizeValue": 20,
        "cuttingSizeUnit": "mm" } ],
  "http://rdf.data-vocabulary.org/#description":
                        "cut into small pieces",
  "http://www.loa-cnr.it/ontologies/
                    FunctionalParticipation#patient": 102,
  "next_steps": [ "_b4" ]},
  ...                 ]
```

Fig. 3. Every process step, expressed as a semi-structured node, has its own ID. The next_steps element denotes a list of edges from the current node to other steps. The attributes of a data object may come from established ontologies.

references (foreign keys). The data model is stored in a polystore based information management system called AWESOME [9] developed at UC San Diego.

1. **Process Model (Semistructured – DAG).** The objective for designing the process model is to enable new scientists find previous material synthesis experiments based on ingredients, instruments, experimental results and subprocesses that might possibly be reused. The process model takes the structure of a directed acyclic graph (DAG) where nodes represent subprocesses, and edges designate a direct transition from one subprocess to the next. The nodes of the graph are typed, semistructured objects implemented as JSON-LD so that one element can reference another element within the same process or to an external object through a hyperlink. Figure 3 shows a node of the Process Model DAG. A schematic of the full process DAG is shown in Fig. 2. Partially inspired by [17], the process DAG illustrates the following features.

   - A process node may belong in one of many system-defined types, e.g., a mechanical process, a chemical process, a computational process, etc. Each process may have subcategories. For example, gas flow synthesis, spark plasma sintering and annealing are chemical processes.
   - For each process type, there are a set of mandatory metadata attributes. For example, a mechanical or a chemical process must record the *environment* in which the process occurs. grinding might be performed in open air and another may require an inert gas environment at a prescribed pressure. Similarly, a measurement process must specify the measuring instrument and must point to the measurement settings.
   - A node attribute may have external references. There are two kinds of references. A *URI reference* is used to point to other information objects

like a PubChem entries, while a *data reference* points to a measurement item or a computational item within our system.

2. **Measurements Model (Relational/Semistructured).** Measurements are primarily outputs of different measuring devices or from computational processes. Figure 2 shows three measurement nodes, namely, X-ray diffraction, Synchrotron diffraction and magnetometry. These nodes point to data files whose formats may be relational or semistructured (XML). In either case, the measurement data has a "settings" component and a "measured values" component. Since one of our goals is to find materials synthesis processes that use similar measurement settings, we maintain both the original and flattened versions of the settings data. The "measured values" component is stored to be primarily consumed by analysis routines, and is transformed to a relational form for querying as well as to a form that the analysis routines expect. When measurements are produced from computational processes, the "settings" component contains the identity of the corresponding computational process and the parameters of its execution.

3. **Computation as Data (Semistructured/Vector).** The final component of the data model are computations that analyze data. These computations can be in a *black box* or *gray box* mode. A proprietary analysis software is considered a black box, while an accessible computation, performed through a Jupyter Notebook, is a gray box because parts of the notebook, including documentation, are expressed in an interpretable form (JSON) and can be analyzed algorithmically, while other parts, like the inner details of libraries called in a notebook cannot. As mentioned before, only the invocation information can be stored for black box computations. Gray box computations, provide significantly more details including (a) a vector of libraries used, (b) document vectors comprising all commentaries, and (c) a list of output items that are stored externally. These three vectors are preserved and can be used in finding similar processes in downstream analysis.

The above description illustrates the cross-pointers between different parts of the data model, allowing us to query for materials synthesis processes through any of the stored parameters and then navigating to find all related information stored in other parts of the AWESOME system, which is designed to store relational, semistructured, graph and text-centric data.

The operations supported by the QFDM are developed based on the intended use. Typically, the actual data values, like an X-Ray diffraction measurement value at a specific angle, are not queried for. Rather, the whole measurement data is consumed by a computation or a visualization process. Similarly, an image produced through an experiment is not queried through content analysis.

At this point, QFDM operations are being designed to support **sample-based queries** and **process-based queries**. In one type of sample-based query, the user knows (or can query the system to determine) the sample, and retrieves a data product derived from the sample by measurement or computation. A more general type of sample-based query locates samples for which general process parameters are specified. For example, "which samples were subjected to the X-

Ray diffraction method but did not get characterized based on magentometry"? Yet as third category of sample-based queries would be on the characterization and experimental settings of the synthesized material. For example, in the use case described in Sect. 2.2 the query can be stated as "Find all samples for which, tin (Sn) is a component material, Single Crystal Neutron Diffraction and Electron Paramagnetic Resonance were measured and Field-dependent magnetization data were collected at a temperature below 5K with magnetic field below 10T". In contrast to sample-based queries, a process-based query retrieves a subgraph of the process DAG based on query conditions. For example, "What mechanical and chemical process steps are executed for synthesizing materials for which electron probe microanalysis are conducted? In which of these process steps do we need to use high-pressure inert atmosphere"? The resulting subgraphs may be edited and extended to create a new synthesis process.

The full extent of the QFDM is designed but it is currently under development. Next, we provide a summary of the progress as a part of the QDH architecture.

3   Quantum Data Hub Architecture

The Quantum Data Hub (QDH) platform provides scientists and researchers a unique combination of virtually unlimited storage space combined with a powerful data analysis environment. To this end, the QDH platform exploits developments in storage and database systems to provide large scale storage capabilities. QDH also leverages developments in computing to create a secure layered architecture. This coupled compute-data structure gives each scientists an agile workshop to build their ideas by leveraging advanced data science and artificial intelligence (AI) libraries to transform large quantum material datasets.

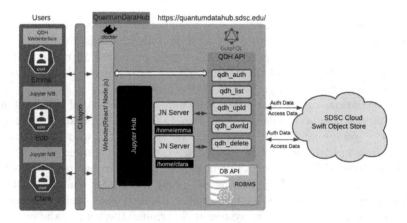

Fig. 4. System architecture of the Quantum Data Hub platform.

The QDH architecture, shown in Fig. 4, consists of 5 main subsystems: (a) a User Authentication and Authorization; (b) a Cloud-based Object Storage back-end and an associated QDH application programming interfaces (APIs); (c) a web-accessible user front-end; (d) a Jupyterhub based analysis environment; and (e) a database system that implements the data model described in Sect. 2.3. The QDH^2 platform enables multiple users to log in, record process and associated metadata, and upload data products related to the material synthesis process. A common usecase involves the material research scientists to log their sequence of steps, various equipment settings and corresponding outcomes of material synthesis process in their logbooks manually as shown in Fig. 2a. The front-end web interface was designed to empower researchers and students to electronically capture a material synthesis process in form of scientific workflow DAG (e.g., the DAG shown in Fig. 2b). The associated data is saved as metadata in a relational database whereas the data and computational products are saved in the cloud object store.

The platform is designed for collaborative research and the environment enables multiple researchers to simultaneously perform complex data analysis using the QDH, and yet store their data securely with required permission management to a unified shared Object Storage system. In order to discover relevant quantum datasets and experiments from unified Cloud Object store, researches can leverage the web-based search interfaces which uses DB queries with metadata attributes in a FAIR-compatible [19] way (e.g., Findable, Accessible, Interoperable, and Reusable). Researchers can connect to the unified Swift Cloud Object store using our QDH APIs and ingest data for analysis. The QDH APIs provide seamless integration of distinct users to shared storage space with precise access control capabilities. The platform leverages JupyterHub [14] as a data analysis platform. The JupyterHub enables each user to launch a dedicated Jupyter Notebook inside a sandboxed Docker [16] container that proxies it back to the user's web browser. The QDH APIs provide simple commands to perform data access operations on the unified Cloud Object storage such as, upload, download, update and delete through the Jupyter notebooks. Integrating the JupyterHub in the QDH amplifies the value of the Quantum Foundry data through data science. Its *workshop-in-a-browser* structure enables users to perform data analysis on the quantum datasets with low overhead.

In the rest of this section, we summarize the main components of the QDH.

Authentication. The QDH platform authenticates users through CILogon [5]. CILogon leverages the OAuth 2.0 standard for token-based authentication to the cyberinfrastructure. Users can gain access to the QDH with their existing university credentials, or other preferred identity providers in few steps. The QDH platform uses a single-sign-on authentication paradigm to allow for navigation and access across its subsystems, such as the QDH front-end, JupyterHub and the Cloud Object Store. Once users' credentials are established at entrypoint to the QDH, users can log their experiments, work in Jupyter notebooks to analysis

---

^{2} The URL for Quantum Data Hub: https://quantumdatahub.sdsc.edu.

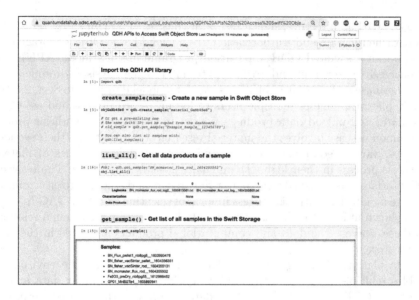

Fig. 5. QDH API calls to access swift storage from Jupyter Notebook

large amounts of data, leverage the APIs to store and access data products using multiple interfaces, and exploit other platform capabilities for their use-cases.

Authorization. The access control model in this system is based on a hierarchical group-based model and represented by the access control triple `<subject, object, permission>`. The subject of the access control model is an individual group, and objects are artifacts, results, or documents produced by the user. Users may have read, write or update permission on the document. A user may be a member of multiple groups. The fine-grained object-level access control will be the future work [11].

Cloud Object Store. We are using the Cloud storage at the San Diego Supercomputer Center using Swift (https://www.sdsc.edu/support/cloud_storage_account.html) to store and archive data products associated with quantum material synthesis process. Swift stores unstructured data in a scalable way to support growth of data over the time, and reliably maintains redundant copies of data, performs error checking, and provides an economical option for research and academic projects.

QDH API. The Python-based QDH API library provides a high-level interface for researchers to interact with the Swift Object storage to store and retrieve data. The API commands use GraphQL queries to call Swift APIs, and can be called from a Jupyter notebook as well as to support the data operation behind the web-interface to perform create, upload, download, delete operations on data objects stored in Swift. Some of the QDH API commands include:

- `qdh.create_sample("sampleName")` - Creates an object with the given sampleName by a user as a top-level object to stores its associated data products.
- `qdh.list_samples()` - Lists all the samples stored in the Swift storage.
- `obj.list_all()` - Lists all the data products associated with a sample, e.g., the data products and files under categories Logbooks and Characterization.
- `sampleName.upload_logbook()` - Submits a logbook into Swift.
- `uploaded_file.delete()` - Removes a file from Swift.
- `uploaded_file.download()` - Creates a local copy of a file.

Data Analysis Platform. JupyterHub provides a platform where multiple users can access a Jupyter Notebook environment to perform data analysis. The Jupyterhub in QDH enables users to perform data analysis, access, upload, and share data. Once the user logs in the QDH, with a click of a button they can start a dedicated Jupyter Notebook single-user application that proxies back to the user's browser. Users can develop or upload their algorithms in Jupyter notebooks and can use simple QDH API commands from Jupyter notebook to access quantum data objects in Swift. Figure 5 shows simple API calls from Jupyter Notebook to create, upload, download, delete operations on this data.

Web Interface. The QDH front-end leverages modern advancements in web development practices in order to provide a feature-rich application for researchers on all platforms. It was engineered with the following goals:

- Create, modify, and collaborate on quantum material sample projects.
- Upload, download, and edit data files and Jupyter Notebooks on Swift.
- Launch JupyterHub to edit and run Jupyter Notebooks.
- Edit procedures for each sample.

The front-end is written using React—an open-source JavaScript state management library that allows for the rapid creation of reusable stateful components, with each encapsulating its own logic. By composing a web front-end with a component-oriented model, development iteration cycles are sped up, and any user-experience issues may be triaged efficiently.

Once authenticated, a dashboard that includes data and notebooks available to the user is presented. The dashboard dynamically caches data retrieved from a GraphQL microservice that aggregates data from the Cloud Object store and Database. We leveraged technical advancements in Kepler scientific workflow management system, provenance and team science [3,4,15] to design QDH Procedure Editor (QDHPE). Users are able to create the procedure associated with each sample using the QDH Procedure Editor (QDHPE), designed to streamline the input of metadata by users. Each vertex displayed in the QDHPE stores associated mutable metadata for each procedure step. For each vertex, users can add custom sets of parameters, and reuse these modifications. The inspector pane for each vertex provides input suggestions based on linked URI references. For instance, a chemical sample vertex may provide a URI reference to a research chemical catalog entry. From there, associated properties (i.e. form, purity) will be offered as an auto-complete suggestion. After each change by a user, the

QDHPE dynamically translates the graph representation into the QFDM (see Sect. 2.3).

4   Related Work

Quantum materials research is a sub-branch of material science research that sits at the convergence point of Quantum Physics and Material Science. Approaching Material Science from a Quantum Physics perspective necessitates a fundamental shift in the process of innovation. The QDH was designed to support such an innovation process in a malleable, scalable, adaptable and collaborative fashion.

In recent times, there has been growing interest in the development of storage and computing platforms that are problem domain sensitive. Some of the storage platforms that allow users to share data for research purposes are CKAN [1], Seedme [7], NoMaD [10], OQMD [13]. These platforms solve a very pertinent problem of data sharing and tackle the problems related to scientific data itself. In contrast, QDH is a unified system that combines computation and data storage technologies to enable quantum material researchers to perform complex analytical tasks. Further, there have been numerous developments in the application of core computer science tools to benefit material science research in recent times. This includes the development of platforms such as Material-Cloud [18], NOMAD [10], AFLOW [8], Material Project [12], CMR [6]. The QDH removes the need to build and maintain separate systems and allows for streamlined research.

Although there are other material science data and collaboration platforms as described above, to the best of our knowledge, The QDH is the only material data platform dedicated to quantum material research for collaborative research that enables multiple researchers to capture a material synthesis process in form of scientific workflow as a Directed Acyclic Graph (DAG). In this DAG, each node saves associated metadata, data, and computational products of the respective synthesis steps and users can perform complex data analysis in JupyterHub using the data porting capability of the system. It offers a searchable database with user-provided metadata that scientists can query to find datasets relevant to their problem domain and combine it with reproducible analysis in a unified platform to accelerate experimentation and streamline the innovation process for synthesis, discovery, storage, and analysis of quantum materials.

5   Conclusions Ad Future Work

This paper presented a new material data and analysis platform for ingestion, management and analysis of quantum material data to couple theory, experimentation and synthesis of quantum materials, built as a part of UC Santa Barbara's NSF Quantum Foundry. Quantum Foundry is the only resource funded by NSF for design and development of materials related to quantum information. The open exchange of data and its organization together with a built in analytical platform within one environment accelerates quantum material design and development as well as enabling new forms of training in this field. It also enables validation, reuse and repurposing of data and analytical products within the Foundry as well as sharing the built in know how with the rest of the world.

In this first description of the QDH, our objective was to describe the vision and progress towards this new resource as an example and in relationship to other related work in scientific computing and material science. While the QDH is fully functional and accessible to a limited group of researchers, the development is ongoing towards the full vision presented in Sect. 2.

As a part of the future work, we would like to link the generated data and insights with other material science data platforms through ontologies and knowledge graphs developed, e.g., the material commons by [2]. In addition, the presented QDH is being extended to enable multiple notebook based analysis process with seamlessly as an analytical workflow with self-reporting capabilities. Future work also includes an evaluation of system performance related to data ingestion, querying efficiency, analytical scalability and collaboration capabilities.

Acknowledgments. This work was supported by the National Science Foundation (NSF) through Enabling Quantum Leap: Convergent Accelerated Discovery Foundries for Quantum Materials Science, Engineering and Information (Q-AMASE-i): Quantum Foundry at UC Santa Barbara (DMR-1906325).

References

1. Ckan. https://ckan.org/
2. Aagesen, L.K., et al.: PRISMS: an integrated, open-source framework for accelerating predictive structural materials science. JOM **70**(10), 2298–2314 (2018). https://doi.org/10.1007/s11837-018-3079-6
3. Altintas, I., Purawat, S., Crawl, D., Singh, A., Marcus, K.: Toward a methodology and framework for workflow-driven team science. Comput. Sci. Eng. **21**(4), 37–48 (2019). https://doi.org/10.1109/MCSE.2019.2919688
4. Altintas, I., Wang, J., Crawl, D., Li, W.: Challenges and approaches for distributed workflow-driven analysis of large-scale biological data: Vision paper. In: Proceedings of the 2012 Joint EDBT/ICDT Workshops, EDBT-ICDT 2012, pp. 73–78. Association for Computing Machinery, New York (2012). https://doi.org/10.1145/2320765.2320791
5. Basney, J., Flanagan, H., Fleury, T., Gaynor, J., Koranda, S., Oshrin, B.: CILogon: enabling federated identity and access management for scientific collaborations. In: Proceedings of International Symposium on Grids & Clouds 2019 — PoS (ISGC2019), vol. 351, p. 031 (2019). https://doi.org/10.22323/1.351.0031
6. Bligaard, T., et al.: The computational materials repository. Comput. Sci. Eng. **14**(06), 51–57 (2012). https://doi.org/10.1109/MCSE.2012.16
7. Chourasia, A., Nadeau, D., Norman, M.: Seedme: data sharing building blocks. In: Proceedings of the Practice and Experience in Advanced Research Computing 2017 on Sustainability, Success and Impact, PEARC 2017, Association for Computing Machinery, New York (2017). https://doi.org/10.1145/3093338.3104153

8. Curtarolo, S., et al.: AFLOW: an automatic framework for high-throughput materials discovery **58**, 218–226. https://doi.org/10.1016/j.commatsci.2012.02.005, https://www.sciencedirect.com/science/article/pii/S0927025612000717

9. Dasgupta, S., Coakley, K., Gupta, A.: Analytics-driven data ingestion and derivation in the AWESOME polystore. In: IEEE International Conference on Big Data, Washington DC, USA, pp. 2555–2564. IEEE Computer Society (December 2016)

10. Draxl, C., Scheffler, M.: NOMAD: the FAIR concept for big data-driven materials science. MRS Bull. **43**(9), 676–682 (2018). https://doi.org/10.1557/mrs.2018.208

11. Gupta, M., Patwa, F., Sandhu, R.: An attribute-based access control model for secure big data processing in hadoop ecosystem. In: Proceedings of the Third ACM Workshop on Attribute-Based Access Control, pp. 13–24 (2018)

12. Jain, A., et al.: The materials project: accelerating materials design through theory-driven data and tools. In: Andreoni, W., Yip, S. (eds.) Handbook of Materials Modeling : Methods: Theory and Modeling, pp. 1–34. Springer International Publishing. https://doi.org/10.1007/978-3-319-42913-7_60-1

13. Kirklin, S., et al.: The open quantum materials database (OQMD): assessing the accuracy of DFT formation energies **1**(1), 1–15. https://doi.org/10.1038/npjcompumats.2015.10, https://www.nature.com/articles/npjcompumats201510

14. Kluyver, T., Ragan-Kelley, B., Pérez, F., et al.: Jupyter notebooks - a publishing format for reproducible computational workflows. In: Loizides, F., Scmidt, B. (eds.) Positioning and Power in Academic Publishing: Players, Agents and Agendas. pp. 87–90. IOS Press, Netherlands (2016). https://eprints.soton.ac.uk/403913/

15. Ludscher, B., et al.: Scientific workflow management and the kepler system. Concurrency Comput. Pract. Experience **18**(10), 1039–1065

16. Merkel, D.: Docker: lightweight linux containers for consistent development and deployment. Linux J. **2014**(239), 2 (2014)

17. Russ, T.A., Ramakrishnan, C., Hovy, E.H., Bota, M., Burns, G.A.: Knowledge engineering tools for reasoning with scientific observations and interpretations: a neural connectivity use case. BMC Bioinform. **12**(1), 1–15 (2011)

18. Talirz, L., et al.: Materials cloud, a platform for open computational science **7**(1), 299 (2020) https://doi.org/10.1038/s41597-020-00637-5, https://www.nature.com/articles/s41597-020-00637-5

19. Wilkinson, M.D., Dumontier, M., et al.: The FAIR guiding principles for scientific data management and stewardship. Sci. Data **3**, 2052–4463 (2016). https://doi.org/10.1038/sdata.2016.18

Hierarchical Analysis of Halo Center in Cosmology

Zichao (Wendy) Di[1]([✉]) [iD], Esteban Rangel[2], Shinjae Yoo[3] [iD],
and Stefan M. Wild[1] [iD]

[1] Mathematics and Computer Science Division, Argonne National Laboratory,
Lemont, IL 60439, USA
wendydi@anl.gov
[2] Argonne Leadership Computing Facility, Argonne National Laboratory,
Lemont, IL 60439, USA
[3] Computational Science Initiative, Brookhaven National Laboratory,
Upton, NY 11973, USA

Abstract. Ever-increasing data size raises many challenges for scientific data analysis. Particularly in cosmological N-body simulation, finding the center of a dark matter halo suffers heavily from the large computational cost associated with the large number of particles (up to 20 million). In this work, we exploit the latent structure embed in a halo, and we propose a hierarchical approach to approximate the exact gravitational potential calculation for each particle in order to more efficiently find the halo center. Tests of our method on data from N-body simulations show that in many cases the hierarchical algorithm performs significantly faster than existing methods with a desirable accuracy.

Keywords: N-body simulation · Hierarchical analysis · Clustering · Halo center

1 Introduction

Cosmological N-body simulation [2,4] is essential for identifying dark matter halos and studying the formation of large-scale structure such as galaxies and clusters of galaxies. For example, the halos in a simulation provide the information needed to analyze structure formation and the galaxy distribution of the universe, which is useful to predict specific models to be compared with observations and therefore understand the physics of gravitational collapse in an expanding universe [17]. One way of identifying halo is through the friends-of-friends (FOF) algorithm [8,9]. Alternatively, one can identify a halo by adopting a certain definition for the halo center and grows spheres around it until given criteria is satisfied [21].

However, one key limitation of these N-body simulations is its rapid increase in computational load with the number of particles, even given the computing power of today's advanced supercomputers. In particular, **the calculation of**

© Springer Nature Switzerland AG 2021
M. Paszynski et al. (Eds.): ICCS 2021, LNCS 12742, pp. 671–684, 2021.
https://doi.org/10.1007/978-3-030-77961-0_53

the bounded potential (BP), i.e., the gravitational force, is the most time consuming task in N-body simulations [4]: in modern cosmological simulations, large halos can comprise tens of millions of particles [13]. Therefore, any improvement for this online analysis is vital.

A number of existing algorithms have been developed to run such large calculations. The most straightforward way to calculate the force is to carry out a direct pairwise summation over all particles, which is a brute-force algorithm and requires $O(N_p^2)$ operations [1,16]. A large halo in a modern simulation may have up to 20 million particles, and thus such a global operation comes with considerable expense and quickly becomes impractical. Alternatively, given the fact that the particles close to each other share common properties, one can group particles that lie close together and treat them as if they are a single source, hence, the force of a distant group of particles is approximated by pseudo particle located at the center of mass of the group. Such methods include the tree method where the particles are arranged in a tree structure [5,14], and the fast multipole method (FMM) [7]. FMM improves the tree method by including higher moments of mass distribution within a group.

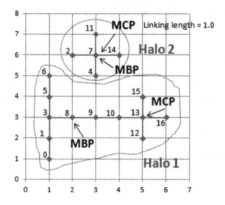

Fig. 1. Illustration of MBP and MCP of two halos with different shapes [19]. Notice that when the halo is approximately spherical, MBP and MCP coincide. Otherwise, they can be far away from each other.

In this work, we focus on the calculation of the halo center, which is a natural byproduct of the BP calculation, and is commonly defined as the "most bounded particle" (**MBP**). MBP is the particle within a halo with the lowest BP. Another type of halo center is the "most connected particle" (MCP), which is the particle within a halo with the most "friends." Figure 1 [19] demonstrates MCP and MBP separately for two different halos and suggests that, given a halo, its MBP and MCP may or may not coincide. Finding the MCP is relatively simple: one sweeps through the virtual edges connecting two particles provided by FOF-based halo finding. However, finding the MBP is much more computationally expensive directly due to the calculation of BP. Some efforts have been made to estimate the exact MBP. The most intuitive and common way is to approximate the MBP location by MCP. However, this approach can provide a reliable estimation only when the halo is roughly spherical, which unfortunately is not always the case. Other recent developments include using a A^* search algorithm to approximate the BPs [10], which is proved eight times faster than the brute-force algorithm. Another approach is to utilize a binning algorithm to rule out the particles that cannot be the MBP, in order to reach a complexity of $O(mN_p)$; however, for most of the time, m can be still very large. High-performance computing also has been used to accelerate the calculation [19]. One common feature of these approaches is that the BP is exactly calculated, meaning that the summation is over every other particle in the halo.

Inspired by the tree-based methods, we propose to further exploit the latent or intrinsic structure within the data, grouping the data into corresponding clusters and then performing operations on the respective groups. We notice that, given the definition of BP (defined below in Sect. 2), the farther the particle is away from another, the less the impact of this particle on the other's BP. This realization can be illustrated by the BP map of a halo. Given a 3D dark matter halo from an N-body simulation with 89,887 particles, we show its 2D projection for better visualization in Fig. 2 on the left side. The right side shows that the BP map is smooth in the sense of small value change around a local neighborhood. Therefore, in this work, instead of exactly calculating every BP to find the MBP, we exploit the local smoothness of the BP map and propose a hierarchical approach to approximate the BP by its so-called local BP, which is defined only on a local neighborhood.

Different from existing methods, our approach provides following benefits:

- It provides a more flexible framework to incorporate domain knowledge, such as different properties of halo in terms of linkage length. Therefore, the accuracy and computational cost can be optimally balanced.
- It is able to recover the local minimum which also contains important physical information for the scientific discovery.
- We provides the theoretical error bounds and complexity analysis of our proposed framework.
- Systematic experiments reveals that our proposed method is significantly faster and yet accurate as the number of particles are increasing, which is critical for halo center finding.

We organize our presentation as follows. In Sect. 2, we introduce our hierarchical framework based on several key components, including construction of the tree structure and an algorithm for finding local extremes. In Sect. 3, we analyze the error bounds and complexity of our proposed hierarchical framework. In Sect. 4, we compare our proposed framework with a brute-force algorithm (as to provide gold standards for accuracy) and FMM on many synthetic halos with different configurations. In Sect. 5, we present our conclusions and discuss ideas for future development.

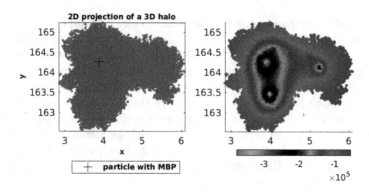

Fig. 2. Left: 2D projection of a 3D halo with 89,887 particles. We can see three major dense areas that can yield at least three local minima of problem (2.2); right: its global BP map calculated by Eq. (2.1). We can see a smooth change of the BP value around a local neighborhood.

2 Method

Given a collection of N_p distinct particles $\mathbf{X} = \{X_i : i = 1, \ldots, N_p\}$ in a halo, we denote a MBP by X_p. The BP of a given particle X_i is computed as the sum over all other particles of the negative of mass divided by the distance,

$$P(X_i) = \sum_{\substack{j \neq i \\ j \in [1, \ldots, N_p]}} \frac{-m_j}{d(X_i, X_j)}, \qquad i = 1, \ldots, N_p, \tag{2.1}$$

where $X_i \in \mathbb{R}^3$ is the ith particle represented by its position coordinates, m_i is its corresponding mass, and $d(X_i, X_j)$ is the Euclidean distance between particles X_i and X_j. Then, we have the following optimization problem,

$$X_p \in \operatorname*{argmin}_{X_i \in \mathbf{X}} \sum_{\substack{j \neq i \\ j \in [1, \ldots, N_p]}} \frac{-m_j}{d(X_i, X_j)}. \tag{2.2}$$

We notice that the BP map from every particle is relatively smooth in its local region. Therefore, our approach is, starting from the whole domain, to approximate the BP map by a few sampled (seed) particles that are sufficiently far away from each other. By comparing the local BP of the sample particles, we gradually narrow the search region to a smaller area where the MBP can be. In other words, at each hierarchical level (as described in Algorithm 1), we approximate the BP map of the corresponding search region by the local BPs of a few sample particles.

Three critical questions need attention. First, how many sample particles should be enough to approximate the whole domain? Second, how far do the sample particles have to be away from each other to guarantee a sufficient coverage? Third, how can one determine the threshold used to calculate the local BP?

Intuitively, the deeper the hierarchical level is, the smaller the search region is; therefore, larger thresholds can approximate the BP better. Thus, an essential step is to perform a range search to allocate the neighboring particles given a threshold. Our approach is to use a kd-tree to first construct the neighborhood among particles. A kd-tree is a data structure that partitions the space through alternative dimensions for organizing points in a k-dimensional space [6]. Since kd-trees divide the range of a domain in half at an alternative dimension at each level of the tree, they are efficient for performing range searches. This structure is constructed only once, in the beginning. Throughout the process, the preconstructed tree structure is used to search for ranges at a given distance threshold. For example, if a tree is storing values corresponding to distance, then a range search looks for all members of the tree in the distance that are smaller than the given threshold. Throughout the paper, we employ the range search algorithm presented in [15].

Once the kd-tree is constructed, on the first level we uniformly sample a few particles from $\mathbf{X}$ and approximate their local BPs (denoted as $\tilde{P}$) by the range search at a given distance threshold $\varepsilon_1 > 0$ as follows:

$$\tilde{P}(X_i) = \sum_{\substack{j \neq i \\ X_j \in B(X_i, \varepsilon_1)}} \frac{-m_j}{d(X_i, X_j)}, \qquad i = 1, \ldots, N_p, \tag{2.3}$$

where $B(X_i, \varepsilon_1)$ is the ball with center X_i and radius ε_1. Accordingly, we have a particle with minimum *local BP* as

$$X_{\tilde{p}} \in \operatorname*{argmin}_{X_i \in \mathbf{X}} \tilde{P}(X_i), \tag{2.4}$$

where $X_{\tilde{p}}$ is the approximation to a global MBP X_p. The next step is to interpolate the coarse BP map for every particle in the halo and use a peak-finding algorithm to locate the particles having the first few smallest BPs while their pairwise distance is bigger than another distance threshold denoted by $\varepsilon > 0$. These particles are then used as the next-level sample particles. We mark the particle with the smallest local BP as a temporary $X_{\tilde{p}}$. Then we start a range search on each newly selected sample particle. From the second level, instead of calculating only local BPs for the sample particles, we calculate local BPs for every particle in the ranges and repeat the steps of interpolating, peak finding, and locating the temporary $X_{\tilde{p}}$ until the current temporary $X_{\tilde{p}}$ coincides with the previous temporary $X_{\tilde{p}}$. We fix the number of sample particles at each level as N_s and denote the collection of the current samples as *Seeds*. The full procedure is described in Algorithm 1, and the step of locating peaks in the BP map is described in Algorithm 2.

We now discuss in detail how we choose the distance thresholds. First, to determine ε_1 that defines the neighborhood, we need to analyze the sensitivity of ε_1 for preserving the pattern of the true BP map. Given the same halo as shown in Fig. 2, we first compare the local $X_{\tilde{p}}$ provided by Eq. 2.4 and the global X_p provided by Eq. 2.2. Given 100 different ε_1 that are equally spaced in the

---

Algorithm 1. $X_{\tilde{p}} = recursive\_localMBP(\mathbf{X}, m)$

1: Given $l_{max} > 0$, and initialize $tree = $ kd_tree$(\mathbf{X})$.
2: **for** $level = 1, \cdots l_{max}$ **do**
3:    **if** $(level = 1)$ **then**
4:       Select N_s uniformly distributed random particles $\{X_{\tilde{i}} : \tilde{i} = 1, \ldots, N_s\}$ from $\mathbf{X}$ as $Seeds$.
5:       For each $X_{\tilde{i}}$, calculate $B(X_{\tilde{i}}, \varepsilon_1) = range\_search(tree, X_{\tilde{i}}, \varepsilon_1)$ and $\tilde{P}(X_{\tilde{i}})$ by Eq. (2.).
6:       Find $X_{\tilde{p}} \in \underset{X_i \in \mathbf{X}}{\mathrm{argmin}} \, \tilde{P}(X_i)$ based on the interpolated BP map from every particle in the halo.
7:       $Seeds = find\_peak(\tilde{P}, N_s, \varepsilon)$ that updates $Seeds$ as the particles with local minimum BP value.
8:    **else**
9:       For each $X_{\tilde{i}}$ in $Seeds$, calculate $B(X_{\tilde{i}}, \varepsilon_1) = range\_search(tree, X_{\tilde{i}}, \varepsilon_1)$.
10:      For every particle $X_i \in \bigcup_{i=1}^{N_s} B(X_{\tilde{i}}, \varepsilon_1)$, calculate $\tilde{P}(X_i)$.
11:      Set $X_{\tilde{p}}^{(0)} = X_{\tilde{p}}$.
12:      Find $X_{\tilde{p}} \in \underset{X_i \in \bigcup_{i=1}^{N_s} B(X_{\tilde{i}}, \varepsilon)}{\mathrm{argmin}} \, \tilde{P}(X_i)$.
13:    **end if**
14:    **if** $\|X_{\tilde{p}}^{(0)} - X_{\tilde{p}}\| = 0$ **then**
15:      break
16:    **end if**
17: **end for**

Algorithm 2. $Seeds = find\_peak(P, m, \varepsilon_1)$

1: Order a given vector P (local BPs of previous $Seeds$) monotonically increasing.
2: Choose the particles corresponding to the first N_s components of ordered P and whose pairwise distance is larger than ε, and assign them as the new $Seeds$.

---

region $[0, 1]$, we generate their corresponding $X_{\tilde{p}}$ as a function of ε_1 and calculate its error compared with X_p, respectively, as $|X_{\tilde{p}}(\varepsilon_1) - X_p|$. The result is shown in Fig. 3a. We can see that as ε_1 increases, $X_{\tilde{p}}$ has a better chance to match X_p. In particular, as $\varepsilon_1 > 0.1$, the difference between $X_{\tilde{p}}$ and X_p is negligible given the sufficient physical distance threshold used to distinguish two particles as 10^{-3}.

Furthermore, we fix $\varepsilon_1 = 0.3$ to calculate the corresponding local BPs and compare them with its global BP, as shown in Fig. 3b. Notice that $\varepsilon_1 = 0.3$ is considered very small given the mean of all pairwise distances as 2.1. The x-axis arranges the particle indices to monotonically increase the global BPs. As we can see, although there is almost a constant offset between the local- and global-potential profiles, the local BP map preserves the shape of the global BP, which suggests that the minimizer of the local BPs can closely approximate the minimizer of the global BPs.

These numerical tests suggest a reasonable choice of $\varepsilon_1 = 0.3$ on the first level, so that the local BP map agrees well with the global BP map. Because

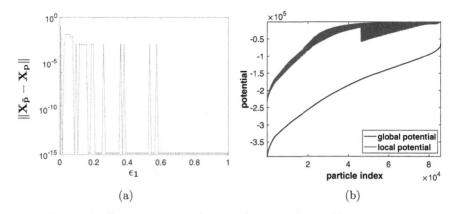

Fig. 3. Left: we compare the relative error between X_p and $X_{\tilde{p}}$ generated by different ε_1 from Eq. (2.4). As ε_1 increases, the local $X_{\tilde{p}}$ has a better chance to coincide with the global X_p; Right: Given $\varepsilon_1 = 0.3$ as the choice for the rest of the tests, the local BP map has a pattern similar to that of the global BP map, which makes them share the same minimizer.

of the common features (e.g., mean of the pairwise distances between particles, linkage length) shared by different dark matter halos, we keep $\varepsilon_1 = 0.3$ for different halos from the same simulation system. Accordingly, the choice of ε_1 suggests that ε, which is the distance threshold to avoid seed particles being too close to each other, should be at least as big as ε_1 in order to guarantee a good coverage of the domain of interest.

3   Error Bounds and Complexity Analysis

In this section, we first provide a preliminary error estimate of the proposed method. Again, since the BP function is a function of inverse Euclidean distance, distance greater than ϵ_1 can be negligible to some extent. Together with the local smooth property of BP function, the error $P(X_i) - \tilde{P}(X_i)$ can be analyzed analogous to the cumulative distribution function error between Gaussian distribution and truncated Gaussian distribution. We assume each term in function $P(X_i)$ follows a Gaussian distribution with respect to $d(X_i, X_j)$, where its mean is 0 and standard deviation is σ, then we have

$$\|P(X_i) - \tilde{P}(X_i)\| = 1 - \frac{1}{2}\left[\text{erf}\left(\frac{\epsilon_1}{\sqrt{2}\sigma}\right) - \text{erf}\left(\frac{-\epsilon_1}{\sqrt{2}\sigma}\right)\right]$$
$$= 1 - \text{erf}\left(\frac{\epsilon_1}{\sqrt{2}\sigma}\right), \tag{3.5}$$

where $\text{erf}\, x = \frac{2}{\sqrt{\pi}} \int_0^x e^{-t^2}\, dt$ is the error function. As shown in Fig. 4, we can see that as ϵ_1 gets larger, the estimation error of local BP is getting smaller and approaching 0.

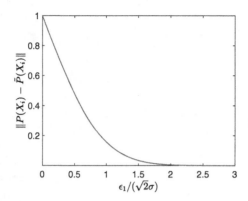

Fig. 4. The estimation error provided by Eq. (3.5). As ϵ_1 gets larger, the estimation error gets smaller and approaches 0.

Next, we estimate the complexity of our proposed hierarchical algorithm. Each level involves the following computations:

- Conduct a range search for each *Seed*: $O(N_p N_s)$.
- After level 1, approximate the local BP for each particle in the neighborhood of the seed particles: $O(n^2 N_s)$, where $n \ll N_p$ is roughly the number of particles in each neighborhood. Figure 5 shows the relationship between n and N_p empirically for various of halos.
- Find the peak: the dominant calculation is sorting, which takes $O(N_s n \log n)$.

On average, three levels are needed in order to converge, and the complexity of a one-time kd-tree construction is $O(N_p \log N_p)$. By summing up these components, the total complexity is

$$O(N_p \log N_p + n^2).$$

4   Numerical Results

We now illustrate the performance of the proposed algorithm on a set of halos from catalogs created by the *in situ* halo finder [18] of the Hybrid Accelerated Cosmology Code (HACC) [12]. All the numerical experiments are implemented in MATLAB and performed on a platform with 32 GB of RAM and two Intel E5430 Xeon CPUs. We first demonstrate the proposed approach on the 2D projection of the 3D halo shown in Fig. 2 where we uniformly downsample only 4,933 particles for a better visualization. This halo is a relatively challenging example since it contains approximately three locally dense areas, which results in three local minima for problem (2.2). We demonstrate the recursive process in Fig. 6, where Fig. 6a shows where the seed particles are on the first level with its corresponding local BP map and Fig. 6b shows similar information for the second level. Figure 7 shows the result where the recursively calculated local $X_{\tilde{p}}$ agrees with the global X_p.

4.1   2D Result

First, we benchmark the performance of the proposed method on 455 2D halos which are projected from the corresponding 3D dark matter halos simulated by HACC. In Fig. 8, we compare the performance of three different methods for finding the MBP: brute force where we explicitly calculate every point-wise distance between particles, the proposed recursive method with $N_s = 10$ and fixed ε, and the FMM using an optimized MATLAB implementation [20]. We can see from the time plot, as the number of particles increases, the proposed recursive approach is outperforming FMM and brute force, while the accuracy in terms of potential error of the returned MBP from our recursive approach is better than FMM in average. Notice here that in order to easily visualize the value, we sort the potential error as shown in y-axis.

4.2   3D Result

Next, we run our algorithm on 455 simulated 3D dark matter halos, using different algorithm configurations. However, due to the difficulty of having optimally implemented MATLAB-based 3D FMM code, in this section, we only compare the performance of our proposed approach against brute force using different parameter settings. For the first test we fix $N_s = 10$, $\varepsilon_1 = 0.3$, and $\varepsilon = \varepsilon_1$ for each level. On the left of Fig. 9, we can see that as the number of particles increases from different halos, our hierarchical method is 100 times faster than the brute-force method. The middle panel shows the location accuracy of our hierarchical method compared with that of the brute-force method. We note that in reality, if two particles are separated by a distance less than 10^{-3}, we do not distinguish these two particles. As a result, 90% of the halos in this test achieve location error smaller than 10^{-3}, which is a desirable accuracy. We also report the relative potential error on the right panel, which is given as $\dfrac{|P(X_{\tilde{p}}) - P(X_p)|}{|P(X_p)|}$.

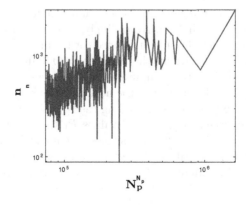

Fig. 5. Given various halos with different numbers of particles, we report the corresponding mean value of n resulting from the proposed Algorithm 1. It suggests that compared with N_p, n is significantly smaller.

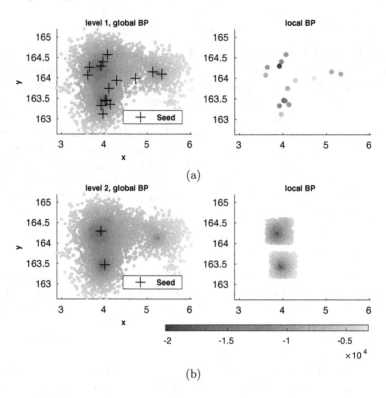

(a)

(b)

Fig. 6. Demonstration of the hierarchical process. Figure 6a illustrates the performance of the first level. The left side shows where the seed particles (denoted as +) are located along with the color-coded global BP map. The right side shows the local BP map provided by the previously chosen seed particles; The second level is shown similarly in Fig. 6b. We can see that the local BP map resembles well the global BP map of the corresponding region.

We also examine our algorithm by gradually reducing the neighborhood radius ε_1 by half along each level. Accordingly, ε is scaled by half as well along each level. Figures 10 and 11 show the effect of having $N_s = 100$ and $N_s = 10$ seed particles, respectively, where the initial $\varepsilon_1 = 0.3$ and scaled by half along each level. We can see that the computational time saved is slightly better than without scaling ε_1. We also note that the accuracy has kept relatively similar to the case where ε_1 is fixed in terms of finding the X_p. Furthermore, if we examine the relative error between the potential values $P(X_p)$ and $P(X_{\tilde{p}})$, we obtain a mean of these errors to be 10^{-3}. This suggests that our hierarchical algorithm can find the local minimum if not the global minimum. This capability is also useful since the local minimum of the MBP optimization problem also obtains

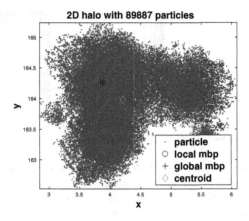

Fig. 7. Result of the recursive process shown in Fig. 6: the approximated local MBP agrees with the global MBP in terms of both location and potential value. Its centroid is labeled to show multiple dense areas in this particular 2D halo.

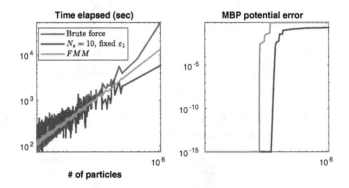

Fig. 8. Performance comparison of finding MBP for various 2D halos using three different algorithms: brute force, our proposed hierarchical method with $N_s = 10$ and fixed ε_1, and FMM. Left: time elapsed; Right: error of potential between the MBP located by proposed method and FMM, respectively.

important features [11]. We note that the oscillatory behavior of the time elapsed and the MBP error on halos with relatively small number of particles are due to the different hierarchical levels required for different structures (such as number of local minima).

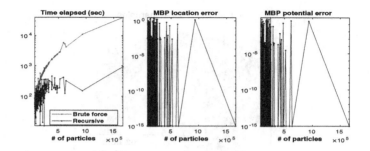

Fig. 9. 3D halo result with fixed ε_1. Left: time elapsed for the hierarchical approach and the brute-force approach. Middle: location accuracy of the MBP approximation for different halos with different structures and numbers of particles. Right: Potential error between the approximated MBP and the global MBP.

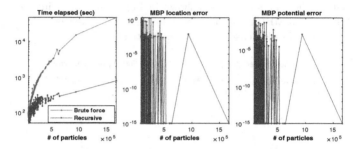

Fig. 10. Result of applying the hierarchical method for the same set of halos with $N_s = 100$, and ε_1 scaled by half along each level.

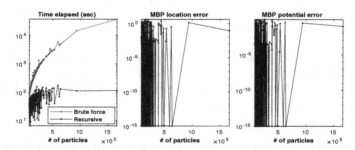

Fig. 11. Result of applying the hierarchical method for the same set of halos with $N_s = 10$, and ε_1 scaled by half along each level.

5   Conclusion and Future Work

We propose a hierarchical framework to accelerate the performance of finding a halo center, in particular, the MBP. Instead of using an exact calculation to find the global MBP, which is extremely expensive because of the large number of particles in a typical halo, we explore the smooth property and the hierarchical structure of the BP map and approximate the global BP only by the local information. The preliminary numerical results suggest that our method is comparable to fast multipole method (even slightly better) in terms of both speed and accuracy, and show dramatic speedup compared with the most common brute-force approach. On the other hand, comparing to existing methods, our approach provides a more flexible framework to incorporate domain knowledge, such as linkage length used to define a halo, to optimally choose the hyper parameters ϵ and ϵ_1 used in the Algorithm 1.

Therefore, opportunities remain to further accelerate the performance. First, one can explore different ways of performing tree data structure, such as R-tree [3], to better exploit the geographical correlations among particles. Furthermore, one can explore ways to shrink the range search parameter along each level guided by the density of the local region. For example, if the density is high, the search parameter ε_1 should not decrease. Another key area of investigation is the initial sampling strategy. We believe the algorithm presented in this work can be generalized to other applications that share the same feature as halo center finding, such as approximation of the pairwise distance matrix.

Acknowledgments. This work was supported by the Exascale Computing Project (17-SC-20-SC), a collaborative effort of two U.S. Department of Energy organizations (Office of Science and the National Nuclear Security Administration). This material was based on work supported by the U.S. Department of Energy, Office of Science, under contract DE-AC02-06CH11357.

References

1. Aarseth, S.J.: From NBODY1 to NBODY6: the growth of an industry. Publ. Astron. Soc. Pac. **111**(765), 1333 (1999)
2. Aarseth, S.J.: Gravitational N-body Simulations: Tools and Algorithms. Cambridge University Press, Cambridge (2003)
3. Arge, L., Berg, M.D., Haverkort, H., Yi, K.: The priority R-tree: a practically efficient and worst-case optimal R-tree. ACM Trans. Algorithms (TALG) **4**(1), 1–30 (2008)
4. Bagla, J.S.: Cosmological N-body simulation: techniques, scope and status. Curr. Sci. **88**, 1088–1100 (2005)
5. Barnes, J., Hut, P.: A hierarchical O (N log N) force-calculation algorithm. Nature **324**(6096), 446–449 (1986)
6. Bentley, J.L.: Multidimensional binary search trees used for associative searching. Commun. ACM **18**(9), 509–517 (1975)
7. Darve, E.: The fast multipole method: numerical implementation. J. Comput. Phys. **160**(1), 195–240 (2000)
8. Davis, M., Efstathiou, G., Frenk, C.S., White, S.D.: The evolution of large-scale structure in a universe dominated by cold dark matter. Astrophys. J. **292**, 371–394 (1985)
9. Einasto, J., Klypin, A.A., Saar, E., Shandarin, S.F.: Structure of superclusters and supercluster formation-III. Quantitative study of the local supercluster. Mon. Not. R. Astron. Soc. **206**(3), 529–558 (1984)

10. Fasel, P.: Cosmology analysis software. Los Alamos National Laboratory Tech Report (2011)
11. Gao, L., Frenk, C., Boylan-Kolchin, M., Jenkins, A., Springel, V., White, S.: The statistics of the subhalo abundance of dark matter haloes. Mon. Not. R. Astron. Soc. **410**(4), 2309–2314 (2011)
12. Habib, S., et al.: HACC: extreme scaling and performance across diverse architectures. Commun. ACM **60**(1), 97–104 (2016)
13. Heitmann, K., et al.: The Q continuum simulation: harnessing the power of GPU accelerated supercomputers. Astrophys. J. Suppl. Ser. **219**(2), 34 (2015)
14. Jernigan, J.G., Porter, D.H.: A tree code with logarithmic reduction of force terms, hierarchical regularization of all variables, and explicit accuracy controls. Astrophys. J. Suppl. Ser. **71**, 871–893 (1989)
15. Kakde, H.M.: Range searching using Kd tree. Florida State University (2005)
16. Makino, J., Hut, P.: Performance analysis of direct N-body calculations. Astrophys. J. Suppl. Ser. **68**, 833–856 (1988)
17. Ross, N.P., et al.: The 2dF-SDSS LRG and QSO survey: the LRG 2-point correlation function and redshift-space distortions. Mon. Not. R. Astron. Soc. **381**(2), 573–588 (2007)
18. Sewell, C., et al.: Large-scale compute-intensive analysis via a combined in-situ and co-scheduling workflow approach. In: Proceedings of the International Conference for High Performance Computing, Networking, Storage and Analysis. p. 50. ACM (2015)
19. Sewell, C., Lo, L.T., Heitmann, K., Habib, S., Ahrens, J.: Utilizing many-core accelerators for halo and center finding within a cosmology simulation. In: 2015 IEEE 5th Symposium on Large Data Analysis and Visualization (LDAV), pp. 91–98. IEEE (2015)
20. Tafuni, A.: A single level fast multipole method solver. https://www.mathworks.com/matlabcentral/fileexchange/55316-a-single-level-fast-multipole-method-solver (2020)
21. White, M.: The mass of a halo. Astron. Astrophys. **367**(1), 27–32 (2001)

Fast Click-Through Rate Estimation Using Data Aggregates

Roman Wiatr[1(✉)], Renata G. Słota[1], and Jacek Kitowski[1,2]

[1] Faculty of Computer Science, Electronics and Telecommunication, Institute of
Computer Science, AGH University of Science and Technology, Mickiewicza 30,
30-059 Krakow, Poland
{rwiatr,rena,kito}@agh.edu.pl

[2] Academic Computer Centre CYFRONET AGH, AGH University of Science and
Technology, Nawojki 11, 30-950 Krakow, Poland

Abstract. Click-Through Rate estimation is a crucial prediction task in
Real-Time Bidding environments prevalent in display advertising. The
estimation provides information on how to trade user visits in various systems. Logistic Regression is a popular choice as the model for this task.
Due to the amount, dimensionality and sparsity of data, it is challenging
to train and evaluate the model. One of the techniques to reduce the
training and evaluation cost is dimensionality reduction. In this work,
we present Aggregate Encoding, a technique for dimensionality reduction using data aggregates. Our approach is to build aggregate-based
estimators and use them as an ensemble of models weighted by logistic
regression. The novelty of our work is the separation of feature values
according to the value frequency, to better utilise regularization. For our
experiments, we use the iPinYou data set, but this approach is universal
and can be applied to other problems requiring dimensionality reduction
of sparse categorical data.

Keywords: Real-Time Bidding, RTB · Click-Through Rate, CTR ·
Dimensionality reduction · Logistic regression

1 Introduction

Online advertising is a ubiquitous form of advertisement that uses the internet to
display ads to the users. Y. Yuan et al. [9] define Real-Time Bidding (RTB) as a
business model for automated online advertising with transaction time constraint
between 10 and 100 ms. There are three key players in the RTB setup: publishers
- offering the internet traffic, generated by sites or applications, aggregated on
Supply Side Platform (SSP); advertisers - running campaigns configured on a
Demand Side Platform (DSP), offering ads to display on sites or applications
provided by the SSPs; Ad Exchanges (AdEx) - platforms for facilitating the
trade between multiple SSPs and DSPs. When a user visits a site with a display
advertisement placement, the visit is being offered on AdEx, where DSP can

© Springer Nature Switzerland AG 2021
M. Paszynski et al. (Eds.): ICCS 2021, LNCS 12742, pp. 685–698, 2021.
https://doi.org/10.1007/978-3-030-77961-0_54

bid on behalf of the advertiser for that particular visit (Fig. 1). The winning
advertiser displays the advertisement to the website visitor. On the DSP side,
the advertiser configures one or more campaigns. Each campaign consists of
preferred targets like age, country, operating system etc. and creatives i.e. images
or videos.

Several event types can be tracked in the RTB environment. Each display
of the advertisement generates an impression event. This event tells nothing
about the user real interest in the advertisement. The next event is a click event
where the user clicks on the displayed advertisement. The relation of clicks to
impressions is called Click Through Rate (CTR). As reported in [1] CTR can be
lower than 1% making the data very unbalanced. The final event is a conversion
event. It is generated when the user takes further actions (like filling out a form or
purchasing a product). The relation of conversions to clicks is called Conversion
Rate (CVR). Typically conversions are very rare and can occur hours after the
click.

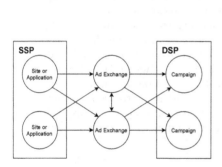

Fig. 1. RTB environment

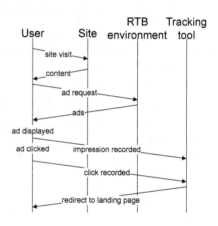

Fig. 2. Tracking events in RTB environment.

RTB environments are highly automated. Because of the number of transactions and time constraints, each transaction has to be made without human
interaction. This forces every entity in the chain to monitor the state of the
traded traffic. Figure 2 shows a typical event chain for sites and how a tracking
tool may be used by an advertiser to monitor the performance of the displayed
ads. When a user visits a site that has a connection to an RTB environment (in
this case SSP), a script is loaded that makes an ad request to the RTB environment. An offer is made on an AdEx and the winners are enabled to present their
advertisements to the user. On advertisement load time an impression event
is generated, indicating an advertisement was presented to the user. A subsequent click and conversion event can be generated by the user actions. All of

these events are tracked using a tracking tool. A similar tool may be used on SSP, AdEx and DSP sides depending on the business model. The traffic buyer is charged by click or by impression. Since the amount of ingested data is huge [1] it is crucial to store the data as an aggregated time series [8] for manual and automated decision making. Raw data are also stored for audit purposes and model building but tend to have a shorter lifespan due to space requirements.

Each time an ad request is done to the RTB environment, an AdEx makes an offer to multiple DSPs (or other AdEx), meaning a single site visit may trigger several DSP queries. Each time a DSP gets an offer, it has to evaluate the CTR/CVR model to decide on how to bid. The DSP does so by evaluating hand written rules as well as various models (Fig. 3) in real time. The amount of bid requests combined with the volatility of traffic and campaigns [1] means the model has to be efficient in both training and evaluation.

In this work, we show how to exploit data aggregates to reduce resources and time required to train and execute the model. We evaluate the model performance on well-known iPinYou data set in a CTR prediction task. The bidding process is complicated and is beyond the scope of this paper. The process can be optimized by several components as stated in [2].

Section 2 presents the state of the art on which this work is based on. Section 3 describes our approach and proposed improvements. In Sect. 4 we compare three types of encodings: Dummy Encoding, Hashing Trick as the baseline and Aggregate Encoding as the proposed improvement. Section 5 describes the experimental setup using iPinYou data set and contains the results of experiments comparing Hashing Trick and Aggregate Encoding. In Sect. 6 we present the conclusions and proposals for future work.

2   State of the Art

A detail statistical analysis of the iPinYou data set as well as a benchmark for predicting CTR are provided by the authors in [11]. The iPinYou data set is available at [10].

In [4] authors present logistic regression as being widely referred to as the state-of-the-art and the most widely used algorithm for CTR prediction. They compare Area under the Receiver Operating Characteristic curve (AuROC) of several methods previously presented in the literature, including [11] on the iPinYou data set. Furthermore, the authors present the impact of certain features, feature generalization and training process parameters. The authors show the high importance of feature generalization and conclude that L2-regularized logistic regression with cross-entropy loss function is the state-of-the-art in RTB CTR estimation. The authors do not consider dimensionality reduction for sparse data that can be useful when feature conjunctions are introduced, leading to an exponential growth of feature vector size.

A scalable and easy to implement framework based on logistic regression is proposed in [1]. The model inputs are categorical variables and categorical variable conjunctions. The dimensionality problem is addressed by the use of hashing trick [6]. Authors present arguments for training one model for multiple campaigns (multi-task learning) instead of separate models for each campaign. Due to class imbalance (positive class is lower than 1%) and huge amounts of data (9 billion events daily), negative class sampling is proposed to reduce the computational cost. Multiple dimensionality reduction techniques are considered in this work and Hashing Trick is considered to be the best one. The authors present practical implications of selecting such a model and propose a feature selection algorithm based on conditional mutual information. The algorithm allows assigning a score to features without retraining the model with each feature. The algorithm is used to select the best conjunction feature and only then the model is retrained. The authors show how fast the model gets outdated and they explain the degradation with the influx of new advertisements. Finally, the authors present an algorithm for large scale model training. This work is a comprehensive guide to build a CTR prediction system and as such it does not attempt to focus on novel techniques for feature encoding.

In [3] the authors focus on predicting Conversion Rate (CVR), a problem similar to CTR prediction with the difference that conversions are much rarer than clicks. The authors model three separate data hierarchies for user, advertiser and publisher. They build a conversion probability estimator for each of these hierarchies and combine the output of these estimators using logistic regression. They reduce the input of the logistic regression to three dimensions which are the success probability based on these estimators. This technique addresses data sparsity but does not address the problem of overfitting due to modelling rare and frequent features together.

An approach focusing on feature engineering is proposed in [5]. The authors present a novel Deep & Cross Network. The architecture consists of two modules: a deep neural network modelling feature interact and a novel cross-network that is better suited to learn cross product of features. This model achieves the state of the art performance and reduction in parameters compared to a deep neural network. Due to input sparsity, a learnable embedding layer is used and no other dimensionality reduction techniques are considered.

3   Research Goals and Approach

Our research focus on CTR prediction, which is a part of a DSP pipeline [2] described by Fig. 3. In our previous work [8] we point out that the marketer has access to aggregated impressions, clicks and conversions via a time series interface. In this work, we show how to reuse the aggregated data stored in a fast database to speed up the training process while preserving the quality of the produced models.

Our approach is to build aggregate-based estimators and use them as an ensemble of models weighted by logistic regression. The novelty of our approach is the separation of feature values of each estimator, to better utilise regularization.

We use the iPinYou data set [10] to conduct our experiments. All features are transformed into categorical variables. In our approach first, we create simple

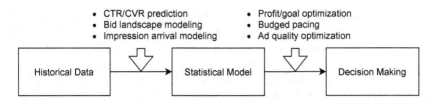

Fig. 3. DSP data pipeline [2]

models based on aggregated features and then we use these aggregates to train a logistic regression model. We compare the baseline to our model in terms of AuROC and parameters used by the model. The code used in the experiments can be accessed via GitHub (see [7]).

4 Modelling Overview

One of the uses of raw data is to train models for CTR (or CVR) prediction. In CTR each impression ends in one of two events: an advertisement can be either clicked or not clicked making it a binary classification problem. The purpose of the model is to support the decision making process by providing it with the probability estimate that a particular user will click the advertisement (see Fig. 3).

The iPinYou data set is divided into test and train sets. The train set contains over 15 million impressions and over 11 thousand clicks. The test set contains over 4 million impressions and over 3 thousand clicks. The data is divided amongst nine advertisers from different industry sectors. For each advertiser, the test data contains events from a later period than the training data. In the iPinYou data set the majority of features are categorical. Meaning that each feature can take one of a predefined set of values. In our model we use the following features: 'weekday', 'hour', 'useragent', 'IP', 'region', 'city', 'adexchange', 'domain', 'url', 'urlid', 'slotid', 'slotwidth', 'slotheight', 'slotvisibility', 'slotformat', 'slotprice', 'creative', 'keypage', 'advertiser'. Feature 'slotprice' is first transformed into a categorical variable and then transformed accordingly.

A simple and efficient way to train a model is described in [1]. The raw data can be encoded using Hashing Trick [6]. Using the encoded feature vectors a logistic regression model is trained. While the model can be improved a new conjunction feature is selected based on conditional mutual information and the model is retrained. We argue that the process can be improved by exploiting existing data aggregate to boost the prediction accuracy of the model as well as reduce the size of feature vectors thus reducing overall training time.

4.1 Feature Encoding

The data set consists of N entries. Each entry has the form of $E = [F_1, F_2, ...]$ where F_k is a distinct categorical feature that can take value f_i where $i \in \{1, 2, ...\}$. Two features F_i and F_j might have different cardinality.

Dummy Encoding in statistics is a standard technique used in regression tasks. The feature F_k is encoded as a $|F_k|$-dimensional vector $\mathbf{k}$. When feature $F_k = f_i$ then it is encoded as $k_i = 1$, e.g. if $|F_k| = 4$ and $F_k = f_2$ then $\mathbf{k} = [0, 1, 0, 0]$. Every categorical feature F_k has to be encoded into its corresponding vector and resulting vectors are concatenated into single vector $\mathbf{x}$. Continuous variables should be first divided into bins and then each bin should be treated as category value. Multi-value categorical variables may be treated as standard categorical variables, with one exception: they may produce vector $\mathbf{k}$ with more than one f_i for which $k_i = 1$. Dummy Encoding produces sparse vectors for each feature. If there are $|E|$ features and k-th feature has $|F_k|$ possible values, then Eq. 1 is the dimensionality of $\mathbf{x}$. The dimensionality, d_{dummy}, can get very large if there is a lot of features with high cardinality.

$$d_{dummy} = \sum_{i=1}^{|E|} |F_i| \qquad (1)$$

Hashing Trick addresses the problem of high dimensionality produced by Dummy Encoding. In [6] the authors outline significant compression of vectors while avoiding costly matrix-vector multiplication amongst the advantages of Hashing Trick. In [1] the authors further state that Hashing Trick is straightforward to implement and effective in terms of resource consumption when used for CTR prediction. They compare the Hashing Trick performance to other methods of parameter reduction stating that the Hashing Trick is slightly better in terms of model performance while being much more effective for real-time computation. Instead of encoding f_i value of F_k as k_i Hashing Trick calculates a hash $h(f_i)$ (Eq. 2).

$$h : f_i \rightarrow \{1, 2, ..., z_{max}\} \qquad (2)$$

$$d_{hash} = z_{max} \qquad (3)$$

If each feature F_k has its own hash space, hash function maps feature value f_i element $k_{h(f_i)}$. If all features share the same space, each feature value is mapped to $x_{h(f_i)}$. Consider $h(F_a = f_a) = 2$ then the encoded vector of the i-th example is $\mathbf{x}^{(i)} = [0, 1, 0, ...]$. In case a collision occurs $h(F_b = f_b) = h(F_a = f_a) = 2$ for $\mathbf{x}^{(i)}$ the second feature may be ignored meaning $\mathbf{x}^{(i)} = [0, 1, 0, ...]$ or increased by 1 meaning $\mathbf{x}^{(i)} = [0, 2, 0, ...]$.

A variation of this method exists. Instead of adding 1, value of $sgn(h_2(f_i))$ is added, where h_2 is an independent hash function. In this case when $h(F_a = f_a) = 2$ the encoded vector is $\mathbf{x}^{(i)} = [0, sgn(h_2(F_a = f_a)), 0, ...]$, and in case of a collision, when $h(F_b = f_b) = h(F_a = f_a) = 2$, the resulting vector is $\mathbf{x}^{(i)} = [0, sgn(h_2(F_a = f_a)) + sgn(h_2(F_b = f_b)), 0, ...]$.

There is a second type of collisions when feature values occupy the same index but for different events. Consider $h(f_i) = h(f_j) = 2$, f_i is a feature value of $\mathbf{x}^{(i)}$ and f_j is a feature value of $\mathbf{x}^{(j)}$. In this case $\mathbf{x}^{(i)} = \mathbf{x}^{(j)} = [0, 1, 0, ...]$ effectively meaning that two features will share regression parameters. As we show later, this can diminish the effect of regularization, and thus may cause overfitting.

It is easy to control the maximum dimensionality of the produced vector by changing the hashing function limit z_{max} (Eq. 2). This causes the dimensionality to be equal to Eq. 3 when the features share the hash space. One of the arguments for Hashing Trick is that it is straight forward to implement and the usage of a hashing function indicates $\mathcal{O}(|E|)$ complexity as it does not require any additional data structures. This is true for offline systems however for online environments such as DSP it may be probed and abused by exploiting hash collisions. In this case, an additional data structure with training set feature values has to be used to prevent unknown values from entering the system, meaning an additional $\mathcal{O}(d_{dummy})$ memory complexity.

4.2   Aggregate Encoding

We propose Aggregate Encoding for dimensionality reduction. It is designed as a set of probability estimators as first suggested in [3]. We propose the use of a single feature estimator but with the possibility of extending to feature conjunctions [1]. The conditional probability of success given that $F_k = f_i$ is given by Eq. 4, where $success_i$ is the amount of successes and $attempts_i$ is the amount of attempts. We encode feature F_k from all events in the training set as vector **k**. Each element of **k** corresponds to the conditional probability (Eq. 4) of that particular event. In our case, the probabilities are very low so we normalize **k** to the range $[0..1]$. Each feature is encoded on a different position of **x** making it a $|F|$-dimensional vector.

$$P(success|f_i) = \frac{success_i}{attempts_i} \qquad (4)$$

First Level Aggregate Encoding preserves the dimensionality of the original data and, as we show later, it behaves better than Hashing Trick with similar dimensionality. This approach, however, causes problems as it contradicts regularization by grouping features with different counts in a single representation. To address this issue we introduce quantile bins as the Second Level of Aggregate Encoding. Each value f_i, of F_k is assigned to a single bin_q where $q \in \{1, 2, ..., Q\}$ based on the quantile that $attempts_k$ belongs to. This causes features with a similar amount of attempts to be grouped in a single bin, reducing the negative effect that the original method has on regularization. This method produces vector **x** with the dimensionality given by Eq. 5.

$$d_{bin} = \sum_{i=1}^{|E|} \min(|F_i|, Q) \quad given \quad \forall_{i,j} attempts_i \neq attempts_j \qquad (5)$$

As in First Level Aggregates, vector **x** is normalized per bin across all features. Aggregate Encoding requires storing a mapping from a feature value f_i to a probability of success $P(success|f_i)$ and a mapping from f_i to a bin_q. Since f_i is a feature from the training set the additional complexity is the same as for Hashing Trick when holding training set feature values in memory.

It is known that in logistic regression the log loss, $LL(\theta)$, is given by Eq. 6 and partial derivative for θ_j by Eq. 7 where θ are the parameters of the model and σ is the sigmoid function. The cost function $J(\theta)$ with L_2 regularization is given by Eq. 8 and partial derivative for θ_j by Eq. 9. By minimizing $J(\theta)$ one can find θ that is optimal under the regularization constraints.

$$LL(\theta) = \sum_{i=1}^{N} y^{(i)} log[\sigma(\theta^T \mathbf{x}^{(i)})] + (1 - y^{(i)}) log[1 - \sigma(\theta^T \mathbf{x}^{(i)})] \qquad (6)$$

$$\frac{\partial LL(\theta)}{\partial \theta_j} = \sum_{i=1}^{N} [y^{(i)} - \sigma(\theta^T \mathbf{x}^{(i)})] x_j^{(i)} \qquad (7)$$

$$J(\theta) = \frac{\lambda}{N} \theta^T \theta - \frac{1}{N} LL(\theta) \qquad (8)$$

$$\frac{\partial J(\theta)}{\partial \theta_j} = \frac{\lambda}{N} \frac{\partial}{\partial \theta_j} \theta^T \theta - \frac{1}{N} \frac{\partial LL(\theta)}{\partial \theta_j} = \frac{\lambda}{N} \theta_j - \frac{1}{N} \sum_{i=1}^{N} [y^{(i)} - \sigma(\theta^T \mathbf{x}^{(i)})] x_j^{(i)} \qquad (9)$$

Let $\mathbf{x}_k$ and $\mathbf{x}_q$ be created by splitting feature F_j with $Q = 2$. Let $x_q^{(i)} = 0$ for $i \in \{1, 2, ..., n\}$ and $x_k^{(j)} = 0$ for $j \in \{n+1, n+2, ..., N\}$. In other words we create two quantile bins bin_k encoded on $\mathbf{x}_k$, where all rows with feature values from bin_q are encoded as $x_k^{(j)} = 0$ and bin_q encoded on $\mathbf{x}_q$, where all rows with feature values from bin_k are encoded as $x_k^{(j)} = 0$ as shown in Fig. 4. Knowing that we split Eq. 9 in to Eq. 10 and Eq. 11.

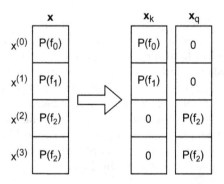

Fig. 4. Vector $\mathbf{x}$ encoding a single feature, as described by First Level Aggregate Encoding (Subsect. 4.2), is splitted into two vectors x_k and x_q, each containing features with similar occurrences. f_0 and f_1 occur once, f_2 occurs two times. $P(f_j)$ is $P(success|f_j)$ as defined in Eq. 4. $x^{(j)}$ is the row index corresponding to the input data.

$$\frac{\partial J(\theta)}{\partial \theta_k} = \frac{\lambda}{N}\theta_k - \frac{1}{N}\left[\sum_{i=1}^{n}[y^{(i)} - \sigma(\theta^T\mathbf{x}^{(i)})]x_k^{(i)} + \sum_{j=n+1}^{N}[y^{(j)} - \sigma(\theta^T\mathbf{x}^{(j)})]\times 0\right]$$
(10)

$$\frac{\partial J(\theta)}{\partial \theta_q} = \frac{\lambda}{N}\theta_q - \frac{1}{N}\left[\sum_{i=1}^{n}[y^{(i)} - \sigma(\theta^T\mathbf{x}^{(i)})]\times 0 + \sum_{j=n+1}^{N}[y^{(j)} - \sigma(\theta^T\mathbf{x}^{(j)})]x_q^{(j)}\right]$$
(11)

We introduce $n_k = n$ and $n_q = N - n$ and we derive Eq. 12 and Eq. 13 by extracting the non-zero components as the gradient expectations when x is drawn from the train set where x_k or x_q is not zero, which are the mean gradient values at the current optimization step.

$$\frac{\partial J(\theta)}{\partial \theta_k} = \frac{\lambda}{N}\theta_k - \frac{n_k}{N}E_{X\sim train_k}[(y - \sigma(\theta^T\mathbf{x}))x_k]$$
(12)

$$\frac{\partial J(\theta)}{\partial \theta_q} = \frac{\lambda}{N}\theta_q - \frac{n_q}{N}E_{X\sim train_q}[(y - \sigma(\theta^T\mathbf{x}))x_q]$$
(13)

We can see that the regularization parameter θ is scaled by a constant that depends on the amount of data and the regularization parameter λ for both θ_k and θ_q. The gradients in both cases are scaled by different values that are corresponding to the element counts, respectively n_k for θ_k and n_q for θ_q. The amount of scaling depends on counts distribution of feature F_j. In our case $n_k < n_q$ meaning that the regularization for $\mathbf{x}_k$ has relatively higher impact than for $\mathbf{x}_q$. A similar effect will take place if Dummy Encoding is used and all features are separated.

If the features are joined randomly, as in Hashing Trick, the effect diminishes, meaning that rare features will be regularized together with frequent features.

The maximum dimensionality of a data set is given by Eq. 1. When d_{hash} approaches the dimensionality of Dummy Encoding (Eq. 3) the amount of conflicts is reduced not entirely to zero due to the nature of the hash function. When $d_{hash} < d_{dummy}$ the conflicts occur, including situations where frequent and rare feature values occupy the same vector positions and might impact regularization. For Aggregate Encoding, if no two features have the same count, d_{bin} given by Eq. 5 is equal at most d_{dummy}. In this case, the methods are identical. If some feature values have the same count, they are encoded in the same bin_q as the probability of success given by Eq. 4 and will most likely be distinguishable by the model. If $d_{bin} < d_{dummy}$ our method groups features with similar counts together, thus preserving the properties of regularization.

5   Experiments and Results

In our experiments, we compare Hashing Trick with Aggregate Encoding. As we argue at the end of Sect. 4 both of the methods degenerate to Dummy Encoding.

Aggregate Encoding exploits the fact that the model uses regularization and Hashing Encoding due to the hashing function fails to do so. In this section we show the impact of this behaviour using the iPinYou [10] data set. We do not use conjunctions of the original variables. We train a separate model for each advertiser. The feature 'usertag' representing a tag given to a user, is a special feature that in sense that a single user can be tagged with none, one or several tags. In our experiments we treat each tag value as a separate feature but we omit the feature in our comparison as it requires special treatment.

5.1   Experiment Setup

Our Hashing Trick implementation encodes all features to single feature space and resolves conflicts by using the sign of an additional hashing function as described in Subsect. 4.1. Our Aggregate Encoding implementation does not resolve conflicts when feature values f_i and f_j have $attempts_i = attempts_j$.

To build training and test sets we sample the negative class without repetition to reduce the amount of time and memory required by the experiments. Negative class sampling was set to 20% for advertisers *1458 and *3386 and 50% for others due to memory constraints. We sample the sets before every iteration to measure how the method behaves depending on the input distributions.

For Hashing Trick we iterate over maximum dimensionality z_{max} from a predefined list. For each z_{max} we sample the test and train data set and set $d_{hash} = z_{max} + z_{delta}$ where z_{delta} is a random. We modify z_{max} by a random number from the range $[-5, 5]$, as we have noticed that the result may vary depending on the selected dimensionality. Next, we encode the data sets using the method described above, train the model and evaluate it using AuROC. For each (advertiser, z_{max}) pair we repeat the experiment five times.

For Aggregate Encoding, we iterate over bin size Q from a predefined list. For each Q we sample the test and train data set. d_{bin} varies between experiments as it depends on the sampled data. As with Hashing Trick, we encode the data sets, train the model and evaluate it using AuROC. For each (advertiser, Q) pair we repeat the experiment also five times. We do not introduce conjunction features but it is possible to do so in both methods.

5.2   Results

Comparison of Hashing Trick and Aggregate Encoding is shown in Table 1. To create the table first a quantile bin size bin_q value was selected amongst the values with the highest average AuROC. Then a hashing dimensionality z_{max} was selected with AuROC close to the corresponding bin_q results. The only exceptions from this rule are advertisers **2259 and **2997 where both bin_q and z_{max} had to be lowered due to a relatively small amount of unique features. The mean gain for AuROC is close to zero as intended. Gain in terms of feature shows almost a two times increase for Aggregate Encoding compared to Hashing Trick. This means that using Aggregate Encoding we were able to significantly reduce the amount of features preserving AuROC results.

Table 1. AuROC and no. of features comparison. Average AuROC gain [A-H] is the difference of average AuROC for Aggregates and Hashing methods. Average number of features gain [H/A] is the quotient of Average number of features for Hashing and Aggregates methods. The average number of features is selected the to minimize the mean average AuROC gain.

Advertiser	Avg. AuROC (%)			Avg. no. of features		
	Hashing	Aggregates	Gain [A-H]	Hashing	Aggregates	Gain [H/A]
*1458	63.75	64.35	0.60	499.00	242.60	2.06
3358	74.55	74.58	0.03	501.00	259.80	1.93
*3386	75.44	75.98	0.54	499.75	255.50	1.96
3427	71.10	70.84	−0.27	499.75	273.00	1.83
3476	62.64	63.08	0.43	498.25	261.00	1.91
**2259	66.75	67.17	0.42	198.40	209.33	0.95
2261	62.34	61.88	−0.46	500.40	275.80	1.81
2821	59.57	59.83	0.26	497.75	250.25	1.99
**2997	57.02	56.04	−0.98	400.60	207.33	1.93
Mean			0.06			1.82

Figure 5 shows AuROC change depending on the number of features produced by Hashing Trick and Aggregate Encoding. We fit an exponential curve to illustrate the trend. AuROC for most advertisers behaves better when Aggregate Encoding is used. In most cases, the improvement is most evident for low feature count and diminishes with the feature increase. This behaviour is expected due to two facts. As shown earlier Aggregate Encoding is expected to exploit the benefits of regularization on low dimensional data compared to a Hashing Trick which is not optimized for this behaviour. With the increase of feature count, as shown in Subsect. 4.2, both of the methods are being reduced to Dummy Encoding with one difference. Error introduced by Aggregate Encoding may be easily reduced to zero opposed to the error introduced by Hashing Trick that is fully dependent on the hashing function used.

For each advertiser, we normalize the feature count and AuROC to the values of Aggregate Encoding as they tend to have less variance. Then we fit exponential curves to both normalized Hashing Trick and normalized Aggregate Encoding data. Figure 6 shows the normalized dependency between AuROC and average no. of features count on combined data. As observed for most advertisers, the most significant gain of AuROC is for low dimensional data, and the difference slowly diminishes as we move towards the dimensionality used by Dummy Encoding.

6 Conclusions and Future Work

In this work, we present Aggregate Encoders - a method for dimensionality reduction with preservation of regularization properties of logistic regression.

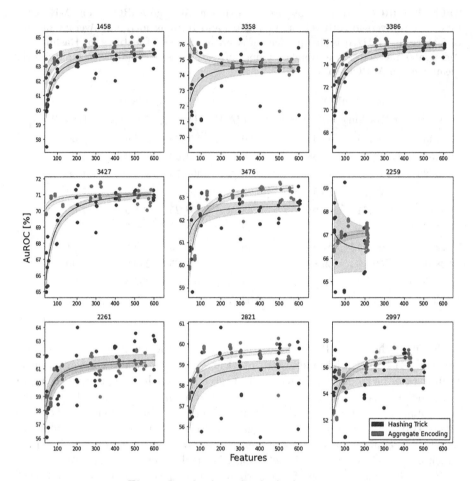

Fig. 5. Results for individual advertisers.

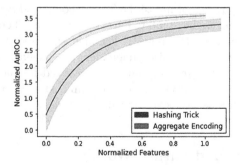

Fig. 6. Combined normalized results for all advertisers.

Using iPinYou data set, we empirically show that it behaves as good as the popular Hashing Trick, producing 42% smaller feature vectors (1.82 mean gain [H/A] in Table 1).

We use this method in a RTB setup but it is universal and can be applied to other problems. Small feature space is crucial for model training and model evaluation. During model training lower feature space reduces memory and time requirements, and during the evaluation, it reduces CPU consumption. Both of these properties are crucial in environments that are processing tens of billions of impressions [1] and potentially hundreds of times more offers daily. In this case, our method can reduce the overall training time and evaluation costs without sacrificing model performance. In [5] the authors present a deep neural network model with better log loss than logistic regression but with more parameters. Given the number of events, logistic regression can be used as the first stage of CTR assessment before a more expensive deep model is evaluated.

We leave three closely related problems for future work. All of them consider the challenge of exploiting cross-feature dependencies. The first one is to measure the effect of conjunction feature aggregates on the AuROC metric. We expect the improvement will be similar than as for Hashing Trick with the exception that Hashing Trick uses a preset vector size, and Aggregate Encoding increases the dimensionality of the vector with each added feature. The second one is using Aggregate Encoding in ensemble methods where each model instead of one value, might use our technique and return a vector divided by frequencies as this could be beneficial for regularization in the meta-classifier. The last proposal for future work is to investigate Aggregate Encoders in conjunction with deep neural networks.

Acknowledgments. We are grateful for support from the subvention of the Polish Ministry of Education and Science assigned to AGH University.

References

1. Chapelle, O., Manavoglu, E., Rosales, R.: Simple and scalable response prediction for display advertising. ACM Trans. Intell. Syst. Technol. (TIST) **5**(4), 1–34 (2014)
2. Grigas, P., Lobos, A., Wen, Z., Lee, K.C.: Profit maximization for online advertising demand-side platforms. In: Proceedings of the 2017 AdKDD and TargetAd Workshop held in conjunction with the ACM SIGKDD'17 Conference on Knowledge Discovery and Data Mining. ACM (2017)
3. Lee, K.C., Orten, B., Dasdan, A., Li, W.: Estimating conversion rate in display advertising from past performance data. In: Proceedings of the 18th ACM SIGKDD International Conference on Knowledge Discovery and Data Mining, pp. 768–776 (2012)
4. Szwabe, A., Misiorek, P., Ciesielczyk, M.: Logistic regression setup for RTB CTR estimation. In: Proceedings of the 9th ICMLC 2017 International Conference on Machine Learning and Computing, pp. 61–70 (2017)
5. Wang, R., Fu, B., Fu, G., Wang, M.: Deep & cross network for ad click predictions. In: Proceedings of the 2017 AdKDD and TargetAd Workshop Held in Conjunction with the ACM SIGKDD 2017 Conference on Knowledge Discovery and Data Mining. ACM (2017)

6. Weinberger, K., Dasgupta, A., Langford, J., Smola, A., Attenberg, J.: Feature hashing for large scale multitask learning. In: Proceedings of the ICML 2009 Annual International Conference on Machine Learning, pp. 1113–1120. ACM (2009)
7. Wiatr, R.: Code repository for this work. Accessed 12 Jan 2021, https://github.com/rwiatr/agge
8. Wiatr, R., Lyutenko, V., Demczuk, M., Słota, R., Kitowski, J.: Click-fraud detection for online advertising. In: Wyrzykowski, R., Deelman, E., Dongarra, J., Karczewski, K. (eds.) PPAM 2019. LNCS, vol. 12043, pp. 261–271. Springer, Cham (2020). https://doi.org/10.1007/978-3-030-43229-4_23
9. Yuan, Y., Wang, F., Li, J., Qin, R.: A survey on real time bidding advertising. In: Proceedings of 2014 IEEE IntThe 8th International Workshop on Data Mining for Online Advertising in conjunction with ACM SIGKDD 2014 International Conference on Service Operations and Logistics, and Informatics, pp. 418–423. IEEE (2014)
10. Zhang, W.: iPinYou Data Set. Accessed 25 Aug 2020, https://github.com/wnzhang/make-ipinyou-data
11. Zhang, W., Yuan, S., Wang, J., Shen, X.: Real-time bidding benchmarking with iPinYou dataset. Tech. rep., UCL (2014), arXiv preprint arXiv:1407.7073

A Model for Predicting n-gram Frequency Distribution in Large *Corpora*

Joaquim F. Silva$^{(\boxtimes)}$ and Jose C. Cunha

NOVA Laboratory for Computer Science and Informatics,
Costa da Caparica, Portugal
{jfs,jcc}@fct.unl.pt

Abstract. The statistical extraction of multiwords (n-grams) from natural language *corpora* is challenged by computationally heavy searching and indexing, which can be improved by low error prediction of the n-gram frequency distributions. For different n-gram sizes ($n \geq 1$), we model the sizes of groups of equal-frequency n-grams, for the low frequencies, $k = 1, 2, \ldots$, by predicting the influence of the *corpus* size upon the Zipf's law exponent and the n-gram group size. The average relative errors of the model predictions, from 1-grams up to 6-grams, are near 4%, for English and French *corpora* from 62 Million to 8.6 Billion words.

Keywords: n-gram frequency distribution · Large corpora

1 Introduction

Relevant Expressions (RE) are semantically meaningful n-grams ($n \geq 1$), as "oceanography", "oil crisis", useful in document classification [15] and n-gram applications. However, most word sequences are not relevant in a *corpus*. Statistical RE extraction from texts, e.g. [7,18], measures the cohesion among the n-grams within each distinct multiword; its performance benefits from predicting the n-gram frequency distributions. Low frequency n-grams are significant proportions of the number of distinct n-grams in a text, as well as of the RE. Assuming, for language L and n-gram size n, a finite vocabulary $V(L, n)$ in each temporal epoch [9,16,17], we model the influence of the corpus size upon the sizes $W(k)$ of equal-frequency (k) n-gram groups, for $n \geq 1$, especially for low frequencies. We present results (and compare to a Poisson-based model), for English and French Wikipedia *corpora* (up to 8.6 Gw), for $1 \leq n \leq 6$. We discuss background, the model, results and conclusions.

2 Background

Zipf's law [20] is a good approximation to word frequency distribution, deviating from real data in high and low frequencies. More accurate approximations pose

Acknowledgements to FCT MCTES, NOVA LINCS UIDB/04516/2020 and Carlos Gonçalves.

M. Paszynski et al. (Eds.): ICCS 2021, LNCS 12742, pp. 699–706, 2021.
https://doi.org/10.1007/978-3-030-77961-0_55

open issues [1,2,6,8,11–14,19]. Low frequency words are often ignored, as well as multiwords. Most studies use truncated *corpora* data [5,8], with some exceptions [17]. In models as [2,3] the probability of a word occurring k times is given by a power law $k^{-\gamma}$ corrected by the *corpus* size influence, but they do not consider other n-gram sizes, unlike e.g. [16].

3   The Model

Successive model refinements are shown: $W_z(k)$, from Zipf's Law; $W_{\alpha_d}(k,C)$ for *corpus* size dependence; and $W^*(k,C)$ for scaling adjustments.

3.1 $W_z(k)$: The Size of the Frequency Levels from Zipf's Law

By Zipf's Law [20], the number of occurrences of the rth most frequent word in a *corpus* with a number of distinct words given by D, is

$$f(r) = f(1) \cdot r^{-\alpha} , \tag{1}$$

α is a constant ~ 1; r is the word rank ($1 \leq r \leq D$). (1) also applies to n-grams of sizes $n > 1$, with α dependent on n (for simplicity α replaces $\alpha(n)$). The relative frequency of the most frequent n-gram ($r = 1$) for each n shows small fluctuations around a value, taken as an approximation to its occurrence probability, p_1. The absolute frequency $f(1) \approx p_1 \cdot C$. So, $\ln(f(r))$ would decrease linearly with slope α as $\ln(r)$ increases. Real distributions deviate from straight lines and show, for their higher ranks, groups of equal-frequency n-grams. $W(k)$ is defined based on Zipf's law [4,16]. For a level with frequency k, with its lowest (r_{l_k}) and highest (r_{h_k}) n-gram ranks: $f(r_{l_k})=f(r_{h_k})=k$; $W_z(k) = r_{h_k} - r_{l_k} + 1$. The model assumes a minimum observed frequency of 1: $f(r_{l_1}) = f(r_{h_1}) = 1$; $r_{h_1}=D$; and only applies to the higher ranks / lower frequencies where adjacent levels ($r_{l_k}=r_{h_{k+1}}+1$) have consecutive integer frequencies: $f(r_{h_{k+1}})=f(r_{h_k})+1$.
    Then, (2) is obtained, with constant α_z.

$$W_z(k) = \left(\frac{1}{D^{\alpha_z}} + \frac{k-1}{f(1)} \right)^{-\frac{1}{\alpha_z}} - \left(\frac{1}{D^{\alpha_z}} + \frac{k}{f(1)} \right)^{-\frac{1}{\alpha_z}} . \tag{2}$$

$$D(C; L, n) = \frac{V(L, n)}{1 + (K_2 \cdot C)^{-K_1}} . \tag{3}$$

For predicting D in a *corpus* of size C, we use (3), following [16] with good agreement with real *corpora*. For language L and n-gram size n, $V(L, n)$ is the finite vocabulary size; K_1, K_2 are positive constants. If V is assumed infinite, (3) equals Heap's law.

3.2 An Analytical Model for the Dependence of α on *Corpus* Size

Empirically, α_z is shown to depend on *corpus* size. So, we consider α in (1) as a function $\alpha(C, r)$ of the *corpus* size and the n-gram rank r:

$$\alpha(C, r) = \frac{\ln(f_c(1)) - \ln(f_c(r))}{\ln(r)} \quad , \tag{4}$$

where $1 \leq r \leq D$, and $f_c(1)$ and $f_c(r)$ are the frequencies, respectively, of the most frequent n-gram and the rth ranked n-gram, in a *corpus* of size C. In (2) α is obtained, for each *corpus* size, by fitting $W_z(1)$ to the empirical level size $W_{obs}(1)$ (for $k = 1$). For that level, $r_{h_1} = D(C, L, n)$ (denoted D or D_c), and $f_c(D_c) = 1$, so $\ln(f_c(r)) = 0$ in (4) for $r = D_c$. Let $\alpha(C, D_c)$ (denoted $\alpha_d(C)$), be the α value at rank D. Let C_1 be the size of a reference *corpus*:

$$\alpha_d(C) - \alpha_d(C_1) = \frac{\ln(f_c(1))}{\ln(D_c)} - Ref_{c_1} \quad . \tag{5}$$

The 2nd term in the right-hand side of (5) (denoted Ref_{c_1}) becomes fixed. It only depends on $f_c(1) = C_1 \cdot p_1$ (p_1 is the occurrence probability of the most frequent n-gram) and D_{c_1} from (3). Using Table 1 (Sect. 4.2) and tuning $\alpha_d(C_1)$ by fitting, for C_1, the $W_z(1)$ from (2) to the observed $W_{obs}(1)$, we find $\alpha_d(C_1)$ and D_{c_1}. Given $\alpha_d(C_1)$ and Ref_{c_1}, then (5) predicts $\alpha_d(C)$ for a size C *corpus*, and $W_z(k)$ (2) leads to $W_{\alpha_d}(k, C)$ (6), where $\alpha_d(C)$ replaces α_z:

$$W_{\alpha_d}(k, C) = \left(\frac{1}{D_c^{\alpha_d(C)}} + \frac{k-1}{f_c(1)} \right)^{-\frac{1}{\alpha_d(C)}} - \left(\frac{1}{D_c^{\alpha_d(C)}} + \frac{k}{f_c(1)} \right)^{-\frac{1}{\alpha_d(C)}} \quad . \tag{6}$$

3.3 $W^*(k, C)$: The Dependence of Level Size on *Corpus* Size

The frequency level size depends on frequency k and *corpus* size C. Firstly, for a *corpus* size C, α_z in (2) is tuned to best fitting $W_z(1)$ to $W_{obs}(1)$. Except for the $W_{obs}(k)$ fluctuations (Fig. 1a), the deviation, closely proportional to $\ln(k)$, between $W_{obs}(k)$ and $W_z(k)$, suggests the improvements due to (7) (Fig. 1a).

$$W_{adjusted}(k) = W_z(k) \cdot k^\beta \quad . \tag{7}$$

β is a constant for each n, obtained from the best fit of $W(k)$ to $W_{obs}(k)$, for a given *corpus*. Secondly, for different *corpus* sizes, Fig. 1b shows $W_{obs}(k)$ curves as a function of k, seeming parallel, but a detailed analysis shows otherwise. If, for each $\ln(W_{obs}(k, C^*))$ for the three smaller *corpora* C^*, an offset equal to $\ln(W_{obs}(1, C)) - \ln(W_{obs}(1, C^*))$ is added ($C = 8.6$ Gw being the largest *corpus*), the resulting curves (omitted due to lack of space) do not coincide, as they should if they were parallel in Fig. 1b. The gap between the curves is proportional to $\ln(k)$. And, for each $\ln(k)$ value, the distance in $\ln(W_{obs}(k))$ for *corpora* of sizes C and C_1 is proportional to $\log_2(C/C_1)$. The distance between the $\ln(W(k))$

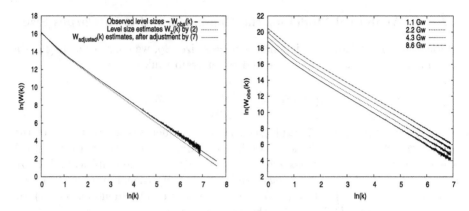

Fig. 1. a) 1-gram equal-frequency level size $W(k)$ *vs* k (log-log scale) – observed and estimates by (2) and (7) from a 1.1 Gw English *corpus*; b) Observed 3-gram level size values, $W_{obs}(k)$ *vs* k (log-log scale), for different English *corpora* sizes.

curves of any two *corpora* C and C_1 is approximated by (8), with δ constant for each n. Joining (6), (7), (8) leads to the final model, $W^*(k, C)$, (9):

$$\Delta = \delta \cdot \ln(\frac{C}{C_1}) \cdot \ln(k) \tag{8}$$

$$W^*(k, C) = W_{\alpha_d}(k, C) \cdot k^{\beta + \delta \cdot \ln(\frac{C}{C_1})} \; . \tag{9}$$

4   Results and Discussion

4.1   The Poisson-Zipf Model

In the $W_P(k, C)$ model of [17] given by (10), an n-gram ranked r occurs, in a size C *corpus*, a number of times following Poisson distribution [10] with $\lambda_r = f(r)$ by Zipf's Law. $W(0) = W_P(0, C)$ is the estimated number of unseen n-grams in the *corpus*. $D = V - W(0)$, for n-gram vocabulary size V.

$$W_P(k, C) = \sum_{r=1}^{r=V} \frac{(p_1 \cdot C \cdot r^{-\alpha})^k \cdot e^{-p_1 \cdot C \cdot r^{-\alpha}}}{k!} \approx \int_1^V \frac{(p_1 \cdot C \cdot r^{-\alpha})^k \cdot e^{-p_1 \cdot C \cdot r^{-\alpha}}}{k!} dr$$

$$\approx \frac{(p_1 \cdot C)^{1/\alpha}}{\alpha \cdot k!} \cdot \left[\Gamma(k - \frac{1}{\alpha}, \frac{p_1 \cdot C}{V^\alpha}) - \Gamma(k - \frac{1}{\alpha}, p_1 \cdot C) \right] \tag{10}$$

4.2   Comparison of Results

Complete *corpora* were built from documents randomly extracted from English and French Wikipedia. For evaluating size dependence, they were doubled successively (Table 2). A space was added between the words and each of the following characters: {!, ?, :, ;, ,, „ (,), [,], <, >, "}. All inflected word forms were kept.

The Model Calculations. (I) To calculate $D(C; L, n)$ in (3), parameters K_1, K_2 and $V(L, n)$ were found for each language L and *n*-gram size n (Table 1, also showing the β and δ values used in (9)). The $V(L, n)$ value is an estimate of the vocabulary size, such that further increasing it, does not significantly reduce the relative error $((E - O)/O) \cdot 100\%$, between an estimated value (E) and the corresponding observed value (O). Pairs (K_1, K_2) were found leading to the lowest possible relative error, for a selected pair of *corpora* with sizes close to the lowest and highest *corpora* sizes in the considered range for each language. (II) To evaluate the relative errors, (9) was applied with k such that the observed level sizes of consecutive frequency levels k and $k+1$ are monotonic decreasing, $W_{obs}(k, C) > W_{obs}(k+1, C)$. This avoids the non-monotony regions of the observed $\ln(W(k))$ curve (Fig. 1a). We considered a basic set of k values, $K = \{1, 2, 3, 4, 5, 6, 7, 8, 16, 32, 64, 128\}$, constrained (to ensure $\ln(W(k))$ monotony) depending on the *corpus* size C: for $C < 250$ Mw, we used $k \leq 16$; for $C < 1$ Gw, $k \leq 32$; the full K set was used only for $C > 4$ Gw. We selected *corpora* of sizes (C_1) 1.1 Gw (English) and 808 Mw (French). (III) The $\alpha_d(C_1)$ values for *n*-gram sizes from 1 to 6 are: (English) 1.1595, 1.02029, 0.88825, 0.82532, 0.8117, 0.8027; (French) 1.158825, 1.0203, 0.86605, 0.84275, 0.80818, 0.7569. The empirical values of p_1 for *n*-gram sizes from 1 to 6: (English) 0.06704, 0.03250, 0.0062557, 0.0023395, 0.0017908, 0.0014424; (French) 0.07818, 0.037976, 0.004685, 0.0036897, 0.001971, 0.00072944. (IV) To run $W_P(K, C)$, α values leading to the lowest relative errors, are, for *n*-gram sizes from 1 to 6: (English) 1.17, 1.02, 0.891, 0.827, 0.814, 0.812; (French) 1.156, 1.01, 0.884, 0.842, 0.806, 0.759.

Table 1. Parameter values K_1, K_2 and vocabulary sizes $(V(L, n))$ to be used in $D(C; L, n)$, (3), and β and δ to $W^*(k, C)$, (9).

English						
	1-grams	2-grams	3-grams	4-grams	5-grams	6-grams
K_1	0.838	0.861	0.885	0.924	0.938	0.955
K_2	3.61 e−11	5.1 e−11	2.66 e−11	1.78 e−11	4.29 e−12	6.5 e−13
V	2.45 e+8	9.9 e+8	4.74 e+9	1.31 e+10	6.83 e+10	5.29 e+11
β	0.044	0.113	0.129	0.135	0.122	0.082
δ	0.0039	0.0118	0.0310	0.0353	0.0339	0.0331
French						
	1-grams	2-grams	3-grams	4-grams	5-grams	6-grams
K_1	0.809	0.794	0.838	0.867	0.903	0.907
K_2	4.501 e−11	3.801 e−11	3.901 e−11	2.501 e−11	2.201 e−11	2.01 e−12
V	2.35 e+8	1.095 e+9	3.1 e+9	8.18 e+9	1.41 e+10	1.45 e+11
β	0.0812	0.120	0.175	0.140	0.160	0.234
δ	0.0061	0.0190	0.0354	0.0491	0.0469	0.0384

Table 2 presents the relative errors for the predictions of the frequency level sizes. For each n-gram size, the left column refers to $W^*(k, C)$ and the right one to $W_P(k, C)$. For each pair (*corpus* size, n-gram size), it shows the *average relative error* for the K set used: $AvgErr(K) = \frac{1}{\|K\|} \sum_{k \in K} Err(k)$, where $Err(k) = |\frac{W(k,C) - W_{obs}(k,C)}{W_{obs}(k,C)}|$. The *average relative errors* for $W^*(k, C)$ are much lower than for $W_P(k, C)$, which assumes an ideal Zipf's Law. The line **Avg** shows the average value of each column over the full range of *corpora* sizes, with errors of the same magnitude across the range of n-gram sizes for $W^*(k, C)$, but having significant variations in the **Avg** values for $W_P(k, C)$. The *global relative error* is the average of the **Avg** values over the range of n-gram sizes, being around 4% for $W^*(k, C)$. Thus, $W^*(k, C)$ curves (omitted due to lack of space) closely follow the $W_{obs}(k, C)$ curves forms of Fig. 1.

Table 2. Average relative error (%) for the predictions of the n-gram frequency level sizes obtained by $W^*(k, C)$, (9), (left col.), and $W_P(k, C)$, (10), (right col.). Each cell in the table gives an average relative error over a subset of k values within the set K considered for that cell, as described in the text.

English											
Corpus	1-grams		2-grams		3-grams		4-grams		5-grams		6-grams	
63 Mw	5.1	29.9	7.5	42.0	3.1	50.1	4.4	53.7	6.3	54.5	5.8	64.7
128 Mw	2.2	23.7	2.8	32.4	4.4	42.5	6.6	45.9	5.7	49.5	7.1	60.3
255 Mw	3.1	22.1	2.0	28.4	2.6	36.3	5.9	37.6	3.7	40.0	5.4	49.8
509 Mw	4.9	15.8	4.7	22.2	3.7	29.4	4.6	30.1	4.7	31.9	6.6	40.3
1.1 Gw	3.1	13.3	2.6	19.8	3.4	26.4	3.9	26.8	5.4	27.9	5.2	32.9
2.2 Gw	5.1	9.5	6.2	23.2	3.9	26.7	3.3	28.5	4.9	31.5	3.7	28.7
4.3 Gw	2.8	10.7	2.7	28.5	2.3	34.8	3.1	37.4	4.2	39.3	4.8	31.4
8.6 Gw	6.1	13.4	6.7	37.6	4.4	47.2	6.0	51.7	5.9	52.4	6.5	40.4
Avg	**4.1**	17.3	**4.4**	29.3	**3.5**	36.7	**4.7**	39.0	**5.1**	40.9	**5.6**	43.6
French											
Corpus	1-grams		2-grams		3-grams		4-grams		5-grams		6-grams	
108 Mw	2.8	22.6	2.4	30.9	2.6	54.9	4.2	40.8	4.9	42.3	5.5	65.4
201 Mw	1.5	18.6	2.0	25.7	2.1	49.8	2.9	33.0	3.2	33.9	3.5	57.7
404 Mw	2.7	15.5	2.9	23.5	4.0	44.4	4.6	28.4	4.9	29.4	5.0	51.6
808 Mw	2.9	12.6	3.0	26.7	3.2	34.8	3.4	23.6	3.6	24.3	3.7	43.8
1.61 Gw	4.6	16.9	3.5	37.2	3.0	29.0	2.9	28.7	3.3	29.6	3.4	41.9
3.2 Gw	4.0	19.2	3.2	48.6	4.2	22.8	5.3	39.5	3.3	41.1	6.6	49.7
Avg	**3.1**	17.6	**2.8**	32.1	**3.7**	39.3	**3.9**	32.3	**3.9**	33.4	**4.6**	51.7

5   Conclusions

Estimating n-gram frequency distributions is useful in statistical-based n-gram applications. The proposed model estimates the sizes $W(k, C)$ of equal-frequency (k) n-gram groups in a *corpus* of size C, for the low frequency n-grams. It applies uniformly to different n-gram sizes $n \geq 1$ and languages, assuming a finite language n-gram vocabulary. It models the dependences of Zipf's Law exponent and $W(k, C)$ on C, agreeing well with n-gram frequency data from unigrams up to hexagrams, from real un-truncated English and French *corpora* with million to billion words. Larger *corpora* evaluation is planned.

References

1. Ausloos, M., Cerqueti, R.: A universal rank-size law. PLoS ONE **11**(11) (2016)
2. Balasubrahmanyan, V.K., Naranan, S.: Algorithmic information, complexity and zipf's law. Glottometrics **4**, 1–26 (2002)
3. Bernhardsson, S., da Rocha, L.E.C., Minnhagen, P.: The meta book and size-dependent properties of written language. New J. Phys. **11**(12), 123015 (2009)
4. Booth, A.D.: A law of occurrences for words of low frequency. Inf. Control **10**(4), 386–393 (1967)
5. Brants, T., Popat, A.C., Xu, P., Och, F.J., Dean, J.: Large language models in machine translation. In: Joint Conference on Empirical Methods in NLP and Computational Natural Language Learning, pp. 858–867. ACL (2007)
6. Cancho, R.F., Solé, R.V.: Two regimes in the frequency of words and the origins of complex lexicons: Zipf's law revisited*. J. Quant. Linguist. **8**(3), 165–173 (2001)
7. Dias, G.: Multiword unit hybrid extraction. In: ACL Workshop on Multiword Expressions, vol. 18, pp. 41–48. ACL (2003)
8. Gerlach, M., Altmann, E.G.: Stochastic model for the vocabulary growth in natural languages. Phys. Rev. X **3**, 021006 (2013)
9. Goncalves, C., Silva, J.F., Cunha, J.C.: n-gram cache performance in statistical extraction of relevant terms in large *Corpora*. In: Rodrigues, J.M.F., et al. (eds.) ICCS 2019. LNCS, vol. 11537, pp. 75–88. Springer, Cham (2019). https://doi.org/10.1007/978-3-030-22741-8_6
10. Haight, F.A.: Handbook of the Poisson Distribution. John Wiley & Sons, New York (1967)
11. Lü, L., Zhang, Z.K., Zhou, T.: Deviation of zipf's and heaps' laws in human languages with limited dictionary sizes. Sci. Rep. **3**(1082), 1–7 (2013)
12. Mandelbrot, B.: On the theory of word frequencies and on related Markovian models of discourse. In: Structural of Language and its Mathematical Aspects (1953)
13. Mitzenmacher, M.: A brief history of generative models for power law and lognormal distributions. Internet Math. **1**(2), 226–251 (2003)
14. Piantadosi, S.T.: Zipf's word frequency law in natural language: a critical review and future directions. Psychonomic Bull. Rev. **21**, 1112–1130 (2014)
15. Silva, J., Mexia, J., Coelho, A., Lopes, G.: Document clustering and cluster topic extraction in multilingual corpora. In: Proceedings 2001 IEEE International Conference on Data Mining, pp. 513–520 (2001)
16. Silva, J.F., Cunha, J.C.: An empirical model for n-gram frequency distribution in large corpora. In: Lauw, H.W., et al. (eds.) PAKDD 2020. LNCS (LNAI), vol. 12085, pp. 840–851. Springer, Cham (2020). https://doi.org/10.1007/978-3-030-47436-2_63

17. Silva, J.F., Gonçalves, C., Cunha, J.C.: A theoretical model for n-gram distribution in big data corpora. In: 2016 IEEE International Conference on Big Data, pp. 134–141 (2016)
18. da Silva, J.F., Dias, G., Guilloré, S., Pereira Lopes, J.G.: Using *LocalMaxs* algorithm for the extraction of contiguous and non-contiguous multiword lexical units. In: Barahona, P., Alferes, J.J. (eds.) EPIA 1999. LNCS (LNAI), vol. 1695, pp. 113–132. Springer, Heidelberg (1999). https://doi.org/10.1007/3-540-48159-1_9
19. Simon, H.: On a class of skew distribution functions. Biometrika **42**(3/4), 425–440 (1955)
20. Zipf, G.K.: Human Behavior and the Principle of Least-Effort. Addison-Wesley, Cambridge (1949)

Exploiting Extensive External Information for Event Detection Through Semantic Networks Word Representation and Attention Map

Zechen Wang[1], Shupeng Wang[2(✉)], Lei Zhang[2(✉)], and Yong Wang[2]

[1] School of Cyber Security, University of Chinese Academy of Sciences, Beijing, China
wangzechen@iie.ac.cn
[2] Institute of Information Engineering, CAS, Beijing, China
{wangshupeng,zhanglei1,wangyong}@iie.ac.cn

Abstract. Event detection is one of the key tasks to construct knowledge graph and reason graph, also a hot and difficult problem in information extraction. Automatic event detection from unstructured natural language text has far-reaching significance for human cognition and intelligent analysis. However, limited by the source and genre, corpora for event detection can not provide enough information to solve the problems of polysemy, synonym association and lack of information. To solve these problems, this paper proposes a brand new Event Detection model based on Extensive External Information (EDEEI). The model employs external corpus, semantic network, part of speech and attention map to extract complete and accurate triggers. Experiments on ACE 2005 benchmark dataset show that the model effectively uses the external knowledge to detect events, and is significantly superior to the state-of-the-art event detection methods.

Keywords: Event detection · External information · Wnet2vec · Semantic network · Attention mechanism

1 Introduction

Event detection mines the necessary information from unstructured data to represent events structurally with 5W1H framework [21]. An event is composed of a trigger which indicates the existence of the event, and several arguments which constitute the detail information [9]. Thus, the target of event detection is to detect whether trigger existing in the sentence, and determine which event type it belongs to.

Event detection is conducive to the storage, retrieval, representation and analysis of event information. Although event detection has been widely used in various areas, [17] such as abnormal public opinion detection [11,20], news recommendation systems [7,19], risk analysis application [1], monitoring system [14] and decision support tools [4], it still faces tremendous difficulties and challenges. There are great obstacles in event detection. Especially, it is difficult to deal with words with multiple meanings on the basis of the information contained in the local context. For example, in the field of financial analysis, an investment event can be detected from the sentence "The charity

© Springer Nature Switzerland AG 2021
M. Paszynski et al. (Eds.): ICCS 2021, LNCS 12742, pp. 707–714, 2021.
https://doi.org/10.1007/978-3-030-77961-0_56

was prospected of giving money from a bank." It can attribute to three problems in event detection tasks as follows: First, **Polysemy**. In different languages, a word may represent different types of entities in different contexts. Thus, the meaning of "bank" can not be judged directly in the sentence from the above example. Consequently, the wrong judgment causes the problem that polysemous word can not be extracted as the correct trigger. Finally, the event element related to investment would be wrongly judged, which may result in the loss of important elements in further analysis and eventually cause unwise investment decision. Second, **Synonym Association**. Some word may only appear once in a corpus, but it may have several synonyms in the same corpus. Hence, it is difficult to establish a relationship between synonyms only through local corpus. For example, the word "giving" in the above example is synonymous with the verb "investing" which expresses investment behavior. Yet these words do not appear in the same context, so the words like "investing" can not be directly associated with the word "giving". The word "giving" dose not always represent a investment-related behavior, but may have other meanings such as "present" and "supply". Therefore, if it does not associate with words like "investing", it is nearly impossible to classify the word as a trigger of investment event. Third, **Lack of Information**. The context usually includes word information, word position information and so on. More seriously, the information does not contain part of speech information. Nonetheless, it is crucial to identify and classify triggers in the event detection task. But if part of speech information is not used sufficiently, it would be difficult to extract triggers which tend to hide in nouns, verbs and adjectives. It will reduce the accuracy of the extraction of triggers.

Therefore, the above three problems need to be solved at the same time to detect events completely and accurately.

2   Related Works

In recent years, with the increasingly wide application of event detection, there are some related research works. The existing studies are in mainly three categories according to the problems they solve: polysemy [3,6], synonym association [8,12] and lack of information [16].

The problem of polysemy is mainly reflected in the stage of word expression. Various word embedding models, such as CBOW and skip-gram, have been proposed. These models can not generate different vectors for different meanings of the same word. At present, the methods to solve the problem of polysemy include: 1) The methods cluster the meanings of the same word corresponding to different contexts to distinguish the meanings of the word [3]; 2) The methods use cross language information. These methods translate different meanings of the same word into other languages to make each meaning of the word corresponding to one word in the target language [6]. But these methods have a disadvantage that a word is fixed to one word vector, no matter how many meanings the word has. Accordingly, the meanings of a word can not be distinguished. As a result, this problem can not be well solved.

The methods aim at solving the problem of synonym association mainly describe the association between synonyms by using rules directly according to dictionaries or through external corpus information. Synonym association plays an important role in event detection task, especially in the event type classification mentioned above. Synonyms can be associated on the basis of event-trigger dictionaries [8] or synonym sets according to semantic networks [12]. However, these methods need to construct a word list in advance to support event detection, and update the word list if corpus changes. Consequently, these methods can only be used in limited situations, and can not solve the problem of synonym association in broad areas.

To solve the problem of lack of information, many features such as vocabulary, syntax and semantics can be used as input. For example, an idea proposed by Liu et al. [16] claims that triggers and arguments should receive more attention than other words, so they construct gold attention vectors which only encode each trigger, argument and context word. Nevertheless, the constitution of gold attention vectors rely heavily on syntax knowledge and domain knowledge which can be used in event detection.

In summary, due to the limitation of the scope of prior knowledge, the studies on event detection till now can partly solve problems in event detection. Furthermore, they also can not solve the problems of polysemy, synonym association and lack of information at once within a single model.

3 Method

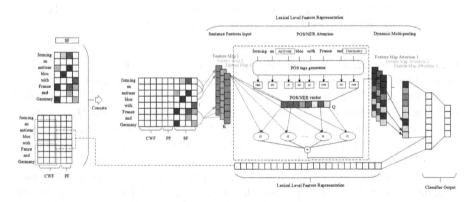

Fig. 1. The EDEEI model includes word vector representation, feature extraction, feature optimization, feature selection and classifier.

In this section, we present our framework for the proposed EDEEI algorithm. The proposed framework is illustrated in Fig. 1.

The word vector representation module generates three kinds of vectors: Content Word Features (CWF), Semantic Features (SF) and Position Features (PF). The newly-generated word vectors are used for feature extraction, feature optimization and feature selection. Feature optimization refined raw features from feature extraction, and provide processed features to feature selection to capture high-quality features. Finally, the features are inputted into the classifier to obtain the triggers and their classification.

Word Vector Representation. We derive a new word vector structure based on BERT [5] and wnet2vec [18]. To solve the problem of polysemy, the proposed framework utilizes BERT to generate the CWF, which identifies different word meanings of one word according to a variable external corpus. Wnet2vec is exploited in our model to generate SFs which can better express the semantic relationship between synonyms. Wnet2vec has the ability of generating word vectors through the transformation from semantic network to semantic space. Finally, the PF is defined, that is, the relative distance between the current word and the candidate trigger. In order to encode the PF, each distance value is represented by an embedded vector. Let the size of CWF be d_{CWF}, the size of SF be d_{SF}, and the size of position code be d_{PF}. Represent word vector of the i-th word in the sentence as $x_i \in R^d$, $d = d_{CWF} + d_{SF} + d_{PF}{}^*2$. Represent sentence of length n as: $x_{1:n} = x_1 \oplus x_2 \oplus \ldots \oplus x_n$, where $\oplus$ is the concatenation operator. The vectors are spliced to form a matrix $X \in R^{n \times d}$.

Feature Extraction. The convolution layer is to capture and compress the semantics of the whole sentence, so as to extract these valuable semantics into feature maps. Each convolution operation contains a convolution kernel $w \in R^{h \times d}$, and the window size is h. For example, from the window size of $x_{i:i+h-1}$, the module generates the feature c_i: $c_i = f(w \cdot x_{i:i+h-1} + b)$, where b is the bias and f is the nonlinear activation function. Apply the calculation process to the sentence $x_{i:n}$ to generate the feature map $C = \{c_1, c_2, \ldots, c_n\}$. Corresponding to m convolution kernels $W = w_1, w_2, \ldots, w_m$, the result of feature maps generated by M is expressed as $\{C_1, C_2, \ldots, C_m\}$.

Feature Optimization. This module uses the POS tag generation tool provided by Stanford CoreNLP to annotate the sentences. One-hot coding is applied to POS tags of different types, and the coding vector is k_1. According to the coding of each POS tag, POS matrix is generated for sentences of length n, and the matrix size is $M_{POS} \in R^{k_1 \times n}$.

The crucial parts corresponding to specific POS tags are emphasised by attention mechanism. The feature optimization module takes POS tags and feature maps as the inputs of attention mechanism. Each feature map generated in feature extraction module is a vector of length $n - h + 1$. The feature maps represented as K are taken as the inputs of attention mechanism. The attention calculation process is as follows. A new random matrix WQ of length w is computed. The product of two vectors is calculated to obtain a new matrix Q. The random matrix WK whose width is w and the length is k_1 is acquired, and the product of the random matrix WK and WQ produces the matrix WV. Calculate the product of WV and feature map to gain the matrix V.

Based on the three generated K, Q, V matrices, an attention matrix Z is calculated by using the following formula: $Z = softmax\left(\frac{Q \times K^T}{\sqrt{n-h+1}}\right)V$. Train Wk, WQ, WV matrices. The scoring function is as follows: $f_{score} = \frac{Q \cdot K^T}{\sqrt{n-h+1}}$. Then the Z matrix is compressed with max pooling to generate a vector z. Based on the updated $WK, WQ and WV$, the product of z and K constructs a new attention map with the size of $n - h + 1$.

Feature Selection. The feature selection module employs dynamic multi-pooling to further extract the valuable features. Furthermore, the features are concatenated to produce lexical level vectors which contain information of what role the current word plays in an event. The calculation process of dynamic multi-pooling is given as follows: $\left[y_{1,p_t}\right]_i = \max\left\{[C_1]_i, \ldots, \left[C_{p_t}\right]_i\right\}$, $\left[y_{p_t+1,p_n}\right]_i = \max\left\{\left[C_{p_t+1}\right]_{i'}, \ldots, [C_n]_i\right\}$, where $[y]_i$ is the $i - th$ value of the vector, p_t is the position of trigger t, and C_i is the $i_t h$ value in the attention map C. We use the maximum pooling result of each segment as the feature vector L at the sentence level.

Classifier. This module concatenates the CWFs of the current word and the words on the left and right of the current one, to obtain the vector P of length $3 * CWF$. The learned sentence level features and word features are connected into a vector F = [L, P]. In order to calculate the confidence of the event type of each trigger, the feature vector is inputted into the classifier $O = W_s F + b_s$. W_s is the transformation matrix of the classifier, b_s is the bias, and O is the final output of the network, where the output type is equal to the total number of event types plus one to include the "not a trigger" tag.

4 Experiment

The ACE 2005 is utilized as the benchmark experimental dataset. The test set used in the experiment contains 40 Newswire articles and 30 other documents randomly selected from different genres. The remaining 529 documents are used as the training set.

Evaluation of Event Detection Methods. To demonstrate how the proposed algorithm improves the performance over the state-of-the-art event detection methods, we compare the following representative methods from the literature:

(1) Li's baseline [13]: Li et al. proposed a feature-based system which used artificially designed lexical features, basic features and syntactic features.
(2) Liao's cross-event [15]: The cross-event detection method proposed by Liao and Grishman used document level information to improve the performance.
(3) Hong's cross-entity [10]: Hong et al. exploited a method to extract events through cross-entity reasoning.
(4) Li's joint model [13]: Li et al. also developed an event extraction method based on event structure prediction.
(5) DMCNN method [2]: A framework based on dynamic multi-pooling convolutional neural network.

In all the methods, EDEEI model has achieved the best performance. Compared with the state-of-the-art methods, F value of trigger identification is significantly improved. The results displayed in Table 1 illustrate three important facts on the method. Firstly, it is necessary to solve the problems of polysemy, synonym association and lack of information at the same time. Secondly, the variable external knowledge can effectively improve the accuracy of event detection. Thirdly, the hierarchical detecting method with feature optimization can make event detection more completely and precisely.

Table 1. Overall performance on the ACE 2005 blind test data

Methods	Trigger identification			Trigger classification		
	P	R	F	P	R	F
Li's baseline	76.2	60.5	67.4	74.5	59.1	65.9
Liao's cross-event	N/A			68.7	68.9	68.8
Hong's cross-entity	N/A			72.9	64.3	68.3
Li's joint model	76.9	65	70.4	73.7	62.3	67.5
DMCNN model	80.4	67.7	73.5	75.6	63.6	69.1
EDEEI model	**77.0**	**72.9**	**74.9**	**77.3**	**63.2**	**69.5**

Analysis of Different Word Vectors. This section presents a detailed comparison of the word vectors generated from Word2vec, BERT, wnet2vec and BERT+wnet2vec respectively. The purpose is to test for advantages and disadvantages of BERT+wnet2vec approaches versus other word vectors under the task of event detection.

Table 2. Performance with different word vectors

Methods	Trigger identification F (%)	Trigger identification + classification F (%)
Word2Vec	72.2	65.9
BERT	72.7	66.7
wnet2vec	67.9	58.8
BERT+wnet2vec	**74.9**	**69.5**

The advantages of using BERT+wnet2vec can be observed visually and quantitatively in Table 2. It can be seen that the combination of BERT and wnet2vec achieves the best performance on both trigger identification and trigger classification.

In conclusion, traditional methods such as word2vec rely on a small corpus to generate word vectors, and can not solve the problem of polysemy and synonym association. Compared with word2vec, the combination of the two methods can obtain the best experimental effect. This proves that BERT+wnet2vec can effectively solve the problem of polysemy and synonym association.

5   Conclusion

This paper addresses three important problems in event detection: polysymy, synonym association and lack of information. In order to solve these problems, we propose a brand new Event Detection model based on Extensive External Information (EDEEI), and give a novel method which involves external corpus, semantic network, part of speech and attention map to detect events completely and accurately. This framework can solve the above three problems at the same time. An attention mechanism with part of speech information is designed to optimize the extracted features and make the features related to triggers easier to capture. The experiments on widely used ACE 2005 benchmark dataset confirm that the proposed method significantly outperforms the existing state-of-the-art methods for event detection. Furthermore, we present numerous qualitative and quantitative analyses about experimental results. In the light of excellent performance and analyses, it is believed that the proposed algorithm can be a useful tool for event detection.

Acknowledgments. This work was supported by National Natural Science Foundation of China (No. 61931019) and National Key Research and Development Program Project (No. 2019QY2404).

References

1. Chau, M.T., Esteves, D., Lehmann, J.: A neural-based model to predict the future natural gas market price through open-domain event extraction. ESWC **2611**, 17–31 (2020)
2. Chen, Y., Xu, L., Liu, K., Zeng, D., Zhao, J.: Event extraction via dynamic multi-pooling convolutional neural networks. In: ACL, pp. 167–176 (2015)
3. Chen, Y., Yang, H., Liu, K.: Collective event detection via a hierarchical and bias tagging networks with gated multi-level attention mechanisms. In: EMNLP, pp. 1267–1276 (2018)
4. Cheng, D., Yang, F., Wang, X.: Knowledge graph-based event embedding framework for financial quantitative investments. In: ACM SIGIR, pp. 2221–2230 (2020)
5. Devlin, J., Chang, M.W., Lee, K., Toutanova, K.: Bert: Pre-training of deep bidirectional transformers for language understanding. In: ACL, pp. 4171–4186 (2019)
6. Ferguson, J., Lockard, C., Weld, D.S., Hajishirzi, H.: Semi-supervised event extraction with paraphrase clusters. In: NAACL-HLT, pp. 359–364 (2018)
7. George, S.K., Jagathy Raj, V.P., Gopalan, S.K.: Personalized news media extraction and archival framework with news ordering and localization. In: Tuba, M., Akashe, S., Joshi, A. (eds.) Information and Communication Technology for Sustainable Development. AISC, vol. 933, pp. 463–471. Springer, Singapore (2020). https://doi.org/10.1007/978-981-13-7166-0_46
8. Han, S., Hao, X., Huang, H.: An event-extraction approach for business analysis from online Chinese news. Electron. Commer. R A **28**, 244–260 (2018)
9. Hogenboom, F., Frasincar, F., Kaymak, U., De Jong, F.: An overview of event extraction from text. In: DeRiVE 2011, vol. 779, pp. 48–57 (2011)
10. Hong, Y., Zhang, J., Ma, B., Yao, J., Zhou, G., Zhu, Q.: Using cross-entity inference to improve event extraction. In: ACL, pp. 1127–1136 (2011)
11. Huang, L., et al.: Liberal event extraction and event schema induction. In: ACL, pp. 258–268 (2016)
12. Iqbal, K., Khan, M.Y., Wasi, S., Mahboob, S., Ahmed, T.: On extraction of event information from social text streams: an unpretentious NLP solution. IJCSNS **19**(9), 1 (2019)
13. Li, Q., Ji, H., Huang, L.: Joint event extraction via structured prediction with global features. In: ACL, pp. 73–82 (2013)

14. Liang, Z., Pan, D., Deng, Y.: Research on the knowledge association reasoning of financial reports based on a graph network. Sustainability **12**(7), 2795 (2020)
15. Liao, S., Grishman, R.: Using document level cross-event inference to improve event extraction. In: ACL, pp. 789–797 (2010)
16. Liu, S., Chen, Y., Liu, K., Zhao, J.: Exploiting argument information to improve event detection via supervised attention mechanisms. In: ACL, pp. 1789–1798 (2017)
17. Mukhina, K., Visheratin, A., Nasonov, D.: Spatiotemporal filtering pipeline for efficient social networks data processing algorithms. In: ICCS, pp. 86–99 (2020)
18. Saedi, C., Branco, A., Rodrigues, J., Silva, J.: Wordnet embeddings. In: RepL4NLP, pp. 122–131 (2018)
19. Sheu, H.S., Li, S.: Context-aware graph embedding for session-based news recommendation. In: RecSys, pp. 657–662 (2020)
20. Wang, Z., Sun, L., Li, X., Wang, L.: Event extraction via dmcnn in open domain public sentiment information. In: ICPCSEE, pp. 90–100 (2020)
21. Xiang, W., Wang, B.: A survey of event extraction from text. IEEE Access **7**, 173111–173137 (2019)

A New Consistency Coefficient in the Multi-criteria Decision Analysis Domain

Wojciech Sałabun$^{(\boxtimes)}$ ⓘ, Andrii Shekhovtsov ⓘ, and Bartłomiej Kizielewicz ⓘ

Research Team on Intelligent Decision Support Systems, Department of Artificial Intelligence and Applied Mathematics, Faculty of Computer Science and Information Technology, West Pomeranian University of Technology in Szczecin, ul. Żołnierska 49, 71-210 Szczecin, Poland
wojciech.salabun@zut.edu.pl

Abstract. The logical consistency of decision making matrices is an important topic in developing each multi-criteria decision analysis (MCDA) method. For instance, many published papers are addressed to the decisional matrix's consistency in the Analytic Hierarchy Process method (AHP), which uses Saaty's seventeen-values scale.

This work proposes a new approach to measuring consistency for using a simple three-value scale (binary with a tie). The paper's main contribution is a proposal of a new consistency coefficient for a decision matrix containing judgments from an expert. We show this consistency coefficient based on an effective MCDA method called the Characteristic Objects METhod (COMET). The new coefficient is explained based on the Matrix of Expert Judgment (MEJ), which is the critical step of the COMET method. The proposed coefficient is based on analysing the relationship between judgments from the MEJ matrix and transitive principles (triads analysis). Four triads classes have been identified and discussed. The proposed coefficient makes it easy to determine the logical consistency and, thus, the expert responses' quality is essential in the reliable decision-making process. Results are presented in some short study cases.

Keywords: Decision analysis · Decision making · Decision theory · Consistency coefficient · Inconsistency coefficient

1 Introduction

Multi-criteria decision analysis (MCDA) is one of the branches of operational research, whose main objective is to support the decision-maker in solving multi-criteria problems [30]. MCDA methods are widely used in many practical decision-making problems, e.g. medicine [14,16], engineering [28], transport [3], energy [17,22,31], management [2], and others [4,10,15]. The quality of the solutions obtained in such cases depends on the specific MCDA method's algorithm and the error of the judgments in pairwise comparison by the expert.

ⓒ Springer Nature Switzerland AG 2021
M. Paszynski et al. (Eds.): ICCS 2021, LNCS 12742, pp. 715–727, 2021.
https://doi.org/10.1007/978-3-030-77961-0_57

The most popular technique, which requires judgments in a pairwise comparison, is the Analytic Hierarchy Process (AHP) [19]. Unfortunately, the most critical AHP method's shortcoming is the ranking reversal phenomenon [23,24], which may strongly discredit the results' reliability. For this approach, Saaty proposed the most popular inconsistency index in [18] and his research was continued in[1,6,8,26]. The AHP uses a multi-valued Saaty scale with a consistency coefficient based on the matrix's eigenvalue, which is not suitable for determining the decisional matrices' logical consistency used three-value scale's. Kendall and Babington [11] proposed their consistency coefficient, which allows the consistency degree of a binary pairwise comparison set. It was also the inspiration to continue the research for many researchers [5,7,12,25].

The rank reversal paradox has become the beginning of proposing a new method called the Characteristic Objects METhod (COMET) [20,29]. Earlier works have shown that it is a method resistant to this paradox, and the obtained solutions are more accurate than with the AHP method [16,21,23]. However, once the pairwise comparison judgments have been received, this method lacks a coefficient to check the matrix's logical consistency.

Both methods use a pairwise comparison to create a comparison matrix. A different scale is used in both cases, i.e., three levels in the COMET and 17 degrees or more in the AHP. Due to an expert's possible mistakes with a series of similar questions, the AHP method uses an inconsistency coefficient based on the eigenvector method. It is helpful for an expert to assess if his answers are sufficiently consistent [13].

In this work, the most significant contribution is a new consistency coefficient for decision matrices with a simple three value scale. This coefficient is presented by using an example of a Matrix of Expert Judgment (MEJ), a crucial step in the COMET method. Currently, there is not possible to verify whether or not the decision matrix received is logically consistent. The proposed coefficient allows us to check how strongly the MEJ matrix is consistent. It is crucial because the decisional model will be as good as an expert is. Therefore, the essential characteristics of this coefficient will be examined in some experiments. The coefficient design will be based on Kendall's work [11] and relate to work [12].

The rest of the paper is organized as follows: In Sect. 2, some preliminary MEJ concepts are presented. Section 3 introduces a new consistency coefficient. In Sect. 4, the discussion on simple experiments shows the most important properties of the presented consistency coefficient. In Sect. 5, we present the summary and conclusions.

2   Matrix of Expert Judgment (MEJ)

The first three steps of the COMET method should be presented to show the whole procedure for creating the MEJ matrix [9,20]. In the first step, we should define the space of the problem. An expert determines the dimensionality of the problem by selecting the number r of criteria, $C_1, C_2, ..., C_r$. Then, the set of fuzzy numbers for each criterion C_i is selected (1):

$$
\begin{aligned}
C_1 &= \{\tilde{C}_{11}, \tilde{C}_{12}, ..., \tilde{C}_{1c_1}\} \\
C_2 &= \{\tilde{C}_{21}, \tilde{C}_{22}, ..., \tilde{C}_{2c_1}\} \\
&\,\,\cdots\cdots\cdots\cdots\cdots \\
C_r &= \{\tilde{C}_{r1}, \tilde{C}_{r2}, ..., \tilde{C}_{rc_r}\}
\end{aligned}
\tag{1}
$$

where $c_1, c_2, ..., c_r$ are numbers of the fuzzy numbers for all criteria.

As a second step, we generate characteristic objects. The characteristic objects (CO) are obtained by using the Cartesian Product of fuzzy numbers cores for each criteria as follows (2):

$$
CO = C(C_1) \times C(C_2) \times ... \times C(C_r)
\tag{2}
$$

As the result, the ordered set of all CO is obtained (3):

$$
\begin{aligned}
CO_1 &= \{C(\tilde{C}_{11}), C(\tilde{C}_{21}), ..., C(\tilde{C}_{r1})\} \\
CO_2 &= \{C(\tilde{C}_{11}), C(\tilde{C}_{21}), ..., C(\tilde{C}_{r2})\} \\
&\,\,\cdots\cdots\cdots\cdots\cdots \\
CO_t &= \{C(\tilde{C}_{1c_1}), C(\tilde{C}_{2c_2}), ..., C(\tilde{C}_{rc_r})\}
\end{aligned}
\tag{3}
$$

where t is the number of COs (4):

$$
t = \prod_{i=1}^{r} c_i
\tag{4}
$$

The third and final step is that we rank the characteristic objects. In the first part of this step, an expert determines the Matrix of Expert Judgment (MEJ). It is a result of pairwise comparison of the COs by the expert. The MEJ structure is presented (5):

$$
\begin{pmatrix}
\alpha_{11} & \alpha_{12} & ... & \alpha_{1t} \\
\alpha_{21} & \alpha_{22} & ... & \alpha_{2t} \\
... & ... & ... & ... \\
\alpha_{t1} & \alpha_{t2} & ... & \alpha_{tt}
\end{pmatrix}
\tag{5}
$$

where α_{ij} is the result of comparing CO_i and CO_j by the expert. The more preferred characteristic object gets one point and the second object gets zero points. If the preferences are balanced, both objects get a half point. It depends solely on the knowledge of the expert and can be presented as (6):

$$
\alpha_{ij} = \begin{cases}
0.0, & f_{exp}(CO_i) < f_{exp}(CO_j) \\
0.5, & f_{exp}(CO_i) = f_{exp}(CO_j) \\
1.0, & f_{exp}(CO_i) > f_{exp}(CO_j)
\end{cases}
\tag{6}
$$

where f_{exp} is the expert mental judgment function. The some interesting properties are described by Eqs. (7) and (8):

$$
\alpha_{ii} = 0.5
\tag{7}
$$

$$\alpha_{ji} = 1 - \alpha_{ij} \tag{8}$$

Based on (7) and (8), the number of comparisons is reduced from t^2 cases to p cases (9):

$$p = \binom{t}{2} = \frac{t(t-1)}{2} \tag{9}$$

3   Consistency Coefficient

This section is divided into two parts. In Sect. 3.1, we analyze characteristic objects in the MEJ and discusses possible triad. It should be noted that the upper triangular matrix is in close relation to the lower triangular matrix. This is important due to the triad analysis, as only the upper triangular matrix will be analyzed. In Sect. 3.2, we propose the consistency coefficient, which is based on the analysis results.

3.1   Triads Analysis

Let suppose that we have four objects which are pairwise comparison, i.e., A, B, C, and D. Based on this pairwise comparison, and we obtain the following judgment matrix (10):

$$
MEG = \begin{array}{c} \\ A \\ B \\ C \\ D \end{array}
\begin{array}{c} A \quad B \quad C \quad D \\
\left(\begin{array}{cccc}
\alpha_{11} & \alpha_{12} & \alpha_{13} & \alpha_{14} \\
\alpha_{21} & \alpha_{22} & \alpha_{23} & \alpha_{24} \\
\alpha_{31} & \alpha_{32} & \alpha_{33} & \alpha_{34} \\
\alpha_{41} & \alpha_{42} & \alpha_{43} & \alpha_{44}
\end{array} \right)
\end{array} \tag{10}
$$

In this case, an expert needs answering to six questions on preferences of the following pairs: (A, B), (A, C), (A, D), (B, C), (B, D), and (C, D). Triad is called a collection consisting of three objects. For this example, we can listed four triads: (A, B, C), (A, B, D), (A, C, D), and (B, C, D). In general, the number of all possible triads (T) from the $t - element$ set can be determined from the formula (11):

$$T = \frac{t!}{(t-3)!3!} \tag{11}$$

Assuming that each characteristic objects have a certain unknown evaluation (constant over time), the expert's preferences must be a transitive relation. If we take the triad (A, B, C) then we can formulate seven rules of transitivity (12):

$$
\begin{array}{ccccccc}
if & A > B & and & B > C & then & A > C \\
if & A > B & and & B = C & then & A > C \\
if & A = B & and & B > C & then & A > C \\
if & A = B & and & B < C & then & A < C \\
if & A < B & and & B < C & then & A < C \\
if & A < B & and & B = C & then & A < C \\
if & A = B & and & B = C & then & A = C
\end{array} \tag{12}
$$

Equation (11) presents the relationship between the number of characteristic objects (t) and the number of all possible triads (T). The number of all possible triads is much higher than the number of all upper triangular matrix elements. However, Eq. (12) presents only seven rules, and we have 27 possible. The term that another 20 rules mean inconsistent triads is not right. Therefore, all 27 rules will be analysed in the next subsection concerning the MEJ matrix.

3.2   Consistency Coefficient

Based on (10) and (12), we are determined a set of consistent triads (CO_i, CO_j, CO_k) for which one of the seven conditions is met (13). The number of all consistent triads is written as T_{con}.

$$
\begin{aligned}
if \quad & \alpha_{ij} = 1.0 \quad and \quad \alpha_{jk} = 1.0 \quad then \quad \alpha_{ik} = 1.0 \\
if \quad & \alpha_{ij} = 1.0 \quad and \quad \alpha_{jk} = 0.5 \quad then \quad \alpha_{ik} = 1.0 \\
if \quad & \alpha_{ij} = 0.5 \quad and \quad \alpha_{jk} = 1.0 \quad then \quad \alpha_{ik} = 1.0 \\
if \quad & \alpha_{ij} = 0.5 \quad and \quad \alpha_{jk} = 0.0 \quad then \quad \alpha_{ik} = 0.0 \\
if \quad & \alpha_{ij} = 0.0 \quad and \quad \alpha_{jk} = 0.0 \quad then \quad \alpha_{ik} = 0.0 \\
if \quad & \alpha_{ij} = 0.0 \quad and \quad \alpha_{jk} = 0.5 \quad then \quad \alpha_{ik} = 0.0 \\
if \quad & \alpha_{ij} = 0.5 \quad and \quad \alpha_{jk} = 0.5 \quad then \quad \alpha_{ik} = 0.5
\end{aligned}
\tag{13}
$$

More interesting are the triads, for which it is impossible to determine whether their relationship is logically consistent. At the same time, their inconsistency cannot be demonstrated. Let us assume that for 3 objects CO_i, CO_j and CO_k we know their preference values as $f_{CO_i} = 0.67$, $f_{CO_j} = 0.47$ $f_{CO_k} = 0.52$. Therefore, we get $\alpha_{ij} = 1$ (0.67 > 0.47), $\alpha_{jk} = 0$ (0.47 < 0.52) and $\alpha_{ik} = 1$ (0.67 > 0.62). For these triads, a binding conclusion cannot be established. Therefore, these triads will be referred to as unknown. It is worth noting that they cannot influence the decrease of the matrix's consistency because, as the example above shows, they may result from real expert knowledge. The number of all unknown triads will be written as T_{unk}, and each unknown triad must be satisfied one of the following rules (14):

$$
\begin{aligned}
if \quad & \alpha_{ij} = 1.0 \quad and \quad \alpha_{jk} = 0.0 \quad then \quad \alpha_{ik} = 0.0 \\
if \quad & \alpha_{ij} = 1.0 \quad and \quad \alpha_{jk} = 0.0 \quad then \quad \alpha_{ik} = 0.5 \\
if \quad & \alpha_{ij} = 1.0 \quad and \quad \alpha_{jk} = 0.0 \quad then \quad \alpha_{ik} = 1.0 \\
if \quad & \alpha_{ij} = 0.0 \quad and \quad \alpha_{jk} = 1.0 \quad then \quad \alpha_{ik} = 0.0 \\
if \quad & \alpha_{ij} = 0.0 \quad and \quad \alpha_{jk} = 1.0 \quad then \quad \alpha_{ik} = 0.5 \\
if \quad & \alpha_{ij} = 0.0 \quad and \quad \alpha_{jk} = 1.0 \quad then \quad \alpha_{ik} = 1.0
\end{aligned}
\tag{14}
$$

The next group of triads is inconsistent triads, which we can divide into two subgroups: weak inconsistent and strong inconsistent triads. One more again, let us assume that for 3 objects CO_i, CO_j and CO_k we know their preference values as $f_{CO_i} = 0.67$, $f_{CO_j} = 0.66$ $f_{CO_k} = 0.65$. Then $\alpha_{ij} = 1$, $\alpha_{jk} = 1$ and $\alpha_{ik} = 1$. Let suppose that the expert gives the answer that $\alpha_{ik} = 0.5$. This answer is inconsistent, but if the expert answers that $\alpha_{ik} = 0$ it will be a bigger mistake. Both situations describe inconsistent triads. The weak inconsistent, we can describe as the following rules (15):

$$
\begin{array}{lllllll}
if & \alpha_{ij} = 1.0 & and & \alpha_{jk} = 1.0 & then & \alpha_{ik} = 0.5 \\
if & \alpha_{ij} = 1.0 & and & \alpha_{jk} = 0.5 & then & \alpha_{ik} = 0.5 \\
if & \alpha_{ij} = 0.5 & and & \alpha_{jk} = 1.0 & then & \alpha_{ik} = 0.5 \\
if & \alpha_{ij} = 0.5 & and & \alpha_{jk} = 0.5 & then & \alpha_{ik} = 1.0 \\
if & \alpha_{ij} = 0.5 & and & \alpha_{jk} = 0.5 & then & \alpha_{ik} = 0.0 \\
if & \alpha_{ij} = 0.5 & and & \alpha_{jk} = 0.0 & then & \alpha_{ik} = 0.5 \\
if & \alpha_{ij} = 0.0 & and & \alpha_{jk} = 0.5 & then & \alpha_{ik} = 0.5 \\
if & \alpha_{ij} = 0.0 & and & \alpha_{jk} = 0.0 & then & \alpha_{ik} = 0.5
\end{array}
\tag{15}
$$

The number of all weak inconsistent triads is called T_{inc}^{weak} (16). Finally, the last group is the strong inconsistent triads, which can be identify by using the following rules (16):

$$
\begin{array}{lllllll}
if & \alpha_{ij} = 1.0 & and & \alpha_{jk} = 1.0 & then & \alpha_{ik} = 0.0 \\
if & \alpha_{ij} = 1.0 & and & \alpha_{jk} = 0.5 & then & \alpha_{ik} = 0.0 \\
if & \alpha_{ij} = 0.5 & and & \alpha_{jk} = 1.0 & then & \alpha_{ik} = 0.0 \\
if & \alpha_{ij} = 0.5 & and & \alpha_{jk} = 0.0 & then & \alpha_{ik} = 1.0 \\
if & \alpha_{ij} = 0.0 & and & \alpha_{jk} = 0.5 & then & \alpha_{ik} = 1.0 \\
if & \alpha_{ij} = 0.0 & and & \alpha_{jk} = 0.0 & then & \alpha_{ik} = 1.0
\end{array}
\tag{16}
$$

The number of all strong inconsistent triads is denoted as T_{inc}^{strong}. Why are we showing two groups of inconsistent triads? It is more likely for very similar assessment values that an error will be classified as weak, inconsistent triads than as strong inconsistent triads. In this work, both groups will be represented as (17):

$$
T_{inc} = T_{inc}^{weak} + T_{inc}^{strong}
\tag{17}
$$

Figure 1 shows triads percentage distributions for random the MEJ matrix, where consistent, unknown, weak inconsistent, and strong inconsistent triads are analyzed. For each of the six cases the MEJ matrix was drawn 10,000 times and then the distributions were determined. The draw was conducted with a uniform probability.

For cases (a) to (f), it can be said with 99% probability that at random selection we keep $[0.2669, 0.2900]$ consistent triads; $[0.1390, 0.1574]$ unknown triads; $[0.3879, 0.4132]$ weak inconsistent triads; and $[0.1631, 0.1827]$ strong inconsistent triads. In general, we can said that randomly obtained matrix has $[0.5510, 0.5959]$ inconsistent triads from the whole number of triads.

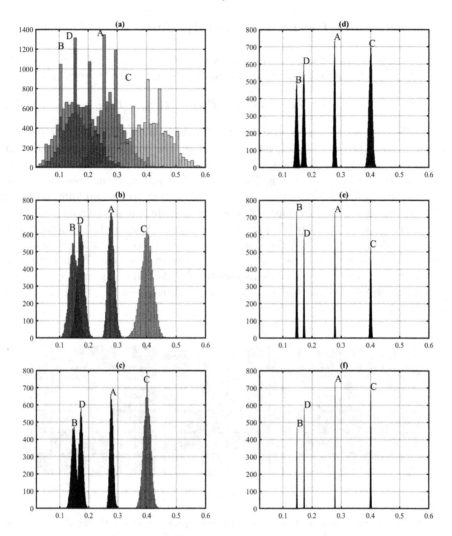

Fig. 1. Triads percentage distributions for 10000 randomly generated samples, where: A logically coherent triads; B triads unknown; C triads are slightly incoherent; D triads are strongly incoherent; (a) for t = 10; (b) for t = 30; (c) for t = 50; (d) for t = 100; (e) for t = 250; (f) for 500.

Finally, we call ξ the coefficient of consistence for the MEJ matrix, and it can be obtained as (18):

$$\xi = 1 - \frac{T_{inc}}{T} \tag{18}$$

4   Consistency Coefficient - Study Cases

Let's analyze a simple experiment for six characteristic objects. First, we will generate a MEJ matrix for elements in the following preference relation:

$$P_{CO_1} < P_{CO_2} < P_{CO_3} < P_{CO_4} < P_{CO_5} < P_{CO_6} \qquad (19)$$

where P_{CO_i} means the preference for CO_i. In that way, we obtained the matrix, which is visualised in Fig. 2a. For this matrix, we get consistency coefficient $\xi = 1.0000$ because all 20 triads are consistent. Let us turn the value into a α_{16} cell from 0 to 1 as it shows in Fig. 2b. As a result, the consistency coefficient will decrease to $\xi = 0.8000$ because we obtain 16 consistent triads and four triads are strongly inconsistent. One more again, let us turn the value into a α_{14} cell from 0 to 1. We get a matrix which is presented in Fig. 2c. As a result, the consistency coefficient decrease to $\xi = 0.7500$, where we have 13 consistent triads and two triads are unknown, and five triads are strongly inconsistent. Each matrix in Fig. 2a-2c provides an order of characteristic objects. The Spearman correlation between reference matrix in Fig. 2a and analyzed matrices Fig. 2b and Fig. 2c is respectively $\rho = 0.9$ and $\rho = 0.8$.

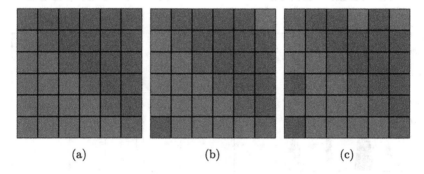

      (a)   (b)   (c)

Fig. 2. The MEJ matrices visualization for exemplary model with six characteristic objects, green 1.0, blue 0.5 and red 0.0 points. (Color figure online)

The second experiment consists of drawing a group of 10 characteristic objects and then calculating several values of the consistency coefficient in response to changes in the original matrix. Table 1 presents the random selected ten characteristic objects with their preference values and an obtained ranking, where one is the highest rank, and ten is the lowest. The MEJ matrix in Fig. 3a was generated based on data from Table 1.

For the matrix in Fig. 3a, the consistency coefficients is $\xi = 1.0000$. More interesting is that only 49 triads were consistent triads, and the rest were unknown triads. This matrix was indeed generated from the data. Thus, despite a large number of unknown triads, it is with all certainty a logically consistent matrix what was also shown by the ξ coefficient. Subsequently, a change was

Table 1. Random ten characteristic objects (CO_i) with their preference values (P_{CO_i}) and a determined ranking rank(CO_i)

CO_i	P_{CO_i}	Rank (CO_i)
CO_1	0.91	1
CO_2	0.22	9
CO_3	0.71	4
CO_4	0.55	6
CO_5	0.72	3
CO_6	0.82	2
CO_7	0.37	8
CO_8	0.63	5
CO_9	0.46	7
CO_{10}	0.10	10

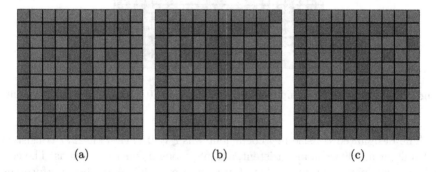

(a)   (b)   (c)

Fig. 3. The MEJ matrices visualization for exemplary model with ten characteristic objects, green 1.0, blue 0.5 and red 0.0 points. (Color figure online)

made to the matrix in the first row and the last column on 0.5, see Fig. 3b. In this case, we obtain $\xi = 0.9333$ with 41 consistent, 71 unknown, and eight weak inconsistent triads. The order of characteristic objects is not changed ($\rho = 1$). Now let us go back to the reference matrix in Fig. 3a and make another change. The two elements have been changed, i.e., $\alpha_{34} = 0$ and $\alpha_{310} = 0$. The new matrix is presented in Fig. 3c.

For matrix on Fig. 3c, we obtain $\xi = 0.9583$ with 44 consistent, 71 unknown, and 5 strong inconsistent triads. The order of characteristic objects is change. Now, the Spearman correlation between reference matrix in Fig. 3a and modified in Fig. 3c is equal $\rho = 0.9573$. Despite the high consistency coefficient, the final result is worse than in the matrix in Fig. 3b. A high degree of consistency in the expert response is a prerequisite for good results, but it is insufficient. In real examples, we rely on the expert's knowledge because data is not available.

Therefore, the expert's answers' consistency is crucial, but only if the expert's knowledge is sufficient and up-to-date.

The last example will come from a previous work where the identified model showed good computational properties. The decision matrix for 27 characteristic objects took the following form Fig. 4.

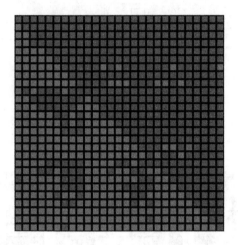

Fig. 4. MEJ matrix for example model, where green 1.0, blue 0.5 and red 0.0 points.

The designated consistency coefficient was $\xi = 0.8711$ with 2277 consistent, 271 unknown, 309 weak inconsistent, and 68 strong inconsistent triads. The coefficient developing is an important step to improve MCDA method performance (in this case, the COMET method). With a series of often similar questions, it is easy to pick a wrong answer, and therefore our task is to identify how consistent the expert-created matrix is. Now with a new coefficient, it is feasible.

5 Conclusions

This study proposes a new consistency coefficient. The proposed ξ coefficient is based on triads analysis. In the study, four groups of 27 presented rules were distinguished, i.e., consistent, unknown, weak inconsistent, and strong inconsistent triads. Triads percentage distributions for 10,000 randomly generated samples were used to determine the confidence interval for which the consistency range of random MEJ matrices was determined with 99% probability.

Three case studies were presented to show how this coefficient works. The consistency is a necessary but not sufficient condition. If we have a strongly consistent matrix created by an inadequate expert, the model identified will still be inadequate. Therefore, the proposed coefficient could not determine the final results' quality, but only the MEJ matrix's consistency. Attempts should now be made to introduce it actively into the COMET procedure and other methods where a three values scale of pairwise comparison matrix is used.

Future directions of research are primarily:

- computer simulations to establish the minimum acceptable level of consistency for further calculation procedure;
- based on erroneously determined triads, the possibility of repairing the matrix by asking the expert again;
- considering the options of improving the proposed formula taking into account the differences between weak and strong inconsistencies;
- considering extended the proposed approach for the incomplete pairwise comparisons matrices [27].

Acknowledgments. The work was supported by the National Science Centre, Decision number UMO-2018/29/B/HS4/02725.

References

1. Aguarón, J., Moreno-Jiménez, J.M.: The geometric consistency index: approximated thresholds. Eur. J. Oper. Res. **147**(1), 137–145 (2003)
2. Al-Harbi, K.M.A.S.: Application of the AHP in project management. Int. J. Project Manage. **19**(1), 19–27 (2001)
3. Baltazar, M.E., Jardim, J., Alves, P., Silva, J.: Air transport performance and efficiency: Mcda vs. dea approaches. Proced.-Soc. Behav. Sci. **111**, 790–799 (2014)
4. Behzadian, M., Otaghsara, S.K., Yazdani, M., Ignatius, J.: A state-of the-art survey of TOPSIS applications. Expert Syst. Appl. **39**(17), 13051–13069 (2012)
5. Bozóki, S., Dezső, L., Poesz, A., Temesi, J.: Analysis of pairwise comparison matrices: an empirical research. Ann. Oper. Res. **211**(1), 511–528 (2013). https://doi.org/10.1007/s10479-013-1328-1
6. Bozóki, S., Fülöp, J., Koczkodaj, W.W.: An LP-based inconsistency monitoring of pairwise comparison matrices. Math. Comput. Model. **54**(1–2), 789–793 (2011)
7. Brunelli, M.: On the conjoint estimation of inconsistency and intransitivity of pairwise comparisons. Oper. Res. Lett. **44**(5), 672–675 (2016)
8. Brunelli, M., Canal, L., Fedrizzi, M.: Inconsistency indices for pairwise comparison matrices: a numerical study. Ann. Oper. Res. **211**(1), 493–509 (2013). https://doi.org/10.1007/s10479-013-1329-0
9. Faizi, S., Sałabun, W., Ullah, S., Rashid, T., Więckowski, J.: A new method to support decision-making in an uncertain environment based on normalized interval-valued triangular fuzzy numbers and COMET technique. Symmetry **12**(4), 516 (2020)
10. Huang, I.B., Keisler, J., Linkov, I.: Multi-criteria decision analysis in environmental sciences: ten years of applications and trends. Sci. Total Environ. **409**(19), 3578–3594 (2011)

11. Kendall, M.G., Smith, B.B.: On the method of paired comparisons. Biometrika **31**(3/4), 324–345 (1940)
12. Kułakowski, K.: Inconsistency in the ordinal pairwise comparisons method with and without ties. Eur. J. Oper. Res. **270**(1), 314–327 (2018)
13. Lane, E.F., Verdini, W.A.: A consistency test for AHP decision makers. Decis. Sci. **20**(3), 575–590 (1989)
14. Liberatore, M.J., Nydick, R.L.: The analytic hierarchy process in medical and health care decision making: a literature review. Eur. J. Oper. Res. **189**(1), 194–207 (2008)
15. Palczewski, K., Sałabun, W.: The fuzzy TOPSIS applications in the last decade. Proced. Comput. Sci. **159**, 2294–2303 (2019)
16. Piegat, A., Sałabun, W.: Comparative analysis of MCDM methods for assessing the severity of chronic liver disease. In: Rutkowski, L., Korytkowski, M., Scherer, R., Tadeusiewicz, R., Zadeh, L.A., Zurada, J.M. (eds.) ICAISC 2015. LNCS (LNAI), vol. 9119, pp. 228–238. Springer, Cham (2015). https://doi.org/10.1007/978-3-319-19324-3_21
17. Riaz, M., Sałabun, W., Farid, H.M.A., Ali, N., Wątróbski, J.: A robust q-Rung orthopair fuzzy information aggregation using Einstein operations with application to sustainable energy planning decision management. Energies **13**(9), 2155 (2020)
18. Saaty, T.L.: A scaling method for priorities in hierarchical structures. J. Math. Psychol. **15**(3), 234–281 (1977)
19. Saaty, T.L.: Decision making with the analytic hierarchy process. Int. J. Serv. Sci. **1**(1), 83–98 (2008)
20. Sałabun, W.: The characteristic objects method: a new distance-based approach to multicriteria decision-making problems. J. Multi-Criteria Dec. Anal. **22**(1–2), 37–50 (2015)
21. Sałabun, W., Piegat, A.: Comparative analysis of MCDM methods for the assessment of mortality in patients with acute coronary syndrome. Artif. Intell. Rev. **48**(4), 557–571 (2017)
22. Sałabun, W., Wątróbski, J., Piegat, A.: Identification of a multi-criteria model of location assessment for renewable energy sources. In: Rutkowski, L., Korytkowski, M., Scherer, R., Tadeusiewicz, R., Zadeh, L.A., Zurada, J.M. (eds.) ICAISC 2016. LNCS (LNAI), vol. 9692, pp. 321–332. Springer, Cham (2016). https://doi.org/10.1007/978-3-319-39378-0_28
23. Sałabun, W., Ziemba, P., Wątróbski, J.: The rank reversals paradox in management decisions: the comparison of the AHP and COMET methods. In: Czarnowski, I., Caballero, A.M., Howlett, R.J., Jain, L.C. (eds.) Intelligent Decision Technologies 2016. SIST, vol. 56, pp. 181–191. Springer, Cham (2016). https://doi.org/10.1007/978-3-319-39630-9_15
24. Schenkerman, S.: Avoiding rank reversal in AHP decision-support models. Eur. J. Oper. Res. **74**(3), 407–419 (1994)
25. Siraj, S., Mikhailov, L., Keane, J.A.: Contribution of individual judgments toward inconsistency in pairwise comparisons. Eur. J. Oper. Res. **242**(2), 557–567 (2015)
26. Stein, W.E., Mizzi, P.J.: The harmonic consistency index for the analytic hierarchy process. Eur. J. Oper. Res. **177**(1), 488–497 (2007)
27. Szybowski, J., Kułakowski, K., Prusak, A.: New inconsistency indicators for incomplete pairwise comparisons matrices. Math. Soc. Sci. **108**, 138–145 (2020)
28. Triantaphyllou, E., Mann, S.H.: Using the analytic hierarchy process for decision making in engineering applications: some challenges. Int. J. Ind. Eng. Appl. Pract. **2**(1), 35–44 (1995)

29. Watróbski, J., Sałabun, W.: The characteristic objects method: a new intelligent decision support tool for sustainable manufacturing. In: Setchi, R., Howlett, R.J., Liu, Y., Theobald, P. (eds.) Sustainable Design and Manufacturing 2016. SIST, vol. 52, pp. 349–359. Springer, Cham (2016). https://doi.org/10.1007/978-3-319-32098-4_30
30. Zavadskas, E.K., Turskis, Z., Kildienė, S.: State of art surveys of overviews on MCDM/MADM methods. Technol. Econ. Dev. Econ. **20**(1), 165–179 (2014)
31. Ziemba, P.: Inter-criteria dependencies-based decision support in the sustainable wind energy management. Energies **12**(4), 749 (2019)

Predicting the Age of Scientific Papers

Pavel Savov[1]([⊠]), Adam Jatowt[2], and Radoslaw Nielek[1]

[1] Polish-Japanese Academy of Information Technology, ul. Koszykowa 86,
02-008 Warszawa, Poland
{pavel.savov,nielek}@pja.edu.pl
[2] University of Innsbruck, Innrain 52, 6020 Innsbruck, Austria
adam.jatowt@uibk.ac.at

Abstract. In this paper we show how the age of scientific papers can be
predicted given a diachronic corpus of papers from a particular domain
published over a certain time period. We first train ordinal regression
models for the task of predicting the age of individual sentences by fine-
tuning series of BERT models for binary classification. We then aggregate
the prediction results on individual sentences into a final result for entire
papers. Using two corpora of publications from the International World
Wide Web Conference and the Journal of Artificial Societies and Social
Simulation, we compare various result aggregation methods, and show
that the sentence-based approach produces better results than the direct
document-level method.

Keywords: Scientometrics · Text age prediction · Embeddings ·
Ordinal regression

1 Introduction

Document dating or timestamping is the process of inferring the age of a document, if it is either unknown or unreliable, based on its textual content. In
the scientific domain, publication dates of documents are usually known, but
the results of document timestamping may be used to complement traditional
scientometric methods in assessing the innovativeness of research papers [20] or
identifying novelty. At a basic level, the larger the difference between the actual
timestamp and the predicted timestamp of a target scientific document, the
higher is its potential innovativeness or novelty of the target paper. This may be
useful to non-expert readers of technical documents, such as potential investors
or decision makers at funding bodies, who wish to know how new or innovative the ideas or methods covered by these documents were at the time of their
creation. Furthermore, in practical scenarios, the timestamping models specialized for scientific corpora can also be applied to other types of documents that
may discuss scientific technology and domain-focused research, or quote content
from scientific papers. Such documents may not have explicit timestamps (e.g.,
web pages) and the determination of their age (as well as the related concept

© Springer Nature Switzerland AG 2021
M. Paszynski et al. (Eds.): ICCS 2021, LNCS 12742, pp. 728–735, 2021.
https://doi.org/10.1007/978-3-030-77961-0_58

of timeliness) can be useful in many cases. Thus, in general, scientific document age prediction can be used for discovering the content parts in a scientific publication that are novel or innovative, or perhaps obsolete/outdated when considering the document publication date [20] as well as for determining the age of science-related content in non-scholarly documents that lack timestamps.

In this paper we focus on improving the accuracy of scientific paper age prediction by using state-of-the-art word embedding models trained on two corpora of papers from related but distinct domains, published at leading publication venues in their respective fields. Typical approaches to automatic document dating are based on modeling language change over time and shifts in word usage. Examples of temporal language models, i.e. time series of statistical language models include [5,9]. Jatowt and Campos [8] have implemented an online visual and interactive system based on n-gram frequency analysis. Garcia-Fernandez et al. [7] used SVM classifiers on feature vectors of word and n-gram frequencies. Ordinal regression models were used for document dating by Niculae et al. [16], or Popescu and Strapparava [18]. Another approach to temporal language modeling are neural language models based on word embeddings such as Word2Vec [15]. Kim et al. [10] studied the shift in word semantics over time by training a model for each time interval and then plotting the words' cosine similarities to their reference points. Soni et al. [21] used diachronic word embeddings to show that scientific papers using words in their newer meanings tend to receive more citations. Vashishth et al. [23] proposed a deep learning approach to document dating, exploiting syntactic and temporal document graph structures. Unlike the above-mentioned methods, which work mainly on news articles or generic documents, we focus on a particular genre of scholarly publications. We also approach the document dating task at a sentence-level, and we test several sentence aggregation approaches.

2   Datasets

We study the following two corpora: (1) *WWW*: 3,896 papers published at the International World Wide Web Conference between 1994 and 2020, containing 1,037,051 sentences, (2) *JASSS*: 884 articles published in the Journal of Artificial Societies and Social Simulation[1] between 1998 and 2020, containing 321,589 sentences. Both corpora contain entire papers. However, we have removed page headers and footers, *References*, *Bibliography* and *Acknowledgments* sections as "noise" irrelevant to the papers' contents. All papers published in the JASSS journal are available in HTML at the journal's website (See footnote 1). Papers from the proceedings of the WWW conference are available at https://thewebconf.org/ in different formats for different years. Most are available in PDF, some in HTML and a small number of older papers in PostScript. We used the *pdftotext* tool[2] to extract plain text from PDF documents. We divided the documents into sentences using the Punkt sentence tokenizer for

---

[1] http://jasss.soc.surrey.ac.uk/.
[2] https://www.xpdfreader.com/pdftotext-man.html.

the English language implemented in the Natural Language Toolkit (NLTK) Python library [4]. Conversion to lower case and tokenization were performed by the BERT tokenizer.

3   Method

We propose to approach the problem of scientific document's age prediction by first predicting the age of its sentences. Thanks to focusing on sentences instead of entire documents we can use more labelled data instances for training, which is quite important for relatively narrow scientific domains with constrained datasets (e.g., proceedings of conferences dedicated to a particular research sub-field). Thus, our approach is composed of two steps: (1) predicting the age of sentences and (2) aggregating sentence age to determine the document age. We describe these two steps below.

3.1   Predicting Sentence Age

As time units are clearly ordinal values, we predict the age of individual sentences by means of Ordinal Regression, a.k.a. Ordinal Classification, based on the framework proposed by Li and Lin [11]. Ordinal Regression was also used by Martin et al. [13] for photograph dating. An N-class ordinal regression model consists of $N - 1$ *before-after* binary classifiers, i.e. for each pair of consecutive years a classifier is trained, which assigns sentences to one of two classes: "year y or before" and "year $y + 1$ or after". Given the class membership probabilities predicted by these classifiers, the overall classifier confidence that sentence s was written in the year Y is then determined, as in [13], by Eqs. 1 and 2:

$$conf(s, Y) = \prod_{y=Y_{min}}^{Y} P(Y_s \leq y) \cdot \prod_{y=Y+1}^{Y_{max}} (1 - P(Y_s \leq y)) \tag{1}$$

where Y_{min} and Y_{max} are the first and last year in the corpus, and Y_s is the publication year of the paper that s comes from.

Thus, the predicted year for the sentence s is:

$$\hat{Y}_s = \underset{y \in [Y_{min}, Y_{max}]}{\operatorname{argmax}} conf(s, y) \tag{2}$$

Unlike the approaches of [11] and [13], we used the Huggingface Transformers[3] [24] Python library to fine-tune SciBERT models [3] for sequence classification in binary *before-after* classification. SciBERT is a BERT [6] model trained on 1.14M scientific papers from the semanticscholar.org corpus. The maximum sequence length supported out-of-the-box is 512, however over 95% of the sentences in our corpora contain up to 64 tokens (see Fig. 1). We have, therefore, decided to cap the maximum sequence length at 64. We have not observed any

---

[3] https://huggingface.co/transformers/.

significant differences in the predictive performance of the models, expressed as Mean Absolute Error, for maximum sequence lengths of 64, 128, and 512 tokens. We trained each model for two epochs, the batch size was 32, and the learning rate: 2e−5. The BERT authors recommend fine-tuning the models for 2 to 4 epochs, but we have found our models to overfit the training data when fine-tuned for more than 2 epochs. In most cases the differences in average loss and accuracy on the validation set for models trained for two epochs vs. one were minimal.

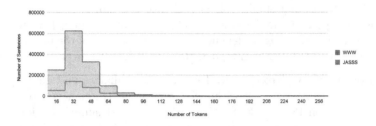

Fig. 1. Number of tokens per sentence

We have made an 80/20 split on the document level so as to make sure of the clean separation of training and testing sentences. Although our approach yielded poor prediction results on the sentence-level (4.49 years for JASSS and 3.56 years for WWW, see Fig. 2), as we will show later, the final prediction of document age produces quite good results.

3.2   Predicting Document Age

As stated above, we predict the age of entire papers by aggregating the results of individual sentence age prediction using various aggregation functions. We have experimented with rejecting sentences for which the model's confidence was below a certain threshold in the range from 0 to 0.5. For values greater than 0.5 in some documents no sentences exceeded that threshold.

Newest Sentence. As a baseline approach we assume the age of the paper p equals the age of its newest sentence. Since most papers contain at least one sentence the most probable age of which is predicted as 0 years, we only take into account the sentence predicted as the newest among those, for which the model's confidence exceeds 0.5. This value was chosen, as it gave the best results.

Topic Distribution Based Classifier. As another baseline approach, which works purely on the document-level, we used a method based on SVM classifier on vectors of latent topic distributions derived from document collections [20].

Arithmetic Mean. In this approach we calculated the predicted age of paper p as the mean predicted age of all its sentences.

Weighted Mean w/Sentence Offset. We assumed that the sooner a sentence appears in the paper, the more important it is. We, therefore, defined the predicted age of paper p as the weighted mean predicted age of its sentences, where the weight of each sentence was its ordinal number within the paper p divided by the number of sentences in p:

$$\hat{Y}_p = \frac{\sum_{s \in p} \hat{Y}_s \cdot \frac{n_s}{|\{s \in p\}|}}{\sum_{s \in p} \frac{n_s}{|\{s \in p\}|}}$$

where n_s is the ordinal number of the sentence s within p.

This concept is a simplified approach to weighted zoning [12], where each sentence is assigned a weight, depending on which section of the paper it appears in, e.g. Abstract: 1, Introduction: 0.8, Related Work: 0.3, everything else: 0.5.

Weighted Mean w/TextRank. TextRank by Mihalcea and Tarau [14] is an unsupervised graph-based algorithm for keyword extraction and text summarization, based on PageRank [17]. Its variant for text summarization finds the most important sentences by running a variation of PageRank on a graph, whose vertices represent the document's sentences. Each edge has a weight corresponding to the similarity of the sentences represented by the vertices connected by that edge. In contrast to PageRank, the graph constructed by TextRank is undirected, since the similarity between sentences is symmetric. Various sentence similarity measures may be used, but Barrios et al. [2] showed that a variation of the Okapi-BM25 [19] ranking function, which is itself a variation of the TF-IDF model using a probabilistic model, yields the best results. We used the implementation of TextRank with the BM25 ranking function from the *gensim*[4] Python library to find importance scores for all sentences in each document. We then used these scores as weights to calculate the predicted publication year of each paper p defined as the weighted mean of the years of its sentences:

$$\hat{Y}_p = \frac{\sum_{s \in p} Imp_p^s \cdot \hat{Y}_s}{\sum_{s \in p} Imp_p^s}$$

where Imp_p^s is the TextRank importance score of s within the paper p.

Citation Removal. In this approach we make the assumption that any sentences citing other papers are unimportant for the content of the paper being analyzed or introduce concepts and ideas from older papers (hence potentially negatively impacting the age detection process). Thus, we remove all sentences containing citations and proceed to calculate the predicted publication year using any of the approaches described above. As shown in Sect. 4, in most cases citation removal improves the prediction results in terms of Mean Absolute Error. Another possible extension could be removing entire *Related Work* sections.

---

[4] https://radimrehurek.com/gensim_3.8.3/index.html.

4   Results

As stated before, the mean absolute age prediction error (MAE) for individual sentences is 4.49 years for the JASSS corpus and 3.56 for WWW. The prediction error distribution is shown in Fig. 2. Although these results are not satisfactory, we obtain much better results for entire documents. As shown in Table 1, the sentence-based approach aggregating individual predictions of many sentences gives much better results in predicting paper publication dates. Except for the naive *newest sentence* baseline, the MAE is always less than 1 year. Also the document level approach proposed in [20] performs much worse.

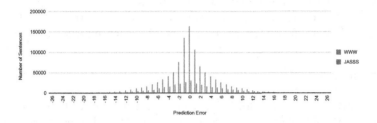

Fig. 2. Sentence prediction error distributions

Weighting the sentence age predictions by sentence offsets performed better on the WWW corpus, while TextRank weights gave better results for JASSS. In all cases, however, removing sentences containing citations improved the document age predictions significantly. This supports our assumption that sentences citing other articles could introduce noise.

Table 1. Results of prediction methods (Mean Absolute Error: #years).

	WWW		JASSS	
Document-level [20]	2.56		3.56	
Sentence-level	All Sentences	Citations Removed	All Sentences	Citations Removed
Newest Sentence	8.959	8.946	8.267	8.33
Arithmetic Mean	0.833	0.816	0.743	0.67
Weighted Mean w/Sentence Offset	0.709	**0.684**	0.738	0.645
Weighted Mean w/TextRank	0.741	0.725	0.67	**0.636**

5   Conclusions and Future Work

In this paper we have shown how the accuracy of scientific paper age prediction can be improved by using state-of-the-art word embedding models at the sentence level, and then aggregating the results. Interestingly, for all aggregation methods except for the most basic baseline approach, i.e. *newest sentence*, increasing the value of the confidence threshold led to worse results. This suggests that unless sentences are rejected based on domain-specific knowledge, e.g. rejecting sentences containing citations, the more predictions are aggregated into the final result the better, similarly to the "wisdom of the crowds" effect, where the aggregated predictions of multiple agents are far closer to the actual value than most of the individual predictions [22]. Finally, we note that as our approach works on the sentence-level, it could also be used to assess the age of text excerpts (e.g., in web pages) about specialized scientific topics, and, therefore, potentially help readers better understand their actual novelty and age.

Having achieved a mean prediction error of less than a year, we plan on experimenting with datasets having narrower time slices, e.g. the Covid-19 dataset from Kaggle[5]. We will also try weighting sentences containing scientific claims [1].

References

1. Achakulvisut, T., Bhagavatula, C., Acuna, D., Kording, K.: Claim extraction in biomedical publications using deep discourse model and transfer learning. arXiv preprint arXiv:1907.00962 (2019)
2. Barrios, F., López, F., Argerich, L., Wachenchauzer, R.: Variations of the similarity function of textrank for automated summarization. arXiv preprint arXiv:1602.03606 (2016)
3. Beltagy, I., Lo, K., Cohan, A.: Scibert: pretrained language model for scientific text. In: EMNLP (2019)
4. Bird, S.: Nltk: the natural language toolkit. In: Proceedings of the COLING/ACL 2006 Interactive Presentation Sessions, pp. 69–72 (2006)
5. De Jong, F., Rode, H., Hiemstra, D.: Temporal language models for the disclosure of historical text. In: International Conference of the Association for History and Computing (AHC 2005), pp. 161–168. Koninklijke Nederlandse Academie van Wetenschappen, Amsterdam, the Netherlands (2005)
6. Devlin, J., Chang, M.W., Lee, K., Toutanova, K.: Bert: pre-training of deep bidirectional transformers for language understanding. arXiv:1810.04805 (2018)
7. Garcia-Fernandez, A., Ligozat, A.L., Dinarelli, M., Bernhard, D.: When was it written? automatically determining publication dates. In: SPIRE, pp. 221–236 (2011)
8. Jatowt, A., Campos, R.: Interactive system for reasoning about document age. In: CIKM 2017, pp. 2471–2474. ACM
9. Kanhabua, N., Nørvåg, K.: Using temporal language models for document dating. In: Buntine, W., Grobelnik, M., Mladenić, D., Shawe-Taylor, J. (eds.) ECML PKDD 2009. LNCS (LNAI), vol. 5782, pp. 738–741. Springer, Heidelberg (2009). https://doi.org/10.1007/978-3-642-04174-7_53
10. Kim, Y., Chiu, Y.I., Hanaki, K., Hegde, D., Petrov, S.: Temporal analysis of language through neural language models. arXiv preprint arXiv:1405.3515 (2014)
11. Li, L., Lin, H.T.: Ordinal regression by extended binary classification. In: Advances in Neural Information Processing Systems, pp. 865–872 (2007)

---

12. Manning, C.D., Raghavan, P., Schütze, H.: Scoring, term weighting and the vector space model. Introduction to information retrieval **100**, 2–4 (2008)
13. Martin, P., Doucet, A., Jurie, F.: Dating color images with ordinal classification. In: ICMR, pp. 447–450 (2014)
14. Mihalcea, R., Tarau, P.: Textrank: bringing order into text. In: EMNLP, pp. 404–411 (2004)
15. Mikolov, T., Chen, K., Corrado, G., Dean, J.: Efficient estimation of word representations in vector space. arXiv preprint arXiv:1301.3781 (2013)
16. Niculae, V., Zampieri, M., Dinu, L.P., Ciobanu, A.M.: Temporal text ranking and automatic dating of texts. In: EACL, pp. 17–21 (2014)
17. Page, L., Brin, S., Motwani, R., Winograd, T.: The pagerank citation ranking: Bringing order to the web. Technical report, Stanford InfoLab (1999)
18. Popescu, O., Strapparava, C.: Semeval 2015, task 7: diachronic text evaluation. In: Proceedings of SemEval 2015, pp. 870–878 (2015)
19. Robertson, S.E., Walker, S., Jones, S., Hancock-Beaulieu, M.M., Gatford, M., et al.: Okapi at trec-3. Nist Special Publication Sp **109**, 109 (1995)
20. Savov, P., Jatowt, A., Nielek, R.: Innovativeness analysis of scholarly publications by age prediction using ordinal regression. In: Krzhizhanovskaya, V.V., Závodszky, G., Lees, M.H., Dongarra, J.J., Sloot, P.M.A., Brissos, S., Teixeira, J. (eds.) ICCS 2020. LNCS, vol. 12138, pp. 646–660. Springer, Cham (2020). https://doi.org/10.1007/978-3-030-50417-5_48
21. Soni, S., Lerman, K., Eisenstein, J.: Follow the leader: Documents on the leading edge of semantic change get more citations. JASIST (2020)
22. Surowiecki, J.: The wisdom of crowds. Anchor (2005)
23. Vashishth, S., Dasgupta, S.S., Ray, S.N., Talukdar, P.: Dating documents using graph convolution networks. In: Proceedings of ACL, pp. 1605–1615 (2018)
24. Wolf, T., et al.: Transformers: state-of-the-art natural language processing. In: EMNLP: System Demonstrations, pp. 38–45 (2020)

Data Augmentation for Copy-Mechanism in Dialogue State Tracking

Xiaohui Song[1,2], Liangjun Zang[1,2(✉)], and Songlin Hu[1,2]

[1] Institute of Information Engineering, Chinese Academy of Sciences, Beijing, China
{songxiaohui,zangliangjun,husonglin}@iie.ac.cn
[2] School of Cyber Security, University of Chinese Academy of Sciences, Beijing, China

Abstract. Traditional dialogue state tracking (DST) approaches need a predefined ontology to provide candidate values for each slot. To handle unseen slot values, the copy-mechanism has been widely used in DST models recently, which copies slot values from user utterance directly. Even though the state-of-the-art approaches have shown a promising performance on several benchmarks, there is still a significant gap between seen slot values (values that occur in both training set and test set) and unseen ones (values that only occur in the test set). In this paper, we aim to find out the factors that influence the generalization capability of the copy-mechanism based DST model. Our key observations include two points: 1) performance on unseen values is positively related to the diversity of slot values in the training set; 2) randomly generated strings can enhance the diversity of slot values as well as real values. Based on these observations, an interactive data augmentation algorithm is proposed to train copy-mechanism models, which augments the input dataset by duplicating user utterances and replacing the real slot values with randomly generated strings. Experimental results on three widely used datasets: WoZ 2.0, DSTC2 and Multi-WoZ demonstrate the effectiveness of our approach.

Keywords: Data augmentation · Dialogue state tracking · Copy-mechanism

1 Introduction

A task-oriented dialogue system interacts with users in natural language to accomplish tasks such as restaurant reservation or flight booking. The goal of dialogue state tracking is to provide a compact representation of the conversation at each dialogue turn, called *dialogue state*, for the system to decide the next action to take. A typical dialogue state consists of goal of user, action of the current user utterance (inform, request etc.) and dialogue history [19]. All of them are defined in a particularly designed *ontology* that restricts which slots the system can handle, and the range of values each slot can take. Tracking

© Springer Nature Switzerland AG 2021
M. Paszynski et al. (Eds.): ICCS 2021, LNCS 12742, pp. 736–749, 2021.
https://doi.org/10.1007/978-3-030-77961-0_59

the goal of user is the focus of this task. To accomplish the tracking task, most DST models take the user's utterance at the current turn, a slot to track and dialogue history as input, and then output the corresponding value if the user triggers the input slot. Considering an example of restaurant reservation, users can *inform* the system some restrictions of their goals (*e.g.*, inform(food = thai)) or *request* further information (*e.g.*, request(phone number)) at each turn.

> **USER**:I'm looking for a restaurant that serves thai food.
> state:{inform(food=thai)}
> **SYSTEM**:There are two, one in the west end and one in the centre of town.
> **USER**:The one on the west end, please. Can I have the phone number?
> state:{inform(food=thai, area=west),request(phone number)}

Having access to an ontology that contains all possible values simplifies the tracking problem in many ways. However, in a real-world dialogue system, it is often impossible to enumerate all possible values for each slot. To reduce the dependence on the ontology, PtrNet [18] uses the Pointer Network [14] to handle the unknown values that are not defined in the ontology. Since then, the attention-based *copy-mechanism* inspired by Pointer Network have been widely used in state-of-the-art DST approaches [4,12,17]. The copy-mechanism based DST models directly copy slot values from the dialogue history, thus reducing the need to pre-define all slot values in the ontology.

Nonetheless, there is still a significant performance gap between seen slot values and unseen ones(*i.e.*, the values that occur in test set but not in train set), we assume the insufficient diversity of values (*i.e.*, the number of unique slot values) is the primary reason for poor generalization capability. We conduct two experiments to prove our assumption and illustrate our observations. In the first experiment, we construct synthetic test sets that only contain unseen values. For WoZ and DSTC2, we replace all values of slot food with food names collected from Wikipedia[1]. For Multi-WoZ dataset, we use 13 slots that contain non-enumerable values. Since it is costly to collect enough values for each slot, we use random strings to construct synthetic test set. After training the baseline model on the original training sets, we observed a huge performance gap between the original test set and the synthetic test set. In the second experiment, we augment the WoZ training sets by duplicating user utterances 10 times and replacing all slot values with collected food names. Results on the synthetic test sets show that the generalization performance is positively related to the diversity of slot values, which further confirms our conjecture.

In a real-world dialogue system, it is impractical to obtain lots of real values for each slot when attempting to improve the diversity. In order to find a more convenient way, we design the third experiment. We first use randomly generated strings as values to enhance the diversity of slot values in the WoZ training set, then train the copy-mechanism based model and test on the test set that contains real slot values. Experimental results illustrate that random strings work as

---

[1] https://en.m.wikipedia.org/wiki/Lists_of_prepared_foods.

well as collected real values and they help a lot to get a better generalization performance.

To design a data augmentation algorithm based on the above observations, a question remains to answer: "How many values we need to generate to obtain a cost efficient result?". It is hard to answer this question in the data pre-processing phase before training (as what traditional data augmentation algorithms do), so we design our algorithm to work in the training phase. The algorithm samples user utterances iteratively from the input dataset, then replaces the real slot values with random strings, and stop training with early-stop mechanism. Experimental results show that our algorithm produces a satisfactory result at a low cost.

The rest of this paper is organized as follows: we first review the recent advances in both DST and the copy mechanism in Sect. 2. Then, we describe the datasets we used and the baseline model in Sect. 3. In Sect. 4, our data analysis process and key observations are described. We propose our data augmentation approach to DST task and evaluate its performance in Sect. 5. Finally, we conclude our work and discuss the future directions.

2 Related Works

2.1 Dialogue State Tracking

Recently, deep-learning has shown its power to the dialogue state tracking challenges [5,6,16]. Some approaches rely on the value set provided by ontology: Neural Belief Tracker (NBT) [9] applied representation learning to learn features as opposed to hand-crafting features; GLAD [20] addressed the problem of extraction of rare slot-value pairs; [10] tried to share parameters across slots, but the model had to iterate all slots and values defined in the ontology at each dialogue turn; [11] generated a fixed set of candidate values using a separate SLU module but suffered from error accumulation. As for those models with generalization capabilities, PtrNet [18] aimed to handle unknown values, which is the first attempt to introduce the copy mechanism into DST; TRADE [17] was a simple copy-augmented generative model that tracked dialogue states without ontology and enabled zero-shot and few-shot DST in a new domain; [4] used the pretrained language model *BERT* [3] and copy-mechanism to predict explicitly expressed values. But the generalization performance of these models still has much room for improvement.

2.2 Copy-Mechanism

Copy-mechanism in deep learning is a generic concept, which means an output is copied from an input sequence. This idea was first proposed in Pointer Network [14]. The Pointer Network is designed to learn the conditional probability of an output sequence with elements that are discrete tokens corresponding to positions in an input sequence. It can address the problems such as sorting variable sized sequences and various combinatorial optimization. As mentioned in the introduction, [18] first reformulated DST problem to take advantage of the flexibility enabled by Pointer Network. Then, the copy-mechanism become a common sub-structure of the DST models, which is used in several state-of-the-art DST models such as TRADE [17], COMER [12], etc. DSTRead [4] used a span prediction method proposed in DrQA [2] for machine reading task, which is also an application of the copy mechanism.

3   Datasets, Baseline Model and Evaluation

In this section, we first describe the datasets we used, then present a baseline model that implements the copy mechanism widely used in DST models, and finally present the evaluation metrics.

3.1   Datasets

Our datasets are extracted from three datasets widely used in DST tasks, *i.e.*, WoZ 2.0 [15], DSTC2 [5] and Multi-WoZ 2.0 [1]. To evaluate the performance of copy mechanism on unseen slot values, we focus on the slots of which the values are non-enumerable. Table 1 lists the slots, the number of slot values, and the number of samples in the following experiments.

Table 1. Slots we use in experiments in WoZ 2.0, DSTC2 and Multi-WoZ datasets. Samples include negative samples for the slot gate (a binary classifier).

Datasets	Slots	Values in train(total)	Data samples in train(test)
WoZ-sub	Food	73(75)	2536(1646)
DSTC2-sub	Food	72(73)	11677(9890)
Multi-WoZ-sub	Hotel-name	35(37)	60027(73720)
	Train-destination	19(20)	
	Train-departure	23(25)	
	Attraction-name	88(95)	
	Taxi-destination	198(214)	
	Taxi-departure	188(210)	
	Restaurant-name	131(139)	
	Restaurant-food	94(95)	
	Bus-departure	1(1)	
	Bus-destination	1(1)	

Each sample corresponds to one dialogue turn, along with a **slot**, its **active** state, and the corresponding **value**. An example is as follows:

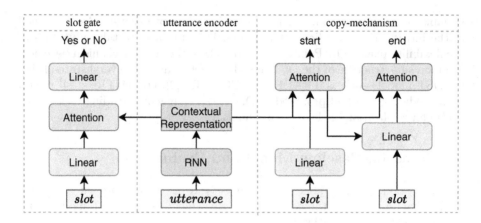

Fig. 1. The architecture of the **baseline** model. The inputs are embedding of slot, word embeddings of user and system utterance$(W_1, W_2, \cdots, W_n)$. It is a common substructure in several SOTA DST models.

> **utterance:**I'm looking for a restaurant that serves thai food.
> **slot:** food, **active:**True, **value:** thai

The active attribute of a slot will be set to `True` (corresponding to a positive sample) if the user triggers the slot, otherwise it will be set to `False` (corresponding to a negative sample). In the test set, each turn of dialogue will be paired with each slot to reach the convincing results.

As shown in Table 1, there are a lot of overlapping values over slots between the training set and test set. To observe the performance on unseen slot values, we construct new develop datasets D'_{woz} and D'_{dstc2}, and new tests set T'_{woz} and T'_{dstc2} by replacing all values with collected real values, and guarantee that these values will not appear in any training set in this paper. As for the Multi-WoZ dataset, it is hard to collect so many values for each slots, so the synthetic develop set D'_{multi} and test set T'_{multi} for Multi-WoZ are constructed using random strings.

3.2 Baseline Model

Without loss of generality, we implement a basic copy-mechanism based model. The model takes an utterance U and a slot s as input, and output whether s is active and the positions of its corresponding value if s is active. The model consists of three important components: **utterance encoder**, **attention-based copy-mechanism**, and **slot gate**. The architecture of the baseline model is shown in Fig. 1.

Utterance Encoder. For utterance U, we simply concatenate user and system utterance (U^{usr} and U^{sys}) by a particular symbol <usr>.

$$U = U^{sys} \oplus <\text{usr}> \oplus U^{usr}, \tag{1}$$

We use a bidirectional LSTM [7] to get the contextual representation H^P of U.

$$H^P = \text{BiLSIM}^P(U_{emb}) \in \mathbb{R}^{n \times d_h}. \tag{2}$$

where n denotes the total number of words in U, $U_{emb} \in \mathbb{R}^{n \times d_{emb}}$ is the word embeddings of U, d_{emb} is the dimension of embeddings, and d_h the dimension of LSTM hidden states.

Attention-Based Copy-Mechanism. We define an attention function $Attn(Q, V)$ to calculate attention scores between query feature $Q \in \mathbb{R}^{d_{emb}}$ and context feature $V \in \mathbb{R}^{n \times d_h}$. The computing process is as follows: 1) do a linear transform for both Q and V, 2) use the dot product result as attention scores, 3) normalize through softmax function.

$$Q' = QW_q, V' = VW_v, \alpha_i = Q'V'_i, \tag{3}$$

$$scores_i = \exp \alpha_i / \sum_j^n \exp \alpha_j, \tag{4}$$

$$contexts = \sum_i^n scores_i V'_i. \tag{5}$$

We use a single linear layer ($\text{Linear}(X) = WX + b$) to encode slots embeddings s_{emb} into s_{enc}, and then use the attention function defined above to calculate both start and end positions of its value.

$$s_{enc} = \text{Linear}_{slot}(s_{emb}), \tag{6}$$

$$p_{enc} = \text{Linear}_p(s_{enc}), \tag{7}$$

$$contexts^p, scores^p = Attn_{span}(p_{enc}, H^P), \tag{8}$$

$$q_{enc} = \text{Linear}_q(s_{enc} \oplus contexts^p), \tag{9}$$

$$contexts^q, scores^q = Attn_{span}(q_{enc}, H^P), \tag{10}$$

$$start = \arg\max_j scores^p_j, \tag{11}$$

$$end = \arg\max_j scores^q_j. \tag{12}$$

Slot Gate. A binary classifier is used to determine whether a slot is triggered by a user or not, where the single linear layer Linear_{cls2} produces a probability over $[0, 1]$ based on attention contexts. A slot is triggered if $cls_{result} > 0.5$.

$$s_{cls} = \text{Linear}_{cls1}(s_{emb}), \tag{13}$$

$$contexts^{cls}, scores^{cls} = Attn_{cls}(s_{cls}, H^P), \tag{14}$$

$$\alpha_{cls} = \text{Linear}_{cls2}(contexts^{cls}), \tag{15}$$

$$cls_{result} = \text{sigmoid}(\alpha_{cls}). \tag{16}$$

Implementation Details. To enhance the effectiveness of the experiments, all experiments in this paper share the same settings. We use randomly initialized word embeddings of dimension 300 with dropout [13] rate 0.5. The model is trained with Adam [8] optimizer and the learning rate is 1e−4. We evaluate the model on the dev set and save the checkpoint and select the best to get the final result on the test set when the training process completes. All experiments are conducted on a single NVIDIA RTX 2080Ti GPU.

3.3   Evaluation Metric

We use F1 scores as the primary metric in all experiments in this paper. Specifically, if a slot is determined to be active by the slot gate and the model predicts the correct value (both the start and end positions) for the slot, then it is a true positive sample. If the slot gate outputs `False`, then the extracted value will be ignored.

4   Data Analysis and Observations

In this Section, we will analyze the factors that influence the generalization performance of the baseline model.

4.1   Original Datasets May Mislead the Model

As shown in TRADE [17], the slots that share similar values or have correlated values have similar embeddings. We suppose that the original datasets incline the attention based copy-mechanism model to memorize values that appear in the train set rather than infer them from contexts, which may lead to a significant performance gap between seen and unseen values. To verify our argument, we train the baseline model and test on original test set and synthetic test set mentioned in Sect. 3.1, which contains all unseen values. The results are shown in Table 2.

Table 2. F1 scores on two datasets. The synthetic test sets of WoZ and DSTC2 contains all unseen values collected from Wikipedia, and that of Multi-WoZ is constructed using random strings.

Datasets	Original test set	Synthetic test set
WoZ-sub	0.9153	0.0820
DSTC2-sub	0.9850	0.0328
Multi-WoZ-sub	0.9297	0.3607

As shown in Table 2, the model performs well on original test sets but poorly on synthetic test sets. The only difference between two kinds of test sets is the values set: slot values in the original test sets and the training sets have a great overlap but the synthetic test sets do not. The low diversity of slot values and strong correlation with the outputs might mislead the model. The model might mistakenly think that the values are the crucial feature and overfits on them, so the model fails when it comes to an unobserved value.

4.2 Greater Diversity of Values Brings Better Generalization

We have argued that the model overfits on the values when there are few values in the training set. In other words, the model pays little attention to context information of slot values, which is a very useful feature for the model to recognize the correct value for a slot. Our goal is to obtain an excellent performance on unseen values. Intuitively, high diversity of slot values make the model more difficult to learn from slot values. Consequently, an interesting question is:

> If we greatly increase the diversity of slot values in the train set, will the model prefer inferring slot values from slot contexts?

Besides the diversity of slot values, we also notice that slot values in the training set obey a power-law distribution, a few values occupy more data samples, which may have a negative effect on the generalization performance of the model.

To clarify the impact of distribution and diversity on generalization performance, we contact experiments as follows. First, duplicate the WoZ training set 10 times to accommodate a large set of values. Second, we adjust the diversity of slot values by controlling the number of unique slot values in the training set. We set the numbers of unique values with $[70, 70 \times \alpha^1, \cdots, 70 \times \alpha^{29} = N]$ (i.e., totally 30 numbers), where 70 and $N = 3000$ is the minimum and maximum number of unique slot values respectively. We sample the values in this way to present a clear trend for increasing number of values. For each number, we replace the values in the training set with collected food names to obtain two constructed training sets, obeying power law distribution and uniform distribution respectively. Third, we train the baseline model using the two constructed training sets and test on the synthetic test set.

As shown in Fig. 2 (left), the F1-scores of the synthetic test set (i.e., all unseen values) increase rapidly with the increasing number of unique slot values and gradually converge. When the number of unique values is relatively small, the performance of the model is slightly worse when the slot values with the power-law distribution than the uniform distribution. But when there are enough slot values in the training set, the difference between the two distributions can be ignored. Hence, we conclude that the generalization capability of the model is positively correlated to the diversity of values, and the uniform distribution is a better choice when constructing a new training set.

4.3 Random Strings Also Work Well

We have known that increasing the diversity of slot values can help the copy-mechanism based DST model to obtain a better generalization capability. But

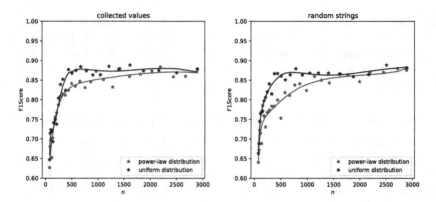

Fig. 2. Performance on T'_{woz} when using the synthetic WoZ training set with different size of value set. In the left part, the datasets are constructed by duplicating the original training set 10 times and replacing slot values with collected food names; in the right part values are replaced with randomly generated strings. The blue lines and points in the figure mean values obey a uniform distribution, and the red lines and points mean values obey a power-law distribution. (Color figure online)

in a real-world task-oriented dialogue system, the ontology often contains many slots that have non-enumerable values. As shown in Fig. 2, the model needs lots of unique values to produce a satisfactory generalization performance, but collecting so many values for each slot cost a lot. Therefore, we wonder if the randomly generated strings can substitute the real slot values.

We design a simple function `randstr(strlen)` to generate random strings. The function takes string length as input and output a randomly generated string, it samples chars from a fix char sequence, we define the char sequence as 26 lowercase letters and 3 spaces[2]. Spaces are used to generate multi-words values, for example, when `strlen` is 10, the probability of generating multi-words values is about $1 - (26/29)^{10} \approx 0.66$.

We use the same experimental setup as that in Sect. 4.2, but use random strings generated by `randstr` function to control the diversity of slot values, the results are shown in Fig. 2 (right). What can be clearly seen in Fig. 2 is that the random strings work well as the collected real slot values. Now we reach a new conclusion: randomly generated strings can be used to enhance the generalization performance of copy-mechanism based DST model.

5   Interactive Data Augmentation for DST

We have shown in Sect. 4 that it is a favorable strategy to employ uniformly distributed random strings to increase the diversity of slot values. In the previous experiments, we duplicate the training set 10 times and set the maximum number

---

[2] in this paper we use 3 spaces and `strlen`=10 to generate more natural multi-words values.

of unique slot values to 3000, but we have no idea about whether we have reached the best performance. So how many data do we need to duplicate and how many random values do we need to generate? To answer these questions, we propose a simple data augmentation algorithm and then evaluate it experimentally.

Algorithm 1. Data Augmentation Algorithm

Input: training set U, synthetic dev set D' and test set T', randomly initialized *Tracker*.
Hyperparameters: sampling bag size s_{bag}, training epochs r at each turn and the patience of early stop mechanism p.
Output: the *Tracker* and its performance on T'
 1: Replace all slot values in U with a string $'[token]'$;
 2: $best\_round=0$, $round=0$, $best\_F1=0$;
 3: **while** $round - best\_round < p$ **do**
 4:     Sample s_{bag} data samples from U;
 5:     Replace all $'[token]'$ with random strings generated by **randstr**, get U_{bag};
 6:     Train the *Tracker* on U_{bag} for r epochs;
 7:     Test the *Tracker* on D', get the F1score $F1$;
 8:     **if** $F1 > best\_F1$ **then**
 9:         $best\_F1 = F1$;
10:         $best\_round = round$;
11:     **end if**
12:     $round = round + 1$;
13: **end while**
14: Test the *Tracker* on T', get the final generalization performance P;
15: **return** *Tracker*, P

5.1   Approach

Traditional data augmentation approaches usually work in the data preprocessing phase, while it is hard in our case to determine how many data we need to duplicate and how many random values we need to generate. Thus, we design a data augmentation algorithm that works in the training phase, as shown in Algorithm 1. Firstly, we replace all slot values in the training set U with a unique token [token]. Secondly, we sample from the training set repeatedly. At each sampling process, we sample s_{bag} data samples and replace all [token] with different random strings to increase the diversity obeying the uniform distribution; re-calculate the start and end positions of random values to update the labels, so we get a smaller training set U_{bag}. We train the model several epochs on U_{bag} and test it on synthetic develop set D'. The entire training process is controlled by an early stop mechanism, after p rounds sampling that fail to produce a better performance, the algorithm will be terminated and report the final performance on synthetic test set T'.

Our algorithm doesn't need to generate all data before training, so it doesn't need to determine how much data to duplicate and how many values to generate. It could reach the satisfactory results at a low cost.

5.2 Performance

We evaluate our data augmentation algorithm on three datasets: WoZ-sub, DSTC2-sub and Multi-WoZ-sub. We focus on slots that have non-enumerable values (see Table 1) and present experimental results in Table 3.

Table 3. F1 scores of our data augmentation (DA) approach on three datasets' test sets and synthetic test sets. * means in which all values in test set are visible to the model. Hyper parameters choice are as follows: bag size $s_{bag} = 1600$ for WoZ-sub, $s_{bag} = 3200$ for DSTC2-sub and $s_{bag} = 8000$ for Multi-WoZ-sub, training epochs at each sampling round $r = 10$, the patience of early stop mechanism $p = 5$.

Dataset	Model	Original test set	Synthetic test set
WoZ-sub	Baseline	0.9241*	0.0820
	+DA	0.9195$_{(\downarrow 0.5\%)}$	0.8819$_{(\uparrow 975.5\%)}$
DSTC2-sub	Baseline	0.9850*	0.0328
	+DA	0.9605$_{(\downarrow 2.5\%)}$	0.9460$_{(\uparrow 2784.1\%)}$
Multi-WoZ-sub	Baseline	0.9297*	0.3607
	+DA	0.9032$_{(\downarrow 2.85\%)}$	0.8847$_{(\uparrow 145.27\%)}$

We can observe from Table 3, with our data augmentation algorithm, the performance of the baseline model on unseen values (synthetic test set) improves significantly, and the performance on seen values (original test set) decrease slightly. We use only one slot in the first two datasets, to further demonstrate the effectiveness of our algorithm on a large-scale data set, we test our algorithm on Multi-WoZ-sub dataset that contains 13 slots, results show that it performs well when there are multiple slots.

5.3 Hyperparametric Analysis

Besides the patience p of the early-stop mechanism, there are two key hyper-parameters in the proposed algorithm, the bag size s_{bag} at each sampling round and the epochs r trained on U_{bag}. We analyzed the impact of these two parameters on the algorithm, and the results are shown in Fig. 3.

From Fig. 3 (left) we can find that given a s_{bag}, a larger r gains a higher performance, but with the s_{bag} increased, the performance gains decreases rapidly. On the other hand, for a certain r, a larger s_{bag} brings a higher performance but it also decreases rapidly with the increasing r. Our goal is to get the best generalization performance with the minimal cost, Fig. 3 (right) presents the training

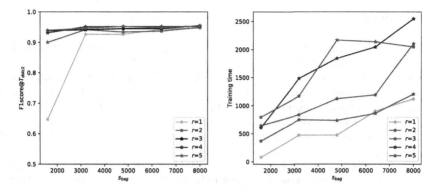

Fig. 3. Performance on T'_{dstc2} of our algorithm. The left part presents the effect of r and s_{bag} on the generalization ability. The right part provides the training time of different r and s_{bag}. We use patience $p = 5$ to produce the results shown in Figure.

time of different r and s_{bag}, Although the time spent on training increases with both r and s_{bag}, we can conclude that a smaller s_{bag} with a larger r is a suitable choice, which can reach the satisfactory results with lower cost.

6   Conclusion and Future Work

This paper focuses on improving the generalization capability of copy-mechanism based models for DST task, especially for the slots that have non-enumerable values. Our conclusions include: Firstly, the copy-mechanism model is easy to over-fit on visible slot values, especially when there are few values, which is the crucial reason for unsatisfactory generalization performance. Secondly, the model's generalization improves dramatically with the increasing diversity of slot values. Thirdly, data augmentation for copy-mechanism models using random strings is feasible and effective in improving the generalization of these models. The interactive data augmentation approach proposed based on these observations shows its effectiveness on three widely used datasets.

In future work, we can study the effect of character-level features of the slot values since the values for the same slot often share somehow similar spelling. In addition, the effect of the diversity of contexts remains unexplored, which may help reduce computational cost further.

Acknowledgements. This research is supported in part by the National Key Research and Development Program of China under Grant 2018YFC0806900 and 2017YFB1010000.

References

1. Budzianowski, P., et al.: Multiwoz-a large-scale multi-domain wizard-of-oz dataset for task-oriented dialogue modelling. arXiv preprint arXiv:1810.00278 (2018)
2. Chen, D., Fisch, A., Weston, J., Bordes, A.: Reading wikipedia to answer open-domain questions. In: Proceedings of the 55th Annual Meeting of the Association for Computational Linguistics (vol. 1: Long Papers), pp. 1870–1879 (2017)
3. Devlin, J., Chang, M.W., Lee, K., Toutanova, K.: Bert: Pre-training of deep bidirectional transformers for language understanding. arXiv preprint arXiv:1810.04805 (2018)
4. Gao, S., Sethi, A., Agarwal, S., Chung, T., Hakkani-Tur, D.: Dialog state tracking: a neural reading comprehension approach. In: Proceedings of the 20th Annual SIGdial Meeting on Discourse and Dialogue, pp. 264–273 (2019)
5. Henderson, M., Thomson, B., Williams, J.D.: The second dialog state tracking challenge. In: Proceedings of the 15th Annual Meeting of the Special Interest Group on Discourse and Dialogue (SIGDIAL), pp. 263–272 (2014)
6. Henderson, M., Thomson, B., Williams, J.D.: The third dialog state tracking challenge. In: 2014 IEEE Spoken Language Technology Workshop (SLT), pp. 324–329. IEEE (2014)
7. Hochreiter, S., Schmidhuber, J.: Long short-term memory. Neural Comput. **9**(8), 1735–1780 (1997)
8. Kingma, D.P., Ba, J.: Adam: A method for stochastic optimization. arXiv preprint arXiv:1412.6980 (2014)
9. Mrkšić, N., Séaghdha, D.O., Wen, T.H., Thomson, B., Young, S.: Neural belief tracker: Data-driven dialogue state tracking. arXiv preprint arXiv:1606.03777 (2016)
10. Ramadan, O., Budzianowski, P., Gašić, M.: Large-scale multi-domain belief tracking with knowledge sharing. arXiv preprint arXiv:1807.06517 (2018)
11. Rastogi, A., Hakkani-Tür, D., Heck, L.: Scalable multi-domain dialogue state tracking. In: 2017 IEEE Automatic Speech Recognition and Understanding Workshop (ASRU), pp. 561–568. IEEE (2017)
12. Ren, L., Ni, J., McAuley, J.: Scalable and accurate dialogue state tracking via hierarchical sequence generation. In: Proceedings of the 2019 Conference on Empirical Methods in Natural Language Processing and the 9th International Joint Conference on Natural Language Processing (EMNLP-IJCNLP), pp. 1876–1885. Association for Computational Linguistics, Hong Kong, China (November 2019). https://doi.org/10.18653/v1/D19-1196, https://www.aclweb.org/anthology/D19-1196
13. Srivastava, N., Hinton, G., Krizhevsky, A., Sutskever, I., Salakhutdinov, R.: Dropout: a simple way to prevent neural networks from overfitting. J. Mach. Learn. Res. **15**, 1929–1958 (2014), http://jmlr.org/papers/v15/srivastava14a.html
14. Vinyals, O., Fortunato, M., Jaitly, N.: Pointer networks. In: Advances in Neural Information Processing Systems, pp. 2692–2700 (2015)
15. Wen, T.H., et al.: A network-based end-to-end trainable task-oriented dialogue system. arXiv preprint arXiv:1604.04562 (2016)
16. Williams, J., Raux, A., Ramachandran, D., Black, A.: The dialog state tracking challenge. In: Proceedings of the SIGDIAL 2013 Conference, pp. 404–413 (2013)
17. Wu, C.S., Madotto, A., Hosseini-Asl, E., Xiong, C., Socher, R., Fung, P.: Transferable multi-domain state generator for task-oriented dialogue systems. In: Proceedings of the 57th Annual Meeting of the Association for Computational Linguistics (vol. 1: Long Papers). Association for Computational Linguistics (2019)

18. Xu, P., Hu, Q.: An end-to-end approach for handling unknown slot values in dialogue state tracking. arXiv preprint arXiv:1805.01555 (2018)
19. Young, S., Gašić, M., Thomson, B., Williams, J.D.: Pomdp-based statistical spoken dialog systems: a review. Proc. IEEE **101**(5), 1160–1179 (2013)
20. Zhong, V., Xiong, C., Socher, R.: Global-locally self-attentive encoder for dialogue state tracking. In: Proceedings of the 56th Annual Meeting of the Association for Computational Linguistics, (vol. 1: Long Papers), pp. 1458–1467 (2018)

Ensemble Labeling Towards Scientific Information Extraction (ELSIE)

Erin Murphy$^{(\boxtimes)}$ ⓘ, Alexander Rasin, Jacob Furst, Daniela Raicu, and Roselyne Tchoua$^{(\boxtimes)}$

DePaul University, Chicago, IL 60604, USA
{emurph35,rtchoua}@depaul.edu

Abstract. Extracting scientific facts from unstructured text is difficult due to challenges specific to the complexity of the scientific named entities and relations to be extracted. This problem is well illustrated through the extraction of polymer names and their properties. Even in the cases where the property is a temperature, identifying the polymer name associated with the temperature may require expertise due to the use of complicated naming conventions and by the fact that new polymer names are being "introduced" into the lexicon as polymer science advances. While domain-specific machine learning toolkits exist that address these challenges, perhaps the greatest challenge is the lack of—time-consuming, error-prone and costly—labeled data to train these machine learning models. This work repurposes Snorkel, a data programming tool, in a novel approach as a way to identify sentences that contain the relation of interest in order to generate training data, and as a first step towards extracting the entities themselves. By achieving 94% recall and an F1 score of 0.92, compared to human experts who achieve 77% recall and an F1 score of 0.87, we show that our system captures sentences missed by both a state-of-the-art domain-aware natural language processing toolkit and human expert labelers. We also demonstrate the importance of identifying the complex sentences prior to extraction by comparing our application to the natural language processing toolkit.

Keywords: Information extraction · Data labeling · Relations extraction · Snorkel · Data programming · Polymers

1 Introduction

Extracting scientific facts from esoteric articles remains an important natural language processing (NLP) research topic due to the particularity of the entities and relations to be extracted. The challenges involved include the fact that entities can be described by multiple referents (synonymy), one word can refer to different concepts depending on context (polysemy), and other domain-specific nuances. These issues arise in many fields as evidenced by NLP tools that rely on domain-specific grammar and ontologies.

© Springer Nature Switzerland AG 2021
M. Paszynski et al. (Eds.): ICCS 2021, LNCS 12742, pp. 750–764, 2021.
https://doi.org/10.1007/978-3-030-77961-0_60

Perhaps the most significant challenge in scientific Information Extraction (IE) is the lack of readily available labeled training data. The process of creating well-balanced, manually-labeled datasets of scientific facts is difficult due in part to the aforementioned challenges, but also due to the scarcity of entities and relations in scientific articles. For instance, it is not uncommon for scientists to write an article about a single newly synthesized material. To annotate sentences in such publications not only requires time and attention, but is also costly as it requires time from domain experts and cannot be easily crowdsourced.

Our ultimate goal is to alleviate the burden of expert annotators and facilitate extraction of scientific facts. Towards achieving this goal, we repurpose a data programming software [14] to identify sentences that contain scientific entities and relations automatically. Typically, data programming relies on existing entity *taggers* in order to identify and label relations. The key novelty of our approach lies in the identification of sentences containing the target entities and relations without identifying the entities through the use of dictionaries nor through complicated hard-coded rules. Instead, we use data programming to describe and combine approximate descriptions of the relations and the entities involved. Not only are we able to identify sentences of interest accurately (94% recall), but our combination of weak, programmed rules retrieves sentences that were missed by human experts and state-of-the-art domain-specific software.

The rest of this paper is organized as follows. In Sect. 2, we briefly discuss the application motivation for this work. Section 3 discusses related work. Section 4 presents the architecture of our system. Section 5 presents the results or our approach, followed by a conclusion in Sect. 6.

2   Motivation

The initial motivation for this work is polymer science. Polymers are large molecules composed of many repeating units, referred to as monomers. Partly due to their large molecular masses, polymers have a variety of useful properties (elasticity, resistance to corrosion and more). Given such properties, polymers are ubiquitous and gathering information about their properties is an essential part of materials design [1]. One specific property with a profound impact on their application, and what this work specifically targets, is the glass transition temperature (T_g): the temperature at which a polymer transitions from a solid, amorphous, glassy state to a rubbery state as the temperature is increased. As the properties between the two states are drastically different, it is crucial to identify polymers with the appropriate T_g for different applications. For example, plexiglass (poly(methyl methacrylate)), a lightweight substitute for glass, has a high T_g of roughly 110 °C, while neoprene (polychloroprene), used for laptop sleeves, has a low T_g of roughly –50 °C [2].

3   Related Work

The medical community has long been invested in applying information extraction methods to medical publications [5,6,9,16]. These tools are designed to extract clinical information from text documents and translate entities and terms to controlled ontologies and vocabularies. Other communities have followed, such as biology, where MedLEE, a tool used to extract clinical information from medical documents [5,6], led to the development of GENIES [7] and BioMedLEE [3] which extract biomolecular substances and phenotypic data from text. However, developing NLP tools for such specialized ontologies can be error-prone, time consuming and hard to maintain, and requires a knowledge of the domain.

Scientific IE models remedy the above challenges by learning from data. Statistical models include Conditional Random Field (CRF), which are graph-based models used in NLP to capture context by learning from sequences of words; long short-term memory (LSTM) networks, which are recurrent neural networks that also capture context by learning relationships between a word and its preceding word; and bidirectional LSTM (Bi-LSTM) networks, which exploit information about the words that come before and after a given word. These models have shown great promise when applied to scientific IE [3,8,10,15,17]. Another example, ChemDataExtractor (CDE)—to which we compare our work and refer to as the state-of-the-art tool—implements an extensible end-to-end text-mining pipeline that can process common publication formats and produces machine-readable structured output data (chemicals and their properties) [17]. While machine learning techniques do not require the implementation of rich domain ontologies and grammars, they do rely heavily on labeled training data to achieve high accuracy, especially when focusing on specialized entities/relations.

While tagging entities and identifying relations between them may be crowd-sourced to the general public for general IE, labeling esoteric scientific articles requires domain knowledge [18–20] and can be costly. Distant supervision [11] circumvents the need for expensive annotation by leveraging available databases or semi-structured text. Deep learning tools like PaleoDeepDive[1] use advanced statistical inference approaches to extract paleontological data from text, tables, and figures in scientific texts by mapping entities and their relations from a large database to text [4,12]. Unfortunately, many fields do not have access to large databases of entities and relations, especially if new data is constantly added.

Snorkel, for example, uses weak programmed rules called labeling functions (LFs) to describe relations between known entities; it can learn and model accuracies and conflicts between LFs to approximately and quickly create labels on unlabeled data [13,14]. However, as mentioned, scientific entities and relations are complex and difficult to extract automatically; while many relations extraction work focuses on relations between two entities, scientific relations may consist of multiple entities and relations or include additional metadata [18,20].

Our work uses Snorkel in a novel manner to address these crucial scientific IE challenges: 1) many NLP tools assume access to costly carefully labeled, balanced datasets, while scientific entities can be scarce in publications; 2) our entities are not always known a priori and are continuously being created or discovered; 3) relations identification is not dependent on first identifying the entities; and 4)

---

[1] PaleoDB at http://paleodb.org.

our relations are complex and may contain entities with multiple relations, hence requiring further expert scrutiny to be extracted.

4   Architecture

Databases that contain information about polymers and their properties are not readily and freely available, thereby creating a need to be able to extract polymers and their properties without relying on external databases to supply known information. A tool is needed to not only extract polymer names from text without knowing them a priori, but can also extract information about the polymer's properties. The particular property this work targets is a polymer's T_g. We have therefore built a tool which aims to identify the three entities of a polymer's T_g, or a polymer-T_g pair: 1) polymers and/or their abbreviations, 2) temperatures and 3) glass transition-mentions.

4.1   Input Dataset

The input dataset was made up of 9,518 unique text sentences (data points) from 31 journal articles containing "Tg" from a keyword search from the journal, *Macromolecules*, a prominent journal in polymer science, during the years 2006–2016 [19]. The full text version of each article was downloaded in HTML format, and split into sentences (Fig. 1) so that each data point was tied to a document (journal article) identifier [19]. The sentences were not preprocessed nor altered in any way prior to this work.

docid	text
acs.macromol.5b01382	A chemically stable and elastomeric triblock copolymer, polystyrene-b-poly(ethylene-co-butylene)-b-polystyrene (SEBS), was functionalized with various benzyl- and alkyl-substituted quaternary ammonium (QA) groups for anion exchange membrane (AEM) fuel cell applications.
acs.macromol.5b01382	Synthetic methods involving transition metal-catalyzed C–H borylation and Suzuki coupling were utilized to incorporate six different QA structures to the polystyrene units of SEBS.
acs.macromol.5b01382	Changes in AEM properties as a result of different QA moieties and chemical stability under alkaline conditions were investigated.
acs.macromol.5b01382	Anion exchange polymers bearing the trimethylammonium pendants, the smallest QA cation moiety, exhibited the most significant changes in water uptake and block copolymer domain spacing to offer the best ion transport properties.

Fig. 1. Example of Input database (Extracted from: Mohanty, Angela D., Chang Y. Ryu, Yu Seung Kim, and Chulsung Bae. "Stable Elastomeric Anion Exchange Membranes Based on Quaternary Ammonium-Tethered Polystyrene-B-Poly (Ethylene-Co-Butylene)-B-Polystyrene Triblock Copolymers." Macromolecules 48, no. 19 (2015): 7085-95.)

4.2   The Snorkel System and Its Built-In Functionalities

Snorkel is a system developed at Stanford University with the objective to "...programmatically [build] and [manage] training datasets without manual labeling" [13]. It applies user-defined programmed rules as weak learners to label data points in a dataset and avoids having to manually assign each data point. The weak learners, or rules programmed in a computer language such as Python, in the Snorkel system are known as labeling functions (LFs). Multiple LFs can be created, and their logic can often be in opposition to each other. After applying LFs to the input data, Snorkel can determine if a data point should be labeled or not. Part of the motivation to use Snorkel was to leverage its speed and ease-of-use of the LFs rather than rely on hard-to-maintain hard-coded rules.

Snorkel Preprocessors and the Uniqueness of Polymer Data. The Snorkel preprocessor [13] allows for each data point to be preprocessed in a user-defined manner. This is important because polymer names do not always follow the same textual rules. For example, it is common for polymer names to be represented throughout polymer texts by abbreviations, consisting largely of uppercase alpha character strings. Applying a preprocessing function to make all text lowercase before applying the LFs would result in missing many abbreviations. On the other hand, there are times when the same sentence containing an abbreviation needs to be made lowercase in order to look for a different entity (i.e. a glass transition-mention). Consider the following sentence:

Bacterial polyhydroxy alkanoates such as poly(3-hydroxybutyrate) (P3HB), poly(3-hydroxyvalerate) (P3HV), or higher hydroxy acids and their copolymers display decreasing melting points from about 180 °C (Tg = 1−4 °C) for P3HB to 112 °C (Tg = −12 °C) for P3HV.[2]

To find a glass transition-mention, a conversion to lowercase and a search for "tg" (a transformation of "TG" or "Tg" or "tg") can be performed. If this conversion were permanent, then polymer abbreviations like P3HB and P3HV in the above sentence would never be identified. Finding different entities may require numerous, impermanent preprocessing applications on the same data point, which are easily accommodated by Snorkel preprocessor functions.

Three preprocessors are built for this work: *makeTextLower()*, *makeCharUniform()* and *removeSpacesInParentheses()*. *makeTextLower()* is self-explanatory: it converts input sentences into lowercase text. *makeCharUniform()* converts special characters, such as dashes and apostrophes (which can appear throughout polymer texts as different characters) to a uniform character. For example, a dash can be represented by the following characters: - − – −. Uniformizing these characters is important, especially if they are used in LF logic. Polymer names can often contain multiple character tokens within parentheses. Consider the polymer name: poly(tetrafluoroethylene). Although this is the common spelling for

---

[2] Extracted from: Petrovic, Zoran S, Jelena Milic, Yijin Xu, and Ivana Cvetkovic. "A Chemical Route to High Molecular Weight Vegetable Oil-Based Polyhydroxyalkanoate." Macromolecules 43, no. 9 (2010): 4120-25.

this polymer, it is possible the polymer could be referred to as: poly(tetrafluoro ethylene). If so, it would be important that a computer program knows that both poly(tetrafluoroethylene) and poly(tetrafluoro ethylene) are the same polymer. Therefore, *removeSpacesInParentheses()* was built to remove spaces only within parentheses.

Labeling Functions (TRUE, JUNK, ABSTAIN). When Snorkel LFs are applied to data points, each LF returns a value of 1, 0 and –1 indicating a TRUE, FALSE (JUNK) or ABSTAIN label, respectively. Attention must be paid to what value is returned since it can greatly impact labeling a data point as TRUE or FALSE. For example, if three LFs are applied to a sentence and their output renders [1, 0, 1], 2 of 3 LFs deemed the sentence to be TRUE (1), if using a majority voting system. If the LF outputs are [1, 0, 0], the sentence would be deemed FALSE (0) since 2 of 3 LFs returned 0. If the LF outputs are [1, 0, –1], this is equivalent to saying that only two LFs voted (1 and 0) and one abstained (–1), resulting in a 50% chance the data point is TRUE or FALSE, and a majority does not exist.

Labeling Functions to Identify Different Entities. For a sentence to be labeled TRUE, it must contain three different entity types: a polymer name or abbreviation, a temperature and a glass transition-mention. Snorkel combines the output of all LFs to label an entire datapoint with a value of 1, 0 or –1. Our work modifies this functionality by first having each LF look for (or lack thereof) one of the three entities within a sentence, where multiple LFs, though expressing different logic, can look for the same type of entity. As a result, each LF is designed to either identify a polymer name or abbreviation entity, a temperature entity or a glass transition-mention entity, thereby grouping LFs by the type of entity for which they are looking. After the LF group determines if the respective entity is present, an ensemble labeler applies a label of 1 (TRUE) or 0 (FALSE) the sentence; if all three entities are present in a sentence, it receives a 1, else it receives a 0. This ensemble labeler will be discussed in Sect. 4.3.

The following code examples illustrate the logic for two different LFs that target temperature entities: *tempUnits()* and *JUNKnoNumbers()*. The rationale for their logic will be discussed in the following sections, but for now we shall illustrate the architecture of LFs.

```
1    @labeling_function()
2    def tempUnits(x):
3        return TG if "°" in x.text else JUNK
```
Listing 1.1. tempUnits() Labeling Function

```
1    @labeling_function(pre=[makeTextLower])
2    def JUNKnoNumbers(x):
3        regexp = re.compile(r"[0 9]")
4        return ABSTAIN if regexp.search(x.text) else JUNK
```
Listing 1.2. JUNKnoNumbers() Labeling Function

In the above examples, *@labeling_function()* signals to Snorkel that a LF is to be defined [13], x refers to the input datapoint which consists of a document

ID (*docid*) and a sentence (*text*) (see Fig. 1), while the variables *TG*, *JUNK* and *ABSTAIN* are assigned values of 1, 0 and –1, respectively. In Listing 1.1, a 1 is returned if a degree sign (°) is found within *text* indicating the LF found a temperature entity within *x.text*, otherwise a 0 is returned indicating that a temperature entity was not found. In Listing 1.2, the regular expressions (re) module [21] allows for a regular expression search to be performed on *x.text* in that if any numeric digits exist, a –1 is returned, otherwise a 0 is returned. It should also be noted that the preprocessor, *makeTextLower()*, is also applied to *x.text* prior to applying *JUNKnoNumbers()*.

The following sections describe the three groupings of LFs, lists the individual LFs for the group, and describes their logic.

Labeling Polymer Entities. Only four LFs are required to identify sentences with polymer entities without a priori knowledge, external reference dictionaries, or writing rules which use extensive REGEX functions.

1. **abbreviation_in_sentence**: This LF looks for a token that consists only of uppercase alpha characters, numbers and special characters. Only 40% or less of the token can consist of special and numeric characters. For example, P3HB is considered an abbreviation, whereas 270 °C is not since 100% of characters in the latter token are numbers and special characters. If the criteria is met, the LF returns 1, otherwise it returns a –1. It would not be appropriate to return a 0 if the logic is not met because polymers do not always have abbreviations, and a sentence should not be penalized for not containing an abbreviation.
2. **keyword_poly**: This LF looks for the character string, "poly" in a sentence. If it exists, a 1 is returned, otherwise a –1 is returned.
3. **keyword_polyParen**: Similar to keyword_poly(), if a sentence contains, "poly(", then a 1 is returned, otherwise a –1 is returned.
4. **keyword_copolymer**: There are naming conventions applied to certain types of polymers known as copolymers. This LF accounts for those rules in that if any of these character strings are found in a sentence, a 1 is returned, else a –1 is returned. Examples of character strings found in copolymers are: "-co-", "-stat-", "-per-", "-ran-", "-grafted-", "-trans-", and "-alt-".

Labeling Temperature Entities. It is simple to identify numbers in a sentence, but it is more difficult to discern what those numbers represent. The below lists the LFs used to identify sentences with and without temperature entities.

5. **tempUnits**: This LF simply looks for a degree (°) symbol. If found, it returns 1, otherwise it returns –1.
6. **tempUnitsAfterNumber**: If a number is followed by a unit of temperature such as C (Celsius), F (Farenheit) or K (Kelvin), then a 1 is returned, otherwise a –1 is returned.
7. **tempUnitsAfterDegree**: If a degree (°) symbol is followed by a C, F, or K, then a 1 is returned, otherwise a –1 is returned.
8. **equalSignBeforeNumber**: If an equal (=) sign exists before numbers (with or without special characters like – or ~), then a 1 is returned, otherwise a –1 is returned.
9. **circaSignBeforeNumberDegree**: If the tokens "circa" or "ca" or "about" precede a number (with or without special characters like – or ~), then a 1 is returned, otherwise a –1 is returned.
10. **tempRange**: Glass transition temperatures can be reported as a temperature range. This LF returns a 1 if more than 40% of a token's characters consists of numbers, such as in the case of "-2 -1" which could read, "negative 2 to negative 1." Otherwise a –1 is returned.
11. **JUNKtempUnitsAfterNumber**: If a number exists and is not followed by a degree (°) symbol, C, F, or K, the number is assumed to not be a temperature and a 0 is returned, otherwise a –1 is returned.
12. **JUNKtempUnitsAfterDegree**: If a degree (°) symbol exists and is not followed by a C, F, or K, it is assumed the sentence does not contain a temperature and a 0 is returned, otherwise a –1 is returned.
13. **JUNKnoNumbers**: If there are no numbers in a sentence, a 0 is returned, otherwise a –1 is returned. A 1 is not returned because that assumes a temperature exists. Since not all numbers represent temperatures, it can only be assumed that a sentence containing numbers is at, best, not a JUNK sentence.

Labeling Glass Transition-Mentions. There are a discrete number of ways that a glass transition mention can be expressed through text, which is either by spelling out "glass transition" (with varying forms of capitalization), shortening it to "glass trans" or "glass-trans," or abbreviating it to simply "tg." Ultimately, this search can be streamlined to searching for: "glass t" or "glass-t" or "tg."

In polymer texts there is a technique called thermogravimetric analysis, which is sometimes abbreviated as, "TGA." Therefore, additional LFs are needed to distinguish sentences that contain "TGA" vs just "TG" to avoid labeling sentences that only refer to TGA as containing a glass transition-mention entity.

14. **keyword_tg**: If the (lowercase) character strings "glass t" or "glass-t" or "tg" are found in a sentence a 1 is returned, otherwise a −1 is returned.
15. **JUNK_tga**: If the character string "TGA" is found in a sentence a 0 is returned, otherwise a −1 is returned.
16. **JUNK_tgAndTGA**: This is considered a "tie-breaker" LF for sentences containing "TGA." If this LF did not exist, sentences with "TGA" would return LF output arrays as [1, 0] and would need to be resolved with a tie-breaker (i.e. randomly assigning the glass transition mention entity as 1 or 0). Therefore, if "TG" is found in a sentence with no other alpha characters following it, a 1 is returned; if the character string "TGA" is found, then a 0 is returned; otherwise a −1 is returned.

4.3   Majority Ensemble Labeler and ELSIE

There are a total of 16 LFs used in this work. The first four LFs, highlighted below in yellow, aim to identify polymer names and abbreviations, the next nine LFs, highlighted in green, aim to identify temperatures, and the last three, in blue, aim to identify glass transition-mentions. Applying LFs is demonstrated in the below sentences and respective output arrays, where characters and/or words are color-coded to indicate the entity identified by a particular LF group, and the LF output values are listed in the corresponding order in the output array as outlined below. The values of the output arrays correspond to the LFs as enumerated in Sect. 4.2—Labeling Polymer Entities, Labeling Temperature Entities and Labeling Glass Transition-Mentions, such that the output array values are designated by the following LFs: [1, 2, 3, 4, 5, 6, 7, 8, 9, 10, 11, 12, 13, 14, 15, 16].

Sentence 1. Bacterial polyhydroxy alkanoates such as poly(3-hydroxybutyrate) (P3HB), poly(3-hydroxyvalerate) (P3HV), or higher hydroxy acids and their copolymers display decreasing melting points from about 180 °C (Tg = 1−4 °C) for P3HB to 112 °C (Tg = -12 °C) for P3HV. See Footnote 2

Output array: [1, 1, 1, −1, 1, 1, 1, 1, −1, 1, −1, −1, −1, 1, −1, 1] → [3/3, 5/5, 2/2]

Of the LFs that did not abstain in Sentence 1 (where the output was either a 1 or 0, but not a −1), all three entities of interest were identified; 3 of 3 LFs found a polymer name or abbreviation (yellow), 5 of 5 LFs found a temperature (green), and 2 of 2 LFs found a glass transition-mention (blue).

Sentence 2. Although the corresponding copolymers were afforded with perfectly alternating nature and excellent regiochemistry control, only glass-transition temperatures of around 8.5 °C were observed in the differential scanning calorimetry (DSC) curve, demonstrating that the polymers are completely amorphous (see Supporting Information Figure S3).[3]

---

[3] Extracted from: Yue, Tian-Jun, Wei-Min Ren, Ye Liu, Zhao-Qian Wan, and Xiao-Bing Lu. "Crystalline Polythiocarbonate from Stereoregular Copolymerization of Carbonyl Sulfide and Epichlorohydrin." Macromolecules 49, no. 8 (2016): 2971-76.

Output array: [1, 1, −1, −1, 1, 1, 1, −1, −1, −1, −1, −1, −1, 1, −1, −1] → [2/2, 3/3, 1/1]

Similar to sentence 1, LFs applied to Sentence 2 also identified all three entities of interest, however, more LFs abstained in this sentence than in Sentence 1.

Sentence 3 (repeated to show how Snorkel can correctly label tricky sentences):
- The TGA scans indicated that APNSi has 5% decomposition in air of 340 °C and in argon of 450 °C (Fig. 1).
- The TGA scans indicated that APNSi has 5% decomposition in air of 340 °C and in argon of 450 °C (Fig. 1).
- The TGA scans indicated that APNSi has 5% decomposition in air of 340 °C and in argon of 450 °C (Fig. 1).[4]

Output array: [1, −1, −1, −1, 1, −1, 1, −1, 1, −1, 0, −1, −1, 1, 0, 0] → [1/1, 3/4, 1/3]

An abbreviation of "TGA" was identified in sentence 3, and polymer LFs cannot determine if the abbreviation represents a polymer or not. The LFs also found the string, "TG" in the sentence. The LFs are able to determine that a glass transition-mention is not present (refer to 4.2 Labeling Glass Transition-Mentions for further clarification), and that only abbreviation and temperature entities were found. As a result, the sentence was labeled as 0. This final example demonstrates the power of LFs and how the combination of weak learners allow the system to carve out entities of interest while ignoring entities not of interest from the sentence by picking up on nuances to discern which sentences to label, even tricky ones.

Once the output arrays of the LFs are generated, the majority ensemble labeler determines which entities exist in a sentence by using a simple majority of LF outputs per entity group; polymer (yellow), temperature (green) and glass transition-mention (blue). The majority ensemble labeler will label a sentence as 1 if and only if all three entities are present in a sentence. This process of considering the output of all LFs per entity group and determining if all entities are present is being called, ensemble labeling toward scientific information extraction, or ELSIE.

5   Results and Analysis

First, we discuss how the initial gold standard dataset—determined by human experts and the state-of-the-art tool—was generated in order to be compared to sentences labeled by ELSIE. Next, we discuss how the initial gold standard dataset was updated after ELSIE identified true positive sentences that were missed by human experts and the state-of-the-art tool. Finally, the state-of-the-art tool's and ELSIE's outputs are both compared to the updated gold standard (hereafter referred to as the "gold standard") dataset. The state-of-the-art tool's performance against the gold standard is discussed as a matter of comparison to ELSIE's performance and labeling abilities.

---

[4] Extracted from: Finkelshtein, E Sh, KL Makovetskii, ML Gringolts, Yu V Rogan, TG Golenko, LE Starannikova, Yu P Yampolskii, VP Shantarovich, and T Suzuki. "Addition-Type Polynorbornenes with Si (Ch3) 3 Side Groups: Synthesis, Gas Permeability, and Free Volume." Macromolecules 39, no. 20 (2006): 7022-29.

5.1   Training Dataset and Its Labels

The intention of the initial gold standard dataset was to label extracted polymer entities and their T_g; this differs from the current intention of ELSIE which aims to label sentences containing the three entities of interest. The motivation behind our approach of labeling sentences before extracting entities is that scientific entities and relations can be too complex to be immediately extracted and may require additional human attention or additional passthroughs of the data. To align the initial gold standard dataset with ELSIE-labeled data, metadata about sentences and polymer-T_g pairs extracted by experts and the state-of-the-art tool was used to determine the sentences from which the entities were extracted. Data extracted by the state-of-the-art tool was previously validated by experts [19]. If the state-of-the-art tool extracted a polymer-T_g pair correctly, the sentence(s) from which the information was obtained by the state-of-the-art tool were labeled as 1; if the state-of-the-art tool extracted an incorrect polymer-T_g pair (i.e. a polymer was paired with an incorrect T_g), sentences containing the correct polymer name/abbreviation and the T_g were both labeled as 0 [19]. Sentences identified by the human experts which contained polymer-T_g pairs were labeled as 1. If a polymer-T_g pair existed in the corpora, and the human experts and/or the state-of-the-art tool did not extract the pair, the sentence was labeled as 0.

5.2   Updated Gold Standard Labels

After running ELSIE on unlabeled data, new polymer-T_g pairs that were not in the initial gold standard dataset (i.e. missed by human experts and/or the state-of-the-art tool) were discovered. We considered these to be false "false positives" from the initial gold standard dataset. More details of these sentences are provided in Sect. 5, but as a result of these findings, the initial gold standard dataset was updated, and the sentences with previously missed polymer-T_g were labeled as 1. It is this updated dataset—the gold standard—to which the state-of-the-art tool and ELSIE are compared.

5.3   Results

The final document corpora contained 9,518 sentences (data points), representing 31 unique scientific journal articles. Overall, the state-of-the-art tool labeled 15 sentences as positive cases, ELSIE labeled 67 sentences, and human experts identified 49; the gold standard dataset contained 64 positive cases. Positive cases represent less than 1% of the data, illustrating the highly unbalanced nature of the dataset, and accuracy alone does not convey each application's performance. Precision and recall results, along with accuracy and F1-scores, are reported in Table 1.

Table 1. Performance compared to the gold standard.

	Gold Standard	Human Experts	State-of-the-Art Tool	ELSIE
Total Cases		9,518		
Total Positive Cases	64	49	15	67
Accuracy		99.84%	99.49%	99.88%
Precision		100%	100%	90%
Recall		77%	23%	94%
F1 score		0.87	0.38	0.92

The analyses were run on a personal laptop using Python 3.8 in Jupyter Notebook. The total processing time to process all 9,518 sentences through ELSIE, including Snorkel preprocessors, was 0:01:03, compared to the state-of-the-art tool's processing time which took approximately 0:26:00 to process 31 documents.

5.4 Analysis

The F1 score of the state-of-the-art's performance compared to the gold standard (0.38) versus the F1 score of ELSIE's performance to the gold standard (0.92) overall demonstrates that ELSIE is better at labeling sentences correctly.

It is more important to capture all true labels than it is to miss true labels, and is therefore acceptable for precision to be compromised in order to obtain high recall. ELSIE identified new polymer-T_g pairs that human experts and the state-of-the-art tool missed (see Table 2). With recall for the human experts (77%) being lower than ELSIE (94%), and the need to update the gold standard dataset to include new polymer-T_g pairs that were previously missed, this demonstrates that a high level of attention is required by humans (even experts) when reading texts, otherwise important information can get missed. This finding also highlights ELSIE's robustness and reliability in labeling scientific [polymer]

Table 2. Sentences missed by human experts, labeled by ELSIE.

Text	Gold Standard	Human Experts	State-of-the-Art Tool	ELSIE
Upon 10 wt % clay loading, the glass transition of the PTMO:MDI – BDO PU nanocomposites shifts slightly from– 44.7 to –46.6 °C. [a]	1	0	0	1
The functionalized polycarbonate exhibited a lower T_g of 89 °Ccompared to its parent(108 °C). [b]	1	0	0	1

[a] Extracted from: James Korley, LaShanda T, Shawna M Liff, Nitin Kumar, GarethH McKinley, and Paula T Hammond. "Preferential Association of Segment Blocks in Polyurethane Nanocomposites." Macromolecules 39, no. 20 (2006): 7030-36.

[b] Extracted from: Darensbourg, Donald J, Wan-Chun Chung, Andrew D Yeung, and Mireya Luna. "Dramatic Behavioral Differences of the Copolymerization Reactions of 1, 4-Cyclohexadiene and 1, 3-Cyclohexadiene Oxides with Carbon Dioxide." Macromolecules 48, no. 6 (2015): 1679-87

sentences for training data over human experts and state-of-the-art tools aiming to perform the same function.

The state-of-the-art tool's recall (23%) is much lower than ELSIE's recall (94%) because the state-of-the-art tool missed labeling more positive cases (n = 49) than ELSIE (n = 4). Given the state-of-the-art tool's objective to extract entities and not label sentences, when the state-of-the-art tool extracted an incorrect polymer-T_g pair, it was penalized and the sentences were not labeled. The state-of-the-art tool would have achieved higher recall (88%) had we focused only on rules-based extraction of the T_g, as opposed to the polymer-T_g pair. However, due to the nuances in complex sentences and complicated polymer naming, it often linked the T_g to an incorrect polymer name [19].

Precision for the state-of-the-art tool was higher than ELSIE's because ELSIE labeled false positive sentences. ELSIE looks for entities within a sentence (even if the entities are not related to one another), whereas the state-of-the-art tool looks for related entities. The number of sentences labeled by the state-of-the-art tool was much smaller (n = 15) than ELSIE (n = 67). The state-of-the-art tool did not label any false positives, whereas ELSIE labeled 7 of the 67 sentences as false positives. An example of a false positive sentence labeled by ELSIE is shown in Table 3; polymer name and glass transition-mention entities were identified, but the temperature entity in the sentence is a melting temperature and not a T_g. Though it is a false positive, reporting this sentence can be beneficial because it could contain important metadata either about the entities of interest, or other polymer characteristics.

ELSIE missed labeling sentences that contained entities of a polymer-T_g pair if all three entities were not contained within a single sentence. This demonstrates how and why the problem of finding polymers and their respective T_g is hard for computers and easier for humans. Table 4 shows two consecutive sentences in a text. The first sentence only contains a polymer entity (which ELSIE identified), but did not contain temperature nor glass transition-mention entities; the human identified this sentence and received credit. The other two entities are found in the next sentence, to which the human experts received credit. The state-of-the-art tool extracted the T_g mention from the second sentence, but paired it

Table 3. False positive sentence (Labeled as TRUE by ELSIE).

Text [a]	Gold Standard	Human Experts	State-of-the-Art Tool	ELSIE
Two or three thermal transitions are expected for SEBS: (1) a low glass transition temperature (Tg1) corresponding to the ethylene-co-butylene block,(2) a high glass transition temperature (Tg 2) corresponding to the styrene block, and (3) a broad endothermic transition at the melting temperature (Tm) near 20 °C depending on the degree of crystallinity of the ethylene-co-butylene block.	0	0	0	1

[a] Extracted from: Mohanty, Angela D., Chang Y. Ryu, Yu Seung Kim, and Chulsung Bae. "Stable Elastomeric Anion Exchange Membranes Based on Quaternary Ammonium-Tethered Polystyrene-B-Poly (Ethylene-Co-Butylene)-B-Polystyrene Triblock Copolymers." Macromolecules 48, no. 19 (2015): 7085-95.

Table 4. True positive sentences missed by LFs.

Text [a]	Gold Standard	Human Experts	State-of-the-Art Tool	ELSIE
The azo- polymer material, poly[4'-[[2-(acryloyloxy)ethyl]ethylamino]-4-nitroazobenzene], often referred to as poly(disperse red 1 acrylate) (here after pdr1a), was synthesized as previously reported.	1	1	0	0
The prepared material was determined to have a molecular weight of 3700g/mol, and a corresponding T_g in the range 95-97 °C.	1	1	0	0

[a] Extracted from: Yager, Kevin G, and Christopher J Barrett. "Photomechanical Surface Patterning in Azo-Polymer Materials." Macromolecules 39, no. 26 (2006): 9320-26.

to the wrong polymer, and did not receive credit for either sentence. Since all three entities were spread among multiple sentences, ELSIE was not able to label either sentence with a 1.

6 Conclusion

This work presented ELSIE, a system that leverages data programming to process scientific articles—specifically in materials science—and identify sentences containing target entities such as polymers, temperatures and glass transition-mentions. We demonstrated that a collection of simple and easy to understand programmed rules are able to detect entity-containing sentences without having to identify the target entities themselves. ELSIE does not use distant supervision—nor a priori known entities—and it does not look for relationship-type words. Instead it determines whether the entities that form the target relationship are present in a sentence. We achieved a recall of 94% when compared to the gold standard, mostly due to an assumption that the three entities are related if they existed in a single sentence. Future work will aim to 1) identify and isolate sentences of interest with their surrounding sentences (e.g., sentences containing 2 out of 3 target entities), and 2) extract polymer entities and their properties.

ELSIE found sentences missed by a best of breed domain-specific toolkit and human experts, whether due to sentences being complicated, such as a sentence containing multiple polymer-T_g pairs, or fatigue/lack of attention paid by human experts. Since ELSIE outperformed a domain-specific toolkit as well as human annotators, this work demonstrates the need for software that can reliably and quickly process polymer texts.

References

1. Audus, D.J., de Pablo, J.J.: Polymer informatics: opportunities and challenges (2017)

2. Brandrup, J., Immergut, E.H., Grulke, E.A., Abe, A., Bloch, D.R.: Polymer Handbook, vol. 89. Wiley, New York (1999)

3. Chen, L., Friedman, C.: Extracting phenotypic information from the literature via natural language processing. In: Medinfo, pp. 758–762. Citeseer (2004)

4. De Sa, C., et al.: Deepdive: declarative knowledge base construction. ACM SIGMOD Rec. **45**(1), 60–67 (2016)

5. Friedman, C., Alderson, P.O., Austin, J.H., Cimino, J.J., Johnson, S.B.: A general natural-language text processor for clinical radiology. J. Am. Med. Inform. Assoc. **1**(2), 161–174 (1994)

6. Friedman, C., Hripcsak, G., Shagina, L., Liu, H.: Representing information in patient reports using natural language processing and the extensible markup language. J. Am. Med. Inform. Assoc. **6**(1), 76–87 (1999)

7. Friedman, C., Kra, P., Yu, H., Krauthammer, M., Rzhetsky, A.: Genies: a natural-language processing system for the extraction of molecular pathways from journal articles. In: ISMB (supplement of bioinformatics), pp. 74–82 (2001)

8. Hong, Z., Tchoua, R., Chard, K., Foster, I.: SciNER: extracting named entities from scientific literature. In: Krzhizhanovskaya, V.V., et al. (eds.) ICCS 2020. LNCS, vol. 12138, pp. 308–321. Springer, Cham (2020). https://doi.org/10.1007/978-3-030-50417-5_23

9. Jagannathan, V., Elmaghraby, A.: Medkat: multiple expert delphi-based knowledge acquisition tool. In: Proceedings of the ACM NE Regional Conference, pp. 103–110 (1985)

10. Jessop, D.M., Adams, S.E., Willighagen, E.L., Hawizy, L., Murray-Rust, P.: Oscar4: a flexible architecture for chemical text-mining. J. Chem. **3**(1), 1–12 (2011)

11. Mintz, M., Bills, S., Snow, R., Jurafsky, D.: Distant supervision for relation extraction without labeled data. In: Proceedings of the Joint Conference of the 47th Annual Meeting of the ACL and the 4th International Joint Conference on Natural Language Processing of the AFNLP, pp. 1003–1011 (2009)

12. Peters, S.E., Zhang, C., Livny, M., Ré, C.: A machine reading system for assembling synthetic paleontological databases. PLoS One **9**(12), e113523 (2014)

13. Ratner, A., Bach, S.H., Ehrenberg, H., Fries, J., Wu, S., Ré, C.: Snorkel: rapid training data creation with weak supervision. In: Proceedings of the VLDB Endowment. International Conference on Very Large Data Bases, vol. 11, p. 269. NIH Public Access (2017)

14. Ratner, A.J., De Sa, C.M., Wu, S., Selsam, D., Ré, C.: Data programming: Creating large training sets, quickly. In: Advances in Neural Information Processing Systems, pp. 3567–3575 (2016)

15. Rocktäschel, T., Weidlich, M., Leser, U.: Chemspot: a hybrid system for chemical named entity recognition. Bioinformatics **28**(12), 1633–1640 (2012)

16. Savova, G.K., et al.: Mayo clinical text analysis and knowledge extraction system (ctakes): architecture, component evaluation and applications. J. Am. Med. Inf. Assoc. **17**(5), 507–513 (2010)

17. Swain, M.C., Cole, J.M.: Chemdataextractor: a toolkit for automated extraction of chemical information from the scientific literature. J. Chem. Inf. Model. **56**(10), 1894–1904 (2016)

18. Tchoua, R.B., Chard, K., Audus, D., Qin, J., de Pablo, J., Foster, I.: A hybrid human-computer approach to the extraction of scientific facts from the literature. Procedia Comput. Sci. **80**, 386–397 (2016)

19. Tchoua, R.B., et al.: Towards a hybrid human-computer scientific information extraction pipeline. In: 2017 IEEE 13th International Conference on e-Science (e-Science), pp. 109–118. IEEE (2017)
20. Tchoua, R.B., Qin, J., Audus, D.J., Chard, K., Foster, I.T., de Pablo, J.: Blending education and polymer science: semiautomated creation of a thermodynamic property database. J. Chem. Educ. **93**(9), 1561–1568 (2016)
21. Van Rossum, G.: The Python Library Reference, release 3.8.6. Python Software Foundation (2020)

Error Estimation and Correction Using the Forward CENA Method

Paul D. Hovland[ORCID] and Jan Hückelheim[(✉)][ORCID]

Mathematics and Computer Science Division, Argonne National Laboratory, Lemont, IL, USA
hovland@mcs.anl.gov, jhueckelheim@anl.gov

Abstract. The increasing use of heterogeneous and more energy-efficient computing systems has led to a renewed demand for reduced- or mixed-precision floating-point arithmetic. In light of this, we present the *forward CENA method* as an efficient roundoff error estimator and corrector. Unlike the previously published CENA method, our forward variant can be easily used in parallel high-performance computing applications. Just like the original variant, its error estimation capabilities can point out code regions where reduced or mixed precision still achieves sufficient accuracy, while the error correction capabilities can increase precision over what is natively supported on a given hardware platform, whenever higher accuracy is needed. CENA methods can also be used to increase the reproducibility of parallel sum reductions.

Keywords: CENA method · Roundoff error · Mixed-precision arithmetic · Reproducibility

1 Introduction

Roundoff error is inevitable in floating-point arithmetic; but rigorous error analysis is difficult even for numerical analysis experts, and such experts are in short supply. This situation leads to two main strategies: perform computations in the highest-available precision, possibly sacrificing time and energy savings available at lower precision, or perform computations in low precision and hope for the best. A third strategy is to employ roundoff error estimation in order to characterize and possibly correct roundoff errors. The correction des erreurs numériques

The submitted manuscript has been created by UChicago Argonne, LLC, Operator of Argonne National Laboratory ('Argonne'). Argonne, a U.S. Department of Energy Office of Science laboratory, is operated under Contract No. DE-AC02-06CH11357. The U.S. Government retains for itself, and others acting on its behalf, a paid-up nonexclusive, irrevocable worldwide license in said article to reproduce, prepare derivative works, distribute copies to the public, and perform publicly and display publicly, by or on behalf of the Government. The Department of Energy will provide public access to these results of federally sponsored research in accordance with the DOE Public Access Plan. http://energy.gov/downloads/doe-public-access-plan.

© Springer Nature Switzerland AG 2021
M. Paszynski et al. (Eds.): ICCS 2021, LNCS 12742, pp. 765–778, 2021.
https://doi.org/10.1007/978-3-030-77961-0_61

d'arrondi method of Langlois [16] (hereafter, original CENA or reverse CENA) is one method for roundoff error estimation and correction but suffers from memory requirements proportional to the number of floating-point operations and an operations count that grows linearly with the number of output variables. We introduce a forward variant of the CENA method (hereafter, forward CENA or CENA) that suffers neither of these deficiencies.

CENA computes local roundoff errors from individual operations and uses automatic, or algorithmic, differentiation (AD) to estimate their cumulative effect on the final output. This estimate is often precise enough to be used as an effective error correction, by subtracting the estimated error from the computed result. The corrected results are not only more accurate than results obtained without the CENA method, but also more reproducible, since they are less affected by the non-associativity of floating-point operators. CENA can easily be implemented and deployed in an existing program by using operator overloading, for example as offered by C++. We believe that CENA can be most useful during the development of numerical software. For example, the error estimates can be used to choose error bounds for regression tests, or the error corrections can be used during the regression tests themselves to remove the nondeterministic effects of parallel sum reductions. Furthermore, CENA could be used to increase the precision of results compared with the best-available precision that is natively supported on a given platform.

The next section summarizes previous work on roundoff error estimation and correction, followed by a brief introduction to AD in Sect. 3. Then, in Sect. 4 we provide a description of the forward CENA and its relationship to reverse CENA, and in Sect. 5 we discuss implementation details. In Sect. 6 we present experimental results, and we conclude in Sect. 7 with a brief summary and discussion of future work.

2   Related Work

Many techniques for estimating or reducing the effects of roundoff error have been developed [4,6,11–13,16–19,21–24]. The forward CENA method builds on the reverse CENA method of Langlois [16]. In contrast to this and related techniques, forward CENA has an operations overhead independent of the number of output variables and a memory overhead independent of the number of operations; it is also much easier to parallelize. Like reverse CENA and in contrast to many other techniques, forward CENA computes deterministic local error estimates and combines them with derivatives to compute a global error correction. Forward CENA requires no source code analysis or transformation and can therefore be implemented as a drop-in replacement numeric type requiring no external tool support. We note that forward CENA can be seen as a way to generalize certain algorithms for accurate summation [1,7,14,20] to other types of computation, and in Sect. 6.1 we compare forward CENA with Kahan's compensated summation algorithm.

3   Brief Introduction to AD

Automatic, or algorithmic, differentiation (AD) is a technique for computing the derivatives of functions defined by algorithms [8]. It computes partial derivatives for each elementary operator and combines them according to the chain rule of differential calculus, based on the control flow of the program used to compute the function. In the so-called forward mode of AD, the derivatives are combined in an order that follows the control flow of the function. In the so-called reverse mode, the derivatives are combined in an order that reverses the control flow of the function. The forward mode computes a Jacobian-matrix product, $JS = \frac{\partial y}{\partial x} S$, at a cost proportional to the number of columns in the so-called seed matrix, S, while the reverse mode computes a matrix-Jacobian product, WJ, at a cost proportional to the number of rows in W. The forward mode is therefore efficient for computing Jacobian-vector products, Jv, and the reverse mode is efficient for computing transposed-Jacobian-vector products, $J^T v$ and the gradients of scalar functions.

4   Forward CENA Method

The forward and reverse CENA methods approximate the error Δ_y in a result y using the formula

$$\Delta_y \approx E_y = \sum_i \frac{\partial y}{\partial x_i} \delta_i, \tag{1}$$

where x_i is the result of each instruction i used in computing y and δ_i is the local round-off error in computing x_i. In the reverse CENA method [16], one computes the derivatives using reverse mode AD. The number of operations is proportional to the number of operations in the function evaluation. Unfortunately, employing the reverse mode also incurs a storage cost proportional to the number of operations in the function evaluation.

Instead of reverse mode AD, we can employ forward mode AD, using a seed matrix (vector) $\delta = [\delta_1 \delta_2 \dots \delta_n]^T$ to directly compute the inner product $\frac{\partial y}{\partial x}^T \delta = \sum_i^n \frac{\partial y}{\partial x_i} \delta_i$. This is most easily comprehended by using the *buddy variable* approach [3]:

```
xibuddy = 0.0
xi = fi(xj,xk) + xibuddy
```

which, after differentiating and initializing the seed matrix for `xibuddy`, yields

```
xibuddy = 0.0
ad_xibuddy = deltai
xi = fi(xj,xk) + xibuddy
ad_xi = (dfidxj*ad_xj + dfidxk*ad_xk) + ad_xibuddy .
```

We note that `deltai` (the roundoff error in computing `xi`) may not in general be available until after the computation of `xi`; however, one can easily simplify the derivative computation to

```
xi = fi(xj,xk)
ad_xi = (dfidxj*ad_xj + dfidxk*ad_xk) + deltai.
```

Theorem. If for each statement $x_i = \phi_i(x_1, x_2, \ldots, x_{i-1})$ we compute

$$E_i = \delta_i + \sum_{j=1}^{i-1} \frac{\partial \phi_i}{\partial x_j} E_j,$$

then E_i satisfies Eq. 1. That is,

$$E_i = \sum_{j=1}^{i} \frac{\partial y}{\partial x_j} \delta_j.$$

Proof. By induction. $E_1 = \delta_1$. Assume that

$$E_i = \delta_i + \sum_{j=1}^{i-1} \frac{\partial \phi_i}{\partial x_j} E_j = \sum_{j=1}^{i} \frac{\partial x_i}{\partial x_j} \delta_j$$

for all $i \leq n$. Then,

$$E_{n+1} = \delta_{n+1} + \sum_{j=1}^{n} \frac{\partial \phi_{n+1}}{\partial x_j} E_j = \delta_{n+1} + \sum_{j=1}^{n} \frac{\partial \phi_{n+1}}{\partial x_j} \sum_{k=1}^{j} \frac{\partial x_j}{\partial x_k} \delta_k$$

$$= \delta_{n+1} + \sum_{j=1}^{n} \frac{\partial \phi_{n+1}}{\partial x_j} \sum_{k=1}^{n} \frac{\partial x_j}{\partial x_k} \delta_k = \delta_{n+1} + \sum_{j=1}^{n} \sum_{k=1}^{n} \frac{\partial \phi_{n+1}}{\partial x_j} \frac{\partial x_j}{\partial x_k} \delta_k$$

$$= \delta_{n+1} + \sum_{k=1}^{n} \sum_{j=1}^{n} \frac{\partial \phi_{n+1}}{\partial x_j} \frac{\partial x_j}{\partial x_k} \delta_k = \delta_{n+1} + \sum_{k=1}^{n} \sum_{j=k}^{n} \frac{\partial \phi_{n+1}}{\partial x_j} \frac{\partial x_j}{\partial x_k} \delta_k$$

$$= \delta_{n+1} + \sum_{k=1}^{n} \frac{\partial \phi_{n+1}}{\partial x_k} \delta_k = \frac{\partial \phi_{n+1}}{\partial \phi_{n+1}} \delta_{n+1} + \sum_{k=1}^{n} \frac{\partial \phi_{n+1}}{\partial x_k} \delta_k = \sum_{k=1}^{n+1} \frac{\partial \phi_{n+1}}{\partial x_k} \delta_k.$$

We note that the proof ignores roundoff errors in the computation of E_i and therefore holds only if the E_i and partial derivatives are computed in real arithmetic. This is sufficient to fulfill our goal of demonstrating equivalence between forward and reverse CENA, which both ignore roundoff errors in the computation of the derivatives.

5   Implementation

We created a C++ type that overloads the standard operators to compute error estimates and corrections using the forward CENA method. They can be used just like any other number type as long as no unsupported operators are used.

Currently, our library supports the usual operators, such as +, -, *, and / (in addition to their compound operators, such as +=); comparison operators, including <, <=, and ==; assignment operators, cast to and from native types, and so on. In addition, we support the sqrt, sin, and cos functions and the << streaming operator to output a textual representation of the number, error estimate, and error correction. One example operator is shown in Fig. 1. Furthermore, we used the OpenMP declare reduction pragma to allow parallel reductions over CENA types.

```
template<typename T>
class freal{
  private:
    T val, err;
    static T addition_error(T a, T b, T x) {
        T corr = x - a;
        return ((x-corr)-a)+(corr-b);
    }

  public:
    freal<T>(T value, T error) : val(value), err(error) { }

    void operator+=(const freal<T> rhs) {
      T value = this->val + rhs.val;
      T localerror = addition_error(this->val,rhs.val,value);
      this->val = value;
      this->err += rhs.err + localerror;
    }
};
```

Fig. 1. Part of the CENA class, showing only the compound addition operator and the internal helper function to compute the local error produced by that operation. The actual implementation used in this work supports many more operators.

In our experiments we use the GNU MPFR library [5] to test the CENA method at arbitrary floating-point precision, in addition to the natively supported single, double, extended double precision, and quad precision as supported by the GNU libquadmath library. All MPFR operations are guaranteed to use the exact precision that was specified, which allows us not only to perform tests at high precision and obtain accurate reference results but also to simulate half, quarter, or more esoteric low-precision number types.

6   Test Cases and Experimental Results

In this section we show the effectiveness of the CENA method on three test cases. The first, shown in Sect. 6.1, uses CENA to obtain reproducible results in parallel sum reductions. Then, in Sect. 6.2 we use CENA to obtain error estimates within an OpenMP-parallel benchmark derived from a cosmology code. Next, in Sect. 6.3 we use CENA to reduce roundoff errors in various implementations of the matrix-matrix-product. Finally, in Sect. 6.4 we apply CENA to a pathological example, the Muller recurrence. All test cases were compiled by GCC 9.2 with flags -O3 -std=c++11 -fopenmp and executed on a 28-core/56-thread Intel® Xeon® Platinum 8180 Processor ("Skylake").

6.1 Sum Reduction

Sum reductions are ubiquitous in numerical programs, for example during the computation of a dot-product, matrix-vector or matrix-matrix product, or numerical quadrature. In this test case we look at reductions in isolation, but two of the subsequent larger test cases also contain sum reductions.

One often wishes to compute large sum reductions in parallel, for example when forming the dot product of two large vectors. This is typically done by accumulating partial sums on each thread in parallel, followed by some strategy to combine these into one overall result. A common problem with parallel sum reductions is the lack of reproducibility because of the non-associativity of the + operator for floating-point numbers. This can result in the same correctly implemented and data-race-free program producing different results every time it is executed, because of the nondeterministic scheduling of the summation. This causes problems, for example, in regression testing, where distinguishing floating-point roundoff from other small errors or race conditions can be difficult.

In this test case we demonstrate how CENA can help reduce nondeterminism caused by a change in summation order. To this end, we initialize an array of 1 million pseudorandom numbers. The same hard-coded seed is always used for the pseudorandom number generator, to ensure that the set of generated numbers remains the same between runs. However, the vector of numbers is then shuffled randomly, using a different seed and thus ensuring a different summation order each time. Any resulting changes are therefore due to roundoff.

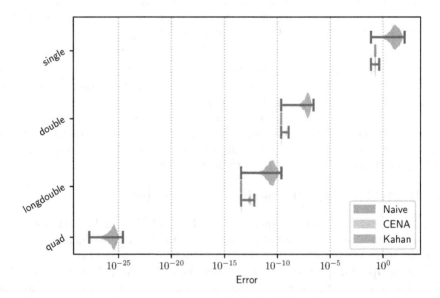

Fig. 2. The result of the sum reduction fluctuates because of the non-associativity of floating-point summations. The CENA-corrected results (and to a slightly lesser extent the Kahan results) are consistent across runs, and more accurate than the uncorrected results. Increasing the working precision has a larger benefit than using Kahan or CENA for accuracy, but not for reproducibility.

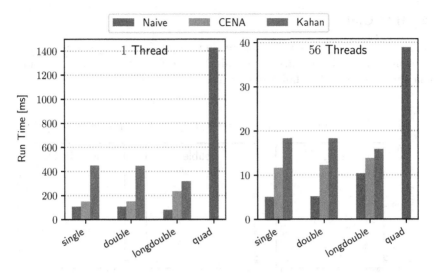

Fig. 3. Run time of conventional, CENA, and Kahan summation on 1 (left) or 56 (right) threads. CENA types are slightly slower than built-in number types but do not affect parallel scaling significantly and still work well on higher thread counts.

We perform this test 1,000 times in IEEE754 single, double, and quadruple precision, as well as 80-bit extended precision, using "naive" summation (adding numbers one by one to an accumulator), Kahan summation, and summation using CENA types. We note that because all of the derivatives in summation are equal to 1, the CENA method (forward or reverse) reduces to Algorithm 4.1 in [20], a form of compensated summation based on Knuth's two-sum algorithm, but without needing to modify the implementation of summation, beyond using our CENA type. The input numbers are generated with a quad-precision mantissa consisting of a uniformly sampled random bit pattern, a random sign bit, and an exponent from a uniform distribution in $[2^{-16}...2^{16}]$ ($\approx [10^{-5}...10^{5}]$). The inputs for the lower-precision tests are obtained by type casting. Reference results are computed by using the aforementioned MPFR library with a mantissa length of 200 bits, well above the 113-bit mantissa of quad precision and almost four times that of double-precision numbers.

Figure 2 shows the errors for each of these settings. CENA and Kahan summation have comparable effects on the mean errors, although CENA is often slightly superior and reduces the variability of errors by many orders of magnitude. In Fig. 3 we show the run times of our tested summation approaches, for sequential or parallel summation on 56 threads. Using CENA instead of built-in number types increases the time by a small factor, typically below 2. Kahan summation is slower, but this is probably a deficiency in our implementation, since Kahan requires fewer operations than CENA. Because of the lack of hardware support, the quad precision summation is very slow.

6.2   HACCmk

The Hardware Accelerated Cosmology Code (HACC) [9] helps in understanding the evolution of the Universe, by simulating the formation of structure and the behavior of collision-less fluids under gravity.

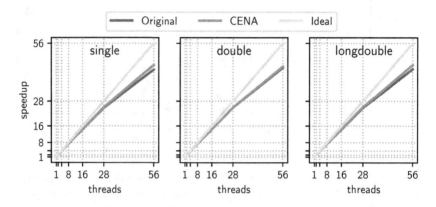

Fig. 4. HACCmk scalability is excellent with and without CENA. The absolute run time between CENA and normal execution differs by a factor of ca. 10 (see Fig. 5).

HACCmk is a compute-intensive kernel routine extracted from HACC that calculates force values for each particle in an OpenMP `parallel for` loop. Our forward CENA method can be used simply by changing the number type through a `typedef`. The only other modification is to replace the pow(·) function by 1/·*sqrt(·), since pow is currently not supported by our implementation.

The HACCmk code scales well on our system with and without CENA, as shown in Fig. 4. Absolute run times in Fig. 5 show that CENA increases run time by a factor of ca. 10× (slightly less for single/double and slightly more for long double precision). CENA estimates the actual error well enough to be able to improve the result by an order of magnitude. Correction is again slightly more effective for even-length mantissas, see Fig. 6.

6.3   Classic and Strassen Matrix Multiplication

In this section we investigate CENA in the context of matrix-matrix multiplications using either Strassen's algorithm or classic multiplication using a triple-nested loop. We briefly summarize results from previous literature showing that Strassen's algorithm produces higher roundoff errors but is faster than classic multiplication for large matrices. We then present numerical experiments.

Strassen's algorithm was the first published way of computing the product

$$C = AB \qquad A, B, C \in \mathbb{R}^{n \times n}$$

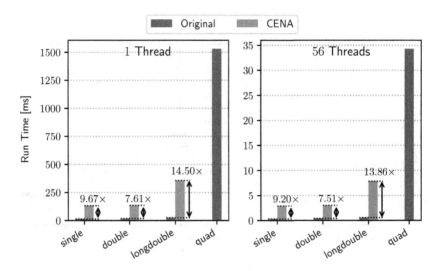

Fig. 5. Absolute run times for HACCmk with and without using CENA. The CENA type slows down execution by one order of magnitude.

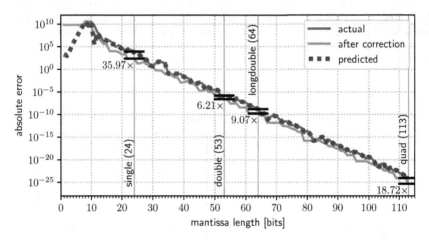

Fig. 6. Errors for HACCmk with and without CENA correction, and CENA error estimate, for various mantissa lengths, compared to a 200-bit mantissa reference. CENA correctly computes the exponent and some mantissa bits of the actual errors.

with a time complexity of less than $\mathcal{O}(n^3)$. Strassen's algorithm and other subsequently discovered subcubic algorithms have been studied extensively, regarding both their run time and their numerical stability. Previous studies have found that Strassen's and related algorithms are generally stable, although their error bounds are slightly worse than those of the classic matrix multiplication [2,10].

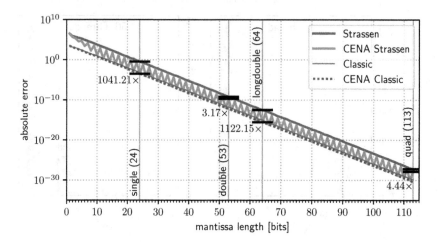

Fig. 7. Absolute error of classic and strassen multiplication for the entire range of mantissa lengths from 1 bit up to 113 bits (IEEE754 quadruple precision), for a 64×64 matrix. Without CENA correction, Strassen multiplication produces an error that is about $500\times$ larger than that of classic multiplication. CENA reduces the error of classic multiplication by a factor of ca. $2\times$ and that of Strassen multiplication by a factor of ca. $3\times$ for odd mantissa lengths and above $1000\times$ for even mantissa lengths.

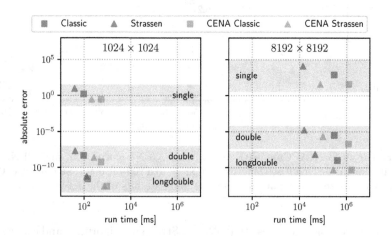

Fig. 8. Error vs run time for classic and strassen multiplication, for various working-precision settings, with and without CENA (left: 1024×1024, right: 8192×8192 matrix). CENA enables new trade-offs between accuracy and run time. For example, in the large test case strassen with CENA offers better run time and accuracy than does classic multiplication without CENA.

We summarize here the error bounds given in [10]. For classic multiplication $\mathbf{C} = \mathbf{AB}$, the error in a computed matrix $\hat{\mathbf{C}}$ is bounded by

$$\left\| \hat{\mathbf{C}} - \mathbf{C} \right\| \leq n^2 \epsilon \|\mathbf{A}\| \|\mathbf{B}\| + \mathcal{O}(\epsilon^2), \tag{2}$$

where ϵ is the unit roundoff error and $\| \cdot \|$ is the maximum norm. Note that [10] also provides a tighter bound for the error in each element in $\mathbf{C}$ that is linear in the matrix size n, while [2] gives an even tighter bound if the summation is performed by using pairwise summation. The authors also remark that no such elementwise error bound can exist for fast (subcubic) matrix multiplication algorithms, whose error is bounded by

$$\left\| \hat{\mathbf{C}} - \mathbf{C} \right\| \leq \left[\left(\frac{n}{n_0} \right)^{\log_2 12} (n_0^2 + 5n_0) - 5n \right] \epsilon \|\mathbf{A}\| \|\mathbf{B}\| + \mathcal{O}(\epsilon^2), \tag{3}$$

where n_0 is the threshold at which the small partition matrices are multiplied using the classic algorithm (the recursion base case).

We note that $\log_2 12 \approx 3.58496$, resulting in a significantly faster growth of roundoff errors than the quadratic growth in the classic algorithm. We also note that for $n_0 = n$ (i.e., a threshold so large that the classic algorithm is used without previous partitioning) the bounds in Eqs. (2) and (3) are identical.

Our experiments use input matrices filled by the same number generator as in Sect. 6.1. After multiplication with either a classic or Strassen's algorithm, we compute the maximum of all elementwise absolute errors by comparing with a reference result obtained through MPFR with a 200-bit mantissa. Our classic implementation is accelerated through the use of OpenMP. Our Strassen implementation supports only matrix sizes that are a power of 2 (a restriction that could be lifted by a better implementation) and is parallelized through the use of the parallized classic multiplication as a recursion base case. We use the best-performing base case size n_0 for our experiments, which we determined to be between 128 and 512 on 56 threads, depending on the working precision and whether CENA was used. We note that the implementation, parallelization, and hardware platform do not change the asymptotic trends and merely affect the break-even point beyond which Strassen's outperforms classic multiplication.

In our experiments, Strassen multiplication reduces run time and increases error, while CENA has the opposite effect. CENA more than offsets the accuracy loss from Strassen multiplication for the tested matrices. For mantissas with even bit length (e.g. single and long double), but not for those with odd length (e.g. double or quad), CENA produces equally good results for either multiplication method. The reason for this even-odd oscillation, shown in Fig. 7, is unclear but can also be observed to a smaller extent in the other test cases.

Since CENA increases run time by a constant factor and Strassen instead reduces the run time complexity class, there is necessarily a break-even point at which Strassen combined with CENA outperforms classic multiplication without CENA. The result is a method that is faster, but at the same time more accurate, than classic non-CENA multiplication. Figure 8 illustrates this effect by showing run time and accuracy for two different problem sizes, one that is smaller and one that is larger than this break-even point.

6.4   Muller Recurrence

Error estimation and correction are not a panacea. We applied CENA to a pathological example from [15], the Muller recurrence

$$x_{n+1} = 108 - (815 - 1500/x_{n-1})/x_n$$

using initial values $x_0 = 4.0$ and $x_1 = 4.25$. This recurrence is carefully designed so that the solution is of the form

$$x_n = (\alpha 3^{n+1} + \beta 5^{n+1} + \gamma 100^{n+1})/(\alpha 3^n + \beta 5^n + \gamma 100^n)$$

where the specific values of α, β, and γ depend on x_0 and x_1. Our initial conditions correspond to $\alpha = 1$, $\beta = 1$, and $\gamma = 0$, Therefore, the recurrence has the solution $x_n = (3^{n+1} + 5^{n+1})/(3^n + 5^n)$ and ought to converge to 5 in the limit. However, the slightest roundoff error causes the recurrence to match the solution for a nonzero γ and the recurrence converges to 100 in the limit.

Figure 9 shows the actual errors as well as the CENA estimate for a number of precisions, for each iteration, compared with the correct value at that iteration. When the sequence first diverges from the true solution with $\gamma = 0$, CENA estimates large roundoff errors up to O(100). However, eventually the nonzero γ terms in the recurrence come to dominate and CENA estimates a very small roundoff error. Thus, consistent with the title of [15], mindless application of the CENA correction would leave roundoff error in the computation of x_{30} undetected. However, monitoring error estimates at each iteration would allow detection of a significant roundoff problem. While real computational science applications are unlikely to exhibit such extreme behavior as the Muller recurrence, it may nonetheless be advisable to monitor error estimates throughout the computation rather than relying exclusively on the final terms.

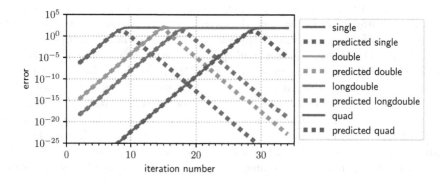

Fig. 9. CENA accurately estimates the roundoff errors up to the point where the recurrence starts to converge to the "wrong" fixed point. Using more precise number types only delays, but does not prevent this problem.

7   Conclusions

We introduced the forward CENA method and an efficient implementation. Using CENA to estimate and to some extent correct roundoff errors in numerical programs can be as easy as replacing the number types with our overloaded CENA number type. We showed that forward CENA does not negatively affect the scalability of parallel codes but has an overhead factor of 2–15. Future work includes analyzing the reasons for the observed superior error correction for even mantissa lengths compared with odd mantissa lengths. We also plan to investigate whether forward-mode AD can be employed in other error estimation methods that have historically relied on reverse-mode AD.

Acknowledgments. We thank Vincent Baudoui for introducing us to the CENA method and constructive conversations about roundoff error. This work was supported by the U.S. Department of Energy, Office of Science, Office of Advanced Scientific Computing Research, Applied Mathematics program under contract number DE-AC02-06CH11357. We gratefully acknowledge the computing resources provided and operated by the Joint Laboratory for System Evaluation (JLSE) at Argonne National Laboratory.

References

1. Ahrens, P., Demmel, J., Nguyen, H.D.: Algorithms for efficient reproducible floating point summation. ACM Trans. Math. Softw. **46**(3), 1–49 (2020)
2. Ballard, G., Benson, A.R., Druinsky, A., Lipshitz, B., Schwartz, O.: Improving the numerical stability of fast matrix multiplication. SIAM J. Matrix Anal. Appl. **37**(4), 1382–1418 (2016)
3. Bischof, C.H., Carle, A., Hovland, P.D., Khademi, P., Mauer, A.: ADIFOR 2.0 user's guide (Revision D). Argonne Technical Memorandum 192 (1998)
4. Christianson, B.: Reverse accumulation and accurate rounding error estimates for Taylor series coefficient. Optim. Methods Softw. **1**(1), 81–94 (1992)
5. Fousse, L., Hanrot, G., Lefèvre, V., Pélissier, P., Zimmermann, P.: MPFR: a multiple-precision binary floating-point library with correct rounding. ACM Trans. Math. Softw. **33**(2), 13es (2007)
6. Garcia, R., Michel, C., Rueher, M.: A branch-and-bound algorithm to rigorously enclose the round-off errors. In: Simonis, H. (ed.) CP 2020. LNCS, vol. 12333, pp. 637–653. Springer, Cham (2020). https://doi.org/10.1007/978-3-030-58475-7_37
7. Graillat, S., Ménissier-Morain, V.: Accurate summation, dot product and polynomial evaluation in complex floating point arithmetic. Inf. Comput. **216**, 57–71 (2012)
8. Griewank, A., Walther, A.: Evaluating Derivatives. Society for Industrial and Applied Mathematics (2008). https://doi.org/10.1137/1.9780898717761
9. Habib, S., et al.: HACC: Simulating sky surveys on state-of-the-art supercomputing architectures. New Astron. **42**, 49–65 (2016)
10. Higham, N.J.: Exploiting fast matrix multiplication within the level 3 BLAS. ACM Trans. Math. Softw. **16**(4), 352–368 (1990)
11. Iri, M., Tsuchiya, T., Hoshi, M.: Automatic computation of partial derivatives and rounding error estimates with applications to large-scale systems of nonlinear equations. J. Comput. Appl. Math. **24**(3), 365–392 (1988)

12. Jézéquel, F., Graillat, S., Mukunoki, D., Imamura, T., Iakymchuk, R.: Can we avoid rounding-error estimation in HPC codes and still get trustworthy results? In: Christakis, M., Polikarpova, N., Duggirala, P.S., Schrammel, P. (eds.) NSV/VSTTE - 2020. LNCS, vol. 12549, pp. 163–177. Springer, Cham (2020). https://doi.org/10.1007/978-3-030-63618-0_10

13. Jézéquel, F., Chesneaux, J.M.: CADNA: a library for estimating round-off error propagation. Comput. Phys. Commun. **178**(12), 933–955 (2008)

14. Kahan, W.: Pracniques: further remarks on reducing truncation errors. Commun. ACM **8**(1), 40 (1965). https://doi.org/10.1145/363707.363723

15. Kahan, W.: How futile are mindless assessments of roundoff in floating-point computation? (2006). http://www.cs.berkeley.edu/~wkahan/Mindless.pdf

16. Langlois, P.: Automatic linear correction of rounding errors. BIT Numer. Math. **41**(3), 515–539 (2001)

17. Linnainmaa, S.: Taylor expansion of the accumulated rounding error. BIT Numer. Math. **16**(2), 146–160 (1976)

18. Martel, M.: Semantics of roundoff error propagation in finite precision calculations. Higher-Order Symbolic Comput. **19**(1), 7–30 (2006)

19. Menon, H., et al.: ADAPT: Algorithmic differentiation applied to floating-point precision tuning. In: Proceedings of SC 2018, pp. 48:1–13. IEEE Press, Piscataway, NJ (2018)

20. Ogita, T., Rump, S.M., Oishi, S.: Accurate sum and dot product. SIAM J. Sci. Comput. **26**(6), 1955–1988 (2005)

21. Solovyev, A., Baranowski, M.S., Briggs, I., Jacobsen, C., Rakamarić, Z., Gopalakrishnan, G.: Rigorous estimation of floating-point round-off errors with symbolic Taylor expansions. ACM Trans. Program. Lang. Syst. **41**(1), 2:1–39 (2018)

22. Tienari, M.: A statistical model of roundoff error for varying length floating-point arithmetic. BIT Numer. Math. **10**(3), 355–365 (1970)

23. Vassiliadis, V., et al.: Towards automatic significance analysis for approximate computing. In: 2016 IEEE/ACM International Symposium on Code Generation and Optimization (CGO), pp. 182–193 (March 2016)

24. Vignes, J.: Discrete stochastic arithmetic for validating results of numerical software. Numer. Algorithms **37**(1–4), 377–390 (2004)

Monte Carlo Approach to the Computational Capacities Analysis of the Computing Continuum

Vladislav Kashansky[1,2]($\boxtimes$) ⓘ, Gleb Radchenko[2] ⓘ, and Radu Prodan[1] ⓘ

[1] Institute of Information Technology, University of Klagenfurt, Klagenfurt, Austria
`vkashansky@acm.org`, `radu.prodan@aau.at`
[2] School of Electronic Engineering and Computer Science, South Ural State University, Chelyabinsk, Russia
`gleb.radchenko@susu.ru`

Abstract. This article proposes an approach to the problem of computational capacities analysis of the computing continuum via theoretical framework of equilibrium phase-transitions and numerical simulations. We introduce the concept of phase transitions in computing continuum and show how this phenomena can be explored in the context of workflow makespan, which we treat as an order parameter. We simulate the behavior of the computational network in the equilibrium regime within the framework of the XY-model defined over complex agent network with Barabasi-Albert topology. More specifically, we define Hamiltonian over complex network topology and sample the resulting spin-orientation distribution with the Metropolis-Hastings technique. The key aspect of the paper is derivation of the bandwidth matrix, as the emergent effect of the "low-level" collective spin interaction. This allows us to study the first order approximation to the makespan of the "high-level" system-wide workflow model in the presence of data-flow anisotropy and phase transitions of the bandwidth matrix controlled by the means of "noise regime" parameter η. For this purpose, we have built a simulation engine in Python 3.6. Simulation results confirm existence of the phase transition, revealing complex transformations in the computational abilities of the agents. Notable feature is that bandwidth distribution undergoes a critical transition from single to multi-mode case. Our simulations generally open new perspectives for reproducible comparative performance analysis of the novel and classic scheduling algorithms.

Keywords: Complex networks · Computing continuum · Phase transitions · Computational model · MCMC · Metropolis-Hastings · XY-model · Equilibrium model

1 Introduction

Recent advancements [3,30] in the field of parallel and distributed computing led to the definition of the computing continuum [6] as the environment comprising highly heterogeneous systems with dynamic spatio-temporal organizational

© Springer Nature Switzerland AG 2021
M. Paszynski et al. (Eds.): ICCS 2021, LNCS 12742, pp. 779–793, 2021.
https://doi.org/10.1007/978-3-030-77961-0_62

structures, varying in-nature workloads [14], complex control hierarchies [8], governing computational clusters with multiple scales of the processing latencies, and diverse sets of the management policies [17,31]. The emergence of these systems is the natural response to the ever-growing variability of computational demands.

This paper investigates how to approach the problem of modeling the computing continuum, considering it as an active part of evolving multi-agent networks, similar to Internet, with complex emergent properties. This problem remains still mostly uncovered in the literature of the scheduling problems. Moreover, since emergence of the concept in 2020 [6], there is a lack of a *reproducible model* of the computing continuum, especially for better understanding scheduling heuristics, as real systems do not preserve this quality and hinder the comparative performance analysis of the novel scheduling approaches.

Our paper aims to fill these gaps through simulation, allowing to study the computing continuum with the preferential attachment topology [1] and given workload, based on a theoretical framework with some initial conditions. We implement our theoretical model as module of a simulator called *Modular Architecture for Complex Computing Systems Analysis (MACS)* [16]. We introduced a parameter related to the noise regime in the system that allows us to study different congested states of the computational network, considered as a graph cut from a global self-organizing network.

The contribution of this paper is as follows:

1. We analyse the definition [6] of the computing continuum and provide a theoretical model of how high-order computational properties emerge from elementary pairwise interactions;
2. We study the behavior of the *fully observable* agent computational network, simulated in the equilibrium regime within the framework of the modified XY-model [15] defined over complex network with Barabasi-Albert topology [1]. We define the Hamiltonian, which encodes dynamical properties of the system over a complex network topology, and sample the resulting Gibbs distribution of spin-orientations with the Metropolis-Hastings technique [15].
3. We derive the bandwidth matrix, as the emergent effect of the "low-level" collective spin interaction, which allows us to study the first order approximation to the behavior of the "high-level" system-wide workflow model in the presence of emergent data-flow anisotropy. We obtained the DAG network structure from the PSLIB project scheduling library [21], namely the state-of-the-art j60-60 benchmark instance consisting of 60 tasks.

The remainder of the paper is structured as follows. Section 2 discusses the related work. Section 2.1 describes the background definitions and introduces reader to our vision of the computing continuum concept, meanwhile Sect. 2.2 and 2.3 devoted to the specific structural aspects. Sections 3 describes proposed model. Section 4 presents an experimental evaluation and Sect. 5 concludes the paper.

2   Related Work and Definitions

2.1   Computing Continuum

To better understand the case of computing continuum, we should recall to complex multi-layer computational networks, that are operated by large corporations and governments. Examples of such networked systems are:

- social platforms that analyse concurrently various motion patterns and opinion dynamics [4,5] related to human behaviour at various spatio-temporal scales;
- self-organizing vehicle fleets and drone swarms [9,24] that receive information from a large group of spatial sensors and need to make decisions accordingly.

Such networks consist of a large number of locally interacting *agents* that form stable emergent flows of data and execute arbitrary workloads. It is important to highlight that we are no longer dealing with machines in the case of computing continuum, but with agents [4] capable of consensus formation of various kind.

A key property typical to all sufficiently complex ecosystems is their tendency to grasp the sub-systems, provide uniformity over heterogeneity and optimize the end-point *workload execution* in the presence of *target criteria with system-wide constraints*. Haken [12] introduced the concept of "order parameter", as lower-dimension projection, governing large-scale dynamics of the complex system as whole, emerged from local interactions. He noted the role of natural language of any kind to synchronize all the individuals, reduce/compress the complexity of mind and maintain the synchronous state across the society, similarly to the order parameter of the social network. The same logic can be seen in the computing systems with the introduction of formal grammars, programming languages and interfaces [13].

Computing continuum can be defined as a *convolution hierarchy* of interacting order parameters from inherently local, associated with tasks to global which are structurally prescribed by the workload network. In our work we interpret global order parameter of whole convolution hierarchy as Makespan.

Finally, we can characterize computing continuum as transfusion of the three components, namely:

- *Computational network*, which provides structural knowledge via statistics about the possible information flows within the system and possible interactions of agents via adjacency matrix;
- *Recursive network* which forms a multi-layer DAG and provides knowledge about non-equilibrium dynamic processes in the computational network. This component is optional and only required if behaviour of the network is considered far from equilibrium. Examples include: Dynamic Bayesian Networks (DBN) and Hidden Markov Models (HMM);
- *Workload network* is the set of tasks, represented as the DAG that governs computational process in the computing continuum, prescribing arrow of the time.

Computational and workload networks are inherently *complex networks*, described in terms of random graphs with corresponding statistical properties [5]. In the computational network, data transfer and local performance rates are emerging at various noise regimes. Tasks are interconnected by a precedence relation and have different data volumes to transmit between each other and computation time, as well as transmission rate, will depend on matching with computational agents. Together tasks form a global collective order parameter called *Makespan*, which does not exist when each task is considered individually. The order parameter for a computational network is the degree of consensus of the agents. As a result, we obtain a hierarchy, which at its highest level characterized by single parameter - Makespan, and at its lowest level is described by the local behavior of the computing agents.

2.2   Computational Network with Scale-Free Topology

In our study, we decided to model computational network topology as a scale-free network, as many of the networks under study fall into the class. This means that they have a power-law distribution with respect to the degree of a node. Scale-free networks are important phenomena in natural networks and human networks (Internet, citation networks, social networks) [1]. Barabasi-Albert theoretical model of scale-free network incorporates two important general concepts: network growth mechanism and preferential attachment principle (PA). Both concepts are widely represented in real-world networks. Growth means that the number of nodes in a network increases over time. Preferential attachment principle prescribes, that the more links a node has, the more likely it is to create new links. Nodes with the highest degree have more opportunities to take over the links added to the network. Intuitively, the principle of preferential attachment can be understood if we think in terms of social networks that connect people together. Strongly connected nodes are represented by known people with a large number of connections. When a newcomer enters a community, it is more preferable to connect with one of the known people than with a relatively unknown person. Similarly, on the World Wide Web, pages link to hubs, for example, well known sites with high Page Rank (PR) [28], than to pages that are not well known. If you choose a new page to link to randomly, then the probability of choosing a particular page will be proportional to its degree. This explains the preferential attachment principle.

The PA principle is an example of positive feedback, where initially random variations in the node number of node links are automatically amplified, thereby greatly increasing the gap and allowing hub formation.

2.3   Modelling Congested States of the Computing Continuum Through Phase Transitions

Natural phase transitions refer to the change between various states of matter, such as the solid, liquid, and gaseous states. The transition from one state to the other depends upon the value of a set of external parameters such as temperature, pressure, and density characterizing the thermodynamic state of the system. In the context of complex networks, various Monte Carlo simulations of Ising model on small-world networks confirmed the existence of a phase transition of mean-field character [5]. Specifically, the critical behavior of the XY-model in small-world networks was studied in [26].

The research on phase transitions of various kinds has a long tradition in physical sciences. Due to the lack of space, we refer to [5,15,18] for general and specific aspects of the problem. For further reading, it is only important to relate "order parameters" with small changes induced by the noise regime. A remarkable connection of the phase transitions and NP-hard problems [27] shows the important influence of the input data distribution on the sensitivity of optimization algorithms. Namely, it is interesting to identify how these small changes will result in dramatic changes in the convolution hierarchy of interacting order parameters due to non-linear effects, from *local* associated with tasks, to *global* measures such as Makespan.

3   Mathematical Model

3.1   Workload Model

We define a *workload* as a directed acyclic graph (DAG) where V is the set of tasks and E is the set of edges between them: $G = (V, E)$. Each edge $(i, j) \in E$ represents a precedence constraint indicating that the task i must complete its execution before the task j starts. Preemption is not allowed.

3.2   Agent Model

We consider a set of independent agents A with different processing speeds and synchronized clocks. Computational properties of the agents are expressed with a non-symmetric positive matrix $\mathcal{B}$. Precisely speaking, off-diagonal entries model bandwidth between agents m and q and diagonal elements correspond to the agent self-performance. All entries in the matrix are normalized and dimensionless.

3.3   Network Model

We model the computational network as a random graph with Barabasi-Albert scale-free topology that provides structural knowledge about its topological properties: $N = (A, Q)$, where A is the set of the agents and $Q \subset A \times A$ is the set of interconnections $(m, q) \in Q$ between them.

For an analytical insight into simulation and phase transitions, we rely on the equilibrium statistical physics framework, which considers that the network N can be in any possible microscopic configuration. Considering XY-model [5,15] defined over the network N we map the two-dimensional unit vector $\vec{S_m}$:

$$\vec{S}_m = (cos(\theta_m),\ sin(\theta_m)) \tag{1}$$

to each vertice in A with angle $\theta_m \in [0; 2\pi]$. Spin vector is inherently local property and has no significant impact in our model when considered separately. In the context of our problem, it accounts on how the noise η numerically impacts the ability of the network to perform computations.

Direct interpretation consists in considering the existence of a centralized policy-maker in the computing continuum. A spin vector θ_m then represents the dynamic scalar degree of agent's belief [5, pp. 216–241] to the "center", prescribing mechanism of the consensus formation. In this scenario, agents assumed to have a continuous opinion amplitude from 0 to 2π. However, this interpretation may not be important, for example when simulating a comparative study of the various scheduling heuristics. The information about the connection between agents is encoded in the adjacency matrix J of the graph N, considering local interaction of the adjacent agents. The Hamiltonian then associates an energy $\mathcal{H}$ to each configuration:

$$\mathcal{H}(\boldsymbol{\theta}) = \sum_{m \in A} \sum_{q \neq m} J_{mq} \cdot [1 - \cos(\theta_m - \theta_q)]. \tag{2}$$

Further, the configuration probability is given by the Boltzmann distribution with noise regime $\eta \geq 0$:

$$P(\boldsymbol{\theta}) = Z^{-1} \cdot e^{-\frac{\mathcal{H}(\boldsymbol{\theta})}{\eta}}; \quad Z = \int_{[0,2\pi]^A} \prod_{m \in A} d\theta_m \, e^{-\frac{\mathcal{H}(\boldsymbol{\theta})}{\eta}}. \tag{3}$$

where Z is the normalization factor. It is important to highlight that we operate with a *fully observable* computational network.

3.4 Order Parameter of the Computational Network

The distribution $P(\boldsymbol{\theta})$ provides important knowledge about network behaviour and importantly, order formation at low noise regime limit. The first important characteristic is the average orientation $\hat{\theta}_m$ computed via the following integral:

$$\hat{\theta}_m = \int \theta_m \cdot f(\theta_m) \, d\theta_m. \tag{4}$$

Further, we require a two-point co-variation function R_{mq} given by the following two-dimensional integral [18]:

$$R_{mq} = \iint (\theta_m - \hat{\theta}_m) \cdot (\theta_q - \hat{\theta}_q) \cdot f(\theta_m \cdot \theta_q) \, d\theta_m \, d\theta_q. \tag{5}$$

Averaging across all spins in the network N leads to the synchronization degree $\mathcal{M} \in [0; 1]$, also known as mean magnetization in the conventional physics literature on the d-dimensional lattices:

$$\mathcal{M} = \frac{1}{|A|^2} \cdot \left\langle \left(\sum_{i=1}^{|A|} \cos \theta_i \right)^2 + \left(\sum_{i=1}^{|A|} \sin \theta_i \right)^2 \right\rangle. \tag{6}$$

This is zero above a critical temperature in many lattice magnetic systems and becomes non-zero spontaneously at low temperatures. Likewise, there is no conventional phase transition present that would be associated with symmetry breaking. However, it is well-known, that system does show signs of a transition from a disordered high-temperature state to a quasi-ordered state below some critical temperature, called the Kosterlitz-Thouless [22] transition.

3.5   Network Bandwidth Model

We propose in our model the definition of bandwidth $\mathcal{B}_{mq}$ between two agents m and q as a non-symmetric positive matrix, based on the following assumptions:

- Mutual spin correlation at a finite noise regime, as stronger correlation results in the better bandwidth;
- Average spin orientation of a agent m projected to the mean-field orientation $\mathcal{M}$ given by the γ-factor, which reflects the synchrony of the agent to the network;
- Inter-node degree relation $\Omega(m, q)$, obtained from the graph structure, which reflects asymmetry in the traffic handling. A larger amount of links forces an agent to split its bandwidth;
- Non-symmetric structure of the matrix, handling the realistic scenario of non-identical bandwidth (m, q) and (q, m) permutation.

Accounting these assumptions results in a bandwidth $\mathcal{B}_{mq}$ defined by the following S-curve equation:

$$\mathcal{B}_{mq} = \Omega(m, q) \cdot \gamma_m \cdot \frac{\min(B_m, B_q)}{1 + e^{-R_{mq}}} > 0 , \; \forall \, (m, q), \tag{7}$$

where the inter-node degree relation is given by:

$$\Omega(m, q) = 1 - \frac{\deg(m)}{\deg(m) + \deg(q)}, \tag{8}$$

and $\deg(m)$ is the degree of the vertex m and γ-factor is given by vector product:

$$\gamma_m = 1 - \frac{1}{4 \cdot \pi^2} \cdot \frac{\mathcal{M} \cdot (\hat{\phi} - \hat{\theta}_m)^2}{1 + e^{-\mu}}; \qquad \phi = \frac{1}{|A|} \cdot \sum_m^{|A|} \theta_m \tag{9}$$

The random variable ϕ expresses the collective field ϕ averaged across the all graph nodes, and μ-factor is the co-variance:

$$\mu = \int (\phi - \hat{\phi}) \cdot (\theta_m - \hat{\theta}_m) \cdot P(\boldsymbol{\theta}) \, d\theta_1 \ldots d\theta_n, \tag{10}$$

which accounts how strong collective field ϕ is "feeling" arbitrary spin vector θ_m orientation. Finally, $\min(B_m, B_q)$ assumed as amplitude of the S-curve function, shows that the channel bandwidth saturates to the minimum of the two possible nominal values defined for each agent. Hats are indicating sample average.

The special case of diagonal elements $\mathcal{B}_{mm}$ corresponding to the dimensionless self-bandwidth of the given agent:

$$\mathcal{B}_{mm} = \gamma_m \cdot \frac{F_m}{e^{\cdot\sqrt{R_{mm}}}}, \tag{11}$$

where $F_m \in (0;1]$ is the nominal normalized dimensionless self-bandwidth, and $\sqrt{R_{mm}}$ is the variance of the local spin orientation, which prescribes decrease of the computational capacity with growing variance.

3.6   Communication and Computation in the Network

We further define a data matrix D_{ij} that indicates the amount of data transmitted from a task i to a task j. Consequently, we obtain the *delay tensor* D^* for transferring data from task i to task j assigned to the agents m and q:

$$d^*_{ijmq} = \overbrace{\mathcal{T}_{mq}}^{\text{Connection Delay}} + \overbrace{D_{ij} \cdot \mathcal{B}^{-1}_{mq}}^{\text{Data Transfer Latency}} . \tag{12}$$

The first term $\mathcal{T}_{mq}$ in Eq. 12 represents a connection estimation delay, assumed as a small constant. This approach models realistic scenarios of synchronized routing information, leading to the fast connection estimation with low delay. The second term results from the time required to transfer the data from task i to j residing on the agents m and q, obtained by dividing the components of data matrix by the corresponding bandwidth. We compute the execution time of the given task $i \in V$ with the following formula:

$$\tau^1_{im} = \max_j\{d^*_{ijmq}; \ \forall j \in \mathcal{P}_i\} + \tau^0_{im} \cdot \mathcal{B}^{-1}_{mm}, \tag{13}$$

where $\mathcal{P}_i$ is the set of predecessors of task i, the upper indices give the order of the approximation, and the term τ^1_{im} corresponds to the execution time perturbed by the transfer times, meanwhile τ^0_{im} corresponds to the execution time measured in the perfect conditions when there is no noise and data transfer delays. We assume first-order approximation reasonable, when we consider a bandwidth independent of the number of simultaneously transferring agents.

4   Experimental Study

4.1   Monte Carlo Simulation of the Computational Network

We obtain the distribution $P(\boldsymbol{\theta})$ via Monte Carlo simulation using the Metropolis algorithm [5, pp. 108–111], [23]. At each time step, we induce a random walk on the graph N, choose one random spin and rotate its angle by some random increment, keeping it in a range $[0; 2\pi]$. States are accepted with the probability:

$$p(\boldsymbol{\theta}_{n+1}|\ \boldsymbol{\theta}_n) = \min\left\{1, e^{-\Delta\mathcal{H}}\right\}, \tag{14}$$

where $\Delta\mathcal{H} = \mathcal{H}_{n+1} - \mathcal{H}_n$, where n defines number of the step in a random walk. The transitions to the lower energies are certainly accepted, while the transitions to slightly higher energies are accepted with the probability $e^{-\Delta\mathcal{H}}$.

4.2 Implementation Details

We implemented the simulation [16] model in `Python` 3.6 with network processing performed by the graph-tool [29] library. We implemented the scheduler using the `SCIP` 6.0.2 optimization suite [11] with the default configuration and `SoPlex` 4.0.2 LP solver. FB-SGS Heuristic [19,20] is implemented as stand alone callable application via Python integration wrapper, that we have developed. We compiled the source files for the model and SCIP using `gcc` version 4.8.5 and handled the experimental data using the `BASH` 4.2.46, `MariaDB` 5.5.64 and native NumPy (.npy) format.

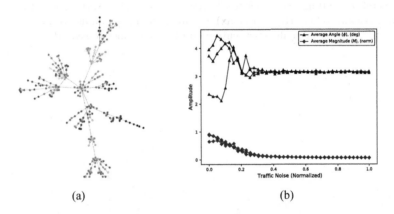

(a)   (b)

Fig. 1. (a) Snapshot after $1.5 \cdot 10^3$ iterations of the Metropolis-Hastings dynamics of the model defined over Barabasi-Albert network ($n = 256$, $\eta = 0.1$) with triangular initial graph; darker colors correspond to values of θ_m close to 0. Visualized with Fruchterman-Reingold layout algorithm and Cairo library. (b) Order parameters ϕ and $\mathcal{M}$ as functions of the noise regime η.

4.3 Numerical Results and Discussion

We considered nominal capacities B_m, F_m of the computing agents equal to 1 without loss of generality. While this brings homogeneity across agents, the heterogeneity is encoded in the network structure and its stochastic behaviour, which depends on the noise regime.

During the first stage of the experiment, we focused on the analysis of the computational network. The topology visualization of the computational network N is depicted on the Fig. 1a. For layout generation we have used force-directed Fruchterman-Reingold algorithm [10] with 2000 iterations. We generated the network structure an from initial triangle graph with three nodes, by

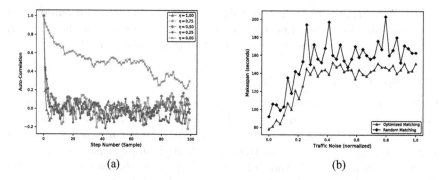

(a)   (b)

Fig. 2. (a) Auto-correlation of the global state-transition of the network graph N, with respect to the 0th step. (b) Makespan of the DAG obtained with a variant of the classical local descent with monotonic improvement in the objective function. Single and several ($i = 20$) iterations agent selection vs noise regime variation. Both cases use Forward-Backward SGS (RCPSP/max) for minimum makespan derivation. Task times obtained by scaling normalized execution and transfer times to seconds.

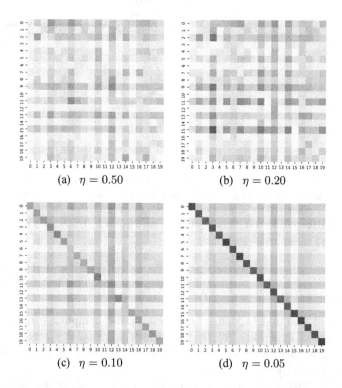

(a) $\eta = 0.50$ (b) $\eta = 0.20$

(c) $\eta = 0.10$ (d) $\eta = 0.05$

Fig. 3. Distribution of the emergent topological excitations $\mathcal{B}_{mq} \in (0; 1]$ on the 20×20 sub-matrix of 256×256 bandwidth matrix, where darker colours correspond to the higher rates of the bandwidth. From left to right, the system evolves from high to low noise influence, exhibiting various computational capabilities of the network N.

selecting one node with probability proportional to the number of its links. We observe heavily linked hubs that tend to quickly accumulate even more links, while nodes with only a few links are unlikely to be chosen and stay peripheral. The new nodes have a "preference" to attach themselves to the already heavily linked nodes. We stopped growth of the network at 256 nodes, since this scales are sufficient [5, pp. 92–115] to study emergent properties. We performed a Monte Carlo simulation with 1500 random walks for each of 100 "statistical snapshots" of N with $\eta \in [0; 1]$, taking 14 min on average. Figure 1b illustrates the existence of the critical transition around $\eta \sim 0.21$. For further reduction of $\eta \leq 0.20$, one can readily see a rapid growth of the $\mathcal{M}$ order parameter to 1, which indicates that the network achieves a synchronous consensus state. At the same time, the average angle demonstrates a bifurcation-like behaviour. Figures 3 and 4 demonstrate a significant difference in the network operation regimes. Even in a relatively simple homogeneous configuration, we are able to achieve a non-trivial transition of the network bandwidth behaviour. As the noise regime η approaches zero, some agents loose ability to transmit the data and become more "computationally" efficient, as reflected by the bright diagonal in Fig. 3d indicating high computational abilities of the agents. Figure 4b shows the distribution of the overall bandwidth in the computational network before and after phase transitions, augmented with non-parametric kernel density estimations. It is remarkable how the bandwidth distribution undergoes a critical transition from single to multi-mode case. These important features allow us to simulate complex behaviour of contemporary computational environments, such as computing continuum. Figure 2a shows an exponential decay of the auto-correlation function. Such ergodic behaviour is natural for Metropolis and Glauber dynamics [5] and indicates that the computational network rapidly looses memory on its initial conditions. The advantage of the ergodic description is that such networks can be described by statistical methods given a sufficient observation time. Depending on the number of transitions in time, this actually rises a question of ergodic hypothesis fairness and suitability for the particular computational network. The discussion of this old issue goes far beyond the scope of this article, however, this approximation is suitable in the low-traffic scenario. We keep the holistic analysis of the more general non-equilibrium scenario for future works in frames of Kinetic schemes and Bayesian's framework.

We further evaluated the behaviour of the DAG network depicted in Fig. 5 for different values of η. It consists of 60 tasks with high inter-dependency and node degrees ranging from 1 to 3. To compute makespan of the DAG network, we use a variant of the classical local descent with monotonic improvements by trying different matching of tasks and agents. Figure 2b depicts a single ($i = 1$) and multiple ($i = 20$) iterations of the random matching. Once the agents are matched with tasks, both cases use Forward-Backward SGS (RCPSP/max) for minimum makespan derivation, with a complexity of $O\left(J^2 \cdot K\right)$, where $J \subseteq V$ and $K \subseteq N$ are the number of tasks and resources required for execution. We obtain the task times by setting $D_{ij} = 1$, scaling the normalized execution and transfer times to seconds and $\mathcal{T}_{mq} = 0$, according to the Eq. 13. This evaluation

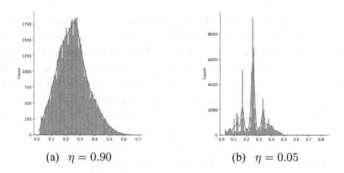

(a) $\eta = 0.90$ (b) $\eta = 0.05$

Fig. 4. Histogram of the bandwidth matrix with non-parametric kernel density estimation before (a) and after (b) phase transition.

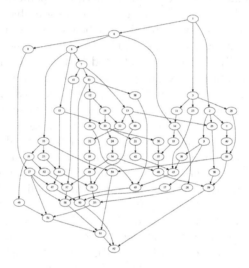

Fig. 5. Directed acyclic graph from PSLIB j60 dataset. Transformed to .dot format and Visualized with GraphViz DOT render engine.

demonstrates how complex precedence networks of the tasks can be analysed in the framework of our simulator. Makespan values from Fig. 2b demonstrate high sensitivity in relation to critical transition at $\eta \leq 0.20$, as the computational efficiency grows. When the amplitude of the noise regime increases, the optimized allocation shows higher makespan values due to the increase in the network disorder. Interestingly, optimized allocation gives not only smaller values of the makespan, but also smaller variance, leading to the more robust estimations of the makespan when the noise regime changes.

5   Discussion and Future Work

In this paper, we analysed the definition of the computing continuum and provided a theoretical model of how high-order computational properties can emerge from elementary pairwise interactions in the context of a topology with preferential attachment policy. We studied the behavior of the computational network by simulating it in the equilibrium regime within the framework of the modified XY-model defined over a complex network with Barabasi-Albert topology. We developed the simulation engine in Python 3.6, which allowed us to study the first order approximation to the behavior of the "high-level" system-wide workflow model in the presence of emergent data-flow anisotropy.

It is important to note that the DAG and scheduling heuristic is non-exhaustive and mainly used for test purposes. We expect to carry out more detailed comparative study of the several heuristics, including Heterogeneous Earliest Finish Time (HEFT), The Dynamic Scaling Consolidation Scheduling (DSCS) [25], Partitioned Balanced Time Scheduling (PBTS) [7] Deadline Constrained Critical Path (DCCP) [2], and Partition Problem-based Dynamic Provisioning and Scheduling (PPDPS) [32]. We hypothesize that Forward-Backward SGS (RCPSP/max) will outperform the aforementioned heuristics in the quality of schedules for a relatively small amount of tasks $|G| \sim 200 - 300$. Particularly interesting is to research this question at a larger scale, where possibly non of these schemes provide satisfactory outcome in terms of convergence speed and quality of the solutions. The only applicable scenario will evolve around multi-agent mapping. We will also continue to work on kinetic schemes and Bayesian's framework to incorporate non-equlibrium phenomena and partial-observations of the network. Integration of the MPI stack to speed up the computations on larger scales and expand our work on new types of network topologies with specific statistics is an important research direction too.

Acknowledgement. This work has been supported by the ASPIDE Project funded by the European Union's Horizon 2020 Research and Innovation Programme under grant agreement No 801091.

References

1. Albert, R., Barabási, A.L.: Statistical mechanics of complex networks. Rev. Mod. Phys. **74**(1), 47 (2002)
2. Arabnejad, V., Bubendorfer, K., Ng, B.: Scheduling deadline constrained scientific workflows on dynamically provisioned cloud resources. Future Gener. Comput. Syst. **75**, 348–364 (2017)
3. Asch, M., et al.: Big data and extreme-scale computing: pathways to convergence-toward a shaping strategy for a future software and data ecosystem for scientific inquiry. Int. J. High Perform. Comput. Appl. **32**(4), 435–479 (2018)
4. Axelrod, R.: The dissemination of culture: a model with local convergence and global polarization. J. Conflict Resolut. **41**(2), 203–226 (1997)
5. Barrat, A., Barthelemy, M., Vespignani, A.: Dynamical Processes on Complex Networks. Cambridge University Press, Cambridge (2008)

6. Beckman, P., et al.: Harnessing the computing continuum for programming our world. In: Fog Computing: Theory and Practice, pp. 215–230 (2020). https://doi.org/10.1002/9781119551713.ch7

7. Byun, E.K., Kee, Y.S., Kim, J.S., Maeng, S.: Cost optimized provisioning of elastic resources for application workflows. Future Gener. Comput. Syst. **27**(8), 1011–1026 (2011)

8. Copil, G., Moldovan, D., Truong, H.-L., Dustdar, S.: Multi-level elasticity control of cloud services. In: Basu, S., Pautasso, C., Zhang, L., Fu, X. (eds.) ICSOC 2013. LNCS, vol. 8274, pp. 429–436. Springer, Heidelberg (2013). https://doi.org/10.1007/978-3-642-45005-1_31

9. D'Andrea, R., Dullerud, G.E.: Distributed control design for spatially interconnected systems. IEEE Trans. Autom. Control **48**(9), 1478–1495 (2003)

10. Fruchterman, T.M., Reingold, E.M.: Graph drawing by force-directed placement. Softw. Pract. Exper. **21**(11), 1129–1164 (1991)

11. Gleixner, A., et al.: The scip optimization suite 6.0. Technical Report, pp. 18–26, ZIB, Takustr. 7, 14195 Berlin (2018)

12. Haken, H.: Synergetics. Phys. Bull. **28**(9), 412 (1977)

13. Hopcroft, J.E., Motwani, R., Ullman, J.D.: Introduction to automata theory, languages, and computation. Acm Sigact News **32**(1), 60–65 (2001)

14. Ilyushkin, A., et al.: An experimental performance evaluation of autoscaling policies for complex workflows. In: Proceedings of the 8th ACM/SPEC on International Conference on Performance Engineering, pp. 75–86 (2017)

15. Kadanoff, L.P.: Statistical Physics: Statics, Dynamics and Renormalization. World Scientific Publishing Company, Singapore (2000)

16. Kashansky, V.: Modular architecture for complex computing systems analysis. http://www.edmware.org/macs/, Accessed 29 Jan 2021

17. Kashansky, V., et al.: M3at: Monitoring agents assignment model for data-intensive applications. In: 2020 28th Euromicro International Conference on Parallel, Distributed and Network-Based Processing (PDP), pp. 72–79. IEEE (2020)

18. Klimontovich, Y.L.: Statistical Theory of Open Systems: Volume 1: A Unified Approach to Kinetic Description of Processes in Active Systems, vol. 67. Springer, Dordrecht (2012)

19. Kolisch, R., Drexl, A.: Local search for nonpreemptive multi-mode resource-constrained project scheduling. IIE Trans. **29**(11), 987–999 (1997)

20. Kolisch, R., Hartmann, S.: Heuristic algorithms for the resource-constrained project scheduling problem: classification and computational analysis. In: Węglarz, J. (eds) Project scheduling, International Series in Operations Research & Management Science, vol. 14, pp. 147–178. Springer, Boston (1999) https://doi.org/10.1007/978-1-4615-5533-9_7

21. Kolisch, R., Sprecher, A.: Psplib-a project scheduling problem library: or software-orsep operations research software exchange program. Eur. J. Oper. Res. **96**(1), 205–216 (1997)

22. Kosterlitz, J.M., Thouless, D.J.: Ordering, metastability and phase transitions in two-dimensional systems. J. Phys. C Solid State Phys. **6**(7), 1181 (1973)

23. Landau, D.P., Binder, K.: A Guide to Monte Carlo Simulations in Statistical Physics. Cambridge University Press, Cambridge (2014)

24. Langbort, C., Chandra, R.S., D'Andrea, R.: Distributed control design for systems interconnected over an arbitrary graph. IEEE Trans. Autom. Control **49**(9), 1502–1519 (2004)

25. Mao, M., Humphrey, M.: Auto-scaling to minimize cost and meet application deadlines in cloud workflows. In: SC 2011: Proceedings of 2011 International Conference for High Performance Computing, Networking, Storage and Analysis, pp. 1–12. IEEE (2011)
26. Medvedyeva, K., Holme, P., Minnhagen, P., Kim, B.J.: Dynamic critical behavior of the xy model in small-world networks. Phys. Rev. E **67**(3), 036118 (2003)
27. Monasson, R., Zecchina, R., Kirkpatrick, S., Selman, B., Troyansky, L.: Determining computational complexity from characteristic 'phase transitions'. Nature **400**(6740), 133–137 (1999)
28. Page, L., Brin, S., Motwani, R., Winograd, T.: The pagerank citation ranking: Bringing order to the web. Technical Report, Stanford InfoLab (1999)
29. Peixoto, T.P.: Graph-tool - efficient network analysis. https://graph-tool.skewed. de/. Accessed: 29 Jan 2021
30. Reed, D.A., Dongarra, J.: Exascale computing and big data. Commun. ACM **58**(7), 56–68 (2015)
31. Reuther, A., et al.: Scalable system scheduling for hpc and big data. J. Parallel Distrib. Comput. **111**, 76–92 (2018)
32. Singh, V., Gupta, I., Jana, P.K.: A novel cost-efficient approach for deadline-constrained workflow scheduling by dynamic provisioning of resources. Future Gener. Comput. Syst. **79**, 95–110 (2018)

Author Index